Rhiannon Bowen (handwritten)

4M 2008

Fourth International Conference on
Multi-Material Micro Manufacture
9th–11th September 2008
Cardiff, UK

Edited by

Stefan Dimov, Cardiff University, UK
and Wolfgang Menz, Germany

Organised by: FP6 4M Network of Excellence
Sponsored by: The European Commission

30.9.2008 (handwritten)

WHITTLES PUBLISHING

Published by
Whittles Publishing,
Dunbeath,
Caithness KW6 6EY,
Scotland, UK
www.whittlespublishing.com

© 2008 Cardiff University
ISBN 978-1904445-76-0

Typeset by
Thomson Digital

Printed by
Dardedze Holography, Riga, Latvia

Preface

It is a pleasure to welcome you to the Fourth Multi-Material Micro Manufacture (4M) Conference in the "birth place" of the 4M Knowledge Community. Four years ago this initiative started in Cardiff with the ambitious aim to create a forum for researchers striving to broaden the range of Micro Systems Technology (MST) based products and their underpinning multi material manufacturing processes. After a flying start in Forschungszentrum Karlsruhe (FZK) 4M2005 was followed by the successful conferences 4M2006 and 4M2007 held at two contrasting venues – the high-tech conference centre at MINATEC and the splendid mountain scenery of Borovets. Now Cardiff, the Capital of Wales, with its mixture of the old and the contemporary in the city centre and its recent waterfront development plays host to 4M2008, providing an inspiring venue for mutually rewarding discussions in micro and nano technologies, and new ideas for joint research and development activities.

Four exciting years have passed since the 4M community was initially formed. This fourth Conference is an excellent opportunity to review the progress in putting the 4M processes and applications on both the European and global science and technology maps. By joining forces we managed to link the fragmented expertise and knowledge in 4M and are on track to achieve our ambitious objective of creating the necessary pre-requisites for establishing a successful micro and nano manufacturing industry in Europe, one of the main priorities of the EC's 7th Research Framework Programme. In particular, to push forward the research, education and innovation agenda that underpins the creation of new knowledge-intensive manufacturing platforms for products based on emerging micro and nano technologies that demand an intelligent mix of materials. The joint activities of the 4M technology and application clusters led to many long-term collaborative programmes, starting with the FP6 R&D project, "Batch Integration of High-quality Materials to Microsystems" (Q2M), and culminating with the latest three FP7 large scale industry-led projects, "Converging technologies for micro systems manufacturing" (COTECH), "Rolled multi material layered 3D shaping technology" (MULTILAYER) and "Development of a highly flexible process for the patterning of large area complex microstructures" (FLEXPAET). All these projects, initiated by the 4M Knowledge Community, target the broadening of the range of MST materials and exploring the convergence of ultra-precision engineering and IC-based technologies, and through them the created 4M capabilities are on track to achieving successful industrial applications. In this context, we believe that 4M2008 will be again a "cradle" for initiating breakthrough technology and application development and an excellent forum for the cross fertilisation of industry and academia driven ideas and research themes. In addition, 4M2008 will be the first gathering of the 4M Association that aims to build upon the success of the 4M Network of Excellence and establish a self–sustaining knowledge community ready to exploit the outstanding potential of the multi-material micro and nanotechnologies for future products in the context of the European Micro and Nano Manufacturing (MINAM) Platform.

Faithful to its goal, the 4th annual conference offers four complementary activities to engage the participants in mutually beneficial discussions. In particular, the conference programme includes three plenary sessions in which six keynote speakers will provide overviews of important developments in micro and nano technologies and their applications. Six technical sessions, complemented by poster sessions, cover important research and development issues in:

- Novel materials: characterisation and processing;
- Process modelling and simulation;
- Process characterisation including process chains;
- Metrology: inspection and characterisation methods;
- Components: fabrication and assembly technologies;
- Systems: novel product and system designs.

The programme is completed by a further two sessions, one presenting the latest results of the FP6 Q2M project. The other is a session dedicated to a new FP7 project for "integrating European research infrastructures for micro-nano fabrication of functional structures and devices out of a knowledge-based multimaterials' repertoire"

(EUMINAfab). This will be followed by a panel discussion to debate industry and academic users' views, needs and benefits associated with the establishment of such a distributed infrastructure for trans-national access and practical assistance in research and development of new products and their underpinning manufacturing platforms.

As were the previous conferences, 4M2008 is an important spreading of excellence activity for the 4M Network. However it is also the first major event organised by the newly formed 4M Association and the EUMINAfab project, bodies that will support the 4M Knowledge Community in the future. We hope that its success will be a testament to the hard work of the Programme Committee, all the staff of the 4M Network Office, and especially Marika Takala, the Network Liaison Officer, and Jeanette Whyte, the Network Secretary. We would also like to thank the Local Organising Committee for their valuable advice and timely assistance in making the conference possible. Last, but not the least, we would like to acknowledge the support of all researchers in the Network without whom it would have been impossible to implement a policy of full peer review for all papers before acceptance for presentation at the conference.

Finally, and most importantly, we would like to thank the authors of all the papers, theme chairs, reviewers, speakers and all attendees for participating in the conference. Once more, we wish you all a very enjoyable and stimulating conference.

4M2008 Conference Committee

4M2008 Programme Committee

S. Dimov, Cardiff University, UK (Conference Chair)

W. Menz, Germany (Conference Co-Chair)

L. Mattsson, KTH, Sweden

E. Jung, The Fraunhofer Institute for Reliability and Microintegration, Berlin, Germany

U. Engel, University of Erlangen-Nuremberg, Germany

P. Johander, IVF Industrial Research and Development Corporation, Sweden

C. Wenzel, Fraunhofer Institute of Production Technology, Aachen, Germany

M. Richter, The Fraunhofer Institute for Reliability and Microintegration, Munich, Germany

P. Kirby, University of Cranfield, UK

A. Schoth, IMTEK, University of Freiburg, Germany

From left to right: S. Dimov and W. Menz

4M2008 Scientific Committee

B. O'Neill, University of Cambridge, UK

K. Boehringer, University of Washington, USA

M. Vellekoop, Vienna University of Technology, Austria

P. Tang, IPU Manufacturing, Denmark

R. Leach, Engineering Measurement Industry & Innovation Division, National Physical Laboratory, UK

T. Wirth, Cardiff University, UK

4M2008 Industrial Advisory Board

C. Hanisch, Festo, Germany

D. Ulieru, Romes - SA, Romania

F. Bartels, Bartels Microtechnologies, Germany

J. Anjeby, Saab Ericsson Space AB, Sweden

M. Knowles, OxfordLasers, UK

M. Ganz, Martin Ganz, Austria

R. Wimberger-Friedl, Phillips Research Laboratories, The Netherlands

U. Ljungblad, ARCAM, Sweden

Sponsors

Llywodraeth Cynulliad Cymru
Welsh Assembly Government

Welsh Assembly Government

SARIX SA

TECAN

Nanotechnology Knowledge
Transfer Network (KTN)

OxfordLasers

MM Micro Manufacturing

EU Sixth Framework Programme

4M Network of Excellence

Cardiff University

MEC, The Manufacturing
Engineering Centre

IMRC, Innovative Manufacturing
Research Centres

MicroBridge: micro and nano engineering
and fabrication in non-silicon materials

Minam, Micro-and NanoManufacturing

IET, The Institution of
Engineering and Technology

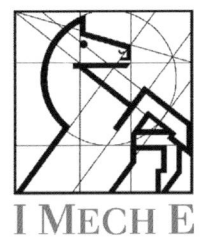

Imeche, Institution of
Mechanical Engineers

I*PROMS, Innovative Production Machines
and Systems

Minos Euronet, Micro-Nanosystems
European Network

Contents

Components: fabrication and assembly technologies

Process characterisation including process chains

Systems: novel product and system designs

4M Network of Excellence, Progress Report 2006–2008

S.S. Dimov[a], C.W. Matthews[a], E.Brousseau[a], S.Bigot[a], A. de Grave[b], H.N.Hansen[b], G. Bissacco[b], G. Tosello[b], B. Fillon[c], P. Bolt[d,] U. Engel[e], P. Johander[f], E. Jung[g], P.B. Kirby[h], L. Mattsson[j], M. Richter[k], H.-J. Ritzhaupt-Kleissl[l], A. Schoth[m], C. Wenzel[n], J. Matovic[o].

[a] *Manufacturing Engineering Centre, Cardiff University, Cardiff CF24 3AA, UK*
[b] *Department of Mechanical Engineering, Technical University of Denmark , Denmark*
[c] *French Atomic Energy Commission, (CEA), Laboratory of Innovation for New Energy Technologies and Nanomaterials (LITEN), 38054 Grenoble, France*
[d] *TNO Science and Industry, 5600 HE Eindhoven, The Netherlands*
[e] *Chair of Manufacturing Technology, University of Erlangen-Nuremberg, D-91058, Erlangen, Germany*
[f] *SWEREA-IVF Industrial Research and Development Corporation, Argongatan 30, S431 53 Molndal, Sweden*
[g] *Fraunhofer Institute for Reliability and Microintegration, IZM, Berlin Division, Germany*
[h] *Nanotechnology Group, School of Industrial and Manufacturing Science, Cranfield University, UK*
[j] *Department of Production Engineering, Royal Institute of Technology, KTH, SE-10044 Stockholm, Sweden*
[k] *Fraunhofer Institute for Reliability and Microintegration, IZM, Munich Division, Germany*
[l] *Forschungszentrum Karlsruhe, Institute for Materials Research III, 76021 Karlsruhe, Germany*
[m] *IMTEK, University of Freiburg, Georges-Koehler-Allee 103, EG-79110, Freiburg, Germany*
[n] *Fraunhofer Institute for Production Technology, IPT, Steinbachstrasse 17, 52074 Aachen, Germany*
[o] *ISAS, Technical University, A1040, Vienna, Austra.*

Abstract

This report follows on from last year's "Progress Report 2004-2006" [1] and gives an update on the continuing activities, such as the 4M Network cross-divisional projects and annual conference, as well as a description of the new activities in its third and forth year, such as the first 4M Summer School and Book Series . Finally, as the end of the funded lifetime of the network approaches the steps being taken to set up a 4M Association, which aims to create the organisational infrastructure to support the 4M Knowledge Community established in the last five years, are described.

Keywords: Micro manufacture, Micro and nano technologies, Network of Excellence, Roadmapping, Summer School

1 Introduction

The 4M Network of Excellence has created a Knowledge Community in Multi-Material Micro Manufacture, comprising over 150 researchers from 30 partner institutions in 15 European countries. Funded by the EC, the Network has brought together expertise in Micro- and Nano- Technology (MNT) for the batch-manufacture of microcomponents and devices in a variety of materials for future microsystems products. The Network acts as a knowledge resource to both the research community and industry in the development of microsystems devices that provide increased functionality in tiny packages, integrating micro and nano scale features and properties into products and systems.

1.1 4M Network scope and overall objectives

Microsystems-based products will be an important part of Europe's industrial and manufacturing future, exploiting the possibilities for functionality, mobility and intelligence in devices that a new generation of materials and components will offer. To benefit from this opportunity, these potential multi-function, multi-material products must advance from the laboratory test bed to low cost volume manufacture. To do so a convergence of technologies must be achieved.

Microsystems Technology (MST) for the first decade of its development dwelt mainly in the neighbourhood of semiconductor technology. Within the last few years MST has started to explore the potential of precision engineering technologies such as milling, drilling and turning, to overcome some of the

limitations of IC based processes. Converging these two strands of technology will require the development of new hybrid methods of manufacture encompassing the suitable elements from both, creating new processing technologies that are capable of manufacturing the miniaturised products of the future.

To address these challenges, and at the same time to capitalise on the opportunities that they represent for European industry, the 4M Network was established in 2004, with the aim of working towards such a technology convergence by the horizontal and vertical integration of processes. Benefiting from the technological and geographical spread of expertise that the Network brought together, it was able to formulate a common set of requirements from different application areas in the form of design/functional features. These have then been compared against the features that can be reliably produced by known and developing technologies. This has enabled discrepancies between the technological requirements of multi-material MST-based products and current and future manufacturing capabilities to be identified.

This 4M approach to achieving integration is illustrated in Figure 1.

1.2 Objectives for years 3 and 4

Year 1 established the operating structures of the network and saw the first conference held. It was also the period when relationships were founded. Year 2 built upon this initial phase, cross divisional projects were introduced (Section 4) and the first phase of the 4M Roadmapping exercise was carried out (Section 7) all of which was described in last year's "Progress Report 2004-2006" [1].

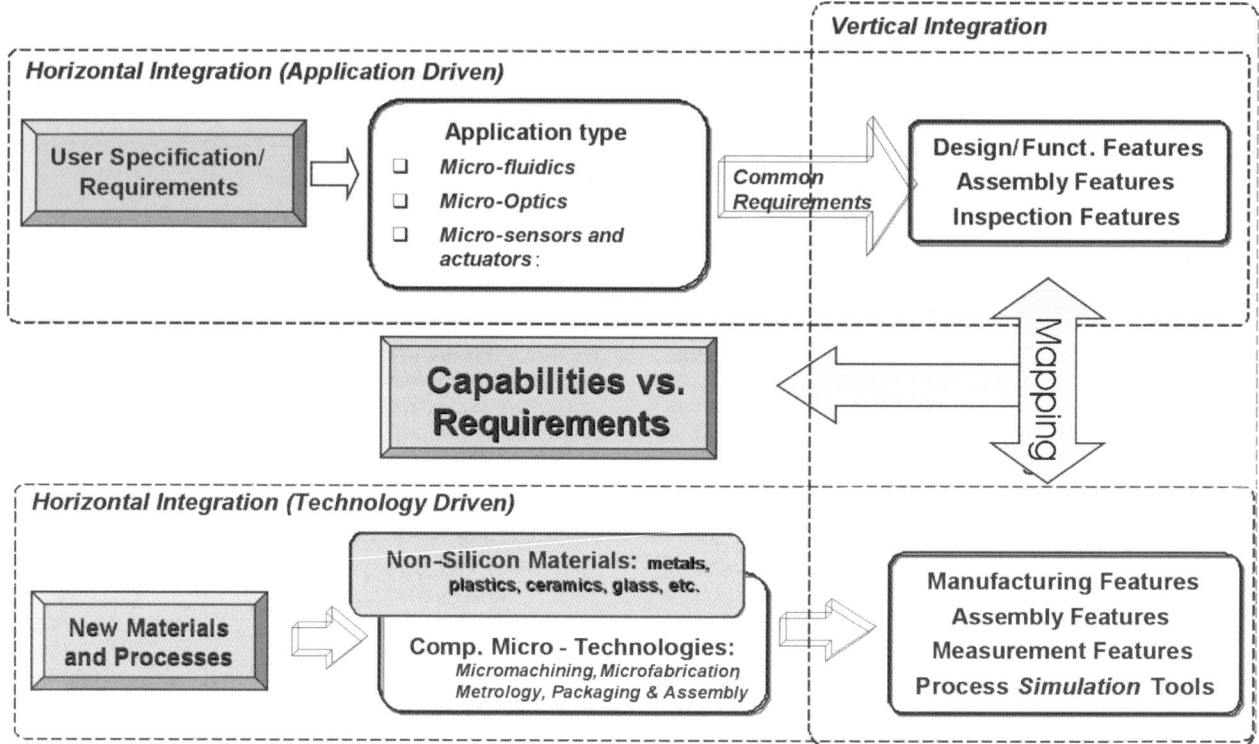

Fig. 1 An approach for achieving horizontal and vertical integration of 4M technologies

In Years 3 and 4, whilst continuing the established activities of the network such as the annual conference (Section 6), special attention has been paid to building a sustainable Knowledge Community in multi-material micro manufacture of 4M researchers and affiliates, one that will continue after the funded period of the Network. In connection with this much effort has been invested this year in consultation with the partners to formulate a mutually acceptable Business Plan, one that will ensure the Knowledge Community continues to provide a focal point and knowledge resource to industry and the wider community long after the funded life of the project. The result of this exercise, described in Section 10, will be the establishment of a 4M Association.

1.3 Main achievements to date

After over 3 years of operation the network has established an active community of researchers and industrialists with a common interest in multi-material micro manufacture. The main achievements of the 4M Network to date are:

- 8 active specialist divisions established
- producing over 50 deliverables
- more than 200 joint publications
- collaborating on 19 cross-divisional projects
- with 119 partner exchanges
- Over 220 research and advisory service (RAS) projects to industry
- 44 workshops with over 2000 attendees
- Leadership of 2 MINAM expert groups
- 4 annual 4M conference and proceedings
- 4M Roadmapping Study 2006
- 4 4M Special issues
- 4M Book series
- 4M website with access to the

- 4M knowledge repository
- 2 Summer Schools

2 4M and MINAM: setting up the research agenda

A new Micro- and NanoManufacturing community is emerging at European level, involving the collaboration of manufacturers of micro- and/or nano-inside-products, equipment suppliers, research organisations and networks such as 4M. From this community a new industrially driven platform has been established called Micro- and NanoManufacturing (MINAM) which is closely associated with the existing European Platform Manu*future*.

One task of MINAM is to identify emerging trends and provide strategic direction for future investment in research and development aimed at sustaining and further enhancing the leading position of European industry in micro and nano manufacturing technologies. Coordinated through the µSAPIENT and IPMMAN projects, MINAM is structured in two organisational groups: an Industrial Management Group (IMG) and an Operational Support Group (OSG). Working together with inputs from industry and other projects such as 4M, they have developed the Vision and Strategic Research Agenda (SRA) [2] for the coming years, a summary of which is given below.

In addition to developing the SRA, early results from the platform were provided as input to the 7th Framework Programme, and more specifically to the NMP programme.

To support this, expert groups were formed, which contribute in their particular field to establishing and continuously updating the MINAM Strategic Research Agenda, including gathering input from industry and, if feasible, indicating the link between products and technologies.

Two of these expert groups are led by members of the 4M Network, the expert group on Manufacturing of Micro Components and the expert group on Nanosurfaces.

2.1 Nanosurfaces expert group

The MINAM nanosurface expert group, led by Bertrand Fillon of CEA, focused on a review of the new trends in surface engineering with an analysis considering some of the key innovations that have emerged during the last decade, both in scientific and technical developments.

Nanosurfaces are structures containing at least one dimensional feature smaller than 100 nm. The manufacturing of nanosurfaces is relevant for both surface functionalisation (nanolayered thin films) and surface structuring (topographical nanofeatures, nanoclustered coatings). During the first year, the roadmap "NanoManufacturing" highlighted the developments in the manufacturing of nanosurfaces in the coming years. A questionnaire was used to identify the opportunities in nanosurface creation. The questions were aimed at gaining information about one of nanosurface's features which might be topographical, thin-film, modified surface areas, or can be a coating (up to the mm size) having phase modulations or crystal sizes in the mentioned range. Such features are created on the surfaces of several solid materials, e.g. metals, ceramics, glasses, semiconductors, polymers. Three main families of nanosurfaces have been identified (nanolayering, nanotexturing and nanoclustering)

This attempt to rank the technologies in nanolayering, nanotexturing and nanoclustering does not constitute a pure scientific approach but has the advantage of connecting more directly surface material structures to manufacturing processes in the perspective of introducing these technologies within the industry.

Some examples of original surface functionalities have already been obtained at labscale (Figure 2) and show that technology coupling offers a significant enlargement of the application fields together with actual breakthroughs.

Fig. 2 SEM image of a teflon-like film including nanosized, micron-long ribbons of crystalline Teflon produced in a single step plasma deposition from C2F4 under power modulation conditions. The film exhibits spectacular super-hydrophobicity (water contact angle of about 170°) – Plasma Processes and Polymers, Wiley-VCH (By courtesy of Bari University)

2.2 MIcrocomponents expert group

MINAM expert group on manufacturing of micro components focuses on components with either overall dimensions and functionalities in the micrometre range or micro structured features (from hundreds of micrometers down to hundreds of nanometers). This is not restricted to shaping of single parts but includes also the integration of functionalities such as interconnect and interfaces in components.

In its first year, this expert group, led by Pieter Bolt of TNO, contributed to the first version of the MINAM SRA by giving input on state of the art and trends, with cooperation of 4M experts in areas such as micro milling and EDM, micro injection moulding and embossing of polymers, micro forming of metals and processing of ceramics. It is also engaged in the formulation of the MINAM roadmap questionnaire, which will be used to update the MINAM SRA.

Manufacturing of microcomponents involves scaled-down ('mechanical') precision working technologies such as micro milling, EDM or laser ablation, or micro injection moulding or embossing for shaping of the components in the micro or sub-micron accuracy range. Next steps involve the application of surface technologies for finishing, such as nanoprecise optical structures and coatings, and assembly steps.

Some developments in micro manufacturing are:

- Higher accuracies and more entwinement of manufacturing processes and material micro or even atomic structure, put new demands on simulation techniques and process control.
- Handling and logistics are improved by faster and more accurate pick-and-place systems, trying to bridge the gap between nanometer accurate slow systems (e.g. atom force based systems) and micron range fast systems (from electronics industry).
- Shaping and finishing processes are integrated (converging technologies), especially for integration of functions such as interconnect and interfacing in one component. This involves new micro-factory concepts (such as poly-fabs based on micro-injection cells), lamination concepts (which suit very well reel-to-reel or on-wafer production), or lithographic or layer built-up methods (printing, jetting) – or combinations of these concepts.
- The enduring challenge is the shaping of 3D components or features as well as selection and development of processes that meet both functional and economical demands. In high-volume production, mould (mask) based replication often remains the most suitable (Figure 3), while in small-volume production environment, processes without specific tooling (masks) are often more favourable. But with new or further process and equipment development, application domains shift.
- Smart microsystems have integrated (on board) functionalities such as power, logics, sensing, communication, which classically would need their own components and packaging. The subsequent costs and size bottlenecks urge their integration in one package. E.g. stacked IC and memory dies without individual packages or embedding bare dies in polymer sensor devices with integrated connect.
- The further entwinement of (multi-)material properties and processes, of processes with different length-scales, the need of in-line inspection in production processes, the merging of precision engineering and IC/MEMS technologies, they all lead to a need for a new "design for manufacture knowledge base" that facilitates a concurrent product and process design and will reduce the product development cycle.

The further entwinement of (multi-)material properties and processes, of inspection and production, of integration, of precision engineering and IC/MEMS technologies requires the establishment of a new "design for manufacture knowledge base" that facilitates a concurrent product and process design that will reduce the product development cycle.

An important aim of the working group is to broaden the inputs to the SRA to include that from industry experts with those from academia and research institutes. Hence the expert group will conduct real live or on-line industrial interviews and organise industrial workshops, among others in cooperation with the 4M divisions, to gather views and needs for the SRA.

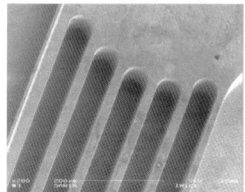

Fig. 3 SEM picture of a detail of a micro structured mould insert (source IMTEK)

2.3 The MINAM SRA – a summary

First presented in January 2008 [2], the MINAM SRA addresses several key areas: manufacturing of nanomaterials, processing of nanosurfaces, micromanufacturing processes and integrated systems and platforms for micro- and nano-manufacturing. The outlined trends and research and development priorities are illustrated with examples from key industrial sectors in Europe.

A key objective of the SRA is to identify emerging trends and provide strategic directions for future investment in research and development. In particular, the SRA outlines key challenges and research priorities with the objective of accel-erating the development of new micro- and nanomanufacturing technologies and their rapid transformation from laboratory based prototypes into volume manufacturing applications.

The SRA concludes that micro- and nanomanufacturing is a highly resource and knowledge intensive sector and that capitalising on the latest technological developments can only be achieved by a concerted coordinated effort of industrial stakeholders, research and academic organisations and public bodies.

2.4 The MINAM Roadmapping Exercise

The development of the MINAM SRA was, and will con-tinue to be, informed by ongoing coordinated MINAM roadmap-ping activities. These are aimed at providing a holistic overview, linking together the major driving factors such as applications, market requirements and technological capabilities of a highly interdisciplinary field.

The outcome of this process leads to a holistic view based on different roadmaps, 4M's for example (see section 7), making the partial results comparable. For this a novel "meta"-methodology has been discussed and agreed upon to allow the combination of results from market and application driven information while focussing on technological aspects. Some of the current results are shown in Figure 4.

The course of MINAM roadmapping activities facilitates a more precise description of the technology-application push-pull link and facilitates the development of a common understanding of needs and barriers faced by the MINAM community. It also strengthens the integration of the different players (end-users, technologists and equipment providers) in the MINAM process.

With a view to creating excellence and enhancing indus-trial competitiveness in all areas which are of crucial relevance for the achievement of the European micro- and nanomanu-facturing vision, a number of key areas and topics have been identified and defined in a global structure for MNMT. These are sectors such as consumer electronics, automotive, healthcare and defence industries. Some of the major challenges facing the European precision manufacturing companies today are increasing demand for a wider variety of microproducts and an increasingly global and distributed supply chain. Microproducts are typically of a high complexity with shorter life cycles, whilst the supply chains and value networks are setting ever more stringent demands.

MINAM roadmapping will continuously analyse in detail the requirements of both, customers and technology providers, and report emerging trends to the MINAM community.

3 The 4M Divisions

Since its launch in October 2004 the Network has, through its joint programme of activities, encouraged Partners to work collaboratively, taking advantage of the range expertise and equipment available across the Network. Partners have integrated their research into eight 4M Divisions. Five are Technology Divisions, three based on materials processing: Polymers, Metals and Ceramics, and two based on process technologies: Metrology and Assembly & Packaging. These five are complemented by three Applications Divisions: Micro-optics, Micro-fluidics and Micro-sensors & Actuators. Each Division, with its own appointed leader and management board, have covered specific topics of research and development, and have

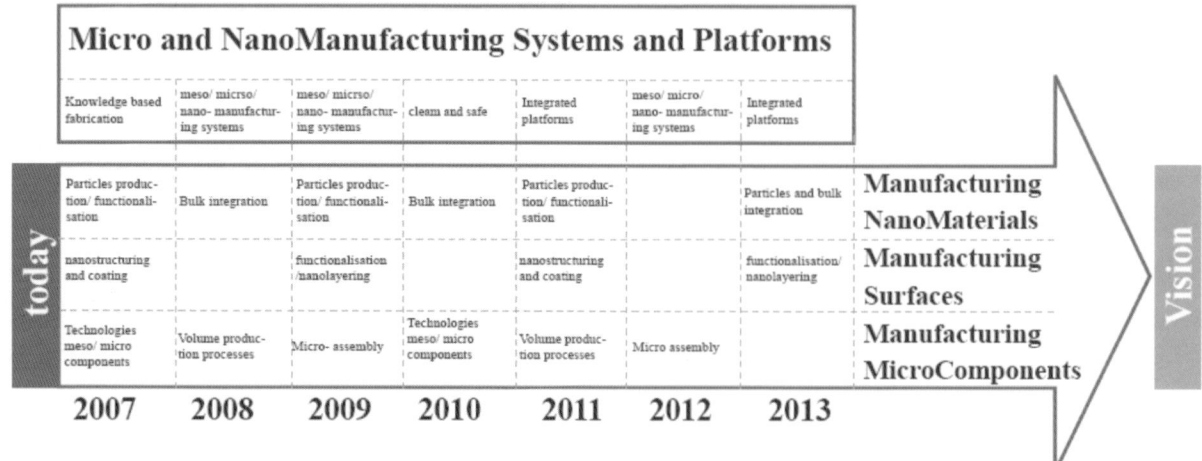

Fig. 4 Roadmapping for the European Micro and Nanomanufacturing vision (from MINAM SRA)

been free to choose their own method of operation, organising their own workshops, training and initiating partner exchanges. The deliverables produced by the Divisions were described as being "of a high quality" at the last project review.

This section gives an overview of the activity in each Division in year 3.

3.1 Polymer Processing Division

The third year saw a big increase in all aspects of the polymer division's activities, not only the interaction between the research partners (through DoE, round robin tests, research exchanges) but also with a big involvement of industrial partners. In fact twelve industrial partners from different European countries were involved directly in the Polymer's 2007 activities. Companies Sarix (CH), UBS (Fr) and Microsystems (UK) have supported with expertise in industrial tooling. Ticona (De), Basell (It), have supplied a few hundred of kilograms of polymers (COC, LCP, PP, PP with nanofillers) free of charge. Battenfeld (Aut), Arburg (De), Jenoptik (De) provided support for the injection moulding and hot embossing machines. Moldflow (Aut) provided support for the simulation of the injection and also for the manufacturing based on design of experiment. Zygo (USA) and Alicona (Aut) provided support for the polymer part and mould insert quality control (aspect ratio up to 100).

As mentioned above, during the third year round robin testing and experiment design were the main activities regardless of task. More than 5000 polymer components were produced to propose different types of demonstrators. These demonstrators were produced through mould insert fabrication, injection moulding and hot embossing. These demonstrators have also been used for different metrology measurements. This offered the identification of the best practise in the polymer process chain. Melt temperature and mould temperature are the main parameters to produce perfect polymer components. A big improvement in hot embossing was achieved – 10 seconds of holding time is now enough to produce good quality plastic parts.

Several man months of personal and equipment exchange took place during the tests. Seven workshops were organised with roughly 500 attendees. More than 40 researchers were trained within the division during the last period.

All these industry and R&D activities highlighted the biggest industrial requirement to offer the new generation of polymer replication processes. In fact assembling is killing the global polymer μ-replication process chains. The answer to this problem could be initiated by the design and development of converging technologies. Such convergence of technologies will accelerate the production and assembly of multi-component devices in order to obtain an important reduction of needed supply chain space and manufacturing costs of next generation polymer based multi-material products. To address this issue, a proposal "COTECH" has been submitted in the seven framework.

The virtual institute will be more and more a real technical sub-platform for European exchange on polymer processing offering the best practice in Micro/Nano-polymer processing technologies to MINAM and MANUFUTURE communities.

This polymer division 2007 success has been highlighted by an invitation from Battenfeld to share their booth at the biggest polymer exhibition in the world K2007 in Dusseldorf (October 2007) free of charge for 4M. 4M flyers and booklets were distributed and exchanged with polymer industrialists. This offered the opportunity to promote, extend and benchmark the 4M expertise all around the world.

3.2 Metrology Division

Since restructuring the way it worked in Year 2, this Division has focussed on collaboration with other Divisions and industry to achieve its goals of establishing a European Metrology Centre and, in particular, investigating the critical area of High Aspect Ratio metrology for micro-components.

The European Metrology Centre is being established by carrying out metrology audits in the partners' laboratories with the purpose of building up a database of the equipment suitable for micro metrology, which will also be a valuable source of input to standardization. Five new laboratories and their equipment have been added to the European Metrology Centre.

There has been continued collaboration between Metrology and Polymers in its Round Robin investigation. Detailed instructions on how to measure surface roughness and dimensions made several of the labs confident that the measurement procedures were good. However, a few labs deviated by far too much from expected results and will be subject to special attention in the final year. In the process of evaluating the high aspect ratio performance of optical profilers we found severe lateral measurement errors among very well known commercial instruments. These results were presented to the instrument manufacturers and after a visit and joint work with one manufacturer we were not only able to reduce the errors but also to improve the performance of the profiler to a world record level of 50:1 in high aspect ratio micro structure measurement. This is an outstanding achievement that all 4M and MEMS manufacturers will benefit from.

3.3 Assembly & Packaging Division

The main focus of this Division is placed on polymeric Microsystems with hybrid integration aspects. Topics were identified (in conjunction with the roadmapping) and are strongly in oriented towards a conclusive process flow and on final reliability over lifetime.

Outside these focal aspects, efforts were shifted into the cross divisional projects (CDP). There is a strong involvement in CDPs with a total of 9 CDPs benefitting from the input of members of the Assembly & Packaging Division with specific tasks drawn from the Division's expertise.

3.4 Metals Division

The Metals Division has harnessed the collective expertise available in the 4M network in the field of micro metal processing to ensure that the advantageous physical, mechanical and chemical properties of metals are exploited to the full.

To do this, new processes, process routes and process combinations have been developed that will meet the ever increasing technical, economic and ecological demands.

Furthermore the Division has considered the new materials available in the field of micro metal processing, for example, nano crystalline materials or shape memory alloys. These new materials also require characterisation e.g. hardness, strength, electrical conductivity. Hence there is a need to develop and improve the processing of metals such that parts and products with smaller feature sizes and higher surface quality can be manufactured with higher and reproducible accuracy (Figure 5).

In the first year a comprehensive survey of partners' capabilities was carried out. In the second year the focus was on two feasibility studies which led to the design and manufacture of two demonstrators.

The results of these studies clearly demonstrated the capabilities of the partners and built cooperation between them.

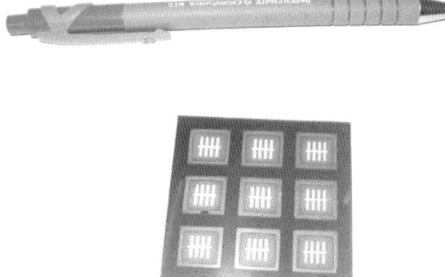

Fig. 5 Solar cell of nine "sub cells" deposited on a stainless steel substrate

In Year 3 a lot of effort has been put into the state of the art report of small quantity and serial micro-manufacturing technology. Based on the already submitted classification report, the description of each technology mentioned there was carefully reviewed and updated information included where necessary. A description of new technologies has also been added to complete the report.

A European workshop on Microforming was also prepared and delivered in Erlangen at the end of November 2007. Several partners from 4M and a large number of industrial participants presented their current activities in Microforming.

Together with partners from the metals division and industry, the kick off meeting of a EUREKA initiative called "Micrometals" (under the FACTORY Umbrella) took place in Copenhagen. Future collaboration was discussed. As a result of this initiative, an FP7 project proposal was prepared and submitted.

In order to create a suitable database and to make the information, collected via the reports on the state of the art, available to a broad range of interested people, several possibilities were investigated and analysed for realisability. Finally the decision was taken to closely link the database to the planned European Information System on Micro Metals Processing which will be finalised in Year 4.

3.5 Ceramics Division

Since its inception the Ceramics Division has worked to find solutions to the problems that hinder the successful application of ceramics such as the limitations of existing manufacturing platforms and the lack of "design for manufacture rules" for ceramic micro components.

Early years saw the Division focus on two tasks and accompanying demonstrators to address these issues – Hybrid Manufacturing Techniques and Integrated Design & Manufacture.

This year work on two further demonstrators has been performed within two cross-divisional projects. The first demonstrator is an inertia sensor, the other a microwave component. Work on the inertia sensor has been completed, and the results published at Eurosensor2006; activities on a MW-component have been performed and the possibilities of various manufacturing processes evaluated.

This Division also has a natural affinity with the Micro-sensors & Actuators Division, ceramics being an important material in many sensors. Consequently they have held a number of bi-lateral meetings and workshops.

Three workshops were held; one in Grenoble on September 19, 2006 on Direct Printing, one together with the sensors &actuators division at IMTEK in Freiburg on November 11, 2006 and another industrial workshop at the Hannover fair on April 18, 2007. At these workshops about 20 – 40 attendees from research institutions as well as from industry could be welcomed. Perhaps most importantly, from these activities a Ceramics and Sensors & Actuators industrial interest group was established.

In the Ceramics Division all the information gathered and the knowledge generated by the above activities is being gathered together into a single document. There is already an agreement with a publisher signed and in place and the book, "Ceramics processing in Micro Technology" should printed in the Autumn of 2008 by Whittles Publishing (See section 8.1). This will be a major achievement for the Division and the book will represent an invaluable and enduring resource for anyone working within the field, consisting of contributions from 28 authors, presenting an overview of the current status and perspectives of the processing of ceramic materials in micro technology.

3.6 Micro-Optics Division

During the first year, a classification of micro optical components was achieved with the identification of corresponding manufacturing technologies and their interaction within entire process chains. Technical drawbacks of the technologies themselves as well as process chains, including process-material interactions, were thoroughly analysed and feasible solutions for long lasting improvements were made during the second year.

In year 2 the main focus was put on the transfer of the identified aspects into industry and also to gather feedback and additional matters from the site of production. Workshops were organised, held and their outcomes summarised. To support European micro optics development, road mapping activities were conducted to give an insight on reachable innovations and necessary research on the base of a detailed time line.

During the third year, the previous analysis has been verified by machining real micro optical components using state-of-the-art micro manufacturing technologies such as ultraprecision machining and lithography. By means of these demonstrator parts the entire process chain in manufacturing micro optical plastic components has been reproduced and analysed.

Support for the European micro optics industry has been advanced by the setting up an internet portal to inform and exchange know-how about state of the art and future developments in micro optics manufacturing. This can be viewed at :
www.Opticsmanufacturing.net

3.7 Micro-Fluidics Division

This year the Division dealt with technological gaps and challenges in a particular microfluidic application (labs-on-a-chip for point of care testing (POCT)) as well as with the problems and challenges typical for a variety of microfluidic applications.

The people involved in the Division have got to know each other well as a result of the 4M activities during the first three years. This created a very good working atmosphere based on trust and the scientific cooperation and integration has consequently been very smooth and successful.

These interactions and networking have produced joint publications and two industrial workshops. Cross divisional projects have been started in order to identify technology gaps,

A report on the concept of "microfluidics on foil" (Figure 6) as well as a report on "New Solutions for Nanofluidic Manufacturing Operations and Interfaces". Finally a Roadmap for closing a gap in the field of POCT has been established. The gap this addressed is the need for a continuous process chain for the replication of microstructured parts, which allows for a high throughput at low production costs.

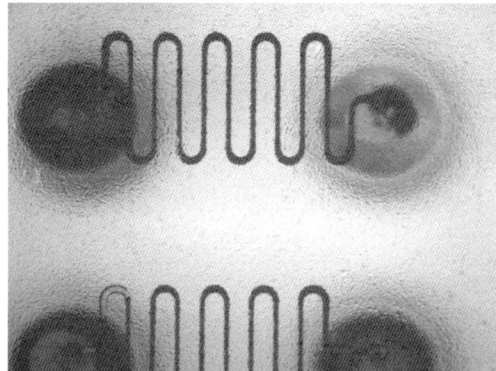

Fig. 6 A microfluidic system on foil. Channels are filled with black ink

One aspect which has been identified for further improvement is the involvement of European Industry in the microfluidic division. Here, the microfluidic division plans to carry out not only roadmaps but also market studies of promising microfluidic market segments. More industrial workshops are also to be established to attract industry, especially local SMEs.

3.8 Micro-Sensors & Actuators Division

As discussed in the second year periodic report, the activities in the m-S&A division required adjustment to address the difficulty of covering the wide ranging nature of the technologies involved in sensors and actuators and the breadth of the application fields for sensors and actuators, taking into account the technologies available within 4M. There was also the issue of the 4M fabrication technology being in other divisions and so the need for mechanisms and structures that could bring the fabrication and sensors and actuator experts together.

Cross-divisional projects were shown in year two to be an excellent vehicle for doing this and this has continued to be the case in year 3, with the study of 'Nanomembranes' which is leading to some blue-sky thinking about potentially important technologies.

This theme of looking to future technologies has been incorporated as one of the main activities in the mS&A cluster with the emphasis moving from the road-mapping needed improvements in well understood sensing and actuating components to the review of emerging micro-sensing and actuators. In detail this was achieved for example by shifting from the review of low cost manufacturing of sensing and actuation devices to functional microsystem components that could be used in new environments. Tasks on Implantable sensors and actuators, and about micro-sensing and actuation technologies for cell transport and measurement systems were introduced and are examples of this recent development. This emphasis on emerging sensors and actuator technologies is proving successful and leading to increased partner participation with improved output. The basic reason for this is probably the improved fit with research projects being undertaken across Europe.

Encouragingly a workshop held on implantable sensors & actuators in this reporting year attracted very high industrial participation and shows that it is possible to combine adventurous research with the interests of future industries. It is hoped that this success will be repeated with the task about micro-sensing and actuation technologies for cell transport and measurement systems and early indications are that industries (especially SMEs) wish to be involved in developments and are willing to map out the future technology needs for cell transport and measurement systems – an important area for pharmaceutical companies.

To further address the diversity of micro-sensing and actuator applications and to make activities relevant to 4M technology, joint activities have been undertaken with the ceramics division. Two workshops were held, one at IMTEK in Freiburg in November 2006, the other, an industrial workshop, at the Hannover Fair in April 2007. As a result several projects were established as a means of assessing promising new technologies. Both the ceramic and mS&A divisions feel that this is an important way of drawing attention to their activities by being able to address future technology needs from a broader base of expertise and capability. It is hoped to be able to continue this association beyond the term of 4M.

4 The 4M Cross-Divisional Projects

In addition to the work of the Divisions, and in order to encourage further collaboration and integration that ensured common research topics were addressed across the divisions, a series of competitive internal calls for cross-divisional project proposals have been held. In the first three years of the network competitive internal calls led to 19 projects being funded and completed and are considered to have been a great success.

The work carried out in both the Divisions and in the cross-divisional projects was recently described by the project reviewer as "of high quality and relevance to industry".

In the first two years of the network 11 projects were funded and completed. In year three a fourth competitive call was held, from which eight proposals were selected for funding and these are described here.

4.1 4M Micro-Wave

The size of the wireless communications market has grown continuously over the last 15 years in the area of mobile and wireless systems including handsets, base stations, RF tagging and wireless interconnections (W-LAN etc.). The system applications include automotive, radar, traffic management, space and satellite communications and aeronautics. At higher terahertz frequencies developments are taking place in security scanning applications and for environmental monitoring. This wireless pull gives a strong demand for high performance, low cost, passive components such as microwave switches, cavity resonators, and metal inductors which is being addressed through the development of RF-MEMS techniques. However, in many cases the required device geometries and materials can be addressed with 4M fabrication techniques creating a need to assess their applicability and prospects.

3-D micro components for microwave applications are a scarcely developed application area for 4M technologies. Microwave engineers generally adopt planar solutions in 2½ D, manufactured with traditional printed circuit board technology. Some 3D microwave components have been manufactured in silicon with MEMS techniques. At terahertz frequencies the exclusion of use of planar circuits due to excessive electrical losses makes the use of precision waveguide structures essential and there is a need to assess 4M replication and machining techniques.

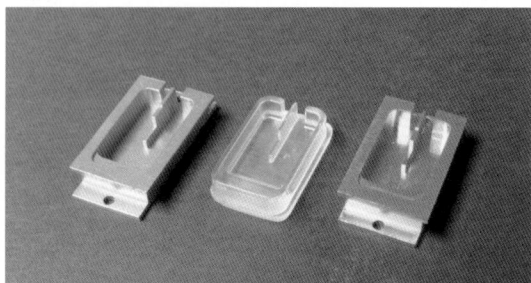

Fig. 7 Microwave components manufactured by layer manufacturing, injection moulding and milling

This project therefore aimed to evaluate multi material micro manufacturing technologies for the manufacture of 3D microwave components for a range of applications. The study was extended to the manufacture of ceramic components, such as filters and packaging structures and also considered the integration of components with antenna structures.

The cross divisional project was run in parallel with the work of a Master student, in collaboration with an industrial partner, IVF and Cranfield University. In the Master student's project a specific microwave component will be evaluated, that was designed and developed by the industrial partner.

The project experimentally evaluated four manufacturing processes: Layer manufacturing with metal, micro-stereo lithography, injection moulding with polymers and injection moulding with ceramics (Figure 7). The geometry and the microwave properties of the parts will be measured and evaluated at the test facilities of the industrial partner.

4.2 Biocompatibility Evaluation of Encapsulation Materials

Microsystems have become a major instrument in the treatment and diagnosis of patients requiring permanent medical care (Alzheimer, Paralysis, Retinal Degradation etc.). However the materials typically used in electronics are, unfortunately, incompatible with the bodily environment. Consequently protective encapsulants are used to protect both the body and the electronics from each other. However, the choice of proven encapsulants is very limited. This is due mostly to the fact that no systematic study dedicated to material selection has been done until now. The manufacturers of systems tend to cling to proven materials, as this approach does not add to their research budget. The resulting solutions are often non-optimal. This cross-divisional project targets this field by bringing together the expertise of various 4M divisions, collecting the findings in a survey to identify white spots and to lay the basis for a fast screening of a wide range of materials being used for encapsulation/protection of medical devices.

The main outputs of this Cross-Divisional Project will be:

• A roadmapping/benchmarking report on biocompatibility tests and on encapsulation materials with proven biocompatibility. This will also cover the aspect of bio-resistivity against bodily fluids.
• Identifying the lack of knowledge (e.g. comparing the modern, much better encapsulation materials used in electronics against the available data for biocompatibility)
• A proposed test setup allowing a quick screening of a multitude of materials.

• Round robin testing of two known encapsulant samples using the established test setup at the partners' sites to allow future cross correlation of other materials tested utilizing this broadened test basis for faster screening.

4.3 Future Tooling: Tooling Technology vs. Application Requirements

The accuracy, geometrical dimensions and physical characteristics of a manufactured part depend strongly on the shape (quality) of the tool. In particular at the micro scale, the comparatively small forces, small nominal dimensions and small tolerances require high-precision tools for the manufacturing of products with high quality. Thus, very precise and accurate manufacturing processes are indispensable for toolmaking at microscale and only few manufacturing processes qualify for this task at present. Each manufacturing process is characterised by the process capability and quantitatively expressed by the achieved dimensional accuracy, accuracy of the shape, roughness of the machined surface, defective layer of the machined surface etc.

If all these processes are to be considered for the purpose of microtooling, the capabilities of each individual process should satisfy the requirements of both, tooling and the replication (fabrication) technology. This means that a certain accuracy and roughness should be achieved firstly for the micro tool to work properly and secondly, to satisfy the requirements of the same criterion of the final part.

The aim of this cross divisional project is to analyse the current capabilities of the different toolmaking processes and to analyse the demands of the replication (fabrication) manufacturing processes. Based on this analysis, the knowledge gaps in micro-tool manufacturing will be identified and recommendations for future tool design and manufacturing will be given.

To reach this goal, two working teams will be established. A first team (tooling team) will analyse the current state-of-the-art in tooling. The second (product team) will characterise the requirements of the micro-replication (fabrication) processes. The results of both teams will be reviewed and recommendations for future tool developments techniques (from design to manufacture) will be given to broaden knowledge and improve micro-replication (fabrication) processes.

4.4 High Throughput Replications Technologies for Organic Electronics and Photonics

Organic LEDs are already appearing in displays for consumer electronics such as mobile phones and MP3 players. In contrast, organic electronics are still under development for applications such as flexible display backpanes, sensors, photovoltaics and RFID tags. One of the major problems of organic semicondcutors, namely the ambient stability, is being overcome and some organic semiconductors can already meet the least demanding of requirements for commercial products. In addition ambipolar, ferroelectric and light emitting OFETs have been demonstrated recently. One remaining major issue is to develop high-throughput, reliability and low-cost manufacturing techniques for producing high-speed organic electronic devices and nanostructured functional surfaces that are the key to most potential applications including RFIDs, driving circuitry of organic displays, smart cards, and biosensors.

Based on the conducted top-level 4M roadmapping study the following generic conclusions can be made: (1) there is no one technology that will prevail – the "breakthroughs" if any will come from an innovative integration of complementary technologies and their implementation in new manufacturing platforms;

(2) process chains incorporating printing, nanoimprinting and reel to reel embossing were identified as very promising for achieving function integration in new emerging products, e.g. organic electronics, photonics and biosensors; (3) there is a mismatch between the perceived future importance of these technologies and the current R&D funding for their development and integration.

In this context, the aim of this cross divisional project is to define the research agenda for developing new manufacturing platforms that combine the capabilities of printing, direct write, projection maskless lithography and nanoimprint techniques to achieve a length scale integration, e.g. micro and nano structuring, in printing/pattering large areas utilising each of these technologies in their most economical processing window. In particular, the objective of this project will be: (1) to study the current state-of-the-art in opto-electronic organic materials and technologies for their processing; (2) to propose innovative processing chains for structuring organic materials that underpin the development of innovative manufacturing platforms for a number of emerging organic electronics and photonics applications; (3) to conduct a feasibility study to demonstrate the viability of the proposed processing chains.

4.5 Micro- and Nanostructured Surfaces for the Liquid and Gas Management in Microstructured Flowfields

This project aims to determine the feasibility of a new generation of passive, capillary driven micro-fluidics. In such enhanced flow structures fluidics are not guided through channels but on surface modified paths (Lotus Microfluidics).

Like the super-hydrophobic effect of the Lotus leaf they will be based on multi scaled micro- and nanostructures in combination with adapted wetting properties of the bulk materials. The introduction of a structural gradient is expected to generate a capillary gradient that will move liquid samples along channel structures. The realization of super hydrophilic surfaces by a combination of micro patterns with hydrophilic materials is expected to increase the mobility of gas bubbles in micro fluidic systems. A systematic patterning may lead to fluidic systems where liquids are confined by potentials of effective surface energy rather than by channel walls.

To realize Lotus–like patterns on large surfaces will require the precise control of the structuring method over a wide range of magnitude. Submicron structures from e-beam writing must cover large areas and be combined to fluidic patterns. To replicate accurately such structures electroplated mould inserts with high aspect ratios must be copied from the resist structure. Finally the replication of high aspect ratio structures into polymers with a very low defect rate, to ensure the function of the Lotus effect, is a challenge.

This approach presents completely new challenges in terms of design and in the manufacture and assembly of such channels. Guidance of liquid movement without the need of confining walls will allow the parallel operation of gases and liquids in a single channel system. This system will require the assembling of structured surfaces with precise alignment, constant gap and leak-tight conditions.

The project will be based on established fuel cell designs. Current designs deal with the passive transport of droplets and bubbles of waste materials and their separation from the embedding flow in the microscale. Lotus Microfluidics should be able to substitute current approaches for the transport of liquids along channels as well as the phase separation inside the channels.

The output of the project will be twofold. Firstly an identification of the advantages of structured channel walls on fluidic flow management versus state of the art solutions based on smooth channels walls. Secondly, a feasibility study of the nanostructuration of micro-channels involving multi-scale manufacturing.

4.6 Micro-milling

Compared to other techniques like LIGA or EDM, micro milling is a cheap and fast way to create a large variety of high aspect ratio microstructures in a wide range of materials. However, below tool diameters of 500 µm the optimization of the micro milling process is often carried out by a very time consuming trial and error approach due to:

- large number of tool suppliers,
- large variety of hard metal substrates for tool manufacturing
- wide range of cutting edge geometries
- different end mill lengths for the same tool diameters
- relatively large deviations of tool diameter and edge profile from the nominal values
- huge variety of materials to be machined
- variability of the delivery conditions of work materials
- different machine equipments in the workshops

The aim of this proposal is therefore to share equipment and knowledge in micro milling for measurement of low cutting forces of small diameter end mills at high spindle speeds, when machining high aspect ratio microstructures. The starting point will be a critical review of the currently available technologies for generation of high aspect ratio micro structures, with particular reference to micro milling and the LIGA process. The comparison of the technologies in terms of freedom of the manufacturability of geometries, cost and accessibility, is expected to emphasize the leading role of micro milling among them.

The activities carried out within the cross divisional project will identify critical micro cutting parameters and lead to the establishment of an initial knowledge database which will serve as a feasibility study for a possible future and more comprehensive work. Furthermore, as a result of first measurements, correlations between cutting forces and several machining parameters like feed per tooth and infeed will be evaluated.

As a prerequisite for the cutting force measurements, micro end mills have to be characterized by diameter variation, edge geometry and length. Afterwards, by high resolution metrology the quality of the micro milled features and thereby the limitations of the machining system will be evaluated. In this way, data for development and validation of predictive analytical models will be made available and prediction of cutting forces allowing a stable micro machining process can be derived.

This cross divisional project constitutes the prerequisite for, and is complementary to, a linked cross divisional project dealing with polymer replication performance of high aspect ratio micro features, focussed on the filling of thin cavities and demoulding of thin protrusions by injection moulding.

4.7 Planar Micromanipulation

Current industrial microfabrication methods produce microparts or microcomponents in batch quantities. These parts need to be properly distinguished, sorted and aligned for the next step of the assembly and packaging of the final MEMS product (e.g. wafer bonding). In many applications an alignment accuracy of below 1 micrometer is already required. Established solutions such as part-feeders and serial/automated-serial assembly with microgrippers are not efficient enough for the constantly increasing desire for mass production of MEMS devices. The main reasons are the lack of flexibility and the time-consuming and quite frequently costly sequential operation (for example optical tweezers can manipulate and assemble around 5 microparts per hour). The need for fast mass sorting, positioning and orienting of microparts, without the usage of robotic microgrippers, micromanipulators or part-feeders, is becoming more and more evident.

This cross-divisional project seeks therefore, new, alternative suggestions for flexible, low cost microhandling of high yield.

4.8 Progress Towards 3D Structured Composite Nanomembranes

Nanomembrane functionalization is an essential step towards a fundamental extension of their applications. Nanomembrane functionalization alters the basic nanomembrane properties whilst at the same time bringing a wide spectra of brand new properties. In particular, the main objective is to modify/enhance their mechanical, electronic, chemical, biological, optical and magnetic properties.

There are three main techniques for nanomembrane functionalization. The first one is the replacement of inactive nanofiller particles with active particles. In future work the fillers could be Pt, Ag or Rh (for a catalytic action), TiO_2 (photocatalysis) and/or SiC, ZnSe (light-emitting semiconductors). The fillers can be also soft and hard magnetic particles, electrets or piezoelectric ceramics, chemically active substances, carbon nanotubes and fullerenes etc.

The second method is based on a modification of nanomembrane surfaces by micro/nano- patterning and micro corrugation. Fabrication of 2D nanohole (nanopore) arrays also belongs to nano-patterning techniques and brings itself another group of functionalities. This method will be verified in the framework of this cross-divisional project.

The third method available for functionalising nanomembranes is their lamination. There are 4 main nanomembrane classes, in particular based on polymer composites, metal composites, diamonds and semiconductor nanomembranes. The nanomembranes belonging to these four classes can be laminated to enhance their properties and thus to broaden further their application area.

In this project we propose to carry out a proof-of-concept study on nanomembrane functionalization, focusing on surface modification and lamination. Nanoengraving of membranes is a technique for fabricating templates for nanopores and nanomasks that are currently application fields of growing interest. Obviously, there is a considerable application potential for such nanosized holes or nanopores. Indeed numerous applications were recently reported aiming to use such membranes as stencils or masks to grow or depose nanostructures, to localise molecular scale electrical junctions, switches and nanotransistors or spin–injection devices. Our goal is to use such patterned membranes for molecular biophysics applications, such as optimized filters for DNA separation or electrical sensors for single biomolecule electrical detection.

5 First 4M Summer School

In the summer of 2007 the first 4M Summer School was hosted by the Department of Manufacturing Engineering and Management (IPL), at the Technical University of Denmark (DTU) in Lyngby, Denmark. The school, which was sponsored by SARIX and by ESPRIT, was considered to be a great success and is being held again in 2008. A description of the content of the two week course is given below.

5.1 Background

The 4M summer school is situated in the context of emerging micro/nano technologies. Micromechanical components play an increasing role in microsystems. The use of metals, polymers and ceramics for miniature components requires product development methods as well as manufacturing technologies. Product dimension will range from micrometre to millimetre. The aim is to give the attendees an overview of the complete product development process, from a list of requirements and technology possibilities to manufacturing, characterisation and testing. The context is also set in and toward an industrial perspective. Indeed it is now well known that micro/nanotechnology is not only a matter of downscaling applications, manufacturing processes and methods.

After the course the attendees will:

- be able to apply product development methodologies to microtechnology in general and to micromechanical products in particular
- be able to choose and apply the most relevant process chains given the requirements of the desired micromechanical system
- be able to select and use supporting technologies such as metrology, handling and assembly in a microtechnology context
- gain understanding of the complete product development, emphasising on collaboration and integration

5.2 Pedagogical Approach

The course covers various elements of product development, within the field of microtechnologies. It couples lectures and activities on micromanufacturing technologies, on products solutions principles and also on design methodology. The schedule tries to balance all aspects between pure manufacturing technologies (such as laser processing, microinjection moulding, etc.) and more product oriented approaches. The lectures are linked to industrial issues, product functionality, fitting into the whole product development scheme.

Two lectures were planned each morning and attendees were asked to actively participate in most of them. Whenever necessary, exercises and practical sessions were held. Lectures normally took place during the morning and practical work was done during afternoons. The lectures in the morning would cover the necessary knowledge needed to continue the product development of the afternoon project, together with some technical workshops for more specific and applied knowledge and practice related to equipment.

Concurrently, the attendees were asked to complete a design project running the whole course length. For this work, the students were split into groups. It started the second day by some functional and process chain design (based on a list of requirements from the teachers, it can be done in accordance with industrial needs) then moved to tooling, production of prototypes (in a "mass" production scheme) and testing of physical implementations of the design. On the last day a presentation of the project acted as a wrap up of the course and evaluation. A 4M partner was asked to act as an external censor. Furthermore, each group completed a report as documentation for their work.

Participants were also asked to prepare a short 15min presentation about their laboratory and Ph.D. topic (to be presented during the course). The presentation was based on a small paper about their Ph.D. subject that they were asked to write prior to the start of the course, to be included in the documents they will retain at the end of the course (slide prints, extra material, exercises, etc.).

5.3 Project work

5.3.1 Assignment

The students were asked to design and manufacture a microfluidic device capable of mixing two fluids and with the ability to visually/optically assess the quality of the mixing. The device should be sealed, i.e. the fluid should be confined in closed structures. Furthermore, the device should be fabricated in polymer materials preferably by mass production technologies. The students were grouped into three groups with the following specific tasks:

- Group 1: Mixing unit 1
- Group 2: Mixing unit 2
- Group 3: Lid including optical assessment

The available technologies included micromilling, micro-EDM milling, laser micromachining and welding, electroforming, microinjection moulding, hot embossing and metrology equipment. Due to available moulds, the overall dimension of the device was restricted to 20x30 mm. Furthermore, the edge of the moulded part as well as the region of the inlet had to be avoided. The groups were asked to fabricate two devices with different mixing units but fitting to the same lid. This resulted in the following common features off all three groups: inlets, outlets, visual assessment window and alignment features.

In order to accomplish this task within the time frame of two weeks, a highly collaborative design approach had to be adapted. The students were teamed up on the first day while performing the so-called Delta Design game [3] in order to get into the team roles quickly.

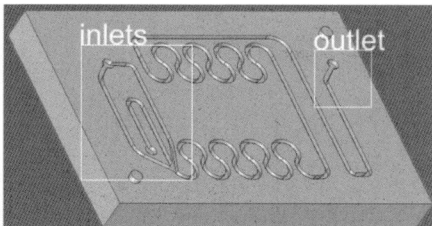

Fig. 8 Design of mixing unit 1

5.3.2. Mixing unit 1

The team related to mixing unit 1 decided to design a mixer based on the recommendations in [4]. The chosen design is shown in Figure 8 and 9. Inlet and outlet areas are also indicated. The group chose an indirect tooling approach consisting of the following steps:

- Micromilling of aluminium substrate
- Electroforming of Ni and Cu
- Fitting of overall dimensions to mould by milling
- Dissolution of aluminium
- Hot embossing

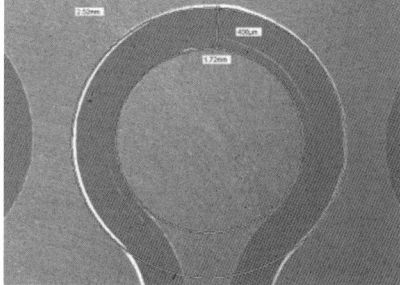

Fig. 9 Detail of mixing unit 1

5.3.3 Mixing unit 2

The team related to mixing unit 2 decided to design a mixer based on the recommendations in [5]. The design is based on microchannels with a microstructure in the bottom of the channels referred to as herringbone structure. With this structure the length of the channels can be shortened compared to the principle in mixing unit 1. The chosen design is shown in Figure 10.

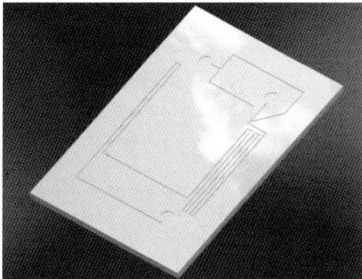

Fig. 10 Design of mixing unit 2

The group chose an indirect tooling approach consisting of the following steps:

- Micromilling of large features
- MicroEDM milling of aluminum substrate
- Electroforming of Ni and Cu
- Fitting of overall dimensions to mould by milling
- Dissolution of aluminium
- Microinjection moulding

It was a challenge to fabricate the small herringbone structures in the bottom of the channels. This was analysed to be a specific challenge related to the CAM programming of the microEDM machine.

5.3.4 Lid

The lid was designed using the same principle as reported in [6]. An indirect tooling approach was used including the integration of optical elements into a master geometry before electroforming. The lid also contained inlets and outlets and alignment features, as seen in figure 11.

The group chose the following process chain:

- Micromilling of aluminum substrate
- Integration of the optical elements (by close fitting)
- PVD of integrated master
- Electroforming of Ni and Cu
- Fitting of overall dimensions to mould by milling
- Dissolution of aluminium
- Microinjection moulding

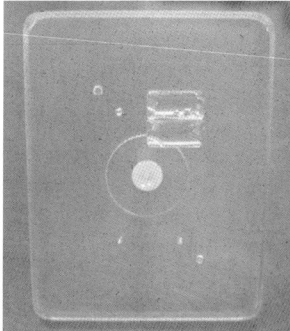

Fig. 11 Injection moulded lid with optical elements, inlets, outlets as well as alignment features

5.3.5 Joining and testing

The lid and the two mixing units were joined by means of laser welding. The final samples were tested using two fluids injected by syringes, figure 12.

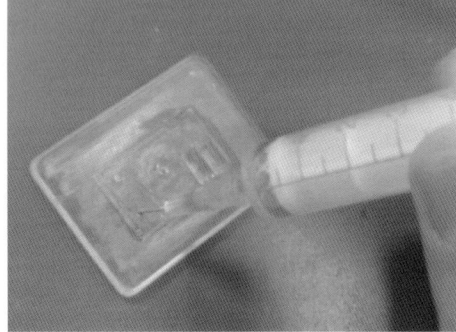

Fig. 12 Final unit during testing

5.4 Summer School Summary

The 4M PhD Summer School 2007 was attended by 16 PhD students from 6 different countries. The response from the students was very positive although the workload during the two weeks was heavy. Their feedback will be used to improve the next version of the summer school. Figure 13 illustrates the three process chains and the challenges experienced during the summer school as identified by the students. The 4M PhD Summer School will be repeated in 2008.

6 The 4M Conference and 4M Special Issues

The 4M Conference is now well established and is one of the main tools for the dissemination of the Network's research findings to the 4M community and beyond. Following on from the successful conferences in 2005 and 2006 which were held in Germany and France respectively, the third conference 4M2007 was held in Bulgaria.

6.1 4M2007

4M2007 was held in the mountain resort of Borovets in Bulgaria, and attracted some 120 delegates. Six fascinating addresses were delivered by the invited speakers to the interested and knowledgeable audience of over 120 attendees. These were complemented by the papers selected for oral presentation in the thematic sessions. A well-presented poster session gave everyone further opportunity to network and discuss each other's work. In total 84 papers were accepted for publication in the proceedings [1] which were published by Whittles Publishing in time for the conference.

6.2 Future conferences

The fourth conference is to be held in Cardiff, UK, from 9th–11th September 2008 and will be the last conference held during the funded life of the project. However there are already plans to continue the 4M series of conferences with a suitable venue for 4M2009 currently being sought.

6.3 4M Special Issues

To further promote the 4M Network of Excellence and its work, selected papers from the conferences have been submitted for Special Issues.

To date three such 4M Special Issues have been published. The first Special Issue, on Applications of Multi-Material Micro-Manufacture, featuring 10 papers from 4M2005 was published as Issue C11 in November 2006 in the Proceedings of the Institution of Mechanical Engineers, Part C: Journal of Mechanical Engineering Science Volume 220, Number 11 / 2006, pages 1609-1705.

A second Special Issue on Multi-Material Micro Manufacture (4M) was published in 2007 in the International Journal of Advanced Manufacturing Technology, featuring seventeen papers from 4M2005.

Thirteen papers from 4M2006 appeared in a third Special Issue on Multi-Material Manufacture and was published in the Proceedings of the IMechE, Part B, the Journal of Engineering

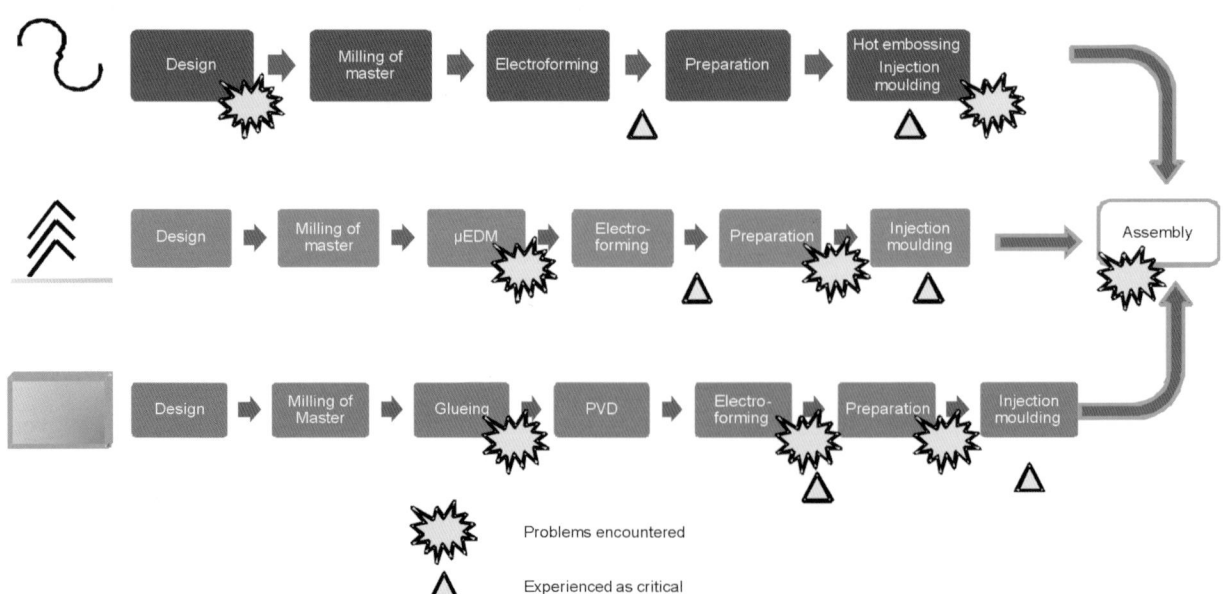

Fig. 13 Process chains and critical steps experienced during the 4M PhD Summer School 2007

Manufacture, Volume 222, Number 1 / 2008, once again enabling the work of 4M to reach a wider audience.

Agreement has now been reached with Springer for a fourth Special Issue to appear, once again in the International Journal of Advanced Manufacturing Technology. This Special Issue is now in preparation and will feature selected papers from 4M2007.

7 The 4M Roadmapping Study – an update

In 2006 the 4M Network of Excellence carried out a roadmapping study to help inform European research and industry about current trends and application requirements in the development of Micro- and Nano-manufacturing Technologies (MNT) for the batch-manufacture of micro- components and devices.

The results of the exercise were based on a roadmapping workshop attended by 30 senior researchers along with parallel questionnaires administered to 38 associated industrialists. The primary application areas addressed were micro-fluidics, micro-sensors & actuators, and micro-optics, while the technologies covered were surface modification and structuring processes, energy assisted and mechanical processes, and

replication processes. The results were first presented at the 4M2006 conference in Grenoble [7].

This section contains update information on the 2006 4M Roadmapping Study. Table 1 shows the results from SWOT analyses with respect to barriers and also technical capabilities for the successful manufacture of micro products, for Europe (barriers) or 4M (capabilities) versus the rest of the world. The perceived relative market importance of the 3 application groups based on 14 industry respondees was: micro-fluidics (23%), micro-sensors & actuators (49%), micro-optics (28%).

Table 2 shows results from the three technology questions including the perceived future importance of micro manufacturing processes. Additional information includes perceived comparative level of characterisation and technique capabilities, including 8 identified in order of importance for the 3 technology groups, for energy-assisted and mechanical processes: accuracy/tolerance (1st), surface properties (2nd), production speed (3rd); for surface processes: repeatability/ reliability (1st), small feature size (2nd), 3D features (3rd); and for replication processes: material capability (1st), low cost (2nd), repeatability/ reliability (3rd).

Table 1 Extended SWOT results for Europe/4M (researchers only)

Category[1]	Micro-fluidics	Micro-sensors & actuators	Micro-optics
A Low cost / volume production	W scale-up	WT cost	W mass production W labour cost
B Interdisciplinary, design, and process/ technology knowledge	S multidisciplinarity S trained people S technical issues O design WT modelling of phenomena	W design for manufacture O modelling	S process knowledge S technology knowledge W design for manufacture W lack of students, education T Export of knowledge
C 3D features, surface properties	W control of surface properties	O 3D	O 3D, surface quality
D Technology / maturity / standards[2]	W lack of technology W knowledge sharing/ awareness T standards	S technology maturity S technology diversity/ new ideas O prototyping O infrastructure	W knowledge sharing with industry T lack of knowledge sharing/ increased development time
E Function and physical integration	O functional interfacing/ integration		O function integration
F Quality/reproducibility/reliability	W		W
G New/improved/multi materials	S smart materials/ systems	S ceramics O combining materials T diversity of materials T material availability	T material development outside Europe
H Integrated process chains		O	O
I Process / machine technologies	S replication	S process technologies	S machine systems S precision engineering SO machine tools
J Assembly & packaging	S bonding, joining, packaging	O T packaging	
K Nano-micro or micro-meso integration	S micro-macro	OT micro-nano interface	O length scale integration
M Markets	W no exemplar products T conservatism, risk/investment	OT silicon competition	T dominance of Far East in key consumer markets
-- Tolerance/accuracy		S (future)	
-- Metrology			W imported (nano) O standards standardisation T US sets standards
-- Other (micro-optics only)	S political and economic stability WO funding focus and efficiency	ST using adapted solutions/ technologies WO product development/ introduction/ timescale	

S, W, O, T = Strength, Weakness, Opportunity, Threat. [1]A-K – common application requirements.[2] In the workshop this included the idea of knowledge sharing.

Table 2 Extended technology results

Workshop group	Technology	Future importance		Capability									Process chain frequency[1]			
		Research	Industry	Characterised	Accuracy/	3D features	Low cost	Material	Speed	Repeatability/	Small features	Surface	All	Fluidics	S&A	Optics
Mechanical and energy-assisted processes	Beam-based														3	1
	E beam														1	1
	Focussed Ion Beam (FIB)													1	2	2
	Laser ablation	+		●	●	●									5	1
	Laser hardening																
	Plasma machining	+		●	●	●						●				
	Projection Mask-Less Patterning (PMLP)	+														1
	Electrical / Chemical															
	Etching	-	-	●					●							2	
	Electrical discharge machining (EDM)	-		●	●	●	●	●		●	●	●		1		4
	Electrochemical machining (ECM)			●	●			●		●	●	●				
	Electrochemical polishing											●				
	Electroforming													1		1
	Mechanical machining													1	1	1
	Abrasive water jet															
	Drilling															
	Milling	+		●	●	●	●	●	●	●	●	●		3	2	8
	Grinding	-	-	●	●				●			●				3
	Lapping															
	Polishing			●	●				●			●		1		3
	Turning / Diamond turning	-	-	●	●	●	●	●	●	●		●				4
	Prototype / layer-based manufacture															
	3D Printing		++											1		7
	3D Lithography		+													
	Selective laser sintering (SLS)	+														
Surface modification and structuring processes	Lithography															
	E beam lithography	+		●												
	Ion beam lithography	-														
	Laser lithography	-		●	●	●			●		●					
	Nanoimprint lithography (NIL)	+		●	●	●			●	●	●	●				1
	Photo / UV lithography	-		●		●				●	●		3	6	1
	X-ray lithography		+											1		
	Coating														10	3
	Physical vapour deposition (PVD)			●		●					●			1		4
	Chemical vapour deposition (CVD)			●		●					●					1
	Electroplating			●										2	1	1
	Spin coating			●						●	●			1		
Replication processes	Replication													2	2	1
	Casting	-		●			●	●		●				1		
	Direct LIGA	+														
	Hot/UV Embossing			●	●	●	●	●	●	●		●		1	2	1
	Injection moulding	-	--	●	●	●	●	●	●	●		●		4	5	9
	Metal Forming			●						●				1	1	4
	Multi-component injection moulding		+	●	●	●	●	●	●	●		●				
	Nanoimprinting		++			●	●	●	●	●	●	●		1		
	Screen printing	-		●	●		●	●	●			●			5	
	Powder injection moulding	+			●		●	●	●	●				1		
	Reel to reel embossing		++											2	3	
	Self assembly	-														
	Assembly & packaging													8	9	1

+/– Increase/decrease over current importance. ● Researchers ● Industry (if at least 3 respondees) ● Industry + Researchers, the larger the symbols the better. [1] 8 micro-fluidic, 15 micro-sensors & actuator (S&A), 14 micro-optic process chains.

Table 2 also includes a further analysis of a total of 37 'promising' process chains were, each with typically 4-5 process steps. A few processes were mentioned that are not contained in the table, e.g. sintering and inspection. A typical optics chain, for polymer lenses, is: mechanical machining (milling, turning), coating, and then replication (injection moulding). An example sensors chain, for a thermo-electric sensor, is: tooling, coating, injection moulding, sintering, and

3D printing. An example fluidics chain, for a blood separator, is: photo/UV lithography, electroplating, hot embossing, and then assembly & packaging.

Industry respondees were asked commercialisation questions, the results of which are summarised below (numbers in brackets indicating numbers of respondees):

- The issues affecting the commercial success of micro components and products are: internally, design capability, development and production cost, lead time, trained staff, supply chain, acceptance of new technology (18), externally, understanding customers' needs/ emerging markets leading to product definition through close collaboration (9), and selling through a good commercial network (3).
- The ways micro technologies and applications are best known to manufacturers are: through cultivating customer contacts (8), through events – seminars, workshops, exhibitions, regional/national networks (14) and new product ideas ideally with functioning prototypes (8), and through publications, website, projects, training (8).

Errata to the 2006 4M Roadmapping Study are as follows. Methodology section: Application area (on which interviewees chose to answer questions) should be micro-fluidics (15), micro-optics (16), micro-sensors & actuators (18), other (10). Table 2: The barriers bars for the micro-fluidics group should be 25% longer than indicated.

8 The 4M Book Series

With the profile of 4M established and a lot of interesting research being produced by the Partners, it was decided to launch a 4M Book Series. This would be an opportunity to further disseminate the 4M knowledge and to leave a legacy of the 4M network. Currently two books are in production.

One mentioned above (Section 3.5) comes from the Ceramics Division and is the culmination of their efforts to assemble a handbook for ceramics processing. It should be published by Whittles Publishing in the autumn of 2008.

The second book, the seed of which was the output of one of the early cross-divisional projects, is to be an overview of the current development of nanomembranes. It is hoped that this will be published in 2009.

8.1 Ceramics Processing in Micro Technology

Due to their outstanding properties ceramics have been used by mankind for millennia, and because of these same properties they are still of great interest and importance in modern technology. Consequently a great variety of ceramic materials are used for high-tech applications worldwide and this holds true for micro systems technology.

The intention of this volume on "Ceramics Processing in Micro Technology" is to present an overview over the current status and perspectives of the processing of ceramic materials in micro technology. Favourable applications of ceramic materials are also considered.

The contributions to this book are written mainly by the scientific partners in the Ceramics Division of 4M but, for the sake of completeness, also by researchers outside the network.

Comprehensive information will be given for materials researchers, process engineers and developers in the field of micro system technology as well as for students. The objective is to give an impression on the whole development and processing chain from design questions, which are very special for micro systems, via material development (synthesis of micro technology adequate materials), powder preparation and conditioning, micro forming by various techniques, thermal processing through to metrology strategies. This includes the characterisation and implementation of adequate quality control with respect to e.g. dimensional stability, reproducibility and the investigation of the dependence of the parts' final properties on the process parameters and on the intrinsic microstructure of the ceramic components.

After an introduction offering a short overview of the 4M network and its Ceramic cluster, an overview on the status, challenges, requirements, chances, perspectives, and applications for micro ceramic components will be presented. As a broad palette of ceramic materials with interesting structural or functional properties exists, so various fields of application can be covered. Their potential for application in micro systems will be presented within this chapter.

In further chapters the modelling and design of micro components and micro systems, material development, various shaping and production processes for ceramic micro manufacturing, processing technologies, tooling, mould fabrication, tape casting, embossing, electrophoretic deposition, high pressure injection moulding, low pressure injection moulding / hot moulding are all covered as well as aspects of prototyping. Subtractive and additive process technologies such as micro milling, laser milling, and electro discharge machining (EDM), layer manufacturing and direct printing of ceramics are discussed. Hybrid manufacturing processes and relevant applications are also covered, mainly the applications and processing of LTCC materials.

Finally, contributions on quality assurance and metrology are presented. These contributions include test methods, suitable equipment for testing micro components as well as considerations on comparability and reproducibility.

By covering most of the important aspects of modelling, design, materials development, processing and quality control the authors hope not only to present a useful guide for students and readers looking for a comprehensive overview on non-silicon micro technology, but also to give an overview on the technologies available and the status of R&D in this field, as performed within the 4M network.

8.2 Nanomembranes

This book summarises recent advances in the field of nano-membranes, a new paradigm in the family of the nanotechnology building blocks. Nanomembranes are defined as artificial or natural structures with a thickness in the range of 100-5 nm and an aspect ratio exceeding 1.000,000 (areas of several square centimetres or larger). Such a low thickness is very close to the fundamental limit for solids, since 5 nm approximates to 15 atomic layers, and makes the nanomembrane structure quasi 2-D. This creates many new and exciting applications, not feasible with other nano or MEMS structures. The importance of the topic is becoming increasingly recognised in the nanoscience community, and with good reason. In every living cell, from bacteria to human beings, nanomembranes divide the cytoplasm from the environment and at the same time enable their active and intelligent interaction with the ambient. Nature created the nanomembrane as one of the most ubiquitous building elements of life.

The book will provide a concise overview of the current development of nanomembranes, as well as a clear vision of their future development toward practical applications. Currently nanomembranes are being developed in several distinct fields including macromolecular chemistry, thin film techniques, bio-engineering and MEMS/NEMS generally. The book represents an attempt to offer the first comprehensive and systematic overview of the most important and pertinent topics in this field of nanotechnologies.

9 4 M Technology Readiness Level Workshops

The 4M Network of Excellence is currently carrying out a study on 4M technologies maturity levels to help inform European and national funding bodies and industry about the maturity of micro and nano technologies. More specifically, the objective is to obtain a picture of the distribution of the research efforts along a maturity scale for the technologies relevant to 4M.

Research efforts supported by EU IP funding

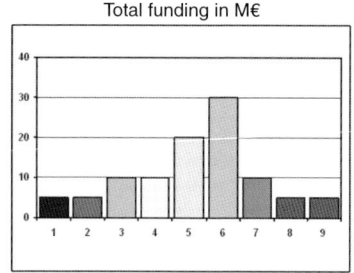

Research efforts targeting the development of micro fluidic products

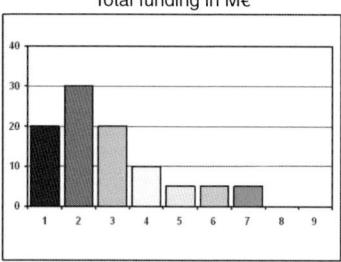

Fig. 14 Examples of envisaged outputs from the technology maturity study

Within the 4M network, more than 300 research projects have been initiated or completed in the field of micro manufacturing over the last 5 years with EU, national or institutional funds. In this study it is proposed to position each of these research projects on a technology maturity scale thus assessing the distribution of research effort in Europe. The graphs in Figure 14 show some examples of the type of information that could be derived from such a study. The presented graphs were prepared using fictional data and assuming a maturity scale composed of 9 levels.

The maturity scale used in this study was inspired by the Technology Readiness Level (TRL) scale developed originally in the 80's by NASA and adopted in the 90's by the United States Air Force [8]. TRL is a measure to assess the maturity of an evolving technology (materials, components, devices, etc.) prior to incorporating it into a system or subsystem. The NASA TRL scale is composed of 9 levels which are grouped into 6 transition phases as shown in Figure 15.

However, this scale needed to be adapted to the specificities of 4M technologies. For example, it is obvious that TRL 9 in Figure 15 does not describe appropriately the highest maturity level of a 4M technology. For this reason, the TRL concept was

presented and discussed at the 4M workshop organised on the 14th February 2008 and hosted by Fraunhofer IZM in Munich. This workshop was facilitated by Cardiff University's Cardiff Business School and Manufacturing Engineering Centre and took place the day before the 4M Governing Council. It regrouped 4M experts from the entire network. In this workshop, the experts were firstly asked to refine the proposed maturity scale in the context of 4M technologies.

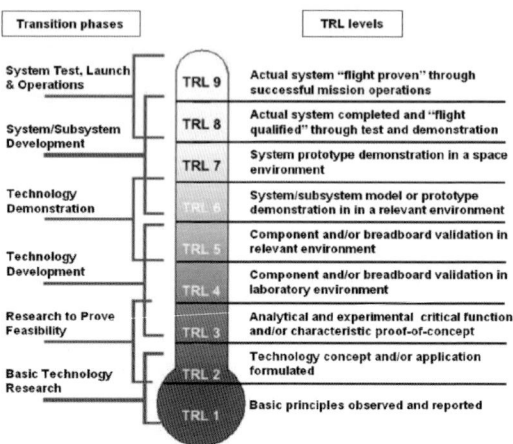

Fig. 15 TRL levels and transition phases
(adapted from NASA and [8])

In order to position a project accurately on the obtained maturity scale, it is clear that the expertise of the researchers actively involved in a project is essential. However, due to the amount of research projects considered in this study and the range of technologies they cover, it is not a straightforward task and different experts with different interpretations of the maturity scale could lead to inconsistent positioning of projects.

Thus during the workshop, a Delphi type of study was also conducted in order to identify key indicators for each phase of a 4M technology development. This was done in order to ensure that the positioning of all projects along a maturity scale could be done in a systematic and consistent way.

Fig. 16 Workshop participants

The workshop participants (Figure 16) were split into 3 groups. The first group focused on identifying maturity indicators for projects that target the development of manufacturing technologies. The second group was asked to focus on projects targeting the development of applications while the third group was dedicated to projects aiming at developing both technologies and applications at the same time.

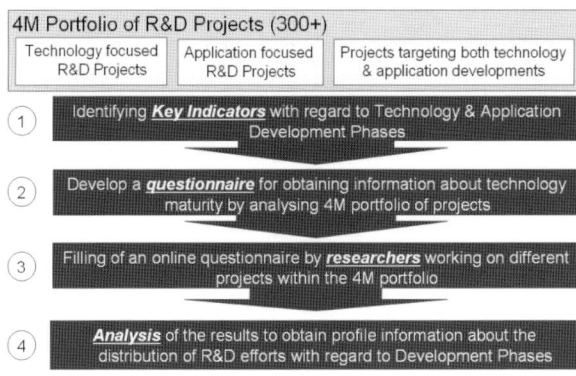

Fig. 17 Methodology

From the answers given by the 4M experts, generic indicators in the form of project triggers and activities typically associated to a specific technology development phase were identified. They were then used to design a questionnaire which allows an expert to answer simple questions about a project while positioning it automatically on the maturity scale. The overall methodology of the 4M technology maturity study is illustrated in Figure 17.

10 Setting up the 4M Association

The 4M Network of Excellence has been a success in generating relationships and links between the research partners in the Network, at both the organisation and personal levels. It has built a "knowledge community" in 4M. An awareness of the skills and expertise of fellow partners has gradually been built up along with a perhaps even more valuable commodity – trust – that is a very important prerequisite for achieving a sustainable integration. Partners report that this has been the principal benefit arising from their participation in 4M.

This is also the aspect that the Partners themselves wish to preserve after the funded period of the Network, in particular:

- engaging jointly with industry in mutually beneficial R&D activities,
- opportunities for joint R&D,
- access to partners' research infrastructure and knowledge,
- jointly publicising its knowledge and technology transfer activities,
- achieving a critical mass in new emerging R&D areas, and
- having a collective "voice" to industry and policy makers at European level.

In fact the Partners see great potential in enjoying the benefits of the Network without having to fulfil the negative aspects (reporting, performance indicators, deliverables etc) associated with any funded project. With this in mind it is proposed to establish a 4M Association.

The 4M Network of Excellence believes that by setting up a 4M Association the relationships formed during the lifetime of the NoE will be converted into a durable and sustainable format, creating an open and permanent forum for the knowledge community, allowing it to grow, spreading the benefits of membership to a wider micro and nano manufacturing constituency than the current 4M NoE, as future membership will be open to all.

In Year 3 and continuing into Year 4, there has been much debate about what the 4M Association should offer in order to achieve its goal of being sustainable through the collection of membership fees. This has not only been a debate amongst Partners, for the coordinator has also conducted an online survey of its industrial

Affiliates in which they were invited to rank a list of proposed services. This has proved to be a very interesting exercise.

Very broadly the results of the survey indicated, perhaps unsurprisingly, that industry is primarily interested in the possibilities for networking and generally building up contacts with other companies and organisations that Association membership might bring. The opportunity to build consortia to bid for EC, national and other funding sources is also attractive.

The least attractive of the proposed services included training, either electronically or face to face, providing a presence on behalf of the membership at conferences and fairs, and assisting start-up companies.

In the middle ground are provision of an annual conference and various website services (guide to funding opportunities, job ads, links, equipment), both of which will almost certainly have to be provided anyway – the continuation of the 4M conference is particularly attractive to the current partners and they are likely to provide the core of any initial membership.

The initial offering of the 4M Association will therefore include the following:

- networking (local/regional/technical contact groups/clusters)
- an interface between research community & industry
- joint working/interest groups addressing specific, but common, topics
- brokerage service offering access to the expert design, prototype and manufacturing services of the other member organisations
- new business opportunities via contacts among membership
- consortium building for industry-lead joint projects in emerging research areas in order to bid jointly for EC, national and industrial funding
- access to 4M research infrastructure and knowledge (4M Knowledge repository)
- an annual 4M Conference

Acknowledgements

The authors and the 4M Knowledge community as a whole would like to thank the European Commission for funding the FP6 Network of Excellence on "Multi-Material Micro Manufacture: Technologies and Applications (4M).

References

[1] S.S.Dimov et al, 4M Network of Excellence, Progress Report 2004-2006, 4M2007, Proc. of the 3rd International Conference on Multi-Material Micro Manufacture, ISBN 978-1-904445-53-1.

[2] Ratchev S., Turitto M., Editors, Micro- and Nano-Manufacturing Strategic Research Agenda, January 2008

[3] Buccarelli L.L., Delta Design game, copyright 1991 MIT, All rights reserved

[4] Jiang F., Drese K.S., Hardt S., Küpper M., Schönfeld F., Helical flow and chaotic mixing in curved micro channels, AIChE Journal, 50(9): 2297–2305, 2004.

[5] Stroock A.D., Dertinger A.D., Chaotic mixer for microchannels, Science, Vol. 295, p. 647–651, 2002.

[6] Tang P.T., Christensen T.R., Simultaneous Replication of both Refractive and Diffractive Optical Components using Electroformed Tools and Injection Moulding, MICRO.tec, Procdings pp. 135–138, München, 13-15 October, 2003.

[7] Dimov, S.S., Matthews, C.W., Glanfield A., and Dorrington, P. (2006), "A roadmapping study in 4M Multi-Material Micro Manufacture". Proceedings of International Conference 4M2006, Elsevier (Oxford).

[8] Mankins J. Technology readiness levels – a white paper, 1995. Accessed on 07th May 2008 at www.hq.nasa.gov/office/codeq/trl/trl.pdf.

Keynote Papers

Multi-Material Micro Manufacture
S. Dimov and W. Menz (Eds.)

3

Traceable measurement of areal surface texture

R.K. Leach, C. Giusca

Industry & Innovation Division, National Physical Laboratory, Teddington TW11 0LW, UK

Abstract

There is a clear need in industry and academia for traceable areal surface texture measurements. To address this need traceable transfer artefacts and primary instrumentation are required. The National Physical Laboratory (NPL) is working on two projects – one to develop areal transfer artefacts and one to develop a traceable areal surface texture measuring instrument. The authors describe the development of the artefacts and instrument, and present some of the challenges that are still required to be able to offer an areal traceability measurement service to industry. The instrument has a working volume of 8 mm x 8 mm x 0.1 mm and uses a co-planar air-bearing slideway to move the sample. It also uses a novel vertical displacement measuring probe, incorporating an air-bearing and an electromagnetic force control mechanism. The motions of the slideway and the probe are measured by laser interferometers thus ensuring traceability of the measurements to the definition of the metre. The artefacts were manufactured using a range of machining technologies and in a range of geometries suitable for stylus and optical based instruments.

Keywords: areal, surface texture, measurement, transfer artefacts, areal standards

1. Introduction

Surface texture plays a vital role in the functionality of modern engineered products. Traditionally, surface texture data is used to monitor changes in a manufacturing process. For this form of monitoring, a two-dimensional, profile measurement is sufficient. Industry often has a need to engineer or structure a surface in three dimensions to impart functionality into the surface and the resulting device. Examples include micro-lens arrays for modern displays, MEMS for sensing applications, and glasses that are patterned in such a way as to make them hydrophobic and hence essentially self-cleaning. Three-dimensional or areal surface texture measurements have a number of advantages over profile measurements including:

- The areal approach comes closer to fully describing a real surface and the derived parameters possess greater functional significance.
- The areal approach allows parameters to be derived relating to area for the first time, for example, texture "strength" and direction, material and void volumes, *etc.*
- The areal approach takes data from an area rather than a profile, therefore, the parameters have greater statistical significance and better repeatability between different parts of the same surface.
- Areal measurements are a better visualisation tool.

The control of complex structured surfaces requires an areal measurement of surface texture. There are many instruments on the market that address this need, for example, coherence scanning interferometers (often referred to as vertical scanning white light interferometers) and scanning stylus instruments, but there is currently no definitive, direct route to traceability for such instruments [1]. At present, traceability is inferred from calibrated artefacts and measurement strategies that were originally designed to calibrate profile measuring stylus instruments. Whilst this method of calibration may be adequate in some circumstances, there are characteristics of an areal instrument that cannot be determined from profile measurements alone.

The UK National Measurement System has recently funded two projects that go a long way to establishing traceability of areal surface texture measurement. Firstly, NPL has collaborated with the Atomic Weapons Establishment (AWE), Rubert & Co. and Taylor Hobson to produce a set of prototype artefacts to address verification and calibration of various performance aspects of areal surface texture measuring instruments [2]. Secondly, NPL has developed a traceable areal surface texture measuring instrument [3]. These two projects are described below. We also discuss the current state of standardisation for areal surface texture and discuss some further work that is still required to fully complete the traceability chain.

2. Areal specification standards

In 2002 the International Organization for Standardization (ISO) Technical Committee 213, dealing with Dimensional and Geometrical Product Specifications and Verifications, formed a working group to address standardisation of areal surface texture measurement methods. The working group is developing a number of draft standards encompassing definitions of terms and parameters, calibration methods, file formats and characteristics of instruments. The first published standards are expected some time in 2009. These standards will finally allow engineers and scientist to start to gain benefit from the deterministic areal structuring of surfaces. However, the change over from profile standards (currently used on engineering drawings to tolerance surface texture) to

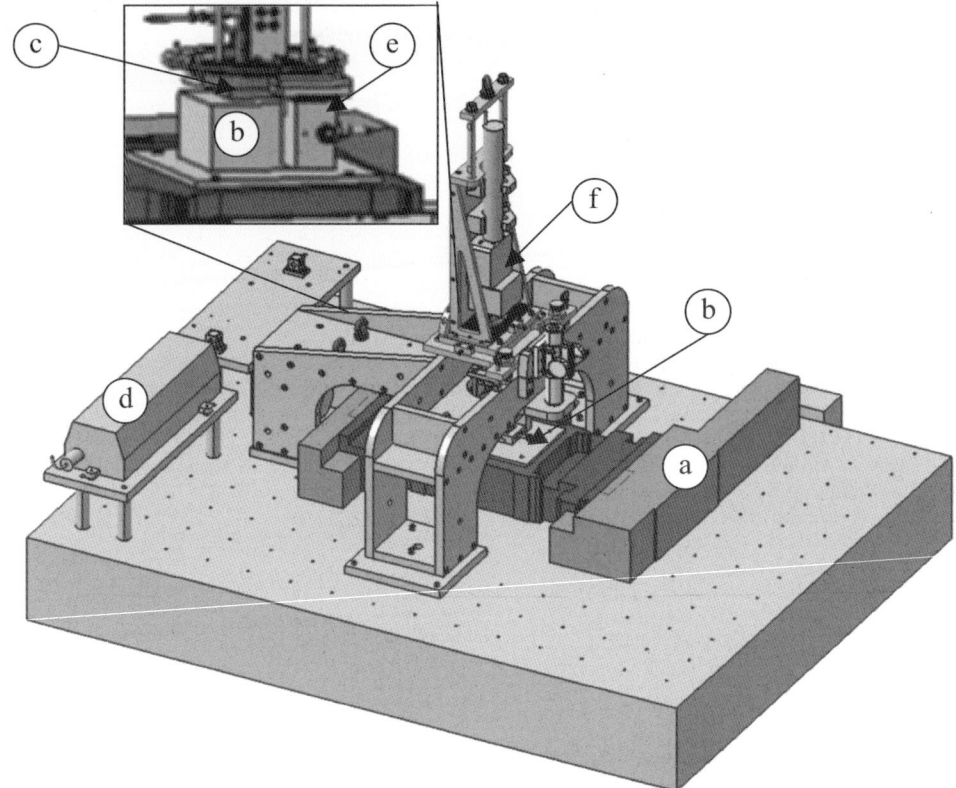

Figure 1 CAD drawing of NPL Areal Instrument: a) co-planar linear air-bearing stage, b) sample holder and Zerodur mirror block, c) reference mirrors, d) laser source, e) linear and angular column-referenced interferometers f) z-axis plane mirror differential interferometer (see also Figure 2)

areal standards (of which profile standards will become a subset) will require a great deal of dissemination and education. Whilst this may be a difficult changeover for some industries, the rewards for embracing areal methods for product design and manufacture will be highly significant.

3. A traceable areal measuring instrument

The NPL Areal Instrument (figure 1) was designed to have a working volume of 8 mm × 8 mm × 0.1 mm and a target uncertainty of 10 nm × 10 nm × 1 nm. The instrument consists of an ABL9000 co-planar linear air-bearing stage (figure 1 a) designed for this application by Aerotech on which is mounted a sample holder and Zerodur mirror block (figure 1 b). The design of the stage is such that pitch, roll, yaw and orthogonality errors are less than two seconds of arc. The mirror block is reflectively coated on three sides and has sub-second of arc orthogonality errors and faces flat to less than 60 nm. Two further reference mirrors (figure 1 c) are mounted on the probe body. The position of the mirror block in the xy-plane is determined using a commercial laser interferometer system utilising two linear and angular column-referenced interferometers (figure 1 e) (Zygo ZMI2000 series). The surface being measured is mounted within the Zerodur block and the motion of a stylus as it is scanned across the surface being measured is detected by the use of a plane mirror differential interferometer (see figure 1 f and figure 2).

The interferometers for measurement of the motion in the xy plane both measure a linear and an angular (yaw) degree of freedom. Therefore, if the mirrors were

perfectly flat and orthogonal, one of the angular interferometers is redundant. The light from a frequency-stabilised laser is input to the interferometers using mirrors and the measurement signals are output to the processing electronics *via* fibre optic cables. The output from the pairs of x and y interferometers (and the z interferometer) are synchronised at the sub-microsecond level using bespoke hardware.

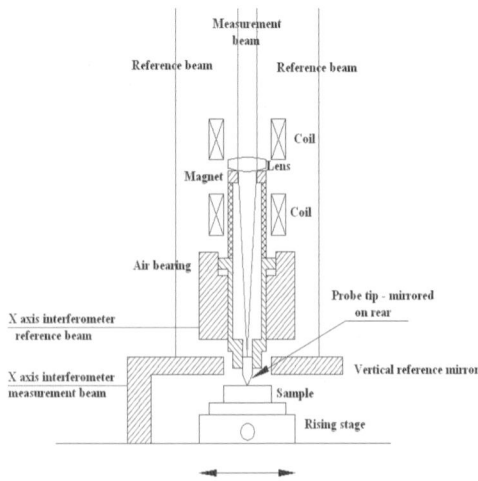

Figure 2 Schematic of z interferometer

At the instrument design stage many types of probe design were considered. When performing areal measurements with a tactile probe, the measurement duration becomes an issue and the use of dry bearings is inappropriate due to their relatively slow motion (as on the NPL NanoSurf IV traceable profile measuring instrument [4]). In an industrial application an optical probe is generally much faster than a tactile probe, but for a stylus-based traceable instrument it is much easier to predict the surface-stylus interaction with a conispherical stylus tip than the interaction of a optical beam with the surface. The probing system (Figure 2) utilises an air bearing (developed by Fluid Film Devices) as a linear guide for a stylus with an electromagnetic force control device, akin to a probe design reported elsewhere [5]. The sample is mounted inside a Zerodur mirror block that is described above, so that it comes into contact with the probe (this is achieved with Zerodur spacers and a height adjustment stage). The stylus is attached to the end of a hollow cylindrical air bearing and consists of a Zerodur rod with a polished and aluminised end face with a conventional diamond stylus on the opposite end. The air bearing is hollow to keep the mass of the probe down and allow the passage of the measurement beams of the z interferometer. The stylus operates through a hole in the vertical reference mirror and contacts the sample.

The static probing force is controlled by an arrangement of two electromagnets and a toroidal permanent magnet. The electromagnet design is that of a Maxwell pair (akin to a Helmholz coil but with the current passing in opposite directions in the two coils). This ensures a constant static probing force with respect to displacement in the z axis [6]. Note that this magnet and coil arrangement requires a current of more than 100 mA and needs to be water-cooled.

The displacement of the probe in response to the surface topography of the sample is measured by a differential plane mirror interferometer [7] where the measurement beams are focused onto the stylus mirror using an aspheric lens. The plane mirror interferometer system is referenced from the vertical reference mirror (see Figure 2). This referencing scheme essentially removes the effect of thermal or mechanical instabilities in the steel metrology frame (although any effects of the spacers and rising stage are not removed). The probe is designed to have a resolution of 0.1 nm, an accuracy of 1 nm, a range of 0.1 mm (with some over travel) and to be capable of responding to structures with wavelengths of 0.001 mm when scanning a surface at 1 mm s^{-1} (*i.e.*, 1 kHz).

At the time of writing only preliminary noise tests for the probe have been carried out. The RMS noise level is less than 3 nm with the probe in contact with a surface and all air-bearings and water cooling running. Comparisons with other traceable instruments [4] and further system performance tests will now be carried out and these results will be presented in a future paper. A full uncertainty analysis is also being developed.

4. Traceable areal transfer artefacts

NPL has collaborated with AWE, Rubert & Co. and Taylor Hobson to produce a set of prototype artefacts to address verification and calibration of various performance aspects of areal surface texture measuring instruments. A primary consideration in the design of the artefacts was the need for compatibility with both contact and non-contact measuring instruments. Compatibility is important to many users,

as it is very common to compare data from non-contact areal instruments and stylus profilometers. The artefacts need to transfer the traceable calibration from the stylus based primary instrument to non-contacting instruments in use in R&D laboratories and industry. The ease of manufacture of the artefacts was also a major consideration at the design phase, since the artefacts need to deliver traceable and accurate calibration at reasonable cost. Artefacts have been manufactured using the following methods: silicon processing, optical lithography, and diamond turning combined with replication in electroformed nickel. The artefacts are designed to address calibration of lateral and vertical scales, verification of lateral resolution, dynamic response and probe condition monitoring.

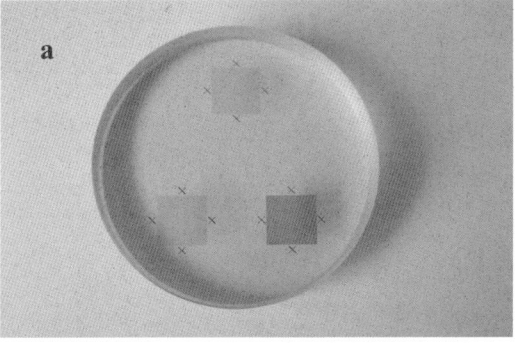

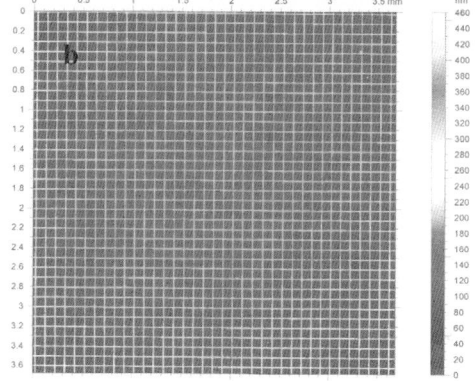

Figure 3 Lateral scale calibration artefact: a) photograph, b) coherence scanning interferometer plot of 100 µm period grid

A set of grid patterns was produced for calibration of lateral scales. The grids are chrome-on-glass patterns with nominal periods of 200 µm, 100 µm and 20 µm with line widths of 20 µm, 10 µm and 5 µm respectively. Each of the three patterns extends over a 12 mm square patch, and all three are on a single, 59 mm diameter, flat glass substrate. Figure 3 a shows a photograph of the artefact and Figure 3 b shows an plot from a coherence scanning interferometer instrument of the 100 µm period grid. The grids were evaluated using two traceable methods of measurement. The average periods of the grids in the two orthogonal directions on four sets of grids were measured by an optical diffraction method [8], and found to be within 0.1% of their nominal values. The uniformity of placement of the lines was measured on one set of grids using a traceable linescale measuring instrument [9]. The cumulative error in line placement over the 10 mm extent of the patterns was measured to

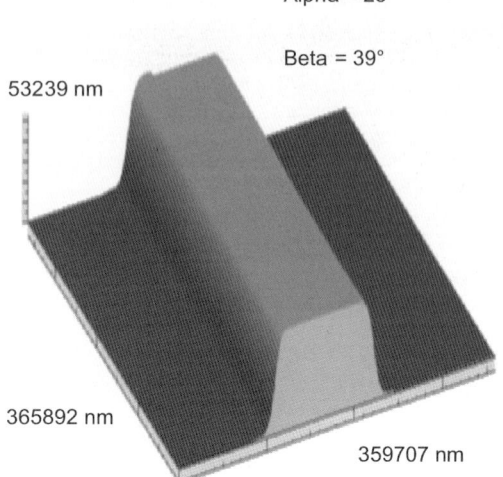

Alpha = 25°

Beta = 39°

53239 nm

365892 nm

359707 nm

Figure 4 Coherence scanning interferometer plot of a 50 μm step

be less than 0.5 μm. The grids may be used to check linearity of scales and their orthogonality.

It is common practice to calibrate the z-axis scale of areal instruments using the method, and standard step height artefact types, described in the stylus profilometry standard ISO 5436-1 (2000) [10]. This practice is satisfactory for calibration, but needs to be supplemented by verification of the uniformity of the response of the probe over the field of view of the instrument. For example, on a coherence scanning interferometer small errors in z-axis measurement may be attributed to imperfections in the reference mirror; these may be corrected for by comparison with a calibrated reference flat. A set of artefacts was produced covering the step height range 10 nm to 50 μm, with step widths at 100 μm and 500 μm, permitting the step height measurement described in ISO 5436 to be fitted within the typical field of view of high and low magnification lenses on optical instruments. The larger step heights, 1 μm, 10 μm and 50 μm were diamond turned in copper and then replicated in nickel. Figure 4 shows a coherence scanning interferometer image of a 50 μm step, 100 μm wide. The smaller steps, 1 μm, 100 nm and 10 nm were produced both in glass and in silicon. In addition waffle step height patterns at 30 μm, 100 μm and 200 μm periods were produced in silicon at the same step heights to enable verification of the z-axis scale over the field of view. Figure 5 shows an image of a 1 μm waffle pattern.

Diamond turning was used to produce various surface profiles useful for verifying dynamic response of scanning instruments, response to slopes for optical probes and for probe condition monitoring. The diamond turned profiles were all turned in copper and then replicated in nickel to give an affordable artefact with a durable surface. Sine wave profiles have been produced with maximum slopes of 20° and 5°, at 25 μm and 8 μm periods.

There is no agreed, specific definition of lateral resolution for areal instruments, but it is important, both for manufacturers when marketing an instrument, and for users when selecting an instrument fit for purpose. Resolution test structures were fabricated on two silicon

samples, approximately 12 mm square, by e-beam lithography. Sample Res A carries nine grating patterns with equal mark/space ratios and periods of 0.6 μm to

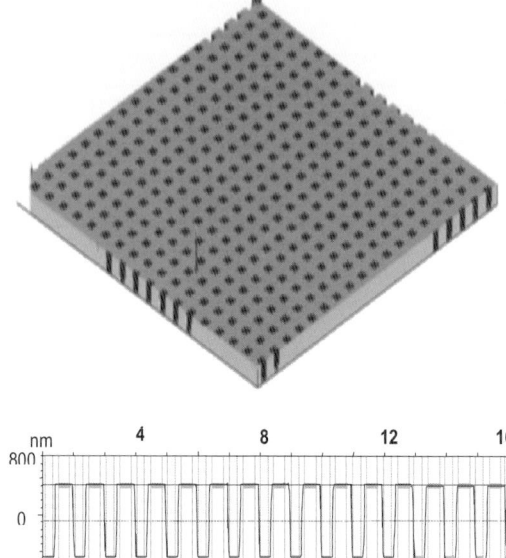

Figure 5 Coherence scanning interferometer plot of a 1 μm waffle pattern

10 μm. Sample Res B carries eight gratings of periods 1 μm to 8 μm and an array of star patterns in the central patch. The periods of the grating patterns have been measured traceably by an optical diffraction method. Figure 6 shows the layout of the two samples with an AFM image of the central grating pattern on sample Res A and a coherence scanning interferometer image of one of the star patterns in the central array on sample Res B. Where the coherence scanning interferometer resolves the pattern on Res B comfortably, the image shows the two levels of the upper and lower surface of the pattern. Near the centre, the pattern may still be distinguished but its true height has not been measured. The measured values agreed with the nominal values to within 0.5 nm in the majority of the patterns.

5. Further work on areal traceability

We have taken the first steps towards a traceable instrument and transfer artefacts for the measurement of areal surface texture. However, there is still a significant amount of research and development required to be able to offer a measurement service to industry. There are many commercially available instruments for measuring areal surface texture, mainly based on stylus or optical methods. ISO 213 is addressing the specification standards for all common types of instruments (at the time of writing of this paper only stylus, coherence scanning interferometry and confocal chromatic instruments are being actively worked on), and research is still required on how to measure the large range of structured surfaces that will become available in the future. Structured surfaces will

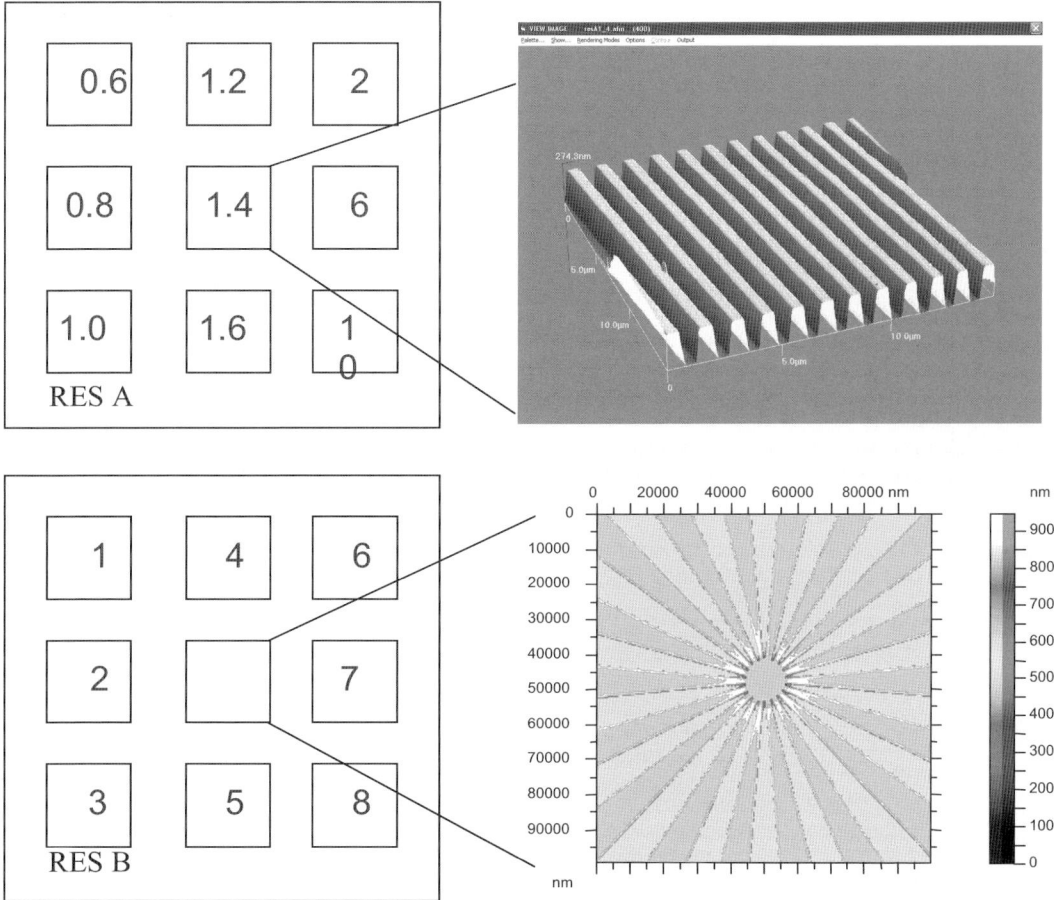

Figure 6 AFM image of the central grating pattern on sample Res A and a coherence scanning interferometer image of one of the star patterns in the central array on sample Res B

present surface bandwidths that may be difficult to measure using some instruments and good practice guidance will be necessary. Coherence scanning interferometers may be very versatile instruments but they can give erroneous results without *a priori* knowledge of the structure of the surface being measured [11]. This is an example where guidance on good practice is required. NPL, the University of Loughborough, the University of Huddersfield, IQE Ltd and Taylor Hobson are producing good practice guides in the use of coherence scanning interferometers (these guides will be published in the first half of 2008).

Once traceable instrumentation and transfer artefacts are in place for areal measurement, software measurement standards will be required to ensure that instrument software for filtering and parameter calculations is correct. New characterisation methods and parameters will also be needed as the number of commercially utilised areal structured surfaces grows.

6. Conclusion

The development of a traceable instrument and transfer artefacts for measuring areal surface texture has been summarised. Future work involves the full characterisation of the instrument and an uncertainty analysis. Once this work is complete the instrument will be fully traceable and used to measure two samples (a waffle plate and a random roughness sample) that have been circulated to and measured by several UK laboratories. Future research will develop software

measurement standards for areal filtering and parameters, new characterisation methods for areal structured surfaces and good practice guidance on the use of stylus and optical instruments.

Acknowledgements

The work was funded under the NMS 2002–2005 Programme for Length and the NMS 2005–2008 Programme for Engineering Measurement. Thanks are due to David Flack, Ben Hughes, Chris Jones, Simon Oldfield, Simon Reilly, Mike Parfitt, Keith Jackson, Alistair Forbes and Dave Bayliss (NPL), to Ivor McDonell and John Garratt (Taylor Hobson), to Julian Lamb and Steve Wheeler (AWE), Paul Rubert (Rubert + Co.) and Ron Wooley (Fluid Film Devices).

References

[1] Leach R K. Some issues of traceability in the field of surface texture measurement. Wear. 257 (2007) 1246-1249.

[2] Haycocks J A, Jackson K, Leach R K, Garratt J, McDonnell I, Rubert P, Lamb J, Wheeler S. Tackling the challenge of traceable surface texture measurement in three dimensions. Proc. 5th Int. euspen Conf., Turin, Italy, May (2005) 253-256.

[3] Leach R K. Traceable measurement of surface texture at the National Physical Laboratory using NanoSurf IV. Meas. Sci. Technol. 11 (2000) 1162-

1172.

[4] Leach R K, Flack D R, Hughes E B, Jones C. Development of a new traceable surface texture measuring instrument. 11th Int. Conf. Metrology & Properties of Engineering Surface, Huddersfield, UK (2007) 75-79.

[5] Thomsen-Schmidt P, Krüger-Sehm R. Development of a new stylus system for roughness measurement. Proc. XIth Int. Colloquium on Surfaces, Chemnitz, Germany, February (2004) 79-85.

[6] Bayliss D, Leach R K, Hall M. Development of an electromagnetic spring for use with a high accuracy surface texture measuring probe. NPL Report DEPC-EM 06 (2006) 1-9.

[7] Downs M, Nunn J. Verification of the sub-nanometric capability of an NPL differential plane mirror interferometer with a capacitance probe. Meas. Sci. Technol. 9 (1998) 1437-1440.

[8] Nunn J. Calibration of 2 dimensional magnification standards for SPMs and SEMs through optical diffraction: method, traceability and uncertainties. Proc. 4th Seminar on Quantitative Microscopy QM 2000, Braunschweig, Germany (2000) 17–24.

[9] McCarthy M B, Gee A T. A 120 mm x 120 mm area photomask metrology standard with absolute position accuracy of 60 nm. Proc. ASPE. 11 (1996) 190–195.

[10] ISO 5436-1. Geometrical product specification (GPS) – Surface texture: Profile method – Measurement standards – Material measures. International Organization for Standarization (2000).

[11] Gao F, Leach R K, Petzing J, Coupland J M. Surface measurement errors using commercial scanning white light interferometers. Meas. Sci. Technol. 19 (2008) 015303.

Multi-Material Micro Manufacture
S. Dimov and W. Menz (Eds.)

Biphasic reactions in microreactors

B. Ahmed-Omer[a,b], D. Barrow[b], T. Wirth[a]

[a] *Cardiff School of Chemistry, Cardiff University, Cardiff, CF10 3AT, UK*
[b] *Laboratory for Applied Microsystems, Cardiff School of Engineering, Cardiff University, Cardiff, CF24 3TF, UK*

Abstract

The contact between immiscible liquids in a microfluidic system creating segmented flow offers great potential in the study of biphasic reactions in organic chemistry with significant advantages with respect to conventional flask techniques. As organic solvents play a key role in many chemical processes within the pharmaceutical and chemical industry, there are many applications of biphasic reactions in different areas of chemistry. For a simple biphasic reactions, we show that the application of various reaction conditions in microreactors using segmented flow can dramatically increase the reaction rate, especially when microwave irradiation, sonication or phase transfer catalysis are combined with segmentation.

Keywords: microreactors, phase transfer catalysis, segmented flow

1. Introduction

The miniaturisation of chemical processes using chip-based microreactors can exhibit significant advantages over existing conventional techniques. The properties and reaction conditions in such microreactors are different to large-scale systems. A high surface-to-volume ratio, short diffusion distances, fast and efficient heat dissipation and mass transfer enable novel and diverse applications.[1] These properties have been advantageously used in organic synthesis.[2]

2. Results and Discussion

A microchip system applied in synthetic chemistry usually consists of an arrangement of microstructures such as capillary scale ducts, sensors and actuators. A combination of such microcomponents and a chip-to-world interface of fluidic, electrical, optical and other interconnects may form a microreactor which can be fabricated in different geometries and from a variety of materials. The majority of chemical reactions in solution carried out in microreactors involve homogeneous reactions at room temperature. Recently, an interest in applying microreactors utilizing multiphase flow (gas/liquid or liquid/liquid biphasic systems) has emerged.[3] In a microchannel, the contact interface between immiscible liquids can follow various flow patterns, due to the forces at the interface generated from the different physical properties of both phases such as viscosity and surface tension. The most common mode of multiphase interface is known as parallel flow in which the respective fluid phases align side-by-side and mixing between them occurs principally via diffusion. Another multiphase mode, segmented flow, can be created in a microchannel when two (or more) fluid phases form serial trains of fluid packets, each phase being separated by the other. Once these fluid packets or segments are formed, an internal fluid vortex is generated which causes rapid mixing within a given segment by continuously refreshing the diffusion interface as shown in Figure 1. The area of this interface is approximately proportional to the cross-sectional area of the microchannel. The cross-section must be smaller than the length of the segments, otherwise emulsions are formed.[4] Furthermore, the constructional material of the microchannel plays a significant role in the formation of segments and influences their shape due to the effects of interfacial tension and surface energies.[5] The comparison of microreactors with conventional processing has become of interest recently as exploitation of their industrial usage is increasing.[6]

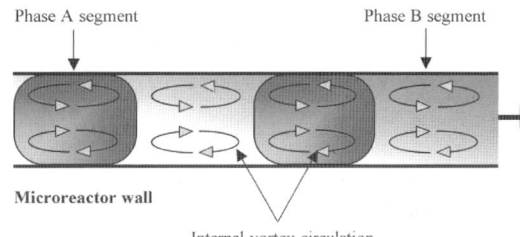

Figure 1. Schematic representation of segmented flow in a microchannel: Rapid mixing within a given fluid segment is caused by the internal vortex fluid flow; mass transfer between contiguous fluid segments is enabled by the continuously refreshing interface.

We investigated the hydrolysis of *p*-nitrophenyl acetate **1** in toluene with 0.5 M aqueous sodium hydroxide as a biphasic system.[7] Under these conditions, the hydrolysis of *p*-nitrophenyl acetate involves a nucleophilic attack by the hydroxide at the carbonyl carbon atom to displace the *p*-nitrophenyl moiety. Once the acetate is hydrolysed, the *p*-nitrophenolate **2** transfers into the aqueous phase resulting in a colour change from colourless to yellow. The reaction progress was monitored by the UV/VIS absorption of phenolate **2** at $\lambda_{max} = 400$ nm in the aqueous layer.

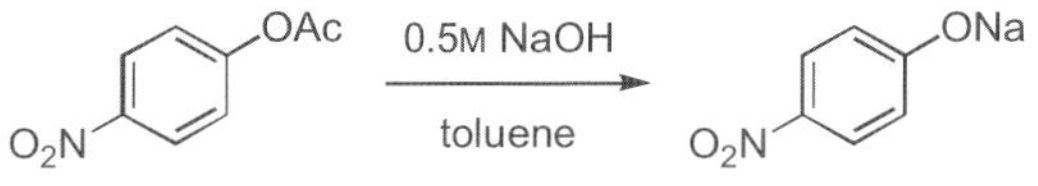

Scheme 1. Hydrolysis of *p*-nitrophenyl acetate **1**.

The yield of the reaction is dependent upon reaction time which is inversely proportional to the flow rate. Additionally, the size of the segments will also have an effect on the reaction rate as the ratio of volume: interfacial area increases with increasing size of the segments.

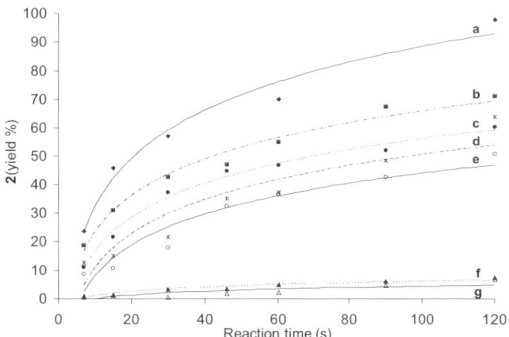

Figure 2. Hydrolysis of **1** using different flow types and reaction times: (**a**) Short segmented flow (approx. 2 mm) under microwave irradiation at 50 °C; (**b**) Long segmented flow (approx. 10 mm) under microwave irradiation at 50 °C; (**c**) Segmented flow in PTFE tubing heated in an oil bath at 50 °C; (**d**) Segmented flow at room temperature in PMMA reactor; (**e**) Segmented flow at room temperature in PTFE tubing; (**f**) Hydrolysis reaction at 50 °C in flask with stirring; (**g**) Hydrolysis reaction at room temperature in flask with stirring.

A solution of substrate **1** in toluene (0.05 M) and an aqueous solution of sodium hydroxide (0.5 M) were passed through the two inlets of the micro-reactor or into T-junction of PTFE tubes using a dual syringe pump.

The hydrolysis of *p*-nitrophenyl acetate **1** was carried out under different reaction conditions in different reactors. It is possible to deduce a number of trends from the results shown in Figure 2. A comparison of the results of reactions labelled (**f**) and (**g**) in Figure 2 with the other results is obvious. The reaction rate of the hydrolysis using a conventional flask is much lower than using microreactors and the difference between hydrolysis at room temperature (**g**) and at 50 °C (**h**) is not significant at that timescale.

Reactions labelled (**d**) and (**e**) were both carried out under similar reaction conditions, but in different reactors. These results show that the reaction in the PMMA (polymethyl methacrylate) microreactor (**d**) performs slightly better than in the PTFE tubing (**e**). The better performance in the PMMA microreactor is probably due to the visually slightly shorter segments and their higher regularity. The different T-junctions for creating the segmented flow might also play a role. As the PMMA microreactor with its metal housing cannot be inserted in the microwave, the experiments at elevated temperature (**a**, **b**, **c**) have been performed using PTFE tubing.

An increase of the reaction rate by heating can be seen by comparison of (**e**) and (**c**). A further increase is observed by microwave irradiation of the microreactor in a water bath (**b**) instead of heating with a conventional oil bath (**c**). By microwave irradiation of the microreactor without a surrounding fluid, insufficient

energy is absorbed to cause significant heating. This is due to the low loss microwave materials (PMMA, PTFE) used in construction and the low absorption characteristics of the fluidic duct geometry. Those are known properties and microwave heating of microreactors with outside deposition of gold metal to increase microwave absorbance are known.[8] Macro scale flow reactors suitable for microwave irradiation have also recently been reported.[9]

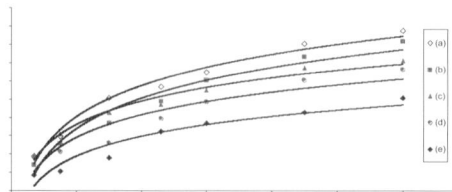

Figure 3. Hydrolysis of **1** in polytetrafluoroethylene (PTFE) tubing (300 μm diameter microchannel, 400 mm length) using segmented flow conditions (organic phase: toluene): (**a**) 10 mol% Bu₄NHSO₄ and sonication; (**b**) Sonication; (**c**) Microwave irradiation at 50 °C; (**d**) 10 mol% Bu₄NHSO₄ at room temperature; (**e**) Room temperature.[10]

We have investigated the effect of phase transfer catalysis and sonochemistry in the hydrolysis of **1** under segmented flow conditions and these results are shown in Figure 3.[11]

Phase transfer catalysis is a common approach used to accelerate a biphasic reaction by ensuring a ready supply of necessary reagent to the phase in which the reaction occurs.[12] When 10 mol% tetrabutylammonium hydrogen sulfate (Bu₄NHSO₄) was used as the phase transfer catalyst (PTC) at room temperature under segmented flow conditions (**d**), an increase in the reaction rate was observed compared to a reaction under segmented flow conditions without phase transfer catalyst (**e**).

Ultrasound irradiation is a transmission of a sound wave through a medium and is considered as a form of energy for the excitation of reactants hence increasing the rate of diffusion.[13] In the sonochemical technique, the microchannel tubing was immersed in the ultrasound bath[14] during the reaction time.[15] As a reasonable amount of heat is generated during sonication, the temperature of the sonicating bath was maintained constant at 25°C during the reaction.

When sonication was used under segmented flow in microreactors, a higher conversion rate was observed in reaction (**b**) than in reactions (**c**), (**d**) and (**e**). During sonication, irregular sized segments (1-10 mm length) are formed together with some emulsions. Increasing the interfacial area during sonication with the help of segmentation led to an enhancement of the reaction rate. This indicates that the reaction rate is now controlled by diffusion. Further increase of the rate of hydrolysis of **1** was obtained (**a**) when sonication, phase transfer catalyst and segmentation methods were all combined together.

3. Conclusions

In conclusion, by utilizing the large specific interfacial area provided by the microreactor under segmented flow, the hydrolysis reaction of *p*-nitrophenyl acetate **1** was found to be much more efficient than parallel flow and flask method.

The combination of microwave irradiation, sonochemistry and segmented flow was found to enhance the rate more than just segmented flow and phase transfer catalyst combined. Further enhancement was obtained when segmented flow was combined with both, sonochemistry and phase transfer catalysis.

Acknowledgements

We thank EPSRC for support of this work and the National Mass Spectrometry Service Centre, Swansea, for mass spectrometric data.

References

[1] (*a*) W. Ehrfeld, V. Hessel, H. Löwe, Microreactors, New Technology for Modern Chemistry. Wiley, **2000**; (*b*) M. Matlosz, W. Ehrfeld, J. P. Baselt, Eds., Microreaction Technology – IMRET 5: Proceedings of the Fifth International Conference on Microreaction Technology, Springer, Berlin, **2001**; (*c*) S. J. Haswell, R. J. Middleton, B. O'Sullivan, V. Skelton, P. Watts, P. Styring. *Chem. Commun.* **2001**, 391–398; (*d*) P. D. I. Fletcher, S. J. Haswell, E. Pombo-Villar, B. H. Warrington, P. Watts, S. Y. F. Wong, X. Zhang, *Tetrahedron* **2002**, *58*, 4735–4757; (*e*) K. Jähnisch, V. Hessel, H. Löwe, M. Baerns, *Angew. Chem.* **2004**, *116*, 410–451; *Angew. Chem. Int. Ed.* **2004**, *43*, 406–446; (*f*) B. Ahmed-Omer, J. C. Brandt, T. Wirth, *Org. Biomol. Chem.* **2007**, *5*, 733–740; (*g*) *Microreactors in Organic Chemistry and Catalysis,* Ed.: T. Wirth, Wiley-VCH, Weinheim, **2008**.

[2] (*a*) P. Watts, S. J. Haswell, *Chem. Eng. Technol.* **2005**, *28*, 290–301; (*b*) P. He, P. Watts, F. Marken, S. J. Haswell, *Electrochem. Commun.* **2005**, *7*, 918–924; (*c*) T. Kawaguchi, H. Miyata, K. Ataka, K. Mae, J. Yoshida, *Angew. Chem.* **2005**, *117*, 2465–2468; *Angew. Chem. Int. Ed.* **2005**, *44*, 2413–2416; (*d*) T. Honda, M. Miyazaki, H. Nakamura, H. Maeda, *Lab Chip* **2005**, *5*, 812–818.

[3] (*a*) V. Hessel, W. Ehrfeld, T. Herweck, V. Havercamp, H. Lowe, J. Sechiewe, C. Wille, *Gas/liquid Microreactors; Hydrodynamics and mass transfer*, Proceedings of the Fourth International Conference on Micoreaction Technology, **2000**; (*b*) B. Zheng, L. S. Roach, R. F. Ismagilov, *J. Am. Chem. Soc.* **2003**, *125*, 11170–11171; (*c*) J. M. Kohler, T. Henkel, A. Grodrian, T. Kirner, M. Roth, K. Martin, J. Metze, *Chem. Eng. J.* **2004**, *101*, 201–216; (*d*) A. Günther, S. A. Khan, M. Thalmann, F. Trachsel, K. F. Jensen, *Lab Chip* **2004**, *4*, 278–286; (*e*) D. Belder, *Angew. Chem.* **2005**, *117*, 3587–3588; *Angew. Chem. Int. Ed.* **2005**, *44*, 3521–3522; (*f*) B. Zheng, R. F. Ismagilov, *Angew. Chem.* **2005**, *117*, 2576–2579; *Angew. Chem. Int. Ed.* **2005**, *44*, 2520–2523; (*g*) B. K. H. Yen, A. Günther, M. A. Schmidt, K. F. Jensen, M. G. Bawendi, *Angew. Chem.* **2005**, *117*, 5583–5587; *Angew. Chem. Int. Ed.* **2005**, *44*, 5447–5451.

[4] (*a*) J. R. Burns, C. Ramshaw, *Lab Chip* **2001**, *1*, 10–15; (*b*) T. Thorsen, R. W. Roberts, F. H. Arnold, S. R. Quake, *Phys. Rev. Lett.* **2001**, *86*, 4163–4166;

[4] (*c*) T. Nisisako, T. Torii, T. Higuchi, *Lab Chip* **2002**, *2*, 24–26; (*d*) S. Sugiura, M. Nakajima, M. Seki, *Langmuir* **2002**, *18*, 5708–5712; (*e*) J. D. Tice, H. Song, A. D. Lyon, R. F. Ismagilov, *Langmuir* **2003**, *19*, 9127–9133; (*f*) S. Okushima, T. Nisisako, T. Torii, T. Higuchi, *Langmuir* **2004**, *20*, 9905–9908.

[5] (*a*) A. Serizawa, Z. Feng, Z. Kawara, *Exp. Therm. Fluid Sci.* **2002**, *26*, 703–714; (*b*) G. M. Greenway, S. J. Haswell, D. O. Morgan, V. Skelton, P. Styring, *Sens. Act. B* **2000**, *63*, 153–158; (*c*) N. Harries, J. R. Burns, D. A. Barrow, C. Ramshaw, *Int. J. Heat Mass Transfer* **2003**, *46*, 3313–3322; (*d*) G. Dummann, U. Quittmann, L. Gtoschel, D. W. Agar, O. Worz, K. Morgenschweis, *Cat. Today* **2003**, *79–80*, 433–439; (*e*) T. J. Jonson, D. Ross, L. E. Locascio, *Anal. Chem.* **2002**, *74*, 45–51.

[6] H. Pennemenn, P. Watts, S. J. Haswell, V. Hessel, H. Löwe, *Org. Proc. Res. Develop.* **2004**, *8*, 422–439.

[7] B. Ahmed, D. Barrow, T. Wirth, *Adv. Synth. Catal.* **2006**, *348*, 1043–1048.

[8] P. He, S. J. Haswell, P. D. I. Fletcher, *Sensors Act. B* **2005**, *105*, 516-520.

[9] M. C. Bagley, R. L. Jenkins, M. C. Lubinu, C. Mason, R. Wood, *J. Org. Chem.* **2005**, *70*, 7003–7006.

[10] Hydrolysis of **1** in polytetrafluoroethylene tubing (300 μm diameter microchannel) using segmented flow conditions (organic phase:toluene): (**a**) y = 24.618ln(x) - 33.32, R^2 = 0.9803; (**b**) y = 24.184ln(x) - 38.617, R^2 = 0.964; (**c**) y = 18.477ln(x) - 19.173, R^2 = 0.9796; (**d**) y = 17.876ln(x) - 24.463, R^2 = 0.8882; (**e**) y = 15.62ln(x) - 27.864, R^2 = 0.9257.

[11] B. Ahmed-Omer, D. Barrow, T. Wirth, *Chem. Eng. J.* **2008**, *135S*, S280–S283.

[12] (*a*) B. Cornils, W. A. Hermann, *Aqueous-phase organometallic catalysis, Concepts and Applications,* Wiley, **2004**; (*b*) J. A. B. Satrio, L. K. Doraiswamy, *Chem. Eng. Sci.* **2002**, *57*, 1355–1377.

[13] (*a*) T. J. Mason, J. P. Lorimer, *Sonochemistry, theory, applications and uses of ultrasound in chemistry,* Ellis Horwood Ltd; J. Wiley and son, **1988**; (*b*) J. L. Luche, *Ultrason. Sonochem.* **1996**, *3*, S215–S221; (*c*) A. Tuulmets, *Ultrason. Sonochem.* **1997**, *4*, 189–193; (*d*) M. H. Entezari, A. Keshavarzi, *Ultrason. Sonochem.* 2001, *8*, 213–216.

[14] Clifton Ultrasonic Bath, 30 – 40 kHz, 80 W.

[15] see also: W. Bock, Ger. Offen., DE 102004059451 (2006).

Multi-Material Micro Manufacture
S. Dimov and W. Menz (Eds.)

Mircofabrication using a Single Mode Yb Fiber Laser

W. O'Neill, K. Li, Q. Hu, P. Chopra, J. Kanghee, A. Buntardjo

Institute for Manufacturing, University of Cambridge, Cambridge, CB2 1RX, UK

Abstract

The advances in design, performance, cost reduction, and brightness for the modern Yb fiber laser have opened up the possibility of redefining the micro processing options for a range of semiconductor materials and micro fabrication production techniques at a wavelength of 1064nm. The usual laser of choice for micro electronics processing is the 532, 355, or 266 nm DPSS system. The provision of a new MOPA high brightness Yb based fiber laser configuration has provided a range of pulse parameters (10-200 ns FWHM), peak powers approaching ~ 2G Wcm^{-2}, and pulse repetition rates up to 500 kHz. These processing parameters offer a broad range of material response characteristics. This paper provides a preliminary analysis of the use of a Yb based fiber laser in the production of Si and Glassy Carbon microstructures and explores the potential of this source for low cost micromachining solutions.

Keywords: Ytterbium, Fiber, Laser, Si, Glassy carbon

1. Background

The response of semiconductors under intense laser illumination and therefore high carrier density has been the subject of much research since the advent of the high power laser. The transient absorption due to laser excitation of free carriers [1] and enhanced reflectivity induced by intense laser pulses has been observed by a number of workers [2, 3, 4]

Energetic (μJ–mJ) optical pulses with short pulse lengths and hence high focussed intensities in the range 10^6 to 10^{99} W cm^{-2} have proved capable of removing a wide range of materials including metals, semi-conductors, polymers and dielectrics. In this intensity regime, the performances of laser machining processes are largely dependent on the absorption of photons by the material which is a strong function of wavelength and temperature. Various laser sources have been employed in machining operations by previous research studies including 1μm, 532nm, 355nm, and 266nmwavelengths, It is generally thought that the absorption by Si of longer wavelengths >532nm is too low to effect accurate machining. In most cases short wavelengths (< or equal to 532nm) are the preferred choice since Si exhibits higher absorption at lower wavelength and the shorter wavelength reduces the negative influence of any surface plasma.

The advent of very low cost 1μm high brightness fiber lasers with high repetition rates, high operational efficiencies, and pulse modulation controls exceeding those of typical Q-switched DPSS lasers, has provided an opportunity to explore the behaviours of Si substrates under various intensity ranges and temporal envelopes. This work examines the performance of a new MOPA based Yb fiber laser when used to process Si and glassy carbon (GC) for the creation of basic microstructures.

2. Optical Properties of Si

Absorption of light by Si at ~1μm wavelength usually exhibits a room temperature absorption coefficient α of ~0.54 cm^{-1}. Lowndes et al [5] studied the time dependent optical properties of Si during nanosecond irradiation from a ruby laser. The results are shown in figure.1 for the transmission of 1152nm wavelength light during the interaction. It was found that above an intensity of E_l = 0.8 Jcm^{-2} (λ=694nm FWHM 14ns) the transmission of the 1152nm line reduced to zero and stayed at zero for a time proportionate to E_l and then recover to its initial value over several hundred nanoseconds.

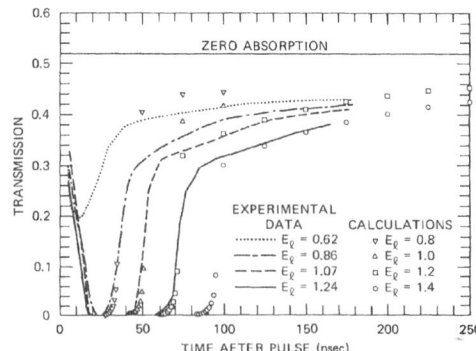

Figure.1 Time dependent transmission of a 1152nm probe laser beam through Si during and after pulsed Ruby irradiation, where E_l is the pulsed Ruby laser energy density [5].

This low transmission behaviour is also accompanied by high reflectivity, which is on the order of 75%, and is attributed to a thin layer of molten Si < 5μm thick. All of the measurements reveal that the time resolved transmission (TRT) recovered slowly with a time constant of ~ 500ns even after the molten layer has solidified. One explanation for long tail behaviour is that there is enhanced absorption in the near surface region of the sample. Aydinli, and Lo et al [6] measured the induced transient absorption in Si at a wavelength of 1152nm. Their results, figure.2, show that under intense laser excitation, the reduced transmission phase is accompanied by a high reflectivity phase that is characterised by high carrier density, high carrier temperature, low vibrational temperature and a self confinement of the plasma region of the order of 0.07μm.

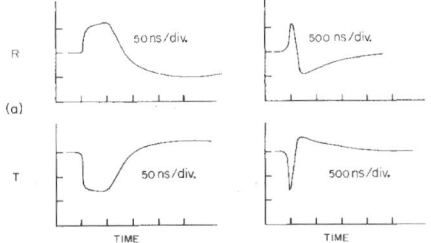

Figure.2 Time dependent transmission T and reflection R of a 1.152mm probe laser incident on Si during and after exposure to a pulsed laser excitation, 1 Jcm^{-2}, 8ns pulsewidth and 455nm wavelength [6].

Given the highly dynamic response of Si under intense illumination and the subsequent reduction in transmission, it is possible to influence the machining response of Si at a wavelength of 1μm (a wavelength usually disregarded in microsystems manufacturing) given sufficient control over the pulsewidth envelope.

A series of experiments were conducted to determine the interaction effects of 1064nm pulses from a Yb fiber laser with varying pulse envelope and peak power. The ability to vary the pulse envelope may lead to the identification of pulse parameters that would take advantage of the optical response of Si at elevated temperatures, namely the elimination of transmission, to produce greater accuracies in Si machining on a par with those obtained with a 355nm DPSS laser.

3. Experimental

The light source was a turnkey pulsed Yb fiber laser , G3 SP-20P-0011 20W, from SPI lasers, based on a Master Oscillator Parametric Amplifier (MOPA) architecture.A directly modulated seed laser (single emitter telecom based laser diode) is used to pump a fiber based Yb amplifier. This seed laser approach is enabled by a high gain and high saturation power from the active Yb fiber. It provides increased pulse energy and peak parameter space with increased control of pulse parameters. The system has the following specifications: peak emission wavelength 1064 +- 10nm; CW-500kHz; maximum pulse energy 0.8mJ; peak power 15kW; PulseTune^{TM} capability (25 preset waveforms for optimised peak pulse power at specified repetition rates); collimated output beam diameter 3.1mm; and an integrated Faraday isolator. The beam intensity spatial profile is near Gaussian

with a M^2 of <2. Output pulse stability is typically ~1% (1s) with a beam divergence of < 1.6 mrad. The laser beam was delivered to the workpiece using a XLR8 x-y scan head from Nutfield Technology Inc. A flat field lens of focal length 160 mm was used which gave a maximum field size of 120x120 mm. Scan patterns were software generated (Waverunner Nutfield Inc) which allowed materials interactions to be investigated whilst allowing full control of all laser parameters. The minimum spot size diameter at the focal plane was found to be 30~55 μm depending on the beam input diameter, 3-9 mm using beam expanders of magnification 1-3.

The output waveforms (WFM) for a number of settings are shown in figure.3. In this case the output pulse parameters have been controlled in order to provide the maximum peak power at higher frequencies.

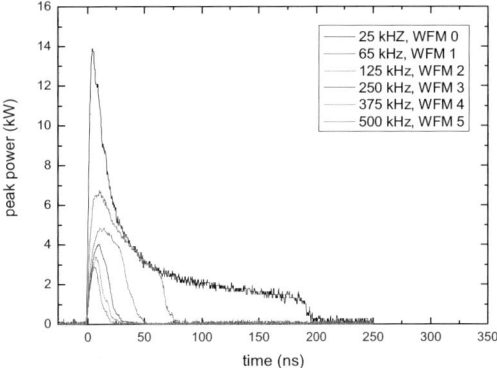

Figure.3 Laser pulse characteristics at varying pulse repetition rates. The PulseTune^{TM} capability enables maximum peak power to be maintained at higher repetition rates.

At higher frequencies the MOPA system is able to modify the pulse envelope in order to high peak power output. Processing with these output pulse waveforms providers much greater control than traditional Q-switched systems, which tend to produce flattened pulse profiles at higher pulse repetition rates. The PulseTune^{TM} system offers a variety of options for processing a host of materials which can provide superior processing results. Table.1 gives the incident beam parameters at focus for a variety of WFM outputs when using a 163 mm focal length objective lens.

Table.1 incident power density for a range of WFM settings an input bean diameters on a 163mm focal length objective lens

Input beam diameter (mm)	Min Spot Diam' (mm)	WaveForm Number (Power density MWcm^{-2})			
		0	1	2	3
6	57	548	274	196	156
9	41	1075	537	384	307

12	31	1878	939	670	536

One can immediately see that extremely large power densities can be achieved for low input powers of 20W, approaching 2 GWcm^{-2} for a spot size of 31 µm, a pulse energy of 0.8 mJ, and a pulse repetition rate of 25kHz. The WFM settings effectively optimise peak power at higher frequencies, so for a WFM of 0, 1, 2, and 3, the corresponding peak pulse powers are 14, 7, 5, and 4 kW respectively. The corresponding optimum pulse repetition rates for WFM of 0, 1, 2, and 3 are 25, 65, 125, and 250 kHz respectively.

4. Si Processing

The results of Si scribing are shown in figure.4.

Figure. 4a Si surface scribes produced at 25kHz (WFM-0), at 1m/s.

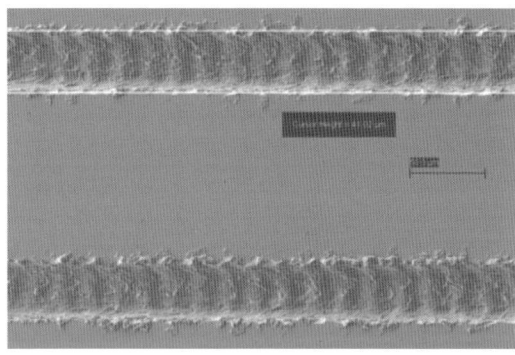

Figure. 4b Si surface scribes produced at 65kHz (WFM-2), at 1m/s.

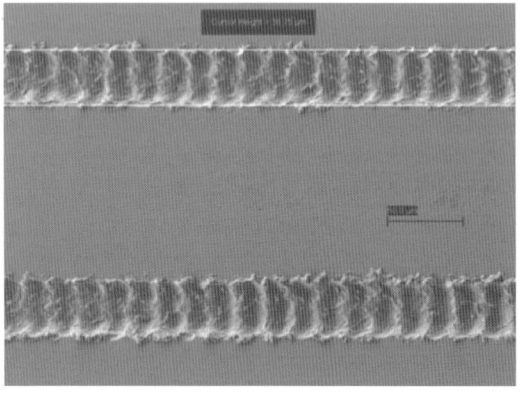

Figure. 4c Si surface scribes produced at 125 kHz (WFM-3), pulse width 30 ns..

All test samples were made at 500m/s, 20W, 163 mm f- theta lens, and a 9 mm input beam diameter. The scribes were produced on a 0.2 mm pitch. Despite the long wavelength, effective surface structuring was demonstrated by producing an array of micro pits on polished Si substrate. These can be seen in figure.5. In this case, the shallow 'pits' are 30-40µm wide and 20-40µm deep, produced using single pulses at 250 kHz repetition rate, and pulse shape of WDM-3 at 20W average power.

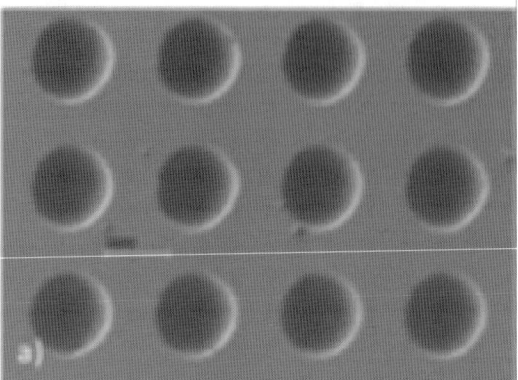

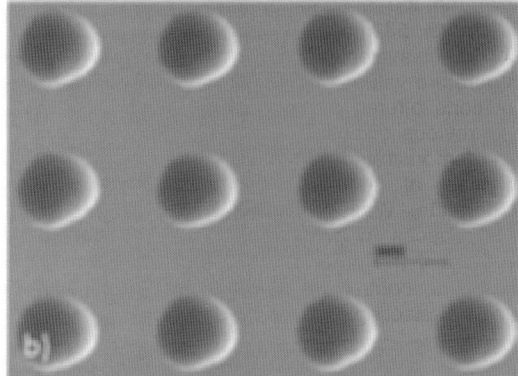

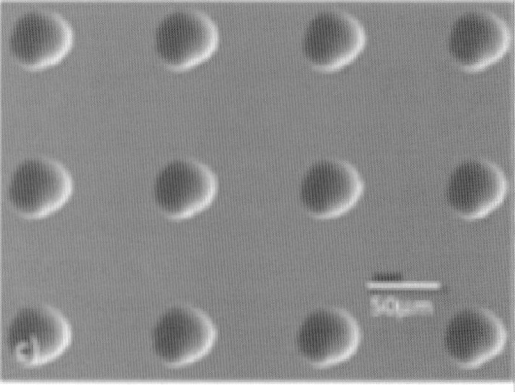

Figure.5 Micro pits ~30µm wide and 2-3 mm high, produced using a single pulse at 25 kHz. 65 and 125kHz repetition rates. (WDM number of 0,1,and 3) 20W average power.

Figure.6 shows the cross section of a blind slot cut into Si at 50kHz repetition rate, 10mm/s scanning speed and a fluence of 7.5Jcm^{-2}. The depth of the channel is ~ 100µm.

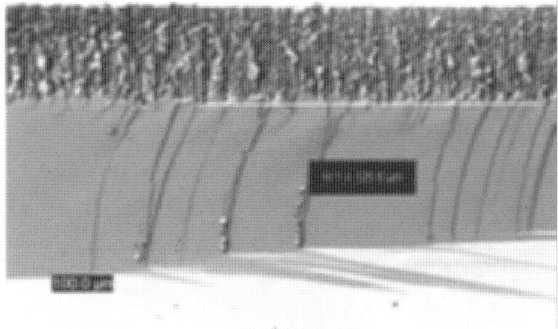

Figure.6 Section of blind slot cut into Si using WDM 23 (50kHz) at 7.5 Jcm^{-2}, 10mm/s scanning speed and seven overscans.

The ability to deliver high peak power pulses at 1µm wavelength can have a significant affect on resultant characteristics of Si processing. The usual laser of choice is the NdYAG DPSS laser operating at 532nm, 355nm , or 266nm. Previous work has shown that the absorption length in Si is of the order of 1µm @ 532nm wavelength and ~200 µm @ 1064nm wavelength [7]. These are room temperature values and as such do not relate to the conditions offered at the melting point and above. The previous discussion on optical properties of Si has demonstrated dramatic changes in optical response at temperatures exceeding the melt temperature. Whilst it is shown that the transmission of Si reduces to a level such that the absorption length is of the order of 0.07µm [6], it is therefore possible to control the nature of the interaction by carefully selecting the basic pulse profile and power density.

5. Glassy Carbon Processing

Glassy carbonis a non-graphitizing carbon which has a combination of glass-like and ceramic-like properties. The properties that are most significant are it's high temperature resistance, extreme resistance to chemical attack and impermeability to gases and liquids. Glassy carbon is widely used as an electrode material, high temperature crucibles, electronic substrates, and is sometimes used in biomedical implants and prosthetics due to its excellent biocompatibility. Glassy carbon is a composite consisting of amorphous carbonand one or more additional materials that possess uniqueproperties and characteristics. Glassy carbon is formed
by carbonizing phenolic resins which are made by reactingphenols with cellulosics, aldehydes and ketones [8]. It is an important material in a range of Microsystems applications although it is very difficult to machine having a shore hardness of 110, Young's modulus of 32,400 MPa, and a flexural strength of

147 MPa. A series of processing experiments were performed on as received GC (Sigradur) from HTW-Germany.

Figure. 7 presents the etch rate data for GC using two waveforms WDM3 and 5, operating at 250kHz and 500kHz respectively, at a range of incident fluences. The pulse energy for WDM3 was 0.08mJ and for WDM5 it was 0.02mJ. Despite having a much lower pulse energy, the WDM5 waveform has a higher etch rate than WDM3. This observation could be due to the reduced beam attenuation effects that occur at lower pulse energies and higher repetition rates, compared to higher pulse energies and lower repetition rates. The ablation rate shown here is an average value over many thousands of pulses, and shows a clear benefit for high frequency low pulse energy interactions. This needs to be investigated further to determine the intensity dependent phenomena occurring during ablation. The ablation process is however quite stable and repeatable, producing features with long range order and regularity. The threshold fluence is found to be ~0.3 Jcm^{-2}.

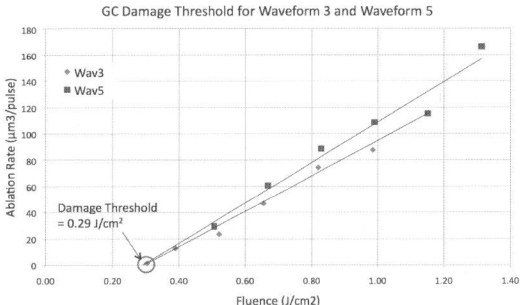

Figure.7 Process interaction data, ablation volume per pulse verses incident fluence on GC using WDM3 and 5 waveforms (250 and 500kHz repetition rates).

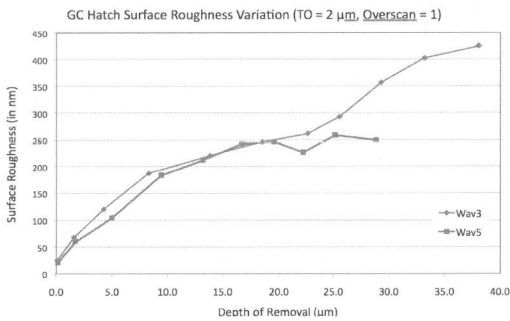

Figure.8 Surface roughness data (Ra) verses depth of removal of GC using WDM3 and 5 waveforms (250 and 500kHz repetition rates), using a transverse overlap of 2µm and a single pass. Speeds varied from 100m/s to 2500mm/s in order to create a variation in trench depth. Fluence = 1Jcm^{-2} in both cases.

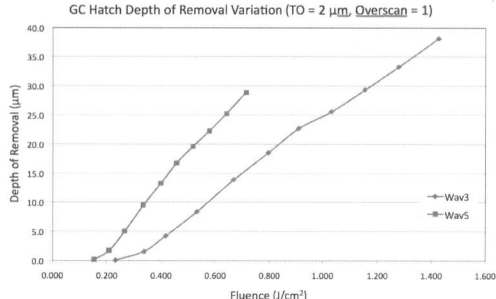

Figure.9   Machined depth verses fluence using WDM3 and 5 waveforms, 250kHz @ 1250mm/s and 500kHz @ 2500mm/s scanning speed. A transverse overlap of 2µm and a single pass was used in both cases.

Figure. 8 presents the surface roughness data for trenches machined in GC using two waveforms WDM3 and 5, operating at 250kHz and 500kHz respectively, as a function of machined depth. The surface roughness is seen to increase with machined depth with a linear dependence. Both waveforms are performing equally well in terms of quality of machined surface.   Although it is worth noting that the surface quality is particularly high for all process parameters compared to those obtained for Si. This is not surprising, since the machining process is dominated by well-controlled ablation effects and the absence of melting. Figure.9 shows machined depth verses fluence for the two waveforms used.   A reasonably linear response is obtained in both cases as long as the process is operated far away from the damage threshold of 0.3 Jcm^{-2}.

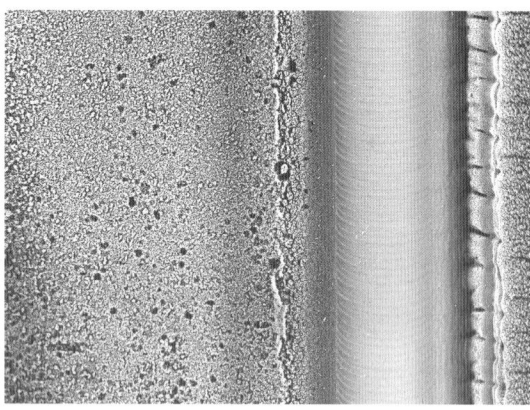

Figure.9 Plan view of a 50µm wide machined trench produced using WDM3 waveform, 250kHz @ 100mm/s, F=1Jcm^{-2}.

Figure.9 shows a plan view of a trench machined in GC. The image shows clear surface features on the base of the which are remnants of the multiple pulse overlaps, and significant debris on the surface bordering the trench.   On closer examination, this debris field is a condensate of carbon that has re-deposited on the neighbouring surface as the trench moves along the sample.   The debris forms a thin coating on the surface as can be seen in Figure.10

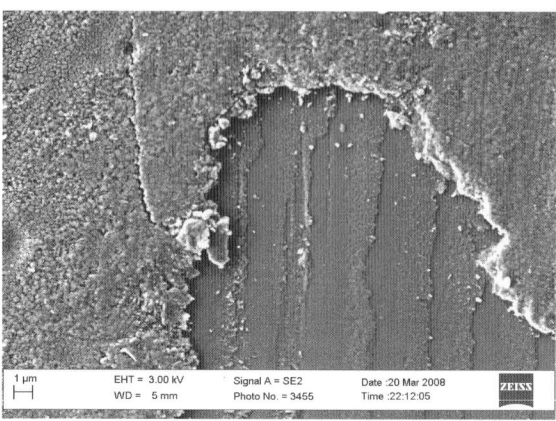

Figure.10   Surface detail of debris field shown in fig.9.

Attempts were made to utilize the discrete machining threshold of GC by employing sub threshold intensities to in-situ laser clean the surface of the newly machined microstructure at a fluence below the threshold value of 0.3Jcm^{-2}. The optimum value was found to be 0.2Jcm^{-2} as shown in figure.11.

Figure.12 shows a periodic surface structure directly machined by Yb fiber laser operating with WDM5 at 100mms/ and F=1Jcm^{-2}.   The long range order and stability are quite clearly seen.   These early results show the promise of a highly effective low cost micromachining solution for GC microstructures.

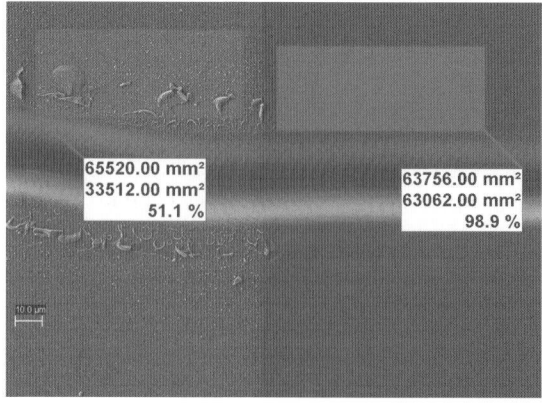

Figure.11 Shows a side by side comparison of two areas of a micro machined trench in GC.   The left hand side shows the newly machined feature with debris field and the right hand shows an area after it has been laser cleaned in-situ at F=0.2Jcm^{-2}.

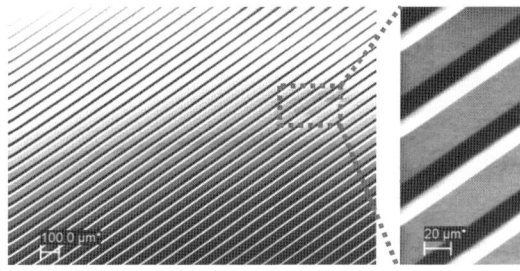

Figure.12 Periodic surface structure, 30μm wide, 15μm deep, machined in GC by at WDM5, 100mm/s, F=1Jcm⁻².

6.0   Summary

The use of a MOPA based Yb fiber laser has provided considerable choice when it comes to selecting pulse profiles, pulse repetition rates, and incident power density. The optical response of Si has shown itself to be complex and time varying, with almost the elimination of transmission at the melting point. Whilst the surface reflectivity increases at this stage, the MOPA laser can deliver sufficient energy to counteract the reflection losses, yet still produce sufficient energy deposition to create a range of surface effects such as re-melting and discrete melt removal, and surface structuring. These results show considerable promise in identifying the capabilities of the new MOPA Yb laser configurations. The markets for Si processing applications are set to increase dramatically as the global production of solar cells ramps up. There are many processing operations that could be carried out with this new low cost high efficiency laser system.

GC has been shown to respond extremely well to Yb fiber laser radiation with a very discrete damage threshold of 0.3Jcm⁻². Repeatable micromachining is easily obtained with operating fluences above this (<3x threshold fluence for best quality and performance). Work is at hand to develop a full 3 dimensional micromachining solution for GC micro-parts.

Exploration of this type of MOPA based Yb fiber laser is at an early stage, further work will develop a full range of interaction maps that will identify the processing domains of interest for a range of materials. Other experiments are in hand to examine the time varying nature of the optical properties of Si and couple this with laser parameters that are designed to meet the optical response of Si.

Ackowledgements

The authors would like to thank the EPSRC for supporting this work.

References.

1.   Blinov L.M., Bobrova E. A, Vvilov V.S, and Galkin G.A, Sov Phys Solid State 9, 2537 1968.

2.   Lampert M O, Koebel J M and Siffert P,"Temperaturedependence of the reflectance of solid and liquid silicon"*J. Appl. Phys.* **52** 4975–6, 1981.

3.   Jellison G. E. Jr and Lowndes D. H. Measurements of theoptical properties of liquid silicon and germanium usingnanosecond time-resolved ellipsometry *Appl. Phys. Lett.***51** 352–4. 1987.

4.   Shvarev K M, Baum B. A. and Gel'd P. V. "Opticalproperties of liquid silicon *Sov. Phys.—Solid State* **16**, 2111–2, 1975.

5.   Lowndes D.H., Jellison G. E. Jnr, Wood R.F., "Time resolved optical studies of silicon during nanosecond pulsed laser irradiation " Phys Rev B, **26**, 12, 6747-6755, 1982.

6.   Aydinli A., Lo H. W., Lee M. C., Compaan A., "Induced absorption of Si under intense laser excitation: evidence for a self-confined plasma", Phys Rev Lett, 46 (25), 1640, 1981.

7.   Bauerle, D., "Laser Processing and Chemistry", 3rd edition, Springer Verlag, 2000.

8.   Kotlenksy, W., Martens, H, Nature**206**, 1246 – 1247, June 1965.

Multi-Material Micro Manufacture
S. Dimov and W. Menz (Eds.)

Utilising Electrochemical Deposition for Micro Manufacturing

Peter T. Tang

IPU Manufacturing, Kemitorvet 204, 2800 Kgs. Lyngby, Denmark (tt@ipu.dk)

Electrochemical deposition, comprising both electroplating and electroless plating, plays an important role as an indispensable process family utilised in many micro manufacturing process chains. Advantages of electrochemical deposition, such as deposition speed, relatively inexpensive equipment, reliability, the large amount of available processes as well as the almost atom-by-atom replication of a given substrate, has given the technology its present position within microelectronics, surface treatment and recently also micro- and nano-manufacturing.

The present paper will briefly describe all the major disciplines of electrochemical deposition, as well as some of the problems and challenges that are usually associated with the different deposition processes. Finally are two applications, an all-nickel AFM cantilever and a new process chain for fabrication of tool inserts for injections moulding, described in some detail.

The chemical abbreviations for the elements (Cu for copper, etc.) are used throughout the paper.

Keywords: Electroplating, Electroforming, Indirect Tooling, Electroless Plating, Pulse Plating

1.0 Introduction

Electrochemical deposition (ECD) is a term used for all types of electrochemical reactions leading to deposition of material. As such it includes electroplating and electroless plating, but also other processes such as electrophoretic polymer processes and anodising of aluminium.

This paper will focus on the use of ECD in micro manufacturing for deposition of pure metals and metallic alloys onto various substrates such as Si, Al, Au, Cu and polymers.

The two most important sub-groups of ECD, electroplating and electroless plating, are both based on the same electrochemical reaction:

$$Me^{z+} - ze^- \leftrightarrow Me^0 \qquad (1)$$

In case of electroplating the part to be plated (called the cathode) is submerged in the electrolyte and connected to a power supply. Another electrode (called the anode) is also submerged in the electrolyte and connected to the power supply, thereby completing the electrical circuit.

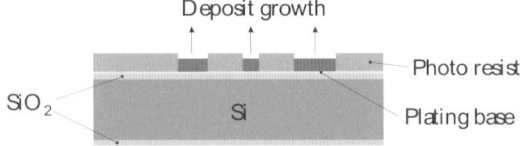

Fig. 1: Electroplating of a Si-wafer with a photoresist pattern and a sputtered plating base

On the surface of the cathode, reaction (1) moves from left to right, while on the anode, the reaction is reversed. In an ideal process the amount of metal deposited on the cathode is equalled by the same amount of metal being dissolved at the anode. However, in many practical processes this is only almost true, but a good electroplating bath can still be used for years with a minimum of maintenance.

In the electrolyte it is relatively easy to control where the deposition will take place, since the electrical resistance of the various cathode surface areas will determine the amount of available electrons for deposition and growth. For instance by applying a

photoresist or other types of masking (see fig. 1), deposition of metallic structures is possible with exceptionally good resolution (primarily governed by the resolution of the lithography process).

1.1 Basic considerations of electroplating

In the following some basic consideration of electroplating will be discussed, using Ni plating as a typical example.

The first useful nickel plating bath was proposed by C.P. Watts in 1916 [1]. This bath is still used extensively since it is relatively easy to establish - both from an economical and a practical point of view. The Watts Ni bath is operated at temperatures from 30 to 60 °C, usually with air agitation and at a pH ranging from 3.5 to 5.0. The current density is typically between 2 and 7 A/dm².

Table 1: Composition of Watts Ni bath [2].

Compound	Formula	Concentration
Nickel Chloride	$NiCl_2 \cdot 6 H_2O$	40 - 60 g/l
Nickel Sulphate	$NiSO_4 \cdot 6 H_2O$	240 - 300 g/l
Boric Acid	H_3BO_3	25 - 40 g/l

The appearance of deposits from the basic Watts Ni bath is not very good. The deposits are coarse and can have pitting holes in the high current density regions (corners, edges, etc.). The pitting holes are made by small hydrogen bobbles that have stayed on the surface long enough to prevent deposition in that point.

The bobbles are formed because a small fraction of the current will be used to reduce H^+ ions in the electrolyte to hydrogen gas:

$$2 H^+ + 2 e^- \rightarrow H_2 (g)$$

This usually undesired side reaction, must be taken into account when the deposition rate is calculated. The amount of coulombs delivered by the power supply is calculated as (I is the current in amperes, t the total plating time in seconds):

$$Q = I \cdot t \qquad (2)$$

and the theoretical amount of Ni deposited by the charge Q is calculated as (F is 96487 C/mole, z=2 for

Ni and M_{Ni}=58.71 g/mole):

$$m_{theory} = \frac{Q \cdot M}{z \cdot F} = t \frac{I \cdot 58.71 \frac{g}{mole}}{2 \cdot 96487 \frac{C}{mole}} \qquad (3)$$

If the current is 1.0 A and the metal is Ni we get 0.304 mg deposited Ni per second (at 100% current efficiency). The current efficiency (θ) is simply calculated as the difference between the theoretical and the actual (measured by weight, m_{Ni}) amount of Ni deposited.

$$\theta = \frac{m_{Ni}}{m_{theory}} 100\% \qquad (4)$$

Since most Ni electroplating processes exhibits a relatively high current efficiency of 96% to almost 100%, one should expect the pitting problem to be very small. For some applications however, due to the geometry and the resulting high local current densities, the problem will be present. For micro applications it is also possible to get pitting problems, especially if the cathode has deep (more than approximately 50 µm) holes or other structures in which the bobbles may be trapped.

Generally the current efficiency for acidic electrolytes is almost 100% (except for Cr plating which is from 6-30%) and for alkaline baths around 60-80% [3]. For the group of acidic baths, to which both sulphamate and Watts Ni baths belong, the current efficiency is highest (closest to 100%) at relatively high current densities. For the group of alkaline baths the tendency is the opposite [3]. Consequently one must be careful not to use very low current densities (from 0.5 A/dm^2 and below) in the acidic baths (or very high in the alkaline baths) as the current efficiency will be significantly reduced.

Whereas there is very little that can be done about the current efficiency (except to avoid extreme current densities) something can be done about the pitting problem. Most commercial plating baths use wetting agents to reduce the surface tension - thereby making it more difficult for the bobbles to attach themselves to the cathode surface. The wetting agents are typically "soap"-like organic chemicals with a long carbon chain connected to a polar group (i.e. sodium lauryl-sulphate, $CH_3(CH_2)_{11}SO_4Na$).

When the current efficiency (θ), the plateable area (A) and the density (δ) are known, it is possible to calculate the thickness of the deposit (x_{Ni}) at a given time (t):

$$x_{Ni} = t \cdot \theta \frac{I \cdot M}{z \cdot F \cdot A \cdot \delta} \qquad (5)$$

As a rule of thumb for Ni electroplating, we get a deposition rate of 0.5 µm/min. at a current density of 2.5 A/dm^2 (which is a very typical current density for Watts Ni baths). The plating rate can be scaled perfectly linearly within a large current density range. Some special Ni electrolytes can reach deposition rates of 5 µm/min. or more, but generally rates from 0.2 to 1.5 µm per minute are normal.

2.0 Electroplating

Deposition of metals using electrochemical deposition (ECD) requires an electrolyte. The electrolyte must fulfil a number of demands specific for the metal involved:

1. suitable salts of the metal can be dissolved in the electrolyte, at a reasonable temperature and in sufficient concentrations,

2. the electrolyte must be a relatively good electrical conductor in order to get an even distribution of the material and to avoid extensive heating of the bath,

3. the pH-value of the bath and the concentration of complexing agents should be kept within a certain range so that reduction of the metal occurs before reduction of hydrogen.

For some metals all these demands can not be fulfilled using a water based electrolyte. Elements such as Ti and Al can only be deposited from organic electrolytes, while other metals such as Mg, Nb, Ta, and W can be plated from molten salt electrolytes (at 700°C and above). For obvious reasons these electrolytes are both difficult to use and expensive, and are normally avoided whenever possible.

The metals that *can* be plated from aqueous solutions can be divided into three groups; electroplateable elements, elements that can (also) be reduced chemically (usually referred to as electroless plating) and elements that can only be co-deposited together with one of the others (see fig. 2).

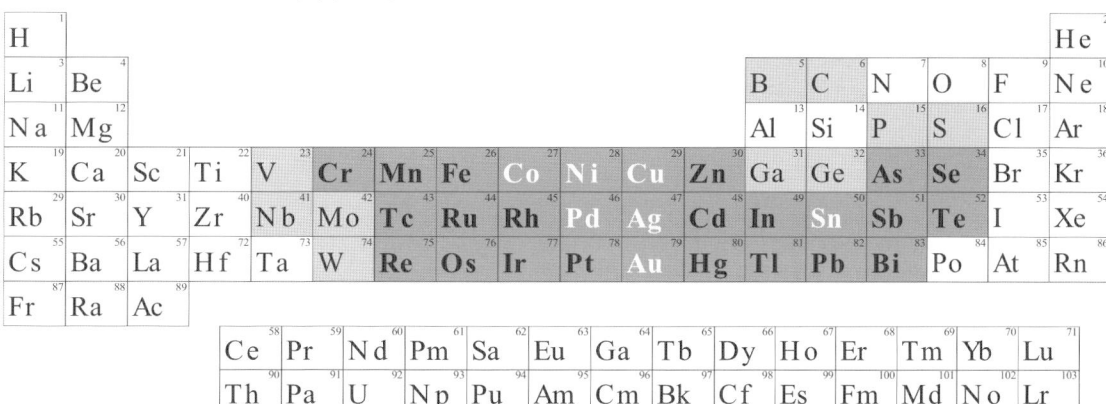

Fig. 2: Elements that can be plated from aqueous solutions (dark grey background) using electrochemical deposition. Elements with light grey background can only be deposited in combination with other elements (alloy plating). White text on dark background indicates elements that can be deposited using both electroless and electrolytic processes (Pt, Pb, Rh and Ru are occasionally considered belonging to this group [4]).

Electroplating is possible with all the elements marked in fig. 2 with a dark grey background.

Elements such as S, various oxides and hydroxides as well as particles (SiC, TiC etc.) can be co-deposited together with one of these elements.

The following pure metals are typically used for microstructures:

Ni is used for most mechanical structures, but the use of Au and Cu has also been reported. Ni is an obvious choice because of the good mechanical properties and relatively high corrosion resistance. Au is used where extreme corrosion resistance is needed, such as for implants or for reliable electrical contact. Cu can be used with advantage where good electric conductivity or high ductility is needed.

In order to create free-standing structures such as bridges or beams, a sacrificial layer is necessary. For this purpose metals such as Cu and Zn may be useful since they can be removed by etching without damaging other metals such as Ni or Au.

Soft metals such as Sn or In are used for soldering or bonding.

2.1 Electroforming

Electroforming is the most commonly used name for a process in which a machined substrate, referred to as the mandrel, is used for the deposition of metal. In some cases the mandrel is separated from the deposit after electroplating (an example of this is the process sequence discussed later called indirect tooling), and in other cases the deposit and the mandrel stay together as one component. Electroforming is in other words a process in which the deposited metal, with or without the mandrel, becomes the finished product itself. This product can then be used as a tool for injection moulding or embossing or for various other applications.

Fig. 3: Scanning electron microscope image of an all-nickel AFM-component fabricated using a combination of micro-etching of Si (pyramidal tip) and two steps of UV-lithography and electroforming - creating the cantilever and the support [5].

Whenever electroforming is utilised, the four most important properties are:

Material distribution

By distributing the material evenly, it is possible to both save time and metal, and to avoid expensive machining after plating. In some applications however, such as tooling for replications of microfluidic systems into polymer, an even distribution may not be sufficient. The so-called high aspect ratio applications require that the electroforming process is capable of filling deep trenches (or other features which are characterized by having a depth to width ratio from 2 up to 5 or even higher) without forming voids inside the structure.

Internal stress

For almost every application a low stress level is important. With too much stress there is a risk that the deposit will change shape when it is released from the mandrel. In extreme cases the accumulated stress can exceed the adhesion forces between deposit and mandrel (and thus create partly or full exfoliation) or deform the mandrel during the deposition process.

Hardness

When electroforming is used to produce tools for injection moulding or hot-embossing, a hard deposit will usually extend the life-time of the tool. Increased wear resistance, by increased hardness, can however lead to brittleness if sulphur containing additives (such as saccharine) are used in the electrolyte [6]. Particularly when Ni tools are repeatedly heated to temperatures above approximately 150 degrees, will re-crystallisation of Ni commence, accompanied by transportation of S and C to the grain boundaries - if these impurities are present in the matrix due to co-deposition.

Re-crystallisation will lead to larger crystals, and consequently lower the hardness, while diffusion of S to the grain boundaries will increase the hardness – typically accompanied by an almost glass-like brittleness.

Corrosion resistance

Even the smallest corrosion attack could be fatal for a micro component. For replication techniques such as injection moulding, corrosion in combination with wear (especially from polymer fillers) must be considered.

Table 2: Typical composition of conventional (Conv.) and high-speed (HS) sulphamate Ni baths [7].

Compound	Formula	Conv.	HS
Nickel sulphamate	$Ni(NH_2SO_3)_2 \cdot 4\,H_2O$	300	600
Nickel chloride	$NiCl_2 \cdot 6\,H_2O$	30	10
Boric acid	H_3BO_3	30	40
		g/l	g/l

Ever since its introduction, shortly after 1960, the sulphamate Ni bath has been the most used process for electroforming. Compared to the Watts solution, the main improvements that can obtained using sulphamate Ni, is a tremendous increase in deposition rate and the possibility to make almost stress-free deposits. The disadvantages are that the sulphamate bath is more expensive, more sensible to pollution and has a high density (especially the high-speed solution). In a high density solution the pitting problem is intensified, because the hydrogen bobbles are more difficult to remove from the cathode surface. Another disadvantage is the risk of hydrolysis according to the following irreversible reaction:

$$Ni(NH_2SO_3)_2 + 2 H_2O \rightarrow NiSO_4 + (NH_4)_2SO_4$$

Hydrolyses will occur if the bath temperature is increased above 70°C or if the pH-values is decreased below 3.0. Hydrolysis will also occur if the potential gets very high because of insoluble anodes (the anodes must be S alloyed to facilitate the dissolution) or small anode surface area. High concentrations of both sulphate (SO_4^{2-}) and ammonium (NH_4^+) will increase the internal stress in the deposits.

The mechanical properties of Ni are in many ways comparable to Si, and the fabrication of an all-Ni AFM cantilever is discussed later in application part, to demonstrate this. The utilisation of this cantilever for AFM lithography on Si is described by Birkelund et. al. [8].

2.2 Electroless plating

Chemical or electroless plating can be divided into four groups:

- immersion plating or ion exchange plating
- substrate catalysed electroless plating
- auto-catalytic electroless plating
- contact plating

The most useful are groups 1 and 3, but all four groups will be defined below.

Immersion plating

In the case of a metal substrate submerged in an electrolyte containing a solution of a more noble metal, a process usually referred to as ion-exchange deposition or immersion plating will start spontaneously. This type of process produces very well defined [9] thin films of the more noble metal (such as Au in the reaction below) while the less noble substrate (Ni) is dissolved (and thus supplies the electrons needed).

$$Ni^0 \rightarrow Ni^{2+} + 2 e^- \text{ and } 2 Au^+ + 2 e^- \rightarrow 2 Au^0$$

Thicknesses of up to 8 µm have been reported [10], but usually much thinner layers are obtained using this method. Commercial baths for immersion gold plating, which is the most widely used immersion plating process, are available. The combination of (auto-catalytic, see below) electroless nickel (EN) and immersion gold (IG) is used extensively by the electronics industry for contacts, soldering pads etc. and is usually referred to as the ENIG process.

The immersion plating processes are very stable and produced thin but well-controlled layers that tend to stop growing once the substrate is fully covered by the more noble metal.

Substrate catalyzed electroless plating

An intermediate between ion-exchange and auto-catalytic electroless plating is substrate catalyzed electroless plating. This process requires a substrate that is catalytic for a specific electroless plating process. The deposited metal however, is not itself catalytic for metal deposition. As a result this type of process leads to very uniform layers since the process will stop when there are no available areas of the original catalytic substrate. A hydrazine based electroless Au process containing cyanide as a complexing agent is an example of a substrate catalyzed electroless plating process that works on substrates such as Ni or Co [11].

Auto-catalytic electroless plating

When referring to an auto-catalytic electroless plating process, it is implied that the deposited metal act as catalyst for the reduction of more metal ions. If the substrate is not a catalyst for the desired process in itself, it must be made catalytic. This can lead to several pre-treatment or activation steps. Electroless plating is nevertheless a widely used process, mainly because of its ability to create metal coatings on non-conducting materials such as polymers and ceramics, but also because of the excellent material distribution that can be obtained due to the lack of an electric field. Electroless plating of Ni on Si wafers has been reported [12], but also electroless plating of Au and AuPd [13] could be very important processes for the fabrication of some microstructures.

Tabel 3:   Simple and relatively stable recipe for an auto-catalytic electroless Ni bath. The bath is operated at 90 °C and deposits around 25 µm/ hour. The pH-value is adjusted to 4.5 with ammonia [from 14].

Compound	Formula	Concentration
Sodium hypophosphite	$NiCl_2 \cdot 6 H_2O$	0.25 M
Nickel Sulphate	$NiSO_4 \cdot 6 H_2O$	0.10 M
Acetic acid	H_3BO_3	0.60 M

Auto-catalytic electroless plating is possible with the elements Co, Ni, Cu, Pd, Ag, Au and Sn [15] (white text on dark grey in fig. 2). However auto-catalytic electroless plating of Pt, Pb, Rh and Ru has also been reported [4].

Electroless plating of Ni (and only Ni) always involves a co-deposition of P or B originating from the reducing agent used in the electroless plating solution (see table 3). Depending on the type of complexing agent used, very different deposition rates (from below 4 to more than 25 µm/hour), internal stress values (from 140 MPa compressive to 70 MPa tensile) and P contents (from 7 to 12 wt%) can be obtained [14].

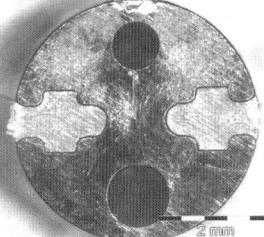 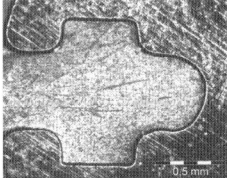

Fig. 4:   Polymer component made by cavity transfer two component (PPE/PEI) injection moulding, before (left) and after (right) selective metallization of PPE with auto-catalytic Cu [16].

The driving force in electroless plating is not applied current but a chemical reducing agent such as sodium hypophosphite (NaH_2PO_2), formaldehyde (HCHO) or hydrazine (NH_2NH_2). When metal is reduced the reducing agent is oxidized and must therefore be replenished frequently. There are no anodes in the auto-catalytic electroless plating bath, and therefore the deposited metal must also be replenished. This gives rise to a build up of oxidized reducing agents, as well as other ions added with the metal as metal salts. Because of this, and because of the nature of most of the reducing agents used, electroless plating baths are usually relatively unstable and have a limited lifetime.

Contact plating

This rather unusual electroless plating process requires three different conducting materials at the same time - a metal in solution (the one that is going to be deposited) and a substrate consisting of two different metals. One metal will be used as sacrificial electron donor and the last metal simply serves as a conducting plating base. The sacrificial metal will dissolve in the plating solution, thereby releasing electrons that can be used for deposition of the metal in solution. A gold plated silicon wafer partly covered with aluminium and submerged into a electroless nickel bath, will be covered by nickel (although gold is not a catalyst for electroless nickel plating) because aluminium is dissolved in the process [17]. After a while, when the wafer is almost covered with nickel, regular auto-catalytic electroless nickel deposition will take over.

2.3 Additives

Electroplating will, for most metals, produce a relatively rough surface that tends to enhance surface irregularities rather than making the surface smoother. This effect can be acceptable for thin deposits (from about 5 µm and below, depending on the application), but it is not acceptable for electroforming of microstructures. As illustrated in fig. 6, additives can occasionally solve this problem by producing surfaces that are even smoother than the substrate, provided that the deposit is allowed to grow to a certain thickness.

Additives are chemical compounds added to the electrolyte in small amounts. The additives are often organic molecules, known to attach themselves to the substrate during plating. When deposition of metal has to take place through this additive film, both texture and properties of the deposit will change. Using additives in Ni electrolytes, which is one of the most important metals for electroforming, properties such as; hardness, internal stress, smoothness, texture and structure (laminar growth versus columnar growth) can be altered considerably [7, 18-19].

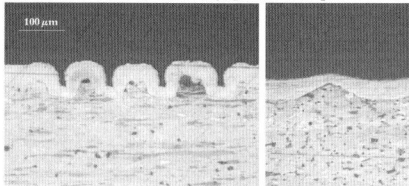

Fig. 5: Cu plating without (left) and with (right) level-ling additives. The average thickness is 50 µm [2].

The electronics industry in particular has encouraged the development of several additive systems. These systems, used in Cu electrolytes for the production of printed circuit boards and for the so-called damascene copper process, contain different kinds of levelling agents and brighteners that, together with high electrolyte conductivity, improve the material distribution considerably.

Levelling effects as illustrated in fig. 5, can also be obtained for electroplating of Ni, Zn and some alloys.

2.4 Pulse Plating

Electrodeposition by means of pulsating current (or

potential) also gives the possibility to change the properties of the deposit, just like the addition of organic compounds. The advantage is, that the pulse patterns or waveforms can be changed easily while the additives are difficult, if not impossible, to remove from an electrolyte once added. Furthermore, some of the improvements that pulse plating implies are not possible by adding organic or other additives to the solution. This is especially true when pulse reversal plating (fig. 6) is applied.

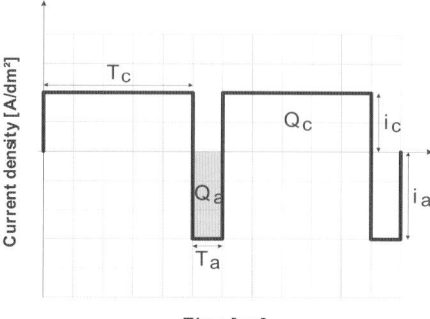

Fig. 6: Current versus time diagram for pulse reversal plating. Typical values for pulse reversal plating of Ni are; T_c = 100 ms, T_a = 20 ms, i_c = 4 A/dm^2 and i_a = 6 A/dm^2 [18].

Using ordinary pulsed current (on/off plating) it is possible to reduce the size of the deposited crystals, thereby improving hardness and tensile strength of the deposit and reducing the number of pores [20-21]. This is only true however, when relatively high frequencies (above 100 Hz) are used - lower frequencies usually have little effect [22]. For some metals, such as Au, on/off plating will change the crystal growth pattern [23], leading to porosity and hardness improvement that are not obtainable using additives.

Pulse reversal plating makes it possible to improve the material distribution, by dissolving "unwanted" crystals during the anodic periods. This type of pulse plating also has a dramatic influence on the texture of the film [20] and is known to be able to reduce internal stress of the deposits [18].

As the size of the holes on the printed circuit boards is decreased and the thickness of the boards is increased, it becomes more and more difficult to deposit enough copper in the holes. Using pulse reversal plating it is possible to improve the material distribution [24], by dissolving previously plated copper in high current density regions during the anodic periods.

2.5 Plateable alloys

Perhaps even more useful and interesting than the pure metals are the numerous alloys that can be plated from aqueous solutions (fig. 8). The special rules that apply for alloy plating, such as the importance of agitation [25] diffusion and limiting current density for both metals, are excellently described by Brenner [10] and will not be discussed here. Deposition of alloys with special properties includes:

Magnetic materials

One of the most important soft magnetic materials is PermAlloy (79% Ni, 21% Fe) which can be plated from sulphate based electrolytes [26], with better current efficiency from citrate electrolytes [27] or for special application from chloride electrolytes [25].

Several commercial plating baths are also available for PermAlloy plating. More sophisticated materials such as the CuCo composition modulated alloy exhibits exceptional properties such as in-plane magnetic anisotropy [28].

Depositions of both NiFe and CoNiFe (magnetic saturation flux density, B_s of 1.1 and 1.8 T respectively) for MEMS application have also been reported [29]. To avoid co-deposition of S (thus avoiding stress reducing additives such as saccharine) while keeping the internal stress at a very low level, pulse reversal plating was used.

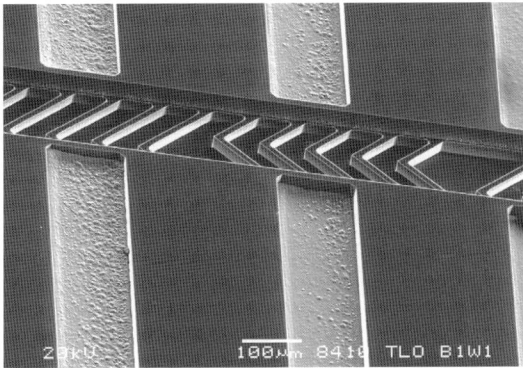

Fig. 7: Microfluidic channel fabricated in Si using reactive ion etching. The herring bone structure is designed to promote turbulent flow in the channel. Electroplated NiFe alloy acts as soft magnetic bars, capable of capturing magnetic beads in the channel when magnetised by an external field [30].

Solderable alloys

Classic solderable alloys such as PbSn (90% Pb) and SnPb can be deposited [10]. Recently a number of lead-free alloys such as SnAg [31], SnAgCu (the so-called SAC solder), AuSn [32] and SnZn have been studied intensively.

Corrosion protection

CuNi alloys such as Monel (30% Cu) or konstantan (60% Cu) are stable in various corrosive environments. Konstantan is also known to have a temperature independent (high) electric resistance. The CuNi alloys can be deposited from several different electrolytes [10]. Brass (CuZn) and bronze (CuSn) are fairly easy to deposit [10], while stainless steel (FeNiCr) can be very difficult to electroplate with an acceptable result.

Mechanical properties

The most important use of alloy plating however, is perhaps the possibility to improve the mechanical properties of metals such as Ni, Au and Cu. The hardness and wear resistance of Cu is increased 75% by alloying with 0.2-0.4% of Sb [26]. The hardness of electroplated Ni can be almost tripled (560 HV) using Co (30-40%) as the alloying element [29].

Temperature resistant alloys such as CoW (59/31) maintain a hardness of almost 700 HV at 600°C [33] and also NiW (12% W) provides a hard deposit with substantial thermal stability [34].

Other important mechanical properties such as internal stress can be improved as well. AuPd alloys containing 20% Pd can be plated absolutely stress-free

[26] whereas pure Au can not. Deposits with low thermal expansion coefficients can be obtained by plating Invar (35% Ni and 65% Fe) [35-36] or super Invar (30% Ni, 65% Fe and 5% Co) [37].

2.6 Multilayered deposits

By combining layers of different metals or layers with different alloy composition, deposits exceeding the properties of the single metals can be obtained.

Composition Modulated Alloy (CMA)

If the thickness of each layer in a multilayered material is low (typically 1000 nm or less), and the total deposit consists of several alternating layers of two different metals or alloy compositions, the deposit is referred to as a Composition Modulated Alloy (CMA).

The properties of CMA materials exceed by far the properties of the individual metals involved. Extreme changes in tensile strength [38] and magnetic properties [39] have been reported.

The reason for the dramatic change in properties is not fully understood, but has to do with the smaller crystals that the metals are forced to create because of the frequent changes and the composite behaviour of the systems. If two carefully selected metals or alloys are mixed in a multilayered material, it is occasionally possible to get a deposit with a combination of the best properties of the two constituents.

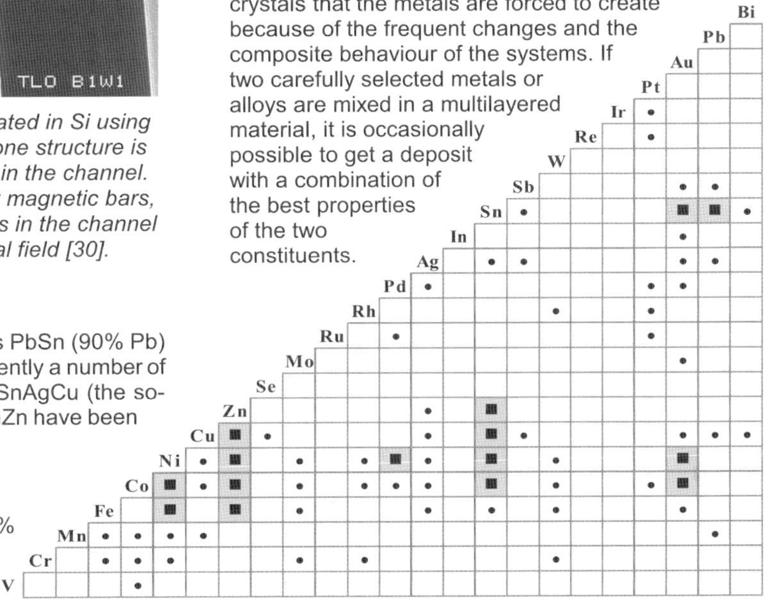

Fig. 8: Binary alloys (•) that can be deposited from aqueous solutions. A box on a grey background (■) indicates that the alloy deposition process is commercially available (adapted from [10]).

CMA electrodeposits can be produced in two different ways; by a dual bath process or by a single bath process. The dual bath system involves two separate electrolytes and a substrate that is moved back and forth. In the single bath system, alternating layers of different alloy composition are created, using pulse plating in an alloy plating bath [9, 32].

3.0 Applications

3.1 All nickel AFM cantilever

When a new technology or material is introduced, it is not unusual to incorporate it in a "high-end" application to prove that it works. If someone invents a new type of batteries, they may be used in an electrical sports car, just to show what they are capable of.

Without any direct comparison to the above, it was decided to fabricate an all-nickel atomic force microscope (AFM) cantilever – simply to demonstrate the possibilities and properties of using Ni as a construction material in a MEMS environment.

The cantilever consists, quite traditionally, of a tip, a cantilever beam and a support structure (for handling).

The mould for the tip is defined in Si using traditional Si micro machining; i.e. anisotropic silicon etching (KOH, 28 wt%, 80°C) using silicon dioxide as etch-mask. This results in a pyramidal tip-mould defined by the (111) crystal planes. This tip-mould is sharpened by growing a non uniform, low temperature (950°C), wet thermal oxide (4000 Å) [5].

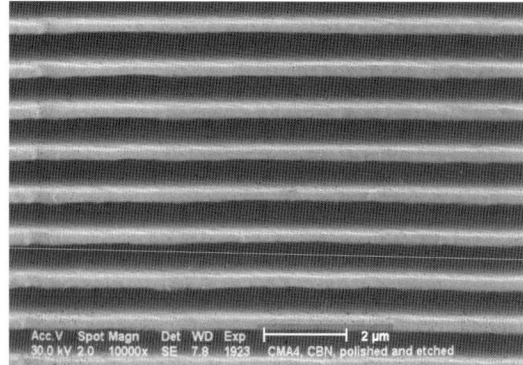

Fig. 9: CMA material consisting of Cu and Ni. Cu has been removed by selective etching. Deposits of more than 1000 layers have been produced [9].

A plating base (adhesion and conducting layer for the plating process) of 50 Å Ti and 1000 Å Au is deposited by physical vapour deposition (PVD). The cantilever and tip are electroplated in Ni using the KOH-etched cavity and a 9.5 µm thick photoresist pattern as a mould. As photoresist the AZ 4562 from Hoechst is used.

In the electroplating process the thickness of each cantilever is typically set to 7 µm. The support structure for the cantilever is electroplated in Ni using a 97 µm thick photoresist pattern as mould. The photoresist is the same as before but the processing is a little different. Four layers of approximately 25 µm resist is stacked on top of each other. This result in a film of 97±2 µm. The resist is exposed in a mask aligner using contact printing. The total thickness of the plated support structure is usually equal to the thickness of the photoresist, but this is not a critical step.

At last the Si wafer is dissolved in a selective wet etch (KOH, 28 wt%, 60°C), and the AFM-probes can now be released one by one with a pair of tweezers and are ready for use.

In order to assess the internal stress in the deposited Ni, a thin-film (2-4 µm) without pattern is electroplated on an entire 4"-wafer. Stoney's expression for stress [see ref. 5 for more details] is then used to calculate the stress in the Ni film. The input parameters are the thickness of the plated thin-film and substrate, the modulus of elasticity of the substrate and the change in curvature of the substrate before and after electroplating. The change in curvature is measured at room temperature using a mechanical profiler. Since the material distribution of the electroplated nickel is not perfect, the calculated stress is an average stress value based on an average thickness calculated from the weight gain. The

uncertainty on the measurement and calculation is in the order of 10 MPa.

The difference in the thermal expansion of the different materials involved will also cause some problems with stress. This is especially true when working with thick metal layers on top of Si. As an example the calculated introduced stress in a Ni thin film on a thick silicon substrate that is cooled down from 43°C (typical plating temperature) to 22°C (room temperature) is approximately 62 MPa or almost 3 MPa/°C.

The plating parameters are chosen so that the Ni deposit contains a small amount of compressive stress (as plated at the bath temperature), balancing the tensile stress occurring as a result of cooling down to room temperature. This is done to minimise stress in the materials. After the last plating step the Si wafer is dissolved, ending up with an array of ready for use all-metal AFM-probes.

The thickness of the cantilevers is controlled in the plating process. The total thickness variation of the cantilevers across the entire 4"-wafer is less than 10%, but it is better than 2% in the central 2"-area. The design of the photo mask for the resist mould and the design for the wafer plating holder are very important parameters for the thickness distribution of the electroplated deposit across the wafer. A special holder with an aperture (a so called "current thief") surrounding the wafer is used [see detailed description in ref. 2]. The aperture is at the same electrical potential as the wafer during plating, which provides a more even thickness distribution. Another problem is the pattern distribution/density across the wafer. Generally the thickness is smaller in areas with a high pattern density (large plating area) than in areas with a small pattern density. This problem is reduced by a rational mask design with a homogenous pattern distribution.

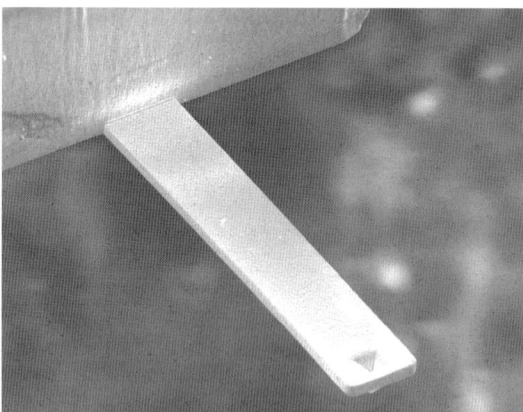

Fig. 10: The back of the AFM cantilever, clearly showing that the pyramidal tip is not filled with nickel and that the thickness of the deposited nickel appears to be very uniform [5].

3.2 Tooling concepts for polymer replication

One of the most important applications of electroforming is tooling. The tools are typically used for injection moulding or stamping of various – mainly polymer based – products, ranging in size and complexity from blue-ray discs to interior decoration panels for passenger aircrafts.

Utilising a strictly systematic approach, the

possible tooling concepts can be divided into four groups or schemes [40]. The first division is made by identifying the most important shaping process, i.e. the process that creates the shape of the finished tool. This process can either place material on a substrate (additive process) or remove material from a substrate (subtractive). The substrate is normally a homogeneous material, typically metallic or ceramic (Si), but it could also be a hybrid material (multi-layered, containing particles or fibres, etc.). The shape of the substrate is typically a flat disc or plate, but other simple shapes (rods, spheres, etc.) are also possible.

The second division is made with regards to the way the final tool is obtained. If the tool is fabricated directly it is understood that the substrate – after additive or subtractive machining – will become the tool. In case the substrate is removed during one of the subsequent steps in the process chain, the tooling concept will be considered an **indirect** one [41].

Additive processes used in micro fabrication include:
- electroforming
- laser sintering
- physical and chemical vapour deposition
- printing

Subtractive processed used in micro fabrication are:
- milling, turning or other machining processes
- electrodischarge machining (EDM)
- chemical etching
- electrochemical machining (ECM)
- laser machining/ablation (alternatively other beam processing technologies)

Some of the processes mentioned above require photolithography or other masking methods to define the areas that will be affected (and thus also the areas that will remain unchanged). To a certain extent, lithography can also be considered as belonging to the group of subtractive processes.

For the indirect process chains, one particular process is in common and has some special demands on the choice of materials. Indirect tooling, additive or subtractive, always requires – at some point – separation of the substrate and the more or less finished tool. A complex tool, with micrometers sized features and high accuracy, is not easy to separate from a substrate using mechanical methods or brute force. Consequently the most gentle way to do the separation is to chemically dissolve the substrate. In this case the substrate material should be one that can be dissolved easily and cleanly, without damaging the surface or structure of the tool.

Another important point when selecting a tooling process chain is the type of features to be realized and their relationship with the rest of the insert. As many micro structuring processes are based on material removal (e.g. subtractive processes such as micro milling, micro EDM, etc.), the smaller the total amount of material to be removed is, the faster the tool production will be completed. Thus when the tool is characterized by small cavities on a relatively large substrate, direct machining of the tool can be an advantage. In contrast, when the tool is characterized by small protrusions relatively isolated on a large substrate, indirect tooling is often the best approach.

3.3 Indirect tooling example for microfluidics

A very good example of a type of tool which is much easier to machine indirectly than directly, are replications tools for microfluidic applications.

Microfluidic systems are characterised by having channels, reservoirs and similar structures taking up a relatively small amount of the surface of the substrate. It is consequently much faster and easier to machine the channels than to remove almost the entire surface leaving only protruding walls (the opposite of a channel).

The term indirect tooling covers the fact that the master structure produced by machining is the positive geometry, which is identical in shape to the final product (i.e., the opposite of what is needed for the actual mould insert). In addition to that, due to minimum micro feature sizes, the EDM-milling process was selected. The material substrate (or master) needed to be suitable for both EDM-milling (to be conductive or semi-conductive) and for selective etching after the electroforming is completed.

Moreover, due to the micro fluidic application, it is preferable to have a flat and smooth substrate surface to ease the sealing during the packaging of the final product (i.e., bonding of a lid on top of the micro fluidic system). For these reasons, an 8'' silicon wafer (1,000 lm thick) was used as a substrate to be machined by EDM-milling.

The process chain was composed by the following steps:
- μEDM of micro structures on the silicon substrate (master).
- Laser cutting of the Si wafer to the specified size and shape to fit into the holder for the electroforming bath.
- Pre-treatments including cleaning and deposition of a thin layer of Ti/Cu by PVD.
- Electroforming of Ni and Cu for the insert.
- Selective etching of Si in NaOH.
- Mechanical machining of the back and side of the insert to fit the external mould.
- Final cleaning and selective etching of the Ti/Cu layer.

After μEDM milling, the Si master was cut out from the wafer in a circular shape using an Nd:YAG laser and the original Au layer on the wafer was removed by selective etching. In the areas affected by the EDM-milling the Au had already been removed, and therefore - in order to avoid problems with different electrochemical behaviour of the various materials - it was decided to remove the Au layer everywhere.

Fig. 11: Cu/Ni tool insert after the final machining using wire EDM to cut the tool insert from the block of deposited metal [43].

In a second step a thin layer of Ti/Cu was applied by PVD DC-magnetron sputter coating. After the deposition of the Ti layer (50 nm) an estimated 300 nm thick Cu layer was deposited as a conduction layer, without breaking the vacuum of the chamber, in order to avoid any oxidation of the Ti. In order to apply the coating to the side walls, the sputtering process was optimized for maximum step coverage. For higher aspect ratios, or negative slopes, problems can occur as sputtering generally is a line-of-sight process.

Electroforming of Ni and, after that, of Cu was then performed. In particular, the Ni deposition was carried out by immersing the master for 5 hours in a specially optimized low-stress nickel bath. The bath was based on nickel sulphamate and operated at 32 °C with a relatively low current density of 1.0 A/dm^2 (corresponding to a deposition rate of 0.18 µm/min), resulting in a deposited thin nickel layer (54 µm) for tool wear resistance. The relatively low temperature of the Ni sulphamate-based bath nearly avoids the introduction of thermal stresses induced by the difference of the thermal expansion coefficient of Si and of Ni. The decrease of the effect of the internal stresses induced by the electrochemical deposition, was one of the reasons for choosing the 8" wafer, as the thickness of a standard Si wafer increases with increasing size. Having a relatively thick substrate made the stresses unable to bend the substrate.

After the deposition and a surface re-activation,

the Ni plated silicon master was immerged in the Cu bath, which was mainly composed by an aqueous solution of $CuSO_4$ and H_2SO_4 at room temperature. With a deposition time of 168 hours, a 2.75 mm thick layer of Cu could be deposited. The advantage of using Cu deposition as the second and thicker layer is that the bath can run at room temperature which means only low stresses were induced on the work-piece. From the point of view of the micro tool, the Cu layer has a higher thermal conductivity than Ni (385 and 60 W/m·K, respectively) which for a mass-production application could turn out to be an important feature since the better conductivity enables fast cycle times in the injection moulding process.

The next step was the Si dissolution, which was performed using a NaOH (60 g/l) solution. Once the Ti/Cu surface appeared (previously deposited by PVD as a plating base) due to the silicon being dissolved, it was removed using a selective etching solution [42], which will only remove Cu and not attack the Ni tool surface in any way.

Eventually the tool insert was used for injection moulding of polypropylene (PP) replicas. Machine settings and other parameters are described in detail by Tosello et al. [43] along with extensive metrology results.

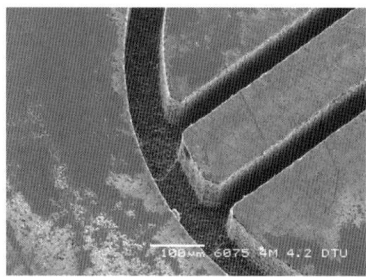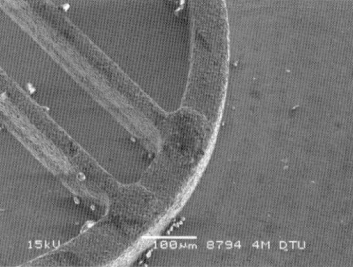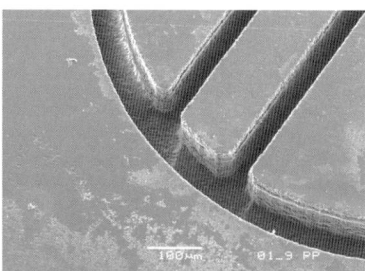

Fig 12: SEM images of the Si master (left) machined by µEDM-milling, the Ni insert (middle) made by electroforming and an injection moulded polypropylene replica (right) [43].

4.0 Acknowledgements

The author wishes to thank the Technical University of Denmark (DTU), and - in particular - the members of the micro and nano manufacturing group of the department of mechanical engineering.

5.0 References

[1] C.P. Watts, Trans. Electrochem. Soc., Vol. 29, pp. 395 (1916)

[2] P.T. Tang, "Fabrication of Micro Components by Electrochemical Deposition", Ph.D.-thesis, Dept. of Manufacturing Engineering, March, KMT 980301-1 (1998)

[3] P. Møller, "Avanceret Overfaldeteknologi", Technical University of Denmark, Compendium 8007 (1994)

[4] T. Watanabe, "Surface Modification Technology", Sangyou Gijutu Service Center (in japanese), pp. 49 (1993)

[5] J.P. Rasmussen, P.T. Tang, C. Sander, O. Hansen & P. Møller, "Fabrication of an All-Metal Atomic Force Microscope Probe", Proceedings, Transducers '97, Chicago (1997)

[6] A. Kubota et al., "The Effect of S Content in Deposits on Low-cycle Fatigue Properties of Ni Film Electroplated from Sulphamate Baths", Iron&Steel (in Japanese), Vol. 86, No. 1, pp. 38-44 (2000)

[7] S.A. Watson, "Compendium on nickel electroplating and electroforming", Nickel Development Inst., Toronto (1989)

[8] K. Birkelund, E.V. Thomsen, J.P. Rasmussen, O. Hansen, P.T. Tang, P. Møller & F. Grey, "New approaches to atomic force microscope lithography on silicon", J. Vac. Sci. Technol. B, Vol. 15, No. 6, pp. 2912-2915 (1997)

[9] P. Leisner, C.B. Nielsen, P.T. Tang, T.C. Dörge & P. Møller, "Methods for electrodepositing composition-modulated alloys", J. of Mat. Proces. Tech., Vol. 58, pp. 39-44 (1996)

[10] A. Brenner, "Electrodeposition of Alloys", Academic Press, New York (1963)

[11] C.D. Lacovangelo & K.P. Zarnoch, "Substrate-Catalyzed Electroless Gold Plating", J. Electrochem. Soc., Vol. 138, No. 4, pp. 983-988 (1991)

[12] M. Parameswaren, D. Xie & P. G. Glavina, "Fabrication of Nickel Micromechanical Structures Using a Simple Low-Temperature Electroless Plating Process", J. Electrochem. Soc., Vol. 140, No. 7, pp. L111-L113 (1993)

[13] E. Quéau, G. Stremsdoerfer, J.R. Martin & P. Cléchet, "Electroless Metal Deposition as a Useful Tool For Microelectronics and Microstructures", Plating and Surface Finishing, Vol. 81, No. 1 (1994)

[14] K. Tashiro, K. Chiba, Y. Fukuda, H. Nakao & T. Watanabe, "Origin of Nodules in Electroless Nickel-Phosphorous Deposition", J. Electrochem. Soc. of Japan, Vol. 47, No. 4, pp. 349-355 (1996)

[15] G.O. Mallory & J.B. Hajdu, "Electroless Plating", AESF publications, Orlando (1990)

[16] A. Islam, "Two component micro injection moulding for moulded interconnect devices", Ph.D.-thesis, Technical University of Denmark (Dept. of Mechanical Engineering), February (2008)

[17] M. Frydendall, "Kemisk udfældning af mikrostrukturer", Internal report Institute of Manufacturing Engineering KOT 960627-17, Technical University of Denmark (1996)

[18] P.T. Tang, H. Dylmer & P. Møller, "Nickel Coatings and Electroforming Using Pulse Reversal Plating", AESF SUR/FIN'95, pp. 529-536, Baltimore Jun. 26-29 (1995)

[19] J. Amblard, I. Epelboin, M. Froment & G. Maurin, "Inhibition and nickel electrocrystallization", J. Appl. Electrochem., No. 9, pp. 233-242 (1979)

[20] P.T. Tang, T. Watanabe, J.E.T. Andersen & G. Bech-Nielsen, "Improved corrosion resistance of pulse plated nickel through crystallisation control", J. of Applied Electrochem., Vol. 25, pp. 347-352 (1995)

[21] W. Paatsch, "Galvanotechnik mit Strompulsen", Metalloberfläche, Vol. 40, No. 9, pp. 387-390 (1986)

[22] P.T. Tang, P. Leisner & P. Møller, "Improvement of Nickel Deposit Characteristics by Pulse Plating", AESF SUR/FIN '93, pp. 249-256, Anaheim Jun. 21-24 (1993)

[23] J.W. Dini, "Electrodeposition - The Materials Science of Coatings and Substrates", Noyes Publ. (1993)

[24] P. Leisner, P. Møller & A. McNelly, "Throwing power and ductility of pulse reversal plated copper for PCB's", Processing of Advanced Mat., Vol. 9, pp. 148-154 (1994)

[25] J. Horkans, "Effect of Plating Parameters on Electrodeposited NiFe", J. Electrochem. Soc., Vol. 128, No. 1, pp. 45-49 (1981)

[26] W.H. Safranek, "The Properties of Electrodeposited Metals and Alloys", AESF, Florida (1996)

[27] H.V. Venkatasetty, "Electrodeposition of Thin Magnetic PermAlloy Films", J. Electrochem. Soc., Vol. 177, No. 3, pp. 403-407 (1970)

[28] R.D. McMichael, U. Atzmony, C. Beauchamp, L.H. Bennett, L.J. Swartzendruber, D.S. Lashmore & L.T. Romankiw, "Fourfold anisotropy of an electrodeposited Co/Cu compositionally modulated alloy", J. Magn. Magn. Mater., Vol. 113, pp. 149-154 (1992)

[29] P.T. Tang, "Pulse Reversal Plating of Nickel and Nickel Alloys for MEMS", Proceedings pp. 224-232, SUR/FIN 2001, Nashville, 25-28 June (2001)

[30] K. Smistrup, T. Lund-Olesen, P.T. Tang & M.F. Hansen, "Microfluidic magnetic separator using an array of soft magnetic elements", J. of Applied Physics, Vol. 99, 08P102 (2006)

[31] S. Arai & T. Watanabe, "The electroplating of Sn-Ag alloy for the use of Pb-free solder film", MRS Symposium, Tokyo May (1996)

[32] A. He, B. Djurfors, S. Akhlaghi & D.G. Ivey, "Pulse Plating of Gold-Tin Alloys for Microelectronic and Optoelectronic Applications", proceedings of AESF Sur/Fin, Chicago, 9 pages, June 24-27 (2002)

[33] G. Ekström (ed.), "Lärebok i Elektrolytisk och Kemisk Ytbehandling", (in Swedish) Ytforum Forläg AB (1990)

[34] I. Mizushima, P.T. Tang, H.N. Hansen & M.A.J. Somers, "Residual stress in Ni-W electrodeposits", Electrochem. Acta, Vol. 51, pp. 6128-6134 (2006)

[35] A.F Bogenschütz, J.L. Jostan & A. Ficker, "Galvanische Abscheidung von Invar-Legierungen (Ni 35/Fe 65) aus Sulfamatbädern", Oberfläche, No. 11, pp. 396-402 (1969)

[36] A.F. Bogenschütz, J.L. Jostan & W. Hemmrich, "Galvanische Abscheidung von FeNi-Legierungen mit niedrigem thermischem Ausdehnungskoeffizienten", Oberfläche, No. 12, pp. 506-514 (1970)

[37] A.F. Bogenschütz & U. Georg, "Galvanische Legierungsabscheidung und Analytik", Eugen G. Leuze Verlag, Saulgau/Germany (1982)

[38] D.M. Tench & J.T. White, "Tensile Properties of Nanostructured Ni-Cu Multilayed Materials Prepared by Electrodeposition", J. Electrochem. Soc., Vol. 138, No. 12, pp. 3757-3758 (1991)

[39] V.M. Fedosyuk, O.I. Kasyutich & N.N. Kozich, "Electrodeposition and Study of Multilayered Co/Cu Structures", J. Mater. Chem., Vol. 1, No. 5, pp. 795-797 (1991)

[40] G. Bissacco, H.N. Hansen, P.T. Tang & J. Fugl, "Precision manufacturing methods of inserts for injection molding of microfluidic systems", Proceedings ASPE Spring Topical Meeting, pp. 57-63, Columbus, April 18-19 (2005)

[41] P.T. Tang, J. Fugl, L. Uriarte, G. Bisacco & H.N. Hansen, "Indirect Tooling Based on Micromilling, Electroforming and Selective Etching", Proceedings, 2nd International conference on Multi Material Micro Manufacture (4M), pp. 183-186, September 20-22, Grenoble (2006)

[42] P.T. Tang, "A method of manufacturing a mould part", Patent application, WO (PCT) 2006/026989 A1 (2004)

[43] G. Tosello, G. Bissacco, P.T. Tang, H.N. Hansen & P.C. Nielsen, "High aspect ratio micro tool manufacturing for polymer replication using μEDM of silicon, selective etching and electroforming", Microsystem Technologies, available on-line from Feb. 26, ISSN 1432-1858 (2008)

Multi-Material Micro Manufacture
S. Dimov and W. Menz (Eds.)

Engineered Self-assembly From Nano to Milli Scales

Karl F. Böhringer

Department of Electrical Engineering, University of Washington, Seattle, WA 98195-2500, USA

Abstract

Self-assembly is the autonomous and spontaneous organization of components into patterns or structures. Self-assembly is ubiquitous in nature, e.g. in the growth of crystals and organisms, but also at macroscopic scales – it is nature's prevalent paradigm for manufacturing. Self-assembly also provides the basis for important new industrial manufacturing techniques, especially for components at the milli, micro, and nano scales: their small sizes and large numbers scale unfavorably for common serial techniques but favorably for a new, massively parallel approach. We believe that self-assembling systems will be able to create complex, heterogeneous, non-periodic, three-dimensional devices in massively parallel production processes. Hence, our research investigates the scientific and engineering foundations of self-assembly processes for integrated micro/nanoelectromechanical systems (MEMS/NEMS).

Keywords: self-assembly, stochastic manufacturing processes, MEMS, NEMS, packaging

1. Stochastic manufacturing by self-assembly

Let us start with three basic observations that are widely applicable to current industrial manufacturing:

(1) Production follows a centralized, top-down approach that tries to impose complete control over the production flow; every part and every assembly has its predetermined place in the overall scheme, and any uncertainty or randomness is considered detrimental.

(2) Microfabrication and nanotechnology allow the production of tiny but highly functional parts and devices at very high volumes and low unit cost.

(3) Historically, revolutionary changes in manufacturing (and simultaneously, arguably, in society) occurred when production shifted from individual, "0-dimensional" manual labor (e.g., blacksmith) to pipelined, "1-dimensional" assembly lines (e.g., Ford Model T), and again at the transition to "2-dimensional" batch-fabrication (e.g., VLSI).

We believe that recent developments have set the stage for another revolutionary change, where decentralized, parallel, stochastic, 2D or 3D manufacturing processes will become superior to the current state of the art. A central concept in this new way of manufacturing will be *engineered self-assembly*: like self-assembly in nature, engineered self-assembly uses energy minimization to organize components in stochastic, parallel processes; but unlike self-assembly in nature, we will design components such that the assembly process proceeds towards a specific desired final outcome.

Products of engineered self-assembly take form stochastically but not arbitrarily. Therefore, we must (a) develop self-assembly processes that minimize variation, and (b) design self-assembling systems that tolerate occasional defects. We can address these imperatives, e.g., by (a) tight control over process conditions, part tolerances and surface properties, and (b) by robust or redundant architectures. Such methods, however, are not unique to engineered self-assembly, but rely on techniques for increasing yields that are common in current manufacturing.

2. Engineered self-assembly: a brief review

Here, we introduce some of the central concepts in engineered self-assembly, and list selected references.

Fluidic self-assembly uses a slurry of parts, which are delivered by the fluid flow to the substrate [1]. Once this random motion brings the parts close to a *binding site*, capture may occur by shape matching and gravity (i.e., the part falls into a cavity with complementary geometry), or by other binding forces, such as, e.g., *capillarity* or *DNA hybridization*. *Capillary-force driven self-assembly* [2] relies on the minimization of interfacial energy: in an aqueous environment, hydrophobic surfaces and hydrocarbon droplets provide strong driving forces; in an air environment, water [3, 4] or molten solder [5] are typically used. *Template-based self-assembly* introduces an alignment template on which the parts are assembled [6, 7]; after self-assembly they may be batch-transferred to their final substrate and the template may be used again. Figure 1 shows examples of this approach.

In general, energy minimization alone is not sufficient for successful self-assembly; such an approach would be prone to errors due to local energy minima, hysteresis, friction and stiction, etc. A controlled amount of global energy input, typically in the form of random agitation (analogous to kT in chemistry), can help overcome these problems and transform the simple "gradient descent" method into "physical annealing."

Understanding and using self-assembly is essential for creating future MEMS/NEMS. For example, at the sub-millimeter scale, Alien Technology (Morgan Hill, CA) is set to produce billions of RFID tags by self-assembly; at the nano-scale, the use of self-assembly for carbon-nanotube-based devices is being discussed as a viable alternative to in situ growth. Micro-scale sensor and transmitter nodes ("smart dust") may consist of disparate device elements—ICs, MEMS transducers, photonic devices, RF components, solar cells or batteries—that are incompatible with a monolithic fabrication process, thus making their production without assembly all but impossible.

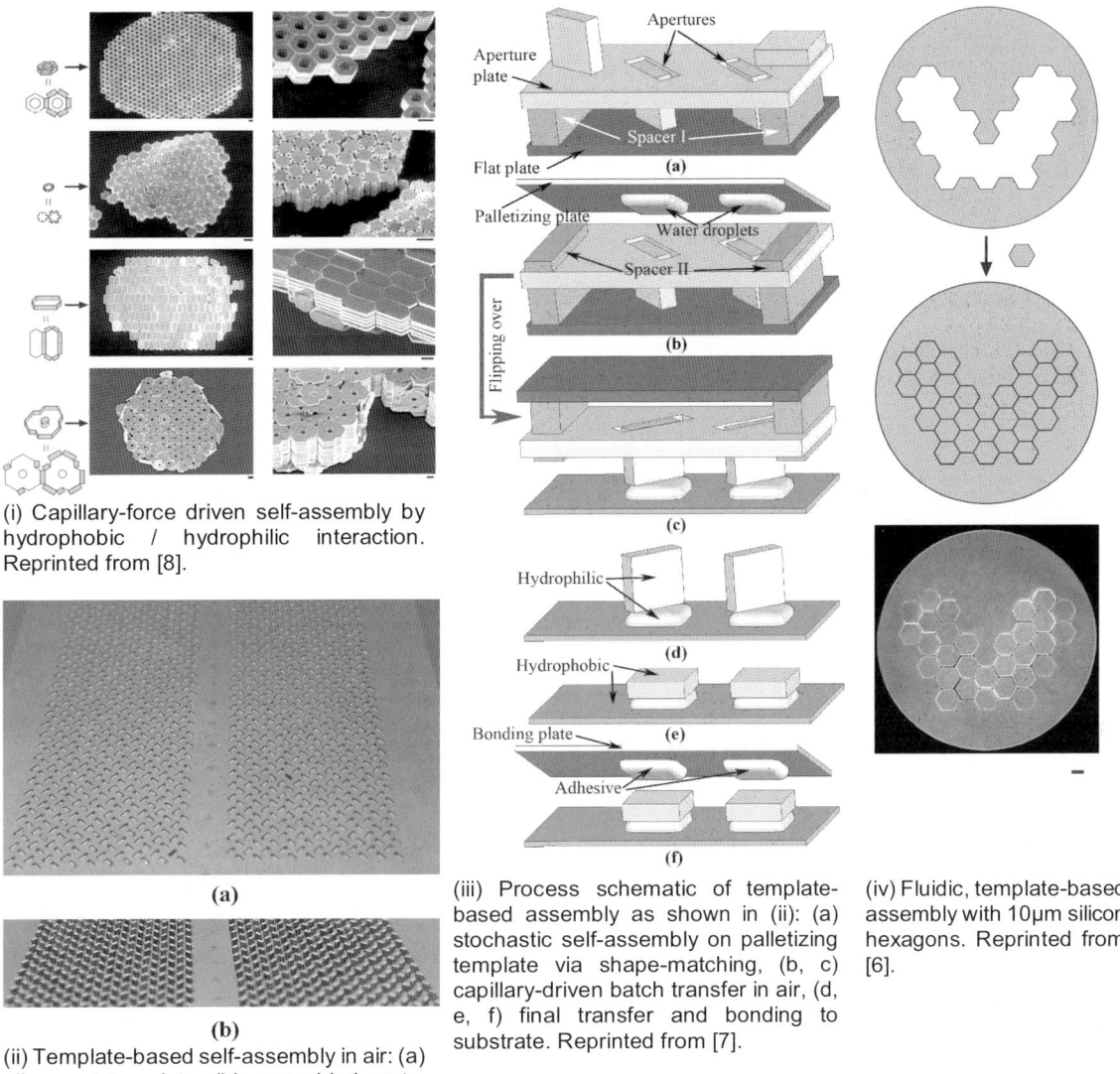

(i) Capillary-force driven self-assembly by hydrophobic / hydrophilic interaction. Reprinted from [8].

(ii) Template-based self-assembly in air: (a) alignment template, (b) assembled parts. Reprinted from [7].

(iii) Process schematic of template-based assembly as shown in (ii): (a) stochastic self-assembly on palletizing template via shape-matching, (b, c) capillary-driven batch transfer in air, (d, e, f) final transfer and bonding to substrate. Reprinted from [7].

(iv) Fluidic, template-based assembly with 10µm silicon hexagons. Reprinted from [6].

Figure 1: Examples of engineered self-assembly at the micro-scale.

A state-of-the-art pick-and-place robot (Datacon, Austria; Assembléon, the Netherlands) for assembly of parts having lateral dimensions of at least 500µm (e.g., for surface mount technology) has a placement accuracy of a few micrometers and a throughput of about 10,000 parts per hour. Given the mature state of robotic assembly automation, we cannot expect dramatic improvements upon these specifications. While the current assembly costs with a robot can be calculated at well below ¢1/unit, the trend toward increasingly small parts is likely to make these low costs unsustainable: smaller parts and higher placement accuracy will likely lead to more costly assembly. More importantly, if higher overall volumes demand installation of multiple robots operating in parallel, then the large capital investment will drive up unit assembly costs substantially.

Our philosophy is to exploit self-assembly to reduce this cost overhead. Self assembly introduces parallelism at several levels: besides the inherently parallel character of self-assembly, the lower cost for an overall setup will allow the addition of multiple

production lines as needed without ballooning capital investment.

3. Research goals in engineered self-assembly

Our goal is to develop the science and technology necessary to create a fundamentally new way for manufacturing microsystems. Conventional microfabrication methods have excelled in producing integrated circuits. Microelectromechanical system (MEMS) technology has extended the range of these technologies to other domains such as micro optical and micro fluidic devices. Although solid-state IC and MEMS manufacturing techniques have been remarkably successful in producing devices, they have been markedly less successful in generating integrated systems. A number of factors including process incompatibility and the need for three-dimensional integration have limited the scope of traditional techniques in parallel production of integrated systems. The goal of further research in engineered self-assembly is to develop the fundamental scientific understanding and the relevant technologies to enable microsystem production via self-assembly. The new production technique allows for cost-effective, readily reconfigurable integration of heterogeneous microsystems in 2D and 3D configurations. Specifically, we identify the following science and engineering goals:

1. Develop *processes for massively parallel integration of heterogeneous materials and structures* into 2-dimensional and 3-dimensional devices, including "agile" manufacturing techniques that allow fast adaptation of product designs in response to evolving specifications.
2. Establish *performance metrics* that permit a systematic and objective evaluation of these processes, including an understanding of their scaling properties.
3. Develop *models and simulators for self-assembly* that can effectively deal with the complex physics and surface chemistry involved in the interactions between possibly huge numbers of independently moving parts.

4. Recent advances in engineered self-assembly

Our group is studying a wide range of self-assembly techniques from nano to milli scales. Here we list some representative work. Xiong et al. [9] introduced multi-batch programmable self-assembly, using binding sites that can be selectively activated and de-activated. Programmable self-assembly requires the ability to modulate interfacial surfaces with good spatial and temporal resolution. An overview of surface modulation techniques is given in [10]. Fang et al. [4, 7] achieve uniquely oriented assemblies driven by capillary forces and gravity. Fang et al. [3] also demonstrate millimeter-scale assembly of piezoelectric micropumps. Saeedi et al. [11] give an overview on molten solder driven self-assembly. Capillary forces can also be exploited to produce assemblies of colloidal nano-particles, as demonstrated by Xiong et al. [12]. Baskaran et al. [13] improves self-assembly performance by introducing "catalysts" that increase mobility of assembly components.

FEM models for capillary force driven self-assembly are described by Lienemann et al. [14]. A simplified model based on geometric convolution of binding sites is developed in [15]. Liang et al. [16] discovered component designs that accomplish unique position and orientation in capillary force driven self-assembly.

Self-assembly is not limited to microelectronic or MEMS components. Cheng et al. [17] demonstrate how proteins and cells are patterned on thermally responsive polymer thin films that switch between a hydrophobic and a hydrophilic state.

Engineered self-assembly is being studied by an increasing number of scientists and engineers. An early paper by Shimoyama et al. [18] showed analogies to chemical reaction kinetics. For recent advances, the reader may consult the proceedings of the *Foundations of Nanoscience: Self-assembled Architectures and Devices* conference series, the review papers by the Whitesides group, e.g. [19, 20], Syms's et al. [2] review on surface tension powered self-assembly, or Pelesko's popular science book on self-assembly [21].

Acknowledgements

We gratefully acknowledge support from the National Science Foundation (NIRT-07-09131, ECS-05-1628, ECS-02-23598, ECS-98-75367), DARPA (FA9550-04-1-0257), the National Institutes of Health (5-P50-HG002360-06), and Intel Corporation.

References

1. Yeh, H.-J.J. and J.S. Smith, *Fluidic Assembly for the Integration of GaAs Light-Emitting Diodes on Si Substrates.* IEEE Photonics Technology Letters, 1994. **6**(6): p. 706-708.
2. Syms, R.R.A., et al., *Surface Tension-Powered Self-Assembly of Microstrutures: The State-of-the-Art.* Journal of Microelectromechanical Systems, 2003. **12**(4): p. 387-416.
3. Fang, J. and K.F. Böhringer, *Self-Assembly of PZT Actuators for Micro Pumps with High Process Repeatability.* ASME/IEEE Journal of Microelectromechancial Systems, 2006. **15**(4): p. 871-878.
4. Fang, J. and K.F. Böhringer, *Wafer Level Packaging Based on Uniquely Orienting Self-Assembly (the DUO-SPASS Processes).* ASME/IEEE Journal of Microelectromechancial Systems, 2006. **15**(3): p. 531-540.
5. Jacobs, H.O., et al., *Fabrication of a Cylindrical Display by Patterned Assembly.* Science, 2002. **296**: p. 323-325.
6. Clark, T.D., et al., *Template-Directed Self-Assembly of 10-micron-Sized Hexagonal Plates.* Journal of the American Chemical Society, 2002. **124**: p. 5419-5426.
7. Fang, J. and K.F. Böhringer, *Parallel micro component-to-substrate assembly with controlled poses and high surface coverage.* IOP Journal of Micromechanics and Microengineering (JMM), 2006. **16**: p. 721-730.
8. Clark, T.D., et al., *Self-Assembly of 10-μm-Sized Objects into Ordered Three-Dimensional Arrays.* Journal of the American Chemical Society, 2001. **123**: p. 7677-7682.
9. Xiong, X., et al., *Controlled multibatch self-assembly of microdevices.* ASME/IEEE Journal of Microelectromechanical Systems, 2003. **12**(2): p. 117-127.
10. Böhringer, K.F., *Surface modification and modulation in microstructures: controlling protein adsorption, monolayer desorption and micro-self-assembly.* Journal of Micromechanics and Microengineering, 2003. **13**(4): p. S1-S10.
11. Saeedi, E., et al., *Molten-Alloy Driven Self-Assembly for Nano and Micro Scale System Integration.* Fluid Dynamics & Materials Processing, 2007. **2**(4): p. 221-246.
12. Xiong, X., K. Wang, and K.F. Böhringer. *From Micro-Patterns to Nano-Structures by Controllable Colloidal Aggregation at Air-Water Interface.* in *IEEE International Conference on Micro Electro Mechanical Systems (MEMS).* 2004. Maastricht, Holland.
13. Baskaran, R., et al. *Catalyst enhanced micro scale batch assembly.* in *IEEE International Conference on Microelectromechanical Systems.* 2008. Tucson, AZ.
14. Lienemann, J., et al., *Modelling, Simulation and Experimentation of a Promising New Packaging Technology - Parallel Fluidic Self-Assembly of Micro Devices.* Sensors Update, 2003. **13**(1): p. 3-43.

15. Böhringer, K.F., U. Srinivasan, and R.T. Howe. *Modeling of Capillary Forces and Binding Sites for Fluidic Self-Assembly.* in *IEEE MEMS'01 - International Conference on Micro Electro Mechanical Systems.* 2001. Interlaken, Switzerland.

16. Liang, S.-H., X. Xiong, and K.F. Böhringer. *Towards Optimal Designs for Self-alignment in Surface-tension Driven Micro-assembly.* in *IEEE Conference on Micro Electro Mechanical Systems (MEMS).* 2004. Maastricht, Holland.

17. Cheng, X., et al., *Novel Cell Patterning Using Microheater Controlled Thermoresponsive Plasma Films.* Journal of Biomedical Materials Research, 2004.

18. Hosokawa, K., I. Shimoyama, and H. Miura, *Dynamics of Self-Assembling Systems - Analogy with Chemical Kinetics.* Artificial Life, 1994. **1**(4): p. 413-427.

19. Boncheva, M. and G.M. Whitesides, *Making things by self-assembly.* MRS Bulletin, 2005. **30**(10): p. 736-742.

20. Whitesides, G.M. and B. Grzybowski, *Self-Assembly at All Scales.* Science, 2002. **295**: p. 2418-2421.

21. Pelesko, J.A., *Self Assembly: The Science of Things That Put Themselves Together.* 2007, Boca Raton, London, New York: Chapman & Hall/CRC.

Q2M Special Session

The integration of mono-crystalline silicon micro-mirrors on CMOS for SLM applications

F. Zimmer[a], M. Friedrichs[a], M. Lapisa[c], F. Niklaus[c], M. Mueller[a], T. Bakke[b],
H. Schenk[a], H. Lakner[a]

[a] Fraunhofer Institute for Photonic Microsystems (IPMS), Maria-Reiche-Str. 2, D-01109 Dresden, Germany
[b] SINTEF Department of Mikrosystems and Nanotechnology, Gaustadalleen 23C, Oslo, Norway
[c] KTH, The Royal Institute of Technology, Stockholm, Sweden

Abstract

Spatial light modulators (SLMs) based on micro-mirrors for use in DUV lithography and adaptive optics need very high mirror planarity as well as mirror stability. We will present results of new micro-mirror arrays, consisting of mono-crystalline silicon, which is a material to fulfil these requirements. As all mirrors of the SLM can be separately activated by an underlying CMOS circuit, the integration of CMOS and MEMS must be achieved, which results in certain restrictions on processing temperatures and the compatibility of materials. Therefore a special low temperature bonding technology has been developed, using an adhesive polymer. This technique provides the transfer of a 300nm thin mono-crystalline silicon layer to the CMOS wafer using only 250°C. First silicon micro-mirrors have been made and characterized using pure adhesive polymer (PMGI), improvements using a mix of an inorganic material with a thin bond-polymer (benzocyclobutene BCB) on top are in development. Both approaches and their results will be discussed and presented in detail.

Keywords: Optical MEMS, spatial light modulator, micro-mirrors, wafer bonding, maskless lithography

1. Introduction

Today, spatial light modulators (SLMs) are used in projection display systems [1], adaptive optics and also in advanced lithography. Especially the last item demands for very accurate micro-mirror arrays [2],[3], which have perfect mechanical and optical properties. The goal of the development presented here has been a matrix, consisting of 1 million separately addressable silicon micro-mirrors with a pitch of 16 x 16 μm^2. As the temperature impact for CMOS-electronics is limited by approximately 450 °C, only low temperature process steps can be used for the mirror fabrication. Thus, the formation of mono-crystalline silicon membranes using high temperature processes, e.g. epitaxy must be excluded. One possibility is the transfer of thin mono-crystalline layers from a donor wafer to the CMOS wafer, using a low temperature adhesive bond process.

The use of mono-crystalline silicon as mirror material has several advantages. Especially for lithography applications (e. g. mask writing) in the DUV (deep ultra violet) region, as described here, the accurate and repeatable positioning of each individual mirror must be perfect. This can be only achieved with mirrors of high elasticity, low internal or compensated stress gradients and no material plasticity. Beyond sputtered materials like SiGe or aluminium alloys, mono-crystalline silicon fulfils almost all specifications perfectly. Thus, highly planar mirrors without material degradations can be expected using mono-crystalline silicon. The reflectance of 64% in the DUV region is sufficient, positive is the surface quality of the silicon membranes.

We will present two fabrication methods of mono-crystalline SLMs on CMOS substrates using adhesive wafer bonding. One is the combination of Polymethylglutarimide (PMGI) as bond polymer, the

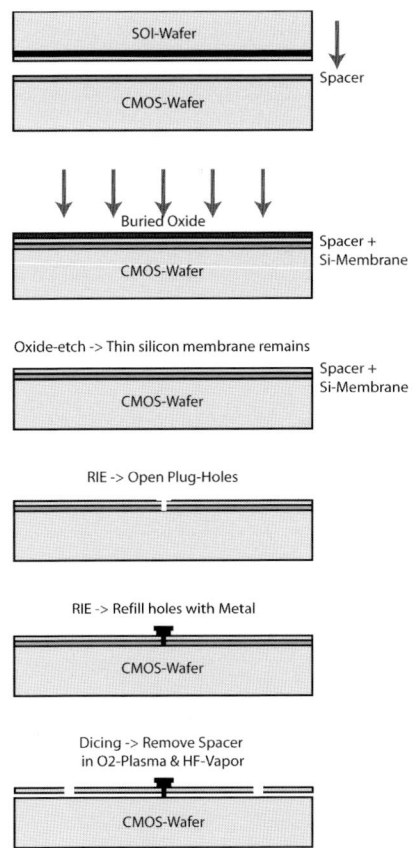

Fig. 1. Fabrication process flow

second BCB.

The technological concept has been already presented by Niklaus et. al. in 2003 [4]. Based on these first results, technology improvements have been attained. This involves the transfer of the basic technology from a 4 x 4 mirror array to a 2048 x 512 mirror array, which can be directly integrated onto a CMOS substrate, the improvement of the bond interface using Benzocyclobutene (BCB) and a technology change from electroplated gold to sputtered aluminium posts, due to its better compatibility to the CMOS process.

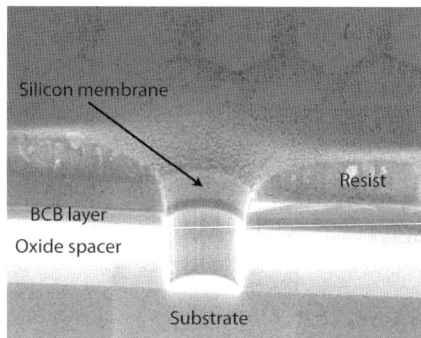

Fig. 2. Post hole in oxide / BCB spacer

2. Technology

2.1 General information

The technology development of MEMS on CMOS substrates is expensive and time consuming. In order to reduce costs and to get fast but comparable results, the CMOS wafer was at first replaced by silicon test wafers containing only the top CMOS metal layer. As CMOS wafers are polished several times during the process, the topology of test- and CMOS wafers, which are the basis for the MEMS technology development, can be regarded as equal.

The fabrication process is illustrated in [Fig.1]. Starting point is a silicon wafer, containing the top CMOS metal layer. First, a 600nm thin spacer layer, with adhesive properties is deposited on the wafer.

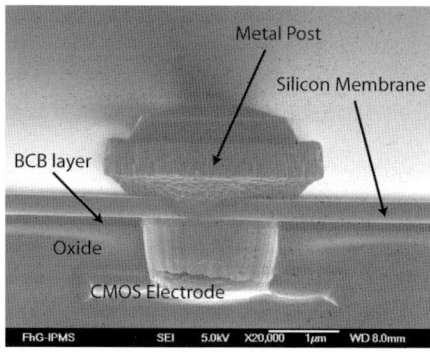

Fig. 3. Metal Post pinning silicon membrane to CMOS-wafer

Two different material combinations have been tested, which will be explained in detail in the next section. Then a SOI-wafer with a 300nm thin silicon membrane and a 400nm thin buried oxide layer is bonded on the adhesive layer. The handle silicon of the SOI-wafer is removed by grinding and a wet spin-etch process using nitric and hydrofluoric acid. The etching process is stopped in the buried oxide, which is removed in a next step in buffered hydrofluoric acid. Thus, the transfer of the mono-crystalline silicon membrane is finished.

In a further step, holes for pinning the silicon membrane to the CMOS wafer are etched using a reactive ion etch process. These holes pass the silicon membrane and the spacer and stop on the underlying CMOS metal layer [Fig.2]. The holes are refilled (sputtering) with an aluminium alloy, providing the mechanical and electrical connection of the silicon membrane [Fig.3]. In a last step, the membrane is etched, resulting in a movable micro-mirror array in which every mirror can be actuated separately. For dicing, the wafer is covered with a photo resist layer in order to avoid sawdust on the silicon mirror surface. The mirrors are released using oxygen plasma respectively oxygen / CF4 plasma and vapour hydrofluoric acid, depending on the chosen spacer material. During this release process, the mirrors are still covered with the photo resist layer, used during the dicing process. This layer avoids an attack of the mirror surface. Combined with a special release step, using only a small amount of CF4 in the plasma, the etch attack on the mirror backside can be reduced to a minimum.

2.2 Full polymer spacer

One approach to fabricate mono-crystalline silicon micro-mirrors was to use a reflowable polymer as spacer material (Polymethylglutarimide, PMGI). The polymer is structured before the bonding process, leaving only material on the actuation electrodes (s. [Fig.4]). The correct spacer height is provided by special oxide stopper, situated between the reflowable polymer regions. The bond of the silicon membrane is then accomplished in combination with a reflow at

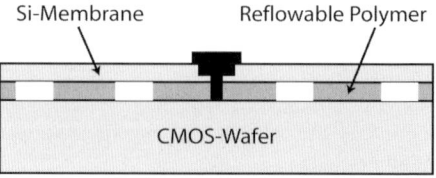

Fig. 4. Spacer technology with reflowable polymer

250°C.

2.3 Inorganic and polymer spacer

As alternative to full polymer spacers, the spacer can be also built up using an inorganic material (e.g. silicon oxide, deposited by a chemical vapour deposition process) and a thin polymer layer on top (s. [Fig.5]). In our approach, a nano-imprint resist and

Benzocyclobutene (BCB) have been investigated. This technique provides the advantage to level the oxide surface in advance with a chemical mechanical polishing step. Thus the polymer layer can be made very thin (several 100nm), maintaining still the possibility of low temperature adhesive wafer transfer

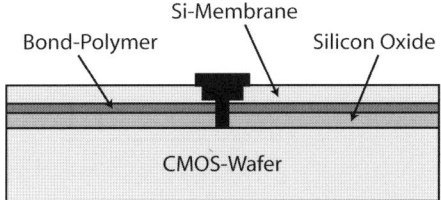

Fig. 5. Spacer technology with silicon oxide and thin polymer

bonding.

3. Measurement and Results

3.1 Bond properties

Both concepts, the full polymer spacer and the combination of silicon oxide with a thin polymer showed perfect bond interfaces (s. [Fig 6]). No voids or defects could be detected. Nevertheless, the nano-imprint resist showed delamination at the edge of the wafer, when exposed to temperatures of approximately 200°C. This is critical as the delamination appeared during the lithography step. This can only be avoided by using lower temperatures or changing to BCB, with which no damage could be detected.

3.2 Micro-Mirrors

First micro-mirror arrays consisting of mono-crystalline silicon are shown in [Fig.7]. They have been made using a complete PMGI polymer spacer. The mirrors had a size of 16 x 16 µm² and an averaged RMS planarity of 0.8nm (more than 2800 mirrors have been measured with white light interferometry).

The actuation characteristic of fabricated micro-mirrors is shown in [Fig.8]. At 15V, a deflection of 48nm has been reached. Looking on the application in DUV lithography, this is enough to generate a black pixel, when modulating light with a wavelength of 193nm. Long-term investigations showed no measurable level of drift (s. [Fig. 9]), when the mirrors were deflected at 48nm for 1 hour.

4. Conclusion and Outlook

We have shown the fabrication of mono-crystalline silicon micro-mirror arrays using low temperature and CMOS-compatible adhesive wafer bonding. This process is well suited for fabrication of SLMs on a CMOS substrate, providing the actuation of each mirror separately. First investigation show, that the mirrors reach the expected deflection of 48nm by applying a voltage of only 15V. A stable deflection could be

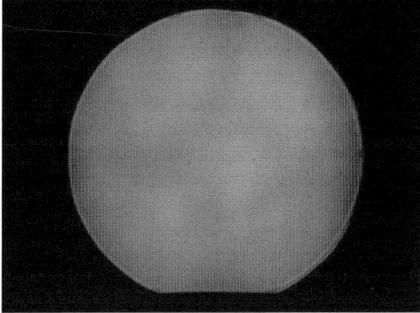

Fig. 6. IR-Image of Bond-interface using BCB thin film

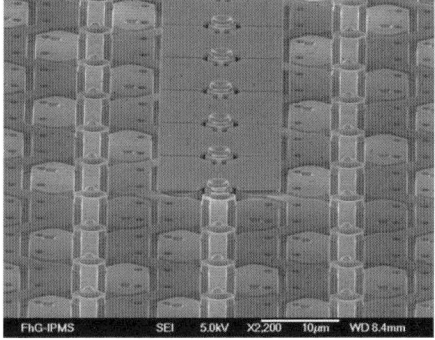

Fig. 7. SLM with mono-silicon micro-mirrors and bottom electrodes

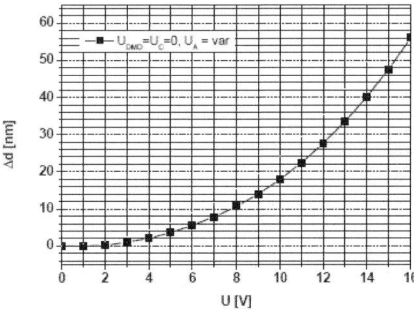

Fig. 8. Actuation characteristics of mono-silicon micro-mirrors on passive substrates

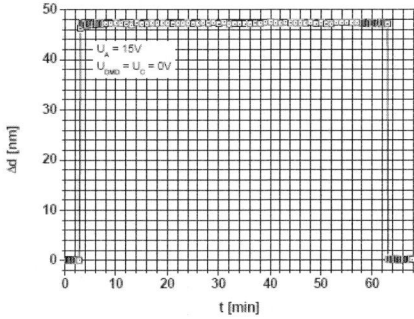

Fig. 9. Long-term stability as indicator for the level of drift

achieved for more than 1 hour, a drift level of deflected mirrors could not be detected.

Further investigations will include the fabrication of micro-mirrors using an inorganic oxide spacer with a thin BCB layer on CMOS substrates. SLMs for adaptive optics are also being developed using two spacer options, the inorganic oxide spacer technology with thin polymer layer and the full organic polymer spacer technology. Results will be presented in the near future.

References

[1] Texas Instruments Technical Journal, vol. 15, 1998, no. 3.

[2] U. B. Ljungblad, P. Askebjer, T. Karlin, T. Sandstrom, H. Sjoeberg, Proc. SPIE, vol. 5721, 2005, pp. 43-52.

[3] U. Dauderstädt, P. Dürr, U. Ljungblad, T. Karlin, H. Schenk, H. Lakner, Proc. SPIE, vol. 5721, 2005, pp. 64-71.

[4] F. Niklaus, S. Haasl, G. Stemme, IEEE/ASME J. of Microelectromechnical Systems, vol. 12, No. 4, 2003, pp. 465-469.

Batch Fabrication Methods for Polymer Based Active Microsystems using Hot Embossing and Transfer Bonding Technologies

T. Grund, M. Heckele and M. Kohl

Forschungszentrum Karlsruhe GmbH, Institute for Microstructure Technology (IMT), Postfach 3640, 76021 Karlsruhe, Germany

Abstract

A batch compatible process flow to overcome the costly piece by piece assembly of hybrid microsystems is shown. Hot embossing is used to fabricate microstructured polymer layers. Wafer scale compatible bonding tasks are carried out by ultrasonic welding and heat activated bonding with micromachined bonding foils. As demonstrator device, a shape memory alloy (SMA) actuated polymer microvalve is introduced. The valve concept, fabrication technologies and device characteristics are discussed.

Keywords: batch fabrication, polymer microvalve, shape memory alloy, TiNi actuator, transfer bonding

1. Introduction

Polymer microsystems offer superior properties for e.g. the field of pharmaceutics, biotechnology and life sciences. They can be fabricated economically and the systems chemical resistance and temperature operation range can be tailored to various applications by selecting the appropriate polymers. The realization of mechanically active polymer microsystems requires the integration of transducer materials like piezoelectric layers or shape memory alloy (SMA) films. However, dissimilar material properties require adapted fabrication technologies. For example high process temperatures for micromachining silicon are a K.O. criterion for most polymers. So far, the only approach to overcome these limitations has been the use of a costly pick-and-place assembly [1], which prevents a broad introduction of the product into the market. In the following, the bonding technologies of heat activated bonding, ultrasonic welding and a process flow for batch fabrication of polymer microvalves driven by SMA microactuators is introduced.

2. Microvalve Layout and Shape Memory Actuator

The microvalve is designed as a seat valve and operation is in normally open mode. It consists of several layers, which can easily been stacked during assembly (see Fig.1). The outer dimensions, disregarding the electrical and fluidic interconnection, are 11 x 6 x 3 mm³. The assembled valve parts are: a polymer valve housing, polyimide membrane, polymer actuator carrier, SMA microactuator and a spherical spacer for prestraining the actuator. The membrane is integrated as a layer in between the valve housing and the actuator carrier. Thus, the valve is divided into two sections, an actuator and a fluidic chamber, both having a diameter of 2 mm. The chosen design provides sufficient thermal isolation between fluid and actuation chamber allowing control of gases as well as temperature-sensitive liquids. The SMA actuator offers large strokes concurrently allowing small device dimensions [1, 2, 3]. It is fabricated by wet etching of a 20 µm thin NiTi cold rolled foil. The microactuator is designed as a circular arrangement of six microbridges, which are joined in the center. The beams have a length and width of 1 and 0.15 mm, respectively. By applying a pressure

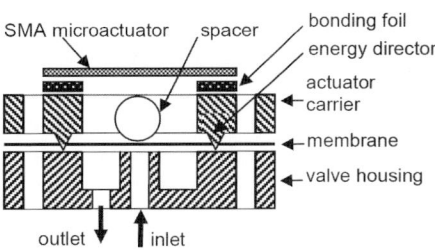

Fig. 1. Schematic of the microvalve.

difference between inlet and outlet, membrane and microactuator are deflected causing the valve to open. By direct electrical heating the microactuator, the membrane is pushed down via the spacer onto the valve seat and the valve closes. For test purposes, a batch of four microvalves has been produced.

3. Double Sided Hot Embossing of the Polymer Valve Parts

For forming the polymer valve parts (housing and actuator carrier), hot embossing is used [4]. Especially the fast and easy change between different polymers, inexpensive mold insert fabrication and possible fabrication of thin layers makes this technology attractive for the production of polymer microsystem parts. Standard hot embossing [5] creates a surface topology into the upper surface of the polymer film. The opposite surface remains normally smooth. Especially in the field of microfluidics and capillary analysis more and more details are required, impossible to realize by simple surface structuring. Typical examples are fluidic connections from the backside to supply the fluidic microstructures on top of the film. Therefore an additional structuring of the backside is necessary, which finally must be linked by through holes to the fluidic structure.

Replacing the flat counter plate by an additional mold insert allows a double sided replication. Only minor changes of the hot embossing machine are necessary to have now two mold inserts to be opposite. Using the double sided hot embossing allows a multitude of new applications but

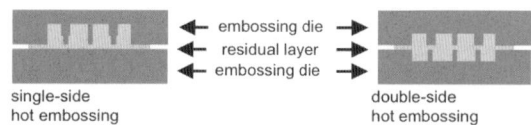

Fig. 2. Scheme of single- and (left) and double-side (right) hot embossing.

creates additional problems compared to the classical single side replication.

The residual layer, unavoidable in the hot embossing process remains also characteristic for the double sided process, because even using two structured mold inserts, a complete closing of the tool is impossible. But now the residual layer is not anymore necessarily located on the backside of the molded part. It is now possible to shift this layer in an intermediate layer, which can be advantageous in subsequent process steps (see Fig. 2). In the present case, the residual layer within the valve seat channel and within the holes for electrical contacting was removed manually with a simple punching tool after the embossing process.

Necessary for the double sided embossing is an alignment, because the molded structures on the front and backside mostly are not independent of each other. The precision of the alignment depends on the application and is typically in the range between 5 and 100 µm. This precision can be obtained with high precision embossing machines (e.g. Wickert WP1000, Jenoptik HEX). Jenoptik supplies with the HEX03 a machine with an active position control. The position of the countertool can be modified from one cycle to the next cycle by a high precision piezo drive. The data for such a position correction are from an image treatment software analyzing the picture coming from a microscope, which can be rotated into the hot embossing machine and realizing in this way an online position control.

Finally, the process creates problems in the demolding step, when the structured foils are separated from the mold insert. In the single sided process the demolding of thin films is solved by a special micro roughness of the stamping plate, which served also as demolding tool. Replacing this stamping plate by a second mold insert results in the lack of a demolding unit. One possibility to enable successful demolding without ejectors or other moving parts is to exploit the difference of the two mold inserts. Normally, one of the two mold inserts requires a higher demolding force due to high microstructures, steep side walls or the high density of microstructures. The other mold insert is characterized by only a few microstructures sometimes even with ejector slopes. To surmount the high demolding force of the first mold insert, the second one can be treated in this way, that is surface is modified in order to increase the adhesion to this second mold insert. Now, the first mold insert will be demolded when opening the hot embossing machine. The subsequent separation of the second mold insert will happen by shrinking when cooling down. In the present case, polymethylmethacrylate (PMMA) is used as a polymer.

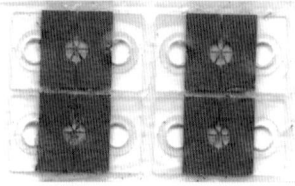

Fig. 3. A batch of four SMA actuators on polymer actuator carriers after adhesive bonding with microstructured bonding foils.

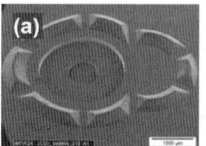 

Fig. 4(a). SEM picture of energy directors with improved design for membrane integration and (b) photograph of the integrated membrane after ultrasonic welding.

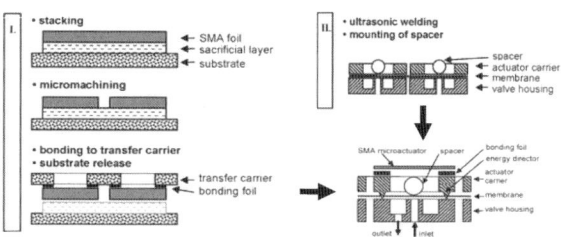

Fig. 5. Process flow for the batch fabrication of SMA actuated polymer microvalves.

4. Bonding Technologies

4.1. Heat Activated Bonding Foils

The used heat activated bonding foil has a thickness of 60 µm and is micromachined by laser cutting. Thus, the bonding areas are well defined and blocking of fluidic channels by the uncontrolled flow of adhesive is prevented. The foil is not adhesive at room temperature and is covered on both sides with a protective layer, which allows for easy handling. The protective layer on one the first side is removed by simple peeling. After aligning, the bonding foil is fixed on the actuator carriers using elevated temperatures and a short pressure step. Thereafter, the remaining protective layer is removed and a substrate with the structured SMA actuators is aligned and applied on the bonding foil, followed by a compression and heating step (see Fig. 3) [6]. As all actuators for the current batch are already contained on the substrate, no costly pick-and-place assembly is necessary.

4.2. Ultrasonic Welding

Welding technologies allow foreign matter free bonds, which is important for chemically inert joints. Preliminary tests show that the membrane in between valve housing and actuator carrier is punched during the welding process. Since the membrane is no longer supported in this case, it folds into the fluidic. In order to provide mechanical support of the membrane after welding, an improved design

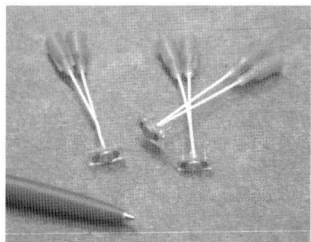

Fig. 6. Batch fabricated polymer microvalves.

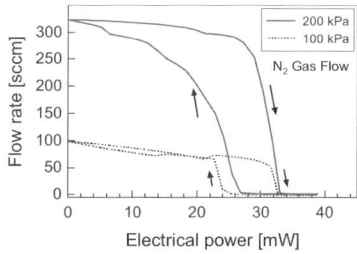

Fig. 7. Typical gas flow characteristic of the fabricated microvalves.

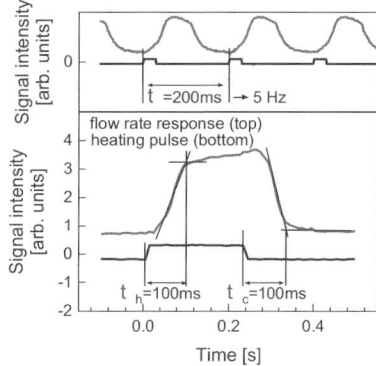

Fig. 8. Typical dynamic characteristic of the fabricated microvalves.

of the energy directors is developed consisting of locally interrupted triangles, see Fig. 4(a). In addition, small slits are provided in the membrane for the energy directors. In this case, small support structures remain after the welding process, keeping the membrane above the fluidic chamber under tension and thus, in flat condition, see Fig. 4(b) [7].

5. Batch Fabrication Process

Both bonding technologies are combined in a novel process flow, shown in Fig. 5. The SMA foil is fixed on a temporary substrate using an sacrificial layer. After standard lithography and wet etching, the SMA actuators are bonded on a transfer carrier using the introduced bonding foils. The sacrificial layer is removed by a solvent. In a separate process, the actuator carrier and valve housing is joined by using ultrasonic welding. The spacer is a spherical ceramic and as no alignment is necessary it is simply dropped into place. An corresponding opening in the actuator carrier centers the spacer passively above the valve seat. In a last step, both components are joined by using again the bonding foils. The transfer carrier can remain on the the actuators (not shown in the last process step in Fig. 5) as it can be used as a heat sink.

6. Characteristics

Figure 6 shows a batch of fabricated microvalves with electrical and fluidic interconnections. The measurement of the flow rate is carried out in quasi static condition for different pressure differences using a flow meter by ramping the heating power up and down. Flow rates upon heating and cooling of the SMA actuator are investigated separately. Figure 7 shows that at zero power, the microvalve is in open state showing a maximum gas flow of about 250 standard ccm at a pressure difference of 200 kPa. Above a critical electrical power of about 30 mW, the valve changes from the open to the closed state demonstrating the expected valve function. Control of liquids is also possible. Depending on the valve design, time constants upon cooling are in the range of 100 ms (see Fig. 8).

7. Conclusions

Batch integration of micromachined polymer layers and of other components like a membrane and SMA foil microactuators is facilitated by the novel technologies of two-step ultrasonic welding and heat-activated bonding with micromachined bonding foils. By combining these technologies, mechanically active polymer microsystems have been batch-

fabricated. Prototypes of SMA-actuated polymer microvalves, fabricated with the presented technologies, reveal performance characteristics similar to specifications of corresponding microvalves fabricated by a pick-and-place process. Reproducibility of the microvalve characteristics will be investigated in future studies, comparing valve characteristics within one batch and between several batches.

Acknowledgements

The presented work is part of the Q2M project and receives research funding from the European Commission through the sixth framework program.

References

[1] Kohl, M.; Liu, Y.; Dittmann, D., A Polymer-Based Microfluidic Controller, 2004, IEEE Catalog No. 04CH37517, pp. 288-291

[2] Kohl, M.; Hürst, I.; Krevet, B., Time Response of Shape Memory Microvalves, Proc. ACTUATOR 2000, pp. 212-215

[3] Kohl, M. et al., Shape memory micromechanisms for microvalve applications, SPIE '04 Vol. 5387

[4] Heckele, M.; Schomburg, W.K; (2004), Review on Micro Molding of Thermoplastic Polymers, Journal of Micromechanics and Microengineering, 14

[5] M. Heckele, et al., Double-sided hot embossing of microstructures. Proc. HARMST'01, pp.137-138

[6] T. Grund, R. Guerre, M. Despont and M. Kohl, EPJ - Special Issue - EMRS Fall Meeting '07

[7] T. Grund, T. Cuntz and M. Kohl, MEMS 2008, Tucson, USA, IEEE Catalog #: CFP08MEM-PRT

Wafer-scale manufacturing of robust trimorph bulk SMA microactuators

N. Sandström[a], S. Braun[a], T. Grund[b], G. Stemme[a],
M. Kohl[b], W. van der Wijngaart[a]

[a]Microsystem Technology Lab, KTH - Royal Institute of Technology, Stockholm, SWEDEN
[b]Institut für Mikrostrukturtechnik, Forschungszentrum Karlsruhe GmbH, Karlsruhe, GERMANY

Abstract

This paper demonstrates the concept of wafer-level fabrication and integration of robust bulk SMA microactuators based on adhesive bonding of cold-rolled SMA sheets to silicon wafers. Contact printing of an adhesive polymer ensures a selective bonding when transferring full SMA sheets to silicon structures on a patterned wafer. The induced stress of a thin dielectric film deposited on top of the SMA sheet ensures a stable and built-in reset mechanism of the actuators. The trimorph microactuators can be actuated by indirect resistive heating through a thin metal film. We report on the successful wafer-scale fabrication of actuator cantilevers and their characteristics. First test cantilevers show a cold-state deflection of 300 μm which, however, is limited by the silicon substrate. Upon heating, the cantilever shows a stroke of approx. 80 μm.

Keywords: SMA, microactuators, wafer-level integration, adhesive bonding

1. Introduction

Shape Memory Alloy (SMA) materials can be easily pseudo-plastically deformed at temperatures below the transformation temperature. Upon heating above the transformation temperature, the material recovers the initial shape and when hindering the recovery, the material generates high forces with the energy density exceeding that of other actuation principles by at least one order of magnitude [1].

In most cases, the SMA is heated by electrical resistive heating – however, the necessary electrical contacting is difficult because of the stable native oxide of the SMA. Furthermore, the power consumption is relatively high [1].

To obtain an actuator structure, the SMA material must be deformed after the heating cycle at temperatures below the transformation temperature by an external bias spring (cold-state reset).

Basically, there are two different approaches to integrate SMA material in microelectromechanical system (MEMS) devices. One method is to fabricate the SMA actuator structure and the MEMS structure separately from each other and integrate them subsequently in a per-component assembly [2]. The cold-state reset is provided by a mechanical obstruction that pre-stresses the SMA component during the assembly. However, the per-component assembly is not batch compatible and therefore results in unacceptable high costs, which outbalances the featured advantages such as the availability of NiTi-foils in a wide thickness range and with reproducible bulk material characteristics.

Another method is to directly sputter deposit SMA material onto the MEMS structures [3]. However, sputter deposition of SMA is complicated and a subsequent annealing at high temperatures is necessary, which potentially causes problems with interdiffusion of SMA into the substrate as well as incompatibility issues with other processing steps. Furthermore, the film thickness and therewith the mechanical performance of the SMA is limited. However, this approach features an integrated cold state reset mechanism by built-in film stresses.

Previously, the authors of the present paper reported on a concept for batch manufacturing of robust trimorph bulk SMA microactuators [4], which allows for a novel integration method circumventing the limitations of the previous methods. In this concept, we utilized thin cold-rolled TiNi foil as the bulk material and added a dielectric layer at an elevated temperature which provides the cold-state reset in form of a stress induced deflection of the actuator in the cold state due to the different thermal expansion coefficients of the TiNi foil and the dielectric layer. Finally a thin metal layer is added to be operated as a thermal resistor to indirectly heat the SMA through the dielectric layer and thereby actuate the structure. However, this concept was realized on a per device level, only.

In the present work we report on a concept for both the wafer-level manufacturing and integration of robust trimorph bulk SMA microactuators into silicon structures.

2. Wafer-scale manufacturing of SMA microactuators

2.1. Principle of trimorph SMA actuators

There are two main technical challenges to be addressed to allow for batch integration of bulk SMA microactuators. The first is to provide a batch manufacturing compatible cold-state reset mechanism; the second is to allow batch manufacturing compatible electrical contacting.

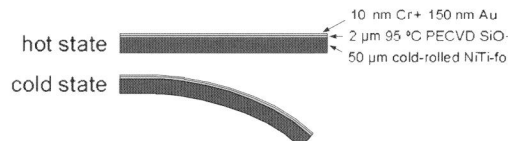

Figure 1: The trimorph microactuator is deformed in the cold state due to the induced stress of the dielectric layer. Upon actuation, which can be accomplished by resistive heating through the top gold layer, the SMA sheet regains its flat shape in the hot state.

In this study, our previously developed actuation concept using a trimorph SMA/dielectric/metal structure [4] was realized on a wafer-scale level. The dielectric layer stress deforms the actuator in the cold-state, see Figure 1. This built-in cold-state reset eliminates the need for mechanical pre-tensioning or additional microsprings. The choice of dielectric material, deposition conditions, layer thickness and thermal treatment [5] allows for tuning the actuator characteristics.

Actuation can be accomplished by an indirect heating scheme, in which the thin metal layer is deposited on top of the oxide. The heater is electrically isolated from the SMA and allows for easy electrical contacting and low actuation current. Moreover, the heater can be patterned prior to the etching of the SMA for optimized heat transfer, reducing thermal gradients along the beam and reducing power consumption. However, for certain geometries of the SMA structures, e.g. u-shaped beams, patterning of the metal layer is not required.

The electrical contacting and the intrinsic cold state reset are the key enablers for batch processing. Three key actuator performance factors, i.e. the shape memory effect, the cold-state reset and the electrical heating, can be optimized independently by each respective layer of the trimorph.

2.2. Principle of wafer-scale integration of SMA actuator structures

To allow for wafer-scale fabrication of the trimorph SMA actuators two key issues must be solved. One is to selectively bond SMA foils to a silicon substrate and the other is to then pattern the foil into desired actuator shapes. Adhesive bonding allows for good adhesion between many materials and can be done in low-temperature conditions in inert atmosphere [6] not risking devastating side effects such as substantial oxidation of the SMA.

Here we present a method to integrate SMA microstructures to silicon on a wafer-level by adhesive bonding followed by patterning into actuator structures. Contact printing of the adhesive to a silicon wafer with topographical structures results in a transferred adhesive layer only on the top surfaces of the silicon. SMA foil covering the whole wafer is applied above the conversion temperature to ensure a flat shape of the foil which enables good conformal contact to the silicon. During curing of the adhesive, the SMA is selectively bonded only to the regions of the silicon to which contact was made with the intermediate adhesive, defined by the topographical pattern on the wafer.

After bonding of the SMA foil to the silicon wafer, actuator structures can be patterned by lithography and standard wet chemical etching of NiTi. However, the

1) KOH-etching of 300 μm deep wells in 525 μm thick oxidized Si-wafer

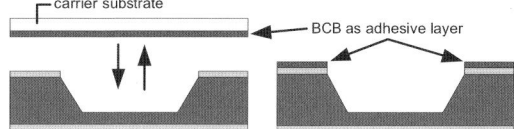

2) Stamping of adhesive layer (BCB) on non-etched top surface, resulting in a patterned adhesive layer.

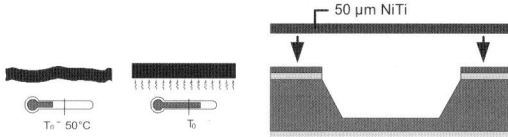

3) Flattening of a 50 μm thick TiNi-foil by heating above the conversion temperature T_0 and then applying the foil on the adhesive layer.

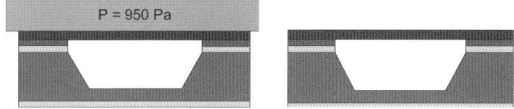

4) Loading the stack with a weight to ensure a uniform bond and hardcure the adhesive layer at 250°C in a nitrogen atmosphere.

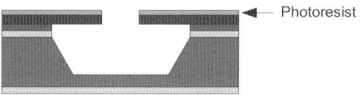

5) Patterning the NiTi foil using lithography and wet etching.

6) PECVD of 2 μm SiO_2 at 300°C and evaporation of 10/150 nm Cr/Au. Finally the cantilevers are bending down into the KOH etched well due to compressive stress in the PECVD oxide.

Figure 2: Cross sectional drawings illustrating the steps (1-6) of the wafer-scale manufacturing of robust trimorph bulk SMA microactuators.

adhesive must be carefully selected to not risk delamination upon exposure to the etchant. A possible way to circumvent this problem is to first pattern the SMA, which is temporarily bonded to a carrier wafer, and subsequently transfer bond the SMA from the carrier to the target silicon wafer. However, this introduces an additional and complicated alignment step of the SMA structures to the silicon during assembly of the two layers.

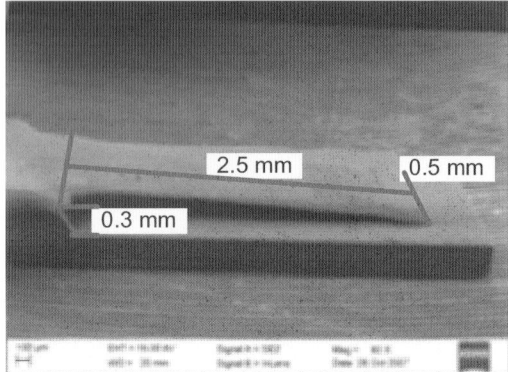

Figure 3: SEM-Picture of a wafer-scale manufactured trimorph microactuator, bending downwards due to the compressive stress in the PECVD oxide, which forms the built-in bias spring.

2.3. Fabrication

Figure 2 illustrates the process flow for the first test structures. We KOH-etched 300 µm deep wells in an oxidized Si-Wafer which define the maximum deflection of the cantilevers during operation (a). Then, we stamped a layer of Benzocyclobutene (BCB) on the unetched parts of the substrate, resulting in a self-aligned adhesive layer pattern (b). Next, we flattened a 50 µm thick commercially available TiNi foil (Johnson-Matthey, USA) by heating above the transition temperature, T_0, and applied it onto the adhesive layer (c). The wafer-SMA-stack was compressed to ensure a uniform bond and the bonding process was subsequently completed by hard-curing the BCB in a nitrogen atmosphere (d). Then, the SMA was patterned using lithography and wet etching (e). Finally, we PECVD deposited 2 µm of SiO_2 at 300 °C, which also further cures the BCB, and evaporated 10/150 nm Cr/Au (f). The resulting actuators are bending downwards into the wells due to the compressive stress in the oxide layer. The thin metal layer is electrically isolated from the SMA and can potentially be used as an electric resistor to heat the actuator with an electrical current, but it must then be deposited and patterned prior to the etching of the SMA or be used on other geometries of the cantilevers.

3. Results and discussion

3.1. Evaluation with cantilever test structures

Test cantilever structures with a width of 0.5 mm and a length of 2.5 mm were fabricated and diced from the wafer into single pieces for evaluation. Figure 3 shows a SEM-picture of a single cantilever and its dimensions.

The current SMA etch procedure results in an unacceptable underetch rate, as can be seen in Figure 4, and is therefore currently being improved.

3.2. Measurements

Figure 4 shows photographs of a single cantilever during operation. In the cold state the compressive stress in the oxide, forming the built-in bias spring, deflects the cantilever-tip downwards until it touches the bottom of the well. In the hot state, the SMA works

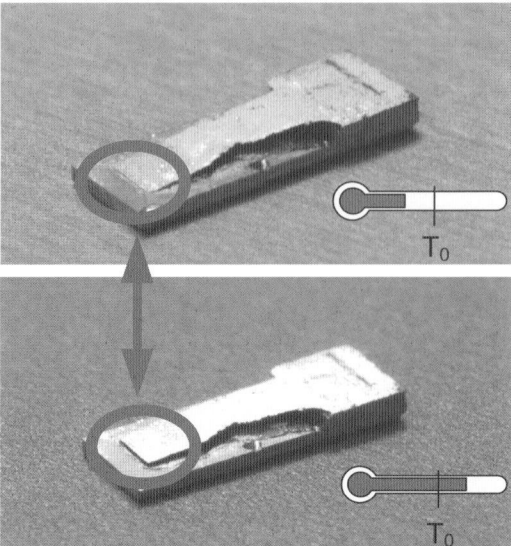

Figure 4: Pictures of a sample taken in the cold state (top) and in the hot state (bottom), showing the stroke range of the cantilever tip. In the cold state the cantilever deflects downwards due to the PECVD SiO2 layer which acts as bias spring. In the hot state the SMA works against the SiO2 bias spring and lifts up the cantilever.

against the bias spring and lifts the cantilever-tip up from the substrate.

Temperature-deflection measurements at quasi-static equilibrium conditions using a thermostat are shown in Figure 5. Starting at temperatures above T_0, the actuator shows simple bimorph behaviour. After decreasing the temperature to the transformation temperature between 60 and 40 °C the SMA phase-change decreases the SMA stiffness and allows for a considerable quasi-plastic deformation, resulting in a rapid deflection of the cantilever. The latter is mechanically limited at 300 µm by the silicon substrate. Hence, the maximal deflection in the cold state is equal to the depth of the silicon well, which is 300 µm.

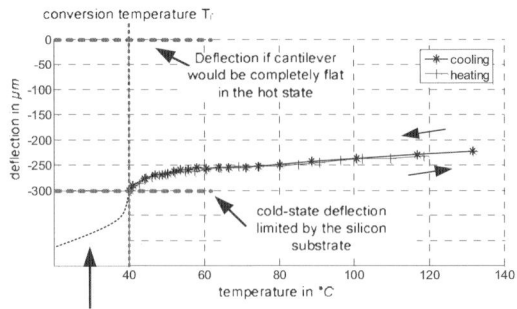

Figure 5: Temperature–deflection measurement of a first test actuator. In the cold state, the measurements show a large deflection of 300 µm. Heating the actuator results in stroke of max. ~80 µm. The dotted line shows the expected shape of the deflection curve without the limitation, in accordance to previous similar work [4].

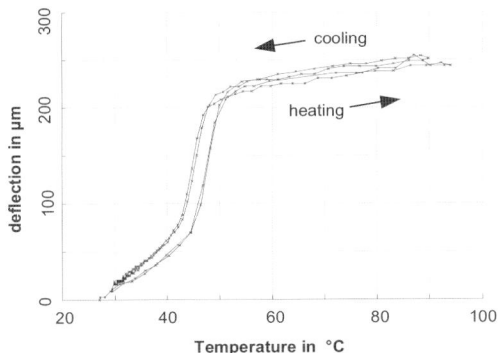

Figure 6: Temperature–deflection measurement of an earlier, similar actuator [4]. The deflection of the cantilever was not mechanically limited and showed a stroke of approx. 250 μm. This device, however, was fabricated on a per-device level.

Heating the actuator results in a relatively small stroke of max. ~80 μm. The actuators in this first batch run cannot completely recover the flat shape due to the high compressive stress in the oxide. However, earlier work of the authors [4] indicates, that the overall deflection and the resulting stroke would be much larger without the deflection-limiting substrate. Figure 6 illustrates the deflection of an earlier presented trimorph cantilever actuator [4], which was not mechanically limited and showed a stroke of approx. 250 μm.

In this work, we incorporated a limit to the deflection stroke due to two advantages: the bias spring is never fully relaxed, which potentially increases the stability of the actuator, and the risk for plastic deformation of both the SMA and the oxide is considerably reduced.

4. Conclusions

We successfully developed and demonstrated a process for the wafer-level integration of SMA actuators to silicon microstructures. The process adapts well to batch fabrication and allows for patterning of the SMA either prior or after the transfer bonding.

The processed materials are only exposed to low-temperature treatment and bonding pressures below 1 kPa which allows for the integration of sensitive components. The concept was demonstrated by the wafer-level fabrication of bulk trimorph SMA microactuators showing a working deflection range of approx. 80 μm.

Acknowledgements

This work is part of the Q2M project and receives research funding from the European Commission through the sixth framework program.

References

[1] M. Kohl, Shape Memory Microactuators. Springer, 2004.

[2] K. Skrobanek, M. Kohl, and S. Miyazaki, "Stress-optimised shape memory microvalves", in *Micro Electro Mechanical Systems, 1997. MEMS'97, Proceedings, IEEE., Tenth Annual International Workshop on*, 1997, pp. 256–261.

[3] P. Krulevitch, A. Lee, P. Ramsey, J. Trevino, J. Hamilton, and M. Northrup, "Thin film shape memory alloy microactuators," *Journal of Microelectromechanical Systems*, vol. 5, no. 4, pp. 270–82, December 1996.

[4] S. Braun, T. Grund, S. Ingvarsdottír, W. van der Wijngaart, M. Kohl, G. Stemme, "Robust trimorph SMA microactuators for batch manufacturing and integration", Proceedings Transducers 2007 in Lyon, pp. 2191-2194

[5] Z. Cao and X. Zhang, "Mechanism of temperature-induced plastic deformation of amorphous dielectric films for MEMS applications," in *Proceedings of the 18th IEEE Inter-national Conference on Micro ElectroMechanical Systems (MEMS)*, 2005, pp. 471–474.

[6] F. Niklaus, G. Stemme, J.-Q. Lu, R.J. Gutmann, "Adhesive wafer bonding", Journal of Applied Physics – Focused Review, vol. 99, no. 1, pp. 031101.1-031101.28, 2006.

Multi-Material Micro Manufacture
S. Dimov and W. Menz (Eds.)

Material aspects for batch integration of PZT thin films using transfer bonding technologies – Q2M development

D. Bhattacharyya[a], R. V. Wright[a], Q. Zhang[a], P.B. Kirby[a], R. Guerre[b], U. Drechsler[b], M. Despont[b], F. Saharil[c], J. Oberhammer[c]

[a]*Materials Department, Cranfield University, Bedford MK43 0AL, UK*
[b]*IBM Research Gmbh, Zurich Research Laboratory, Rueschlikon, Switzerland*
[c]*Microsystem Technology Lab, KTH – Royal Institute of Technology, Stockholm, Sweden*

Abstract

Transfer bonding is a reliable cost-efficient and low-temperature CMOS compatible technique which allows batch integration of materials whose incompatibility with Si makes them unsuitable for monolithic integration. In this heterogeneous device integration method the material and process incompatibilities inherent in Si IC technology are overcome by fabricating devices on separate substrates and then transferring them onto target (e.g. CMOS) wafers. Transfer bonding has great potential for integrating RF-MEMS devices incorporating, for example, high thermal budget materials such as PZT and PST or non-ferroelectric piezoelectrics such as AlN and ZnO into microwave ICs for enhanced systems performance. This paper presents an overview of technology developments within the EU sponsored project Q2M for the realization of transfer bonded piezoelectrically actuated RF MEMS switches and other components focusing in particular on material factors relating to growth of the piezoelectric films, in this case sol-gel deposited PZT, that restricts the choice of device layers and impact on PZT properties such as microstructure, film orientation and piezoelectric coefficients. New process developments such as hard masking of PZT pattern during RIE etching and its compatibility with polymer transfer bonding are discussed.

Keywords: PZT, transfer bonding, RF MEMS switches, sol-gel

1. Introduction

If the recent drive to incorporate sensing and actuation functions into electronic circuits continues it will revolutionize the fields of microsystems and RF technology, particularly wireless communications and automotive electronics. The potential for integration of RF-MEMS components with well-established IC technology has already been demonstrated with realization of thin film bulk acoustic resonators (FBAR) duplexer filters for mobile phones [1]. There is now a current technological quest to develop reliable and cost-efficient batch integration techniques that provide for incorporation of high thermal budget materials such as $PbZr_xTi_{(1-x)}O_3$ (PZT) and $Pb_xSr_{(1-x)}TiO_3$ (PST) or non-ferroelectric piezoelectrics such as AlN and ZnO and so enable increased functionality and superior systems performance. Piezoelectric sensing/actuation offers advantages over alternatives such as high actuation force at low voltage and linear response with wide dynamic range [2] and among thin film piezoelectrics PZT has received most attention for MEMS applications as it has the highest piezoelectric coefficients and hence exhibits the best electromechanical performance. However, for some applications PZT's high dielectric constant is a limitation as is its CMOS incompatible growth temperature and in these cases, AlN, in particular, is an attractive alternative which in addition to CMOS direct integration offers improved signal-to-noise ratio and high power efficiency [2].

As monolithic integration of PZT continues to prove elusive, new and alternative CMOS-compatible heterogeneous device integration methods have received great attention recently with the focus on novel transfer bonding techniques to overcome the material and process incompatibilities that arise when combining dissimilar materials. For example within the EU FP6 sponsored project Q2M [3], a consortium is currently investigating the suitability of a range of transfer bonding techniques for application specific RF-MEMS device integration incorporating PZT thin film actuators. This article gives an overview of the batch integration of PZT thin film devices onto dissimilar substrates, and consequent materials and process constraints applicable to two of the transfer techniques recently developed within the Q2M: wafer level microdevice distribution technology using selective transfer and adhesive full wafer transfer bonding [4,5]. The separation of the device processing onto different substrates, i.e. an auxiliary substrate and target RF substrates, and the transfer bonding between these substrates allow new possibilities to develop innovative actuator designs and fabrication processes. As an example, hard masking for RIE patterning of PZT is described in this paper.

2. PZT transducer stack – Microstructure design

The ideal PZT device structure consists of a PZT thin film sandwiched between two electrodes in direct contact with the substrate as shown in Fig.1. However, PZT film quality depends largely on control of interdiffusion and surface chemical reactions involving the layers that form the PZT composite [2,6] and for the sol-gel deposited films used in the present work it has been established that Pb from the sol diffuses through the bottom electrode into the substrate during the nucleation and growth of the perovskite PZT, and if for example a standard Si substrate is used, forms lead silicates (PbSiO$_4$) that can cause delamination of the overlying electrode [6,7]. To avoid this, a barrier layer such as SiO$_2$, Si$_3$N$_4$ or TiO$_2$ must be inserted between the bottom electrode and the substrate to limit the diffusion of Pb and also to prevent formation of PtSi which would also have a deleterious effect.

Fig. 1. Schematic - an ideal PZT transducer stack

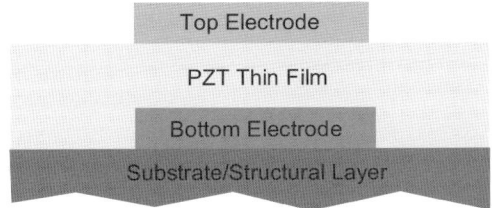

2.1. Device layer – Standard structure

Until recently most research studies have been concerned with obtaining high quality PZT films on Si with a standard structure composed of Pt/Ti bottom electrodes on top of a thermally grown SiO$_2$ barrier layer. In addition to Pb diffusion, which in fact, both these layers together prevent, other processes occur during PZT crystallization and these are depicted schematically in Fig. 2. The Ti which diffuses along grain boundaries to the Pt surface provides nucleation sites and the fleeting formation of an intermetallic phase, Pt$_3$Pb also influences perovskite phase transformation with the preferred PZT grain orientation [7]. As evident, material constraints imposed by the layers available for a specific device design means that the actual layer structures has to be more complex for practical applications. To make progress on novel device development, it is interesting to understand the material factors that influence the choice of layers and the consequent effect on PZT actuation properties.

Fig. 2. PZT device structures - schematic diagram of chemical interactions at layer interfaces

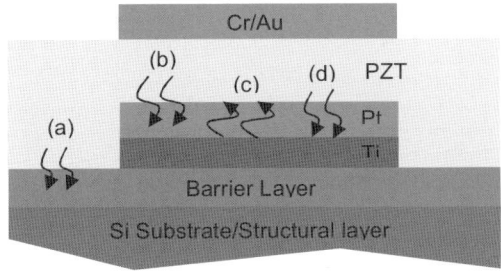

(a) Pb diffusion into the barrier causes cracks; (b) Pb diffusion into Pt and alloying (Pt$_3$Pb); (c) Ti diffusion to Pt surface provides nucleation sites; (d) diffusion of O$_2$ to Ti adhesion layer forming an oxide layer

3. Transfer bonding process – Layer design

As proposed within the Q2M project, two routes for transfer bonding technologies, already described in detail in our previous publications [4, 5], for realization of piezo-actuated RF switches are under development which requires PZT stacks with different types of layer structures for the different design configurations and process layouts under consideration. The choice of materials and process parameters is crucial to the successful implementation of transfer bonding with no degradation of PZT film quality.

While pursuing process layouts and overall device design, different types of substrates and barrier layer combinations were considered as shown in Table 1 and schematic switch configurations with various PZT stack designs are depicted in Fig.3. In the selective transfer process LPCVD grown Si$_3$N$_4$ was chosen as a structural layer whereas for full wafer transfer both Si and PECVD Si$_3$N$_4$ are being investigated. In principle, the choice of a Si$_3$N$_4$/Si substrate offers better tunability of stress-induced pre-bending of PZT cantilever and is also cost-effective as compared to SOI wafers.

Table 1. List of materials for PZT thin film devices

Device Layer	Materials (wafer/oxides/metal)	
	Wafer level microdevice distribution	Adhesive full wafer transfer
Substrate	Si	Si and SOI
Structural Layer	Si$_3$N$_4$ (LPCVD)	Si and Si$_3$N$_4$ (PECVD)
Diffusion Barrier	TiO$_2$ (RF/DC sputtered & annealed)	TiO$_2$ and SiO$_2$ (Thermal & PECVD)
Bottom Electrode	Pt/Ti	Pt/Ti
Functional Layer	PZT, AlN	PZT, AlN
Top Electrode	Pt/Ti, Au/Cr	Pt/Ti, Au/Cr

Fig. 3. Schematic device design layout

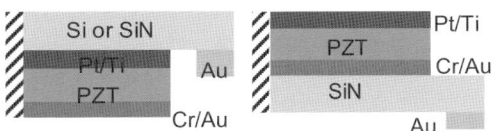

(a) Structural layer (Si/Si$_3$N$_4$) above and below PZT

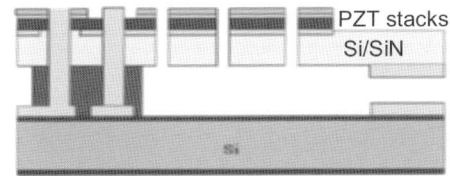

(b) Cross-sectional layout of a full-wafer transfer-bonded RF switch on the target RF substrate

4. Fabrication - PZT thin film devices

Focusing on devices in which Si$_3$N$_4$ was used as the structural layer, a TiO$_2$ layer was selected as the barrier material - the layer thicknesses used for these were 1.5µm and 30nm respectively. The TiO$_2$ was prepared by RF sputtering of Ti and thermally oxidizing at 700°C to obtain the stable rutile phase over the metastable anatase and brookite phases. The RF sputtered bottom Pt/Ti electrodes were patterned or left unpatterned, depending on the design layout for particular transfer bonding scheme, before sol-gel PZT thin films (with Zr/Ti compositions of 30/70) were deposited. The PZT was, therefore, spin-coated onto either a continuous Pt surface or a non-planar mixed surface of TiO_2 and Pt. Following spinning PZT films underwent pyrolisis and crystallization hotplate bakes at 200°C and 530°C respectively.

Finally RF actuators were fabricated by further processing of wafers with PZT microstructure that had been produced by either dry etching (full wafer transfer) or wet etching (selective transfer). The dry etching of PZT stacks is a critical step, since the physical etching of the PZT layer results in sidewall deposits, commonly called "fences". The problem of fencing could affect the device functionality or fabrication yield and was resolved as described below in section 4.1. PZT was wet-etched in two steps using a solution of $(BHF:HCl:NH_4Cl)$ and followed by HNO_3. Fig.4 shows successful distribution of a cell of piezo-actuated RF switches using selective transfer bonding technique.

Fig. 4. Distribution of a cell of RF switches

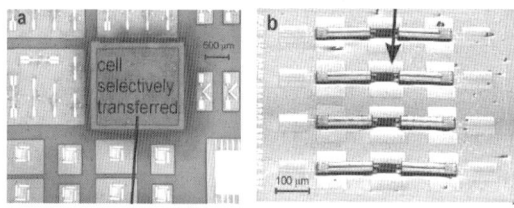

(a) Optical micrograph of the source wafer
(b) SEM image - Transferred devices after release on the receiver using selective transfer bonding

4.1. Plasma etching of PZT microstructure

Although reactive ion etching (RIE) is used extensively for producing fine patterns in PZT e.g. for FeRAMs a limitation has been that with all the currently used etch chemistries (generally F or Cl based) the involatility of some of the etch products often leads to excessive sidewall deposits when a conventional photoresist mask is used. It has been found, however, that fence problems can be alleviated by hard masking and good results have been achieved recently in the fabrication of FeRAM devices, for example, using TiN and TiO_2 masks [8]. The sputtered Ni has also been a popular choice due to its extremely low RIE etch rate but a low film stress, another key requirement in view of the often observed low adhesion of layers deposited on top of PZT, is difficult to obtain with this material. Consequently, following the work of Subasinghe on bulk PZT [9], a low stress electroplated Ni hard-mask process has been developed for the PZT thin films. A critical factor in this is the use of a sulphamate bath which has the advantage that under carefully controlled conditions it produces extremely low stress Ni films.

To facilitate electroplating a Cr/Au seed layer is deposited on top of the PZT and Ni is then selectively plated through a photoresist mask. Following this photoresist is stripped and the seed layer is removed by dry etching using the Ni itself as a mask. For etching the PZT a low power CHF_3/Ar plasma process has

been developed to provide good stopping on the underlying thin Pt electrode. An example of the heavy sidewall deposits produced by this process when conventional photoresist masking is used is shown in Fig.5 (a). Ni etches at a much lower rate than the PZT (etch selectivity ~8:1) making it an ideal etch mask. Ni thicknesses in the range 0.5-1μm are used to etch up to 1μm thick PZT stacks although thinner layers could be used. Following PZT etching the Ni is removed in a $FeCl_3$ solution to leave fence free features with steep side walls (~63°) as shown in Fig.5 (b).

Fig. 5. SEM micrographs of etched PZT stacks

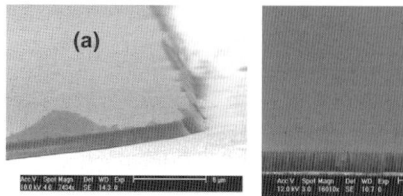

(a) Sidewall deposits during dry etching when using a photoresist mask; (b) Fence-free etched profile when electroplated Ni hard mask used instead.

4.2. PZT microstructure evaluation

The surface morphology of PZT microstructures was analyzed using AFM method. The state of the surface underneath the Pt/Ti electrodes and TiO_2 barrier layer strongly influences PZT thin film growth and so for example Pt hillock formation during PZT crystallization must be avoided to ensure a smooth texture with lower surface roughness [10]. A thermally stable electrode and barrier interface must be achieved prior to depositing PZT and key to this is complete oxidation of the Ti so as to prevent delamination by Pt hillock formation a well known effect that occurs due to increased compressive stress during high temperature annealing. Fig.6 (a) exhibits AFM image of a high quality PZT film with dense texture (surface roughness ≈ 0.7nm). In contrast Fig.6 (b) shows a blistered PZT surface as evidence of Pt hillocks formed on the relatively flat PZT grains.

Fig. 6. Surface morphology of PZT on $Pt/TiO_2/Si_3N_4$

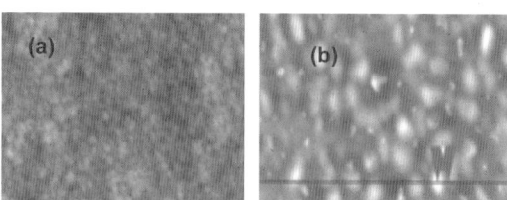

AFM images illustrate (a) high quality PZT thin film and (b) Pt hillocks on a blistered PZT surface

It has been found with all the different layer structures investigated that the (111) PZT perovskite grain orientation dominates. This can be seen clearly for example in the XRD spectra depicted in Fig. 7 revealing the fact that the PZT on $Pt/Ti/TiO_2/Si_3N_4$ structures also exhibit a less preferred (110) orientation unlike $PZT/Pt/Ti/SiO_2$ structures. However the higher peak intensity of the preferred perovskite (111) phase compared to (110) orientation in all such PZT stacks comprising various device layer combinations indicates that the piezoelectric properties in the films will be dominated by the perovskite (111) phase as expected with PZT thin films of 30/70 compositions [7].

Fig. 7. X-ray diffraction spectra of PZT thin film (Zr/Ti = 30/70 compositions) grown on different materials

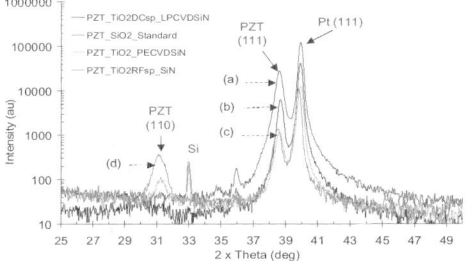

(a) Standard PZT/Pt/Ti/SiO$_2$/Si stack;
(b) PZT stack on TiO$_2$/SiN$_{LPCVD}$ (DC sputtered TiO$_2$);
(c) PZT on TiO$_2$/SiN$_{LPCVD}$ (RF sputtered TiO$_2$);
(d) PZT on TiO$_2$/SiN$_{PECVD}$ (RF sputtered TiO$_2$)

4.3. Piezoelectric thin film properties

Electrical characterization was carried out to assess the dielectric properties and polarization hysteresis of the PZT films. The measurements were conducted on PZT thin film capacitors composed of various layer designs and Cr/Au evaporated dots as top electrode. The measured values of the dielectric constant were 378 and 314 and the loss tangent as 0.02 and 0.025 at 100 kHz, before and after poling respectively. The parameters obtained from different materials exhibit only ~10% variation in capacitance values and have very low loss compared to standard PZT stacks on SiO$_2$/Si. Fig. 8 shows hysteresis loops of a PZT composite on TiO$_2$/Si$_3$N$_4$. The saturation polarization, remnant polarization and coercive field in poled devices were 41μC/cm^2, 29μC/cm^2 and 78kV/cm respectively, establishing that the films had fully crystallized in the perovskite phase.

Fig. 8. Hysteresis loops of PZT/Pt/Ti/TiO$_2$/Si$_3$N$_4$ stacks before and after poling

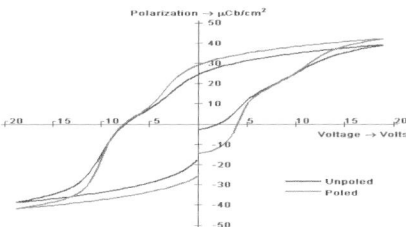

The actuation characteristics were evaluated using laser scanning vibrometry and the effective longitudinal piezoelectric coefficient, d$_{33,f}$ of the PZT films was determined after poling by applying an AC sinusoidal signal at 5kHz, i.e. well below the resonant frequency. The optimum value of d$_{33,f}$ was 30pm/V which is consistent with results obtained from standard PZT composites. The experimental results established that high quality PZT thin films can be grown on top of

a range of different layer structures thus demonstrating their suitability for wafer-scale transfer. Work is in progress to characterize RF switches fabricated using two different transfer bonding techniques and comparison of piezo-actuated switching performance under different excitation conditions subsequently.

5. Conclusions

Layer designs for devices incorporating PZT thin films have been formulated which feature thermally stable barrier/electrode interfaces combined with suitable process layouts that are compatible with transfer bonding techniques (developed within the Q2M project) for wafer-scale integration of piezo-actuated RF MEMS switches and components. The surface morphology and the electrical properties of PZT microstructures were assessed to understand material constraints imposed by the choice of different layers. A new low stress electroplated Ni hard mask process was developed to establish fence-free patterning of PZT stacks during plasma etching of full wafer transferred cantilever beams. The first successful distribution of PZT-based RF actuators using selective transfer bonding has been demonstrated.

Acknowledgements

This work is part of the Q2M project and receives research funding from the European commission through sixth framework program.

References

1. Aigner R, "High performance RF-filters suitable for above IC integration: Film bulk acoustic resonators (FBAR) on silicon", Proc. of the custom integrated circuits conf., IEEE, 2003, p141-146.
2. Trolier-Mckinstry S. and Muralt P, "Thin film piezoelectrics for MEMS", Jr. Electroceramics, Vol.12, 2004, p7-17.
3. http://q2m.4m-net.org
4. Niklaus F, Stemme G, Lu J, Gutmann R, "Adhesive wafer bonding", Journal of Applied Physics, Vol.99, 2006, p 031101.1-031101.28.
5. Guerre R, Drechsler U, Jubin D, Despont M, "Selective transfer technology for microdevice distribution" IEEE Jr. Microelectromechanical systems, Vol.17 (1), 2008, p157-165.
6. Kaewchinda D, Chairaungsri T, Naksata M et al, "TEM characterization of PZT films prepared by a diol route on platinised silicon substrates" Jr. European Ceramic Soc., Vol 20(9), 2000, p.1277-88.
7. Huang Z, Zhang Q, Whatmore R, "Structural development in the early stages of annealing of sol-gel prepared lead zirconate titanate thin films", Jr of Applied Physics, Vol 86 (3), 1999, p 1662-69.
8. Chung C, Chung, I. "Etch behavior of Pb(Zr$_x$Ti$_{1-x}$)O$_3$ films using a TiO$_2$ hard mask", Jr. Electrochem. Soc., Vol 148 (5), 2001, p. C353-6.
9. Subasinghe R, Srimath S., "High aspect ratio plasma etching of bulk Lead Zirconate Titanate" Proc. SPIE, Int. Soc. Opt. Eng., Vol. 6109, 2006, p. 61090D.
10. Jung W, Choi S, Kweon S, Yeom S, "Platinum (110) hillock growth in a Pt/Ti electrode stack for ferroelectric random access memory" Applied Physics Letters, Vol.83, 11, 2003, p 2160-62.

Multi-Material Micro Manufacture
S. Dimov and W. Menz (Eds.)

Towards Batch Integration of SMA into Microsystems: An Actuator Prototype

D. Clausi, J. Peirs, D. Reynaerts

Katholieke Universiteit Leuven, Department of Mechanical Engineering, Division PMA

Abstract

Shape Memory Alloys have a considerable potential for integration into microsystems, where scaling down of their size allows favorable exploitation of the intrinsic adaptive capabilities, providing an actuation mechanism for applications (e.g. micropneumatics) requiring large force control and large actuator stroke. However, the implementation of these materials into actual structures is rather complex and mostly confined to depositing thin NiTi films onto certain target substrates, resulting in devices having a relatively high cost-per-piece.

This paper is aimed at investigating a novel approach for batch integration of SMA to microactuators, which might provide a cost-effective alternative to thin film technology while enhancing functional properties and design flexibility. Indicative requirements for the actuator design have been drawn from typical microvalve applications. In order to evaluate the actuator performance, brass microcantilevers have been produced, with prestrained SMA thin wires bonded on top of them, eccentrically with respect to the cantilever's neutral plane. The activation of SMA element is obtained by direct heating through electrical current. The bending actuation of the cantilever leads to large strokes, expected to match the requirements of a wide range of applications.

Keywords: SMA, actuator, Microsystems, Micro-electro-mechanical-system (MEMS), micro-fluidics, microvalves

1. Introduction

In the recent past, the field of Microsystems has been subjected to growing attention from both industry and research community. Microsystems have been recognized as having the potential to revolutionize the performance of a wide range of products by merging silicon-based microelectronics with micromachining technologies, thus enabling complete systems-on-a-chip to be realized and allowing novel functionalities at reduced costs.

Certain classes of applications, namely microvalves, have been found particularly attractive when produced with micromachining technologies, having the potential to achieve control of large gas flows at relatively large pressure differences, with the rapid response time and low power consumption offered by microsystems. Furthermore, the perspective of batch production promises a cost-efficient approach to the production of single microvalves and opens up the integration of many devices on a common fluidic plate, enabling a modular approach in designing microfluidic systems.

Several microvalves, using different actuation mechanisms, have been investigated by many research groups in the last years. Despite the wide variety of devices that have been built and tested, the basic configuration of microvalves can be divided in two main groups: seat valves an gate valves (Fig.1).

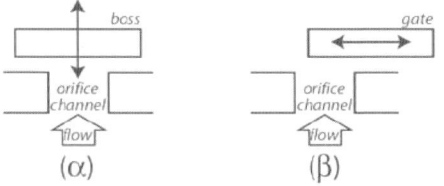

Fig.1: Working principle of seat valve (α) and gate valve (β)

Seat microvalves are also known as "membrane-type" and are characterized by an actuation mechanism that counteracts the large pressure forces controlled by the valve. In order to allow for a large gas flow, a large valve stroke is needed.

An interesting approach in designing such microvalves is presented in [1], where multiple circular nozzles are expected to increase the flow capacity when compared to a single orifice of the same area, while keeping the required actuation force constant.

In gate microvalves, the static pressure and valve actuation are perpendicular ("cross-flow") and therefore do not counteract each other. This valve design removes the requirement on the actuator's high force output. Nevertheless, large stroke actuators are needed to obtain large gas flows. This configuration lacks contact between gate and orifice, which means that some leakage cannot be avoided.

As clear from the above described basic valve configurations, a common requirement for effective control of gas flows at micro-level is a robust high force/high stroke actuation mechanism. Another crucial factor is the availability of such microactuators at low unit cost. For this reasons, gas microvalves have not fulfilled their maximum commercial potential yet, primarily for the cost-per-performance ratio of today's devices, whose miniaturization is limited by technological constraints and low energy densities of conventional actuation principles at small scale.

In this context, the application of shape memory alloys for actuation of micropneumatic devices might bring a relevant technological breakthrough.

SMA materials exhibit the highest energy density amongst current MEMS compatible materials and, importantly, as size is reduced towards the micro-scale, they benefit from improved heat transport, which increases their response speed. Other remarkable advantages of SMAs for micro-applications are their simplicity, clean operation and low voltage requirement, which make them suited for CMOS processing [2, 3, 4].

Most of the current efforts for the implementation of

SMAs into microsystems are based on thin film technology [5,6]. Despite their attractiveness in terms of batch fabrication, the techniques used for depositing thin NiTi films onto target substrates (e.g. vacuum vapour deposition and sputtering) require expensive equipment and present several problems in terms of composition control and prestraining of the deposited material[3,6,7].

An alternative approach is based on cold-rolled SMA sheets, which has been proven successful to a certain extent and rather promising.

Kohl et al.[8] used SMA rolled sheets of 95μm thickness operated in bending mode to actuate microvalves, obtaining controlled pressure differences and gas flows respectively of 100kPa and 1200 Standard ccm.

More recently, S. Braun et al.[9] reported on the fabrication of a NiTi-SiO$_2$-gold trimorph microactuator. The NiTi was etched out of a cold-rolled sheet 20μm thick, pre-stressed by the dielectric layer consisting in 2μm thick SiO$_2$ PECVD deposited, and actuated by resistive heating through the 150nm thick gold layer evaporated on top of the structure. The observed stroke of the unloaded 3mm long actuator was about 730μm, with a power requirement of 20 mW.

One potential limit of rolled sheets is the detrimental effect of bending actuation on the efficiency of the actuator, being the energy efficiency of SMA in bending mode lower than the one in tension mode.

Cost-effective integration of SMAs into microsystems is believed to be viable by the authors by using prestrained thin fibres in combination with Si micro-cantilevers, which would serve the role of bias spring. With SMA wires placed eccentrically with respect to the neutral plane of each cantilever and fixed at their ends, the actuation mechanism could exploit the high recoverable strains of the SMA, while keeping the stresses into the Si cantilever well below the elastic limit of the material.

The approach envisioned to fabricate such actuators at wafer level starts from a SOI wafer and is based on standard IC processing for patterning of the Si structures: front side DRIE to fabricate the cantilevers and obtain the overlap for out-of-plane placement of SMA wires and back side DRIE to release the cantilevers. The actual integration of the active elements is achieved by spinning the adhesive that bonds the SMA fibers onto the Si substrate, deposit the prestrained fibers and finally cure the adhesive locally at the fixing points by exposing it with UV light. Contact pads ensure electrical contact between the active material and the power source via electrical wires contacted by conventional wire bonding.

As feasibility study for implementation of SMA fibers onto the cantilevers and in order to evaluate the performance of the actuators, some prototype structures have been fabricated. Conventional production techniques have been used in order to make the process easier and faster and avoid dealing with mask design and fabrication and with setting up of a complete photolithographic process, which at this stage would have been premature.

The fabrication process of the actuators and some preliminary results are reported in the next sections.

2. Fabrication of SMA microactuator prototype

The main effort required in the production of prototypes of SMA microactuators involves the fabrication of the cantilevers. Brass has been selected as substrate material for the production of these structures, in view of its easy machinability and favorable mechanical (E modulus fairly close to the one of Si) and electrical properties (good conductivity, favorable for the subsequent spark erosion process). A sheet of 600μm thickness was cut in rectangular shape of 60x30mm size by spark erosion and eventually machined by micro milling to obtain the desired thickness for the cantilever and the supports for the SMA wire. For this purpose, groups of 2 pockets (respectively 3x1x0.53mm^3 and 1x2.5x0.53mm^3 size) were milled into the brass sheet. Reference holes were also made, to serve as alignment marks for the SMA wire and starting holes for the subsequent sparking process.

A Φ37.5μm Flexinol™ wire, previously cleaned in Acetone, was clamped to a vertical support, with a mass of 17g crimped at its free end. Electrical contacts were made at the two ends of the wire and connected to a DC source. The actuator wire was subsequently cycled under constant load, with a resulting stress of 150MPa.

The brass plate was eventually positioned with respect to the SMA wire, using the holes previously machined as reference. Strain gauge glue (X60) was used to fix the wire onto the substrate at discrete locations (see Fig.2). UV curing glue was used in other samples instead of X60, displaying good bonding performance.

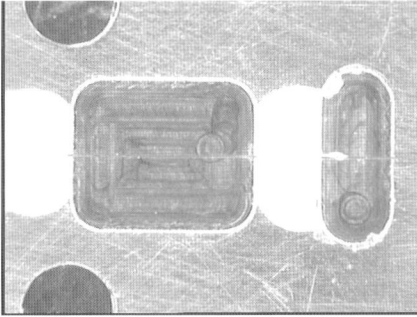

Fig.2. Detailed view of SMA wire glued onto brass substrate. It is possible to see the two reference holes and the pockets milled to provide out of plane placement of the SMA element.

The following step consisted in spark eroding the brass plate by wire EDM to fabricate and release the cantilever structures. The resulting actuators were 3mm long and 0.5mm wide between the droplets of glue, with an anchor of 2x2.5mm hosting the glue and providing room for one of the electrical contacts. The cantilever thickness was 70μm, according to design. The final configuration of the microactuator is depicted in Fig.3.

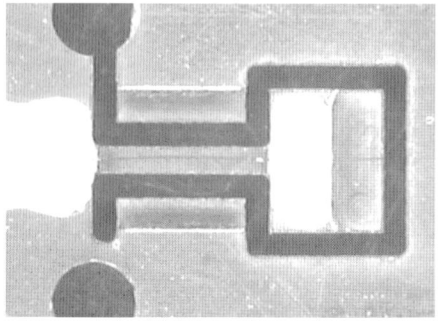

Fig.3. Microactuator released from substrate. The actuation length of the SMA wire, between the two fixed ends, is 3mm.

3. Testing procedure and results

Resistive heating was used to induce the phase transformation of the SMA material, for which reason electrical contacts were needed. Since the brass substrate has a much lower electrical resistance compared to the SMA wire, contacting the latter directly onto the cantilever would result in a parallel arrangement that would draw most of the current to the cantilever instead of the SMA wire, leading to highly inefficient actuation. For this reason, an open circuit has been made with respect to the brass cantilever: at its free end, the SMA fiber and the electrical wire were contacted directly onto the metal, while on the clamped side they were bonded onto the insulating layer of glue. The resulting electrical circuit was then connected to a DC source to provide current for resistive heating of the SMA element.

Objective of the testing phase has been the assessment of the general behavior of the fabricated device, leaving the thorough investigation of its performance to a later stage, when silicon microstructures designed for specific applications and with strictly defined requirements will be available. Despite the non optimal design and production process used to fabricate the actuators, the observed performance under actuation featured a stroke in the order of 0.5mm (theoretical value) at a frequency of about 1.5Hz and power consumption of 120-140mW. In particular, the high stroke displayed is expected to match the requirements of knife gate valves, being the typical orifice high in the order of 250 – 300μm [10]. The actuator is shown in closed and open configuration in Fig.4

Fig.4. Actuator in rest position (upper image) and actuated configuration (lower image).

The measured values can be sensibly improved by reducing the droplet size of the glue and using electrical insulating material as substrate with contact pads patterned onto its surface. The actuator could then be contacted on the clamped side only, leaving the free end of the cantilever unloaded. Using silicon as structural material will also allow higher maximum stresses in the cantilevers, thus providing a more effective reset mechanism.

4. Conclusions

Integration of SMA material into microsystems has been pursued in the present work. Actuators based on thin SMA fibers bonded on cantilever structures were designed, produced and tested, showing attractive performances. The work has served the purpose of demonstrating the feasibility of this approach for applications where high work outputs are required.

The concept hereby described has several advantages when compared to conventional implementation strategies of SMA into microstructures: it is based on off-the-shelf components (SMA wires and commercial adhesives), therefore resulting into cheaper devices; It is readily implemented into Si microstructures, by depositing the wires automatically in place and selectively curing the adhesive; it allows a wide design flexibility, in terms of prestrain of SMA wire, number of wires on a single actuator, positioning with respect to the neutral plane of the cantilever. The latter point makes it possible to tailor the performance of the actuator, in terms of force-displacement characteristics, on the application's needs.

Future work will focus on fabrication of actuators on Si wafers, using conventional clean room technologies to etch the cantilevers and spinning or automatic dispensing techniques to deposit the glue for bonding the wires onto the substrate.

Acknowledgements

The authors wish to acknowledge the EU for financial support through the FP6 Project "Q2M: Batch Integration of High Quality Materials to Microsystems".

References

[1] W. Van der Wijngaart et al., "A Seat Microvalve Nozzle for Optimal Gas Flow Capacity at Large Pressures". JMEMS, vol.14, nr.1, February 2005.

[2] J. Peirs et al., "Scale effects and thermal considerations for micro-actuators". Proceedings of the 1998 IEEE Internation Conference on Robotics & Automation, Leuven, Belgium, May 1998.

[3] J. Peirs, D. Reynaerts, H. Van Brussel, "Design of micromechatronics systems: scale laws, technologies and medical applications". D.Phil. Thesis, Katholieke Universiteit Leuven, Belgium, 2001.

[4] S. A. Wilson et al., "New Materials for Micro-scale Sensors and Actuators - an Engineering Review". Materials Science and Engineering R 56 (2007) 1–129.

[5] B.-K. Lai, G. Hahm, L. You, C.-L. Shih, H. Kahn, S. M. Phillips and A. H. Heuer, "The Characterization of TiNi Shape-Memory Actuated Microvalves". Mat. Res. Soc. Symp. Proc. Vol. 657, 2001.

[6] Yongqing Fu, Hejun Du, Weimin Huang, Sam Zhang, Min Hu, "TiNi-based thin films in MEMS

applications: a review". Sensors and Actuators A 112 (2004) 395–408.

[7] H. Kahn, M. A. Huff and A. H. Heuer, "The TiNi shape-memory alloy and its applications for MEMS". J. Micromechanics and Microengineering. **8** (1998) 213–221.

[8] M. Kohl, D. Dittmann, E. Quandt, B. Winzek, S. Miyazaki, D.M. Allen, "Shape memory microvalves based on thin films or rolled sheets". Materials Science and Engineering A273–275 (1999) 784–788.

[9] S. Braun, T. Grund, S. Ingvarsdottír, W. van der Wijngaart, M. Kohl and G. Stemme, "Robust Trimorph SMA Microactuators for Batch Manufacturing and Integration", The 14th IEEE conference on solid-state Sensors, Actuators and Microsystems (Transducers 2007), Lyon, France, 2007, June 10-14.

[10] S. Haasl, S. Braun, S. Sadoon, A. S. Ridgeway, W. van der Wijngaart, G. Stemme, "Out-of-plane knife-gate microvalves for controlling large gas flows". IEEE Journal of MicroElectroMechanical Systems (JMEMS), vol.15, nr.5, October 2006, 1281-1288.

Multi-Material Micro Manufacture
S. Dimov and W. Menz (Eds.)

Fabrication of piezoelectric thick-film bimorph micro-actuators from bulk ceramics using batch-scale methods

R.P.Jourdain and S.A.Wilson

Materials Department, School of Applied Sciences, Cranfield University, Cranfield, Bedfordshire, MK43 0AL United Kingdom

Abstract

Piezoelectric ceramic films in the 20-60 micron thickness range are rarely employed today in commercial micro-mechanical devices, even though their expected force capability suggests that they are well suited to many micro-fluidic and micro-pneumatic applications. Some examples would be micro-scale fuel cells and micro-combustors. Head sliders, radio-frequency (RF) micro-switches and powered micro-optics are further potential application areas. These are only a few and the barriers in bringing them into reality are those of processing compatibility rather than commercial desirability. Such issues are being addressed in the EU Framework 6 Project 'Q2M', which focuses on batch-scale fabrication issues for high quality new micro-mechanical devices that are cost-effective and which have extended capabilities.

This paper discusses a potential batch-scale production route for piezoelectric thick-film bimorph micro-actuators that combines ultra-precision grinding of ceramics and femto-second laser machining, along with standard micro-fabrication techniques such as wafer bonding. This new method has the key advantage that many different shapes and thicknesses of actuator can be made with only minor process changes, meaning that actuators can be designed to suit their intended application. It contrasts with current practice whereby micro-actuators are often designed around a limited range of standard components, with consequent reduction in their achievable performance. The examples used are a 6mm diameter plane-spiral bimorph actuator for integration into a polymeric micro-valve and 2-5mm long bimorph cantilevers intended for use in a new type of silicon 'house' micro-valve, with pneumatic applications.

Keywords Bimorph, PZT, micro-actuator, wafer bonding, femto-second, ultra-precision grinding, MEMS

1. Introduction

Commercial manufacture of piezoelectric ceramic thick films in the thickness range 20-60 microns currently presents a significant technological challenge [1]. Traditional mixed-oxide, high temperature sintering routes tend to result in ceramics that are deformed or cracked at this level. Alternative bottom-up deposition methods involve spinning nanometre-scale layers of a sol-gel ceramic precursor onto a substrate, followed by rapid thermal annealing and sintering. Film thickness is increased by successive deposition, layer-by-layer. The resultant ceramic is effectively 'clamped' to its substrate preventing its free microstructural development. In practice the ceramic tends to evolve under tensile stress, which becomes more severe as layer thicknesses increase. The ceramic is then prone to cracking and the achievable electro-active properties are around 30-50% of the bulk values, when taking the example of the most commonly used lead zirconate titanate (PZT) family of ceramics. This is perfectly adequate for some applications, however the full potential of the materials is not realised using the bottom-up technique.

For improved performance bulk ceramics are sometimes thinned down to the required dimensions by lapping and polishing. This method can indeed produce higher performing thick films, but the procedure tends to be painstaking and slow. In this paper we describe a method that extends this basic idea, using ultra-precision grinding techniques to increase the material removal rate, and combines it with standard micro-fabrication procedures [2]. The new route is designed to produce actuators that can operate at their full potential and it provides the flexibility to design micro-actuators that are tailored to their intended application. In the opinion of the authors many designs of micro-fluidic actuators presented in literature are compromised in some way by the need to build around the availability of standard electro-ceramic components. Frequently these are thicker than required, precluding the optimum mechanical design solution from being achieved; and, significantly, failing to address the wider systems requirements, as relatively large electronic components must still be used. Typical drive voltages for bulk PZT ceramics are around $1V/\mu m$. Hence, by adopting 20-60μm thick-film PZT micro-actuators, direct integration with CMOS electronics can be said to be achievable. Progress towards this goal is an objective of the EU 'Q2M' project consortium, our partners in this work [3-4].

2. Process

2.1 Assembly of the multi-material stacks

The key feature that distinguishes the new process from previous work is the adoption of ultra-precision grinding of bulk ceramics in combination with standard micro-fabrication procedures. This enables the PZT ceramic components to be fully integrated at the wafer scale with huge time savings over conventional lapping and polishing techniques and with excellent layer-thickness control [5-6].

Standard 50mm diameter PZ26 discs (Ferroperm Piezoceramics A/S, Kvistgard, DK) are first machined to ensure their flatness, nominally +/-1µm form, and surface roughness, nominally <30nm, using an 8-inch wafer face-grinder, designed and built by Cranfield Precision Ltd UK (Figure 1). Figure 2 illustrates the four major steps that follow. By first preparing the ceramic surfaces it is possible to achieve much thinner bonds than would normally be the case [7]. Adhesive (BCB) wafer-bonding the ceramic components together with a titanium shim, using a SUSS Substrate Bonder SB6 VAC/SKM, introduces a compressive stress to the ceramic in the multi-layer structure. This is a consequence of the thermal expansion mismatch in the metal and the ceramic components. In Stage 2 the structural layers remain flat and parallel owing to the stiffness of the stack. In Stage 3 a glass carrier wafer is introduced and temporarily bonded to the stack using a UV adhesive. The bond is quick to make at room temperature and robust. Importantly for complex applications such as this, it does not impart extra stress to the structure so that the devices can be released cleanly and without distortion. The UV curable adhesive is Delo-Photobond4464 (Delo Inset) spun at 2000rpm onto a glass wafer carrier at room temperature and exposed to 320-450nm wavelength light for 40s. The carrier wafer is required to maintain the parallelism of the layers during the second machining step, Stage 4. Figure 3 shows the bonding set-up adopted. It was found that the PZT ceramic can be sensitive to a local temperature variation associated with the central push-down pin of the wafer bonder. This affects the bonding process, as revealed by the second machining step. In Figure 4 a very slight 'orange peel' effect can be seem which indcates an imperfect bond. The twin wafer stack configuration (schematic) in Figure 3 was used to avoid this issue. In Figure 5 the multi-material make-up of the wafer stack is shown. Polyimide (PI) layers have been used for planarization and to prevent surface defects such as pores and cracks from developing into sites for localised electrical breakdown when the ceramic is subsequently poled under a strong electric field.

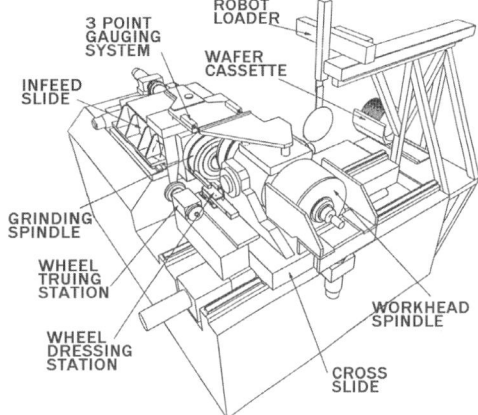

Figure 1: Schematic – 8-inch wafer face-grinder (Cranfield Precision Ltd, UK)

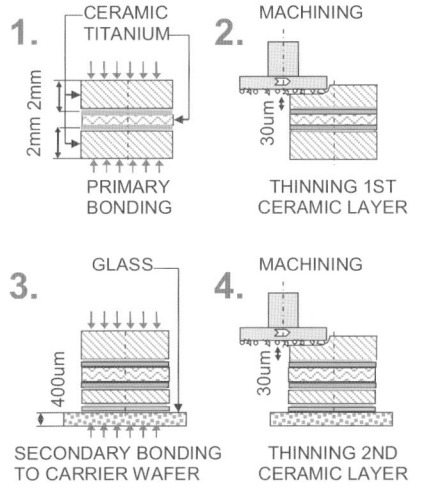

Figure 2: Four key steps in fabricating the multi-material sandwiches

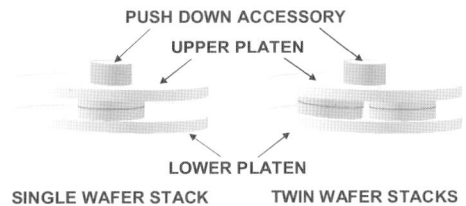

Figure 3: Bonding the multi-material sandwiches - schematic

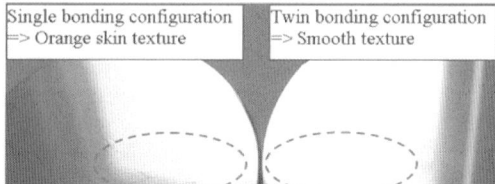

Figure 4: Effect of the bonding method on the PZT thick films

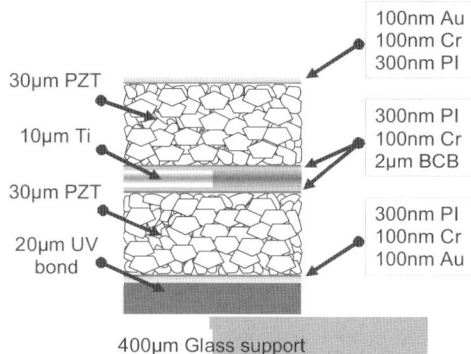

Figure 5: (top) Schematic diagram of a wafer stack – not to scale

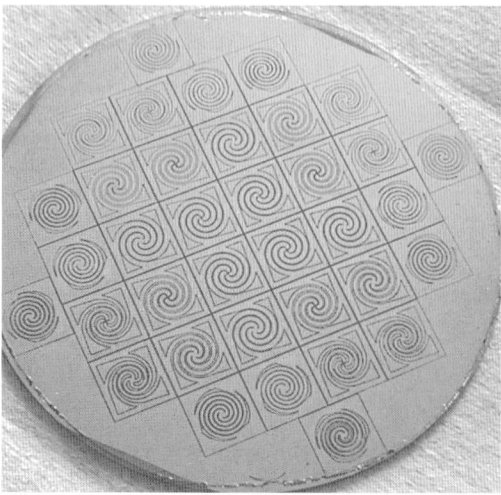

Figure 5: A wafer stack after laser dicing. The total thickness of the electro-active components is ~70 microns

Figure 6: An individual 6mm plane-spiral bimorph

2.2 Laser Machining

The multilayer sample was subsequently diced using a femto-second laser by Micreon GmbH Hannover, using a 30-micron diameter beam. Variations to the basic plane-spiral actuator designs were implemented at this stage and hence 9 different designs and 36 actuators were fabricated on a test wafer (Figure 5). The minimum width of cut was not fully explored in this test, however it is comfortably below 100µm. Some re-deposition of vaporised material onto the cut surface is to be expected using this technique. This is potentially harmful to the operation of the devices and it was removed using a nitric acid etch before the devices were poled. The versatility of the laser machining process (Figure 6) in producing complex designs is illustrated here and, in the opinion of the authors, its inclusion in this process is justified, despite its relatively high cost. Laser machining is a very gentle process in contrast to wafer saw-dicing, which could be viewed as a possible alternative, albeit with more limited scope for design variations. Saw-dicing trials revealed some cracking of the pre-stressed multi-layer structure and it may well prove to be unsuitable for these devices.

2.2 Electrode Design

Following successful fabrication of the multi-material stacks the attention has turned to device operation and integration with ancillary components. Three electrical connections are required in order to give the actuators bi-directional capability. Blanket electrodes to the top and bottom surfaces are insufficient to enable this to be achieved. Under high-field conditions, such as those applied during poling, it is possible for short circuiting to occur. This is most commonly initiated by surface inhomogeneity at the exposed edges. Patterned electrodes on the top and bottom surfaces have been used to increase the inter-electrode gaps substantially and this has resulted in consistently high impedances between the layers. Figure 7 shows a range of cantilevers of different lengths from a second test wafer. Figure 8 shows the 6mm spiral configuration redesign. The laser machining process has been exploited to create access holes for the centre electrode.

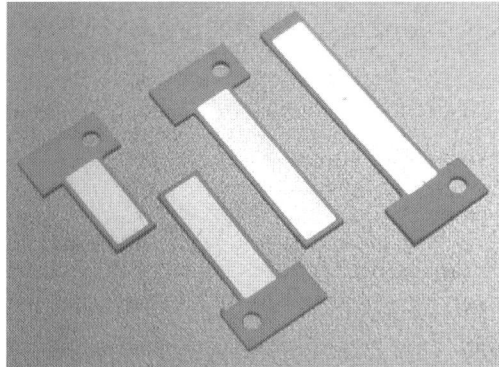

Figure 7: Cantilever bimorphs with patterned electrodes in the length range 2-5mm

Figure 8: A 6mm bimorph spiral actuator showing redesigned electrodes

Conclusion

A range of pre-stressed PZT thick-film bimorph micro-actuators have been created that are suitable for a variety of micro-valve applications. The process sequence used has been demonstrated to provide more flexibility for device design than has previously been the case and the wafer-scale production route is compatible with selective transfer bonding methods for the creation of more complex micro-electromechanical systems. The actuators are intended to operate with a 30V drive signal, making integration with CMOS electronics a possibility.

Acknowledgements

This work was carried out in collaboration with the EU 'Q2M' Consortium under Framework 6 STREP No. 027926 *Batch Integration of High-Quality Materials to Microsystems* - Information Society Technologies. The authors wish to thank Dr Frank Niklaus and colleagues at ELAB/KTH for their assistance and their expertise.

References

[1]   Wilson SA, Jourdain RP, Kohl M, van der Wijngaart W et al. New materials for micro-scale sensors and actuators - an engineering review (2007) Materials Science and Engineering Reports-R 56 1-6 1-129

[2]   Wilson SA, Jourdain RP, Whatmore RW, Morantz P, Corbett J, Hucker MJ, Warsop C. Ultra-precision machining of 30 micron PZT-on-silicon laminates for piezoelectric MEMS Proc. of ACTUATOR 2006 10th Int. Conf. on New Actuators pp 748-751.

[3]   Guerre R, Drechsler U, Jubin D and Despont M. CMOS-compatible wafer-level micro-device distribution technology. The 14th IEEE conference on solid-sate Sensors, Actuators and Microsystems (Transducers 2007), Lyon, France, 2007, June 10-14, pp.2087-2090, 2007.

[4]   Guerre R, Drechsler U, Jubin D, and Despont M. Selective transfer technology for micro-device distribution. IEEE Journal of Microelectromechanical Systems 17 (1) Feb 2008 157-165

[5]   Arai S, Corbett J, Whatmore RW, Wilson SA and Hedge J. Surface Integrity Control of Piezoelectric Materials in Ultra-precision Grinding. Proc. 4th International Conference of EUSPEN the European Society for Precision Engineering and Nanotechnology, 201-3, Glasgow 2004

[6]   Jourdain RP, Wilson SA, Whatmore RW, Morantz P, Corbett J, Hucker MJ, Warsop C. Precision Multi-point Grinding of Commercial PZT Wafers for Piezoelectric Micro-actuators. Proc. 4th International Conference of EUSPEN the European Society for Precision Engineering and Nanotechnology, 137-9, Glasgow 2004

[7]   Jourdain R and Wilson SA. Thermally induced stresses in an adhesively bonded multilayer structure with 30-micron thick piezoelectric and ceramic components. 4M2006 2nd International Conference on Multi-Material Micro Manufacture, 259-262, Grenoble 2006

Components: fabrication and assembly technologies

A new tool for aligned micro-embossing and nano-imprinting

T.Rogers & I.Malmros

Applied Microengineering Limited, Unit 8 Library Avenue, Didcot, Oxon.,OX11 0SG, UK

Abstract

A new multi-purpose MEMS fabrication tool is described. The tool enables in-situ aligned embossing and nano-imprinting, in addition to surface activation and aligned wafer bonding. De-embossing is also included in-situ via the use of vacuum chucks and chamber pressurisation. The multi-purpose tool enables the fabrication of bonded, embossed, multi-layer, micro-fluidic devices, for example PDMS structures on silicon, including the alignment of the embossed structure to any pre-existing patterning on the silicon. Examples are presented of various structures that have been made using the tool along with a description of the principles of operation.

Keywords: hot embossing, surface activation, MEMS, wafer bonding

1. Introduction

One of the factors that has inhibited the widespread commercialisation of MEMS has been the failure of silicon based manufacture to deliver the low costs that the early MEMS pioneers predicted. In particular, devices that require high aspect ratio micro-machining can incur high processing costs when solutions like LIGA and DRIE are used.

The use of polymers for MEMS fabrication offers cost reductions, but issues such as the machining of high aspect ratio structures, and alignment for the integration of other processes then need to be resolved.

This paper describes a tool that addresses the aligned micro-embossing / imprinting of polymer substrates. This tool is based on the AML AWB aligner –bonder platform and has similar specifications to that aligned wafer bonding equipment. In fact with suitable tooling the machines can be used in an in-situ aligned wafer bonding mode, or aligned embossing mode.

By utilising emboss stamps that have been fabricated using a high aspect ratio MEMS process such as silicon DRIE or LIGA, the high aspect ratio structures can be faithfully transferred into polymer substrates at low cost.

2. Emboss Tool – Description and Specifications

Hot embossing [1, 2] is a process in which a pattern on a stamp is replicated into, mainly polymer, substrates. In order to perform the replication process, both substrate and stamp are heated to a temperature at or above the glass transition temperature (Tg) of the substrate. The stamp is then pressed into the substrate (embossing force). The stamp and substrate are cooled down to a temperature below the Tg, the stamp is then pulled out of the substrate; this is the de-embossing step.

There are many commercial systems available for carrying out hot embossing of microstructures, but not with the in-situ ability for precise alignment of the embossed structure with pre-existing features on the substrate. Alignment is present on the tools that have been designed for nano-imprint lithography, but these machines tend to be much more expensive than the aligner-emboss / imprint tool described here.

The AML emboss tool is shown in Figure 1 and schematically in Figure 2. The tool is also a fully functional wafer bonder so all processes that can be carried out on a stand-alone wafer bonder are also possible on the embosser.

The embosser comes equipped with an alignment feature that makes it easy to align the embossing to structures that already exist on the substrate. If the embossed polymers require a subsequent bonding step (eg for encapsulating microfluidic structures, then surface activation of the polymer can also be carried out in–situ. This process uses a novel RAD-tool [3], that works in a similar manner to plasma activation [4], to activate the substrate during the process. The RADical activation process is a feature that uses free radicals to activate surfaces and is also being investigated as a means of improving the anti-stiction properties of the stamp, thereby making the de-emboss process easier to perform.

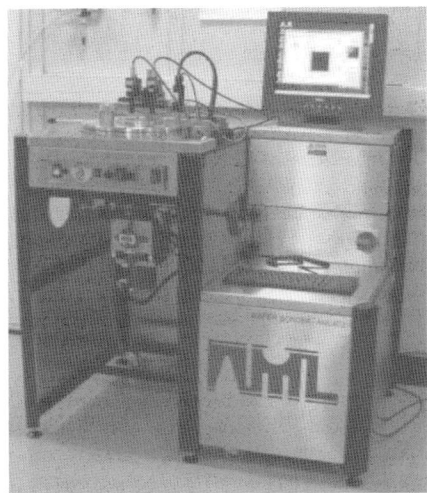

Fig 1. Basic machine for in-situ aligned bonding and / or aligned embossing / imprinting

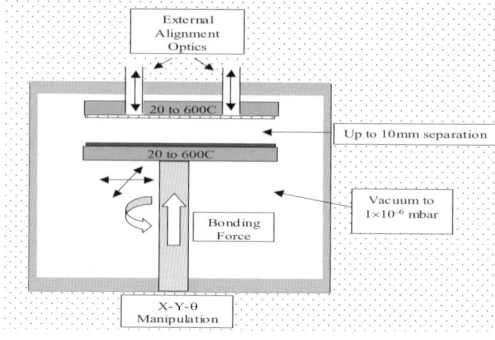

Fig 2. Schematic of process chamber

The tool utilises a process chamber that can either be evacuated or pressurised, a lower platen that is movable in X,Y,Z and Theta, and a fixed upper platen that includes apertures that allow optical split-field alignment via externally mounted lenses / cameras. In addition both platens can be independently heated, and a force of up to 15kN can be applied between them. For de-embossing the vacuum can be switched from the chamber to the platens (ie they then behave as vacuum chucks) whilst pressurising the chamber.

The vacuum chucks, and the ability to switch the applied vacuum from the chamber to the chucks, are two of the functions that have been added to convert the existing commercial tool, the AML Aligner Bonder, into a dual aligner bonder / embosser. One of the other important additions has been the modification of the Z-drive to reduce XY movement over a long stroke. For an aligner bonder, the "true Z" stroke only has to be ~10(m in order to enable the movement from the proximity separation, needed for the alignment process, to contact in order to enable wafer bonding. However, in order to be able to emboss high aspect ration structures, for example for a microfluidic device, much longer strokes are needed, and if micron scale precision is needed over the lateral dimensions, for the whole depth of the embossed structure, then the stroke has to be "true" to micron accuracy. The specifications for the tool are shown in Figure 3.

Wafer Sizes	3", 100mm, 125mm, 150mm
Max emboss force	15kN
De-emboss force	630N
Max temperature of platens	560°C
Platen temperature uniformity	+/-2.5°C (100mm wafers) +/-3.5°C (150mm wafers)
Z stroke	2mm
XY error over full Z stroke	1.4µm (NB XY error for shorter strokes is pro rata, ie 1.4nm for a 2µm stroke
Alignment accuracy	+/- 2µm

Fig 3. Emboss Tool Specifications

The emboss / print procedure is as follows:

- Mount wafers on platens
- Close lid
- Pump down chamber and simultaneously heat to process temperature (for embossing

this is typically just above the glass transition temperature (Tg).
- Align stamp wafer with workpiece (for embossing this is typically a polymer wafer or polymer coating on carrier wafer)
- Align wafers
- Apply force for defined time
- Cool (using nitrogen) to defined de-emboss temperature (typically just below Tg)
- Separate (de-emboss) wafers – for this process the chamber is vented and pressurised to 2 Bar absolute. This enables the wafers to be secured via vacuum chucks.
- Vent to atmosphere, open lid and remove embossed wafer

In order to perform the alignment operation, the lower platen can be moved in X,Y,Z and θ, to enable precision contacting of the stamp with the workpiece. The alignment is achieved in a similar fashion to mask aligners, and the stamp / workpiece need to have alignment marks processed onto them. These marks are then imaged using externally mounted cameras. Either visible reflected light, or transmitted IR can be used depending on the optical characteristics of the materials being used.

The de-embossing process can be performed in-situ and uses the same precision z movement that is utilised to perform the high aspect ratio embossing. Thus there is no possibility of the precision of the embossed reproduction becoming degraded by the withdrawal of the stamp from the polymer. To perform this de-emboss function, the machine operator can switch the vacuum from the chamber and apply it to the platens on which the stamp and substrate (workpiece) are mounted. This then enables a separation force to be applied, and this force is amplified by pressurisation of the chamber. Note that during the embossing step the stamp is held by a special edge clamp thus removing the need for a vacuum chuck at that stage, and therefore allowing the embossing to be performed within a vacuum – which is beneficial in preventing any air trapping as the stamp is pressed into the polymer. The workpiece, during the emboss stage, sits on the lower platen under gravity.

An actual process run showing chamber pressure, platen temperature and applied force is shown below in Figure 4.

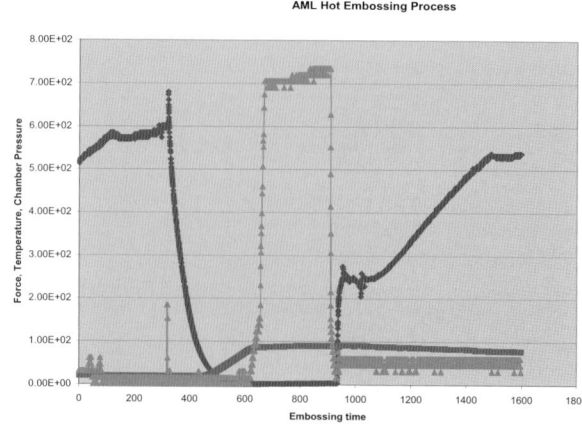

Emboss Force (N)
Substrate Temp. (°C) (100°C / div)
Chamber Pressure (mBar)

Fig 4. Data for a typical emboss process

3. Results

To date the tool has been used for embossing the following polymers, but many others are possible:

PDMS, PEEK, Polycarbonate, PMMA

Some examples of typical embossings, in this case in a spun-on layer of PDMS on a silicon substrate are shown below in figures 5 and 6.

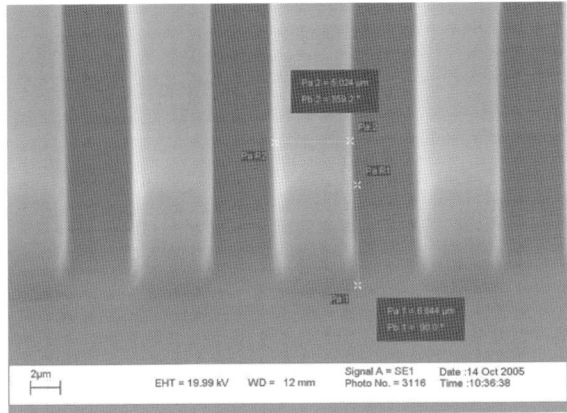

Fig 5. Vertical wall structures in PDMS on silicon

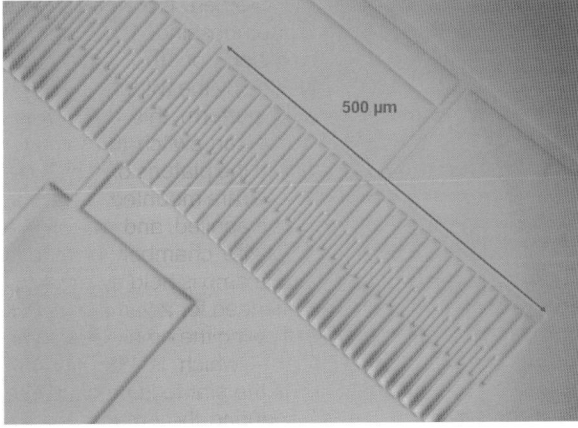

Fig 6. Embossing in PDMS layer on silicon

In addition to hot embossing, the tool has also been used to demonstrate nano-imprinting. Figure 7, below, shows an example of a nano-imprinted material in which 100nm lines are completely resolved.

Fig 7. 100nm printed lines in surface layer

If the stamp includes holes, that align with the apertures for the two split-field cameras that are used for the alignment process, then reflected, through-the-lens, visible light can be used for alignment. For standard silicon stamps, transmissive IR light can be used. This avoids having to machine through holes in the silicon. Figure 8 shows an image of the substrate wafer as seen through the in-situ alignment optics, when using transmissive IR illumination.

By including alignment marks on the stamp and on the substrate it is then possible to accurately align the embossed structure with pre-defined structure on the substrate. As with standard photolithography the overlay of a cross within another is the preferred alignment technique -See figure 9. A typical example whereby this alignment procedure would be used is the embossing of flow channels in a PDMS layer that has been spun onto a silicon wafer that has already been metallised and photolithographically patterned to include electrodes and bond pads.

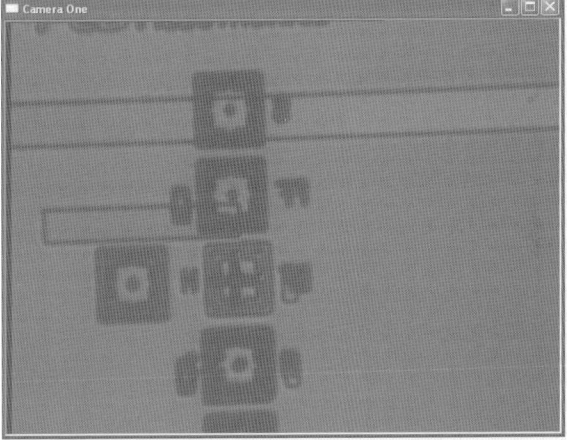

Figure 8. IR image of patterned substrate as seen through the in-situ alignment optics.

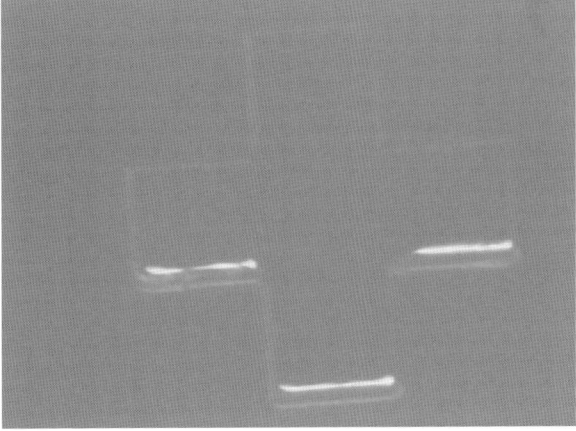

Fig 9. Overlaid crosses used for alignment of the embossing to pre-existing structures on the substrate. Magnification: 200X

The combined capabilities of embossing and bonding in the same machine have recently been exploited for the fabrication of a three-layer microfluidic polycarbonate structure, in which the through-embossing of the central layer is used to define flow channels. Two other substrates are also embossed such that when the three layers are subsequently bonded together, the embossings in the outer two layers form structured walls for the central flow channels. The bonded structure is shown schematically in Figure 10, and Figure 11 shows a photo of the structured wall at the end of one of the 100 micron deep flow channels.

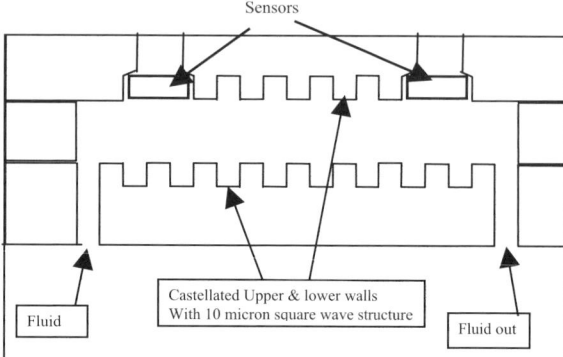

Fig 10. Schematic of three layer, embossed / bonded micro-fluidic device

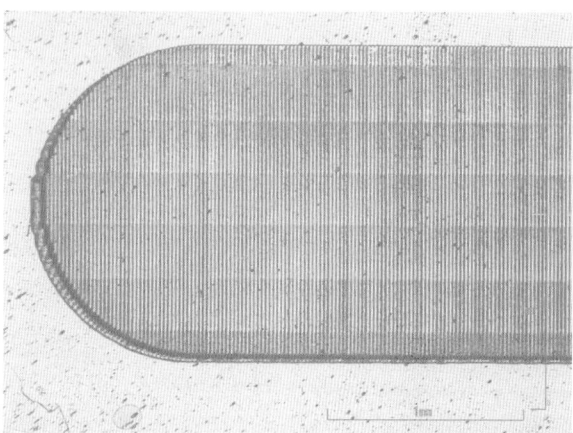

Fig 11. View of Structured polycarbonate surface forming the wall of a flow channel.

Figure 12 shows a close up of the 10 micron castellated structure that forms the channel upper and lower walls. The undulating edges are an artefact of the low cost acetate mask that was used for masking the silicon stamp during the DRIE fabrication. The undulating edges were then subsequently faithfully transferred in to the polycarbonate during the emboss step, thereby demonstrating the accuracy of the reproduction.

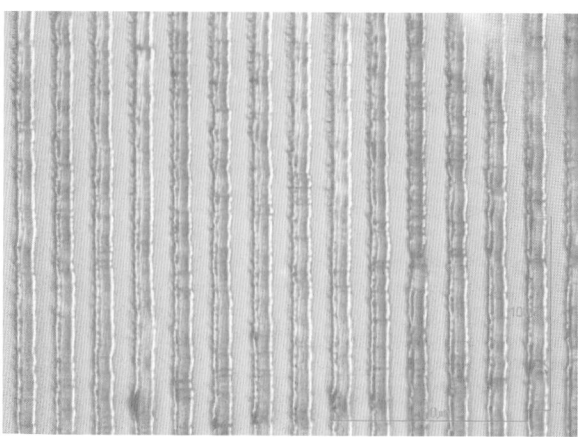

Fig 12. Close up of structured walls of the channels of a micro-fluidic device

4. Conclusions

The multi-purpose tool has been demonstrated to perform the following:

> Embossing of 100 micron high features
> Micron scale features with vertical walls and an aspect ratio of ~5:1
> 100nm printed lines / spaces
> In-situ alignment using transmissive infra-read illumination

These results are examples of structures that have been produced to date rather than representing the limit of the tool.

In addition to the above emboss / imprint processes, the tool has been demonstrated as being capable of a wide range of standard wafer bonding processes, also featuring in-situ alignment. In addition, in-situ surface activation using oxygen radicals can be used for polymer bonding.

The tool therefore provides a multi-purpose fabrication capability for MEMS devices.

5. References

[1] Chen C L and Jen F, Fabrication of Polymer Splitter by Micro Hot Embossing Technique. Tamkang Journal of Science and Engineering. Vol. 7, No. 1 pp 5–9, 2004.
[2] Malmros I, Hot Embossing of Polymers. MME2001. pp 7-10.
[3] Rogers T, Aitken N and Kowal J, Activation, Alignment and Bonding using Radical Activation. Ninth International Symposium on Semiconductor Wafer Bonding. 2006.
[4] Katzenberg F, Plasma-bonding of poly-(dimethylsiloxane) to glass. E-polymers 2005, No. 060. pp 1-5. ISSN 1618-7229.

Ultrasonic welding of micro plastic parts

W. Michaeli[a], E. Haberstroh[b], W.-M. Hoffmann[a]

[a] Institute of Plastics Processing (IKV), RWTH Aachen University, Aachen, Germany
[b] Lectureship and Research Field of Rubber Technology, RWTH Aachen University, Germany

Abstract

Due to the ongoing miniaturisation in many industrial branches plastics are increasingly applied in microsystems technology. To guarantee the functionality of the system suitable joining processes must be applied to join separate components. Most of the welding processes commonly used for series production are not suitable for welding micro parts made from plastics, since either the mechanical or the thermal load of the joining partners during the welding process are too high. Only laser transmission welding and ultrasonic welding are applicable for welding complex micro components. Since with ultrasonic welding a certain frictional load of the components cannot be avoided totally with standard welding equipment, specially adapted machinery has to be used as it could be shown at the Institute of Plastics Processing (IKV) at RWTH Aachen University. While micro parts with two-dimensional weld seams have already been successfully welded in previous investigations, recent research deals with the ultrasonic welding of micro parts with a more complex three-dimensional weld seam geometry. It could be shown that for appropiate welding parameters this can be accomplished, whereby the mechanical load of the parts has to be kept as small as possible.

Keywords: welding, ultrasonic welding, adapted welding equipment, tensile strength, weld seam morphology

1. Introduction

Due to low material costs and the great variety in design plastics are becoming more and more important in microsystems technology. By means of micro injection moulding and micro hot embossing, micro parts can be produced in high numbers and short cycle times.

However, microsystems often consist of several components which have to be joined together in order to ensure the functionality of the system. In general, adhesive bonds of plastic parts can be realised by welding or glueing [1, 2]. Welding has the advantage that there is no need for additional material and that the achievable bond strength are higher compared to glueing. Moreover, there is no need for curing times.

There are several welding processes applicable for welding plastics in the macro-range which use different mechanisms of energy input. However, the micro technology has special requirements to a suitable welding process. Therefore, only few welding processes can be applied for the welding of micro plastics parts.

A welding process for plastics which is suitable for the application in microtechnology has to meet the following demands [3, 4]:

- precisely controllable energy input
- low mechanical load on the parts
- low thermal load on the parts
- single-stage process
- small flash and no abrasion
- high positioning accuracy during welding

On account of these requirements most welding processes for plastics are not suitable for microtechnology. With the heated tool welding, for example, the energy input cannot be controlled exactly enough, so that filigree structures would be destroyed. Besides, all processes, which melt the joining parts by means of friction, i.e. moving both parts relatively to each other, are not applicable in the micro-range, because the mechanical load of the components is too high. So sensitive structures would be demolished [4].

Solely the laser transmission welding as well as the ultrasonic welding are suited for the application in microtechnology. Laser radiation can be focused very well so that the energy can be put locally and precisely metered into the joining area. Moreover, it is a contactless process, i.e. there is no mechanical load of the parts during the welding process. Although the ultrasonic welding uses friction as an energy source, it is nevertheless a possible joining process in microtechnology, since the material is plasticised mainly by internal, dissipative friction between the polymer molecules. The polymer is heated inside so that the mechanical load of the components due to boundary friction is relatively low [1].

At the Institute of Plastics Processing (IKV) at RWTH Aachen University complex micro parts with two-dimensional weld seam geometries could successfully be welded by ultrasound in previous investigations [5]. In the following it is indicated that this process can be also applied to the welding of micro parts with a three-dimensional seam geometry.

2. Ultrasonic welding

Ultrasonic welding is a joining process for thermoplastics which is often applied in series production. The reason is that this process features very short cycle times in the range of 0.1 to 1.0 seconds. Since there are physical restrictions limiting the maximum size of the welding tool, also called 'horn', the process is constrained to small and middle-sized components with a weld seam length up to 300 mm. If the joining surface or the weld length are too big, a homogeneous oscillation of the horn and thus a homogeneous energy input cannot be ensured [2].

With ultrasonic welding the joint is realised by melting the polymer due to dissipation, i.e. the

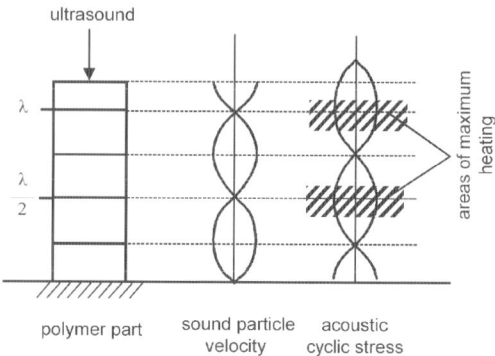

Fig. 1 Principle of ultrasonic welding

transformation of mechanical oscillation energy into heat [2, 6]. A longitudinal ultrasound wave forms a standing wave within the component between the horn and the bottom of the component by reflection of the wave. The areas of maximum heating of the polymer are located where the acoustic cyclic stress has its maximum (Fig. 1).

Their location depends on the wavelength of the sound wave amounting to approximately 5 cm to 10 cm depending on the polymer. So ideally, the parts should be designed in such a way that the joining zone is located in these areas of maximum cyclic stress. Generally, this is not possible in practice, since this would limit the part design extremely. However, a cross section narrowing can be realised by constructive measures, which locally increases the mechanical cyclic stress and therefore the energy conversion [3]. This can be achieved in form of an energy director or a shear joint. Thus the location of the joining area can be chosen freely.

Apart from the longitudinal oscillation, also transversal oscillations occur which originate among other things from the flexural oscillation of the horn. Due to the transversal oscillation there is to a relative movement between the joining components and thus to boundary friction. This leads to a not negligible mechanical load of the components which would damage filigree stuctures. That is why without modification standard ultrasonic welding equipment cannot be applied in microtechnology.

3. Ultrasonic welding equipment for micro parts

On account of the small component and joining area dimensions in microtechnology there are certain demands which the ultrasonic welding equipment and the joining process have to fulfill:

- A sufficient energy input into the joining area has to be possible.
- Small joining areas (<1 mm ^{2}) must be realisable.
- The reproducibility and the positioning accuracy have to be very high.
- Flash should be as small as possible in order to ensure the functionality of the microsystem.
- the mechanical load of the joining parts during the welding process must be as low as possible.
- the energy input should be adjustable by variation of the welding parameters.

Because there is no standard ultrasonic welding

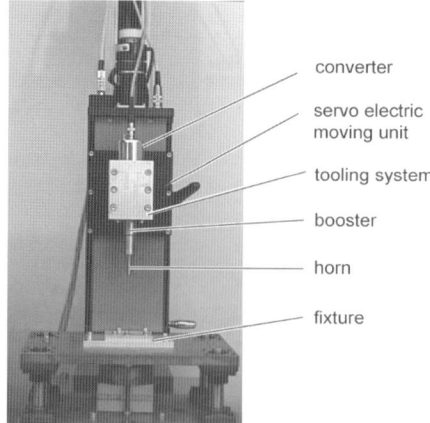

Fig. 2 Modified welding equipment

machine which fulfills these requirements, an ultrasonic welding unit was developed at IKV which has been adapted to the abovementioned requirements, see Fig 2.

One of the most important aspects of the modification of the welding unit is the frequency of the ultrasound. Instead of a frequency of 20 kHz which is often used for standard welding machines a frequency of 40 kHz is applied. So lower amplitudes can be chosen in order to bring nearly the same amount of energy into the joining area. The lower the amplitude the lower is the mechanical load of the components during the welding process. To allow for an exact movement of the oscillation system, a servo electric moving unit is used. It features a traversing range of 100 mm with an accuracy of ± 1 μm. The oscillation system is mounted to this moving unit. Furthermore, the horn and the fixture are adapted to the sample geometry.

4. Methodology

Ultrasonic welding has already been applied successfully for joining plastic micro parts [1, 5]. However, the components had only two-dimensional weld seam geometries so far. The two-dimensionality of the weld seam is according to the state of the art of the ultrasonic welding technology a precondition for the successful joining of macro-ranged parts. The welding surfaces should be parallel with the front surface of the horn and in one plane to allow for a favorable and homogeneous ultrasound energy input.

Components in microsystem technology have very small dimensions. So even if the weld seam geometry has three-dimensional properties, this will have a

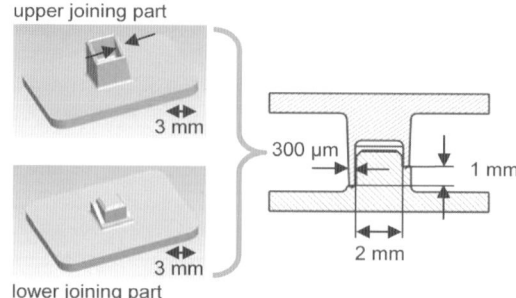

Fig. 3 Sample part with three-dimensional weld seam geometry

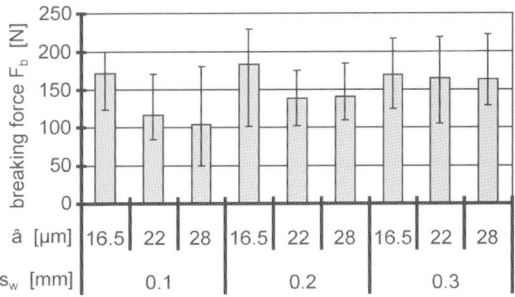

Fig. 4 Breaking forces depending on the welding parameters

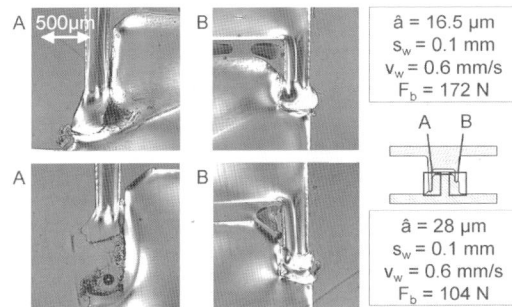

Fig. 5 Weld seam morphology for different amplitudes

smaller influence on welding process compared to welding parts of the macro-range.

Therefore, within the scope of the investigations carried out at the IKV Aachen it was examined if it is possible to join micro parts with a three-dimensional weld seam geometry by ultrasound. For this purpose, the sample part shown in Fig. 3 was developed. The square-shaped sample part features a wall thickness of only 300 µm with a step of 1 mm in height at a length of 2 mm. The joining surfaces of the upper joining component are equipped with energy directors.

The sample parts were made from the amorphous polymer polycarbonate (PC) whose weldability by ultrasonic welding is rather good. The sample parts were welded with the parameter settings of the amplitude of the ultrasonic wave â, the joining displacement s_w and the welding velocity v_w indicated in Table 1. With each set of parameters 10 parts are welded out of which 8 samples are taken for the determination of the breaking force by tensile tests. The results are statistically evaluated within the scope of an analysis of variance (ANOVA) [7]. Besides, the weld seam morphology was determined by microscopic analyses.

Since the welding velocity had no significant influence on both the breaking force and the weld seam morphology only the results regarding the parameters amplitude and joining displacement are considered in this paper. The welding velocity amounts to v_w = 0.6 mm/s for the following discussion. At this velocity the cycle time ranges from approximately 0.2 s up to 0.5 s.

Table 1 Welding parameters

amplitude â (µm)	16.5	22	28
joining displacement s_w (mm)	0.1	0.2	0.3
welding velocity v_w (mm/s)	0.1	0.6	

5. Results

Fig. 4 shows the results of the tensile tests of the welded samples. Average breaking forces up to F_b = 180 N can be realised depending on the amplitude and the joining displacement. Thereby, the amplitude has the biggest influence on the breaking force. An amplitude of â = 16.5 µm leads to a breaking force of F_b = 174 N averaged over all settings of the joining displacement. An increase of the amplitude to â = 28 µm causes a decrease of the breaking force to an average of F_b = 136 N. Thereby, the level of significance exceeds 99 %. This indicates that the energy input is too high during the ultrasonic welding process, so that the polymer decomposes and

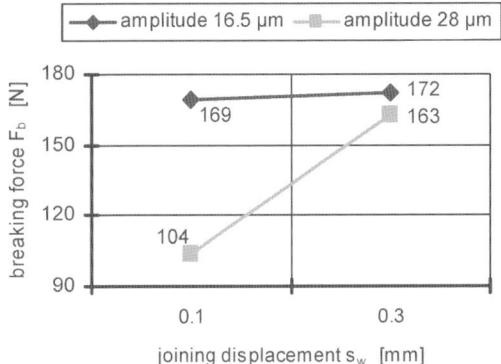

Fig. 6 Significant interdependences between the welding parameters amplitude and joining displacement

weakens the weld.

These findings can be confirmed by microscopic analyses. In Fig. 5 the weld seam morphologies for two parts are shown which were welded with the same joining displacement of s_w = 0.1 mm, but with different amplitude settings of â = 16.5 µm and â = 28 µm, respectively. It can be recognised that the left web of the upper component - here the distance between the welding zone and the horn is the largest - is decomposed at an amplitude of â = 28 µm (on the left side at the bottom of Fig. 5). This decomposition occurs for all components which are welded at this amplitude. So there is no circumferential weld seam resulting in a decrease of the breaking forces. An explanation for this could be an elevated mechanical load for higher amplitudes originating from boundary friction.

For small joining displacements the influence of the amplitude is significantly higher than for s_w = 0.3 mm. At a joining displacement of s_w = 0.1 µm the breaking force is decreased from F_b = 163 N down to F_b = 104 N when increasing the amplitude from â = 16.5 µm to â = 28 µm. At a joining displacement of s_w = 0.3 mm the amplitude has no influence any more, see Fig. 6. As microscopic analyses indicate the reason is that in this case the components are not only welded in the intended weld area, but also between the positioning structures, see Fig. 7. Even though the upper welding component is partly decomposed, the welded area contributing to the breaking force is larger and the breaking force increases accordingly.

Regarding the joining displacement, an increase from s_w = 0.1 mm to s_w = 0.3 mm results in an increase of the average breaking force from F_b = 131 N to

68

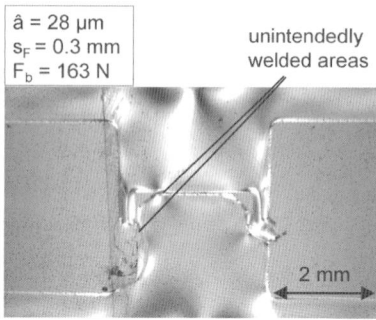

â = 28 μm
s_F = 0.3 mm
F_b = 163 N

unintendedly welded areas

2 mm

Fig. 7 Enlargement of welding area at a joining displacement of s_w = 0.3 mm

F_b = 166 N averaged over all settings of the amplitude. The reason is that the meltflow is improved since the joining parts are moved towards each other. Thus the molecules can entangle themselves better, so that the mechanical properties of the weld get better. Moreover, as mentioned above, the parts do not get in contact only in the intended welding area, but also in adjacent regions.

6. Conclusion

To sum up, good results can be realised with PC. This is reflected in the high breaking forces. However, the three-dimensional weld geometry leads to an inhomogeneous ultrasound energy input and thus to an irregular plasticising of the weld area. This results in partial decomposition at high amplitudes which could be shown in microscopic analyses. At an amplitude of â = 16.5 μm, a joining displacement of s_w = 0.1 mm and a welding velocity of v_w = 0.6 mm/s the best result could be achieved for PC, both regarding mechanical properties and the outer appearance of the weld.

Welding trials have also been carried out with the semicrystalline material polyoxymethylene (POM). Normally, POM features a good weldability with ultrasonic welding. However, it was not possible to prevent the partial decomposition recognised at high amplitude with PC when welding the depicted sample part geometry with the three-dimensional weld seam geometry. Even for welding parameters with low amplitude and joining displacement the inhomogeneous ultrasound energy input due to the three-dimensionality of the weld interferes with the welding process. On the one hand, this leads to lower breaking forces since no circumferential weld line could be realised. On the other hand, the outer appearance is deteriorated by a relatively large amount of decomposed material.

This problem does not occur when welding samples made from POM which have a more suited weld seam geometry, i.e. the weld seam is in one plane with the same distance to the horn along the whole weld contour. In this case, good weld results are possible also for POM. Figure 8 depicts the results achieved with a similar sample geometry without the step in height of 1 mm. As it can be seen, there is no decomposition and thus no imperfection in the weld. So the breaking force is quite high, as well. Depending on the welding parameters the formation of flash can be totally avoided which is an important aspect concerning the possible application of this welding process in microtechnology, see Fig. 8. at the top.

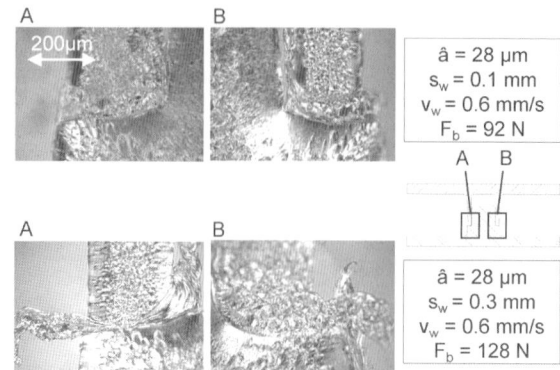

A   B

200μm

â = 28 μm
s_w = 0.1 mm
v_w = 0.6 mm/s
F_b = 92 N

A   B

A   B

â = 28 μm
s_w = 0.3 mm
v_w = 0.6 mm/s
F_b = 128 N

Fig. 8 Welding results for POM using a sample part with two-dimensional weld seam geometry

Acknowledgements

The authors gratefully acknowledge the financial support of the Deutsche Forschungsgemeinschaft (DFG) within the Collaborative Research Centre SFB 440 "Assembly of Hybrid Microsystems". Moreover, we gratefully thank Bayer MaterialScience AG and Ticona GmbH for providing the required material for the investigations presented in this paper.

References

[1]   Haberstroh E. and Lützeler R. In: Dilthey U. and Brandenburg, A. (Ed.) Montage hybrider Mikrosysteme. Springer, Heidelberg, Berlin, 2005, pp. 137-148

[2]   Rotheiser J. Joining of Plastics. Carl Hanser Verlag, Munich, 2004, pp. 462-505

[3]   Haberstroh E. and Hoffmann W.-M. Laser transmission welding of micro plastics parts. Proceedings of the Second International Conference on Multi-Material Micro Manufacture, 2006, pp. 71-74

[4]   Klein H. Laserschweißen von Kunststoffen in der Mikrotechnik. Dissertation, RWTH Aachen, Germany, 2001

[5]   Haberstroh E. and Hoffmann W.-M. Sanft zu zarten Bauteilen. Plastverarbeiter 58 (2007) 8, pp. 44-46

[6]   Grewell, D., Benatar, A. and Park, J. B. (Ed.) Plastics and Composites Welding Handbook. Carl Hanser Verlag, Munich, 2003, pp.141-188

[7]   Turner, J.R. and Tahyer, J. F. Introduction to Analysis of Variance: Design, Analysis & Interpretation. SAGE Publications Inc, London, 2001

Manufacturing of Versatile Ceramic or Metal Micro Components by Powder Injection Moulding

V. Piotter, K. Plewa, J. Prokop, A. Ruh, H.-J. Ritzhaupt-Kleissl, J. Hausselt

Forschungszentrum Karlsruhe, Institute for Materials Research III
P.O. Box 3640, 76021 Karlsruhe, Germany

Abstract

Although microsystems technologies products have been steadily launched worldwide markets the development and improvement of manufacturing processes suitable for medium or large-scale production is still one of the most important prerequisites.

A well-known technology to meet such demands is micro injection moulding which has already reached an industrial viable status for polymeric materials. Nevertheless, there is still a lack of methods for the processing of materials with a wider range of properties.

A promising option to close this gap, development of the so-called MicroPIM process to facilitate the fabrication of metal and ceramic micro components was started.

Presently, the smallest dimensions achievable are 25-50µm of part thickness or minimum structural details of less than 5µm. Theoretical densities of up to 99% were achieved depending on the particular powder applied. As further improvement, the technology to produce rotational-symmetric parts by making use of a special head spindle system has been developed.

To enlarge the application possibilities of MicroPIM further, micro two-component injection moulding enables, for example, the fabrication of micro components consisting of two ceramic or metal materials with different physical properties and, not less important, significantly minimises assembly expenditure.

Keywords: micro injection moulding, powder injection moulding, two-component injection moulding

1. Introduction

Miniaturization clearly represents a significant and remarkable trend in the worldwide progress of manufacturing technology. Injection moulding well-established in macroscopic production presents no exception: First experiments were carried out in the eighties whereas the technological basics for the so-called micro injection moulding process were accomplished in the nineties [1]. In the last years a considerable number of – mainly polymeric material - products were developed or even entered the market.

Two promising sub-variants presently followed by micro injection moulding will be described in the following sections, i.e. the production of micro devices with higher complexity and precision and the attempts to realize parts consisting of more than one material. Examples for both aspects will be described in the following chapters.

2. Micro Powder Injection Moulding

2.1 The MicroPIM Process in general

At present, powder injection moulding (PIM) using metal- (MIM) and ceramic-filled (CIM) feedstocks is a well-established technique in industrial practice. Worldwide, more than 500 enterprises and institutions are working in this production sector.

This wide and rapidly increasing use of powder injection moulding is mainly a result of the high economic efficiency reached when producing medium and large series. Furthermore, it is possible to manufacture complex device geometries with the contours desired, i.e. hardly any finishing work is required. As practically all known metal and ceramic materials are also available in powder form, a large variety of materials is in principle available for PIM.

Fields of use of PIM technology extend over components for industrial machines, automotive engineering, household applications, microelectronics, and medical engineering. In the field of microelectronics, for instance, nozzles for bonding wires are produced with high accuracy and small tolerances.

Therefore, it is not surprising that many attempts have been undertaken to use PIM process technology for the fabrication of micro components, as well.

Presently, various institutes and first industrial enterprises are working on the practical applicability of MicroPIM [2, 3]. Due to the limited space, however, they cannot be presented in detail.

At Forschungszentrum Karlsruhe, development work related to microPIM started in 1995, at first with the use of commercially available feedstocks. Work mainly focused on determining process limits depending on the feedstocks used.

Investigations of the manufactured micro samples revealed theoretical densities of up to 97% for metals and aluminum oxide, whereas the values for zirconium oxide reached 99%. The linear shrinkages varied in a range of 15-22% depending on the composition of the feedstocks. Very important for the industrial applicability of the process, nominal sizes of the micro parts after sintering could be obtained within a tolerance of ±0.3-0.7%, under certain conditions replication accuracy could be improved further down to ±0.1% only [4]. Due to the smaller particle sizes of ceramic powders which are usually in the range of 300 – 600nm, the moulded samples achieved better surface qualities (R_a=0.02μm) compared to the metal micro parts (R_a=0.5-0.8μm).

While first experiments were performed with macroscopic templates carrying microstructures on their surface, subsequent development was concentrated on the fabrication of easily separable micro parts with outer dimensions securely below 1 mm. For fabrication, a facility of the type Microsystem 50 by the Battenfeld company is applied among others. Additionally, for powder injection moulding of e.g. micro specimens and gearwheels of less than 1mm in diameter, novel tool concepts had to be developed and implemented. Examples of such micro parts are shown in Fig. 1.

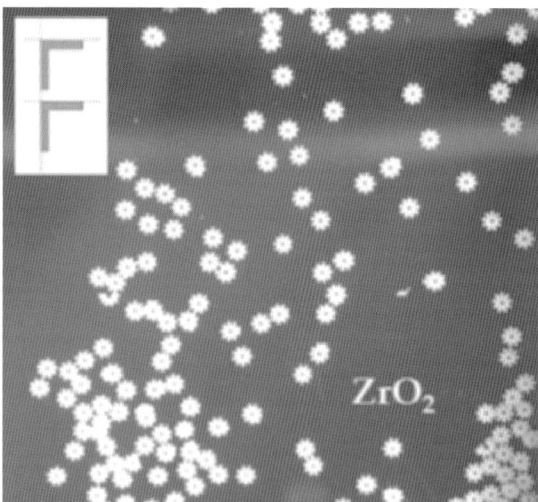

Fig. 1: Singular micro gear wheels made of zirconia with outer diameters of approx. 850μm.

2.2 Increasing Shaping Versatility

To tap the full application potential of one-component MicroPIM attempts were started within the Collaborative Research Project SFB 499 to manufacture more complex shaped devices. As a demonstrator a nearly free-formed feed screw for a micro-dispenser unit was chosen. Production of the tool inserts was carried out by milling and EDM. For replication a modified micro injection moulding process was inaugurated: usually, such screws are manufactured in tools containing two or more clamping

units to release the cut-back structure of the screw. However, the formation of burrs can never be completely avoided. Therefore, a special spindle construction was developed to screw out the moulded green body before opening the tool (Fig. 1). Although the principle looks easy at the first glance the real implementation of the outscrew technique including features of micro replication lead to a quite complex tool design (Fig. 3).

First trials were carried out with zirconia feedstocks, and the samples could be debindered and sintered without major problems (Fig. 4). Continued miniaturization of samples geometries is currently going on.

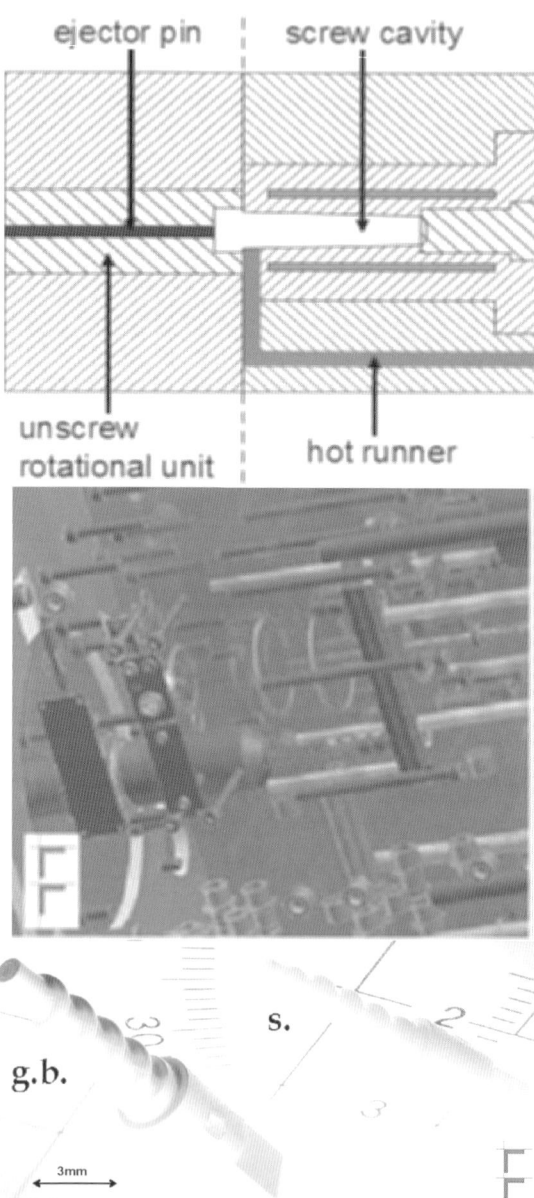

Figs. 2 - 4: Schematic drawing of the unscrew tool. Elevation of the tool for PIM dispenser screws (above). Green bodies (g.b.) and sintered (s.) zirconia samples of the screws (bottom).

3. Micro Two-Component Powder Injection Moulding (2C-MicroPIM)

Typical MST products are not only characterized by smallest dimensions of their detached pieces, but they also integrate a number of different functions in smallest space. The resulting assembly and bonding expenditure might be reduced by using material compounds. Two-component micro injection moulding is considered to be a viable process for the series production of micro components from multi-functional materials in the long term [5].

To combine multi-component injection moulding with micro manufacturing technology, new machinery and tooling equipment had to be developed at Forschungszentrum Karlsruhe. For example, typical micro manufacturing features like tool venting or variothermal temperization had to be implemented twice in the moulding cycle. The final result was a worldwide unique combination of three special subvariants: two-component micro powder injection moulding.

In two-component-PIM, (at least) two feedstocks filled with different powders are injected into a thermostated tool. The procedure can be performed simultaneously or successively. In the latter case a barrier unit has to be implemented in the tool. The feedstocks used for two-component micro powder injection moulding have various, partly opposite functions. Complete filling of micro structured mould areas shall be ensured by their low viscosity, whereas deformation-free demoulding requires a high mechanical strength: Distortion-free sintering under isotropic shrinkage has to be achieved by homogeneous powder distribution in the green compacts.

Apart from their high mechanical and thermal stability, such compounds produced from multi-functional materials may combine materials of different or even contradictory physical properties like e.g. electrically conductive and insulating ceramics or magnetic and non-magnetic metals.

Two examples for multi-component MicroPIM parts shall be presented here. The first one utilizes ceramic feedstocks based on a mixture of insulating aluminum oxide and electrically conductive titanium nitride. In this case, aluminum oxide is used as matrix material. Electric conductivity of the ceramic part produced may be varied within certain limits by changing the fraction of percolated titanium nitride in the matrix. Based on feedstocks processed in a twin-screw extruder, micro heater samples have been moulded by a two-component micro injection moulding process. Two fractions of different ceramic compositions and thus of different electrical conductivities were used [6]. The position of the joint line could either be adjusted by the movable slide incorporated in the tool or by a suitable selection of the injection moulding parameters, e.g. injection rate.

The second example is the 2C moulding of a shaft-gear wheel combination. For realization, a new 2C-tool with a rotating index plate plus complete micro replication features had to be developed (Fig. 5). Each subproduct can be moulded using a particular material so that combinations of two ceramics, two metals or even metal and ceramic are feasible within one injection moulding cycle [7].

The procedure in general can be configured to obtain both mobile and immobile bonds. With respect to mobility, the composition of the feedstocks and here especially the powder-binder ratio is of great importance: For immobile bonds the powder contents have to be nearly identical to achieve equal shrinkage rates. For the mobile bonds, however, the shaft has to have a significantly higher shrinkage rate than the surrounding gear wheel. There are a few other aspects, e.g. the sintering temperatures and sintering rates, which have to be adjusted too.

Experiments started with the aim of realizing immobile bonds of a combination of alumina as shaft material and zirconia for the gear wheel. Using feedstocks with nearly the same powder content (approx. 55-vol.%), tight and stable connections were achieved (Fig. 6).

At present, trials to obtain mobile bonds are being carried out. Additionally, experiments using metal feedstocks, i.e. filled with fine grain 17-4PH steel powders, are running. Although steel combinations were already tested for immobile and mobile bonds of macroscopic samples [8], transformation to micro technology leads to additional challenges: Due to the very small distancies between the two green/brown portionss it is quite difficult to prevent them from sintering together so that e.g. novel sinter bearings have to be developed.

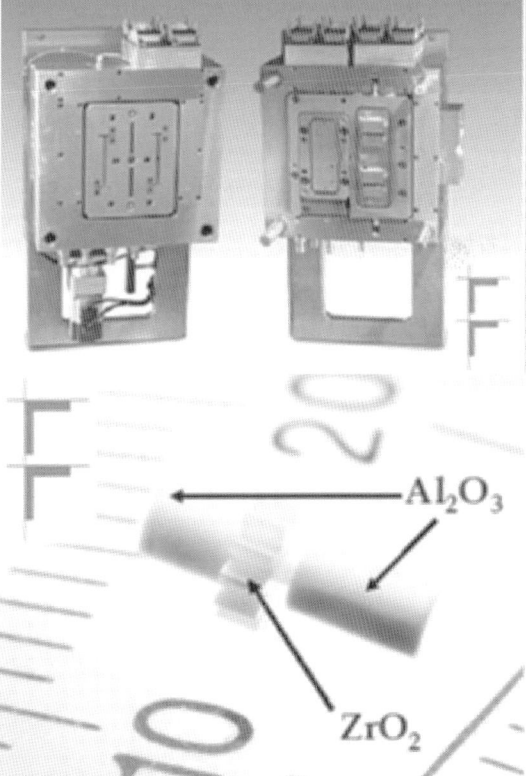

Figs. 5, 6: 2C-Micro injection moulding of shaft-gear wheel combinations. New 2C-tool with turnable index plate (above); two-component zirconia/alumina sample (bottom), outer diameter of the gear wheel: 2.932mm.

4. Outlook

Of course, development of the above-mentioned variants of micro powder injection moulding will not rest on it´s laurels.

In order to enhance the performance capability of Micro-PIM, further material and process development is going on. These experiments will deal mainly with the utilization of very fine metal powders as well as near-nano powders in case of ceramic materials.

As explained above, entirely new perspectives are opened up by multi-component micro injection moulding. A promising technological level has been reached but, of course, a lot of research work remains to be done, especially concerning the adjustment of feedstock formulations and the optimization of sintering procedures.

It has to be emphasized that 2C-MicroPIM is not only bound to result in reduced assembly expenditure for micro systems, but also in the production of new functional units. Therefore, two-component micro-injection moulding offers a clear economic and technical potential for current and future applications.

For all variants mentioned above, an enhanced development and use of simulation tools is expected for the future. New approaches with innovative material models and software tools will be utilized for micro- and powder-specific aspects for the first time.

Acknowledgements

The authors of the present article expressly thank the Deutsche Forschungsgemeinschaft (SFB 499) and the Federal Ministry of Education and Research (BMBF) for their support.

Additionally, this work was carried out within the framework of the EU Network of Excellence "Multi-Material Micro Manufacture: Technologies and Applications (4M)".

Further thanks go to the cooperating industrial companies and all colleagues at Forschungszentrum Karlsruhe as well as to the Universities of Freiburg and Karlsruhe.

References

[1] Piotter, V., Beck, M., Plewa, K., Ruprecht, R., Haußelt, J.: Micro Injection Moulding – Status and Perspectives; Proceedings of the 7th euspen International Conference, Bremen 2007; euspen ISBN 978-0-9553082-2-2 (2007); pp. 328-331.

[2] Piotter, V., Oerlygsson, G., Ruprecht, R., Hausselt, J., Nishiyabu, K.: New Developments in Micro Powder Injection Moulding; Proceedings of PM 2004 Powder Metallurgy World Congress, Vol. 1, ISBN 1899072 15 2, pp. 473-480.

[3] Piotter, V.; Finnah, G.; Zeep, B.; Ruprecht, R.; Hausselt, J.: Metal and Ceramic Micro Components Made by Powder Injection Moulding; Materials Science Forum, Vols. 534-536; Trans Tech Publications (2007); pp. 373-376.

[4] Beck, M.; Piotter, V.; Ruprecht, R.; Haußelt, J.: Dimensional tolerances of micro precision parts made by ceramic injection moulding. In: Dimov, S., ed., 4M 2006 : Proc. of the 2nd Internat. Conf. on Multi-Material Micro Manufacture, Grenoble 2006, Amsterdam: Elsevier (2006); pp.135-38.

[5] Piotter, V., Finnah, G., Prokop, J., Ruprecht, R., Haußelt, J.: Multi-component micro injection moulding – trends and developments; Proceedings of the 3rd International Conference on Multi-Material Micro Manufacture, 3rd-5th October 2007, Borovets, Bulgaria; Whittles Publishing, ISBN 978-1904445-53-1 (2007); pp. 107-110.

[6] Oerlygsson, G.; Piotter, V.; Finnah, G.; Ruprecht, R.; Hausselt, J.: Two-Component Ceramic Parts by Micro Powder Injection Moulding; Proceedings of the Euro PM 2003 Conference, 20.-22.10. 2003, Valencia, Spain, pp. 149-154.

[7] Ruh, A.; Dieckmann, A-M.; Heldele, R.; Munzinger, C.; Piotter, V.; Ruprecht, R.; Fleischer, J.; Haußelt, J.: Herstellung zweikomponentiger Micro-PIM-Baugruppen mittels 2-Komponenten-Pulverspritzgießen; Proceedings of the Kolloquium „Mikroproduktion", 22./23.11.2007, Karlsruhe; Editors: O. Kraft, B. Emmerich; ISBN 978-3-923704-61-3; pp. 129-134.

[8] Maetzig, M.; Walcher, H.: Assembly moulding of MIM materials; Proceedings of EPMA 2006 Conference, Ghent, 23. - 25. Oktober 2006.

High Density Interconnections Fabrication by UV Lasers Microprocessing of Microvias and Microstructures

D. Ulieru[a], Alina Matei[b], Elena Ulieru[c], A. Tantau[c], Florin Babarada[d]

[a] ROMES S.A., 126A, Iancu Nicolae Str., Bucharest, 72996, ROMANIA
[b] National Institute for Research and Development in Microtechnologies, 32B,
Erou Iancu Nicolae Str., Bucharest, 077190, ROMANIA
[c] SITEX 45 SRL, 114, Ghica Tei Blvd., bl. 40, ap. 2, Dept. 2, Bucharest 72235, ROMANIA
[d] "Politehnica" University of Bucharest, Splaiul Independentei Str., No. 313, Bucharest 060042, ROMANIA

Abstract

The strong evolution of electronic packaging in the field of high performance hand held electronic products involves, from the fabrication point of view, to manufacture small, lightweight, reliable and, very important too, cost effective electronic modules. In the last years new techniques and technologies for production of rigid/flexible MCM-type multilayer were introduced. The manufacturers of laminate substrates are being challenged to realize boards with very good electrical and mechanical properties. In the past the biggest issues regarding vias and via capture pad sizes were only solderability and manufacturability. Today the vias density is also an important electrical issue. The more vias on a board are presented, the more discontinuities into PCB/MCM passive interconnection structure are placed. For High Density Interconnection (HDI) circuits design one solution is to reduce the via hole and the via capture pad, but still maintain manufacturability at board fabrication stage.The most indicated solution is to use UV laser microprocessing for HDI production.

Keywords: UV lasers, microsystems, electronic products, microvia, microstructures

1. Introduction

In the frame of electronic packaging domain some components, as BGA (Ball Grid Array), µBGA, DCA (Direct Chip Attach), CSP (Chip Scale Package or Chip Size Package) or even FC (Flip Chip, which is a necked die but one can be estimated as a package), overcame the capability of the standard PCB?MCM industry to offer high performance substrates/boards from the W-S-V (Width of tracks-Spacing between tracks-Via parameter) criterion point of view. For example, due to many advantages of array packages against QFPs, BGAs will set a new standard in the packaging of high-end ICs. Specialists from many companies have studied the specific requirements for the manufacture of BGA substrates, mainly for IC with high I/Q counts and have identified a number of difficulties with traditional manufacturing technologies. In particular, when four layers have to be used, the mechanical drilling of the through holes is not only costly, but also adversely operation. Even when build-up structures are used with micro-vias from layer 1 to 2 and 4 to 3 with buried vias incorporated in the core, these buried vias have to be THT.

A as result of these considerations, the new developed UV technology for PCB/MCM modules have to be characterized by much finer lines/spacings and much small vias (through-hole, blind or even buried). Unfortunately, the conventional technology with mechanical manufactured holes by punching and/or drilling has several inconveniences regarding the W-S-V criterion. For this reason, during the last years a new technology was developed and tested with excellent results in practice. This is so named High Density Interconnect (HDI) technology, which permits a dramatically increasing of circuit routability and decreasing of PCB/MCM size using microvias and very fine lines/spacings by laser microprocessing.

2. The new technological challenges

Today PCB producers have to take into account the growing demand for "High Density Interconnect boards". The leading companies have to produce HDI boards with microvias and fine and dense circuitry for large series of hand held devices, smaller PCB shops face now the demand on HDI boards from many parts of the electronics industry.

But the production of these complex and expensive boards call for specific capital equipment not necessary for the production of "standard" multi-layer boards.

One the specific elements of HDI boards are interconnections between two or three layers by blind and buried vias. Conventional mechanical drilling machines are not suitable for the production of reliable blind vias at reasonable costs. Laser drilling machine will become are the common tool for this job on the future.

In some cases also the usual photochemical processes are inadequate for achieving ultrafine circuitry as required to route flip chip and micro packages. Even solder mask structures become an issue with small packages on the board.

The pressure the market exerts on smaller pcb manufactures to step up their processing ability confronts them with a return-on-investment dilemma. With the technology currently available, they would have to purchase a whole range of high-priced cutting-edge machines. On the other hand, the small batch sizes they usually have to live with leave them few chances to reach the degree of utilization needed to achieve sufficient profitability from the investment.

The PCB manufacturing requests could be as follows: micro via forming, circuitry direct structuring, resist and solder mask structuring, flex and rigid board cutting, cover foil opening.

3. Experiments and results

3.1. Microvia Formation

An increasing number of reliable interconnections between the layers of HDI boards has led to smaller microvias and to higher quality standards for the forming process of microvias.

For reliable interconnection the main requirements to the formation of microvias are: clean vias, no residue, no delamination, tapered sidewalls, no undercut, large via diameter on the land for robust interconnection, no perforation on the inner layer.

There are primarily three methods of microvia formation: mechanical drilling vias, photo formed vias, laser ablated vias.

Mechanical drilling is well known to the PCB community and the machine providers try hard to control the drilling depth in order to qualify their machines for blind via formation.

Nevertheless this method is limited by throughput and efficiency as well by its accuracy especially in reinforced materials.

It is not surprising that laser ablation has become the leading method for microvia forming. CO_2 lasers generating infrared light and solid state lasers generating ultraviolet light are used on laser drilling machines.

CO_2 lasers are primarily used to drill bare substrate due to their inability to cut through copper. The outer copper layer has to be opened by etching, which causes a number of additional alignment problems. After CO_2 laser formation, the via has to be cleaned from residue.

UV laser on the other hand are able to cut copper as well as organic substrate materials and glass. This way a very simple microvia formation process is feasible – one laser is able to form the complete microvia through the outer copper layer and the substrate material. The short wavelength and the high density of the pulse power of these result in an almost cold ablation process.

Though the ablation rate of different PCB materials depends on certain material properties like optical absorption, glass transition temperature, inner structure, etc. the UV laser formation leaves almost ideal shaped microvias in a wide variety of PCB materials including glass reinforced once like FR4. The so formed blind vias show a clean, textured copper land ready for plating without desmearing.

Other benefits of UV laser formation of microvias are:

- The extended limit for the minimum diameter at 25 um

- The excellent alignment to the inner layer circuitry by using fiducials on the inner layer allows reduction of the land pads.

- The capability to form stacked vias, connecting three layers reduces the Sequential Build-Up steps.

The automotive and computer industry – with their requirements for PCBs offering extended temperature range and mechanical stability – require HDI boards, the greater the demand for microvias formed in reinforced materials becomes.

Woven glass fibers bedded in epoxy provide the necessary material properties but they also cause a general problem for microvia drilling due to its inhomogeneous absorption of laser light.

In the past most of the UV laser drilling machines were by their laser output power.

Resin coated copper (RCC) and similar non reinforced materials were the materials of choice and a lot of results of laser via formation in these materials were reported.

New laser sources especially developed under the requirements of the microvia formation process ensure excellent results in glass reinforced materials and accelerate the formation process.

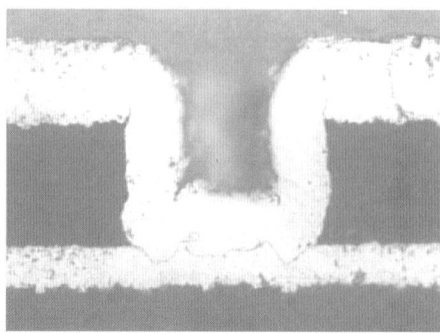

Fig. 1 Microvia in FRA 1x1080 with 18µm copper

Figure 1 shows the cross section of a 100 um diameter microvia of ideal shape. The outer copper layer is of 28 um thickness, the substrate is FR4, 1 x 1080, the inner layer is of 18 um copper.

The walls are well tapered, the via is circular and the glass bundle does not disturb the plating.

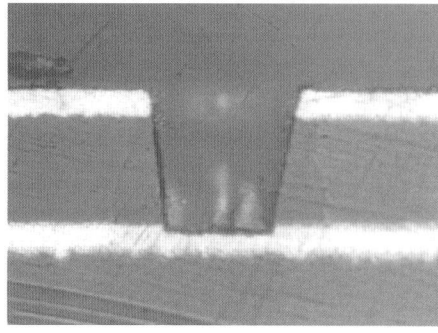

Fig. 2 Perfect interconnection over the whole via bottom

Figure 2 shows a magnified bottom zone of the same microvia. Perfect connection to the inner copper layer over the whole bottom area is provided by UV laser formation.

Boards with 60000 vias of different diameters from 50 um to 300 um were formed by our experiments result, plated without desmearing and subjected to cycles of thermal shock. The electrical test revealed excellent resistance results and yield.

3.2. Laser direct structuring

Structures on PCBs primarily mean conductive lines from copper separated by isolating channels or areas. Photo chemical processes are used to build these structures. Photosensitive resist of a certain thickness is applied to the PCB and exposed.

A film contains the layout of the board which becomes transferred to the resist by exposition. With laser direct imaging, a laser exposes the structure into the resist. The resist thickness at least is 20 µm and us based on chemical etching of the copper layer.

For the production of prototype PCBs, mechanical milling is a known process. Lines and spaces of 100µm minimum are produced by direct removal of the copper and formation of the isolation channels avoiding the etching problems. The minimum space dimension of this sequential method is limited by the stability of the milling bit.

With laser direct structuring shown on the figure 3 it is possible to produce lines and spaces as small as the copper thickness with a minimum defined by the size of the focused laser beam. Structures of 20 µm line/space width are possible in 18 um copper as well as 65 um line/space in 70 µm copper.

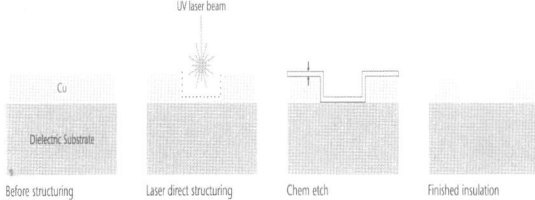

Fig. 3 Direct Circuit Structuring

Copper layers of rigid and flexible boards can be directly structured with UV laser.

With this method the insulation will be created into the cooper clad and clad and thin Cu layer remains to protect the substrate underneath. This remaining copper will be removed with a short clean etch step afterwards. The UV laser removes a high percentage of the copper thickness, leaving only few micrometers of copper to protect the substrate material. The etching process that follows does not last as long as the one with the photochemical method.

As a result, the underetch problem is reduced significantly. The best results were obtained with a metal resist on top of the copper layer, which saves the copper on top of the lines from the etch.

Figure 4 shows the SEM picture of a typical structure produced by laser direct structuring.

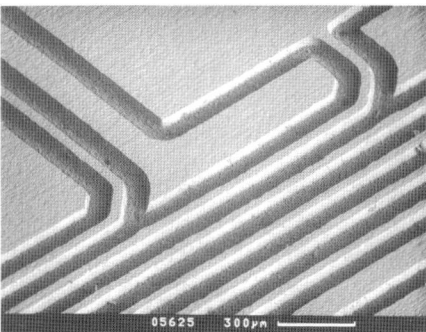

Fig. 4 Copper direct structuring

3.3. Fine pitch solder mask

Wafer level packages and CSP require ultra fine structures not only in copper, but also in different types of solder masks. The vector scanned UV laser can be used to "skive", that is to remove a layer of polyimide cover foil or applied solder mask from the copper layer beneath. The fine spot of the UV laser and the big difference in ablation thresholds between the organic materials and copper allows for the precise and rapid removal of solder mask. Figure 5 shows a part of the solder mask for a CSP.

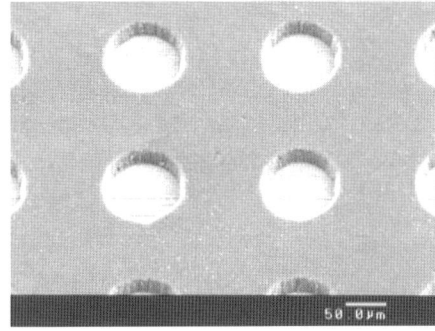

Fig. 5 Solder mask with 50µm diameter openings

3.4. Drilling

Although methods such as mechanical drilling, punching and plasma etching are still used, laser drilling is the most widely used microvia formation technique for flex. The primary reasons for the widespread use of laser drilling are productivity, flexibility and maximum optime.

Mechanical drilling and punching machines use precision drill bits and dies to generate holes to approximately 250 mm in diameter in conventional flex circuits. These precision tools are expensive and have relatively short lifetimes. Because high-density flex requires via diameters less than 250µm, mechanical drilling is not considered a viable option.

Plasma etching can produce microvias with diameters smaller than 100 mm on 50 mm thick polyimide substrates, but the capital equipment costs associated with both processes are extremely high. Plasma etch is an expensive process to maintain, too, especially in regards to chemical waste treatment and the cost of consumables. Furthermore, it makes a significant amount of time to develop new processes that consistently yield reliable microvias. The advantages of plasma technology are higher reliability and reported yields of 98 percent in microvias formations. This finds niche applications in medical and avionics.

In contrast, laser drilling microvia holes is a straightforward and comparatively inexpensive process. The capital equipment costs are far lower, and because a laser is a noncontact tool, there is no expensive hard tooling to replace, as in mechanical drilling. Furthermore, modern sealed CO_2 and UV-DPSS lasers are maintenance-free, which minimizes downtime and maximizes productivity.

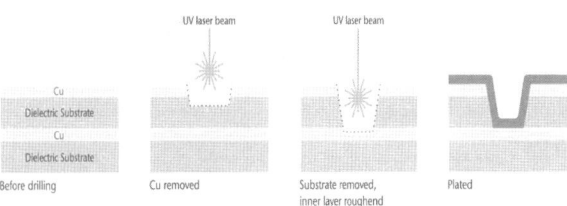

Fig. 5 Drilling of Blind Vias

The process of forming blind vias in PCB is a two step process is could be seen on the figure 5.

In step 1 a laser beam with a power density higher than the ablation threshold of cooper structures the cooper and removes a part of the substrate. In step 2 the fluence of the laser beam is adjusted to remove the remained substrate and to remove the remained substrate and to leave the via land rough, clean and ready for plating.

3.5. UV-DPSS Processing

Both dielectrics and copper easily absorb the 355nm output wavelength of UV – DPSS lasers. In dielectric processing applications, because of an inherently small focal spot size and lower output power compared with CO_2 lasers, UV – DPSS lasers are usually used to process features with small (<50 μm) dimensions. As a result, UV light is ideally suited for drilling smaller than 50μm diameter microvias in high density flex substrates. As new, more powerful UV – DPSS lasers become available, there will be a potential increase in UV dielectric machining and drilling speeds.

An advantage of UV – DPSS lasers is than their high-energy UV photons directly break molecular bonds at the surface layer of many nonmetals in a "cold" photo-ablation process that produces features with smooth edges and minimal thermal damage or charring. Thus, UV micro-machining is preferable in more demanding applications where post-processing is either impossible or undesirable.

Materials with high UV ablation thresholds such as cooper are processed at high energy with low repetition rates. Low – threshold dielectric materials like polyimide dielectrics are processed at lower energy and higher repetition rates, both to avoid damage to the copper pad and to maximize throughput. To increase throughput, most large-diameter microvias are drilled in a two-step process, with UV – DPSS lasers direct-drilling the copper surface and CO_2 lasers drilling the exposed dielectric material. The table 3 shows typical processing speeds for a UV-DPSS laser. Comparing Table 2 and Table 3 clearly shows the advantages of each laser.

Table 2. Polyimide Processing Speeds for a Typical 9.4 μm Wavelength CO_2 Laser

Process	Polyimide Thickness	Nominal Processing Speed
Machining/cutting	50 μm	500 mm/sec.
Microvia drilling	100 μm	100 μm diameter via 450 holes/sec.
Microvia drilling	50 μm	25 to 50 μm diameter via 400 holes/sec.

Table 3. Processing Speeds for a 3 watt, 355 nm Wavelength UV-DPSS Laser

Process	Material Thickness	Processing Speed
Machining/cutting polyimide	50 μm	500 mm/sec.
Microvia drilling	50 μm	30 μm diameter via 100 holes/sec. (copper+dielectric)
Direct copper drilling	50 μm	30 μm diameter via 100 holes/sec.

Conclusion

The UV laser developed technology for PCB wide range microprocessing technologies especially for HDI/microvias has the main advantage to use only a single THC laser. This allows clean and accurate process. General of both the copper clad on top as well as the substrate underneath. Te heating effects of side walls from CO_2 lasers completely avoided and an extracleaning of blind vias no necessary. The short wavelength of 355nm allows a very small focus and extremely small holes diameter. The best ratio pure/performances are a very important advantage.

The increasing demand for high HDI/microvias for mobile phones portable media like for nano-ceramic IC's package substrates reach a total HIS/microvias value to exceed 25 billion by 2008 will represent more than 25% percent of combined PCB sand packaging substrate world wide market.

This will confirm us the big potential of our technology and the interest of producers for it expend their market shoes by innovative technology applications.

Acknowledgements

The authors thanks for kind technical and material support of company SITEX 45 to realize the experiments to develop the UV laser technology as Eng. Alina Matei/IMT Bucharest for paper issuing on the best conditions and results processing.

References

[1] D. Ulieru, Alina Matei, Elena Ulieru, Application of laser microprocessing for high density interconnections of microsystems manufacturing, ICOMM Proceedings 2007, pp.299-302
[2] D. Ulieru, Alina Ciuciumis, A. Tantau, Elena Ulieru, „Sensors microprocessing by laser direct patterning (IDP) for industrial production", CAS 2006 Proceedings, p. 379-382, vol II, ISBN
[3] IPC, National Technology Roadmao for Electronic Interconnections 2000
[4] K, Vollrath, B. Lange: A versatile laser machine for PCB production, to be published
[5] S. Raman, UV laser drilling of multilayer blind vias, technical paper at IPC 1998, S 17-1

Single- and multi-layer conductive patterns fabricated using M³D technology

B. Obliers-Hommrich[a], A. Fischer[b], H. Willeck[a], W. Eberhardt[a], H. Kück[b]

[a] Hahn-Schickard-Institute of Microassembly Technology HSG-IMAT, Stuttgart, Germany
[b] University of Stuttgart, Institute of Micro and Precision Engineering, Germany

Abstract

The continual trend of miniaturization and increasing complexity in the field of microelectronic devices pose a challenge for today's manufacturing technology. The novel Maskless Mesoscale Material Deposition (M³D) manufacturing technology offers the potential for printing superfine circuitry as well as for the building up of multi-layer systems. Therefore it could be an interesting technology to meet the requirements of miniaturized systems. The M³D process depends on aerosol formation and uses aerodynamic focusing of aerosol streams for a high resolution deposition of colloidal suspensions and liquid raw material. Since M³D is a contactless and maskless Direct Write Technology, it also offers new possibilities for 3D devices.

This paper will report on first results of depositing conductive and non conductive materials onto glass substrates as well as onto typical MID (Moulded Interconnect Devices) substrates. Furthermore it will present first multi-layer systems that have been fabricated using the M³D technology.

Keywords: direct write technology, M³D, aerosol, microsystem, 3D device, conductive pattern, multi-layer

1. M³D Technology

By means of M³D it is possible to directly write on a variety of substrate materials such as plastics, metals, ceramics, glass or silicon. Since the M³D process depends on aerosol formation, of the material to be deposited, it has to be available as a suitable liquid or suspension with a viscosity less than 1000 cP [1, 2]. Accordingly, raw materials like metals, insulators, ceramics, polymers or biological materials can be deposited by M³D as long as the particle size of the raw materials does not exceed a size of 500 µm [2]. Currently, feature sizes down to 10 µm have been realized with M³D [1, 3, 4]

The aerosol is produced inside an atomiser chamber and is led into a special deposition head (figure 1). Inside the deposition head the aerosol stream is focussed aerodynamically by a second gas stream (sheath gas) which encloses the aerosol in an annular way. Thus the M³D deposition head can reduce the aerosol stream to a diameter which is as small as a tenth of the size of the nozzle orifice. At present, feature definitions down to 10 µm can be realized. During the printing process the stand off point of the deposition head to the substrate is about 3 mm. Moreover, the material beam emitting the nozzle is characterised by a long focal length. Thus M³D has the potential to precisely deposit materials even on non-planar substrates without any readjustment of the deposition head. The curing of the deposited layers can be done by either conventional thermal sintering or by laser. The laser module is integrated within the M³D system and can be used to sinter deposited layers on substrates that are not suitable for thermal sintering since it only heats up the substrate locally without affecting the surrounding material. At the moment the print velocity of the M³D process is in the range between 1 mm/s and 20 mm/s for good printing results. Although the resulting process time is high compared to other processes like ink jet printing or LPKF-LDS® process, M³D offers the advantage to deposit very small structures down to 10 µm onto planar and non-

planar substrates. Up to now neither ink jet printing nor

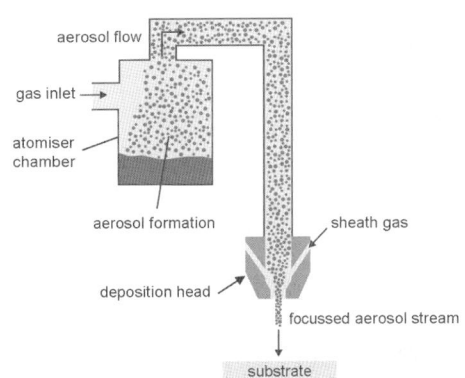

Figure 1: Scheme of the M³D process

LPKF-LDS technology can be used for such purposes.

Currently M³D is already used for 2D electronics applications, high density interconnects, embedded passives, micro antennas, gates and interconnects in displays [4].

2. Substrates and Deposited Materials

For the build-up of single- and multi-layer conductive patterns several substrates were used. Most experiments were performed on glass substrates. Furthermore printing has been done on typical MID substrates like LCP Vectra E840i LDS, PA6/6T Ultramid T 4381 LDS, PET+PBT Pocan DP T7140 LDS.

Both the conductors and the isolators have been built-up by M³D. Materials with a viscosity of less than 5 cP can be easily atomised with the ultrasonic atomiser of the M³D system whereas for viscosities up to 1000 cP a pneumatic atomiser is used for aerosol formation. For the deposition of the conductors a silver suspension in ethylene glycol has been used. The suspension has been atomised with the ultrasonic module of the M³D system. In order to adapt the viscosity the silver suspension has been diluted with water before use.

The insulating layers of the multi-layer conductive pattern were made from a soluble polyimide. The polyimide was solved in N-methyl-2-pyrrolidone (NMP) for the use with the pneumatic atomiser of the M³D system. The second insulating material which has been tested for building up a single-layer is an epoxy resin typically used as an encapsulant. For use with the pneumatic atomiser the material was diluted with dipropylene glycol monomethyl ether (DPGME).

3. Single-Layers

The aim of the work presented in this paper is to show that multi-layer systems can be built-up by M³D. Therefore first conducting and nonconducting materials are deposited on the relevant substrates in order to determine suitable process parameters for dense, pinhole-free layers with a good edge definition. The parameter determination for all deposited materials was done on glass substrates. Important process parameters are gas flows (gas inlet, sheath gas) and the amount of generated aerosol.

3.1. Printing on glass substrates

Figure 2 shows the printing result on a glass substrate. The line width of the filigree conductors is about (30 ± 3) µm and the layer thickness is approx. 0.2 µm.

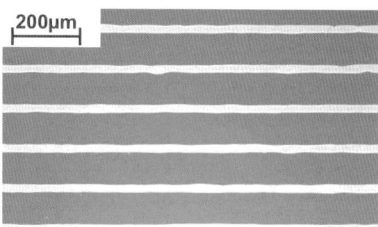

Figure 2: Silver lines on glass

The layer thickness can be easily increased by repeating the deposition process. Since the wetting of silver on silver differs from the wetting of silver on glass, the layer thickness of the second layer is larger than that of the first layer. After sintering the silver lines passed the standard tape test.

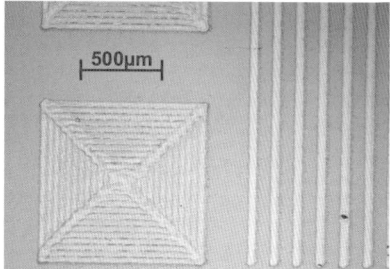

Figure 3: Epoxy resin layers on glass

The deposition of two nonconducting materials on glass is presented in the figures 3 and 4. Figure 3 shows the printing result of the epoxy resin. The isolator lines have a width of about 60 µm. To build-up the insulating area shown left in the picture a square of

1 mm² was filled like a serpentine. In this print example the minimum layer thickness after a single deposition is approx. 0.1 µm.

A comparable printing result is achieved for a single deposition of polyimide on glass (figure 4). The insulation area has the same dimension of 1 mm² and was printed the same way as before. In this example the minimum layer thickness is approx. 0.2 µm and the width of the insulator lines is about 70 µm.

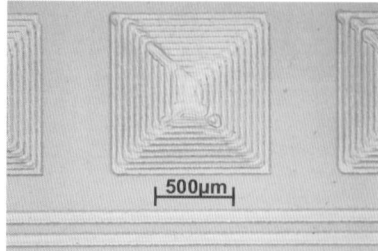

Figure 4: Polyimide layers on glass

Both polymers did not pass the tape test after curing in the oven. Therefore it would be necessary to modify this polyimide material to obtain a better adhesive strength on glass. In this case the primary aim was to determinate suitable process parameters for the deposition of this material. Hence no improvement of the adhesion of this material on glass was carried out.

3.2. Printing on thermoplastic MID substrates

To investigate the potential of M³D for patterning of thermoplastic substrates as used for multifunctional 3D-Packages printing tests were carried out on typical substrates.

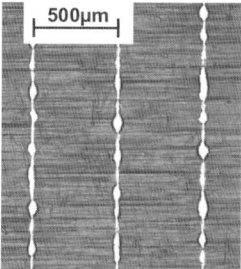

Figure 5: Silver lines on untreated PET+PBT

The silver lines shown in figure 5 are directly written onto PET+PBT without any pre-treatment of the substrate. Obviously the wetting of the silver ink on PET+PBT is very poor. On the untreated surface the line width of the conductors is about (21 ± 9) µm.

To improve the wetting of the PET+PBT substrates a pre-treatment with oxygen plasma was done. In figure 6 the result of the plasma treatment is demonstrated. The wetting behaviour is enhanced since the appearance of the silver lines is much more evenly. As an effect of the improved wetting the line width has increased up to (91 ± 11) µm using the same process parameters as before. After sintering the silver lines passed the tape test.

In case of silver deposition on LCP and PA6/6T exactly the same behaviour has been observed. The wetting of the silver ink is poor for the untreated substrates and can be strongly improved by a pre-treatment with oxygen plasma. All silver lines on the different substrates passed the standard tape test after the sintering process.

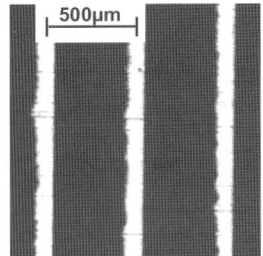

Figure 6: Silver lines on pre-treated PET+PBT

In the same way polyimide was deposited onto the three MID substrates. The wetting of the untreated surface has already been sufficient and could not be improved by a pre-treatment with oxygen plasma. But the insulation patterns only passed the tape test on untreated PA6/6T and PET+PBT substrates. However, on oxygen plasma pre-treated substrates the tape test was passed on all three MID substrates. Figure 7 shows the polyimide deposition on an untreated PET+PBT substrate.

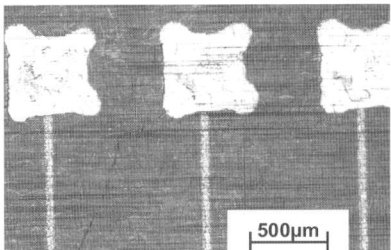

Figure 7: Polyimide on PET+PBT

4. Multi-layer systems

Based on this results first simple multi-layer conductive pattern were realised. Figure 8 shows a crossing of two conductors on a glass substrate.

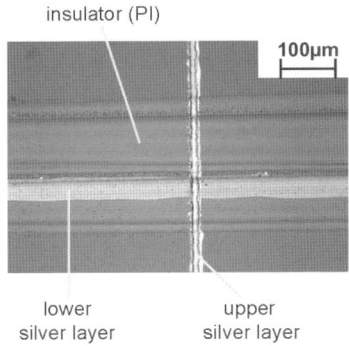

Figure 8: Crossing of two conductors on a glass substrate

The crossing was built-up in three steps. First silver was deposited onto the glass substrate. In a second step a polyimide layer was deposited over the silver layer. The minimum layer thickness of the insulating layer is approx. 1 µm. In the last step a further silver line was printed across the previous deposited layers. First characterisation results show that all of the deposited silver lines are fully conductive and a complete isolation between the two silver lines of the crossing was accomplished.

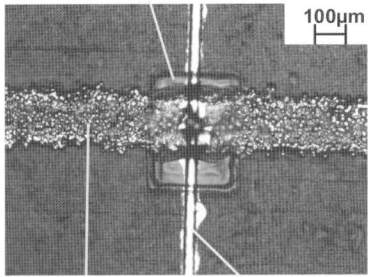

Figure 9: Crossing of two conductors on MID substrate

Furthermore a similar crossing was built-up on a MID substrate. The substrate was made from LCP Vectra E840i-LDS using the LPKF-LDS® technology [5]. After laser patterning of the thermoplastic substrate a layer system of copper, nickel and gold with a total thickness of approx. 10 µm was deposited by electroless plating. On these electroless plated substrates an insulating layer of polyimide was deposited by M³D technology. Afterwards a silver layer was deposited with M³D across the insulating layer. Due to the comparatively high thickness and roughness of the first layer deposited by electroless plating, the patterning process with the silver lines was done for five times to get continuous conductors. On the basis of the results in paragraph 3.2. the MID substrate was pre-treated with oxygen plasma before silver deposition. Figure 9 presents the result of the crossing of the two conductors. First electrical measurements have shown that electrical isolation between the two metal lines was accomplished.

5. Conclusion

The novel Maskless Mesoscale Material Deposition Technology M³D shows high potential for printing superfine circuitry and multi-layer conductive patterns. Because of the diversity of the usable materials printing technologies such as M³D are a promising enhancement for the manufacturing of thermoplastic multifunctional 3D packages.

Acknowledgements

The authors would like to acknowledge the support of the "Bundesministerium für Bildung und Forschung" (BMBF Förderkennzeichen 16SV3545) within the research project "Anwendung nanoskaliger Dispersionen zur maskenlosen Strukturierung von mehrlagigen 3D-Leiterbahnsystemen zur Erhöhung der Integrationsdichte bei multifunktionalen thermoplastischen 3D- Packages für die Nano-Mikro-Integration" (NADIMASMEL).

References

[1] Hedges, M., Kardos, M., King, B., Renn, M., "Aerosol-Jet Printing for 3D Interconnects, Flexible Substrates and Embedded Passives", Proceedings of the 3rd International Wafer Level Packaging

Congress IWLCP 2006, San Jose, 1.-3. November 2006.

[2] Optomec®, M^{3}D®, Aerosol Jet™ Deposition, System Manual, 2007.

[3] King, B., "Maskless Mesoscale Materials Materials Deposition", EP&P, Electronic Packaging & Production, 2003, vol. 43, issue 2, pp 18-20.

[4] Hedges, M., Kardos, M., King, B., Renn, M., "3D Direct Writing via M^{3}D™", pp 157-166, Proceedings of the 7th International Congress on Molded Interconnected Devices MID 2006, Fürth, 27.-28. September 2006.

[5] Schlüter, R., Rösener, B., Kickelhain, J., Naundorf, G.: Completely additive laser-based process for the production of 3D MIDs – The LPKF LDS Process, 5th International Congress Molded Interconnect Devices MID 2002, Erlangen, Germany, 2002

Wafer-scale transfer of nanoimprinted pattern into silicon substrates

G. Hubbard[1], S.J. Abbott[1], Q. Chen[2], D.W.E. Allsopp[2], W.N. Wang[2], C.R. Bowen[2],
R. Stevens[2], A. Satka[3] D. Hasko[3], F, Uherek[3] and J. Kovac[3]

[1] MacDermid Autotype Ltd, Grove Road, Wantage OX12 7BZ, England
[2] Faculty of Engineering and Design, University of Bath BA2 7AY, England
[3] International Laser Center, Ilkovicova 3, Bratislava 812 19, Slovakia

Abstract

Nanoimprinting provides a low cost alternative to Deep Ultra-Violet and electron beam lithography for producing deeply sub-micron features in semiconductor device fabrication. This paper presents a flexible nanoimprint process capable of wafer-scale pattern transfer into Si substrates. The technique is based on the novel concept of a disposable soft master with matching imprint resist formulations. Both of the resists developed enable the transfer of vertical sided, nearly flat bottomed features in Si substrates. The technique lends itself to large-scale low cost roll-to-roll processing based on the concept of a disposable master. The process is a promising method for low-cost formation of photonic crystal structures in hard substrates and is potentially suitable for high volume production.

Keywords: nanoimprint, silicon, wafer, photonic crystal structures

1. Introduction

Nanoimprinting provides a low cost alternative to Deep Ultra-Violet (DUV) and electron beam lithography for producing deeply sub-micron features in semiconductor device fabrication [1]. Nanoimprint techniques frequently involve the formation of either a pre-patterned die of limited size, from either a hard [2] or soft material [3] that is then pressed into a curable lacquer spin-coated on to the target substrate. Wafer-scale or large area nanoimprinting arises from stepping and repeating the impression. Frequently encountered issues during large scale nanoimprinting include release of the master, stitching errors in the step and repeat process, damage to bowed wafers, especially in high-pressure imprint processes, and clearing the inevitable residual layer when device fabrication involves pattern transfer into the underlying substrate.

In many applications of nanoimprinting, the formation of a pattern in a polymer film coating is merely a stage in a device manufacturing processing. Often the imprinted polymer coating must also withstand aggressive chemicals used to replicate the nano-scale pattern in a hard substrate to a depth determined by the functionality of the device. For example, the formation of a photonic crystal structure may require etching a highly regular array of holes with critically controlled section and depth into a hard substrate that has a high refractive index.

The aim of this paper is to present a novel nanoimprint process for wafer-scale pattern replication in hard substrates. The technique lends itself to large-scale low cost roll-to-roll processing based on the concept of a disposable master. Further, the method is shown to enable the transfer of a nano-scale pattern into a thin silicon dioxide layer that then acts as a hard etch mask for subsequent pattern transfer into a silicon substrate by reactive ion etching.

2. Methods

Master structures were made on a 250mm scale

via laser interference photolithography [4] in photoresist and then transferred to nickel replicas by standard electroforming techniques. The pattern on the nickel was transferred to the surface of a polyethylene terephthalate (PET) film using a roll-to-roll UV replication process. This provides a large supply (100's of metres) of "disposable masters" for nanoimprinting. The resulting master structure, shown in Figure 1, mimics a motheye structure and comprises a pseudo-hexagonal array of sine[2] shaped features, used as an anti-reflection film for displays [5]. The peaks and depressions in the pattern have a near elliptical rather than circular cross-section.

Fig. 1 SEM microphotograph of a motheye structure roll-to-roll replicated onto a polymer substrate (a disposable master) by MacDermid-Autotype UV nanoreplication.

Two types of UV-crosslinking material were used in the replication process. The first was an epoxy-silicone, the second a conventional acrylate system containing a small percentage of a fluoro-acrylate. Both resists were designed to have reduced surface energy enabling clean release of the master without additional processing steps. The disposable masters are flexible and their stiffness is determined by the combination of the PET backing film and the formulation of the lacquer containing the relief. Two different masters having different period and relief height were used in this work.

The first step in the process is to spin coat the substrate with a thin film of UV sensitive imprint resist, followed by a short low temperature pre-bake to evaporate excess solvent and to promote adhesion of the polymer film to the wafer. The imprint step involves using a roller to press the disposable master into the resist with only light force. Figure 2 shows a schematic of the nanoimprinting roller and disposable master. Unlike conventional imprinting, the line-contact of the flexible master makes contact pressure virtually irrelevant. Experiments showed that the degree of filling and the thickness of the residual layer are controlled by the thickness of the spin-coated layer rather than pressure. Nanoimprinting is followed by a cure step follows which involves a short exposure to UV light and an optional thermal cure to complete the cross-linking of the imprint resist. Finally the disposable master is released by peeling it from the imprinted substrate.

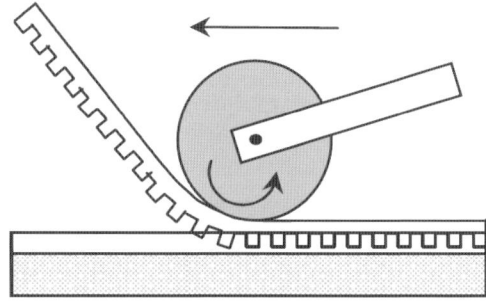

Fig. 2 Nanoimprinting roller and disposable master.

3. Results

Two novel imprint resist have been developed and are presented here. One is based on Oxetanyl Silsesquioxane (OXSQ) with an iodonium photo-initiator. This was designed so that the Si atoms of the OXSQ (Figure 3) oxidised during CHF_3/O_2 reactive ion etching (RIE) which were used to clear the residual layer after imprinting. The aim was to create a resist more able to withstand the etching required for pattern transfer into a hard substrate.

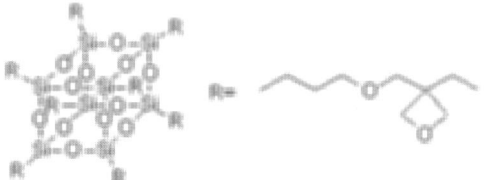

Fig.3 Oxetanyl Silsesquioxane, showing the effect of Si cross-linking atoms.

The second imprint resist is based on the UV sensitive acrylate used to make the disposable master. The purpose of developing the acrylate based resist was to ease its removal and descumming, after pattern transfer into the substrate. Flow enhancers were added the formulation to improve wetting and spin coating of the substrates. Both types of resist were dispersed in toluene for spin coating.

While 4" diameter Si wafers have been imprinted with 100% successful coverage using the above technique, 2"diameter Si substrates with or without a 100 nm thick layer of SiO_2 deposited by CVD were used in the examples described here. Figure 4 shows a Scanning Electron Microscopy (SEM) microphotograph of an imprinted OXSQ film of original thickness 165 nm.

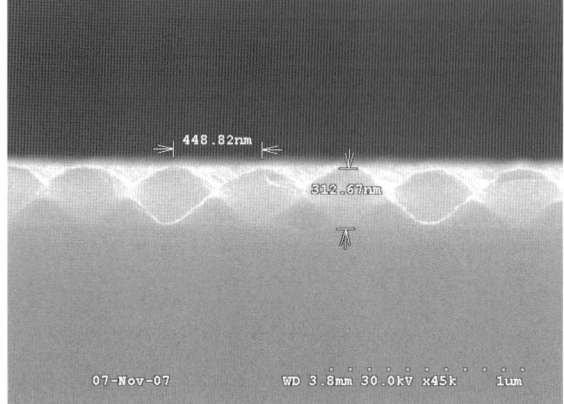

Fig. 4. Nanoimprinted OXSQ. Pattern period = 450 nm, relief height = 313 nm.

At first sight the profile of the imprinted resist looks unpromising for an etch mask. However, the 9nm/min etch rate of OXSQ in the SF_6/CHF_3 plasmas used to reactive ion etch SiO_2 and Si was sufficiently low to enable pattern transfer into both these materials. In the process developed here the removal of the residual layer (typically ≤20 nm) is the critical, but controllable step. Figure 5 (a) shows the impact of removing a sufficient amount of the OXSQ residual layer to form a pseudo-hexagonal array of holes in the resist for pattern transfer into the silicon by Ar+ RIE. Figure 5 (b) shows the impact of more extensive CHF_3/O_2 RIE that leaves isolated islands of OXSQ. In the case a SF_6/CHF_3 RIE was used to create the structure in the Si in a way that yields an undercut edge, demonstrating the resilience of the OXSQ to SF_6/CHF_3 RIE.

The performance of the acrylate based nanoimprint resist is shown in Figs 6 (a) and (b). An array of holes has been created in the acrylate by removing the residual layer using O_2 RIE (no CHF_3 is needed). The substrate was coated with a ~100 nm thick SiO_2 layer. Fig. 6 (b) shows the effect of sequential CHF_3 RIE, to transfer the pattern into the SiO_2 layer, and then SF_6/CHF_3 RIE to transfer the pattern in the underlying Si. Whilst the etching of the SiO_2 is not anisotropic, RIE of the Si results in the formation of vertical sided, flat bottom holes on a highly regular lattice. Such structures are of interest for photonic crystal structure applications. Vertically-sided holes in the Si as deep as 250 nm have been achieved when longer etch times have been used with the acrylate resist mask.

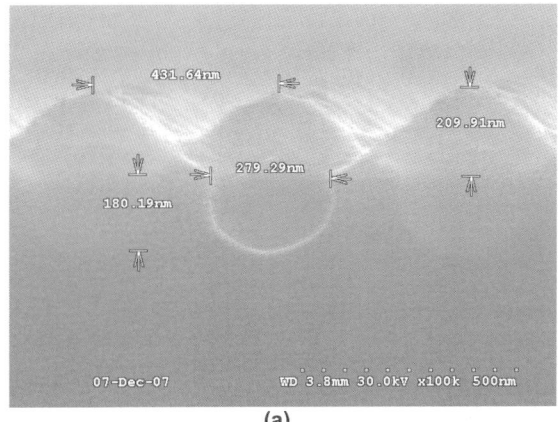

(a)

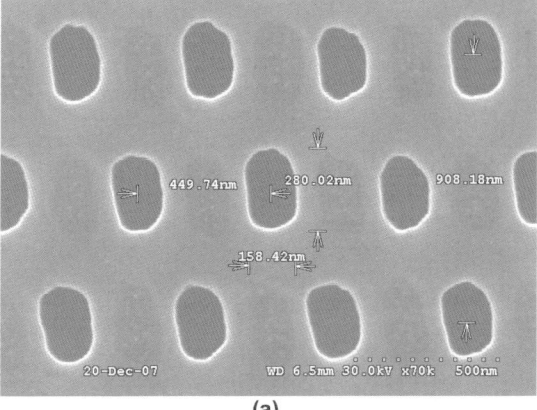

(a)

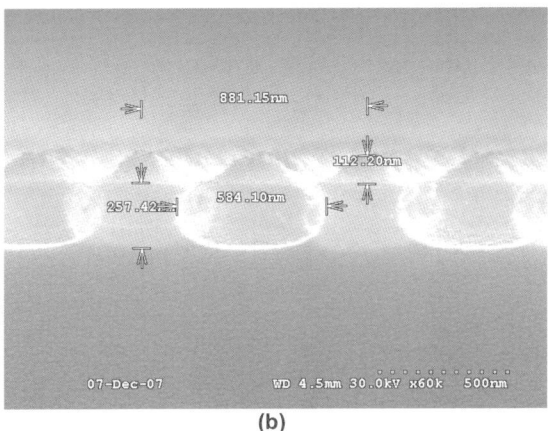

(b)

(b)

Fig. 5 (a) Edge profile of 2D photonic structure etched into Si using SF_6/CHF_3 RIE with nanoimprinted OXSQ as the etch mask; (b) The effect of over widening the holes during removal of the residual layer by CHF_3/O_2 RIE.

Fig. 6(a) surface view of nanoimprinted acrylate resist after removing ~20 nm thick residual layer and (b) side view after subsequent CHF3 etching to clear the SiO2 and 3 min SF6/CHF3 RIE to transfer the pattern into Si.

4. Conclusions

In summary, a flexible nanoimprint process capable of wafer-scale pattern transfer into Si substrates has been demonstrated. The technique is based on the novel concept of a disposable soft master with matching imprint resist formulations. Both the resists developed enable the transfer of vertical sided, nearly flat bottomed features in Si substrates. The process is a promising method for low-cost formation of photonic crystal structures in hard substrates and is potentially suitable for high volume production.

Acknowledgement:

The Authors wish to acknowledge support from the European Union under Framework 6 contract number 017481, STREP "N2T2" and partial support from the Slovak Research and Development Agency grant APVV-RPEU-0005-6.

References

[1] "Alternative lithography: unleashing the potentials of nanotechnology," Ed: C.M.Sotomayor Torres, Kluwer Academic/Plenum Publishers, ISBN-10: 0306478587 (2003).

[2] K.-I. Nakamatsu and S. Matsui, Japan. J. Appl. Phys. vol. 45, L546-L548 (2006).

[3] L.-R. Bao et al, J. Vac. Sci. Technol. B vol. 20, 2881-2886 (2002).

[4] www.holotools.de (accessed 5th Feb 2008)

[5]http://www.autotype.com/autotype.nsf/webfamilieseurope/AUTOFLEX%20OPTICAL%20FILMS (accessed 5th Feb 2008)

Multi-Material Micro Manufacture
S. Dimov and W. Menz (Eds.)

Large-area metal-coated dielectric nanopillar array for excitation of surface plasmon resonance

X. Chen, K. Jiang

Micro Engineering and Nanotechnology Group, Department of Mechanical Engineering, University of Birmingham, B15 2TT, UK

Abstract

Many of current techniques are not suitable for the fabrication of metallic nanostructure on the scale of usual optical coatings at reasonable fabrication cost and time. A fabrication process for producing large-area metal-coated periodic nanopillars is presented. A hybrid metallic nanostructure array was obtained by depositing a silver film with a thickness of ~40 nm on the fused silica nanopillars with an in-plane diameter of ~140 nm and out-of-plane height of ~130 nm, which was fabricated by a combination of interference lithography, metal deposition and etching. There are two peaks in the extinction spectrum of the p-polarized incident light, one at 585.3 nm and the other 493.6 nm. The shift of the higher peak is 32.9 nm (a red-shift), while that of the lower peak is 42.3 nm (a blue-shift) with the addition of absolute ethanol on the sample surface. Such structure was used to monitor the evaporation process of the absolute ethanol on the sample surface. It was found that narrowest extinction peak appears at normal incidence, while the polarization of the incident light does not affect the experimental result due to the symmetrical distribution of the nanostructures. The fabrication process and unique optical properties of the structure array are expected to be suitable for the development of high-throughput ultrasensitive chemical sensor arrays.

Keywords: Large-area, Hybrid metallic nanostructure, Interference lithography, Surface plasmon resonance.

1. Introduction

Localized surface plasmon resonance (LSPR) refers to the ability of the conduction electrons in the nanoparticle to oscillate collectively, which can concentrate and enhance electromagnetic energy surrounding the nanoparticle. Recent technical developments in nanolithography and the chemical synthesis of metal nanostructures have allowed the production of metallic nanostructures with various shapes such as dots, triangles[1-4], shells [5,6], rings [7], rods [8], disks [9] and cups [10]. Unfortunately, many of these techniques are not suitable for the fabrication of metallic nanostructure on the scale of usual optical coatings (i.e., on a cm^2 scale) at reasonable fabrication cost and time. Nanosphere lithography is a powerful technique and has been used to inexpensively produce nanoparticle arrays with controlled shape, size and interparticle spacing, but the typical defect-free domain sizes are in the 10-100 μm range [1,2,4]. Wang et al.[11] reported a one-step electron-beam lithography process to fabricate a nanopin array, in which each nanopin is constituted by a metal-capped dielectric pillar sitting on a ring-shaped metallic disc. However, the exposure dose they used is much higher than normal dose to enable the centre cross-linked PMMA to survive during the subsequent lift-off process. Interference lithography, as an inexpensive and versatile technique, has already successfully been used on the fabrication of 1D metallic photonic crystal slabs [12], of magnetic metamaterials [13], and of negative-index metamaterials at infrared range [14]. Here an interference lithography based fabrication technique is presented, which is capable of producing large-area hybrid metallic nanostructure array for excitation of surface plasmon resonance [15]. The structure consists of a metal capped dielectric nanopillar array and a metallic hole array.

2. Experimental details

The large-area periodic metallic nanostructures were fabricated by a combination of interference lithography, metal deposition and etching. Fused silica substrates with a size of 2 cm by 2 cm were cleaned in isopropyl alcohol for 10 minutes, assisted with an ultrasonic processor, to remove dust and then prepared by washing in running deionised water for 1 minute, dipping in a 1:1 mixture of concentrated sulfuric acid and hydrogen peroxide for 10 minutes to remove organic contaminants, and washing again for 1 minute in running deionised water. Finally substrates were dried with a nitrogen gun and baked on a hotplate for 1 hour at 170 °C. A chromium film with a thickness of ~20 nm was deposited on the newly cleaned substrate by thermal evaporation. On top of the chromium, a layer of AZ3100MI(20cp) (positive resist) with a thickness of 120nm was spun, followed by a post application bake of 100 °C for 15 minutes. The resist was patterned by IL and periodic resist nanostructures were obtained after development in Microposit 303A (Shipley Co.). Oxygen reactive ion etching (RIE) was used to remove resist residual in the patterned regions (O_2 gas flow=10 sccm, pressure=1 Pa, power=30 W). Chromium patterns were obtained by dipping the substrate into a chromium etchant. The resist on chromium was removed by sonication in acetone. CHF_3 RIE was used to etch the newly exposed fused silica (CHF_3 gas flow=30 sccm, O_2 gas flow=1 sccm, pressure=1 Pa, power=80 W), transferring the patterns to the oxide layer. After the remaining chromium is removed in chromium etchant, a new chromium film with a thickness of 5 nm (to improve adhesion) and a silver film with a thickness of ~40 nm are deposited onto the periodic oxide nanopillars. Finally a hybrid metallic nanostructure array was obtained.

An IL system based on Lloyd's mirror configuration [16, 17] was built up, in which a mirror is placed normal to the substrate and illuminated with an s-polarized laser beam with a wavelength of 442nm. A part of the incident laser beam is reflected by the mirror and interferes with the un-reflected part of the beam to form interference patterns. Nanopillar arrays were fabricated by rotating the sample and double exposure. In order to obtain a two-dimensional nanopillar array in the resist, the exposure dose must be high enough otherwise that the resist only at the crossings of the grid lines (after first and second exposure) receives enough energy to be removed after development. The intensity of the incoming light in the exposed area was measured to be 0.65 mW/cm2 for normal incidence. The resist was exposed to the interference line pattern for 8 seconds. After rotating the substrate to 90° the exposure was repeated. The resist was developed in Microposit 303A (Shipley Co.) for ~30 seconds and rinsed in deionized water.

The extinction spectra of the structure were measured to study the optical characteristics of this structure. A beam of white light (300nm-800nm) was sent to a collimating lens via an input fibre. The well-collimated light beam was incident to the sample surface at an angle, which is controlled by a rotating stage. The probe diameter was approximately 3 mm, which is controlled by a diaphragm. The output fibre carried the light from the sample to the spectrometer (Ocean Optics USB4000) connected to a computer. All spectra in this study are the results of macroscopic measurements performed with polarized light. The measurement was performed at room temperature with a relative humidity of 47%. A drop of absolute ethanol was added on the surface of the metallic nanostructures to change the surrounding refractive index. The difference between the extinction maximum before and after the addition of absolute ethanol is the wavelength shift response. To measure the evaporation process of this drop of absolute ethanol on the surface of this structure, a series of extinction spectra were collected every 1 minute. Effects of the incident angle and light polarization on extinction spectra were studied by independently changing the rotating stage and the polarizer.

3. Results and discussion

Figure 1(a) shows SEM images of fused silica nanopillar array with a pitch of 321 nm, an in-plane diameter of ~140 nm and out-of-plane height of ~130 nm. Although some residues are observed in the nanostructures, they may originate from the chemical reactions in pattern transfer from chromium to fused silica by RIE and could be eliminated by prolonged sonication in acetone. Figure 1(b) is the SEM image showing the final hybrid metallic nanostructure array obtained through this approach. In comparison with figures 1(a), it can be observed that there is some silver extending from the top of the oxide pillar and partly down the side wall. The variation of oxide pillar size between its top and bottom and evaporative

edge effects due to the minor change of evaporation direction are the possible causes. The metallic structure can be improved with sharper oxide nanopillars by optimizing etching conditions during the pattern transfer processes.

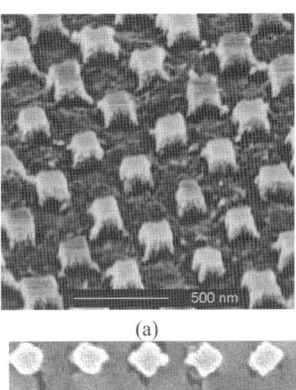

(a)

(b)

Fig.1 (a) An SEM image of the pillars taken with a tilt angle of 45°. (b) An SEM image of the metallic nanostructure after silver deposition.

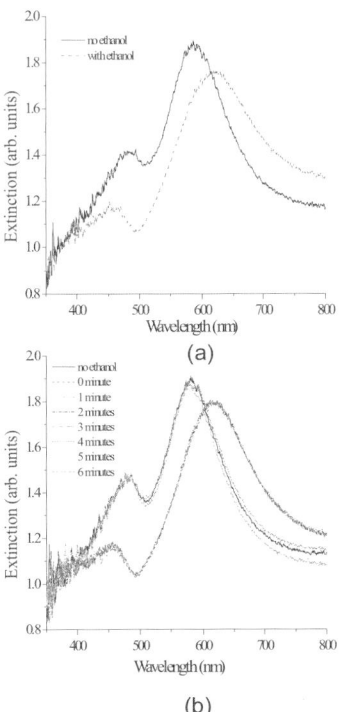

Fig.2 The spectra with p-polarized incident light obtained in experiments. (a) Experimental comparison with and without addition of absolute ethanol. (b) Evaporation process of absolute ethanol.

A well-collimated white light beam was incident normally to the surface of the hybrid metallic nanostructure array. Figures 2(a) and 2(b) show the extinction spectra for the periodic nanostructures. There are two peaks in the extinction spectrum of the p-polarized incident light, one at 585.3 nm and the other 493.6 nm. The peaks are shifted to 451.3nm and 618.2 nm with the addition of absolute ethanol on the sample surface. The shift of the higher peak is 32.9 nm (a red-shift), while that of the lower peak is 42.3 nm (a blue-shift). Such structure was used to monitor the evaporation process of the ethanol on the sample surface. Fig. 2(b) shows the extinction spectrum change during the evaporation process. The initial spectrum in the experiment was obtained without absolute ethanol. The spectrum turned into a plot very much similar to the spectrum shown in Fig. 2(a) shortly after the addition of absolute ethanol. This spectrum lasted for 3 minutes. One further minute later, the curve turned back to the initial spectrum, indicating that the whole evaporation process took 3-4 minutes.

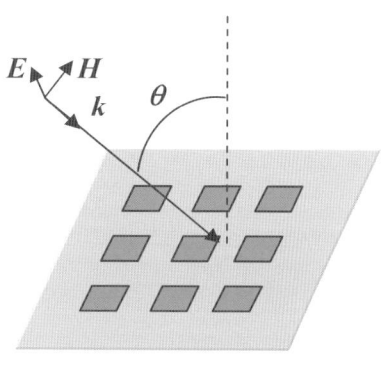

(a)

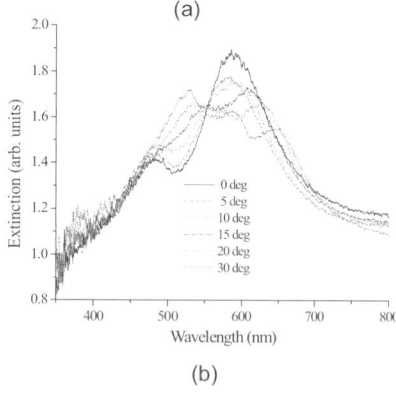

(b)

Fig.3 (a) Illustration of the experimental configuration, in which p-polarized light is incident at an angle in order to illuminate the sensor chip, (b) Effect of incident angle on extinction spectra.

Effects of the incident angle and light polarization on extinction spectra were studied in further experiments. Figure 3(a) shows the experiment configuration where the incident angle and polarization can be controlled by independently changing the rotating stage and the polarizer. Figure. 3(b) shows the experimental results when the incident angle was changed. It can be observed from this figure that the narrowest extinction peak appears

at normal incidence. It was found that the polarization of the incident light does not affect the experimental result due to the symmetrical distribution of the nanostructures.

The geometry-dependent plasmon response can be considered as an interaction between the essentially fixed-frequency plasmon response to a periodic array of metallic holes and an array of metal capped dielectric nanopillars. At the same time, if the distance between metal piece on the dielectric pillar and the metal piece at the bottom is small enough, the hybridization of two plasmons will be formed due to their interaction [13]. This research is only focused on the experimental study and theoretical models can be found in the literatures [1,6,13,14]. There is some potential for further optimization of parameters to obtain stronger field enhancement. The structure can be used on a chemical sensor chip, which can be divided into many pieces for experimental comparison and reused for many times after the silver layer is removed and deposited again at a certain thickness. It would be straightforward to further increase the resulting sample area to many square inches by scaling up the apertures of the optics. Future work is ongoing aiming at increasing the exposure area to enlarge the area of the sensor chip.

4. Conclusions

A low-cost, high-throughput fabrication process has been developed to produce hybrid metallic nanostructure array over large areas without the need for e-beam lithography. The pillar size and pitch of the nanostructures can be independently adjusted by selecting different exposure doses and by changing the incident angle of the laser beam in the IL step, respectively. The height of the dielectric pillar can be controlled by modifying the RIE etching condition during the pattern transfer step. The plasmon response of the hybrid metallic nanostructures can be viewed as the collection of plasmons arising from two geometries to form an interacting system, which makes it possible for the structure to possess unique optical properties. Hence, it is possible to tune the optical properties by tailoring the geometries of the hybrid metallic nanostructures in order to achieve specific extinction spectra on demand. The large-area fabrication process, unique optical properties, and further improvement of the hybrid metallic structure array are expected to be suitable for the development of high-throughput ultrasensitive chemical sensor arrays, nanobiosensor array etc.

Acknowledgements

The authors acknowledge the financial support to the FP6 RaSP project from European Union and assistance from Dr J. Teng in SEM imaging.

References

[1] Haes A.J., Hall W. P., Chang L., Klein W. L., Duyne R.P.V., A Localized Surface Plasmon Resonance Biosensor: First Steps toward an Assay for Alzheimer's Disease, Nano Lett. 2004, 4, 1029-1034.
[2] Zhang X., Whitney A. V., Zhao J., Hicks E.M.,

Duyne R.P.V., Advances in contemporary nanosphere lithographic techniques, J. Nanosci. Nanotechnol. 2006, 6, 1-15.

[3] Vlasov Y.A., Bo X.Z., Sturm J.C., Norris D.J., On-chip natural assembly of silicon photonic bandgap crystals, Nature 2001, 414, 289-293.

[4] Kuo C., Shiu J., Cho Y., Chen P., Fabrication of large-area periodic nanopillar arrays for nanoimprint lithography using polymer colloid masks, Adv. Mater. 2003, 15, 1065-1068.

[5] Prodan E., Radloff C., Halas N.J., Nordlander P., A Hybridization Model for the Plasmon Response of Complex Nanostructures, Science 2003, 302, 419 – 422.

[6] Oldenburg S. J., Averitt R. D., Westcott S., Halas N. J., Nanoengineering of optical resonances, Chem. Phys. Lett. 1998, 288, 243-247.

[7] Aizpurua J., Hanarp P., Sutherland D.S., Käll M., Bryant Garnett W., García de Abajo F. J., Optical Properties of Gold Nanorings, Phys. Rev. Lett. 2003,90, 057401/1-4,

[8] Jana N. R., Gearheart L., Murphy C. J., ,Wet Chemical Synthesis of High Aspect Ratio Cylindrical Gold Nanorods, J. Phys. Chem. B. 2001,105, 4065-4067.

[9] Maillard M., Giorgio S., Pileni M. P., Tuning the Size of Silver Nanodisks with Similar Aspect Ratios: Synthesis and Optical Properties, J. Phys. Chem. B. 2003,107, 2466-2470.

[10] Charnay C., Lee A., Man S., Moran C.E., Radloff C., Bradley R. K., Halas N.J., Reduced Symmetry Metallodielectric Nanoparticles: Chemical Synthesis and Plasmonic Properties, Phys. Chem. B 2003, 107, 7327 -7333,

[11] Wang S., Pile D.F.P., Sun C., Zhang X., Nanopin plasmonic resonator array and its optical properties" Nano Lett. 2007, 7, 1076-1080.

[12] Guo H.C., Nau D., Radke A., Zhang X.P., Stodolka J., Yang X.L., Tikhodeev S.G., Gippius N.A., and Giessen H., Large-area metallic photonic crystal fabrication with interference lithography and dry etching, Appl. Phys. B 2005 **81**, 271-275.

[13] Zhang S., Fan W., Panoiu N.C., Malloy K.J., Osgood R.M., and Brueck S.R.J., Experimental demonstration of near-infrared negative-index metamaterials, Phys. Rev. Lett. 2005, **95**, 137404/1-4.

[14] Fan W., Zhang S., Malloy K.J., and Brueck S.R.J., Large-area, infrared anophotonic materials fabricated using interferometric lithography, J. Vac. Sci. Tech. B 2005, **23**, 2700-2704.

[15] Chen X., Jiang K., Large-area hybrid metallic nanostructure array and its optical properties, to be published.

[16] Rijn Cees J. M. van, Laser interference as a lithographic nanopatterning tool", J. Microlith., Microfab., Microsyst. 2006, 5, 011012/1-6.

[17] Solak Harun H, Nanolithography with coherent extreme ultraviolet light, J. Phys. D: Appl. Phys. 2006, 39, R171–R188.

Multi-Material Micro Manufacture
S. Dimov and W. Menz (Eds.)

Fabrication of stainless steel micro components using soft lithography

Mohamed Imbaby[a], Kyle Jiang[*a], Isaac Chang[b]

[a] School of Mechanical engineering, University of Birmingham, Edgbaston, Birmingham, UK
[b] School of Metallurgy and Materials, University of Birmingham, Edgbaston, Birmingham, UK

Abstract

316-L stainless steel has good mechanical properties and has been widely employed for making different devices. This paper presents a study for making micro 316-L stainless steel components by soft lithography in combination with powder metallurgical processes. The process involves producing deep and solid micro moulds using SU-8 photo resist, making soft replica of the moulds using silicon rubber (PDMS), forming green patterns by filling stainless steel slurry into the PDMS moulds. The green parts are de-moulded, de-bound, and finally sintered in tube furnace including nitrogen atmosphere to obtain the final micro parts. The resultant micro components show good quality micro parts with complex geometry. The density of the sintered parts reaches 91.5% of the theoretical one and the linear shrinkage of the micro components after sintering is investigated and it is found to be dependent on the percentage of the solid loading in the green patterns. The fabrication process is described in detail and the results of characterization in shrinkage and density have been analysed.

Keywords: SU-8 master mould, PDMS, Duramax D-3005, 316L stainless steel, micro components

1. Introduction

Micro electro mechanical systems (MEMS) were used in various applications such as pressure sensors, biomedical sensors, drug delivery systems, fluid management processing devices etc. in which the fabrication technology was evolved from silicon based integrated circuits techniques [1]. The demands for multiple material micro components encourages researchers to develop new fabrication techniques that allow metals, ceramics, alloys, and polymers be used. In terms of good mechanical properties and medical applications, stainless steel components are good materials for these purposes. 316L stainless steel grade that contains chromium-nickel and molybdenum provides higher resistance to pitting and crevice corrosion in chloride environments. These properties in addition to low carbon contents make it the best candidate for the implanted applications because of the decreasing in vivo corrosion [2].

Different methods have been used in micro components fabrication, such as LIGA process based on combining synchrotron radiation lithography and galvanoforming [3], focus ion beam (FIB) [4], laser micro machining [5] and micro electro discharge machining (MicroEDM) [6]. These methods are either very expensive, such as LIGA, FIB and MicroEDM, or of low quality and resolution, such as laser. Micro injection moulding (MIM) is another fabrication technique suitable for polymers [7], metals [8], alloys [9], and ceramics [10]. However, micro injection moulding relies on precise metal mould that is fabricated by electroformed process [11] which can increase the overall cost of micro fabrication.

This paper presents a study on fabrication of 316-L stainless steel micro parts with high shape retention and complex shapes using softlithography and powder metallurgy processes. Softlithography is a relatively new process that relies on soft mould inserts. The process includes fabrication of SU-8 master moulds and replication with PDMS negative replicas moulds. Powder metallurgy process includes preparation of stainless steel slurry, filling the PDMS moulds, de-moulding, de-binding, and sintering to obtain the final micro part. The research work is based on previous successful experience in fabrication of micro components [12, 13] and the techniques have been developed further. The fabrication process is investigated in details. The linear shrinkage and the density of the sintered micro parts are discussed.

2. Fabrication Process

2.1. Fabrication of SU-8 master mould

SU-8 is a negative tone resist. It can be fabricated to a thickness higher than 1mm because of its very low optical absorption in the UV range 360-420 nm. Thus, a relatively constant exposure dose can go through the entire resist thickness without losing much energy and excellent vertical sidewalls are obtained. Moreover, SU-8 is highly resistant to solvents and acids when cured [14]. Deep x-ray lithography is one of the fabrication techniques used for fabrication ultra thick layers of SU-8 [15]. Due to higher cost of x-ray source, UV lithography was successfully developed to fabricate ultra thick layers of SU-8 micro engine parts [16].

The fabrication process of ultra thick SU-8 master moulds used in this research was based on [12, 13, 16], and modification was made due to the change of the SU-8 type. SU-8 2075 [MicroChem, USA] is the type of photo resist used in this work in which 1mm thick is fabricated. The SU-8 mould fabrication procedure started with casting SU-8 on the 4 inch wafer and soft baking at 65C° for 2 hours followed by 95C° for 34 hours. Exposure was done in Canon PLA-501 FA UV-mask aligner. As discussed in [16], the exposure time was 17.5 units, but in this work PL-350 filter was used for reducing the overall exposure time to 13 units. Afterwards, the wafer went through post exposure bake

* For correspondence: k.c.jiang@bham.ac.uk

and developing in EC solvent. The SU-8 master moulds achieved from the fabrication process are shown in Figure 1 in which uniform and straight sidewalls are obtained.

2.2. Replication with PDMS mould

PDMS is viscoelastic, non toxic, non flammable, optically clear. Moreover, very accurate impressions of micro structures can be obtained. The steps needed to replicate PDMS mould inserts are as follows. Firstly, PDMS raw material (DOW Sylgard Silicone) and curing agent were added in 10:1 in weight then the mixture was vacuumed in order to remove all bubbles formed during the mixing. Secondly, the mixture is purred on SU-8 moulds and vacuumed again. Finally the PDMS on top of the moulds was cured in an oven at 90 C° for 2 hours. The cured PDMS moulds were peeled off from SU-8 master moulds with the help of cutter and tweezer. Figure 2 shows PDMS replication mould. It was observed that very thick PDMS layers damaged the SU-8 master moulds during the peeling off, on the other hand very thin PDMS layer was damaged during the peeling off.

2.3. Preparation of 316-L stainless steel slurry

Stainless steel 316L powder supplied by Sandvik Osprey Ltd., UK is used in this research. The particle size distributions and the chemical compositions of the powder delivered by the supplier are presented in Table 1. The powder was examined in SEM (JOEL 6060) and its image is shown in Figure 3. The powder shape is spherical with different size distributions. In softlithography where soft mould (PDMS) insert is used, no injection pressure or heating temperature are needed, thus, the binder must have low viscosity at room temperature during the moulding and high viscosity during the drying processes. In this work, Duramax D-3005 delivered by Chesham Speciality Ingredients Limited, UK was used as a binder. Duramax D-3005 is the ammonium salt of an acrylic homo-polymer and it has been used as a dispersant in the ceramic and nickel fabrications [17, 18]. In this work, stainless steel slurry containing 80% and 85% weight of stainless steel powder were prepared. Duramax D-3005 and de-ionized water were mixed in beaker 30:70 by weight respectively, then stainless steel powder is added. The mixture was homogenized by mechanical stirrer for one hour and put under vacuum to remove the bubbles formed during mixing.

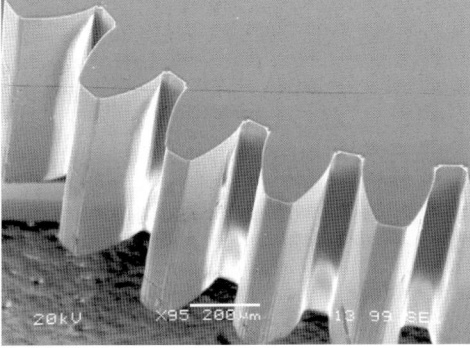

Fig. 1. SU-8 master moulds

Fig. 2. PDMS moulds insert

Table 1
Chemical compositions and size distributions

Chemical composition %					size distribution		
Cr	Ni	Mo	Mn	C	D10	D50	D90
18.5	11.6	2.3	1.4	.048	1.1 um	1.8 um	3.6 um
P	Si	S	Fe				
.027	0.65	.008	Bal.				

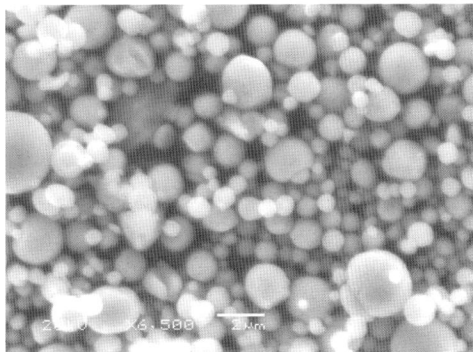

Fig. 3. 316L Stainless steel powder

2.4. Mould filling and preparation of green parts

The slurry was poured in the PDMS mould cavities under gravity. It was observed that the moulds showed incomplete filling because air trapped in small cavities such as gear teeth is difficult to be removed. Therefore, vacuum was applied to let the trapped air to escape from the mould. After that, the filled mould was left to dry at room temperature for 1-2 hours before the green parts were successfully de-moulded with more than 80% defect free samples. Figure 3 shows defect free green part.

2.5. De-binding and sintering

De-binding and sintering were carried out in the same heating cycle in a tube furnace (ELITE thermal system limited) in nitrogen atmosphere. In the de-binding stage, the temperature was slowly ramped up at 3 °C/Min until it reached 600 °C, and it was maintained at that temperature for one hour. Then it came to the sintering stage. The temperature was ramped to 1200 °C at a rate of 6 °C/min and maintained at this temperature for 90 minutes.

90

Fig. 4. Green part

3. Component characterization

3.1. Dimensional shrinkage

The parts were shirked after sintering and the linear shrinkage was calculated based on the diameters D of the gear as follows:

$$\text{Shrinkage}\% = \frac{D_{SU\text{-}8\,gear} - D_{sintered\,gear}}{D_{SU\text{-}8\,gear}} \times 100$$

4. Results and discussions

The fabrication of SU-8 mould is optimized with little difference from the previous work [16] due to the change of SU-8 type. On the other hand, vacuum has the greatest effect of degassing the bubbles formed during the PDMS mould replication and mould filling. Moreover, the sintered gear is examined under SEM and its image was shown in Figure 5. It has been shown that the gear retained all features in which homogenous shrinkage was obtained. The linear shrinkage% of sintered gears containing 80% and 85% solid loading was measured and found to be 17% and 15% respectively. Increasing solid loading during the green part preparation decreases the shrinkage after sintering because of decreasing the amount of the binder that burned out during the heating (de-binding and sintering). The density of the sintered part was measured according to Archimedes principal and found to be 7.23 g/cm3 around 91.5% of the theoretical one.

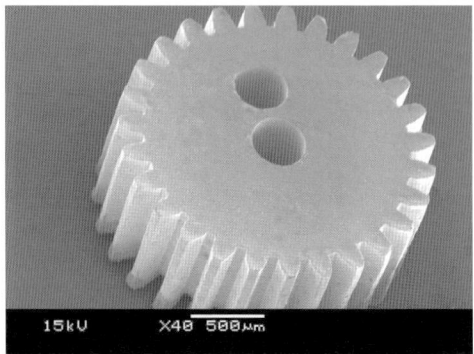

Fig. 5 Sintered part

5. Conclusion

Good shape retention with complex shape stainless steel micro parts were successfully fabricated using softlithography in combination with powder metallurgy processes. Also, Duramax D-3005 was successfully used as a binder in the preparation of stainless steel slurry. The linear shrinkage of the sintered parts was investigated and found to be inversely proportional to the solid loading of the powder in the preparations of green pattern.

Acknowledgements

The authors would like to thank Sandvik Osprey Ltd. for the supply of superfine stainless steel 316L powders. The research was supported by the Ministry of Higher Education, Egypt.

References

[1] Nadim M, Kirt W. An Introduction to Microelectromechanical systems Engineering. London, Artech House, Inc., 2004.

[2] Ranter DB, Hoffman AS, Schoen FJ, Lemons JE. Biomaterial Science. London, Academic Press, 1996.

[3] Becker EW, Ehrfeld W, Hagmann P, et al. Fabrication of microstructures with high aspect ratios and great structural heights by synchrotron radiation lithography, galvanoforming, and plastic moulding (LIGA process). Microelectron. Eng. 4 (1986) 35-56.

[4] Walker JF, Moore DF, Whitney JT. Focused ion beam processing for microscale fabrication. Microelectron. Eng. 30 (1996) 517-522.

[5] Xiaoshan Z, Jin-Woo C, Cole R, et al. A new laser micromachining technique using a mixed-mode ablation approach. IEEE (2002) 152-515.

[6] Liu HS, Yan BH, Huang FY, et al. A study on the characterization of high nickel alloy micro-holes using micro-EDM and their applications. J. Mater. Process. Technol. 169 (2005) 418-426.

[7] Liou AC, Chen RH. Injection molding of polymer micro- and sub-micron structures with high-aspect ratios. Int. J. Adv. Manuf. Technol. 28 (2006) 1097-1103.

[8] Nishiyabu K, Matsuzaki S, Tanaka S. Net-shape manufacturing of micro porous metal components by powder injection molding. Mater. Sci. Forum 534-536 (2007) 981-984.

[9] Gerling R, Aust E, Limberg W, et al. Metal injection moulding of gamma titanium aluminide alloy powder. Mater. Sci. Eng. A 432 (2006) 262-268.

[10] Wu C, Atre SV, Park SJ. Material homogeneity in ceramic micro arrays fabricated by powder injection molding. IIE Annual Conf. Exp. US, (2006).

[11] Su-dong M, Namsuk L, Shinill K. Fabrication of a microlens array using micro-compression molding with an electroformed mold insert. J. Micromech. Microeng. 13 (2003) 98-103.

[12] Jung-Sik K, Jiang K, Chang I. A net shape process for metallic microcomponent fabrication using Al and Cu micro-nano powders. J. Micromech. Microeng. 16 (2006) 48-52.

[13] Zhigang Z, Xueyong W, Kyle J. A net-shape fabrication process of alumina micro-components

using a soft lithography technique. J. Micromech. Microeng. 17 (2007) 193-198.

[14] website: http:/www.microchem.com..

[15] Jian L, Loechel B, Scheunemann HU, et al. Fabrication of ultra thick, ultra high aspect ratio microcomponents by deep and ultra deep X-ray lithography. IEEE Comput. Soc, (2003) 10-14.

[16] Jin P, Jiang K, Sun N, Ultra-thick SU-8 Fabrication for Micro Reciprocating Engines, J. Microlith. Microfab. Microsys. 3 (2004) 569-573.

[17] Tsetsekou A, Agrafiotis C, Leon I, et al. Optimization of the rheological properties of alumina slurries for ceramic processing applications. J. Europ. Ceram. Soc. 21 (2001) 493-506.

[18] Sanchez-Herencia AJ, Millan AJ, Nieto MI, et al. Aqueous colloidal processing of nickel powder. Acta Mater. 49 (2001) 645-651.

Multi-Material Micro Manufacture
S. Dimov and W. Menz (Eds.)

Concept for Packaging of a Silicon based Biochip

T. Velten[a], M. Biehl[a], T. Knoll[a], W. Haberer[a]

[a] *Fraunhofer Institute for Biomedical Engineering, Ensheimer Strasse 48, 66386 Sankt Ingbert, Germany*

Abstract

We report on a concept for packaging of a silicon-based biochip for integration with a fluidic cartridge, thus forming a lab-on-chip (LOC). The biochip, which has dimensions of 2 mm x 2 mm, comprises a central membrane having a diameter of 200 µm, and 20 bond pads with metal tracks leading to the membrane. The packaged biochip provides a fluidic interface to the cartridge as well as electrical interfaces to the biochip electronics being located in a readout instrument. The packaging method ensures the strict separation between the wet sensing area and the electrical contacts. The challenge is that the biochip has a freely moving membrane, additionally with a delicate biological coating, and this membrane is positioned on the same side of the silicon chip as the bond pads for the electrical interconnection. For packaging, the biochip is mounted into a recess of a rigid printed circuit board (PCB). The biochip is electrically connected with the PCB using a proprietary MicroFlex interconnection (MFI) technology, thus resulting in a flat surface towards the reaction chamber of the fluid cartridge. After the realization of the electrical contacts between the sensor chip and the PCB, the entire chip is encapsulated with an epoxy layer, leaving the membrane of the biochip uncovered. To protect the membrane against the fluidic epoxy, a specially shaped silicone casting-mould is used. In a last step, the biochip with the epoxy layer is glued on the bottom side of the cartridge.

Keywords: biochip packaging, lab-on-chip

1. Introduction

Here, we report on the packaging concept for a silicon biochip fabricated by MEMS (micro electro mechanical system) technology. The biochip's sensing element is a freely moving membrane, which is coated with a special polymer in order to bind the biological species to be detected. Both, excitation and detection of the membrane oscillation is done electrically. This so-called circular disc resonator (CDR) sensor has been developed by Newcastle University, UK, and will be described elsewhere. The packaging technologies as described in this paper represent important steps in developing a user-friendly, practically usable LOC device from a bare biochip. This development aims at a prototype device that can be applied for the evaluation of the biochip's analytical properties and the implementation of suitable assays in the field of medical diagnostic testing.

Normally, for the electrical connection of silicon dies in semiconductor device fabrication, silicon micro technology offers a number of well-established solutions, the most common thereof being wire bonding technologies and flip-chip-technology. However, the problem-free application of these technologies presumes, that the only integrated functional unit is electronic circuitry – in case that there are parts on the chip, which potentially have to undergo mechanical, chemical or biological interaction with the environment, electrical interconnection and encapsulation tend to become a challenge. This difficulty in finding suitable interconnection and encapsulation solutions is one of the main reasons, why the majority of biochip prototypes, which have been developed in scientific laboratories during the last decade, did not come up to capture the market, yet.

Similar considerations apply to the CDR-sensor: The challenge is that it has a freely moving membrane, additionally with a delicate biological coating, and this membrane is positioned on the same side of the silicon chip as the bond pads for the electrical interconnection. Standard interconnection technologies, as listed above, cannot be used, as the bond connections and their encapsulations, respectively, would hinder the flow of the analyte towards the sensor membrane.

2. Packaging and integration concept

2.1. Packaging of the biochip

An overview of the packaging concept is depicted in Fig. 1. The CDR-sensor will be mounted into a

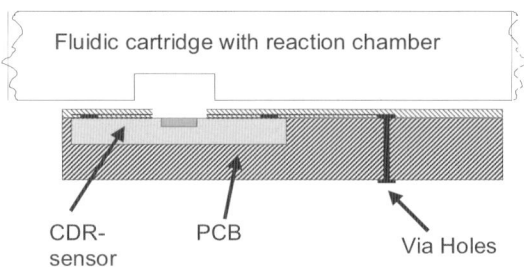

Fig. 1. Packaging concept. The packaged CDR-sensor will finally be bonded against the fluidic cartridge.

recess of a rigid printed circuit board (PCB), with its sensor surface being flush with the surface of this PCB. The CDR-sensor is electrically connected with the PCB using IBMT's MicroFlex interconnection technology, thus resulting in a flat surface towards the sensor chamber of the fluid cartridge. Via holes in the PCB provide electrical feed through to the backside of the PCB, thus allowing for an electrical connection between contact pads on the biochip's front side to those on the backside of the PCB. After the realization of the electrical contacts between the sensor chip and the PCB, the entire chip is encapsulated with an epoxy layer, leaving the membrane of the CDR-sensor uncovered. In a last step, the CDR-chip with the epoxy layer is glued against the bottom side of the cartridge. In this way all electrical connections are reliably separated from the fluidic interface.

A detailed description of the MFI technology can be found in [1]. In brief, gold balls produced by a standard ball bonder are used to mechanically and electrically connect the bond pads of the chip to the bond pads of an extremely thin micromechanically produced polyimide substrate. The polyimide substrate is bonded to the CDR chip first, and to the rigid PCB afterwards. The substrate is composed of two individual polyimide layers, each having a thickness of 5 μm. In-between, a structured metallic layer electrically connects the contact pads, which will be bonded to the sensor chip, to those bond pads provided for connection to the PCB. A very low overall height of the assembly can thus be achieved. Furthermore, this method is very robust because no mechanically fragile bond wires are used.

The described concept is illustrated in Fig. 2. The depicted design of the PCB supports the possibility for the integration of three CDR-dies in total.

We produced a tailored support for gluing CDR chip and PCB and we developed a special silicone mould for the casting of the CDR chip – PCB assembly with a suitable biocompatible epoxy coating (Fig. 3). The purpose of this mould is to guarantee the conformity of the coated PCB with the CDR test cartridge, which has been produced by MiniFAB Ltd, Victoria, Australia, and to prevent the CDR chip membranes from being wetted by the epoxy coating. This is achieved by pressure sealing of the CDR membrane surrounding area with the specially shaped silicone mould (Fig. 4). One part of the mould is made of transparent material in order to allow for optical control of the casting process.

To validate the feasibility of the packaging process, a PCB insert was assembled with special CDR dummy chips having the same dimension as the future CDR-chip. They are completely made of silicon with gold bond pads and a gold structure marking the position of the CDR membrane – however, they do not feature a real membrane or any functionality. The chips were integrated and electrically connected like described above. The resulting packed CDR-sensors are shown in Fig. 5. Microscopic inspection showed that the coating did not contain any imperfections and that the

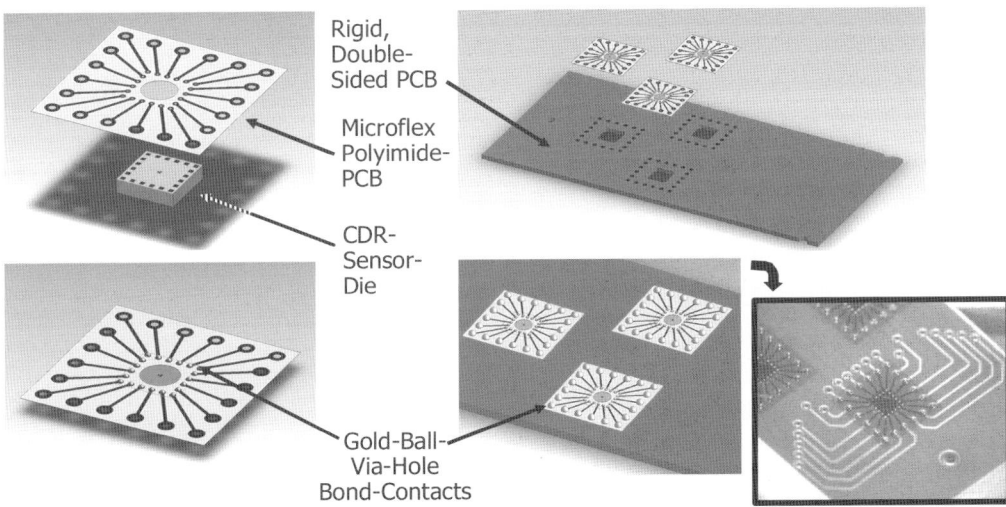

Fig. 2: Electrical interconnection and assembly of the CDR sensors to the PCB

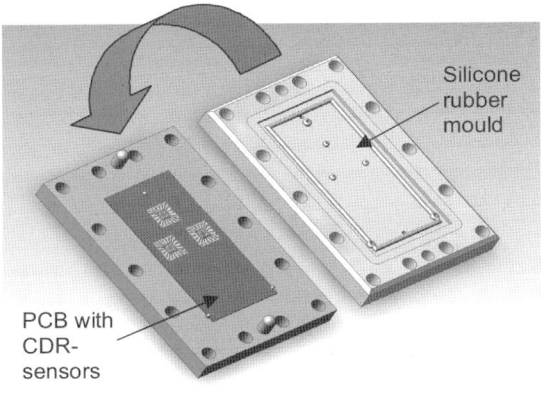
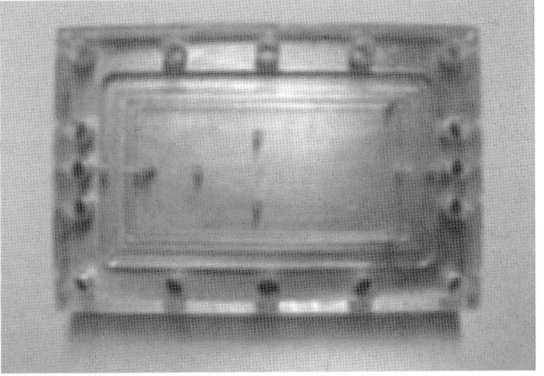

Fig. 3: Device for the coating of the CDR PCB carrier with epoxy, while leaving the CDR sensor membranes uncovered. Left: schematics. Right: Photograph of the mould. One part of the mould is transparent in order to allow checking for air bubbles

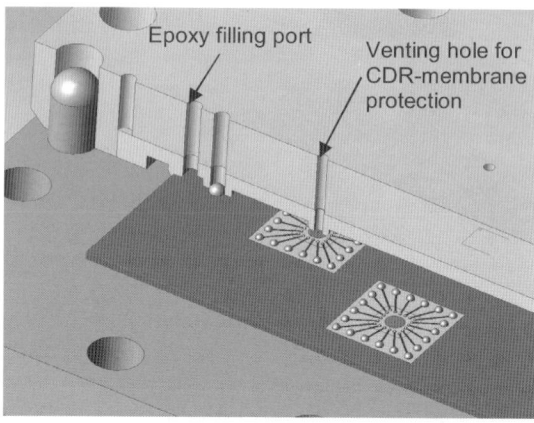 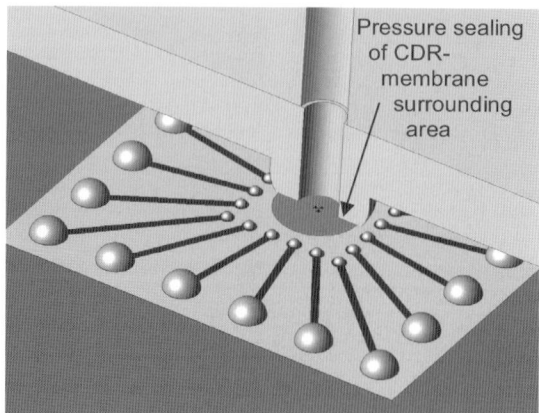

Fig.4: Details of the casting concept and device

membrane area, as desired, actually could be kept free of coating material.

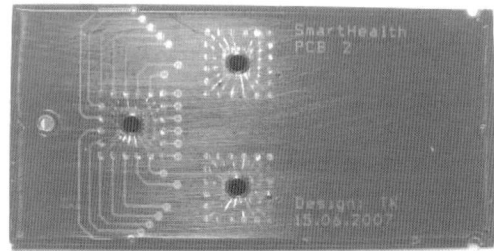

Fig. 5: Photograph of a coated PCB containing three CDR-chips.

2.2. Electrical connection to readout instrument

All electrical contacts on the backside of the PCB are led to one edge of the PCB. Once inserted into the readout instrument, the electrical contact between the PCB and the instrument is established by a pogo-pin connector (Fig. 6).

Fig. 6: Photograph of the pogo-pin connector.

The 48 Pogo-pins have a pitch of 0.8 mm. They are double-side spring loaded, thus allowing a pressure-initiated electrical connection to the PCB as well as to the signal processing board of the readout instrument. The alignment of the connector towards the respective PCBs is guaranteed by semi-circular grooves in its front side and by dowel pins at the back. The elaborate, multi-component design of the connector and its precise implementation allows adhesive connections to be avoided. Thus, single pogo-pins, which are delicate parts, can be replaced in case of damage.

The whole connector arrangement between the PCB, which has been integrated in the fluidic cartridge, and the electronics board of the readout instrument is depicted in Fig. 7.

Electrical measurements on the CDR test assembly have been performed in order to validate the quality of the electrical connection achieved with the MicroFlex method. For this purpose, some of the CDR dummy chips were completely covered with a gold layer on their surface, and then integrated and connected to the PCB as described above.

After having made the MicroFlex electrical interconnections between the chip surface and the PCB, the resistance of each of the 20 bond pad connections was measured. This was achieved by measuring the resistance between the gold surface in the centre of the dummy chip towards the pogo-pin-connection pad of the PCB, which is connected to the respective bond pad. The results showed that the resistance for a successful bond connection is less than 7 Ω, with an average of 5.27 Ω.

3. Summary and Outlook

A concept for packaging of a MEMS biochip as well as a method for its integration to a LOC has been presented. The packaged biochip provides a fluidic interface to the fluidic cartridge as well as electrical interfaces to the sensor electronics located in a readout instrument. The packaging method ensures the strict separation between the wet sensing area and the electrical interconnection area. Dummy sensor chips have been packaged in order to demonstrate the feasibility of the described packaging method. A pogo-pin connector has been fabricated and tested with a packaged dummy sensor chip.

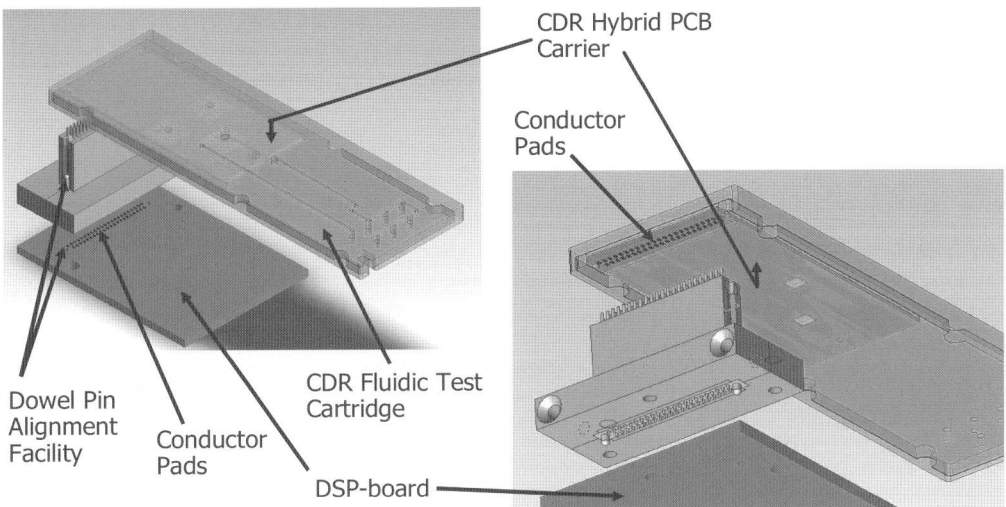

Fig. 7: Connector arrangement between PCB in fluidic cartridge and electronics board of the readout instrument

It remains to be proven that the described casting procedure ensures a robust and reliable protection of the electrical interconnects against the liquids, which will be in contact with the sensor's membrane area.

Acknowledgements

The authors would like to thank the teams of Professor Calum McNeil, University of Newcastle upon Tyne, UK, and Dr. Andrew Campitelli, MiniFAB (Aust) Pty Ltd., Victoria, Australia for the good collaboration as well as for the design of the CDR-sensor and the fluidic test cartridge, respectively. The authors are also grateful to the European Commission for funding the presented work within the framework of the SmartHEALTH research project (FP6-2004-IST-NMP-2-016817).

References

[1] Meyer J.U., Stieglitz T., Scholz O., Haberer W., and Beutel H., High Density Interconnects and Flexible Hybrid Assemblies for Active Biomedical Implants, IEEE Trans. On Advanced Packaging Vol. 24, NO. 3 (2001), pp 366-374.

Multi-Material Micro Manufacture
S. Dimov and W. Menz (Eds.)

Flexible microfluidics based on commercial SU8 foils

Chantal Khan Malek and Laurent Robert

Institute FEMTO-ST/Dpt. MN2S, CNRS UMR 6174,
32 Av. de l'Observatoire, 25044 Besançon, FRANCE

Abstract

Polymer-based microfabrication technologies are gaining momentum as they enable low cost fabrication of a variety of microsystems, with major developments in optical and microfluidic systems. The use of dry film resist for microsystem applications is briefly reviewed. A method based on the lamination of commercial SU8 dry films and photolithography for the formation of flexible thin film micro-devices is presented. Fast prototyping of multi-layer microfluidic simple chips with embedded channels is reported.

Keywords: SU8, lamination, microfluidics

1. Introduction

Polymer-based microfabrication technologies are gaining momentum as they enable low cost fabrication of a variety of microsystems, with major applications in optical and microfluidic systems. Along replication techniques like hot embossing, injection moulding and casting, fabrication methods like lithography and laser ablation are amenable for prototyping microfluidic devices.

Photolithography at the microscale range is a standard fabrication technique well adapted for laboratory needs. It also involves moderate price of fabrication. Its drawbacks include the need of a mask and the limited number of suitable photostructurable materials. UV lithography using SU8 resist has demonstrated its attractiveness for producing Micro-Electro-Mechanical Systems (MEMS), Micro-Opto-Electrical-Mechanical Systems (MOEMS) or micro-Total-Analysis Systems (µTAS), in particular high aspect ratio microstructures.

Dry film technology is well established in the printed circuit board industry. Use of laminated dry film resist in microsystem engineering applications is less common, despite the multiple advantages that it can offer. In microfluidics, it is mainly used for sealing microscale components.

We report here on the use of commercially available SU8-based dry thin film for the fabrication of flexible thin-film microfluidic devices using UV photolithography and lamination technique.

2. Lamination of dry films

In a thermally-activated lamination process, a lamination sheet containing a thin layer of glue is heated and pressed onto the polymer (microstructured) workpiece to form a joint after cooling. This simple method presents several advantages like a fast implementation and low-cost, the ease of generating 3-D microstructures via stacking in a batch process and the possibility of bonding different materials and incorporating hybrid functional elements into the design such as electrodes, filter membranes, sensors, etc. In addition to just joining parts, the adhesive films can also

be structured to become functional.

Dry film resists for fast and low-cost fabrication derived from printed circuit board technologies provide several advantages over liquid resists such as:
- No liquid handling, hence no wet chemistry nor disposal problem;
- Simplicity of forming controllable film thickness, good planarity of the resist, over-wafer thickness uniformity, and inherent planarization for multilevel 3D processes as well as coating capability for substrates that are either non-planar or present complex geometries, as an alternative to spin-coating or spray-coating;
- Good adhesion to almost any substrate;
- Simple and fast fabrication process by lamination technique, not necessarily requiring a cleanroom environment;
- It is also compatible with large surface and roll-to-roll process for high volume, so it lends itself well to upscaling;
- It can also be patterned or structured with a number of techniques such as printing or imprinting techniques [1].

Several materials were laminated and used in microsystem technologies:
- Riston® for example is a photo-patternable resist in dry film form that can be thermally laminated [2, 3];
- Polyimide has also been used, and its thermal imidization enabled bonding of substrates with incorporated channel structures [4]. Polyimide in the form of a Kapton® foil covered with a very thin film of Teflon® allows for joining parts of different materials when heated. This process is now used in fluidic bio-MEMS [5, 6];
- A two-layer poly (ethylene) terphtalate / poly (ethylene) (PET-PE) tape, with PE acting as a thermal adhesive, was used for sealing PET microanalytical devices [7]. Laminate fabrication has been used to build 3-D microfluidic disposable circuits [8] as well as polymeric active parts [[9];
- Ordyl SY300 dry resist was also used for prototyping fluidic microdevices [10];
- Epoxy materials (e.g. SU-8 resin) can also be laminated [11];

- Etertec HQ-6100 laminated [12].

Other examples include:
- The double-sided adhesive transfer tape (VHBTM, 3M) tested in microsystems with multiple fluidic and electrical connections for joining an electrofluidic chip to an electromicrofluidic dual inline package [13];
- The joining with punched adhesive film (VHB acrylate, 3M) in a batch process at low temperature [14] for six level polymer microball valves composed of bonded microstructured layers (three embossed polysulfone and three FeNiCr metal layers) [15].
- Thin Zeonor film lamination.

3. SU-8 resist

SU-8 is an epoxy-based near-UV chemically amplified negative-tone photoresist which has been specifically developed by IBM and EPFL for applications requiring high aspect ratios in the microsystem area. SU-8 films span a large range of thickness in a single coat (from micrometer thick to several hundreds of micrometers) and can be patterned with nearly vertical wall profiles and high aspect ratio using standard ultraviolet exposure tools [typically, UV broadband near UV aligner (320 nm-420 nm)]. The technology is relatively cost-effective for the production of accurate structures in polymer compared with other techniques such X-ray lithography or dry etching techniques using inductively coupled plasma.

SU-8 has become a popular structural material employed in the construction of microfluidic-based and BioMEMS devices. It has been used as an adhesive bonding material for systems built in SU-8 or other materials such as silicon or glass. It has also been employed as a mould material for several replication techniques [16] and in particular for casting for which it is frequently used to produce a master in the fabrication of microfluidic chips in PDMS [17].

Most applications require enclosure of microchannels and cavities. Most methods rely on bonding a cover to predefined microchannels. The main requirements are to prevent the microfluidic features from leaking, provide sufficient mechanical strength for the device and not interfere with the process taking place in the microfluidic system. Each group employs customized techniques to form enclosed channels based on SU-8, such as bonding. Closed or embedded channels have been fabricated by a number of techniques:
- Adhesive bonding between SU-8 layers using rigid substrates without intermediate layers [18, 19, 20] or with an intermediate layer, or using a flexible substrate such as Kapton® as a carrier to SU8 [21, 22] or lamination of SU-8/PET films [23];
- Diffraction-based effects in proximity-mode photolithography [24];
- Curing duration and dose control [25];
- Using an embedded mask [26];
- Using inclined exposure UV-photolithography [27];
- Using a sacrificial filler material to form channel space [11, 28];
- Laminating Riston® on SU-8 to form channel space [3];
- Using a direct-write proton beam to partially expose SU8 to form buried channels [29];
- Using a UV-laser system [30];
- Controlling the UV exposure depth while using anti-reflection coating to absorb the reflection UV light on SU-8 [18];
- Enclosed microchannels could also be produced by low-energy e-beam exposure [31].

Tuomikoski and Franssila (2005) [22] demonstrated the manufacture of free-standing SU-8 chips with enclosed microchannels in a three-layer process involving SU-8 to SU-8 adhesive bonding and sacrificial etching. Fluidic inlets were made by adding one lithography step, eliminating through-wafer etching or drilling. These free-standing SU-8 chips were mechanically strong and show consistent wetting and capillary filling with aqueous fluids.

4. Experimental

The principle of the lamination process is illustrated in Fig. 1. In these first experiments, no dedicated equipment was used: a PDMS cylinder was manually rolled over the substrate instead of using a roll mill.

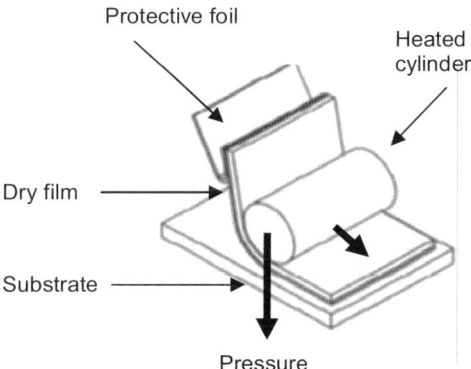

Fig. 1: Principle sketch of the roll lamination process.

Foils of 20 µm thickness (XP Microform™ 1000DF20 from Microchem) were used. The protective foil on the face to adhere to the substrate was removed prior to lamination of resist against the substrate. The dry SU-8 resist film was applied manually on the substrate heated at 65°C by a hot plate. The temperature was controlled but the pressure could not be controlled. It was then baked like standard SU-8.

The resist was exposed by UV contact exposure through a mask. The exposures could be performed with the protective layer in place but removing it before exposing the SU8 film yielded a better definition of the microstructures.

The exposure of one SU-8 foil was performed with 210 mJ/cm^2. The resist was then baked at 65°C for 4 min and at 95°C for 4 min following the technical datasheet provided by MicroChem using a heating and cooling ramp. A post-exposure bake was performed in the same conditions as the prebake step. The resist film was developed by immersion in the standard of developer of the SU-8 resist, propylene glycol monomethyl ether acetate (PGMEA) for a few minutes. Removal of the protective sheet was required before development. Fig. 2 shows a one-layer SU-8 fluidic circuit on a silicon substrate produced by this method.

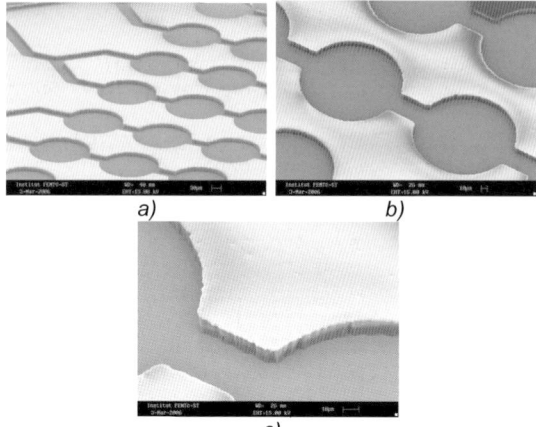

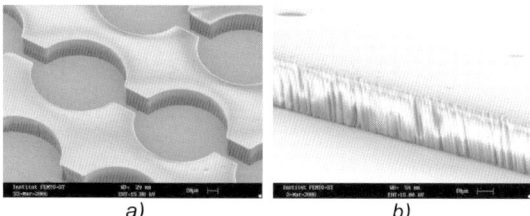

*Fig. 2 a-c): SEM pictures of one 18 µm thick SU-8 foil
after exposure and development;
b)-c) close-up views.*

Thicker resist layers could also be prepared by repeated application of the SU8 foils. Fig. 3-a) and 3-b) show such a stack of two foils. The foils were laminated on top of each other at 65°C with the top protecting layer removed. The two-foil system was exposed at 230 mJ/cm². This resulted in a seamless interface between the sheets as can be seen in the picture of Fig. 3-b).

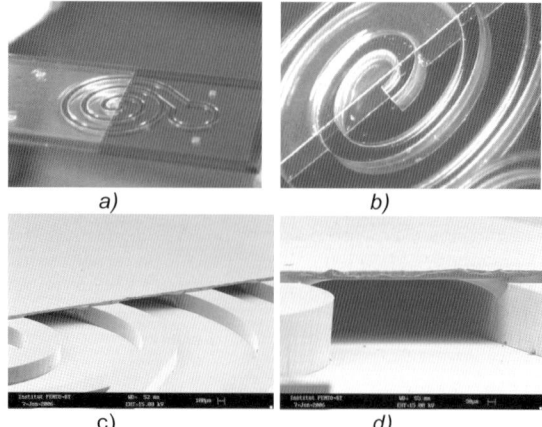

*Fig. 3: a) SEM picture of a microstructured stack of 2
layers of SU-8 (36 µm thick);
b) detail of the sidewall after development.*

The sidewall aspect with the SU-8 one- or two-foil process shows a somewhat columnar structure in contrast to what is obtained with spin-coated SU8 film (from SU8 conventional liquid resist) (Fig. 3-b). Small dips are also observed on the surface and seem to be linked to the intrinsic granularity of the SU-8 foil material which is visible on virgin foils.

A device consisting of a first fluidic level of channels and reservoirs and a second level corresponding to a cover was produced (Fig. 4). The fluidic level was produced by spin-coating SU-8 2075 liquid resist followed by a photolithography step using a standard process. The cover was formed by laminating a 20 µm thick SU-8 foil while keeping the protection layer on top of the fluidic level. The sealing of cover to channels results from the flow of SU-8 as illustrated in Fig. 4-d).

The fabrication of multilayer devices for more complex fluidic structures involving several stacked layers is challenging.

*Fig. 4: Microfluidic device in SU-8:
height of fluidic level: 275 µm, width 1.2 mm;
cover: 20 µm thick SU-8;
a)-b) optical pictures of the device (a)
and close-up view of a channel (b);
c)-d) SEM picture of the channel with cover (c)
and close-up view (d).*

A self-supported two-layer chip was also produced (Fig. 5).

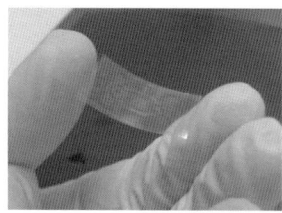

*Fig. 5: Optical picture of a self-supported two-layer
SU-8 microfluidic chip with enclosed channels.*

The temperature of lamination was reduced to 50°C for 2 min for the formation of a multiple stack of foils. Each lithographic level with microfluidic channels were formed by exposure with 200 mJ/cm², then baked at 50°C-55°C for 10 min and developed in the PGMEA developer for 3 min. The protective layer was removed by a brief pulling gesture to avoid cracks.

Two levels of embedded microchannels could be produced in a stack of four layers of SU-8 foils (72 µm). A total of 13 SU-8 layers were laminated on top of each other.

Fig. 6 illustrates this device and the two levels of channels (Fig. 6-b and c). One issue is to prevent void formation at the layer interface, as can be seen on Fig. 6-c). Another issue is to avoid filling of the channels, especially with small channels, in particular with multiple-layer stacks.

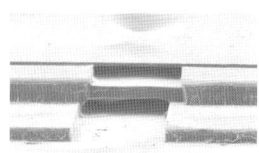

c)

Fig.6: a) SEM picture of a four-layer microfluidic device in SU-8 consisting of two levels of micro-channels built by lamination and photolithography. The channels are 100 µm wide, and the thickness of each layer 36 µm (two foils);
b) and c) close up-view of the channel levels.

5. Conclusion

We have investigated the use of commercial laminated SU8 dry films for both the formation of a microfluidic circuit and its cover. Two- and four-layer microfluidic chips with embedded channels were produced successfully using successive steps of lamination and photolithography. This method enables the fabrication of monolithic multilayer microfluidic-based chips in SU-8: it uses simple tools, is compatible with IC-based processes and makes the fabrication process easier and faster than previous methods used by the various groups (as recalled in paragraph 2).

Acknowledgements

Guillaume Pécoud is thanked for his help in developing the process during his engineering training.

This work was carried out within the framework of the EC Network Of Excellence "Multi-Material Micro Manufacture: Technology and Applications (4M)" (EC funding FP6-500274-1; www.4m-net.org).

References

[1] Velten T., Schuck H., Richter M., Klink G.. Bock K., Khan Malek C., Polster S., and Bolt P., "Microfluidics on foil", 2nd Int. Conf. on Multi-Material Micro Manufacture (4M), Grenoble, 20-22 sept. 2006. Proc. W. Menz, S. Dimov and B. Fillon (Eds.), Elsevier (2006) pp. 313-317.
[2] Zhu et al. 1996   Zhu J., Holmes A. S., Arnold J., Lawes R. A., Prewett P. D., "Laminated dry film resist for microengineering applications", Microelectronic Engineering 30 (1996) pp. 365-368.
[3] Heuschkel M. O., Guerin L., Buisson B., Bertrand D., and Renaud P., "Buried microchannels in photopolymer for delivering of solutions to neurons in a network," J. Sens. Actuators B 48(1–3) (1998) pp. 356–361.
[4] Metz S., Holzer R., and Renaud P., "Polyimide-based microfluidic devices," Lab on a Chip 1(1) (2001) pp. 29–34.
[5] Paul] B. K., and Peterson R. B., "Microlamination for microtechnology-based energy, chemical, and biological systems", Proc. ASME Int. Mechanical Eng. Congress and exposition, Nashville, Tennessee, Nov. 15-20 (1999), AES 39 (1999) pp. 45-52,.
[6] Bargiel S., Walczak R., Knapkiewicz P., Gorecka-Drzazga A., and Dziuban J. A., "Micromachined silicon-glass dosing device with built-in conductivity detector", Eurosensors XIX, Barcelona, Sept. 11-14thCatalonia,Spain (2005).
[7] Roberts M. A., Rossier J. S., Bercier P., and Girault H., "UV laser machined polymer substrates for the development of microdiagnostic systems," Anal. Chem., 69 (1997) pp. 2035–2042.
[8] Weigl B.H., Bardell R.L., and Cabrera C.R., "Lab-on-a-chip development," Adv. Drug Delivery Rev., 55(3) (2003) pp. 349–377.
[9] Truongand T.Q. and Nguyen N.-T.,"A polymeric piezoelectric micropump based on lamination technology,"J. Micromech.Microeng. 14 (2004) pp. 632–638
[10] Vulto P., Glade N., Altomarre L., Bablet J., Del Tin L., Medoro G., Chartier I., Manaresi N., Tartagni M., and Guerrieri R., "Microfluidic channel fabrication in dry film resist for production and prototyping of hybrid chips," Lab on a Chip 5 (2) (2005) pp. 158–162.
[11] Guérin L. J., Bossel M., Demierre M., Calmes S., and Renaud Ph., "Simple and low cost fabrication of embedded micro-channels by using a new thick-film photoplastic," in Proc. IEEE Transducers (1997) pp. 1419–1421.
[12] Stephan K., Renaud L., Kleinmann P., Pitet P., Morin P., and Ferrigno R., "Fast prototyping using a dry film photoresist: microfabrication of soft-lithography masters for microfluidic structures", J. Micromech. Microeng. 17 (2007) N69-N74.
[13] Galambos P., Benavides G.L., Okandan M., Jenkins M.W., and Hetherington D., "Precision alignment packaging for microsystems with multiple fluid connections," in Proc. 2001 ASME Conf., New York, Nov. 11–16 (2001) pp. 1–8.
[14] Harrison C., Cabral J. T., Stafford C. M., Karim A., and Amis E. J., "A rapid prototyping technique for the fabrication of solvent-resistant structures," J. Micromech. Microeng. 14 (2004) pp.153–156.
[15] Fu C., Rummler Z. and Schomburg W., "Magnetically driven microball valves fabricated by multilayer adhesive film bonding," J. Micromech. Microeng. 13 (2003) pp. 96–102.
[16] Edwards T. L., Mohanty S. K., Edwards R. K., Thomas C., and Frazier A. B., "Rapid tooling using SU-8 for injection molding microfluidic components", SPIE Micro Fluidic Devices and Systems Conference, Santa Clara, CA, September (2000) pp. 82-89.
[17] Duffy D. C., McDonald J. C., Schueller O. J. A., and Whitesides G. M., "Rapid prototyping of microfluidic systems in poly(dimethylsiloxane", Anal. Chem., 70 (1998) pp. 4974-4984.
[18] Chuang Y.-J., Tseng F.-G., Cheng J.-H. and Lin W.-K., "A novel fabrication method of embedded micro channels by using SU-8 thick films photoresists", J. Sens. Actuators A 103 (2003) pp. 64–69 .
[19] Blanco F. J., Agirregabiria M., Garcia J., Berganzo J., Tijero M., Arroyo M. T., Ruano J. M., Aramburu I. and Mayora K., "Novel three-dimensional embedded SU-8 microchannnels fabricated using a low temperature full wafer adhesive bonding", J. Micromech. Microeng., 14 (2004) pp. 1047-56
[20] Tuomikoski S., Tikanen T., Ketola R., Kostiainen R., Kotiaho T., and Fransssila S., "Fabrication and optimization of enclosed SU-8 tip structures by electrospray ionization mass spectrometry",. 9th Int. Conf. on Miniaturized Systems for Chemistry and Life Sciences, Boston, Oct. 9-13, 2005, Proc. µTAS (2005) pp. 982-984.
[21] Song Y., Kumar C. S. S. R. and Hormes J., "Fabrication of an SU-8 based microfluidic reactor on a PEEK substrate sealed by a 'flexible semi-solid transfer' (FST) process, J. Micromech. Microeng., 14 (2004) pp. 932-940.

[22] Tuomikoski S. and Franssila S., "Free-standing SU-8 microfluidic chips by adhesive bonding and release etching", J. Sensors and Actuators A 120 (2005) pp. 408-415.

[23] Abgrall P., Lattes C., Conédéra V., Dollat X., Colin S., and Gué A. M., "A novel fabrication method of flexible and monolithic 3D microfluidic structures using lamination of SU-8 films", J. Micromech. Microeng. 16 (2006) pp. 113-121.

[24] Gaudet M., Arscott S., Camart J. C., and Buchaillot L., " SU-8 based arch-like microfluidic microchannels using single mask/single step photolithography", 9th Int. Conf. on Miniaturized Systems for Chemistry and Life Sciences, Boston, Oct. 9-13, 2005, Proc. µTAS (2005) pp. 1189-1191.

[25] Gracias A., Xu B., and Castracane J., "Fabrication of three dimensional microchannels in SU-8", 9th Int. Conf. on Miniaturized Systems for Chemistry and Life Sciences, microTAS 2005 , Boston, Oct. 9-13, 2005, Proc. µTAS (2005) pp. 663-665

[26] Alderman B. E. J., Mann C. M., Steenson D. P., and Chamberlain J. M., "Microfabrication of channels using an embedded mask in negative resist", J. Micromech. Microeng. 11 (2001) pp. 703-705.

[27] Yang R., Williams J. D., and Wang W., "A rapid-micromixer/reactor based on arrays of spatially impinging micro-jets", J. Micromech.Microeng. 14, (2004) pp. 1345-1351.

[28] Chung C. and Allen M., "Uncrosslinked SU-8 as a sacrificial material", J. Micromech. Microeng. 15 (2005) N1-N5.

[29] Tay F. E. H., van Kan J. A., Watt F., and Chong W. O., "A novel micro-machining method for the fabrication of thick-film SU-8 embedded micro-channels", J. Micromech. Microeng. 11 (2001) pp. 27-32.

[30] Gueit A., Sharon A. and Li B., "Scanning laser produces functional microfluidic structures at a single SU-8 layer", 9th Int. Conf. on Miniaturized Systems for Chemistry and Life Sciences, Boston, Oct. 9-13, 2005, Proc. µTAS (2005) pp. 199-201.

[31] Kudryashov V., Yua X. C., Cheong W. C., and Radhakrishnan K., "Grey scale structures formation in SU-8 with E-beam and UV" J. Microelectronic Eng. 67-68 (2003) pp. 306-311.

Microfabrication of Components for a Novel Biomimetic Neurological Endoscope

A. Schneider[a], L. Frasson[b], T. Parittotokkaporn[c], F. M. Rodriguez Y Baena[b], B. L. Davies[b], and S. E. Huq[a]

[a]*Science and Technology Facilities Council, Rutherford Appleton Laboratory, Technology – Central Microstructure Facility, Harwell Science and Innovation Campus, Didcot, OX11 0QX, UK.*
[b]*Mechatronics in Medicine Lab., Depart. of Mechanical Engineering., Imperial College, London, SW7 2AZ, UK*
[c]*Institute of Biomedical Engineering, B 422 Bessemer Building, Imperial College, London SW7 2AZ, UK.*

Abstract

The development of a novel biomimetic neurosurgical probe is inspired by nature. Some insects have spines with a unique surface texture which enables them to penetrate tissue more easily. This surface texture consists of cutting teeth and fin-like pockets on the spine. Instead of drilling, the insect slides its spine into the fibre through the reciprocating motion of independent segments. Applying the same or similar microtexture to a miniaturized neurosurgical endoscope could improve existing tools for brain surgery and brain biopsy. The development of such endoscope could minimize the damage caused by inserting the probe whilst avoiding the risk of buckling, which is a common occurrence when thin flexible probes are axially loaded.

To replicate the surface microtexture, teeth and fin-like high-aspect-ratio microstructures were fabricated. Different geometries of these fins and teeth were studied for insertion into tissue so that the texture could be characterized for friction and tribological interaction with tissue. For these tests, free-standing long and narrow strips with microstructures in up to 525 μm thick SU-8 were designed, fabricated, and mounted onto prototypes made by stereolithography. This paper focuses on the fabrication of the microtextured strips. The required geometry of these strips can cause considerable bending. The structures were investigated regarding fabrication and stress conditions.

Keywords: SU-8, microfabrication, high-aspect-ratio, biomedical device, endoscope, prototype, neurology

1. Introduction

In neurosurgery, brain biopsy is a diagnostic procedure that involves the collection of a small quantity of "undefined" (benign or malignant) tissue from a specified region (usually a lesion) of the brain. Such biopsies are carried out in vivo by using endoscopes. For various reasons, flexible and steerable endoscopes have advantages over conventional rigid endoscopes. Currently commonly used endoscopes for brain biopsy consist of a stainless steel tube, which is several centimetres long and comprises up to four channels including one to accommodate optical fibres (for illumination and viewing) and a working channel for the surgeon's minimally invasive instruments (e.g. forceps, cutter). Conventional endoscopes typically come in a range of outer diameters between 3 mm and 6 mm [1,2].

A novel flexible endoscope, which is currently being developed in collaboration between Imperial College London and Rutherford Appleton Laboratory in the UK, will improve the manipulation of the probe so that the affected brain tissue area can be better targeted. Investigations towards a miniaturized probe are currently underway. In addition to the flexible mechanical properties of the probe, the microtexturing of the probe's surface is currently been investigated in an attempt to ascertain whether small patterns on the probe's outer walls could aid the insertion process whilst minimising tissue damage. Inspired by nature, the surface texture consists of cutting teeth and fin-like microstructures mimicking the spine of a wood boring insect. Such insects have spines with surface textures which enable them to penetrate tissue more easily. Instead of drilling, the insect slides its spine into the fibre by a reciprocating motion. Various shapes of the microstructured probe surfaces have been fabricated for testing by dynamic indention into cadaveric brain tissue and dummy tissue samples (gelatine-based), which have close properties to brain tissue. This work focuses on the elastic and stress properties of several prototype microtextured surfaces. Further work is also underway to investigate the effect of different surface topographies on the friction and tribological properties of a prototype probe during indentation into tissue and gelatine.

2. Microfabrication of samples

2.1. Mircotexture of SU-8 test strips

Test strips 75 mm long and 0.9 mm to 1.4 mm in width with high-aspect-ratio (HAR) microstructures of SU-8 were fabricated using UV-lithography. Microstructures of tooth and fin structures with characteristic lateral size of 500 μm, 250 μm, 100 μm, 50 μm, and 10 μm (Fig. 1) were exposed and developed to create a texture of the side wall along these strips. HAR up to 50:1 of these microstructures was chosen due to the requirements of the subsequent penetration experiments by the neurology team. Such microstructures at the sidewall emulate a typical microtexturing of the ovipositor of the wood wasp *Sirex noctilio* [3].

For the exposure of the micropattern by UV photolithography, the photoresist SU-8 2150 (MicroChem Ltd.) was applied and a Karl Suss MA6 mask aligner was used. The photoresist was either spincoated by a Suss Gyrset spinner or directly poured / cast on 4 inch Si wafer substrates. Due to the high viscosity of the SU-8 2150 resist, the direct

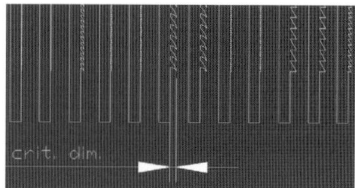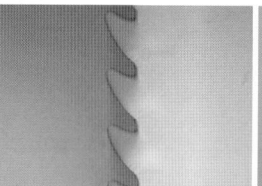

Fig. 1: Detail of fin and tooth structure design (left). Arrows indicate addendum between strip and fin tip – the typical critical dimension. Centre pictures show typical fin and tooth profile of SU-8 structures of critical dimension larger than 10 μm (top view in optical microscope). Profile of structures with 10 μm size on the right.

casting of SU-8 is the preferred method to achieve uniform coatings on the whole wafer [4]. Hence the required resist thickness could be achieved in a one-step process without the necessity of multi-coating and alignment. Process parameter such as exposure dose, soft bake, post exposure bake (PEB), resist development in EC solvent, and hard bake were varied to customise process conditions for the test strips (in particular to control the stress in the strips). After developing exposed SU-8, the microstructures were release from the substrate using chemical etching processes. A sacrificial layer was applied to the substrate in advance to facilitate this release process. SU-8 photoresist is based on a multifunctional bisphenol A novolak epoxy resin, which is considered to be biocompatible [5].

2.2. Assembly of strips in stereolithography core probe

As an initial prototype of the neurological endoscope, SU-8 strips were manually inserted and fixed to a cylindrical needle so that the SU-8 microtexture is protruding from the needle surface. This needle – a core with grooves in axial direction along the cylinder surface (Fig. 2) – was made by microstereolithography (MSTL) using an *Envisiontec PERFACTORY SXGA+W/ERM Mini Multi Lens System* by applying high resolution MSTL acrylic-based resin R11 [6]. Such MSTL system would also be capable to fabricate similar textured needles as one entire integral object but this object would be inappropriate for future endoscopes because the cured material is very rigid and not biocompatible. For initial in vitro tests on cadaveric pig brain or dummy material (gelatine), probes assembled from SU-8 strips in a MSTL core needle are better suitable to investigate the rheological properties for the small geometry of the surface microtexture for a potential endoscope. Alternative flexible MSTL resins with shore hardness 75A to 100A,

Fig. 2: SEM micrograph of tip of MSTL core needle with insertion grooves.

which would be basically more apt for a functional prototype of the neurosurgical endoscope, do not achieve such high optical resolution as R11 or R5 resin [6] so that features much smaller than 100 μm cannot

be fabricated. Furthermore, comparable biocompatible MSTL resin has a low optical resolution as well. It may be difficult to fabricate undercut structures like fins in this material.

For manufacturing of such endoscopes in future, injection moulding would be an interesting option but again flexible structures with dimensions on the micro scale with undercutting features would be difficult to use for rapid demoulding from the mould insert, in particular for undercut features like fins.

3. Results and Discussion

3.2. SU-8 test strips

Several SU-8 strips, with thickness ranging from 125 μm to 525 μm, were fabricated with a side-wall texture according to the process outlined in chapter 2.1. The side-wall profile of a typical strip - after it was released from the Si substrate - is shown in a photo (Fig. 3) taken by a scanning electron microscopy (SEM). Suitable tooth and fin structures with an addendum (distance between strip and tooth/fin tip) as small as 50 μm related to HAR of 10:1 were accomplish whereas structures with 10 μm characteristic feature size were clearly observed but the shape of fin and tooth microstructures became indistinct. This is caused by a) imperfections in the photomask, which are transferred into the structures, and b) the limiting resist properties (UV exposure of structures well beyond HAR 10:1).

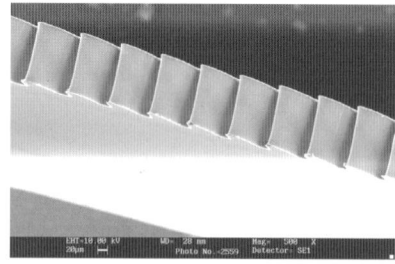

Fig. 3: SEM micrograph of side-wall profile of 150 μm thick SU-8 strip after released from substrate; addendum size of fin structures is 50 μm.

3.2. Stress condition in SU-8 test strips

Because of the required length (75 mm) of a single SU-8 strip, any stress built up in SU-8 causes strain, which leads to a considerable and visible bending of the entire strip. Such bent strips are shown in Fig. 4. Extreme bending can make the assembly of strips into the core MSTL needle enormously difficult. Basically such stress and strain in the SU-8 resist is caused by a difference in quantity of cross-linked SU-8 epoxy groups throughout the resist layer from top to bottom. This can be attributed to the exponential decay of the exposure dose from the resist surface to the interface of resist / substrate. An attempt for a theoretical description of an exposure model as well as stress model in HAR SU-8 structures was made for UV lithography [7,8]. For evaluating the degree of cross-linking across the SU-8 layer thickness, also the entire UV photocuring mechanism (acid formation, diffusion, temperature dependency of polymerization) has to be taken into account. The concentration of an acid, which is created by photons during exposure and which acts as catalyst for cross-linking during the PEB, depends on the local exposure dose in the resist. Hence more bonds occur in areas with higher acid concentration (for instance at the top of the resist) and cause a strong epoxy polymerization and higher stress level. The PEB was carried out on a hot plate, with ramping of the

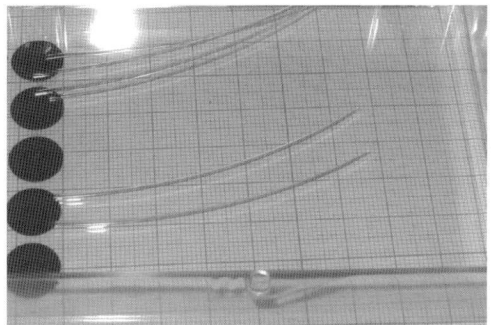

Fig. 4: Several 75 mm long strips of SU-8 structures with extreme bending.

temperature from room temperature to 90°C within a few minutes and after 25 min again cooled down slowly until the sample temperature fell at least below 50°C. So a potential temperature gradient from top to bottom in the resist layer could also occur during cross-linking process. This must be considered to understand the resulting stress gradient and the bending of SU-8 structures fully.

After exposure and development, the SU-8 strips were removed from the substrate. The bending of the strips was measured and the bending radius was calculated. The diagram, shown in Fig. 5, puts the reciprocal value of the bending radius in relation to thickness of strips and exposure dose. A small reciprocal value indicates a minute bending, whereas a

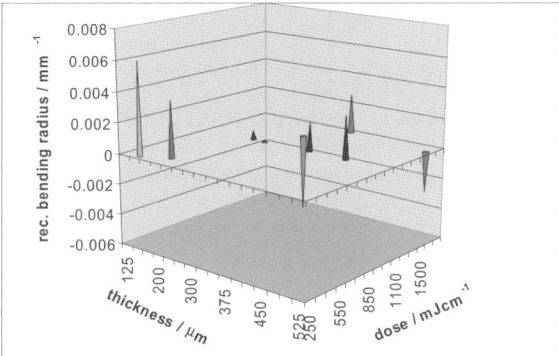

Fig. 5: Graph shows the reciprocal bending radius for 9 SU-8 strips fabricated under different conditions. Samples from the same exposure series (same dose and similar PEB condition) are indicated by the same colour of cones. Magenta and turquoise cones specify samples exposed to a hard bake at 150°C after development and hence strips bend downwards.

large figure (small radius) specifies a large distortion of the strip. Most of the samples have an upwards bending (away from the substrate); but also two samples, which were exposed to a hard bake on a hot plate at 150°C for 15 min after development, bend downwards instead. Samples with a thickness between 250 µm and 275 µm even at a high exposure dose of 1000 mJ/cm^2 - but omitted a hard bake - have the lowest bending. Thicker SU-8 strips also exposed to 1000 mJ/cm^2 show stronger bending, which could be explained by a higher gradient of cross-linked SU-8 bonds consistent with the exposure dose (higher dose difference between resist surface and bottom compared with thinner SU-8 layers). However, other samples, exposed by lower dose, exhibit larger bending than those flat samples. From the present sample series no obvious tendency in bending characteristic in relation to the exposure dose can be concluded. Interesting is the point that hard baking at high temperature reverses the bending. However, suitable process conditions were found for the fabrication of SU-8 strips, which can be assembled to the MSTL needle now. Fully assembled probes are ready for penetration tests at Imperial College.

3.3. Initial tests with assemble probe in gelatine

Twelve SU-8 strips each with the same side-wall profile were assembled into a penetration probe. Currently tests on gelatine, which has nearly identical elasticity properties as brain tissue, are on the way. Fig. 6a-c show details of a MSTL needle with inserted SU-8 strips and the MSTL needle mounted into the test rig at Imperial College. Tests will investigate the different frictional resistance depending on the surface profile and texture.

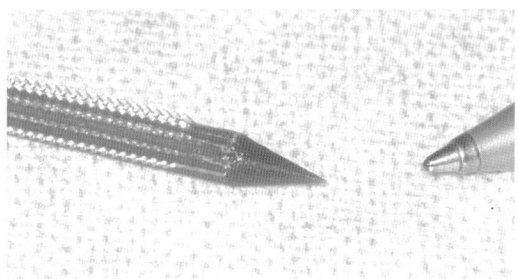

Fig. 6a: On the left SU-8 strips with 500 µm fin structures inserted and glued to a MSTL core needle – for size comparison, tip of a ball pen on the right side.

Fig. 6b: Detail of the SU-8 fin structures mounted into the MSTL needle above.

4. Conclusions

This paper outlines the fabrication of SU-8 test strips, which could eventually be used for a biomimetic neurosurgery probe. A fin or tooth-like outer-surface topography can improve both the friction and tribological properties of the probe during penetration of the endoscope into tissue. This will, in turn, result in reduced damage of brain tissue during e.g. a biopsy procedure. This paper focused on the fabrication of the microstructures out of thick SU-8 photoresist. Generally, in neurosurgery target areas in the brain may lie several centimetres deep with respect to the key-hole aperture in the skull. Thus, the probe consists of long narrow strips of HAR SU-8. Side-wall profiles with detailed geometries down to 50 µm in size and HAR 10:1 were fabricated for test probes. Smaller structures of approx. 10 µm in size (HAR up to 50:1) were still feasible but a clear distinction between tooth and fin structures was no longer possible.

Fig. 6c: Assembled probe mounted into test rig at Imperial College.

The fabrication parameters have a considerable effect on the bending of the probe's strips and the stress conditions in the strips need to be optimised. With regards to the lithography process, critical parameters to be considered are, PEB and hard bake temperature, whereas the exposure dose of the resist does not show conclusive results. The resulting SU-8 test strips were assembled in a prototype probe made by stereolithography, which was subsequently used for indentation tests.

Acknowledgements

The authors would like to acknowledge the financial support of STFC for the Technology Partnership programme TP/07/07 "BIOLOGICALLY INSPIRED MICROTEXTURING". This work was also carried out within the framework of the EC Network of Excellence "Multi-Material Micro Manufacture: Technologies and Applications (4M).

References

[1] Kelliher S., Minimally invasive endoscopic surgery can safely remove deep brain tumors, Cornell University News Service (Sept. 30, 2005) http://www.news.cornell.edu/stories/Sept05/neurosurgery.html

[2] Perneczky A., Aesculap Neurosurgery (2006) 1-60,http://www.tmml.com/Catalogue/SellSheets/A19_INFO_MINOP_SYSTEM.PDF

[3] Vincent J. F. V. and King M. J., "The Mechanism of Drilling by Wood Wasp Ovipositors." Biomimetics 3(4) (1995) 187-201.

[4] Cui Z., Jenkins D. W. K., Schneider A., and McBride G., "Profile Control of SU-8 photoresist Using Different Radiation Sources", Proc. SPIE 4407 (2001) 119-125.

[5] Voskerician G., Shive M. S., Shawgo R. S., von Recum H., Anderson J. M., Cima M. J., Langer R., Biocompatibility and biofouling of MEMS drug delivery devices, Biomaterials 24 (2003) 1959-1967.

[6] envisionTEC GmbH (Al Siblani, Managing Director), http://www.envisiontec.de/05mater.htm

[7] Lawes R., Manufacturing tolerances for UV LIGA using SU-8, J. Micromech. Microeng. 15 (2005) 2198-2203.

[8] Feng R., Farris R. J.. Influence of processing conditions on the thermal and mechanical properties of SU8 negative photoresist coatings. J. Micromech. Microeng. 13 (2003) 80–88.

Investigation of Material Compatibility for Embedding Stereolithography

T. Rechtenwald[a], A. Kopczynska[b], E. Schmachtenberg[b], M. Devrient[a], T. Frick[a], M. Schmidt[a]

[a] *Bayerisches Laserzentrum GmbH, Konrad-Zuse-Str. 2-6, 91052 Erlangen, Germany*
[b] *Lehrstuhl für Kunststofftechnik, Friedrich-Alexander Universität, Erlangen, Germany*

Abstract

Decreasing sales figures and increasing demand of different variants at the same time, as well as the desire of a short time to market pose a challenge for nowadays manufacturing technology. In this context, a new flexible production technology for mechatronical devices, which include also optical functions, called Embedding Stereolithography (eSLA) is introduced. eSLA combines the flexibility to automatically generated inner and outer complex geometries of conventional SLA with embedded functional components, which are conducted by generative manufactured electrical and/or optical structures. To form a rugged mechatronical device out of mechanic and/or electronic parts by eSLA a sufficient wetting of the used components by the processed liquid photopolymer is needed. Therefore the surface tension and viscosity of different photopolymers is measured and compared to surface energies and surface roughness of relevant component materials. Afterwards the characteristics of wetting of the chosen photopolymers on these materials are discussed.

Keywords: Rapid Prototyping, mechatronics

1. Introduction

The increasing number of distributed intelligent sub-systems e. g. in mobile systems like cars and the upward usage of mechatronical modules leads to more and more complex systems. This trend is accompanied by the proliferation of variants and a decreasing order quantity per variant [1].

In contrast to this, the conventional production technologies are perfectly fit for mass production but not for the challenges of the future. These conventional production technologies like injection moulding for technical enclosures or stamping for electrical lead frames as well as the extensive procedure for printed circuit boards involve the need for complex and mainly cost-intensive tools. Due to this, they are strongly restricted in flexibility.

Generative production technologies on the other side have already shown a high potential to fabricate individual but also small series of products [2]. Especially the possibility to generate very complex geometries can help to save costs. This can be established e. g. by reducing the number of parts in the mechanical design and thus reducing the number of tools as well as the costs [3]. But this focuses only on the geometry of parts.

The selection of conventional Stereolithography (SLA) as the basis process out of the variety of generative processes was made because SLA is not dealing with a thermal phase change, the accuracy of the parts is good and the system technology is flexible for modifications.

1.1. Conventional Stereolithography

Like all other types of generative processes SLA enables the physical creation of a part from a designed CAD model. The digital data of this model has to be placed and orientated in the virtual building space of the stereolithographic machine. Further a filigree load carrying construction is generated automatically by the data preprocessing software. This construction supports in the next step the hardened and thus densified overhanging parts of the working piece. As the building operation is performed layer wise the geometrical data of the part and the support construction has to be sliced in adequate digital layers with a height of typically 0.1 mm. One of these layers is fabricated by spreading a liquid photo-resin over the building platform with a spreading knife. Afterwards the radiation of an UV laser is used to cure the material on the surface of the resin bath according to the given layer information. Finally the building platform is lowered. By repeating this procedure the complete work piece is generated layer by layer. After the generative process the part is cleaned from some adherent liquid resin and irradiated with artificial UV light in a post curing apparatus (PCA) for a certain time [4].

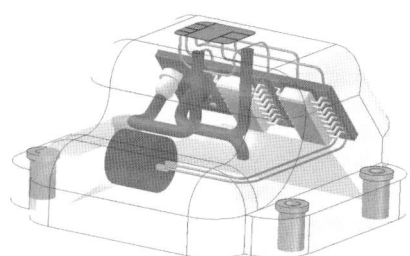

Fig. 1. Example of an aimed eSLA module

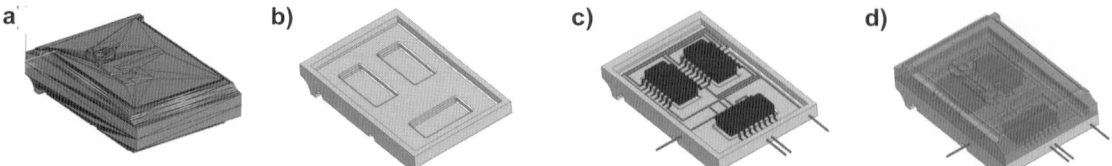

Fig. 2. Starting point is a complete digital design (a). The sup-processes are: Generation of the lower housing (b), placing of functional components and generation of electrical/optical conductive structures (c), generation of the upper housing (d).

1.2. The Idea of Embedding Stereolithography

The basic idea of eSLA is to extend the flexible generation of complex geometries to mechanical, electrical and optical functionality for single and small series products. On basis of conventional SLA a new hybrid production technology is investigated, which enables the production of 3D component assemblies within one generative production chain. Fig. 1 gives an impression what kind of modules is targeted. Fig. 2 illustrates the general procedure. In a first step a lower housing with cavities is generated by conventional SLA. In these cavities functional components like lead frame assemblies or printed circuit board assemblies as well as peripheral components e. g. capacitors or optical transceivers can be placed. As a further extension to conventional SLA electrically and/or optically conductive structures shall be generatively manufactured. Finally the upper housing is fabricated. This 2.5 dimensional approach should be further extended to a real three-dimensional one in a future project phase.

Considering the described process chain the first question on material compatibility appear for the first new application of resin on a placed component. If the resin is forming drops instead of a homogeneous flat layer on the surface material of the component, the light of the laser will cure the drops and thus the recoating of the next layer will be prevented. The second question addresses the adhesiveness, which can be established between the matrix and the components. Here we can distinguish between two cases. The first case is the adhesiveness on the side and bottom surfaces of a component, where the resin is only passively cured by PCA or by UV portion within daylight over time. The second case is the adhesiveness on the top surface, where it is assumed, that the adhesive strength can be controlled by the energy dose, which is brought in by the UV laser. Also the influences of electrically and/or optically-conductive-structure generating processes on the matrix material have to be considered.

All these questions on material compatibility are based on the basic effect of wetting. Thus this work investigates the most important parameters of wetting: surface tension, polar and disperse components of the surface tension, viscosity of liquid SLA resins and surface roughness of solid substrates. Wetting itself is quantified by determining the actually resulted contact angles of interesting liquid SLA resins on a selection of possible component materials.

3. Materials and Methods

3.1. Selection of component and matrix materials

The selection of component materials for this work is driven by the common materials in electronic production. For all important basic elements like conductors, circuit boards, packaging, corrosion protection and also blankly used semi conductor material, representative materials are chosen, see Table 2 materials column.

For the matrix materials the selection was done by considering epoxy-based SLA resins, which tend to imitate the mechanical properties of technical plastics in electronic production, like Vantico SL 5240 (Huntsman Advanced Materials, Basel, Switzerland) for PP or Somos 11110 (DSM , Elgin, USA) for ABS. For exact generated cavities AccuGen 100 (3D Systems, Rock Hill, USA) is chosen. Besides this, Proto Tool 20L (DSM, Elgin, USA) is well known for its very high temperature resistance up to 200 °C.

3.2. Wetting qualities and parameters

Surface theory distinguishes different quality of wetting behaviour depending on the contact angle α [5], see Fig. 3. The case, that no contact angle can be measured ($\alpha \cong 0$), is called spreading. Contact angles up to about 30 ° are presumed to be a good wetting state. For a defined or sufficient wetting contact angles between 30 ° and 90 ° are found. Contact angles of 90 ° and above are considered as partial or insufficient wetting. The lesser the contact angle, the better the liquid resin wets the solid surface.

This qualitative statement can be correlated with the relation of the surface tensions of the involved materials. It is stated that, the higher the surface tension of the substrate is above the surface tension of the liquid, the wetting is expected to be better [6]. Metals are known as materials with the highest surface tensions beyond 1000 mN/m, followed by semi

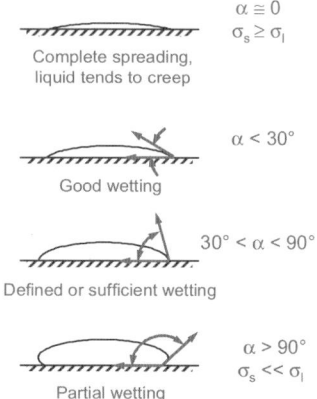

Fig. 3. Characteristic wetting angles (α)
σ_l – surface tension of liquid resin
σ_s – surface tension of solid surface

conductors and ceramics. Polymers have typically surface tensions below 60 mN/m. From these facts it is expected, that liquid SLA resins shall show a very good wetting behaviour on metals, semi conductors and ceramics. Only for certain polymers, which show a smaller surface tension, the wetting behaviour is seen critical. A low viscosity as well as a high surface roughness is expected to support a good wetting behaviour.

3.3. Determination of wetting parameters

Surface tension and wetting angle are determined by using the temperature controlled surface tension measurement equipment OCA 20 from Data Physics, Filderstadt, Germany. The surface tensions of the liquid SLA resins is measured by using a standardised gold stamp (sessile drop method) and analysing the geometry of a drop of the sample fluid by the Young-Laplace-Equation [7]. This method is selected due to its suitability for high viscous liquids like SLA resins. The contact angles between the solid surface and the SLA resin drop are measured directly with a tangent method at a constant temperature of 21.5 °C by placing a drop of the sample liquid on a sample substrate. The measurement is carried out over 120 s with a sampling rate of 2.0 Hz. The surface tensions of component materials are taken from literature [7, 8]. Surface roughness is measured by the confocale multi-pinhole-method using a µsurf apparatus from Nanofocus, Oberhausen, Germany, which allows the analysis of roughness parameters not only regarding one surface profile line (R_a values) but also average over surface elements (S_a values). All values on viscosity of the liquid resins are given in the data sheets of the manufacturers.

4. Results and Discussion

4.1. Wetting parameter of resins and components

Table 1 shows relevant data on the liquid SLA resins. All SLA resins show a surface tension between 35 mN/m and 42 mN/m, also the high viscous Proto Tool as well as the low viscous Somos 11110.

Table 2 shows important data of the solid component materials. The regarded metals show surface tensions above 1000 mN/m, the silicon is expected to have about 750 mN/m from literature and the polymer materials showing values between about 20 and 60 mN/m.

4.2. Contact angles of liquids on solid substrates

Table 3 shows the actually resulting contact angles of the liquid SLA resins on the considered solid component materials after 10 s. This time interval is assumed between application of a new resin layer and the start of laser curing.

Regarding the conductor materials first, contact angles between 35 ° and 70 ° are found over the considered SLA resins. Also the value for the high viscous Proto Tool is within this range. Single exception is shown for the low viscous Somos 11110. The copper sample shows despite its higher value of surface tension a worse wetting behaviour. But both metals feature sufficient wetting behaviour. In contrast, the materials, which are as critically classified according to

Table 1
Surface tension and viscosity of liquid matrix materials

SLA resins	Surface tension in mN/m	Viscosity at 30°C in mPa•s
AccuGen100HC	42.4	500
Accura Si45 HC	35.6	475
SL 5240	38.6	375
Somos 8110	40.5	600
Somos 11110	36.9	92
Proto Tool 20L	40.8	2500

their surface tension e. g. PP or solder resist, show also sufficient wetting. Similar as seen for the metals, the wetting with the high viscous Proto Tool is not worse but for the low viscous Somos 11110 it is slightly better compared with the other SLA resins. Also for the circuit board materials defined or good wetting angles are found only exceeded by the semi conductor material. Versus the viscosity of the liquid resins the wetting behaviour is qualitatively similar as already described.

With this approach surface quality of the substrate materials is still not reflected. But surface quality is also expected to effect the wetting behaviour as already mentioned. For example, the sufficient wetting behaviour of solder resist can be explained by the significant roughness of its surface. In contrast to this, the relatively high contact angles of aluminium can not be understood by its relatively rough surface. As well as by comparing copper and the circuit board material FR4, whose showing a very similar surface roughness, this model of explanation is not sufficient.

4.3. Time dependent development of contact angle

The wetting behaviour of a liquid over time is not the same on all substrate materials. Fig. 4 shows the time dependent characteristics of one SLA resin on different component materials. Despite fast wetting of the metals, the resulting contact angle after 120 s keeps in the range of just sufficient or insufficient wetting. Also PP shows only sufficient wetting. Apart from solder resist all other materials show a very similar time dependent characteristic.

Table 2
Surface tension and roughness of component materials

Function	Materials	Surface tension in mN/m	Roughness S_z in µm
Conductor	Aluminum (AlMg3)	approx. 1200	0.551
	Copper	approx. 1850	0.216
Circuit board	PEEK (foil)	46	0.028
	PI (foil)	50	0.021
	FR4	47	0.215
Packaging	PP	29	0.119
Corrosion protection	Solder resist	approx. 30	0.496
Semi conductor	Silicon	approx. 750	0.011

5. Conclusion

Liquid SLA resins show similar values of surface tension but differ significantly in viscosity. The component materials, which are a selection out of different types of materials used in electronic production, show consequently the complete range of possible surface tension. This range is positioned quite above and slightly below the surface tension of SLA resins.

The common approach to expect a good wetting behaviour for substrates with a much higher surface tension could confirmed e. g. for silicon with a surface tension of 750 mN/m. All SLA resins with surface tensions of about 40 mN/m show a good wetting behaviour on silicon. But this approach could not be fully confirmed for all of the substrate materials, even by considering different surface qualities and viscosities of the SLA resins. A possible explanation for this could be the influence of oxidation or contamination with foreign material, which could be left on the surface of e. g. copper by the chemical lamination manufacturing process of the used unstructured conventional laminated circuit boards.

Despite this, the quality of wetting behaviour of the SLA resins on most of all component materials is between spreading and sufficient wetting. Also for component materials, which were expected to be critical, due to a lower surface tension than the SLA resins. So it can be concluded, that the application of a layer of SLA resin on components can be carried out sufficiently even to permit generally the embedding of functional components within the process of Embedding Stereolithography and thus the effort to adapting the surface of components with additives to adjust the surface tension for a better wetting behaviour is not necessary. The different time dependent wetting behaviour, which affects the necessary time between application of the resin layer on an embedded component, was not found to be critical.

Acknowledgements

This research was supported by Deutsche Forschungsgemeinschaft (DFG) through the Sonder Forschungsbereich (SFB) 694 "Integration elektronischer Komponenten in mobile Systeme" (Integration of electronical Components in mobile Systems).

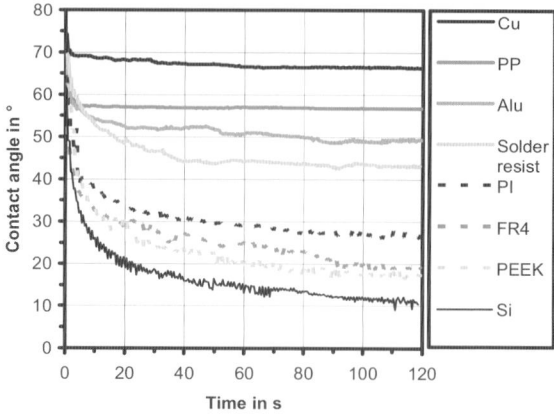

Fig. 4. Time dependent characteristic of AccuGen 100 on selected component materials

References

[1] Runge, W.: Electronics inside Transmission – Chances and signes of Mecha-tronic Control moduls. In: Deutsches IMAPS Seminar, Göppingen, Februar 2000.

[2] Ayers, K.; Hilbrandt, F.J.: The usage of Stereolithographic Parts as the Final Product. In: 3D Systems North American Stereolithography User Group Meet-ing, San Antonio, Texas, March 1–5, 1998.

[3] Wohlers, T. (Editor): Wohlers Report 2003. Fort Collins, Colorado 80525 USA

[4] Jacobs P.F.: Rapid Prototyping & Manufacturing, Fundamentals of Stereo-Lithography. SME, Dearborn, 1992, ISBN 0-87263-425-6.

[5] Comyn, J.: Contact Angles and Adhesive Bonding. Int. J. Adhesion a. Adhesives 12 (1992) 145–149.

[6] G. Habenicht: Kleben - erfolgreich und fehlerfrei; Vieweg Verlag, Wiesbaden, 2006.

[7] Kopczynska, A; Ehrenstein, G. W.: True Surface Tension In Solids and Melts. In The Journal of Materials Education, 29 (2007) 3-4, p. 325-340

[8] Janocha, B.: Change to the wetting and adsorption of plastics-water interfaces under the influence of external electrical fields and plastic surface polarity. Doctoral thesis, University of Tübingen, Fraunhofer IRB Publisher, 1998

Table 3

Contact angles of SLA resins on selected component materials, angles < 30 ° are highlighted grey

Contact angle in °	Proto Tool	AccuGen 100	Vantico SL5240	Accura Si45 HC	Somos 8110	Somos 11110
Aluminium	46,4	54,2	39,7	36,1	52,3	24,9
Copper	60,3	68,7	55,3	45,6	73,3	39,9
PEEK (foil)	28,9	32,7	33,8	27,3	29,5	18,3
PI (foil)	34,5	38,1	24,5	24,5	27,4	17,4
FR4	28,6	32,9	29,3	24,9	32,7	14,0
PP	56,8	57,3	57,0	53,4	54,5	42,5
Solder resist	42,0	52,8	37,6	35,9	46,5	26,5
Silicon	19,8	25,9	18,4	23,0	26,0	13,0

Electric fields in a hybrid batch fluidic micromanipulation concept

P. Lazarou[a], N.A. Aspragathos[a], E. Jung[b]

[a] Robotics Group, Department of Mechanical Engineering and Aeronautics,
University of Patras, Patras T.K. 26500, Greece
[b] Chip Interconnection Technologies, Fraunhofer IZM Berlin, Germany

Abstract

Micromanipulation is a very important issue in several fields of technology (microelectronics, optoelectronics & MEMS device packaging). Current implementations do not provide both sub-micron accuracy and movement of parts over centimeter-scale to a ~100µm final alignment precision. A micropart-inside-a-liquid-droplet manipulation concept that manages to bridge the gap from meso via the micro to the sub-micron scale in a fully contained process has been previously introduced by integrating the phenomena of electrowetting, dielectrophoresis and fluidic self-assembly. In this paper, an investigation of the electric fields that drive the manipulation of the droplet and micropart during the stages of electrowettng and dielectrophoresis is presented. Information for critical factors such as electrostatic force, Maxwell stress and surface charge density distribution is provided. Their effect on the manipulation process is verified, in accordance to theory.

Keywords: MEMS, micropart manipulation, simulation, dielectrophoresis, electrowetting, self-assembly

1. Introduction

Micromanipulation holds great significance in many modern technological fields and especially in microelectronic packaging [1], optoelectronic packaging [2] as well as MEMS device packaging [3]. Also, new emerging fields such as biotechnology and genomics development involve biological sample manipulation [4]. A number of techniques have been developed, dedicated to their specific target applications and with unique benefits in their specific area as well as limitations to their suitability in other fields (e.g. laser based tweezers [4], surface tension assisted microassembly [5], ligand binding assisted microassembly [6]). These techniques offer either alignment precision in the sub-micron area or movement of parts over centimeter-scale to a ~100 µm final alignment precision, but fail to bridge the gap from meso via the micro to the sub-micron scale in a fully contained process.

A micromanipulation concept that allows the filling of this gap has been described in [7], by joint effects of electrowetting [8,9], dielectrophoresis [10] and fluidic self assembly effects [11]. Microparts inside liquid droplets are manipulated from a macro scale delivery container down to sub-micron post alignment precision onto a temporary carrier. Batch microassembly using various templates with different parts to form a full microsystem is then performed.

This paper investigates and analyses the electric fields that drive the droplet/micropart manipulation during the electrowetting and dielectrophoresis stages of the proposed micromanipulation process. The magnitude of the electrostatic force acting on the droplet, electric potential, Maxwell stress and surface charge density distribution as well as their dependence on the applied electrode voltage are determined. These results verify the way the two fields affect the droplet manipulation, in accordance to the electrowetting and dielectrophoresis theory.

2. Concept description

A 4M internal concept *for fluid assisted batch self-assembly of microparts* (Fig.1) has been presented in our previous work [7,12]. This concept relies on the manipulation of a liquid droplet containing only one micropart, instead of the microparts themselves. It involves three stages: 1) an electrowetting based system for fast transportation of the droplets-microparts close to the final position of microassembly, 2) electric fields induced by additional electrode configurations that centre each droplet in their final assembly position providing precise placement (dielectrophoresis stage) and 3) the fluidic self-assembly procedure that takes place due to the tendency of minimization of the interfacial free energy of the system of droplet, micropart and binding site surfaces.

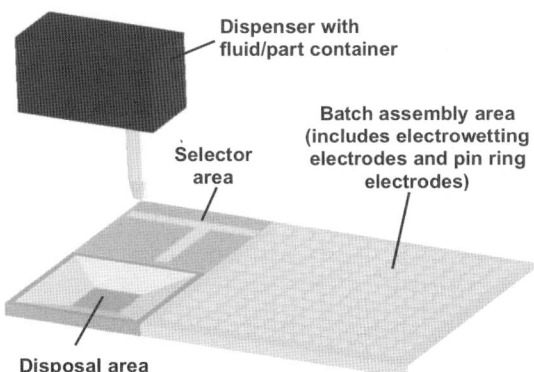

Fig. 1. Schematic of a manipulation platform for industrial use, blue: dispenser with fluid/part container, green: selector area, red: disposal area, gray: assembly area.

A conventional drop dispenser [13] is used to dispense from a suspension of microparts in deionized water (providing both high surface tension and dielectric constant) a drop of water containing one single micropart. Camera vision is used to determine, if the droplet contains one part, several parts or no part at all. Only the droplets containing one part will be guided by the subsequent control of the electrowetting electrodes to the assembly area. The other droplets are guided into a dispose or reuse area and can be collected back into the fluidic container. Indicative CAD model images of the perceived concept are shown in Fig. 2 and 3.

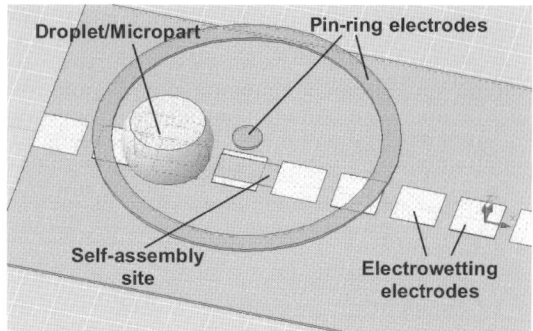

Fig. 2. 3D CAD model of one row of electrowetting electrodes at the bottom layer (square electrodes), pin-ring electrodes at the top layer (transparent except for the pin-ring) and droplet containing one micropart sandwiched between the bottom and top layers. The self-assembly binding site is the rectangle on the bottom layer.

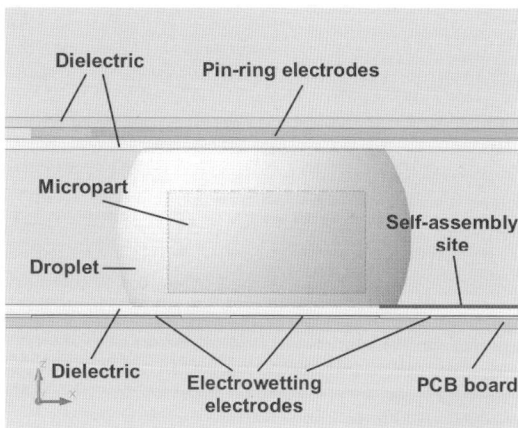

Fig. 3. Cross-section view with the various layers of the CAD model, droplet and micropart.

In the electrowetting stage, the motion of the droplet-part takes place by controlling the voltage of the transport electrodes over a planar hydrophobic surface. The droplet is transported with velocities of up to several cm/s inside the pin-ring electrodes. These generate a radial electric field that accurately positions the droplet on the desired self-assembly position (vertically coinciding with the pin electrode).

After the droplet has been centered on the pin, passive alignment of the part to the binding site by surface energy minimisation takes place (self-assembly). Successive droplet vaporization assists the process by minimizing of the "free" fluidic volume and constraining the micropart more tightly to the self assembly zone.

The objective of this concept is to *position and properly align microparts in parallel on a wafer*. The process can be repeated for other wafers and with subsequent layer by layer bonding the final batch of microsystems can be created (Fig. 4). '1' and '2' are wafers with aligned microparts from the described concept. Using wafer align/bonding equipment the final system is created.

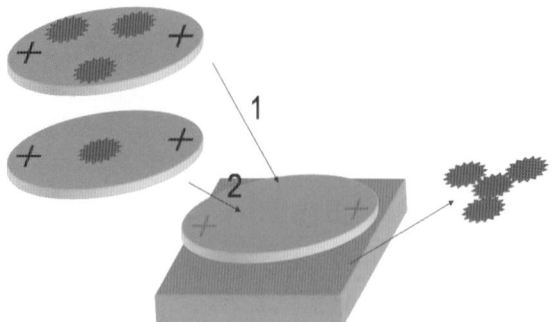

Fig. 4. Transfer and bonding of wafers with microparts in order to form a microsystem.

3. Simulations of electrowetting and dielectrophoresis manipulation stages

In these simulations, an approximately 150 μm diameter-sized water droplet is considered sandwiched (as seen in Fig. 2 and 3) between two Parylene-C dielectric insulator layers, coated with a hydrophobic/superhydrophobic film (e.g. Teflon) in order to reduce the wettability of the surface. The droplet's volume is about 2 nl, its density ρ = 998,23 kg/m^3 and carries a total net charge Q = 0 C. Two FEM models have been created, one for each of the manipulation stages.

3.1 Electrowetting model

In this model, underneath the bottom dielectric layer lie square-shaped copper electrodes of 75 μm length with a spacing of 25 μm (current PCB manufacturing technology allows 75 μm sized pads and 25 μm spacings by laser cutting). One electrode is energized, while the rest are grounded. The voltage applied to the energized electrode is used as a parameter with a step of 20V, up to 100V. The vertical distance between the two dielectric layers is h=100 μm.

Fig. 5 shows a cross-section view of the model, with the electric field created by the energized electrode (the second one from the right) for a voltage value V=100V. The electrowetting effect causes an accumulation of charges in the droplet/insulator interface near the energized electrode, resulting in an interfacial tension gradient across the gap between the adjacent electrodes, which consequently causes the transportation of the droplet.

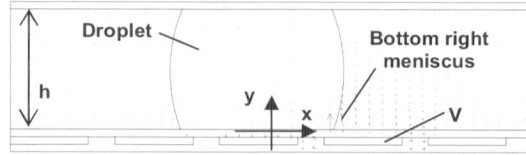

Fig. 5. Cross-section view of the electrowetting droplet manipulation stage and the electric field of the energized electrode with voltage V.

Since the droplet does not have any free charges, the charge accumulation is in reality the appearance of bound charges because of the polarization due to the electric field (Fig. 6). The normal component of the electric displacement **D** is continuous across the interface of water/dielectric (top, bottom) and water/air (left, right). The density of surface bound charge at these *interfaces* is given by the formula [14]:

$$\sigma_b = (P_{1n} - P_{2n}) \cdot n \qquad (1)$$

where P_{1n} is the normal polarization vector at the interface inside the water droplet, P_{2n} is the normal polarization vector at the interface in air or dielectric (outside) and **n** is the normal vector at the interface pointing outside the droplet. Eq. 1 effectively describes the discontinuity of the normal polarization vector at a boundary by the amount σ_b. Fig. 7 shows these vectors at a random point at the bottom right water/dielectric interface.

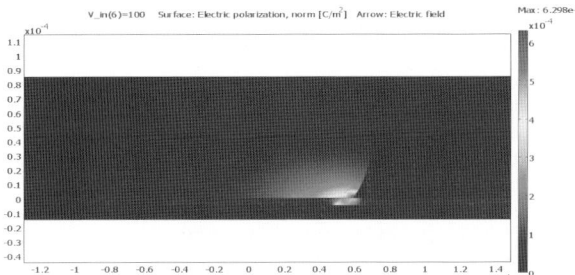

Fig. 6. Droplet polarization due to the induced electric field from the energized electrode.

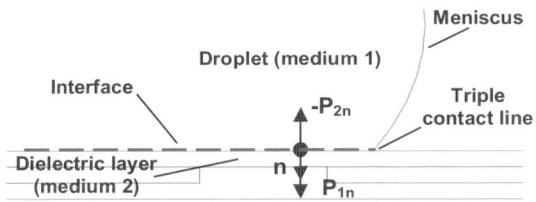

Fig. 7. Normal polarization vectors at a random point at the bottom right water-dielectric interface.

The distribution of the normal polarization vectors along the x axis near the energized electrode for V=100V is shown in Fig. 8. At the point where x=5.9 µm (y=0) lies the triple contact line, where the bound charge density is the greatest with a value of $(6.703 - 2.442) \times 10^{-4} = 4.261 \times 10^{-4}$ C/m^2 according to Eq. 1. This is expected as it is the point closest to the energized electrode. At positions nearer to the beginning of the coordinate system (shown in Fig. 5) the polarization diminishes and so does the surface charge density. The Maxwell stress at the triple contact line reaches 1.05 KPa for the voltage of 100V and rapidly decreases as we move upwards along the meniscus boundary (Fig. 9).

Integrating the Maxwell stress tensor over the two menisci boundaries, the values 57.85×10^{-4} N/m (direction to the right) and -1.49×10^{-4} N/m (direction to the left) are calculated for the right and left menisci respectively, showing a considerable electric pressure difference between the two sides of the droplet. Integration of the electric energy density over the droplet area results in a value of 5.7 J/m. This electrical energy is added to the total surface energy of the droplet. It is understandable that the droplet is bound to move rightwards, in the position of the energized electrode, due to the tendency of total energy (surface and electrical) minimization, in accordance to the existing work [9].

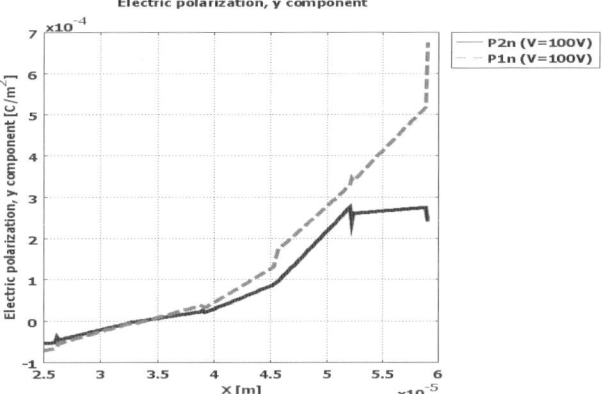

Fig. 8. Normal polarization vectors along the x axis at the bottom right interface of water/dielectric layer.

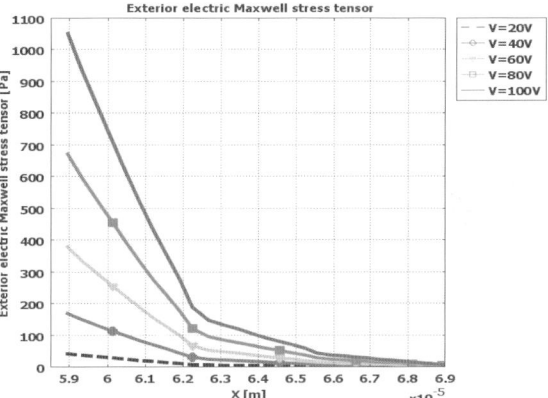

Fig. 9. Maxwell stress on the bottom right meniscus along the x axis, for various voltage values.

3.2 Dielectrophoresis model

In the dielectrophoresis model (2D- top down view), the droplet lies inside a pin-ring copper electrode configuration. The pin electrode has a radius of 25 µm and the ring has an inner radius of 220 µm and an outer radius of 250 µm. The applied voltage to the electrodes is used as a parameter, ranging from 100V to 300V (dc), with a 50V step. The pin and ring electrodes have a negative and a positive voltage V respectively.

Fig. 10 shows this electrode configuration, the droplet – assumed to have been transported with the electrowetting effect- inside the pin-ring region, the distribution of the potential and the electric field's streamlines. It has to be noted that the field is radial (non-homogeneous), directed towards the pin.

As mentioned, the droplet is neutral in terms of charge. However, induced surface and space charges appear due to the non-homogeneous radial field of the electrodes [7]. A re-orientation of the molecular dipoles according to the field takes place and the droplet gets polarized.

The bound surface charge density can be calculated at the points of the droplet closest to the pin and the ring by using Eq. 1. The distribution of the total polarization along a cross-section cut of the droplet along the x axis can be seen in Fig. 11. Due

112

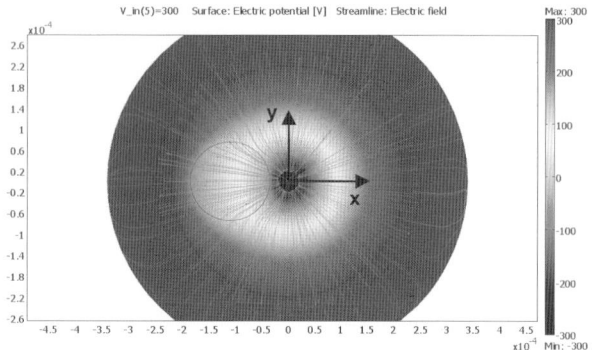

Fig. 10. Top-down view of the pin-ring electrodes, droplet, potential distribution and streamlines of electric field for V=300V.

to the fact that the droplet is surrounded by air, the polarization outside the droplet is zero and thus $P_2=0$. The charge density at *these two points* can be calculated directly from the plot (due to symmetry in the specific droplet position the tangential component $P_{1t}=0$ and $P_{1n}=P_1$). Thus the values of bound charge density at the point nearer the ring is -0.713×10^{-3} C/m^2 (**n** points to the left) and at the point nearer to the pin 3.299×10^{-3} C/m^2 (**n** points to the right) for V=300V.

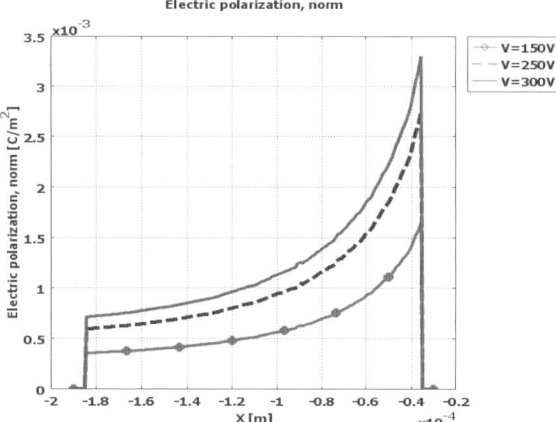

Fig. 11. Polarization vector along the x axis, for various voltages.

As a result of the non-homogeneity of the electric field, there is a significant difference in the electrostatic pressure on each side of the droplet, as seen in Fig. 12. For V=300V on the left side of the droplet (nearer to the ring) a pressure of -7.31×10^{-4} atm (~74.1 Pa) is exerted, whereas on the right side (nearer to the pin), the pressure is 1.644×10^{-2} atm (~1.666 KPa). An integration of the Maxwell tensor over the upper half boundary of the droplet yields a total of $+0.019874$ N/m (direction towards the right).

This asymmetry in the electric pressure distribution is the driving mechanism of the droplet's movement. Alternatively, the sum of electrostatic forces acting on the charges nearer to the pin is greater than the sum of the forces acting on the charges nearer the ring. The total electrostatic force acting on the droplet in the x direction is 3.996×10^{-2} N and 5.903×10^{-6} N in the y direction. It is obvious that the droplet is bound to move towards the pin electrode, where it will rest, due to the symmetry of the electric field.

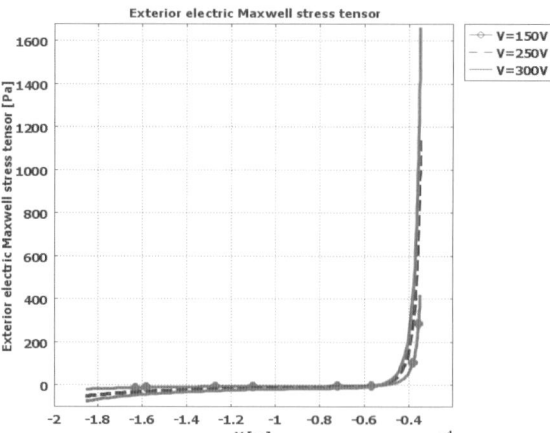

Fig. 12. Maxwell stress of the upper half of the droplet over position x, for various voltage values.

One issue that should be investigated is under what electrode voltage the droplet is able to overcome the water-hydrophobic film surface tension components that tend to resist and retard any motion of the droplet. According to [15], there is a threshold pressure necessary to overcome this frictional ("sticky") tension and has been calculated to be in the order of ~10^{-2} atm for a horizontal capillary tube of height h=100 µm. The report however is based on experimental data and does not explicitly correlate this threshold to contact angles or surface roughness.

From the presented results, the Maxwell pressure difference in the advancing and receding side of the droplet can overcome the threshold of 10^{-2} atm only when the electrode voltage is higher than 240 V. Another issue that arises with this range of voltages is that in the case of a real implementation where the electrodes are planarized and the droplet moves on top of the dielectric layer, the material of the dielectric as well as its thickness should be considered, in order to avoid a dielectric break-down. However, increasing the dielectric's thickness or selecting a stronger one directly affects the intensity of the electric field and thus the droplet actuation. More research and actual, experimental data with the pin-ring configuration are needed in order to see if this pressure threshold is indeed of the afore-mentioned order and under which voltages the required droplet actuation is provided.

4. Conclusions & future work

In this paper, the results of an investigation of the electric fields that drive the droplet/micropart manipulation during the stages of electrowetting and dielectrophoresis are presented. Surface charge distribution, Maxwell stress, polarization, electrostatic force, and pressure difference in the advancing and receding sides of the droplet are calculated. The way the induced electric fields provide actuation to the droplet is explained and verified, in accordance to the existing theoretical work.

Future work will include simulations with a joint 3D model for both stages, investigation of the actual conditions that provide droplet actuation as well as the experimental verification of the concept with the implementation of a functional demonstrator.

A successful implementation will allow high speed, high volume placement of micro-components, on wafers, performing the above mentioned process flow in parallel with a multitude of droplets and controlling these by switching electric fields instead of moving heavy mechanical placement heads.

Acknowledgments

This work was carried out within the framework of the EC Network of Excellence "Multi-Material Micro Manufacture: Technologies and Applications (4M)".

References

[1] G. Gengle, "Fluid self assembly as applied to electronics packaging", Workshop on "Flexible Electronics: Materials, Characterization, Application", Halle, Germany, 2.-3. June 2003.

[2] Zheng, Jacobs, „Fabrication of Multicomponent Microsystems by direct three dimensional self assembly", Adv. Funct. Mater. 2005, 15, No. 5, May 2005, pp. 732-738.

[3] Srinivasan, Helmbrecht, Muller, Howe, "MEMS-Some Selfassembly Required", Optics & Photonics News, November 2002, pp. 20-24.

[4] Berns, Michael, "Laser Scissors and Tweezers", Scientific American. April 1998 V. 278, Issue 4.

[5] Boehringer, "Modeling and Controlling Parallel Tasks in Droplet-Based Microfluidic Systems", IEEE TRANSACTIONS ON COMPUTER-AIDED DESIGN OF INTEGRATED CIRCUITS AND SYSTEMS, VOL. 25, NO. 2, FEBRUARY 2006.

[6] Bock, Bleier, Koethe, Landesberger, "New Manufacturing Concepts for Ultra-Thin Silicon and Gallium Arsenide Substrate", 2003 GaAsMANTECH, Inc. 2003 International Conference on Compound Semiconductor Mfg.

[7] P. Lazarou, N. Aspragathos, E. Jung, "Micropart Manipulation by Electric Fields and Batch Self-assembly", IWMF 2006, 5th International Workshop on Microfactories, Besancon, France, October 25-27 2006.

[8] I. Moon, J. Kim, "Using EWOD (electrowetting on dielectric) actuation on a microconveyor system, Sensors and Actuators A: Physical, Volumes 130-131, 14 August 2006, Pages 537-544.

[9] Lee J., Moon H., Fowler J., Schoellhammer T., Kim C.-J., "Electrowetting and electrowetting-on-dielectric for microscale liquid handling", Sensors and Actuators A: Physical, Volume 95, Number 2, 1 January 2002 , pp. 259-268(10).

[10] A DEP Primer [Online]. Availiable: http://www.dielectrophoresis.org

[11] J. Fang, S. Liang, K.Wang, X. Xiong, K. Boehringer, "Self-Assembly of Flat Micro Components by Capillary Forces and Shape Recognition." 2nd Annual Conference on Foundations of Nanoscience: Self-assembled Architectures and Devices (FNANO), Snowbird, UT, April 24-28, 2005.

[12] P. Lazarou, N. Aspragathos, E. Jung, "Micropart manipulation by electrical fields for highly parallel batch assembly", 4M Conference 2006, 20-22 September 2006, Grenoble, France.

[13] Jean-Maxime Roux1, Yves Fouillet1 and Jean-Luc Achard, "Handling droplets in 3 Dimensions for Lab-on-Chip Applications", μTAS 2004, Malmö, Sweden, 26.-30.Sept. 2004.

[14] G. Pollack, D. Stump, "Electromagnetism", Addison-Wesley / Prentice Hall Publishing, 2002.

[15] G. Beni and M. A. Tenan, "Dynamics of electrowetting displays", J. Appl. Phys. 52, 6011 (1981).

Multi-Material Micro Manufacture
S. Dimov and W. Menz (Eds.)

Concept for Fluidic Self-Assembly of Micro-Parts Using Electro-Static Forces

J. Dalin[a], J. Wilde[a], A. Synodinos[b], P. Lazarou[b], N. Aspragathos[b],

[a] *University of Freiburg – IMTEK, Department of Microsystems Engineering, Georges-Köhler Allee 103, 79110, Freiburg, Germany, contact: Johan.Dalin@imtek.uni-freiburg.de*
[b] *Robotics Group, Department of Mechanical Engineering and Aeronautics, University of Patras, Greece, contact: lazarou@mech.upatras.gr*

Abstract

Self-assembly is relatively unused in industrial micro-fabrication, although it offers opportunities to simplify processes and to lower manufacturing costs. A variety of self-assembly procedures have been introduced that take advantage of various forces, e.g. capillary, gravitational, electro-static. In this paper a concept for the alignment of micro-parts on a substrate using fluidic-self-assembly with electro-static attraction is presented. Further, FEM-simulations for the electro-static alignment force are performed and its dependence on several geometric parameters, e.g. the width of the binding sites and the distance between micro-part and substrate at the binding sites, is investigated. Based on results an analytic model is extracted. Furthermore, simulations are also performed to estimate capillary alignment forces, acting on micro-parts that are self-aligned. Finally, the magnitude of electro-static and capillary forces is compared. This novel assembly concept, where the alignment of the component at the binding site is achieved due to electro-static energy minimisation and, optionally, in combination with capillary alignment, could be beneficial in the manufacturing of heterogeneously integrated MEMS, such as optical and RF micro-systems.

Keywords: MEMS, self-assembly, capillary forces, electro-static forces, simulation

1. Introduction

Self-assembly is the process of spontaneous generation of order in systems of components [1]. It is relatively unused in micro-fabrication, although it offers opportunities to simplify processes, lower costs, develop new processes, use components too small to be manipulated robotically, integrate components provided by incompatible technologies, and generate structures in three dimensions and on curved surfaces. Proof-of-concept experiments in meso-scale and micro-scale self-assembly demonstrate that this technique possesses fascinating scientific and technical challenges and offers the potential to provide access to hard-to-fabricate structures.

Fluidic self-assembly of micro-parts is a relatively new process in the field of micro-electronic fabrication. It involves the manipulation and alignment of micro-parts onto a substrate without the help of additional machinery and makes it possible to assemble a large number of small devices on e.g. a planar surface in parallel. The most commonly used technique for self-assembly utilises capillary [2,3,4] (Fig. 1) and gravitational forces.

(b) A substrate with lubricant droplets, exclusively bonded on the binding sites in water. (c-d) After the parts are dispensed into water, the parts are assembled and aligned.

Electro-static forces are also used as driving forces for self-assembly. In this method the micro-part is subjected to either coulomb forces that act on a part having a net charge in an electric field (electrophoresis) or to forces that act on an uncharged particle in an electric field due to induced dipoles (dielectrophoresis) [5]. Research on self-assembly, utilising so called electro-static "traps" (Fig. 2) has been presented by [5,6,7]. The motion and alignment of parts in these experimental setups were more directed and defined in comparison to common fluidic self-assembly techniques, which are more random.

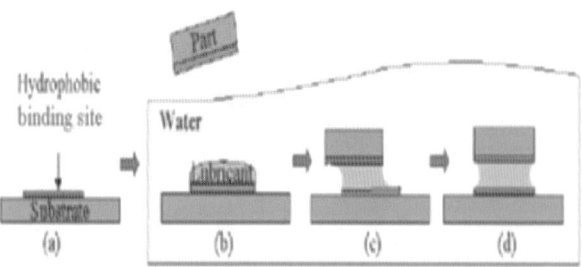

Fig. 1. Fluidic self-assembly steps [2]: (a) A fabricated substrate with hydrophobic binding sites.

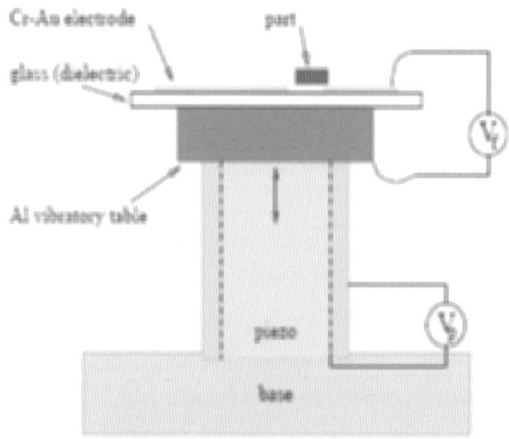

Fig. 2. Experimental apparatus for self-assembly with electro-static attraction [6].

2. Concept description

Consider several complementary patterned areas that are structured on a substrate (serving as binding sites for alignment) and a micro-part. These areas (pads) are marked with a 'B' in Fig. 3a. Between these complementary patterns a voltage potential V is applied. As a result, the patterned areas are attracted to each other due to electro-static forces. The vertical force $\overline{F}_y$ can be estimated, similarly to a parallel plate capacitor, as:

$$| \overline{F}_y | = \frac{\varepsilon_0 \varepsilon_r V^2 A}{2h^2} \qquad (1)$$

where ε_0 is the permittivity in free space, ε_r the relative permittivity, V the voltage potential, A the area of the patterns and h the dielectric gap. It is assumed that fringe effects are negligible.

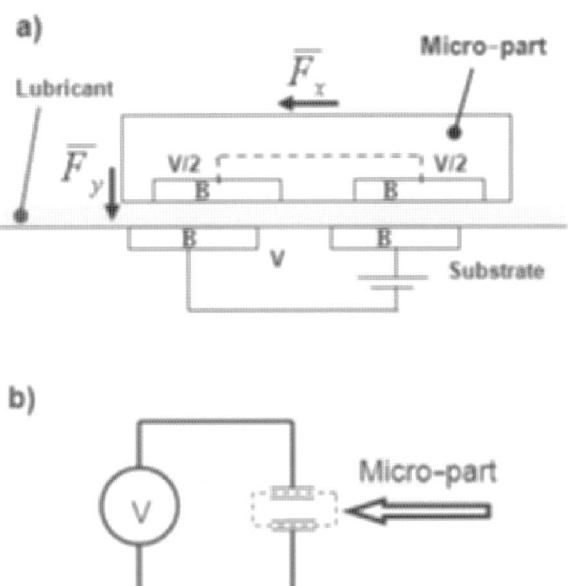

Fig. 3. Fluidic self-assembly using electro-static forces: a) Concept of utilising electro-static forces for self-assembly, b) Electric equivalent of the concept.

In case of a misalignment of the complementary patterns a force vector $\overline{F}_x$, acting on the micro-part, aligns the micro-part to the binding sites on the substrate (Fig. 3a). Since the micro-part has to move freely in the fluidic self-assembly process with some sort of agitation the voltage has to be induced capacitively. The corresponding equivalent electrical circuit is shown in Fig. 3b.

Although the configuration of Fig. 3 relies solely on the alignment due to electro-static forces, capillary alignment in conjunction with the existing electro-static alignment can be achieved by realising lubricant droplets upon the substrate's binding sites. The lubricant in Fig. 3a is a low-viscous film between the micro-part and binding sites on the substrate.

In the following sections modeling and simulations are performed in order to estimate the aligning electro-static and capillary force. The derived results are presented, along with an analytical electro-static force model. The following notations are used in the rest of the paper:

F_x, F_y: electro-static force in x and y direction
F'_x, F'_y: capillary force in x and y direction
W: width of the binding site (pad) of the micro-part / substrate.
h: distance along the y axis, i.e. vertical separation, between the pads of the micro-part and substrate.
O: offset, i.e. misalignment, between the pads in x direction.
V: applied voltage.
γ: surface tension.

3. Modeling of the electro-static alignment force

A three-dimensional model is used for the FEM-simulations in order to compute the electro-static force F_x in Ansys, where the third axis is constrained. The model consists of two blocks that represent the complementary square pads and a surrounding cylinder as the voltage ground level (Fig. 4). The voltage potential is applied at the two pads and the relative permittivity of the dielectric medium is set to 5.0.

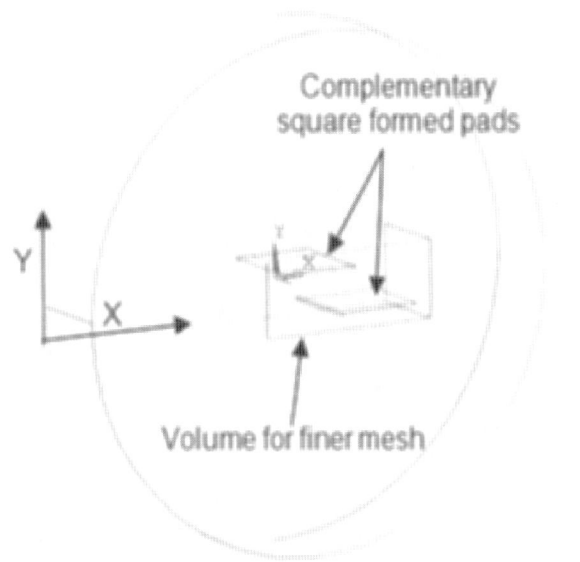

Fig. 4. FE-model in Ansys for electro-static analysis.

In a systematic way the alignment force F_X is analysed and its dependence on the various geometric parameters as well as the applied voltage is investigated. The results of FEM-simulations are presented in Fig. 5 - 8. As shown in Fig. 5, the magnitude of the resulting force increases approximately linearly with the pad's width W. Additionally, F_x is inversely proportional to h (Fig. 6), i.e. a larger separation between the binding sites results in weaker electro-static force. Its dependence on the applied voltage V can be seen in Fig. 7. It can be approximated that F_x is proportional to V^2. Finally, a Gaussian plot can describe the relation between F_x and offset O, as seen in Fig. 8. As O increases F_x increases non-linearly. At the point where $O \approx W$ the magnitude of F_x decreases non-linearly. For large offsets there is practically no alignment force.

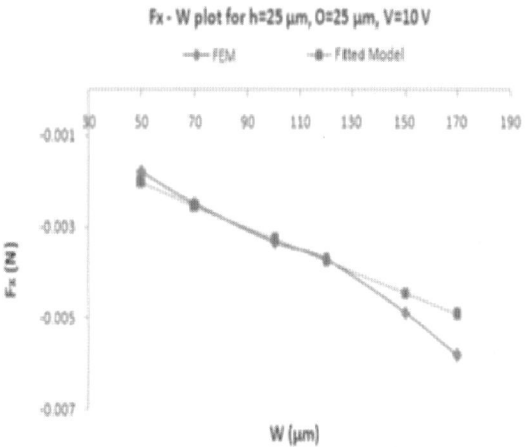

Fig. 5. Dependence of F_X on the width of the pad W.

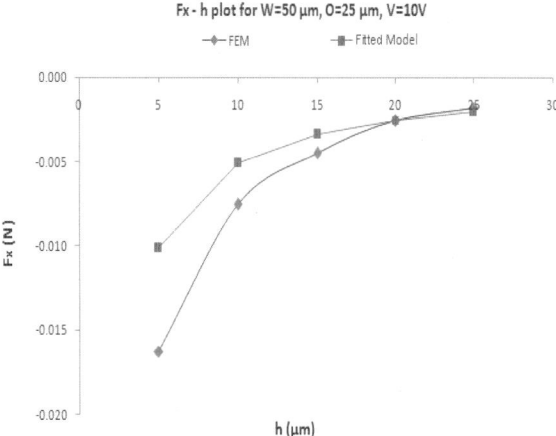

Fig. 6. Dependence of F_X on the separation of the complementary pads h.

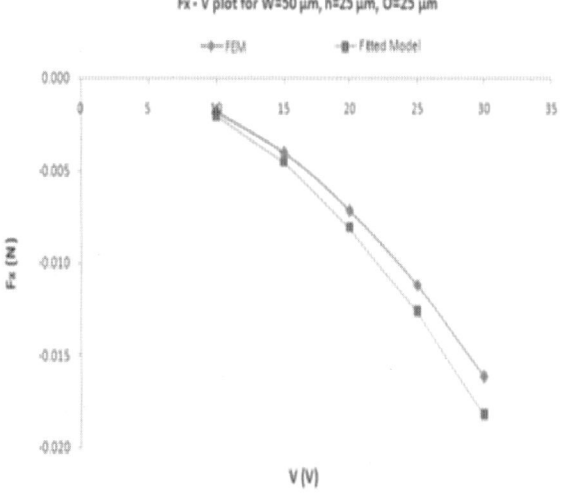

Fig. 7. Dependence of F_X on applied voltage potential V.

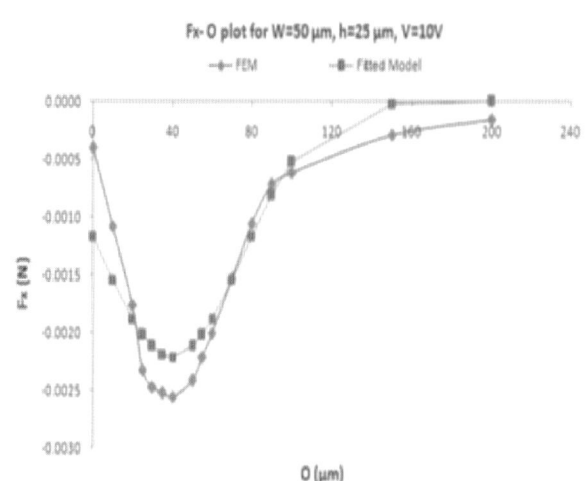

Fig. 8. Dependence of F_X on the misalignment O.

The above results were transformed into dimensionless parameters and fitted to the following model:

$$F_X = -\frac{W\varepsilon_0\varepsilon_r V^2}{4h} \cdot e^{-(-0.8+\frac{W}{O})^2} \qquad (2)$$

By comparing the results from the FEM-simulations and the analytic model several deviations can be observed. These can be explained by the simplifications that were inevitable in the FE-modeling (Fig 4). For example, the perimeter of the cylinder is set to ground voltage, which affects the results. Furthermore, when both the misalignment of the micro-part to the substrate and the width of the pad were large the FE-mesh size had to be increased in order to avoid too many mesh nodes in the FE-model.

4. Modeling of capillary alignment forces

The software Surface Evolver is used to model the capillary force in a fluid, in physical contact with two parallel pads, which are binding sites of a micro-part and a substrate. For this model the pads are assumed to be wetted completely by the fluid (Fig. 9). An algorithm based on the principle of virtual work calculates the alignment force.

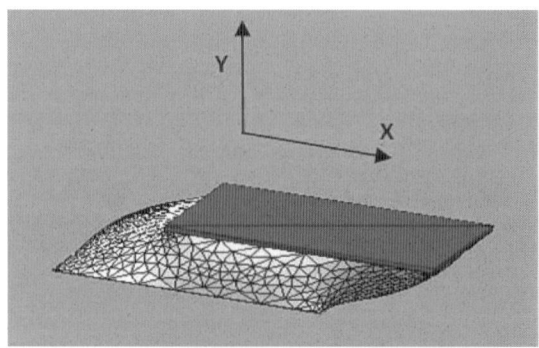

Fig. 9. Model in Surface Evolver to estimate capillary alignment forces.

The computed force F'_X as function of the misalignment of the binding sites, O, for different widths of the pads, W, and separation between the pads, h, is shown in Fig. 10. As long as the top pad is in contact with the lubricant droplet it is clear that for significant misalignments the capillary force is strong and tends to pull the part towards the equilibrium position, i.e where $O = 0$ and the surface energy of the droplet is minimized. The parameters for surface tensions in the model correspond to a self-assembly setup for micro-parts with triethylenglycol-dimethacrylat glue in a water bath [2].

When the distance of the micro-part to the substrate is small enough compared to the width of the pad the capillary force can also be estimated [2]:

$$F'_X = -\frac{\gamma \cdot WO}{\sqrt{O^2 + h^2}} \quad (3)$$

In this model the curvature of the lubricant droplet is neglected. Computed forces with Eq. 3 result in values that are approximately half as large as computed forces in the model in Surface Evolver.

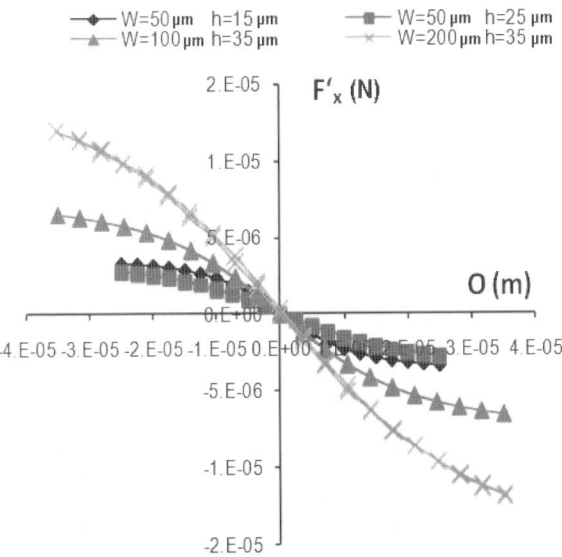

Fig. 10. Results from simulations in Surface Evolver.

5. Conclusions & future work

Fluidic self-assembly of micro-parts utilising capillary or gravitational forces for the alignment within micrometer precision has been demonstrated by several research groups [2,3,4]. However, very little experimental work on self-assembly with electro-static forces has been presented [5,6,7]. A concept of how electro-static forces could be used for self-assembly of micro-parts by inducing voltage potentials on the micro-part capacitively has been investigated in this paper. The alignment at the binding site is achieved due to electro-static energy minimisation. Furthermore, this electro-static alignment process could be used in conjunction with capillary alignment by realising lubricant droplets upon the substrate's binding sites.

Electro-static FEM-simulations and simulations of how surface shapes arise due to surface tension effects have been used to compute electro-static and capillary alignment forces. The electro-static forces are in the range of 10^{-3} to 10^{-2} N, whereas the capillary forces are in the range of 10^{-6} to 10^{-5} N for binding sites in range from 50 x 50 μm^2 to 200 x 200 μm^2. Hence, the electro-static forces are much stronger. Future research will focus towards experimental work, based on the presented concepts.

Acknowledgments

This work was carried out within the framework of the EC Network of Excellence "Multi-Material Micro Manufacture: Technologies and Applications (4M)".

Experimental work on this topic is also performed within the MNI-mst program from the German Federal Ministry of Education and Research (Nanopad) and the University of Patras' internal project "Microassembly Automation".

The Surface Evolver model was implemented by Mr. Jan Lienemann, University of Freiburg – IMTEK.

References

[1] M. Boncheva, G. Whiteshides, "Making things by self-assembly", MRS bulletin, Volume 30, October 2005.

[2] J. Lienemann, A. Greiner, J. G. Korvink, X. Xiong, Y. Hanein, K. F. Böhringer, "Modelling, Simulation and Experimentation of a Promising New Packaging Technology – Parallel Fluidic Self-Assembly of Micro Devices", Sensors Update 13, March 2004, WILEY-VCH Verlag, Weinheim,Germany, 2004, pp. 3-43.

[3] Uthara Srinivasan, Michael A. Helmbrect, Christian Rembe, Richard S. Muller, Roger T. Howe, "Fluidic Self-Assembly of Micromirrors Onto Microactuators Using Capillary Forces." IEEE Journal on Selected Topics in Quantum Electronics Vol. 8, No. 1, January/February 2002, pp. 4-11.

[4] Uthara Srinivasan, Dorian Liepmann, Roger T. Howe, "Microstructure to Substrate Self-Assembly Using Capillary Forces." IEEE Journal of Microelectro-mechanical Systems, Vol. 10, No. 1, March 2001, pp. 17-24.

[5] Theresa S. Mayer, Thomas N. Jackson, Christopher D. Nordquist, "Electro-Fluidic Assembly Process for Integration of Electronic Devices Onto a Substrate", United States Patent 6 687 987 B2, Feb. 10, 2004.

[6] K. Bohringer, M. Cohn, K. Goldberg, R. Howe, A. Pisano, "Parallel microassembly with electrostatic force fields", In Proc. IEEE Int. Conf. on Robotics and Automation (ICRA), Leuven, Belgium, May, 1998.

[7] Christopher D. Nordquist, Peter a. Smith, Theresa S. Mayer, "An Electro-Fluidic Assembly Technique for Integration of III-V Devices onto silicon". Compound Semiconductors, 2000 IEEE International Symposium on, pp. 137-142.

[8] J. Lienemann, "Modeling and Simulation of the Fluidic Controlled Self- Assembly of Micro Parts" Diploma Thesis, IMTEK, University of Freiburg, Chair for Simulation, Freiburg, Germany, 2002.

Multi-Material Micro Manufacture
S. Dimov and W. Menz (Eds.)

Towards automation in AFM based nanomanipulation and electron beam induced deposition for microstructuring

F. Krohs[a], T. Luttermann[a], C. Stolle[a], S. Fatikow[a], E. Brousseau[b], S. Dimov[b]

[a] *Div. Microrobotics and Control Engin., Univ. of Oldenburg, Germany;*
[b] *The Manufacturing Engineering Centre (MEC), Cardiff University, Wales*

Abstract

To move towards complex assemblies at the micro- and nanoscale, manipulation processes have to be automated to increase throughput and accuracy. First, this paper addresses manipulation at the nanoscale by an AFM and second, automated electron beam induced deposition as a method for structuring at the microscale is presented. Nowadays, AFM based nanomanipulation still requires frequent user interaction and remains a very labor intensive task. Spatial uncertainties are identified as a major problem that prevents reliable automation of AFM based manipulation. Results of a novel particle filter based method for measuring thermal drift in an AFM system is presented and future applications for probabilistic methods are discussed.

The automation of electron beam induced deposition (EBiD) for microstructuring purposes builds a multifunctional tool for additive structuring and also bonding inside an SEM. The presented system has the ability to create EBiD depositions from two different precursor materials by automatically executing predefined sequences. The automation includes the precursor flux control with the possibility to alternate between two materials, the deposition of points and lines at defined positions, as well as the ability to find and track already deposited structures with the use of digital image processing. This assures precise positioning of depositions relative to others even in cases of thermal or electrostatic drifting of the specimen substrate or the electron beam.

Keywords: atomic force microscope, electron beam induced deposition, automation, image processing

1. Automation of AFM based nanomanipulation

1.1. Introduction

Although the atomic force microscope (AFM) is primarily used for imaging, utilizing it as a nanomanipulation device has become of increasing interest in the last decade. Especially due to its high resolution and its flexibility against different types of samples and ambient conditions, it is applicable for a variety of nanomanipulation tasks. Possible applications range from prototyping nanoscale devices [1] to the characterisation and handling of biological samples (i.e. dissection of DNA [2]). However, due to several problems, AFM based nanomanipulation still requires frequent user interaction and remains a labor intensive task. Therefore, automation has been identified as an important research goal to increase throughput of AFM based nanomanipulation.

1.2. Difficulties in AFM based nanomanipulation

Manipulations by an AFM always have to be performed in a "blind" way, because the AFM tip can only be utilized either as the nanomanipulator or the imaging device at the same time. Positioning inaccuracies and the physical characteristics in the nanoworld yield high error rates in the manipulation results. Hence, due to the lack of a real-time capable visual sensor, the result of the manipulation process has to be verified afterwards by scanning the relevant area. Manipulation and image acquisition are therefore mostly performed alternately, resulting in very low throughput.

In AFM based nanomanipulation, and especially for its automation, spatial uncertainties constitute one major cause for erroneous results:

These uncertainties are partially caused by hysteresis,

creep and other nonlinearities of the piezo scanning stage. The most common way is to compensate these positioning inaccuracies by measuring the actual displacement of the scanner using position sensors and using these data as feedback signal for operating the scanning stage in a closed-loop. However, position sensors are afflicted with noise and for small scan areas this often leads to oscillations in the closed-loop control. Modern approaches exist that are using feedforward strategies by modelling the scanning stage characteristics [3][4].

More critical and less straightforward to counteract are spatial uncertainties that are induced by thermal drift. Even small changes in temperature cause all AFM's components to vary slightly in size (due to thermal expansion and contraction) which result in an unknown displacement between AFM probe and sample. By operating the AFM under homogeneous environmental conditions, the effect of thermal drift can be reduced, but even in highly temperature stable conditions thermal drift is still observable and amounts from 0.01 to 0.2 nm/s [5]. Even though this motion is very slow, it can obviously be detrimental to the success of nanomanipulation in the long run.

Especially when dealing with objects in the order of magnitude of a few nanometres (like shown in Fig. 1), this effect becomes a crucial issue for the success of manipulation (i.e. pushing a certain object).

Unfortunately, the spatial displacement between AFM tip and sample cannot be observed directly due to the lack of a real-time-capable and high-resolution visual sensor. Hence, the only means to measure this displacement is the AFM tip itself.

A very common approach is based on cross-correlating successively recorded AFM images to measure the drift inside the AFM system. However, this technique is too slow to be applicable for nanomanipulation, since the

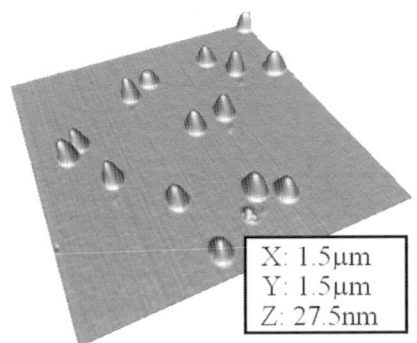

Fig. 1. Topography image of 30nm-Au-nanoparticles on mica.

Fig. 2. Topography image (10x10 µm) of Au coated Si substrate used as a sample for measuring drift.

AFM tip (which is also needed for the manipulation itself) is occupied for a couple of minutes to acquire images, and drift velocity and direction may change even in a small time frame. More sophisticated approaches try to track certain features on the sample (i.e. the center of a nanoparticle [6]). Even though these techniques are able to measure drift reliably with update rates in the order of seconds, knowledge about the sample (i.e. the shape of a particle) is a necessity.

1.3. Sample independent drift estimation

To allow for robust drift compensation even with unknown and sparsely structured surfaces, a novel algorithm to measure drift was developed. Instead of scanning an area or tracking certain features on the sample to gain information about the true lateral position of the AFM probe, the developed algorithm periodically records short height profiles that can be recorded in a small time-frame (~0.1s). These scanned lines are considered as sensor data containing some information about the tip position in relation to the sample. Although it seems obvious that the tip position cannot be extracted directly from a single height profile, the developed algorithm use these data to update its belief about the current system state, namely drift.

This is performed by comparison of the height profiles recorded at arbitrary positions with line scans that are extrapolated from a previously recorded topography image.

Due to a low signal-to-noise ratio and sporadic faulty measurements the recorded line scans only represent the true height profile of the sample roughly. Additionally, depending on the type of sample used, these line scans may also contain only little information. To account for these perceptional uncertainties, a particle filter based algorithm was developed for drift estimation.

Since several years, particle filter based localization has successfully been applied in the domain of mobile, autonomous, macroscale robotics [7]. It is part of the superordinate concept known as *probabilistic robotics*. The key idea behind probabilistic robotics is to explicitly deal with the uncertainties that exist in a robotic system using the calculus of probability theory. It has been shown that probabilistic algorithms are very robust against noisy or faulty sensor data and that they perform well even when the system's behavior can only be poorly modeled [8]. According to algorithms used in mobile robotics, the task of estimating drift can be reduced to a limited global localization problem: In

terms of mobile robotics, the sample surface can be considered as an environment in which a moving robot, in the discussed case the AFM tip, has to be localized. To validate the algorithm, drift measurements were conducted using several types of samples. However, since the algorithm was mainly developed to work even with almost unstructured sample surfaces, most interesting results were obtained with gold coated silicon substrate, which roughness parameters were determined to be $R_{RMS} = 0.765$ nm and $R_a = 0.592$ nm. A topography image of the Au/Si substrate is shown in Fig. 2.

The experiments (see Fig. 3) have shown that even with the highly unstructured substrate used as a sample, drift could be reliably measured over a period of several hours. To prove correctness of the obtained results, a topography image taken before the measurement was cross correlated with an image taken directly after the measurements. The results obtained from this cross correlation have shown to be identical to the drift induced displacement measured by the newly developed algorithm.

Compared to existing state-of-the-art algorithms for drift measurements, the developed algorithm is both fast (with update rates of 0.1Hz) and independent of the type of sample.

1.4. *Probabilistic approaches for automation of AFM based nanomanipulation*

The results discussed above have shown exemplarily the strength of probabilistic approaches and their applicability for nanoscale applications. Especially for

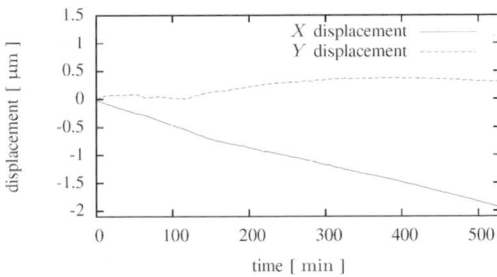

Fig. 3. Drift induced displacement measured over a period of 9 hours using Au/Si substrate as a sample.

systems that have partly unknown behavior and that suffer greatly from noisy and faulty sensor data – which is the case for AFM applications – these algorithms tend to be more robust than existing algorithms. Therefore, more applications for probabilistic methods are discussed:

The AFM drift estimation problem could successfully be solved by using a particle filter based localization algorithm. Moreover, setups are planned that combine a conventional AFM with a nanorobotic platform equipped with an AFM tip as an endeffector. This will allow for more sophisticated measurements (i.e. electrical measurements in the sample plane). However, to achieve accurate results with this setup, both AFM tip and the secondary robot have to be positioned with high accuracy in relation to the sample. Using a probabilistic localization approach could also be feasible for finding the exact positions of AFM probe as well as the robot's endeffector.

Another important issue, which is mostly neglected in literature, is that thermal drift always even occurs during image acquisition. This leads to slightly distorted AFM images, which can affect AFM based nanomanipulation and particularly automation negatively. Path planning algorithms that are based on previously acquired AFM images can therefore fail and lead to collisions of the tip and an object on the sample surface. To correct for drift induced distortions in AFM images, a probabilistic approach called *simultaneous localization and mapping* (SLAM) [8] could be feasible by continuously incorporating sensor data (the AFM tip height) to update the knowledge about the environment (the sample's topography).

1.5. Conclusion and outlook

In AFM based manipulation, a reliable compensation of spatial uncertainties, induced by thermal drift or the piezo based scanning unit, has been identified as a precondition for automation. Particularly when scaling down to objects of a few nanometers, this issue becomes critical. Current research focuses on the development of such algorithms as shown above. However, for fully automated nanomanipulation by an AFM, also path planning and collision avoidance have to be considered.

Further development will focus on a framework to allow automated nanohandling by an AFM without constraining to a certain problem. To account for the high level of uncertainty existing in nanoscale applications, applicability of probabilistic methods will be analysed.

2. Automated nanostructuring with EBiD

2.1. Introduction

Electron beam induced deposition (EBiD) is a technique that allows creating three dimensional structures with sizes down to a few nanometers by directly "writing" them with an electron beam onto a substrate material, inside the vacuum chamber of a scanning electron microscope (SEM). The secondary electrons released by the electron beam of the SEM when hitting a substrate cause a chemical decomposition and partial deposition of a precursor gas, which is evaporated into the vacuum chamber.

In today's technical applications the EBiD process is

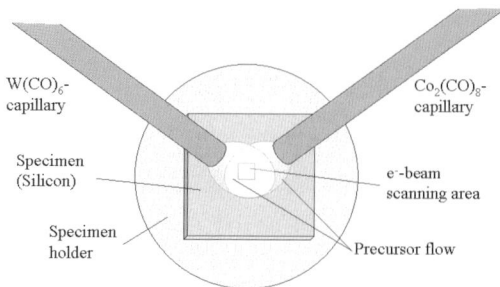

Fig.4 capillary positions and broadening of the precursor (schematic top view)

generally used for mask repair [9], for production of AFM supertips [10] or as a bonding technique in the nano- and micrometer scale [11].

The EBiD process is also comparable with focused ion beam (FIB) deposition, where accelerated ions cause the release of secondary electrons from the substrate material, again inducing the chemical decomposition.

The setup developed in this work uses two self-made precursor evaporation systems inside the SEM chamber with advanced control software. This makes it possible to create depositions from two different precursor materials without any setup changes.

The description of work contains the description of the setup with its precursor evaporation systems and its automation software. Additionally some applications for this setup are explained and evaluated.

2.3. Experimental setup

The deposition setup basically consists of a Zeiss DSM950 scanning electron microscope with two different precursor evaporation systems inside its vacuum chamber and software which controls the precursor flux. Each evaporation setup offers a closed precursor reservoir which can either be heated or cooled by a peltier element to force respectively to avoid sublimation of the precursor in dependence of its vapor pressure. The gaseous precursor gets to the substrate through a capillary with an inner diameter of 0.6 mm and a length of approx. 40 mm. The precursor flux is controlled by a PC with special software, which keeps the chamber pressure on a desired value between $1*10^{-5}$ mbar and $4*10^{-5}$ mbar by adjusting the supply voltage of the peltier element.

The mass flow of the precursor is essentially dependent on its vapor pressure inside the reservoir of the evaporation system, the pressure inside the vacuum chamber and the geometry of the capillary. The

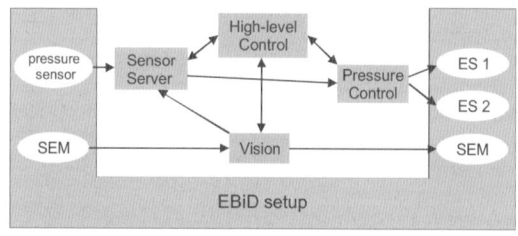

Fig.5 software setup consisting of SensorServer, Vision, PressureControl and High-Level Control (HiLeC).

depositions for the experiments in this paper were made at a controlled vacuum chamber pressure of $2 \cdot 10^{-5}$ mbar which corresponds to a molecular flow of about $1.8 \cdot 10^{21}$ molecules per m² and second at a mean distance of the electron beam to the capillary of 400 µm (Fig. 4).

The electron beam was accelerated with a voltage of 20 kV and a beam current of 90 nA.

2.4. Software architecture

The software control architecture is based on the *Distributed Control Architecture for Automated Nanohandling* (DCAAN) presented in [12, 13]. The flexible architecture is based on several server programs written in C++ connected via Ethernet using the platform independent Common Object Request Broker Architecture (CORBA) as communication framework, which is an object oriented middleware defined by the Object Management Group (OMG).

The programs can be distributed onto several PCs, however in our setup all servers were running on the same machine. For the task of automating the EBiD process the following software servers where involved (Fig. 5):

Several sensor programs (*e.g.* pressure sensor program and vision) are pushing data to the SensorServer at independent update rates. Each sensor data carries the timestamp of the reading. All servers do have a synchronized clock. This way a server can decide whether or not a datum is outdated. The SensorServer collects the most current sensor data and provides them on request to other programs (*e.g.* PressureControl and HiLeC). The collection of all SensorServer data makes up the current world state of the system.

HiLeC coordinates the different parts for different automation tasks. Set values are sent to actuator specific low-level controllers (*e.g.* set pressure) and timing between different system parts gets controlled. HiLeC is capable of tele-, semi and full automation depending on the task requirements. For semi- or full automation all servers connected to HiLeC have a common interface defined in CORBA's interface definition language (IDL). At connection time HiLeC queries the command set of each server. This command set can be directly used at command line or to write automation sequences.

2.4.1 Vision

The vision software [14] tracks objects in visual sensor data (*e.g.* the SEM life images) and publishes the objects' positions to the sensor server. In many applications this is a key ability since most actuators do not have integrated position sensors or the accuracy of the integrated ones is not high enough for nanopart manipulation.

Even though this key ability is used in our application too, Vision does have a second function. For fast position updates it is necessary to set a region of interest (RoI), such that only this section is scanned by the electron beam. This ability of controlling the scan area is exploited in our setup such that Vision acts as a sensor and as an actuator. For point depositions a RoI of width and height one is used and for horizontal or vertical lines a rectangular RoI of width one ore height one is used. This way all 1024x768 points of the displayed scan field can be addressed.

2.4.2. Pressure control software

Deposition rate of Electron Beam induced Deposition is mostly dependent on electron beam parameters and the precursor density and thus the precursor flux through the ES's capillaries [15]. To assure a constant precursor flux while making an electron beam induced deposition it is necessary to control the vaporization of each precursor continuously. In the presented desktop station this is provided by control software where two PID-controllers are implemented, one for each ES. For each controller the control variable is selectable between gas pressure inside the vacuum chamber, which is the most suitable variable to assure a constant precursor flux and temperature of the ES precursor reservoir, which is necessary when both precursors should be vaporized at the same time.

2.5. Automation

The major automation objective for this setup is to be able to grow pins and vertical and horizontal lines in a defined and reproducible manner. In addition it should be possible to change the precursor and start a deposition again on top of an existing pin. This allows *i.e.* deposition of combined pin structures of two different materials in one production step.

Before starting the automation sequence a few preparation steps have to be performed. First a default pin model has to be created in Vision for tracking of the pins. Brightness, contrast, focus and magnification of the SEM have to be adjusted properly. The magnification level in our experiments was fixed to a level of 2000 which corresponds to a displayed image width of 25 µm. With an image resolution of 1024x768 this corresponds to an addressable resolution of about 24.4 nm a pixel. So the process resolution depends on

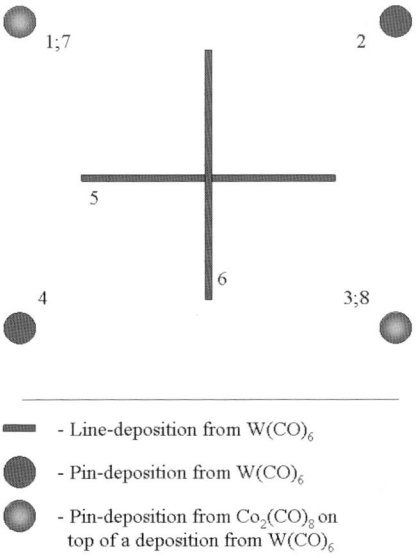

- Line-deposition from $W(CO)_6$

- Pin-deposition from $W(CO)_6$

- Pin-deposition from $Co_2(CO)_8$ on top of a deposition from $W(CO)_6$

Fig.6 schematic view of automation sequence as top view (SEM view). The numbers are the order of depositions steps.

122

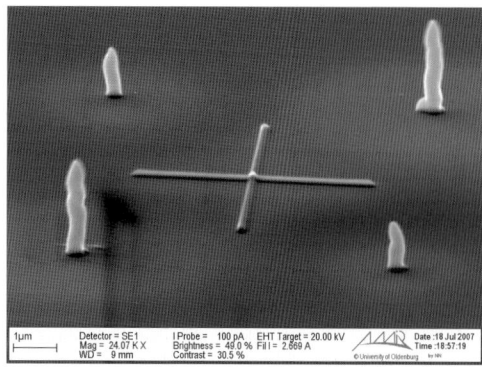

Fig.7 SEM image of the result of a complete automated sequence described in "AUTOMATION", tilted to an angle of 70°

image resolution and the magnification level.

Fig. 6 illustrates the single automation steps used for our experiments. The sequence starts with heating up ES 1 (filled with $W(CO)_6$ precursor) for this HiLeC sends a start signal to PressureControl with the predefined set pressure of $2 \cdot 10^{-5}$ mbar. At the time the goal pressure is stable PressureControl sends a signal to HiLeC.

Up to that time the electron beam is blanked to avoid unnecessary contaminations of the Si substrate. At that time HiLeC signals Vision to set the RoI to the upper left deposition point for 7.5 min before switching to the next point. The points 2-4 have been done accordingly. Afterwards the two lines are grown in step 5 and 6 by setting a rectangular RoI with height one for 10 min and the same with width one for the vertical line. Now HiLeC stops the closed-loop control of PressureControl for ES 1 and blanks the beam again. HiLeC waits for 3 min to let the peltier element cool down and restabilize the standard high vacuum pressure of the chamber. After that time HiLeC signals PressureControl to start closed-loop control of ES 2 with the same set pressure as before ES 1.

In steps 7 and 8 two pins get grown on top of the pins 1 and 3. Before starting the deposition step drift effects have to be compensated. This is done by setting a RoI at the expected position of the pin that should be modified. The RoI needs to be large enough to enclose the real position of the point and small enough to ensure that only one pin is in the RoI. With the predefined default pin model the exact position of the pin gets tracked and the deposition with ES 2 starts. The pin deposition is performed as the pins 1-4 before just having the drift compensation step before.

The whole sequence has been written as part of the HiLeC program using the commands provided by the connected servers. The execution time of the experiments is about 75 min. Deposition time is about 65 min. Most of the residual time is used for heating and cooling.

2.6. Experimental results

The results of the automated deposition sequence were analyzed by an SEM image in a side view by tilting the specimen substrate to an angle of 70°. Fig. 7 shows The SEM image of one complete sequence. Because of the tilt direction of the SEM stage the pattern is turned 90° clockwise so that the first deposited pin is in the upper right corner. As one can see, the second

deposition on top of the first and third pin with the $Co_2(CO)_8$ precursor are exactly matching with the first deposited part so that there is no displacement in x or y direction.

Experiments with different timings during the automated sequence indicated the importance of a constant temperature of the activated evaporation system before starting the deposition. Alternating temperature during a single pin deposition causes drift effects of the substrate material which leads to a displacement of the deposition and thus to sloping pins. Also the absolute positioning accuracy suffers from these drifts during the process. To avoid these problems it is necessary to wait after activating a precursor evaporating system until the temperature is on a constant level before starting the deposition sequence.

2.7. Applications

The aimed application of the presented desktop station in general is the automated electron beam induced deposition of micro and nano scale pin and line structures from different precursors in predefined geometries. Several applications are possible to be automated with this setup. Due to the visual feedback by the SEM image and the implemented object tracking algorithms, one spot on the specimen surface can be approached several times even in cases of thermal or electrostatic drift effects in the SEM image.

In the field of nanorobotics this setup is applicable to build free-standing mechanical structures where different precursors cause various mechanical or electrical properties of the structure. To achieve pin-like structures from different precursor materials it is essential to control the deposition spot because even the changing precursor flow while "switching" the precursor can cause thermal drift of the specimen of about ten times of the diameter of a deposition. The resulting depositions can be cantilevers where a small section is made of a material which is more elastic than the rest of the structure whereby it acts as a solid hinge. As an alternative the different thermal expansion coefficient of depositions from two different precursors can be used for actuation applications where the deposited structures are heated by an electric current.

Another aimed application for the automated desktop station is the EBiD based bonding of micro and nanostructures, i.e. carbon nanotubes or nanowires onto surfaces or STM tips. To do this, the automated sequence will start after positioning the objects that have to be bonded in contact to each other in the SEM image. The sequence itself contains the heating of the evaporation system, automatically finding the couple point of joined objects and than the deposition of a short line across this couple point.

2.8. Conclusions and outlook

We proved the functionality of the automated desktop station with a deposition sequence using the precursors $W(CO)_6$ and $Co_2(CO)_8$. The experimental results have demonstrated, that a precise positioning of depositions relative to each other is possible without manual interaction. Temperature changes due to alternating precursor supply caused by thermal drifts of the specimen substrate have been compensated.

With the developed desktop station it is possible to

deposit point structures and horizontal and vertical lines by electron beam induced deposition. When doing this conventionally, it is essential to set up every line and point manually and to examine the correct time for each deposition process. Additionally the change of the precursor is very time consuming. The main advantage of our automated system is the possibility of predefining complex deposition sequences that can be executed several times without manual interaction.

3. Acknowledgements

We want to thank the German Federal Ministry of Education and Research (Bundesministerium für Bildung und Forschung, BMBF) for funding parts of this work within the research project "ZuNaMi" (promotional referrence: 16SV2276). Additionally, we gratefully acknowledge the support of the german DAAD and the British Council Germany through the British German Academic Research Collaboration (ARC) programme.

References

[1] D. M. Schaefer, R. Reifenberger, A. Patil, and R. P. Andres, "Fabrication of two-dimensional arrays of nanometer-size clusters with the atomic force microscope," Applied Physics Letters, vol. 66, no. 8, pp. 1012 – 1014, 1995.

[2] F. J. Rubio-Sierra, W. M. Heckl, and R. W. Stark, "Nanomanipulation by Atomic Force Microscopy," Advanced Engineering Materials, vol. 7, no. 4, pp. 193–196, 2005.

[3] G. Schitter, F. Allgöwer, and A. Stemmer, "A new control strategy for high-speed atomic force microscopy", Nanotechnology, 2004, 15, pp. 108–114.

[4] Y. Li, J. Bechhoefer, "Feedforward control of a closed-loop piezoelectric translation stage for atomic force microscope", Review of Scientific Instruments, AIP, 2007, 78, 013702.

[5] B. Mokaberi, J. Yun, M. Wang, and A. A. G. Requicha, "Automated nanomanipulation with atomic force microscopes", Proc. IEEE Int'l Conf. on Robotics & Automation (ICRA '07), Rome, Italy, pp. 1406–1412, April 10-14, 2007.

[6] B. Mokaberi and A. A. G. Requicha, "Drift compensation for automatic nanomanipulation with scanning probe microscopes", *Automation Science and Engineering, IEEE Transactions on*, no. 3, pp. 199–207, July 2006.

[7] D. Fox, W. Burgard, F. Dellaert, and S. Thrun, "Monte Carlo localization: Efficient position estimation for mobile robots," in Proceedings of the Sixteenth National Conference on Artificial Intelligence (AAAI'99)., July 1999.

[8] S. Thrun, W. Burgard, and D. Fox, "Probabilistic robotics". MIT Press, 2005.

[9] K. Edinger, H. Becht, J. Bihr et. al., „Electron-beam-based photomask repair," J. Vac. Sci. Technol. B vol. 22(6), 2004, p2902-2906.

[10] H. W. P. Koops, C. Schossler, A. Kaya and M. Weber, "Conductive dots, wires and supertips for field electron emitters produced by electron beam induced deposition on samples having increased temperature," J. Vac. Sci. Technol. B vol. 14(6), 1996, p4105-4109.

[11] T. Sievers, T. Wich, "Assembly inside a Scanning Electron Microscope using Electron Beam induced Deposition," in Proceedings of IEEE/RSJInternational Conference on Robots and Intelligent Systems, 2006.

[12] S. Fatikow, V. Eichhorn, C. Stolle, T. Sievers and M. Jähnisch, "Development and Control of a Versatile Nanohandling Robot Cell", Journal of Mechatronics, 2007, submitted

[13] C. Stolle, "Distributed control architecture for automated nanohandling," in Conference on Informatics in Control, Automation and Robotics (ICINCO'07), p. 127-132, 2007.

[14] T. Sievers and S. Fatikow, "Real-time object tracking for the robot-based nanohandling in a scanning electron microscope," Journal of Micromechatronics - Special Issue on

[15] V. Scheuer, H. Koops, and T. Tschudi. "Electron beam decomposition of carbonyls on silicon", Microelectronic Engineering vol. 5, 1986, p423–430.

Manufacturing and replication of cell aligning micro structures

C. Brecher[a,b], R. Klar[b], F. Pretzsch[b], C. Wenzel[b]

[a] Werkzeugmaschinenlabor (WZL), RWTH Aachen University, Germany
[b] Fraunhofer-Institute for Production Technology Aachen, Germany

Abstract

Biotechnology is becoming more and more important and is influencing our everyday life. One of the most important advantages will become the manufacturing of mass customized cells also known as tissue engineering. To optimize and shorten the cell growth, micro structures have shown an important impact on the cell division, alignment and cell differentiation. This paper deals with the manufacturing of micro structured moulds by diamond machining and the successive replication process to create bio functional surfaces provided with micro structures. While the fabrication of micro structures by diamond machining is an expensive and long lasting process the replication of bio functional surfaces by hot embossing allows the cost efficient production of these surfaces. Compared to lithography processes, diamond machining is very flexible and offers next to 2D or 2 1/2D geometries real 3D forms to copy in-vivo conditions for in-vitro cell replication processes. One of the main advantages of diamond machining is the flexibility to diversify the shape of micro structures rapidly to investigate the cell - substrate - interaction down to the micrometer range.

Keywords: micro structure, bio technology, diamond machining, cell alignment

1. Introduction

The possibilities of bio technology applications are becoming more and more a part of everyday life. One of the most interesting possibilities is the tissue engineering which allows the manufacturing of autogenic implants like skin and cartilage. In that coherence it was discovered that cell adhesion and attachment is mainly influenced by molecular surface interactions. But morphology and orientation of cells are affected by topographic cues of the extracellular matrix.

It is well known that nanometre-scale structures and structured surfaces have a positive influence on cell growth and proliferation (see Figure 1). This is due to the simulation of Extra Cellular Matrix (ECM) by these structures, providing in-vivo topographical conditions for in vitro-cell cultures. Structures on a micrometre-scale on the other hand are expected to have sizeable impact on cell orientation as well as separation and isolation of individual cell types from tissue probes. This can be used in the efficient isolation of important cell types such as adult stem cells or epithelial cells but also to separate different cell types such as Melanocytes, Keratinocytes and Fibroblasts from skin tissue. [1], [2], [3]

In order to maximize the effects of nanometre-scale structuring on the growth of individual cell types, in-vivo ECM topographies can be used as a model for the generation of artificial structures. Different manufacturing processes such as for instance Electro Spinning or Solid Freeform Fabrication (such as 3D-Printing or Fused deposition Modeling) are being developed and optimized for the specific challenges of manufacturing in-vitro ECM structures.[4] [5]

For topographical structures on the micrometre scale however, a comparable model to the ECM doesn't exist and extensive research is required including cell biology and manufacturing technology. Aiming at the systematic approach to the classification and high volume production of cell type specific micro structured surfaces for use in bio reactors, an interdisciplinary team of Fraunhofer researchers in the fields of biology, micro technology and precision manufacturing are therefore working on the development of bio-functionalized surface structures and their mass replication. Micro structured surfaces are being developed, produced and systematically analyzed for their applicability with different cell types. Structure geometries include parallel grooves of various cross sections (square, round, v-shaped) as well as stub like surface textures of different geometries and spatial distribution. Structure sizes range from multiple micro meters down to sub micro meter values.

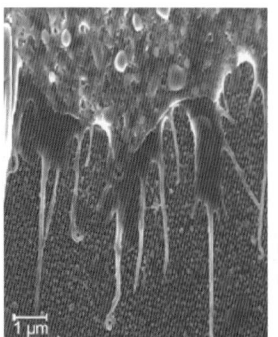 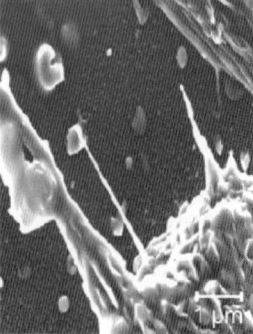

Figure 1: Cell interaction with nano structured surfaces

Multi-Material Micro Manufacture
S. Dimov and W. Menz (Eds.)

X-ray pattern analysis of electroplated two-component moulds used for the production of micro gear wheels

J. Prokop[a,b], J. Lorenz[a], V. Piotter[a], H.-J. Ritzhaupt-Kleissl[a], A. Roch[a], and J. Haußelt[a,b]

[a] *Forschungszentrum Karlsruhe GmbH, Institute for Materials Research III (IMF III) Hermann-von-Helmholtz-Platz 1, 76344 Eggenstein-Leopoldshafen, Germany*

[b] *Department of Microsystems Engineering - IMTEK, University of Freiburg, Germany*

Abstract

A process for the fabrication of metal micro components by combining 2-component injection moulding with metal deposition by electroforming will be presented. To produce these 2-component polymer templates, an electrically conductive base plate is generated by injection moulding of electrically conductive carbon black-filled polymers. In a second injection moulding step microstructures consisting of insulating polymers are mounted onto these plates. The quasi-infinite conductivity gradient of such 2-component templates allows controlled electroplating to start from the base plate only, such that defect-free metal micro components can be achieved. The parameter set of the injection moulding process has been investigated by using an experimental method with an x-ray pattern analysis. Nearly defect-free electroplated micro parts could be fabricated by this process so far.

Keywords: micro injection moulding, two-component injection moulding, electroplating, galvanoforming

1. Introduction

Micro gear wheels have been used in several applications so far. A method to produce gear wheels with very high surface qualities is the LIGA method. Even though this process is very expensive, LIGA-products have already been used in several marketable applications [1-3]. A promising method to replicate these high-quality products is the combination of two-component injection moulding of an electric and an electrically insulating material with subsequent electroforming into micro cavities which are produced during this two-component process [Fig. 1 | 4-6].

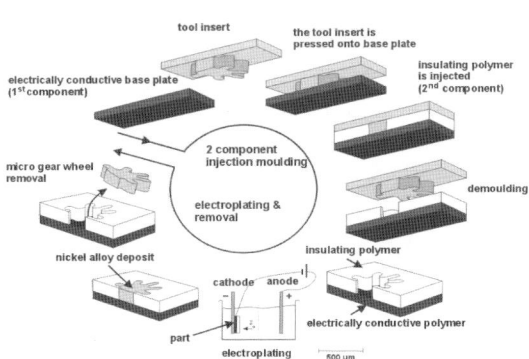

Fig. 1. Principle of the process cycle

The project faces several challenges. One challenge is the homogeneity of the conductive component; it can be achieved by using optimum injection parameters. A staged injection rate during injection moulding results in an improved homogeneity of the moulded parts with respect to their electric conductivity [7]. Another challenge is the microscopic gap which develops between the first and the second component during the process. This gap develops during demoulding, if the adherence between mould insert and insulating polymer is higher than the adherence between the electrically conductive and the insulating polymer. This gap leads to a burr at the end of the micro gear wheel during the electroforming process (Fig. 2).

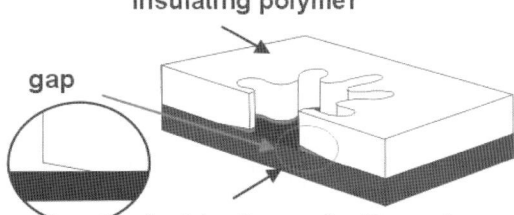

Fig. 2. Gap developed during the process

It is difficult to detect this gap. Neither microscopical analysis of the finished parts nor micrographical examination provides well-defined information. In order to find out how and when gaps are formed, a new detection method was applied using x-ray pattern analysis.

2. Design of the experimental method with x-ray pattern analysis

Processing optimisation for injection moulding can be achieved by using the design of experiment (DOE) method. DOE considers two-way interactions and is increasingly applied for process design tasks. In order to determine optimal process parameters, experiments were performed under various process conditions with the DOE scheduling. This contributed to a better understanding of the process characteristics. Statistical

Figure 7 shows the embossing chart for the depth controlled process strategy. The following process parameters have been found most suitable for the replication of the micro structures into Polystyrol plates:

Tool temperature: 142°C – 144°C
Substrate temperature: 125°C – 128°C
Depth of tool penetration: 100 µm
Substrate thickness: 3 mm

At these values, the embossing force was about 300N. It shall be stated that the temperatures mentioned above and in the embossing chart in Figure 7 do not reflect the actual temperatures in the contact zone but are interpolated values by the embossing system. Careful process development is required to determine the best temperature range for a given embossing set up.

Figure 8 shows examples of structures replicated in PS plates, indicating the distance between grooves and groove width (numbers in the top left corner) as well the cross section of the surface topography. These structures are currently being used in cell culture experiments at the Fraunhofer IGB in Stuttgart, Germany and the Fraunhofer IBMT in St Ingbert, Germany. After successful evaluation, structures with a positive effect on cell growth and proliferation will be machined into the structured rings of the roller set up shown in Figure 4, where they can then be used for continuous roll-to-roll embossing into thermoplastic film materials.

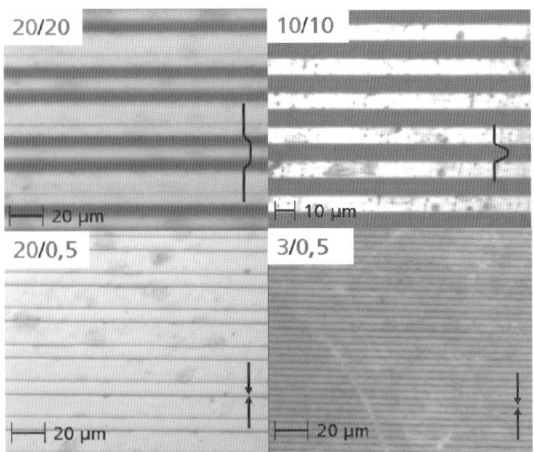

Figure 8: Structured surface examples

4. Conclusions

The combination of diamond machining and high volume polymer replication technologies provide a highly capable process chain for the manufacturing of functionalized surfaces for cell biology applications. Ongoing research efforts are being focused on the classification of surface topographies and their precise and efficient manufacturing. Current investigations show that diamond machining processes are offering high structural flexibility in the micrometre range but must be further improved towards the precise manufacturing of surface structures in the sub micron range. Based on the results of the continuing biological investigations, the manufacturing process chain will be further improved following a holistic approach including both surface structuring and high volume replication.

Acknowledgements

This work is being funded by the Fraunhofer Gesellschaft in Munich, Germany. Their support is gratefully acknowledged.

References

[1] Flemming RG, Murphy CJ, Abrams GA, Goldman SL, Nealy PF. Effects of synthetic micro-and nano-structured surfaces on cell behaviour. Biomaterials 1999; 20; 573-585

[2] Crainghead HG, James CD, Turner AMP. Chemical and topographical patterning for directed ell attachment. Curr Opin Solid Stm 2001; 5; 177-184

[3] Wilkinson CDW, Riehle M, Wood M, Gallagher J, Curtis ASG. The use of materials patterned on a nano- and micrometric scale in cellular engineering. Mater Sci Eng C 2002; 19; 263-269

[4].Sachlos E, Czernuszka JT. Making tissue engineering scaffolds work. Review on the application of Solid Freeform Fabrication technology to the production of tissue engineering scaffolds. European Cells and Materials 2003; 5; 29-40

[5] Boland ED, Wnek GE, Simpson DG, Pawlowski KJ, Bowlin GL. Tailoring tissue engineering scaffolds using electrostatic processing techniques: a study of Poly(Glycolic Acid) Electrospinning. J. Macromol. Sci.—Pure Appl. Chem., A38(12), 1231–1243 (2001)

[6] Brinksmeier E, Preuß, W, Schmütz, J. Manufacture of Microstructures by Diamond Machining Progress in Precision Engineering and Nanotechnology, Proceedings of 9th IPES/UME, Braunschweig, 26. - 30. Mai 1997, pp. 503 – 507

[7] Datta P, Goettert J. Method for polymer hot embossing process development. Microsyst Technol 2006

Figure 4: Roller with replaceable ring elements

While the continuous embossing process is being developed in parallel, for the current investigation and analysis of different surface structures a discontinuous embossing process is being used to replicate the diamond machined structures. Thus a large variety of structures can be investigated simultaneously for different cell types under similar environmental conditions. Furthermore, the intended process chain of master structure manufacturing and successive replication is being investigated throughout the entire project, following a holistic development approach.

Development of this discontinuous embossing process is being performed on a hot embossing system Hex02 by Jenoptic. Featuring a high resolution motion system, precision temperature control and a vacuum embossing chamber for a clean process environment, the system is highly suitable for the replication of micro structures. Figure 5 shows the vacuum chamber and tool holder of a Hex02 hot embossing system, including a structured embossing tool.

Hot embossing uses a quasi-solid state of the polymer at the glass transformation temperature to imprint the structure of the embossing tool into the substrate surface. Ideally the substrate is being heated just below glass transformation while the tool temperature is slightly above glass transformation in order to allow the substrate material to soften flow into the micro structures and precisely replicate the embossing tool topography. [7]

Figure 5: Hex02 hot embossing system

3. Manufacturing of micro structured surfaces

In this first phase of the project the research focuses on the effects of parallel micro grooves with different pitches and groove widths. A matrix of 49 combinations including groove widths and distances between grooves from 0.5 to 20 µm has been chosen to investigate cell behaviour, proliferation and growth effects. A master tool, comprising of 49 micro structured "stamps" of 4 X 4 mm^2 each has been designed to fit the tool holder of the hot embossing system. The structures are being embossed onto 40 X 40 mm^2 clear Polystyrol plates allowing analysis using standard petri-dishes and inverted microscopy which are commonly used in cell biology laboratories.

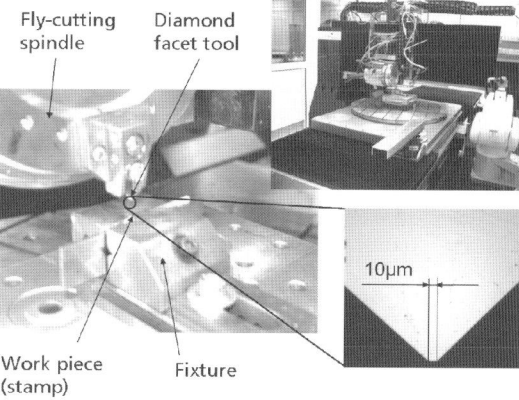

Figure 6: Manufacturing of micro embossing stamps

The micro structured surfaces for the embossing tools have been manufactured using facetted diamond tools thus allowing the use of the same tool for a number of different surface topographies. Figure 6 shows the manufacturing set up for the individually structured stamps of the micro embossing master tool. The stamp material is brass.

For the hot embossing, two process strategies are being analysed and evaluated, a force controlled process using relatively high embossing forces at relatively low temperatures and a depth controlled process, using higher temperatures and less embossing force by precisely controlling the depth of the tool penetration (see Figure 7).

The latter requires stricter temperature control as it works very close to the glass transformation temperature and can lead to uncontrollable melting of the polymer substrate. However, considering the small structure size replication aimed at in this project and the resulting fragile tool geometries, the depth conrolled approach is currently being favoured and has been successfully used in the replication of a first set of structures.

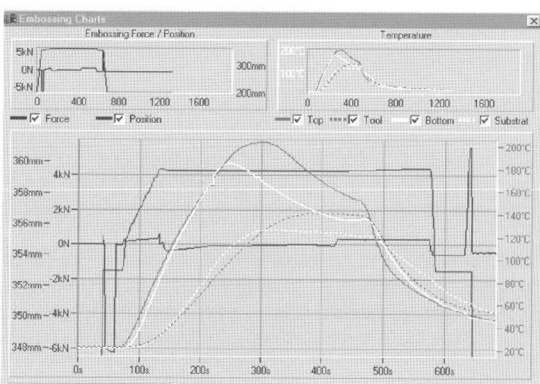

Figure 7: Embossing chart of depth controlled process

2. Micro structuring technologies

Beside the commonly used lithographic methods, diamond cutting has been established as a highly capable alternative in micro manufacturing, mostly due to advanced machine technology. Especially for surface structuring applications diamond machining is offering much higher geometrical flexibility, enabling for instance the manufacturing of structures such as v-grooves and blazed gratings at any given angle or grooves with semi circular cross sections. However, neither lithography nor diamond machining processes alone are suitable for cost effective mass production of structured surfaces which are required for the development of marketable products for laboratory and industrial applications in life science and bio technology. This can only be achieved utilizing high volume replication technologies and biocompatible thermoplastic materials. Therefore, the research scope of the described project includes diamond machining as well as replication technologies such as hot embossing and injection moulding.

2.1. Micro structuring using diamond tools

Microcutting operations using diamond tools are currently being performed on non-ferrous metals, plastics and brittle crystalline materials. The two most important factors determining the manufacturing accuracy are the accuracy of the machine system and the diamond tool accuracy.

Figure 2 shows a precision machining center used for diamond cutting processes featuring a granite machine base for vibration isolation, hydrostatic bearings for friction motion and a high resolution motion system using direct drive technology. This machine can be equipped either with a high frequency milling spindle, a fly-cutting spindle (as seen in Figure 2), a fast tool servo system or a tool holder for planing tools.

Figure 2: Diamond machining center

Diamond machining research in the current project focuses on fly cutting and planing processes, both suitable for the manufacturing of parallel micro grooves at different widths and pitches as used in bio-functionalized surface structures. Fly cutting, also referred to as single tooth milling, combines translatory motion of the machine axes with a rotating cutting tool, thus providing high cutting speeds at a feed rate dependent depth of cut, allowing the economical production of high aspect ratio surface features. Planing uses only the translatory motion of the diamond tool for material removal. This on the one hand significantly shortens machining times especially for low

aspect ratio structures but also requires high machine dynamics in order to provide adequate cutting speeds.

The range of structure geometries which can be manufactured using diamond machining are mostly dependent on the manufacturable tool geometries. The diamond tool cuts a groove with a cross section resembling the shape of the cutting edge of the tool at the applied depth of cut. Figure 3 shows an example of a radius tool and the corresponding groove cross section. Diamond machining offers a large variety of cross section geometries including v-grooves at virtually any opening angle between 180° and 0, semi-circular cross sections including radii down to 1-2 µm, square cross sections and combinations of v-shaped and square sections with the use of faceted tools. Special lapping processes give the geometrically defined cutting edge of the synthetically manufactured or natural diamonds, cutting radii of less than 20 nm.[6]

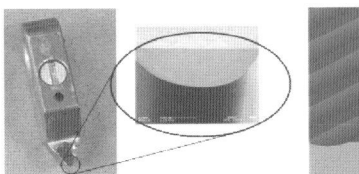

Figure 3: Radius tool and corresponding structure geometry

2.2. Replication using hot embossing technology

Considering high volume fabrication and flexible application of the functionalized surfaces to be developed within this project, structured film materials have been identified as a product with a highly promising marketability. These can be structured using a continuous roll-to-roll embossing process in which the pre-heated film surface is being pressed onto a heated, micro structured roller, thereby replicating a negative of the structure machined into the roller surface onto the surface of the film.

For the development of the continuous embossing process, a roll-to-roll embossing system by Saueressig is being used which has been modified to provide control of the necessary process parameters. In order to achieve higher flexibility with respect to the micro structures to be embossed, the normally directly structured heated roller has been replaced by a smaller sized roller which can accommodate replaceable structured ring elements (see Figure 4). These ring elements are being structured using diamond machining. Thereafter they are assembled, ensuring a close fit to the roller at embossing temperature, in order to provide full heat conductivity from the heated roller. Aiming at a further increase of the system's flexibility, a mounting system for structured metal film as well ultra thin structured silicon wafers is also being developed.

The combination of a continuous roll-to-roll process and flexibly interchangeable master structures, including both diamond machined non ferrous metal foils as well as ultra thin silicon wafers, which are structured using lithographic technologies, provides a high degree of flexibility while simultaneously enabling high production volumes.

Multi-Material Micro Manufacture
S. Dimov and W. Menz (Eds.)

Manufacturing and replication of cell aligning micro structures

C. Brecher[a,b], R. Klar[b], F. Pretzsch[b], C. Wenzel[b]

[a] Werkzeugmaschinenlabor (WZL), RWTH Aachen University, Germany
[b] Fraunhofer-Institute for Production Technology Aachen, Germany

Abstract

Biotechnology is becoming more and more important and is influencing our everyday life. One of the most important advantages will become the manufacturing of mass customized cells also known as tissue engineering. To optimize and shorten the cell growth, micro structures have shown an important impact on the cell division, alignment and cell differentiation. This paper deals with the manufacturing of micro structured moulds by diamond machining and the successive replication process to create bio functional surfaces provided with micro structures. While the fabrication of micro structures by diamond machining is an expensive and long lasting process the replication of bio functional surfaces by hot embossing allows the cost efficient production of these surfaces. Compared to lithography processes, diamond machining is very flexible and offers next to 2D or 2 1/2D geometries real 3D forms to copy in-vivo conditions for in-vitro cell replication processes. One of the main advantages of diamond machining is the flexibility to diversify the shape of micro structures rapidly to investigate the cell - substrate - interaction down to the micrometer range.

Keywords: micro structure, bio technology, diamond machining, cell alignment

1. Introduction

The possibilities of bio technology applications are becoming more and more a part of everyday life. One of the most interesting possibilities is the tissue engineering which allows the manufacturing of autogenic implants like skin and cartilage. In that coherence it was discovered that cell adhesion and attachment is mainly influenced by molecular surface interactions. But morphology and orientation of cells are affected by topographic cues of the extracellular matrix.

It is well known that nanometre-scale structures and structured surfaces have a positive influence on cell growth and proliferation (see Figure 1). This is due to the simulation of Extra Cellular Matrix (ECM) by these structures, providing in-vivo topographical conditions for in vitro-cell cultures. Structures on a micrometre-scale on the other hand are expected to have sizeable impact on cell orientation as well as separation and isolation of individual cell types from tissue probes. This can be used in the efficient isolation of important cell types such as adult stem cells or epithelial cells but also to separate different cell types such as Melanocytes, Keratinocytes and Fibroblasts from skin tissue. [1], [2], [3]

In order to maximize the effects of nanometre-scale structuring on the growth of individual cell types, in-vivo ECM topographies can be used as a model for the generation of artificial structures. Different manufacturing processes such as for instance Electro Spinning or Solid Freeform Fabrication (such as 3D-Printing or Fused deposition Modeling) are being developed and optimized for the specific challenges of manufacturing in-vitro ECM structures.[4] [5]

For topographical structures on the micrometre scale however, a comparable model to the ECM doesn't exist and extensive research is required including cell biology and manufacturing technology. Aiming at the systematic approach to the classification and high volume production of cell type specific micro structured surfaces for use in bio reactors, an interdisciplinary team of Fraunhofer researchers in the fields of biology, micro technology and precision manufacturing are therefore working on the development of bio-functionalized surface structures and their mass replication. Micro structured surfaces are being developed, produced and systematically analyzed for their applicability with different cell types. Structure geometries include parallel grooves of various cross sections (square, round, v-shaped) as well as stub like surface textures of different geometries and spatial distribution. Structure sizes range from multiple micro meters down to sub micro meter values.

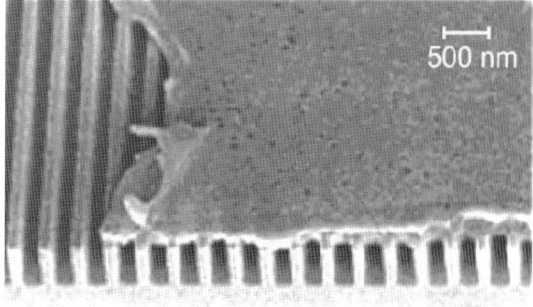

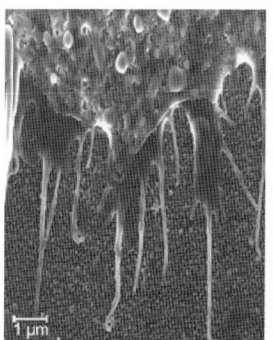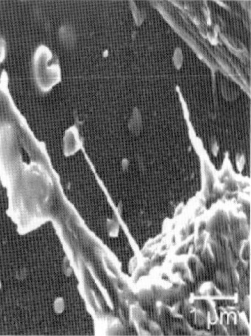

Figure 1: Cell interaction with nano structured surfaces

deposit point structures and horizontal and vertical lines by electron beam induced deposition. When doing this conventionally, it is essential to set up every line and point manually and to examine the correct time for each deposition process. Additionally the change of the precursor is very time consuming. The main advantage of our automated system is the possibility of predefining complex deposition sequences that can be executed several times without manual interaction.

3. Acknowledgements

We want to thank the German Federal Ministry of Education and Research (Bundesministerium für Bildung und Forschung, BMBF) for funding parts of this work within the research project "ZuNaMi" (promotional referrence: 16SV2276). Additionally, we gratefully acknowledge the support of the german DAAD and the British Council Germany through the British German Academic Research Collaboration (ARC) programme.

References

[1] D. M. Schaefer, R. Reifenberger, A. Patil, and R. P. Andres, "Fabrication of two-dimensional arrays of nanometer-size clusters with the atomic force microscope," Applied Physics Letters, vol. 66, no. 8, pp. 1012 – 1014, 1995.

[2] F. J. Rubio-Sierra, W. M. Heckl, and R. W. Stark, "Nanomanipulation by Atomic Force Microscopy," Advanced Engineering Materials, vol. 7, no. 4, pp. 193–196, 2005.

[3] G. Schitter, F. Allgöwer, and A. Stemmer, "A new control strategy for high-speed atomic force microscopy", Nanotechnology, 2004, 15, pp. 108–114.

[4] Y. Li, J. Bechhoefer, "Feedforward control of a closed-loop piezoelectric translation stage for atomic force microscope", Review of Scientific Instruments, AIP, 2007, 78, 013702.

[5] B. Mokaberi, J. Yun, M. Wang, and A. A. G. Requicha, "Automated nanomanipulation with atomic force microscopes", Proc. IEEE Int'l Conf. on Robotics & Automation (ICRA '07), Rome, Italy, pp. 1406–1412, April 10-14, 2007.

[6] B. Mokaberi and A. A. G. Requicha, "Drift compensation for automatic nanomanipulation with scanning probe microscopes", Automation Science and Engineering, IEEE Transactions on, no. 3, pp. 199–207, July 2006.

[7] D. Fox, W. Burgard, F. Dellaert, and S. Thrun, "Monte Carlo localization: Efficient position estimation for mobile robots," in Proceedings of the Sixteenth National Conference on Artificial Intelligence (AAAI'99)., July 1999.

[8] S. Thrun, W. Burgard, and D. Fox, "Probabilistic robotics". MIT Press, 2005.

[9] K. Edinger, H. Becht, J. Bihr et. al., „Electron-beam-based photomask repair," J. Vac. Sci. Technol. B vol. 22(6), 2004, p2902-2906.

[10] H. W. P. Koops, C. Schossler, A. Kaya and M. Weber, "Conductive dots, wires and supertips for field electron emitters produced by electron beam induced deposition on samples having increased temperature," J. Vac. Sci. Technol. B vol. 14(6), 1996, p4105-4109.

[11] T. Sievers, T. Wich, "Assembly inside a Scanning Electron Microscope using Electron Beam induced Deposition," in Proceedings of IEEE/RSJInternational Conference on Robots and Intelligent Systems, 2006.

[12] S. Fatikow, V. Eichhorn, C. Stolle, T. Sievers and M. Jähnisch, "Development and Control of a Versatile Nanohandling Robot Cell", Journal of Mechatronics, 2007, submitted

[13] C. Stolle, "Distributed control architecture for automated nanohandling," in Conference on Informatics in Control, Automation and Robotics (ICINCO'07), p. 127-132, 2007.

[14] T. Sievers and S. Fatikow, "Real-time object tracking for the robot-based nanohandling in a scanning electron microscope," Journal of Micromechatronics - Special Issue on

[15] V. Scheuer, H. Koops, and T. Tschudi. "Electron beam decomposition of carbonyls on silicon", Microelectronic Engineering vol. 5, 1986, p423–430.

methods allow to determine how inputs affect responses. One of the main advantages of this method is that it can be used to systematically determine the optimal process parameters with a reduced number of test runs. In our case, full factorial design with a central point was used. This is the most comprehensive test plan. It can be applied with a large number of parameters and all relationships can be shown. The number of trials depends on the chosen number of parameters.

$$\text{Number of trials} = n^k + c \qquad (1)$$

- n = number of variable steps
- k = number of parameters
- c = central point value

Five parameters were selected and used for the method (Fig. 3). This led to $2^5 + 1 = 33$ individual trials. 330 parts were made with these injection moulding parameter sets.

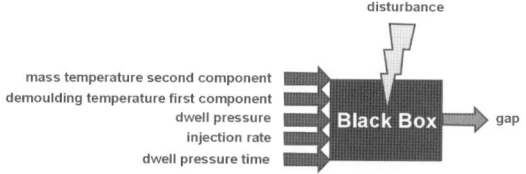

Fig. 3. Design of experiments (DOE) method

To obtain promising values for these parameter sets, a first study was carried out to find out which data would result to a reasonable outcome. The mass temperature of the second component was adjusted to 205°C and 220°C. This is the upper temperature limit for the material. The demoulding temperature of the first component was set to 60°C and 80°C, respectively, in order to find out whether either a higher contact temperature was necessary or a lower temperature was sufficient for the two-component process. To reduce the shrinkage high dwell pressures (250 and 420 bar) were chosen. The injection rate of the second component was set to 40 mm/s and 70 mm/s, respectively, since this injection rate was practicable in first test studies. The dwell pressure time was set to a very high value of 5 to 15 seconds to reduce the shrinkage of the second material before demoulding. Table 1 shows the parameter sets and gives the parameter set for the central point as well.

Table 1
Parameter sets for the DOE method

Parameter		[-1]	Central point [0]	[+1]
Mass temperature second component	°C	205	212	220
Demoulding temperature first component	°C	60	70	80
Dwell pressure	bar	250	330	420
Injection rate	mm/s	40	55	70
Dwell pressure time	s	5	10	15

These parameter sets were integrated in a DOE scheduling which was carried out with a Ferromatic Milacron K50 device and a special two-component tool built for this process development. This tool is additionally equipped with a moveable index plate for demoulding the first component and placing it into the second cavity. In addition the tool contains of a unit for conducting a variothermal process, and a unit to evacuate the cavity. These special features are important to micro injection moulding in general and the moulding of these structures in particular. Each of the parts produced has 12 micro gear wheel cavities. Since every micro gear wheel has a different position on the template every position must be considered separately. This makes a detailed examination of every micro gear wheel necessary. In the DOE presented here, three of the micro gear wheels are analysed in more detail. They are marked in Fig. 4.

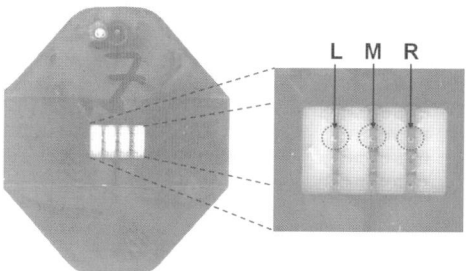

Fig. 4. Two-component template with the micro gear wheels used for detailed investigation

The micro cavities of the two-component parts removed are immersed into an electroplating bath. By using a standard electrolyte for electroforming, the gear wheel cavities are filled with a nickel alloy using the following parameter set (Table 2).

Table 2
Electroplating parameters

Electrolyte	Nickel sulfamate solution with boric acid (buffer) and flourtensid (for better wettability)
pH value	3.3-3.5
Electric current	The deposition is controlled by the current density $i = I / A = \text{const.}$
Voltage	The voltage is adjusted to the resistance of the cathode and varies during deposition
Growth	Approximately 12 µm / h
Period	Approximately 45 min $\rightarrow$ ~ 8 µm

During the electroplating process, the gap between the first and second component is filled completely with a maximum of 8 µm nickel alloy. After this, the part is dried and, in a next step, is analysed using x-rays. The whole part is placed into a Micro-Focus X-ray device (YXLON International X-Ray GmbH, Germany) to analyse the pattern area of the produced gap (Fig. 5). The size and the contour of this area are the targets needed for the DOE analysis. They are required for an interpretation of the design of experiment method.

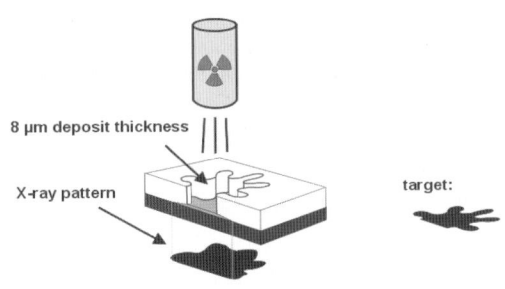

Fig. 5. Scheme of x-ray pattern analysis

As can be seen, the higher mass temperature of the second component show significantly better results. In order to reduce the measuring effort, the parts produced with the lower mass temperature were not analysed in detail. Parts based on the other 17 parameter sets were analysed. The produced gap that reflects the quality of the moulding process was clearly visible. First, as shown in Fig. 6, the contrast of the x-ray pattern is very weak.

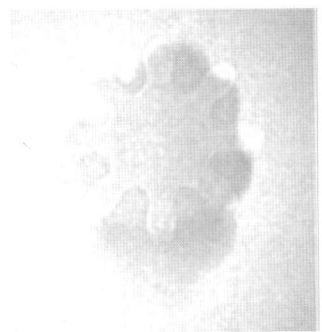

Fig. 6. Concrete example of x-ray pattern analysis

With suitable software and by using filters and generating a binary image with the "Analysis" programme (Software Imaging GmbH, Germany), however, it is possible to measure the area of the gap with high accuracy. This result is the input value needed for the DOE analysis. From all parameter sets, the second line was analysed taking into account the three different positions of the micro gear wheels.

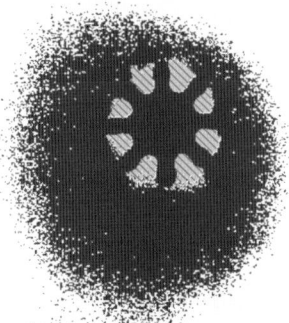

Fig. 7. Part 25-9R | measured value of the gap area: 0.0766305 mm²

The calculated values were processed using Visual-XSel® 10.0 by CRGRAPH. After the input of the parameter set, the programme recommends a defined experimental design sequence. With this sequence, further experiments are carried out. After the import of the measured target value, the programme calculates the best parameter set (Table 3).

Table 3
Parameter sets for "optimum"

Parameter		Value
Mass temperature second component	°C	220
Demoulding temperature first component	°C	60
Dwell pressure	bar	22
Injection rate	mm/s	70
Dwell pressure time	s	5

The detailed analysis of the four values shows a significant characteristic influence of the injection rate and the dwell pressure on product quality.

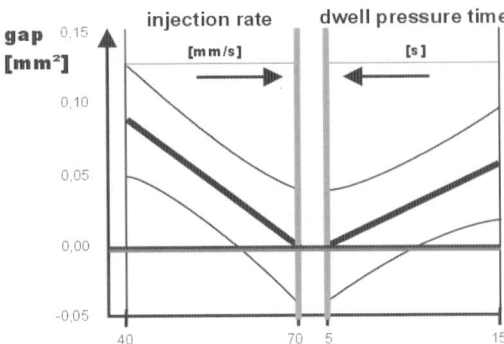

Fig. 8. Influence of the parameters of "injection rate" and "dwell pressure time" on the gap area

In contrast to this, the analysis reveals an insignificant influence of the demoulding temperature of the first component and of the dwell pressure itself.

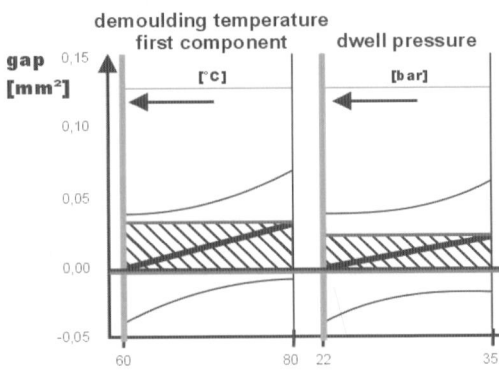

Fig. 9. Influence of the parameters of "demoulding

temperature" and dwell pressure" on the gap area

Nevertheless, investigations of other two-component products revealed similar effects of the demoulding temperature [8].

3. Results

Using the optimum parameter set for the two-component micro injection moulding process, new gear wheels were produced by electroplating. Now, the whole micro gear wheel was filled. The electroplating process took 24 hours. Fig. 10 shows the result of this production cycle. Micro gear wheels with hardly any burrs at the end were fabricated.

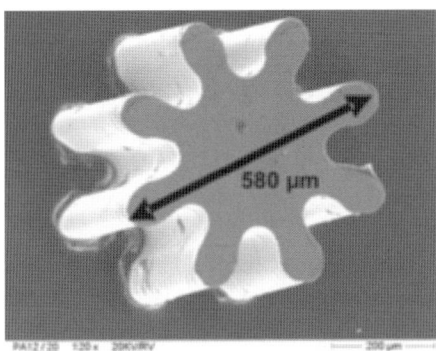

Fig. 10. Replicated gear wheel based on an optimal DOE parameter set

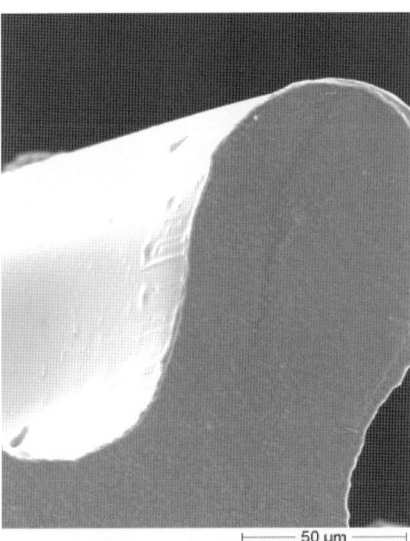

Fig. 11. Detail of a replicated micro gear wheel after optimisation of process parameters

Further improvement efforts will be taken. In a next step hardness, surface quality and microstructure of the electroformed gear wheels will be investigated. In addition different process parameters will be studied to further optimise the properties. The accuracy of the replicated gear wheels has to be investigated and characterised in comparison to that of the mould insert. Finally, the process shall be optimised by adjusting some details of the mould and of the mould insert as well.

A further step will consist in producing a functional demonstrator. This task is planned to be carried out by the research group FOR 702. Additional tests will be aimed at evaluating the geometrical limits of the entire process chain. This will lead to recommendations for the future fabrication of micro components by two-component injection moulding followed by electroplating.

Acknowledgements

This project is supported by the German Research Foundation (DFG) within the Research Group FOR702 "Machine, tool, and process development of new methods to produce micro parts by liquid phases". We also thank all colleagues of IMF III involved and the members of the research group for their assistance. Finally, we would like to thank Evonik GmbH for providing material and CRGraph for the permission to use XSel® 10.0.

References

[1] Dambrowsky, N.; Schulz, J.: Goldgalvanik in der Mikrosystemtechnik. Herausforderungen durch neue Anwendungen. Wissenschaftliche Berichte, FZKA-7308 (April 2007), Dissertation (N. Dambrowsky), Universität Karlsruhe 2006, Germany

[2] Degen, R. ; Kirsch, U.: Extrem leichte und schnelle Mikroantriebe für hochpräzise Montageanwendungen. Galvanotechnik 4/2007, Eugen G. Leuze Verlag, Germany

[3] O'Neil, S.; Thuerigen, C.: Ultraminiature gearmotors break the 2-mm barrier, Machine Design, May 20, 1999, Cleveland, USA

[4] Prokop J.; Finnah G.; Lorenz J.; Piotter V.; Ruprecht R.; Haußelt J. : Metallic Microparts Made by Electroplating on Two-component Injection Molded Templates, MiNaT Congress, Stuttgart, 11 - 15 May 2007, Germany

[5] Piotter, V.; Holstein, N.; Oskotski, E.; Schanz, G.; Ruprecht, R.; Haußelt, J.: Metal micro parts made by electroforming on two-component lost polymer moulds. 2003: Proceedings of 3rd Euspen international conference, Eindhoven, pp. 367–370

[6] Finnah, G.; Naumann, K.; Holstein, N.; Piotter, V.; Ruprecht, R.; Haußelt, J.: Herstellung von metallischen Mikrokomponenten durch Einlegespritzgießen und anschließende Galvanoformung. Galvanotechnik 95 (2004): pp. 2776–2780, Eugen G. Leuze Verlag, Germany

[7] Prokop J.; Finnah G.; Lorenz J.; Piotter V.; Ruprecht R.; Haußelt : Manufacturing Process for High Aspect Ratio Metallic Micro Parts Made by Electroplating on Partial Conductive Templates, Microsystems Technologies, online Publication, 12 February 2008, Springer, Germany

[8] Kühnert, I.: Grenzflächen beim Mehrkunststoff-spritzgießen, Chemnitz, Technische Universität, Dissertation 2005, Germany

Metrology: inspection and characterisation methods

Measurement of frequency response of the bone ossicles in the sheep middle ear by the fiber-optic microphone

Z.V. Djinovic[a,b], R. Pavelka[c], L. Manojlovic[b], D. Vujanic[b], M.C. Tomic[d]

[a] Institute of Sensor and Actuator Systems, Vienna University of Technology, Vienna 1040, Austria
[b] Integrated Microsystems Austria, Wiener Neustadt 2700, Austria
[c] Schwerpunktkrankenhaus, 2700 Wr.Neustadt, Austria
[d] Institut Bezbednosti, Belgrade 11000, Serbia

Abstract

In this paper we present a contactless technique for measurement of acoustic vibrations and frequency response of the bone ossicles in the sheep middle ear. This technique is based on high-coherence interferometry performed by a single mode fiber-optic sensing configuration with just one sensing fiber directed against the target. High-coherence light of 1310 nm wavelength has been impinging a retroreflective target that is firmly fixed upon the incus of the middle ear. We measured frequency response of the middle ear causing vibration of the incus by generation of air pressure from a well calibrated acoustic source in the range of 50 to 90 dB SPL and frequency from 250 Hz to 6 kHz.

Keywords: fiber-optic sensors, index of refraction measurement, interferometry, fiber optic sensor

1. Introduction

Hearing loss is today one of the largest problem affecting modern society. There is estimation that about 10 percent of American population has some kind of hearing loss. This is of about 30 million people. In Europe the situation is very similar; approximately one third of all seniors older than 75 years have a significant hearing impartment. About 14 percent of younger people aged between 45 and 65 years, has hearing dysfunctions too.

Most of the hearing troubles can be solved by some kind of hearing aids. The most popular are conventional hearing devices that can be applied outside in close proximity of the ear pinea. However, there is a list of disadvantages of such devices, e.g. repeated inflammation of the auditory canal, stigmatization, feedback noise, limited speech comprehension, etc. [1].

Implantable hearing devices are able to overcome these obstacles. There are two approaches in developing; partially and totally implantable hearing devices (TIHA) [1,2] aimed for cochlear or middle ear implants. However, TIHA devices are still far from the matured state because of several drawbacks that have to be solved. The largest one seems to be the lack of a reliable microphone and further, an invasive connection between the sensing probe and bone ossicles in the middle ear.

Our group has been for more than ten years involved in the investigation of an optical technique as a basic principle for a miniature and reliable totally implantable microphone [3-5]. A great deal of our effort has been dedicated to the low-coherence interferometry as a physical sensing principle [6,7]. However, we met some drawbacks (e.g. low budget of optical power, small distance between the sensing fiber and target, etc.) that caused to turn on the sensing principle toward high-coherence interferometry. It is well known that this technique can provide high accuracy and resolution in distance measurement. Some authors [8-10] use laser vibrometry for non-invasive characterisation of the middle- and inner ear based on vibrations of the ear drum because of very good sensitivity till to 1 pm. However, they do not use this technique for manufacturing of an implantable microphone or vibrometer as a possible part of TIHA.

In this paper we present that high-coherence interferometry can be used as a contactless technique for measurement of acoustic vibrations and frequency response of the bone ossicles directly in the middle ear of a sheep. We performed measurements by a single mode fiber-optic sensing configuration with just one sensing fiber directed against the target. High-coherence light of 1310 nm wavelength has been impinging a retroreflective target that is firmly fixed upon the incus of the middle ear. We measured frequency response of the middle ear causing vibration of the incus by generation of air pressure from a well calibrated acoustic source in the range of 50 to 90 dB SPL and frequency from 250 Hz to 6 kHz.

2. Principle of Operation

The basic principle of operation is high-coherence interferometry (HCI) performed by "all-in-fiber" interferometric sensing configuration. In Fig. 1 we present a high-coherence interferometric set up composed of the two main parts: optoelectronic interface and sensing head. Central part of the optoelectronic unit is a 3×3 fiber-optic directional coupler made of single mode fiber-optic waveguides of 9/125 μm in diameter and optimized to transmit 1310 nm of light wavelength. Such a configuration provides passive stabilization of the signal by combination of the two high-coherence interferograms mutually shifted by $2\pi/3$.

Light source-laser diode (HCS) is coupled into the middle input arm of the 3×3 coupler, while two receiving photodiodes (PD1 and PD2) are connected to the two other input arms. One output arm of the coupler is the sensing fiber directed to the retroreflective target firmly clamped to the incus of the middle ear. The second arm is the referencing fiber directed to Al mirror. The middle output arm is a passive fiber immersed into the index matching gel (IMG) in order to avoid the back reflection from the fiber tip. Radiations have been back-reflected from the target, i.e. incus and Al mirror and have been recombined into the 3×3 coupler, forming quadrature interferometric signals. These signals are captured by two photodiodes. Signal processing has been made on line in an optoelectronic interface in order to convert vibrations of the ossicle into an audio signal.

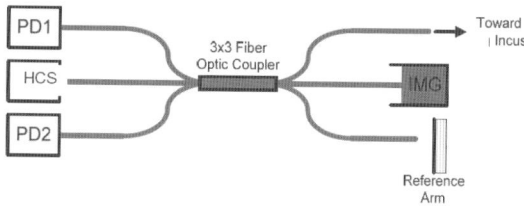

Fig. 1 Schematic presentation of "all-in-fiber" interferometeric sensing configuration, PD1, PD2-photodiodes, HCS-high-coherence source, IMG-index matching gel, Sensing fiber (upper outlet arm) is incorporated into the middle ear

The current signal i, captured by photodiode has the following shape:

$$i = I_0 \left\{ 1 + V \exp \left[-(2\Delta x/L_C)^2 \right] \cos(k\Delta x) \right\}$$ (1)

where I_0 is the maximum photodiode current, V is the fringe visibility, Δx is the optical path difference, L_C is coherence length of the used light source (HCS) and $k=2\pi/\lambda$, where λ is the wavelength of HCS.

3. Experiment

Ex vivo experiments have been done on the incus of the middle ear of a sheep. In Fig. 2 we present the overall view of the experimental set up.

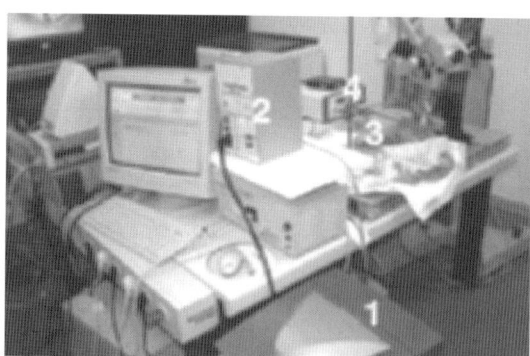

Fig. 2 Overall view of the experimental set up for measurement of frequency response of the bone ossicles of the sheep middle ear , 1-"Afinity" calibrated sound generator, 2-loudspeaker, 3-sheep head, 4-sensing optical fiber; A close up view of the middle ear is given in Fig. 4

We used sensing fiber ended with a collimator packed in a glass tube 1,8 mm in diameter and about 8 mm in length. Working distance between the target and collimator was about 5 mm. Some experiments have been done by the usage of an external x-y-z holder for the fine adjustment of the sensing fibber.   Later we

used a special holder presented in Fig. 3 for easy targeting and fixation of the sensing fiber in the right position with optimal back reflected signal. In order to have better visibility in phase of aiming of sensing fiber we used red laser light through the sensing configuration. It was very convenient to pick up retroreflective target fixed upon the incus. A special care was taken to leave the ear channel and tympanic membrane intact as well as original state of the whole auditory chain.

In Fig. 3 we present 3D design of a fiber-optic holder that provides easy way of aiming and adjusting of sensing fiber-collimator (1) against the retroreflective target. Sensing fiber goes through the ball sitting at the bottom of the extension of the basic holder part (2). After adjusting the ball and fiber are locked by pressing the middle part (3) of the holder. The top part (4) serves for protection of the outlet sensing fiber against overloading of fat, muscles, skin and connective tissue.

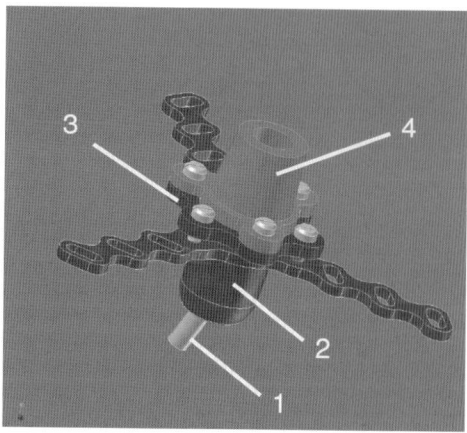

Fig. 3 3D design of a holder for aiming, adjusting and fixation of the sensing fiber against retroreflective target, 1) collimator, 2) basic part 3) middle part, 4) protection, overall size 20 mm

In Fig. 4 we present top view of the middle ear with characteristic bone ossicless; incus (1), stapes footplate (2) and malleus (3). We used joining place (indicated by arrow) between the stapedius and incus for clamping of the omega-ring with the top oriented retroreflective target. Overall dimensions of the incus is about $1,5 \times 1,5$ mm^2.

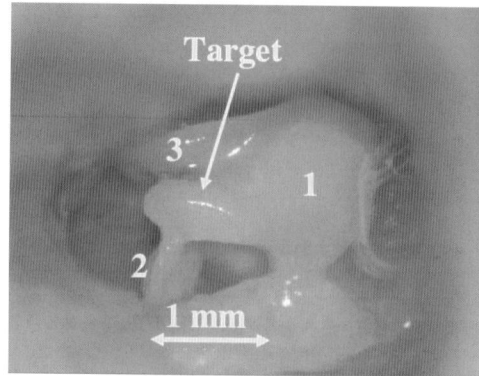

Fig. 4 Top view of the middle ear of a sheep, 1) incus, arrow shows an attaching place of the omega-ring, 2) stapes footplate, 3) malleus

In Fig. 5 we present the retroreflective targets encapsulated in silicone resin for the sake of biocompatibility and firmly fixed to the omega-shape ring. We used to kind of biocompatible materials for the omega-ring manufacturing, titanium and gold. Overall diameter of the rings was about 600 µm and width of about 300 µm. Retroreflective target is 3M special tape made of glass balls of about 50 µm in diameter and overall dimensions of about 250x250 µm^2. This is just one active retroreflector that we used. Overall mass of Ti omega-ring of about 1 mg has no significant influenece on alternation of the sensitivity of the ear.

a)   b)

Fig. 5 Omega-shape rings with encapsulated retroreflective target made of a) titanium and b) gold

We stimulated incus vibration by generating of sound pressure from a calibrated audiometer, with trade name "Affinity". We applied a sinusoidal pure tone with frequency varying between 125 Hz and 6 kHz at a sound level between 50 and 90 dB SPL. Previously we measured the frequency response of the ear-channel itself by placing of miniature microphone till to the eardrum. Sound pressure has been emitted from the loud speaker that was 1 m far away from the subjected middle ear.

4. Results and discussion

In Fig. 6 we depict a signal recorded by "Affinity" audiometer when sinusoidal wave of sound pressure of 70 dB attacked the eardrum of a sheep. This signal presents the frequency response of the ear channel alone. It is obvious that about 2 kHz appears strong amplification of the signal (resonance frequency) as a consequence of the acoustic impendence of the ear-channel.

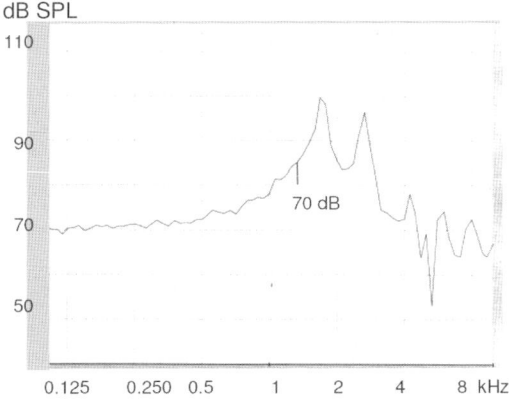

Fig. 6 Frequency response of the free ear-channel of a sheep

In Fig. 7 we present frequency response of the middle ear obtained by high-coherence light source. It was pig-tailed laser diode that delivered of about 2 mW of optical power into the sensing configuration. We measured at the end of the collimator of about 700 µW that was order of magnitude higher intensity than in the case of low-coherence source.

The obtained frequency response has a typical shape characteristic for other animals as well as for human beings. We can see the occurrence of resonant frequency somewhere around 2000 Hz. Our system works well in frequency range that is of the most important for a totally implantable hearing aid. Upper frequency range of 6000 Hz is limited by the bandwidth of the electronic circuit. However, for sound with frequencies between 1000 Hz and 4000 Hz we have very good signal regardless of the applied sound pressure.

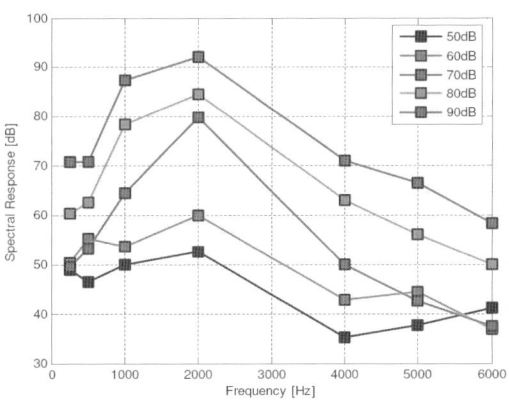

Fig. 8 Frequency response of the middle ear obtained by high-coherence source

The above results show that our system has very high resolution that is even in subnananometer range. Authors in Ref. [1] found that sound pressure of about 100 dB causes of amplitude of vibrations of bone ossicles of about 60 nm at 1000 Hz. We also found similar results that we published Elsevier [3,5]. Based on these findings we can measure amplitude of vibrations of about 50 pm/√Hz. It corresponds with noise floor of about 35 dB as we can see in Fig. 7.

5. Conclusion

In this paper we presented a contact-less optical technique based on high-coherence interferometry that is applicable for manufacturing of an implantable microphone as a part of totally implantable hearing aids. For this purpose we performed a fiber-optic sensing configuration that was used for *ex-vivo* measurement of frequency response of the middle ear of cadaver sheep. We applied a sinusoidal pure tone with frequency varying between 250Hz and 6000 Hz at a sound level between 50 and 90 dB SPL. We obtained a characteristic shape of the frequency response with resonance frequency of about 2000 Hz appears due to amplification effect of the ear-channel.

Acknowledgements

The authors would like to thank the Austrian Science Fund (FWF) for funding this research under the Project L139-N02 "Nanoscale measurement

138

of physical parameters" and the Integrated Microsystems Austria, IMA GmbH that partially supported the research activities in this paper. Also, this work was carried out within the framework of the EC Network of Excellence "Multi-Material Micro Manufacture: Technologies and Applications (4M)".

References

[1] H.P. Zenner, TICA totally implantable system for treatment of high-frequency sensorineural hearing loss, Ear. Nose Throat J. 79 (2000) 770-777

[2] E. Dalhoff, R. Gartnerr, U. Hofbauer, H. Tiziani, H. P. Zenner, A. W. Gummer, Low-coherence fibre heterodyne interferometer for both dc and high-frequency vibration measurements in the inner ear, J. Mod. Optics, 45 (1998) 765-775

[3] Z. Djinovic, M. Tomic, R. Pavelka, D. Vujanic, M. Cordes, Investigation and development of a fiber-optic vibrometer for use in totally implantable hearing aid", J. of Mech. Eng., 52 (2006) 7

[4] Z. Djinovic, L. Manojlovic, D. Vujanic, R. Pavelka, A. Vujanic, M. Tomic, Acoustical vibration measurement of the pig's middle ear ossicles by fiber-optic vibrometer, Eurosensors XX, Technical Digest of the 20th European Conference on Solid-State Transducers, Göteborg, Sweden, September 17-20, 2006

[5] R. Pavelka, A. Vujanic, Z. Djinovic, S. Mitic, D. Vujanic, Ch. Kment, M. Tomic, Animal Experiments with a Fiber-Optic Vibrometer to be used as a Microphone for Totally Implantable Cochlear and Middle Ear Implants, Proceedings of the 4 th International Symposium on Electronic Implants in Otology & Conventional Hearing Aids, 5-7 June 2003, Toulouse, France

[6] Z. Djinović, S. Mitić, Ch. Kment, A. Vujanić, D. Vujanić, R. Pavelka, Middle Ear Investigation by Low Coherence Interferometry, ETRAN 2003, Proceedings of the XLVII Conference, vol. IV, pp. 137-140, 8-13 Juny 2003, Herceg Novi, Serbia & Montenegro

[7] M. Tomić, J. Elazar, Z. Djinović, Low-coherence interferometric method for measurement of displacement based on a 3x3 fibre-optic directional coupler, J. Opt. A: Pure Appl. Opt. 4, pp. S381-S386, Nov. 2002

[8] D. Turcanu, E. Dalhoff, H. P. Zenner, A. W. Gummer, Laser Doppler vibrometric measurements of DPOAE in humans : Eardrum vibrations reflect middle- and inner-ear characteristics, HNO, 55 (2007) 930

[9] D. Turcanu, D. Martu, E. Dalhoff, A. W. Gummer, Laser Doppler vibrometry: a new tool for diagnosing hearing loss with an intact eardrum, Rev Med Chir Soc Med Nat Iasi, 110(2006)357

[10] E. Dalhoff, R. Gaertner, H. P. Zenner, H. J. Tiziani, A. W. Gummer, Remarks about the depth resolution of heterodyne interferometers in cochlear investigations, J. Acoust. Soc. Am., 110 (2001) 1725

How reliable are surface roughness measurements of micro-features?
- Experiences of a Round Robin test within nine 4M laboratories

L. Mattsson[a], P. J. Bolt[b] , S. Azcarate[c], E. Brousseau[d], B. Fillon[e], C. Fowler[f], E. Gelink[b], C. Griffiths[d], C. Khan Malek[g], S. Marson[h], A. Retolaza[c], A. Schneider[f], A. Schoth[i], A. Temun[a], P. Tiquet[e], and G. Tosello[k]

[a] KTH – the Royal Institute of Technology, Department of Production Engineering,
School of Industrial Engineering and Management, SE-10044 Stockholm, Sweden
[b] TNO Science and Industry, 5600 HE Eindhoven , The Netherlands
[c] Tekniker Technological Center, 20600 Eibar, Spain
[d] Cardiff University, Manufacturing Engineering Center (MEC), Cardiff CF 24 3AA, United Kingdom
[e] French Atomic Energy Commission (CEA),
Laboratory of Innovation for New Energy Technologies and Nanomaterials (LITEN), 38054, Grenoble, France
[f] Science and Technology Facilities Council, Rutherford Appleton Laboratory (RAL), Technology – Central
Microstructure Facility, Harewell Science and Innovation Campus, Didcot, Oxfordshire, OX11 0QX, UK
[g] FEMTO-ST Institute, CNRS UMR 6174, LPMO Department, 25044 Besancon Cedex, France
[h] School of Applied Sciences, Cranfield University, Cranfield, Beds, MK43 0AL, UK
[i] University of Freiburg, Institute of Microsystem Technology (IMTEK), 79110 Freiburg, Germany
[k] Technical University of Denmark (DTU),
Department of Manufacturing Engineering and Management (IPL), 2800 Kgs. Lyngby, Denmark

Abstract
Surface roughness of tiny micro machined features is not easy to verify. The statistical variation of the surface itself can be the limiting factor that hampers tolerance verification. In this paper we have studied this effect and we also test the performance of 10 different surface profilers over a very well specified surface area. For this area 6 profilers yielded the same result within a standard deviation window of ±6%. For other areas, on top of narrow bars and in narrow and deep channels, a much larger spread in the Round Robin results was found.

Keywords: Surface roughness, Micro metrology, Round robin, Surface profiler, Ra, Pa

1. Introduction

The 4 M Multi Material Micro Manufacture Network of Excellence has proven to be an excellent forum for testing procedures and throwing light on non-standardised metrology issues of micro components. In a first test round we found that large variations in surface roughness measurements existed among the partners in the Polymer and Metrology divisions. Also, the dimensional measurements of the micro ridges of four metal inserts were far from a complete overlap when the results were analysed. This first test was quite open and gave no detailed instructions to the instrument operator on how to perform the measurements, as the intention was to get an idea of the measurement praxis at the different laboratories and if it is good enough to provide reliable and accurate data. The result was negative. Even on a simple ground metal surface the spread in surface roughness values was astonishing, and the width and height measures of the micro structures departed significantly depending on what lab did the measurement.

As already pointed out in the review paper on micro metrology challenges[1] severe problems arise when surface roughness is to be assessed on tiny surfaces, as the entire concept of ISO standardised line profile based Ra, Rz, Rq, Rp, Rv parameters rely on long traces – by default seven times the sampling length including a start and a stop length. For a surface roughness of about 1 µm the standardised sampling length is 0.8 mm, which means a need to access a default line profile of 5.6 mm. This is far from acceptable and contradictory to the reality of many micro components. The updated ISO standards 4287 [2] and 4288 [3] give the opportunity to specify primary profile based parameters Pa, Pz, Pq, Pp, Pv – values instead, where sampling length, i.e. the measured length along the surface, can be set to an arbitrarily determined length adapted to the surface feature being measured, as long as it is stated with the measured values. The primary profile is by default only filtered by the short wavelength cut off at 2.5 µm, i.e. it is not suitable for sub-micron feature roughness determination like measurements with atomic force microscopes.

If a non-skilled operator just put the sample on the measurement table and press the Ra measurement button the risk is obvious that totally wrong results will come out of the instrument. It is therefore absolutely necessary to have well educated instrument operators when micrometer sized features are to be measured.

After the initial metrology Round Robin on several injection mold inserts, it was obvious that a much more strict measurement description was needed and only one insert, milled at MEC Cardiff, was chosen for the Metrology Round Robin presented in this paper.

2. Equipment and artefact

2.1. Equipment
The instruments used for roughness

measurements in the Round Robin were all commercial

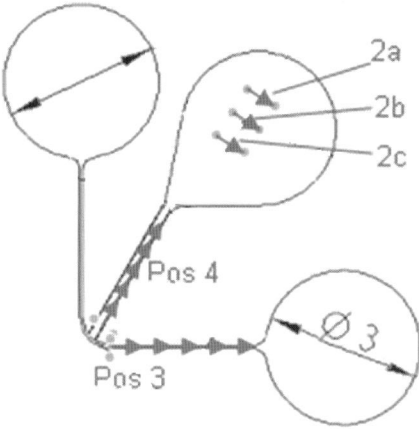

Fig. 1 Layout of the artefact and the measurement positions of the surface roughness. Dots indicate laser drilled holes for alignment purposes and arrows show the individual profiles to be recorded at top of the 75 µm wide ridge (Pos 3) and between two 100 um high ridges separated by 150 µm (Pos 4).

instruments; four optical profilers based on white light interferometry (WLI), two confocal microscopes, one autofocusing profiler, one chromatic focusing profiler and two stylus profilers. Each instruments have its own calibration gauge delivered with the instrument for traceable verification of its performance. Nine different laboratories participated in the surface roughness Round Robin, while only five managed to fulfil the dimensional measurements.

2.2. Artefact

For the Round Robin testing of surface roughness on true 3D micro-features we have to rely on our own artefact, as ISO certified roughness gauges are meant to yield the same value all over the surface, and would not reveal the lateral micro feature size influence that might exist. The artefact, shown in Fig.1, is a steel insert made by mechanical micro milling. The structure contains three 100 µm high Ø 3 mm plateaus connected with 100 µm high ridges of nominally 20, 50 and 75 µm widths. The ground surface of the middle plateau was selected as test area (2a – 2c) for as-good-as-possible surface roughness inter-comparison

Fig. 2 Micrograph of primary area for surface roughness measurements with 6 laser drilled alignment holes 300 µm apart. Pos. 2a is at the top

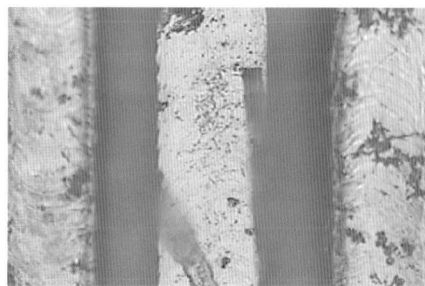

Fig. 3 Micrograph of bottom structure between the 20 um ridges. (Pos 4 in Fig.1) Note the burrs at the bottom that cause measurement problems, in particular for stylus profilers.

measurement. Pos 3 and 4 reveal the roughness reproducibility on top of the 75 um ridge and at the bottom of the 150 um wide and 100 um deep channel respectively. Six alignment holes were laser drilled 300 µm apart as shown in Fig.2 for x,y localization of measurement traces. The arrows in Fig. 1 show the profiling directions.

At position 2a, 2b and 2c the surface profile was recorded from hole to hole and the 250 µm sampling length centred between the holes is used for calculating the P-parameter values for each profile. At position 3, five consecutive 250 µm long profiles were recorded along the nominally 75 µm wide ridge starting at the laser drilled hole pair to the left. The same procedure was done at position 4, at the nominally 150 µm wide bottom level surrounded by the two 100 µm high ridges. For the stylus profilers this structure was critical as the cone angle of the stylus limits the positioning tolerances to some ten microns sideways. It was also evident that there were some burrs left at the bottom from the milling operation, as seen in Fig.3.

3. Measurement procedure

3.1. Parameter selection

As discussed in the introduction the surface roughness parameters to be preferred in micro-metrology are the P-parameters based on primary profile data introduced in ISO 4287:1997[2] By default P parameters are evaluated over the feature size, but P-values can just as well be stated with a clearly defined measurement length that can be set in accordance with the feature to be measured.

Based on these facts the Round Robin parameters to be measured were stated as Pa – arithmetic mean deviation of the assessed profile, Pq – root mean square deviation of the assessed profile (equivalent to the standard deviation of the surface structure), Pv – maximum profile valley depth, Pp – maximum profile peak height and Pz – maximum height of profile.

However, in order to compare the output of instruments having the options of both P and R parameters, Ra, Rq, Rv, Rp and Rz were also requested, but just for one single sampling length of 250 um. In most figures to follow Pa is used as it corresponds to the common Ra values. Pq which measures the standard deviation of the surface structure is a bit more sensitive than Pa to larger pits and bumps.

3.2. Round Robin Measurement instruction

A thorough checklist and measurement protocol was provided with the metal insert sample for the measurements. The checklist, from which excerpts are given here, was aimed to standardise the Round Robin measurements between the different metrology labs, and to eliminate possible mistakes.

General sample handling remarks:
Keep the sample in the delivery container as long as possible to prevent particle contamination. Do not touch the insert with your fingers!! Wear clean dust free gloves! After measurement secure the insert safely in the delivery container, for transportation to next partner.

Sampling length and filtering:
For the analysis we make use of a single sampling length of 250 μm. The measured and stored profile length might be larger, e.g. to cover the laser drilled holes. For terminology and parameters refer to EN ISO Standard 4287:1998.

If possible operate the instrument with minimum filtering in the measurements, and perform the analysis of the 250 μm sampling lengths for the P-parameters and R-parameters as advised in the table.

Roughness measurement at position 2
Alignment marks have been laser drilled into the surface of the flat at the pos. 2 area according to figures 1 and 2. They are positioned in pairs, referred to as 2a, 2b and 2c, with a centre-centre distance of approximately 300 um. Surface profiles must be measured between these indentations from left to right in Fig.2. i.e. close to perpendicular to the grinding marks of the surface. A full description of procedure was then presented.

Analysis of measurements at position 2
Load the measured profile, 2a_1FL, and locate the centre point between the two indentations. Go 125 um towards the beginning of the profile from the center point and select that as the start point of your analysis. From this point let the instrument make a calculation of primary profile parameters Pa, Pq, Pz, Pv, Pp for 250 um (try also Ra1, Rq1, Rz1, Rv1, Rp1 for a single cut off length of 250 um). Note the values in the enclosed table. Repeat the analysis process for the other measurement positions.

4. Measurement results and discussion

4.1. Surface roughness at Position 2
A typical surface profile, recorded at position 2b by the KTH Talystep stylus profiler [4] is shown in Fig. 4. The entire profile length is 500 μm and the two laser drilled holes appear well in the ± 3 μm height scale.
From the profile in Fig. 4 the mid point between the laser drilled holes is determined and a 250 μm long profile centred at this midpoint is saved for parameter evaluation. By repeating the trace three times with a

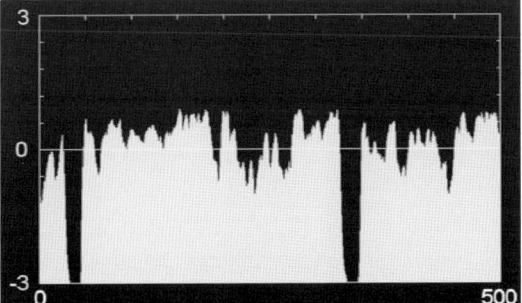

Fig. 4. Surface profile at pos. 2b. Scale in μm.

side-shift of ± 8 μm, i.e. perpendicular to the trace of the stylus, the variation in Pa and Pq were found to be about ±1% at position 2a and 2b, while 2c was close to 4%. The reason for the latter was a pit at the beginning of the profile and a shift of just a few micrometers in and out of this pit changed the Pa and Pq values by 24 nm and 33 nm respectively. For the Pz values measuring the maximum peak to valley height over the entire 250 μm the variations in 2c were up to 9% while 2a and 2b were less than 4%.

In Fig. 5 the Pa results are shown with the standard deviations of 0.9 - 3,8 % obtained at position 2a-2c. This is the variation we can expect in a very well localised area, in this case within 16 μm x 250 μm. By taking all profiles obtained from the three positions 2a, 2b and 2c we get a standard deviation of 14% (see Fig 5) over the area 600 μm x 250 μm, i.e. almost one order of magnitude larger variation than when measurements were kept within a width of 8 μm.

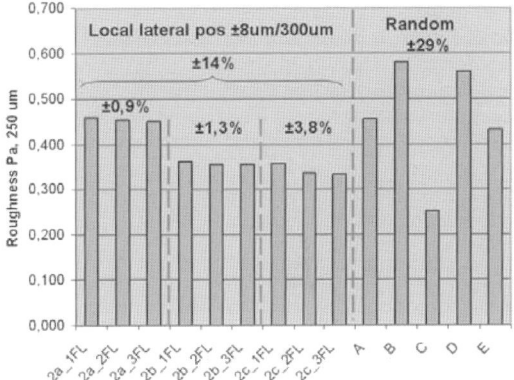

Fig. 5. Pa surface roughness at pos. 2a -2c and at random places.

By making the measurement more random but still aligned with the measurement direction of 2a -2c at five different places within the 3mm feature size, the standard deviation is increased by a factor of two to 29%. Still, the average Pa values for the random (0,39 μm) and 600 μm x 250 μm areas (0,38 μm) agree very well.

This shows the strong influence of surface structure variation when different areas of a surface are measured within short sampling lengths and it is the reason for the ISO-standard recommendation [3] of measuring over 800 μm sampling length for the Ra roughness interval of 0.1 – 2 μm. This exercise shows in a nutshell that inter-comparison measurements for short sampling lengths on rough surfaces are bound to be very uncertain due to the statistical nature of the roughness. That is also to say that a roughness specification on a rough but laterally tiny surface is almost impossible to verify with a reasonable uncertainty.

In Fig. 6 we present the Pa roughness values over 250 μm obtained with ten different profilers at position 2a. The spread of the measurements for each instrument represents the repeatability and is indicated by the line marks at the top of each bar. They are quite small, in agreement with what was obtained in Fig. 5 for the 2.7 μm stylus radius. The variation among the different instruments can not be attributed to the surface variation, but has to be related to the differences of profile interpretation among the different

142

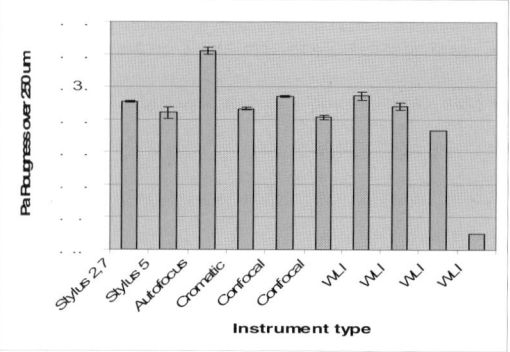

Fig. 6. Round Robin results for Pa roughness values obtained with 10 different profilers over a sampling length of 250 µm at position 2a.

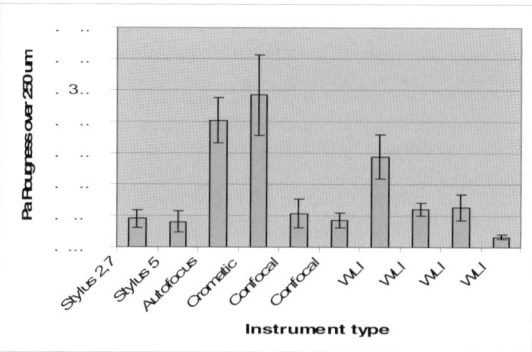

Fig. 7. Round Robin results for Pa roughness values obtained with 10 different profilers over 5 sampling lengths of 250 µm at position 3.

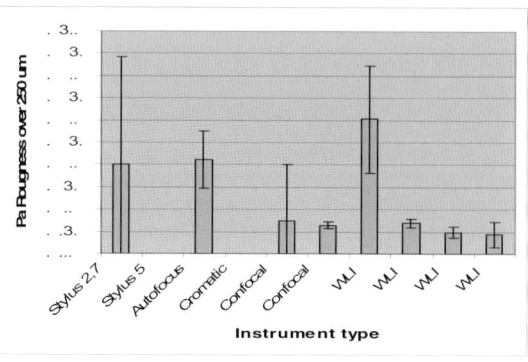

Fig. 8. Round Robin results for Pa roughness values obtained with 8 different profilers over 5 sampling lengths of 250 µm at position 4.

instruments. The 5 µm radius mechanical stylus will for sure resolve less fine structure than the 2.7 µm stylus and can explain the slightly lower value of the former instrument. Auto-focusing profilers are known to boost surface features [5] and the almost 50% higher Pa value is therefore not surprising. The two confocal microscopy results are from instruments of the same brand, but different models. The white light interferometers (WLI) are all of different brands. The extremely low value of the last WLI profiler is a mystery that needs to be solved. Skipping the two main outliers and also the third WLI which is off by a considerable amount, we get an average Pa value of 0,442 µm ± 0,025 µm for the seven remaining profilers. Six of the results fall within this one standard deviation of ±6%, which is reasonably good. For the Pq roughness results, the bar distribution of Fig. 6 was almost identical except for the first WLI which rendered an almost twice as high value as expected. Therefore that instrument needs to be further evaluated.

The Round Robin results obtained at position 3 and 4 are shown in Figs. 7 and 8. At position 3, on top of the nominally 75 µm bar the most striking features are the outliers of the chromatic and the first WLI profilers in addition to the autofocus and the 4:th WLI. No obvious explanation has been found so far for the two new outliers. The Pa measurements at position 4 between the two 20 µm wide and 100 µm high bars separated by nominally 150 µm challenged the profilers as seen in Fig. 8. The 5 µm radius stylus and the chromatic profiler did not work it out, and the variation from one 250 µm profile to the other were very large. The reason has already been shown in Fig. 3 and is caused by remaining burrs in the deep channel. Some operators of the WLI and confocal microscope profilers have been able to avoid the burrs while e.g. the 2,7 µm stylus with 60° cone angle will inevitable hit it.

5. Conclusion

In conclusion we have found that good agreement of surface roughness values Pa were found for 60% of the profilers used in the 4M laboratories. To achieve this level of conformity the measured area had to be well localized and positioned to micrometer accuracy. Some instrument results deviate by large amounts and needs further investigations and probably interaction with the instrument manufacturers to sort out the problems.

Acknowledgements

This paper was compiled with support from EC FP6 NoE on 4M, Metrology and Polymer Divisions.

References

[1] L., Mattsson, "Metrology of micro-components – a real challenge for the future," in the Proceedings of the 5th International Seminar on Intelligent Computation in Manufacturing Engineering (CIRP ISME '06), 25 – 28 July 2006, Ischia, Italy. 547 -552
[2] ISO 4287:1997, "Geometrical Product Specifications (GPS) -- Surface texture: Profile method -- Terms, definitions and surface texture parameters, " International Organization for Standardization, 1, ch. de la Voie-Creuse, Case postale 56, CH-1211 Geneva 20, Switzerland
[3] ISO 4288:1996, "Geometrical Product Specifications (GPS) -- Surface texture: Profile method – Rules and procedures for the assessment of surface texture"
[4] J. M. Bennett and L. Mattsson, Introduction to Surface Roughness and Scattering, 2nd ed. Optical Society of America, Washington (1999)
[5] L. Mattsson and P. Wågberg, "Assessment of surface finish on bulk scattering materials. A comparison between optical laser stylus and mechanical stylus profilometers," Precision Engineering, 15, (1993), 141 – 149

Micro-ultrasonic metrology of multi-material electronic devices

R. Teti, P. De Santo

Department of Materials and Production Engineering, University of Naples Federico II, Naples, Italy

Abstract

The main objective of this work is the investigation on micro-nondestructive evaluation (micro-NDE) metrology for dimensional measurement and quality control of multi-material electronic devices consisting of chipset tablet assemblies. The micro-NDE approach is based on ultrasonic (US) sensors in pulse-echo testing mode applied according to the full-volume immersion scan method that provides for the US axial tomography of the chipset tablet. The thickness of the multi-material chipset tablet assembly layers was evaluated through micro-US 2½ D geometrical measurements and the chipset tablet inter-layer integrity was critically assessed via micro feature US image analysis.

Keywords: Micro-NDE, Ultrasonics, Multi-Material Assembly, 3D Metrology

1. Introduction

An ultrasonic (US) micro-nondestructive evaluation (micro-NDE) procedure was applied to multi-material electronic devices consisting of chipset tablet assemblies. Chipset tablets, and in particular their inter-layer soldered joints, are stressed during service by temperature oscillations for a very high number of cycles [1, 2]. Thus, it is of paramount importance to employ joint materials possessing high temperature resistance and similar thermal expansion coefficient to avoid chipset structure delamination and component cracking that ultimately result in catastrophic device failure [3]. The inspection philosophy usually adopted for chipset tablet devices relies on thermal resistance measurements that are complex and expensive, requiring major electric equipment, and provide for low accuracy [4]. Ultrasonic (US) testing in the pulse-echo mode offers the advantage of using a single transmitter/receiver probe so that only one surface of the assembly structure needs to be accessible [5]. Moreover, by resorting to US 2½ D immersion scanning with full waveform acquisition, it becomes possible to inspect the entire volume of the assembly [6]. This method appears, therefore, particularly suited for testing and inspection of chipset tablet devices to evaluate and assess the geometry and integrity of the chipset multi-material sub-millimetre layered structure based on micro-NDE metrology [7].

2. Chipset tablet electronic assembly structure

The structure of the chipset tablet under consideration is made of various layers of different materials and thicknesses in the micro manufacture range (see Fig. 1). An aluminium nitride (AlN) ceramic layer is sandwiched between two copper (Cu) layers. The upper Cu layer is soldered to the silicon (Si) chip and the bottom Cu layer can be soldered to any metal baseplate [8]. The electrical emitter and gate contacts are made of Al pads wire-bonded onto the chipset.

3. US NDE system and experimental procedures

3.1. US NDE system

The US system for 2½ D NDE of the chipset tablet is composed of a purposely designed hardware configuration and a custom made software code: RoboTest v. 2.0© developed under LabView© [9, 10].

Hardware configuration
The system hardware configuration (see Fig. 2) is:
- an oscillator/detector, generating electrical pulses for US probe excitation and receiving the returning signal
- a transmitter/receiver focused high frequency US immersion probe for pulse-echo testing
- a digital oscilloscope allowing for the acquisition, visualisation and digitisation of the US signal
- a PC for US waveform acquisition and processing as well as US probe displacement control
- a mechanical system, consisting of a tank containing the US coupling medium (water), US probe and test specimen, and a robotic arm providing for US probe displacement in 2D or 3D space

Custom made software code
RoboTest v. 2.0© is a custom made software code, developed in the LabView© environment, to control the US probe displacement in 2D or 3D space and provide for US signal detection, storage and analysis (see Fig. 3) [9, 10]. It contains various test mode options: in this work, the Full Volume Ultrasonic (FV-US) Immersion Scan procedure [6] was used, consisting in the digital detection of the whole US waveform for each position of the transducer during the x-y raster scan of the chipset tablet. US data are stored in a 2½ D volumetric file containing the whole set of complete digitised US waveforms for each material interrogation point. From the volumetric file, US images for any selected segment of the US signal, i.e. to any thickness portion of the chipset tablet multi-material layered structure, can be obtained for analysis, allowing for the US axial tomography of the part

3.2. US NDE experimental procedures

The utilised FV-US Immersion Scan technique had a twofold micro-NDE scope: micro-US measurement of chipset tablet layers thicknesses and simultaneous delamination damage detection at chipset tablet inter-layer joints. A FV-US immersion scan was performed in water as coupling medium between probe and test material; the US signal entered from the chipset tablet lower Cu layer side to obtain a 2½ D volumetric US file for signal processing and image generation.

A 2 inch focused, 15 MHz high frequency, high damping, immersion US probe was used for scanning with the settings reported in Table 1. Each detected US waveform was digitised at 100 MS/s.

3.3. Micrometer contact measurements

Besides US non-contact thickness measurements, micrometer contact measurements were carried out on the chipset tablet layered structure to evaluate the thickness of each layer in the assembly for comparison. It is worth noting that micrometer contact measurement is a robust direct thickness measuring method but highly time consuming and, therefore, not suitable for chipset tablet industrial quality control.

4. Micro-US thickness measurements

The 6-chip chipset tablet and its multi-material layered assembly structure are shown in Figure 1.

In Figure 4a, the typical US waveform retrieved from the chipset tablet 2½ D US volumetric file is shown. The US waveform entered the chipset tablet from the bottom Cu layer side after travelling in water. Thus, the first peak from the left (contained in time window # 1) is the front echo reflected by the free surface of the bottom Cu layer (water - bottom Cu layer interface: first interface encountered by the US signal). The first peak occurred at time t_{fp}, taken as t = 0 for US waveform time measurements. The total time distance measured between the US waveform front echo and back echo is ΔT_{tot} = 0.637 µs: this time-of-flight is the time it takes the US signal to traverse to and fro the full multi-material layered chipset tablet thickness.

To evaluate the thickness of each material layer in the chipset tablet, the US time-of-flight between each couple of consecutive interfaces was measured and the related layer thickness was calculated using the US velocities for the traversed layer materials: the results of the US non-contact thickness measurements are reported in Table 2 with the relevant US parameters, nominal layer thicknesses, and micrometer contact thickness measurements for comparison.

From Figure 4a, it can be seen that the interface between bottom Cu layer and AlN ceramic layer is contained in time window # 2 (0.085 - 0.145 µs from t_{fp}). It is the second interface encountered by the US signal: its exact time position, calculated using the US velocity in the Cu layer, is 0.120 µs from t_{fp}.

The interface between AlN ceramic layer and upper Cu layer is contained in time window # 3 (0.285 - 0.347 µs from t_{fp}). It is the third interface encountered by the US signal: its exact time position, calculated using the US velocity in the AlN layer, is 0.189 µs from the previous interface and 0.309 µs from t_{fp}.

The interface between upper Cu layer and solder layer is contained in time window # 4 (0.400 - 0.473 µs from t_{fp}). It is the fourth interface encountered by the US signal: its exact time position, calculated using the US velocity in the Cu layer, is 0.119 µs from the previous interface and 0.428 µs from t_{fp}.

The interface between solder layer and Si chip is contained in time window # 5 (0.473 - 0.535 µs from t_{fp}). It is the fifth interface encountered by the US signal: its exact time position, calculated using the US velocity in the solder layer, is 0.060 µs from the previous interface and 0.488 µs from t_{fp}.

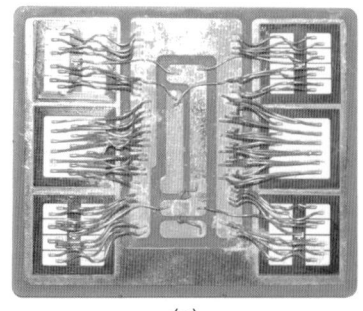

(a)

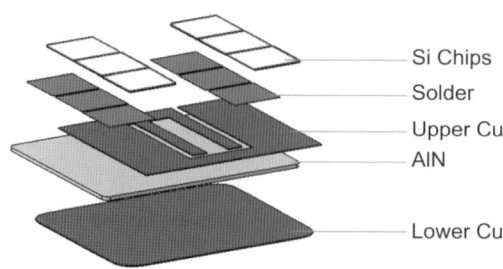

Si Chips
Solder
Upper Cu
AlN
Lower Cu

(b)

Fig. 1. (a) Chipset electronic assembly (view from top); (b) chipset electronic assembly multi-layered structure. AlN = 1000 µm aluminium nitride ceramic layer; Cu = 270 µm copper layers on top and bottom of the AlN ceramic layer; Solder = 60 µm solder layer; Si = 640 µm silicon chip.

Table 1
FV-US immersion scan settings

	Step (µm)	# of steps	Real dimension (mm)
X	200	315	60
Y	200	268	51

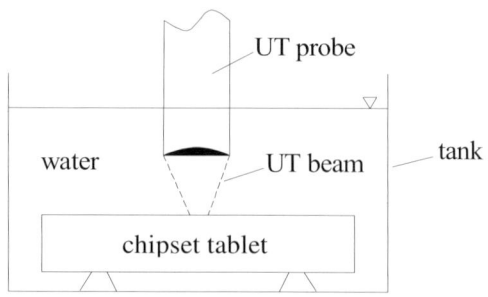

Fig. 2. Ultrasonic NDE system.

Fig. 3. Pulse-echo US immersion testing.

The last peak of the US waveform contained in time window # 6 (0.611 - 0.685 μs from t_{fp}) is the back echo reflected by the back surface of the Si chips (Si chip - water interface). It is the last interface met by the US signal: its exact time position, calculated using the US velocity in the Si chips, is 0.149 μs from the previous interface and 0.637 μs from t_f.

The excellent match between US non-contact measurements and micrometer contact measurements of the chipset tablet multi-material layers (Table 2) confirms the effectiveness and high accuracy of the micro-US metrology approach.

By examining the table, it is worth noting that, although the lower Cu layer and the upper Cu layer display different nominal thicknesses, the US time-of-flight measurements indicated an equal thickness for both Cu layers. This result was confirmed by subsequent micrometer contact measurements, proving the reliability of the micro-US non-contact metrology in identifying non-conformities in the chipset tablet assembly stacking sequence.

5. Micro feature US image analysis

In order to carry out the micro feature US image analysis for micro-NDE of chipset tablet inter-layer integrity, the 0.664 μs long US waveform was divided into twelve equal time windows of 0.055 μs width.

Multiple US images were generated for each of the 12 time windows and, after image examination, selected US images for time windows # 1 to # 6 (see Figs. 4b – 4g) were analyzed and critically assessed.

The US image in Figure 4b shows the front surface of the bottom Cu layer, related to the interface between water and bottom Cu layer (entrance of the US signal in the chipset tablet). The rounded corners of the bottom Cu layer are clearly evidenced on top of the AlN ceramic rectangular layer with sharp corners.

Figure 4c illustrates the interface between bottom Cu layer and AlN ceramic layer. The two layers can be assessed as perfectly bonded since no solution of continuity is noted in the US image of their interface.

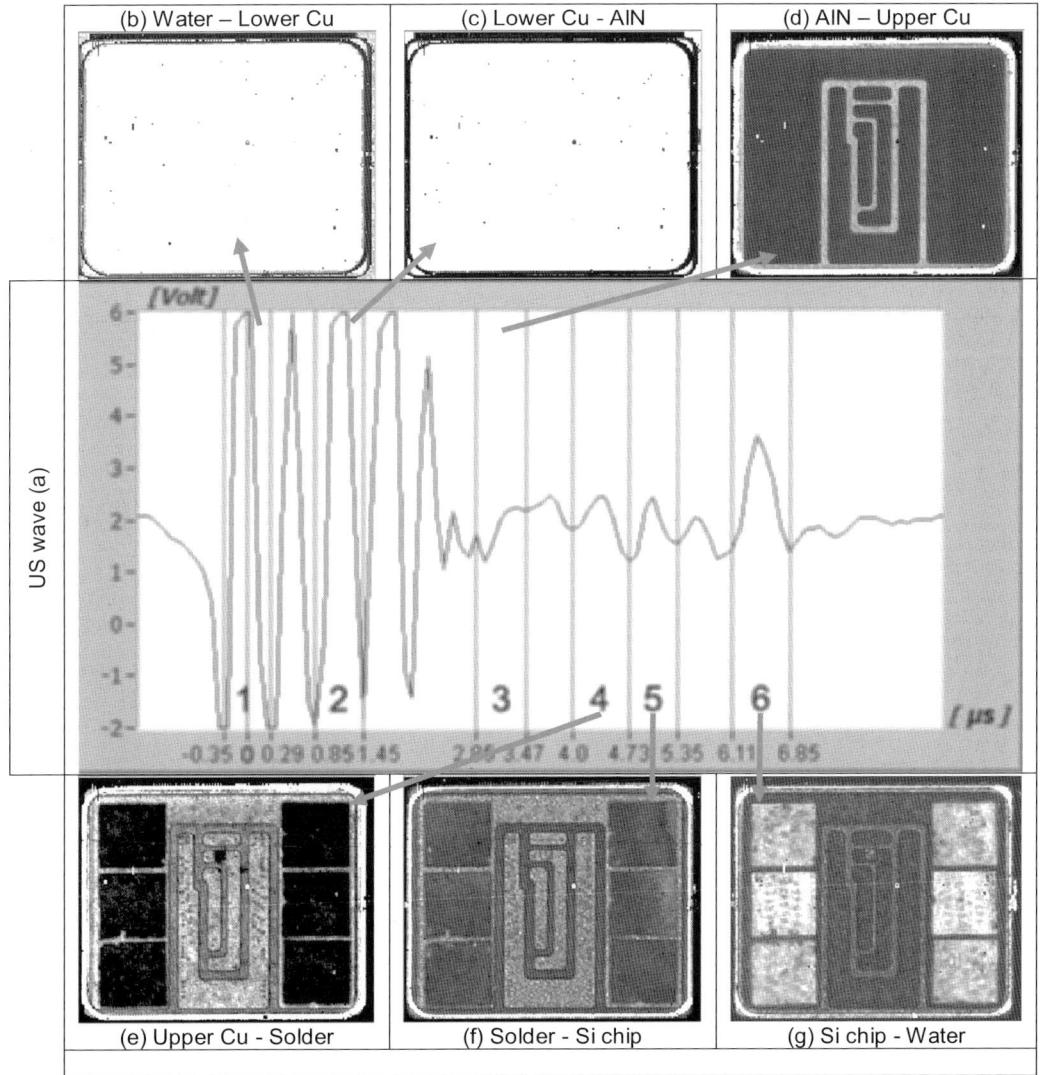

Fig. 4. (a) Typical US waveform retrieved from the chipset tablet US volumetric file.
US images: (b) water – lower Cu interface; (c) lower Cu – AlN interface; (d) AlN – upper Cu interface; (e) upper Cu – solder interface; (f) solder – Si chip interface; (g) Si chip – water interface (C-scan image [5]).

Table 2
Micro-US thickness measurements of the chipset tablet multi-material layered structure
* Time-of-flight between two subsequent interfaces encountered by the US signal in the chipset tablet

	Layer	# Points in the US waveform	Sampling frequency (MS/s)	Time of flight * (µs)	US velocity (m/s)	US measured thickness (µm)	Micrometer thickness (µm)	Nominal thickness (µm)
1	Lower Cu	12	100	0,120	4650	279	270	320
2	AlN	19	100	0,189	10600	1002	1000	1000
3	Upper Cu	12	100	0,119	4650	276	270	270
4	Solder	6	100	0,060	2000	57	60	60
5	Si chip	15	100	0,149	8500	633	640	640
	Assembly	**64**	**100**	**0,637**	-	**2246**	**2240**	**2290**

Figure 4d depicts the interface between AlN ceramic layer and upper Cu layer. The geometrical pattern of the Cu layer is accurately described, indicating a faultless bond between the two layers. No traces of other layer materials are visible in the Cu pattern, witnessing the perfect identification of the 280 µm thick upper Cu layer position in the chipset tablet multi-material layered structure obtained by micro-US thickness measurements through accurate time-of-flight calculations (Table 2).

Figures 4e and 4f portray, respectively, the upper Cu layer - solder layer interface and the solder layer - Si chip interface. In both images, the 6 Si chips are perfectly delineated, including their sharp corners, indicating the flawless soldering of the AlN/Cu/Si chipset joint. Moreover, Figure 4e shows the location of the micro-electrical wire-bonded contacts, on top of the upper Cu layer, providing for the electrical connection with the Si chips. The wire-bonded contacts are hardly visible in Figure 4f because the corresponding time window is displaced by 0.062 µs in the right direction of the US waveform (see Fig. 4a), i.e. slightly away from the upper Cu layer surface.

Finally, Figure 4g reports the US image obtained from time window # 6, set on the chipset tablet back echo, corresponding to the Si chip – water interface. All features of the chipset tablet are clearly shown in the figure, confirming the flawlessness of the chipset tablet multi-material layered assembly structure. The US image is particularly detailed and even the traces of the micro-electrical wire-bonded contacts soldered on top of the Si chips can be easily detected (small dots on the Si chip images).

To sum up, micro feature US image analysis allowed for the assessment of the excellent quality of the multi-material chipset tablet under examination.

6. Conclusions

Micro-nondestructive evaluation (micro-NDE) metrology, based on ultrasonic (US) pulse-echo testing under the 2½ D full-volume immersion scan method, was applied to the multi-material layered assembly structure of a chipset tablet. The scope of the study was twofold: micro-US measurement of the chipset tablet layers thicknesses and simultaneous micro-NDE of the chipset tablet inter-layer soundness.

The results of the micro-US time-of-flight thickness measurements confirmed the accuracy and reliability of the US non-contact micro-metrology. The micro feature US image analysis proved to be highly effective in assessing the quality and integrity of the multi-material chipset tablet assembly.

Acknowledgements

The research activity illustrated in this work was carried out within the framework of the EC FP6 Network of Excellence "Multi-material Micro Manufacture: Technologies and Application (4M)".

References

[1] Coquery G et al., Reliability of the 400 A IGBT modules for traction converters: contribution of the power thermal fatigue influence on life expectancy, EPE 95, 6th Eur. Conf. on Power Electronics & Application, Sevilla, (1995) 60-65.

[2] Reliability of Advanced High Power Semiconductor Devices for Railway Traction Application, Brite-EuRam Project N. 95-2105.

[3] Occhionero MA, Adams RW, Fennessy KP and Hay RA. Cost-Effective Manufacturing of Aluminum Silicon Carbide (AlSiC) Electronic Packages, Advanced Packaging Materials Symp. - IMAPS, Braselton, March 14-17 (1999).

[4] Sumi S et al., Thermal fatigue failures of large scale package type power transistor modules, ISTFA '89 Symposium (1989) 309-322.

[5] Teti R. Ultrasonic Identification and Measurement of Defects in Composite Material Laminates, Annals of CIRP, 39/1 (1990) 527-530

[6] Teti R, Lopresto V, Buonadonna P and Caprino G. Ultrasonic Non-Destructive Evaluation of Impact Damaged CFRP Laminates, 10th Eur. Conf. Comp. Mat. (ECCM-10), Brugge, 3-7 June: (2002) paper n. 346.

[7] Teti R, Mattsson L, Lebar A and Junkar M. Metrology Applications of Two-Dimensional Frequency Analysis for Micro-Features Characterisation, 1st Int. Conf. on Multi-Material Micro Manufacture – 4M 2005, Karlsruhe, 29 June – 1 July (2005) 259-262.

[8] Teti R and Fratelli L. Ultrasonic NDE of IGBT Metal Matrix Baseplate Modules, Advancing with Composites '03, Milan, 7-9 May (2003) 157-169.

[9] Teti R and Salvo F. Programming of Robotic System for Non Contact Ultrasonic Reverse Engineering of Complex Shapes, 4th CIRP Int. Sem. on Intelligent Computation in Manufacturing Engineering – CIRP ICME '04, Sorrento, 30 June - 2 July (2004) 149-152.

[10] Teti R and Buonadonna P. 3D Surface Profiling through Ultrasonic Reverse Engineering, 3rd CIRP Int. Sem. on Intelligent Computation in Manufacturing Engineering – CIRP ICME 2002, Ischia, 3-5 July (2002) 387-394.

Approaching a sub-micron capability index using a Werth *Fibre Probe* System WFP

Richard Thelen[a], Joachim Schulz[a], Pascal Meyer[a], Volker Saile[a]

[a] *Institute for Microstructure Technology, Research Centre Karlsruhe, 76646 Eggenstein, Germany*

Abstract

Reproducibility and precision of LIGA structures has been claimed in many publications, founded mainly on brilliant pictures. Because of the poor accessibility to the sidewalls many publications are based on surface measurements without including information about z depending aspects [1] and focus on reproducibility as measured close to the top.
Often this neglects operator's influence, short time and long time reproducibility, environmental effects on the CMM and others. Tactile optical metrology might help to overcome 2D measurements. Repeatability of tactile optical metrology at IMT was proven to be less than 0,3 µm over some months using ultra fine probes with less than 25 µm diameter. In addition DoE was used to determine the minimum deviation for best possible machine settings. Standard Deviation between 50 and 30 nm was measured. Compared to that, uncertainty remains about 1-2 µm for 3D measurements even with z maximum restricted to 1 mm [2]. Not enough to measure sub-µm product variation that is a typical benefit of LIGA products.
Investigations were started at the Research Centre Karlsruhe to find out more about the effects influencing the measurements to explain why repeatability and capability do not match. Interaction between sample and sensor was the main reason. This was simulated and the results were used to reduce the uncertainty of the system. IMT elaborated a new strategy that improves the capability of a coordinate measurement machine CMM with tactile optical sensor for LIGA parts with sub µm variation.

Keywords: LIGA, tactile optical measurements, sub-µm dimensional variation, DoE for metrology settings

1   Metrology for LIGA structures

1.1   *3D-Measurements using picture data*

A high performance coordinate measurement machine CMM setting with precision table, HD camera and full air conditioning allows sub-µm-precise measurements on 2D structures like Chrome on Glass. Automated measure-ments on 3D structures are very reproducible but include errors of some µm. They are wide spread, but quality of LIGA production is ahead of metrological solutions offered.
The main reason for the extended uncertainty when using 3D structures are surface related effects. It deviates from an ideal geometrical element like a cylinder or cube by having rounded edges, roughness, and waviness as well as surface defects. In addition there are effects by material, setting or users like:

- reflections at/on object edges or surfaces
- focus setting,
- trigger level at grey scale picture interpretation,
- scattering or reflecting light from neighbouring structures,
- light reflections from ground.

Therefore the uncertainty is in the µm range. If product properties like surface gloss, roughness, waviness, colour etc. are not extremely stable, incorrect measurements are performed based on picture interpretation. The reason is that variation of properties definitely influences the metrological result in the µm range. Data from LIGA parts inspection shows this behaviour. Additionally there is a lack of opportunities to adopt the metrology system to parts from different sources with individual production properties.

1.2   *3D-Measurements using a tactile optical sensor*

Picture recognition of a tactile optical sensor illuminated by a white LED is much less sensitive to environmentally influenced variation of light distribution or related effects. Figure 1 shows the basic setup of such a system as offered by the German company Werth. The sphere at the tip of a glass fibre is placed in the focus of a camera within a CMM. The tip acts as a sensor if the sidewall of any structure is touched and the tip is moved relatively to the center of the picture. That is why tactile optical measurements are less sensitive to the above mentioned variation. Focus level or trigger level for picture recognition do not influence the determination of the centre point of the sphere. That enables highly accurate CMM measurements with only ±2 µm uncertainty [2].
The main limitation is that for high aspect ratio and highly dense packed struc-tures, no non des-tructive reference

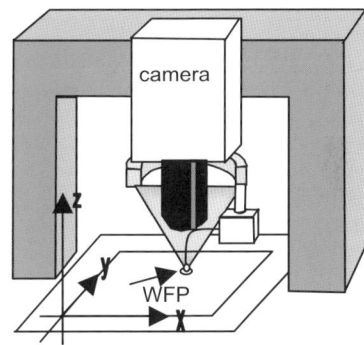

Fig 1: Combining CMM and WFP

metrology system is available that is capable of measuring inside sidewall profiles of 100 µm deep trenches only 30 µm wide [2, 3].

1.3   *3D metrology systems for MST*

As an alternative to tactile optical measurements there is a rigid so called Boss micro sensor commercially available for CMMs [3]. At the Institute for Microstructure Technology IMT at the Research Centre Karlsruhe a confocal micro optical sensor was successfully miniaturized [4]. A semi commercialized highly precise tactile sensor element of MECARTEX is offered [3]. These systems have higher resolution and lower uncertainty but the sensor tips are too big. A new sensor for sidewall measurements is available since 2007 [5], with a design based on AFM tips. But their sensor tip size is too big for the inspection of very small structures with high AR like fuel injection nozzles with multiple injection channels. As alternative volumetric metrology systems like tomography do not attain the resolution claimed as commercially available systems stick to minimum resolution of some µm voxel size [3].

1.4 Strategy to optimize the measurement

To guarantee adequate use of a CMM with multiple parameters to set in a wide range, a DoE approach was used first. By fully factorial tests several sets of parameters were assessed to minimize noise and optimize settings. For validation a class A ceramic gauge with 30 nm thickness variation was used. The statistical deviation is now in the range of 0.15 µm for +-2S, see figure 2. The long term variation adds up to 0.2 µm to this which is still good enough to detect systematic sub µm variation. It includes the chance to measure a single µm tolerance.

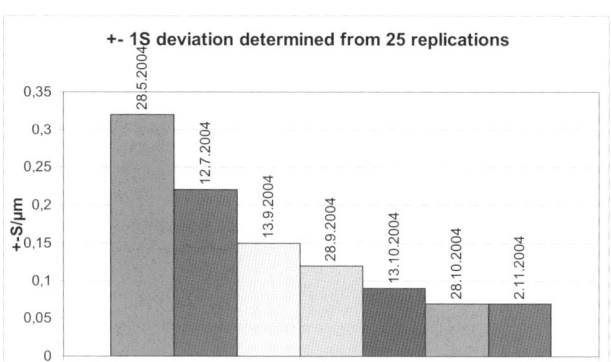

Fig. 2: Graph showing the optimization effect of DoE. All -25 replications measured semi automated.

There is no „one setting fits all". That means that for every relevant change, a new DoE to optimize the possible settings has to be made. That might be acceptable as the number of typical LIGA geometries and materials used are very limited and the measurement procedures are performed fully automated. Settings for geometries measured for the first time can be approached using an interpolation between settings for well known structures.

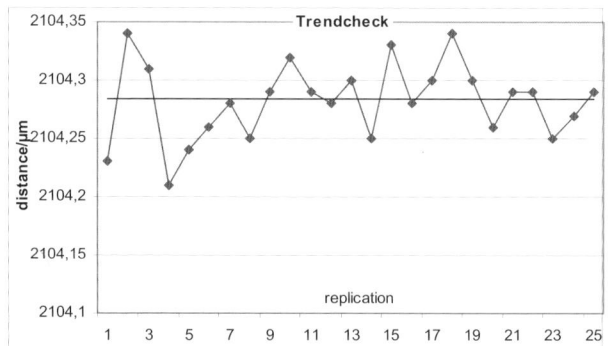

Fig. 3: Repeatability measured with a 2mm gauge with a maximum thickness variation of 30 nm.

Calibration with gauges or internal standards shows a repeatability of 0.12 µm up to 0.2 µm (± 2Sigma, see figure 3). Nevertheless the uncertainty remains in the range of 2-5 µm because all changes depicted are regarded as part of the statistical variation and therefore are added to the system uncertainty. That makes the CMM appear less powerful than it could be. And that is why repeatability and uncertainty do not match.
Figure 4 shows the calibrated WFP diameter as calculated from the IMT standard procedure. As the graph shows, the diameter determined by 25 replications is influenced by similar structures in the vicinity of the 2 mm thick ceramic gauge. The only parameter varying is an additional structure at 160 µm distance. It turns out that the standard calibration procedure -with a relatively lower z-depending deviation from the calibrated value at the surface level- with no structure in

close vicinity, does not contribute to the maximum possible deviation. One way would be to change the calibration strategy to include structures in the vicinity of the gauge. But that would broaden the uncertainty to a level were sub-µm precise measurements would definitely fail.

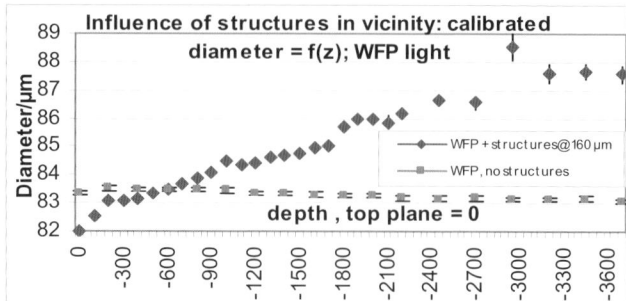

Fig. 4: Structures in vicinity influence the measurement. After the first run the structures parallel to the gauge in 160 mm distance has been removed. Every dot represents 30 replications.

All possible variation could be regarded as a part of the statistical effects that has to be summarized as a global uncertainty. But they could also been caused by systematic influences based on a system-sample interaction [6].

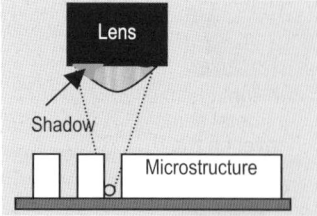

Fig. 5: Aperture related shadow.

Two major effects are the key: first, the increased sensitivity of the CMM to the optical properties of a sample when not measuring on top; and second, the aperture related shadow of the sample rim on the CMM detector when performing measurements not at top level. This is shown with figure 5.

1.5 Sensitivity to optical properties of the sample and aperture related shadow

Figure 6 shows a WFP with 106 µm diameter, illuminated by the diode feeding light into the fibre that the sphere is attached on. The sphere is in contact with a 90 ° hook of a PMMA substrate at 1600 µm depth. The sphere is showing up four times.
The optical sensitivity of the measurement results to the material properties was simulated, see figures 7 and 8. One parameter was the z position of a spherical light source relative to the structure top surface. Another parameter was the wall material, one chosen was PMMA and one a metal

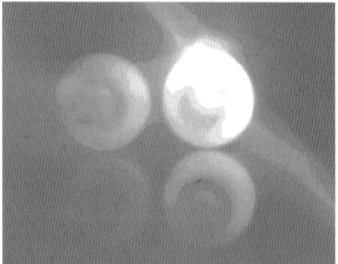

Fig. 6: WFP in contact with a PMMA structure including a 90° angle at 1600 µm depth.

surface. Figure 8 shows simulation from ray tracing. The centre point as calculated by the assessment of the picture generated by ray tracing simulation is shifted about 0.5 µm when placed at 500 µm depth. The results from figure 8 match the results from figure 4 when looking at the graph with no second structure nearby the gauge.

One thing to note that the gauge is ceramic and not a metallic structure as simulated for figure 8. Reflections from the ground do affect the detected centre point of the sphere, as well as neighbouring structures. As micro structures are usually processed on carriers like silicon wafers, the effect is a widespread problem. Single micro structures are difficult to handle, multiple microstructures on a wafer are hard to access by metrology systems.

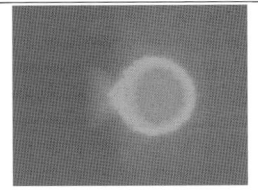

Figure 7: Simulated light distribution of a plane mainly light absorbing side wall (PMMA sample). At 500 µm below top level on the left and at top level on the right. Sphere diameter is 80 µm.

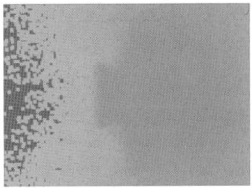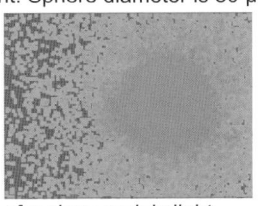

Figure 8: Simulated light distribution of a plane mainly light reflecting side wall (metal sample). At 500 µm below top level on the left and at top level on the right. Sphere diameter is 80 µm

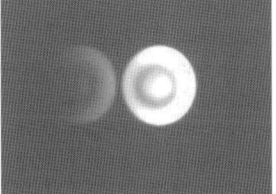

Figure 9: True pictures of a 102 µm WFP. On the right with no structure in vicinity, on the left some µm away from a 30 nm Ra smooth and transparent PMMA side wall at 250 µm depth.

Lenses for optical metrology systems usually go together with high aperture to enable auto focus measurements based on a very limited focus depth. The lower the measurement is performed, compared to the top level of the sample, the more the depiction of the sphere is shadowed by the rim of the sample. That generates an asymmetric distortion of the camera picture. Instead of a true circle the sphere is represented by a flattened ellipsoid with the effect that the centre point representing the physical centre is miscalculated. If the sidewall is mainly absorbing, the centre point is moved towards the sidewall, if the sidewall is mainly reflecting, the centre point is shifted in opposite direction.

A set of measurements along the sidewall of a thickness gauge shows the shadow effect as a function of the depth compared to top level. Figure 10 points out this effect using two sets of illumination. One is backlight that is less typical for MST metrology.

The second is measured with a diode illuminated sphere. The correction of +-1.5 µm, compared to calibration at top level, correlates with some known technical parameters like aperture and light intensity.

Every dot represents 25 replications. The absolute maximum of the dots graph is at that point where the depth is equal to the reciprocal of the NA multiplied by the sphere diameter. For backlight, shown by blue squares, it is similar. But because the light intensity is maximal at negative infinity it is

reversed. For a depth running towards infinity the calculated diameter follows a logarithmic function representing the damping of the light intensity. These effects are systematic and therefore the graph can be used to calculate a correction for similar objects.

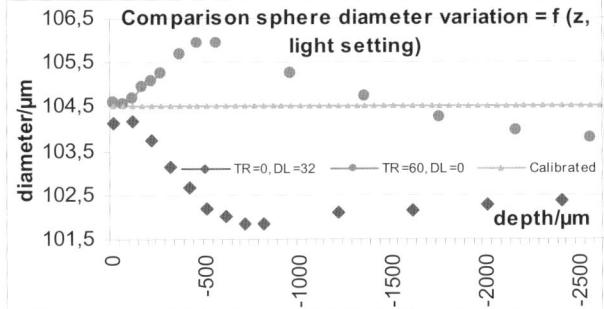

Figure 10: z-depending wall distance measurement using light variation and a precision ceramic gauge class a as a reference.

This is important if the structure height is in the mm range or the aspect ratio is very high while the sphere diameter used is very close to the distance to structures in the vicinity.

2 Application to LIGA structures

2.1 Production reproducibility

2.1.1 Diameter variation of PMMA columns

With the method described, LIGA made columns from PMMA 200/400 µm high PMMA were measured. The diameter deviation compared to the 1000 µm specification is given. The reproducibility is less than 0,2 µm because the main difference is a thickness related change in size as shown more obvious in figure 13. Moreover it can be shown that the exterior columns have a systematically lower diameter compared to those which are more in the centre of the structured field. These are important results as they help to optimize production processes by defining layout rules. Even a weak development process retaining more residues on the sidewalls could be identified. It leads to higher variation of the diameter as seen by figure 12.

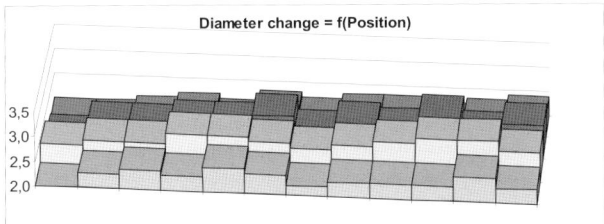

Figure 11: Measured diameter deviation of PMMA columns with 400 µm thickness (lot 2726). Specification was 1000 ±2 µm

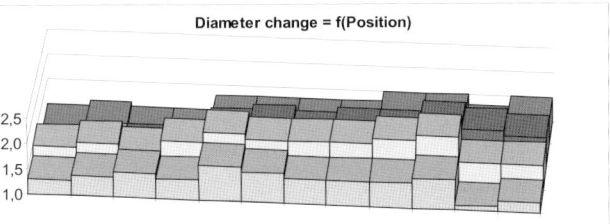

Figure 12: Measured diameter deviation of PMMA columns with 200 µm thickness (lot 2727). Specification was 1000 ±2 µm

After the PMMA development the measurements – as shown by figure 11 and 12 - indicate a systematic offset that exceeds 2 µm. But the effect of water absorption during electroplating nearly compensates this effect and customer specifications were finally fulfilled.

2.1.2 Side wall profiles of 2300 µm high LIGA gears

A LIGA made gear, see figure 13, was measured twice using the WFP sensor. Once from top to bottom and then, after reverting the gear upside down, measured once again. Thereafter measurements with backlight were used to define the reference profile, this time putting the gear on its side.

Figure 13: LIGA made NiCo gear wheels 2300 µm high.

This method is correct if the number of teeth is even and the focus is set properly. Results are shown in figure 14. Systematic z depending deviation of the sphere position was not compensated but added to system uncertainty. The deviation from the ideal orthogonal sidewalls is about 4.5 µm per mm and this is proven by both types of measurements.

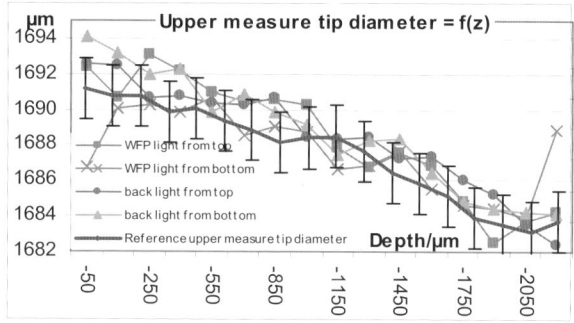

Figure 14: z depending WFP measurement of an upper measure tip diameter using a LIGA gear with 2.3 mm height. Deviation from verticality is 4.5 µm/mm .

3 Conclusions and outlook

With the new strategy developed and patented by IMT [7] gage capability for tactile optical measurements of 3D microstructures could significantly be improved from +-2 µm down to +- 0.2 µm (2 Sigma). That is close to the technical needs for measuring a single µm tolerance of 3D structures.

Systematic influences affecting the capability where simulated using the optical properties of the samples and combined with a ray tracing system to create comparable light distribution. High coincidence between simulated and true light distribution was attained.

That is the future base to integrate light migration along the sidewall, aperture related shadow and material depending light absorption or reflection into a software module combining this information to generate the best possible centre point determination for the metrology system.

Moreover automated routines are possible filtering the most critical details from 3D CAD data, generating the

light distribution prior to the measurements to perform and reduce the operators influence by matching simulated light distribution and life picture to assure that the assumptions the simulation was based on were correct. A concept is shown in figure 15.

For the first time applying this strategy on LIGA parts enabled the IMT to show the sub-µm precision of LIGA

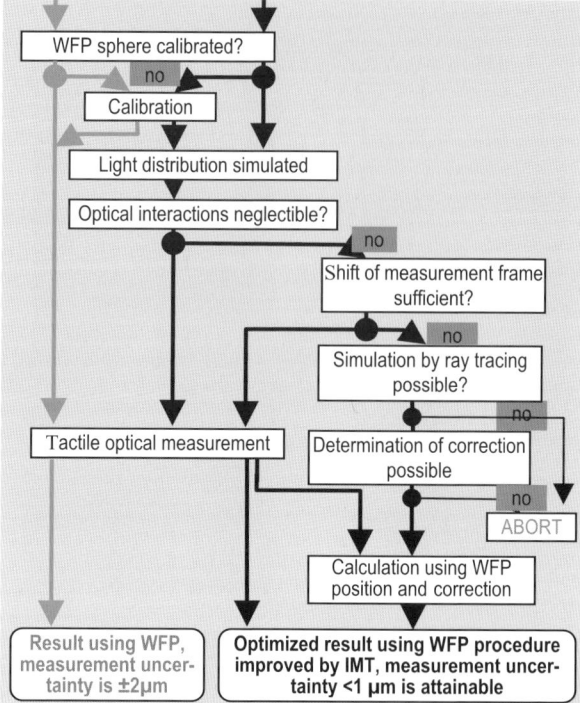

Fig. 15: IMT strategy for optimized WFP measurements

technology. Even for multiple runs most of the layouts with 3D structures vary much less than 1 µm. This is very important to perform systematic process improvements measuring sub-µm deviation to address LIGA related production effects like heat transfer, proximity-effects, swelling of the polymer and others.

4 References

[1] H. Schwenke, F. Härtig, K. Wendt, F. Wäldele: Future challenges in Co ordinate metrology: Addressing metrological Problems for very small and very large Parts. IDW Conference Knoxville (2003)

[2] K. Hasche: Metrological Contributions of PTB for Modern Production Technologies Tm Technisches Messen 1, (2004) 7-8, Page 441

[3] E.J.C. Bos, Fl.M. Delbressine, H. Haitjema: High Accuracy CMM Metrology for Micro Systems: Imeko, Erlangen (2004)

[4] Lücke, P.; Last, A.; Mohr, J.; Ruprecht, A.K.; Pruss, C.; Tiziani, H.J.; Osten, W.; Lehmann, P.; Schönfelder, S. Confocal microoptical distance sensor: realization and results. SPIE Internat. Symp.Optical Metrology, München, June 13-17, (2005)

[5] Institute for Microtechnology Mainz, Germany, IMM handout (2008)

[6] T. Hashimoto, Y. Takaya, T. Miyoshi, R. Nakajima: Fundamental Analysis of the Novel 3-D Probing Technique for Microparts using the Optical Fiber Trapping. CIRP Annals 2006, STC S Vol 55/1 613 ff

[7] PTC/EP 2006/001640

Novel materials: characterisation and processing

Micromachined silicon electrodes for electrochemical micromachining

C. Blattert[a], C. Müller[b], H. Reinecke[a,b]

[a] *Hahn-Schickard-Gesellschaft e. V. Institute for Micromachining and Information Technology (HSG-IMIT),*
Villingen-Schwenningen, Germany
[b] *Laboratory for Process Technology, Department of Microsystems Engineering (IMTEK),*
University of Freiburg, Germany

Abstract

Piracy and counterfeiting as well as retraceability demands of products such as plastic parts or tablets require new and innovative methods for unique product identification. An opportunity is the placement of microstructured codes in moulding tools. These tools are often made from materials that do not allow for highly precise micromachining by traditional technologies. Electrochemical machining (ECM) is a method for structuring construction materials such as steel or titanium. The current paper presents a new technology for the fabrication of microstructured tool electrodes for electrochemical machining by using highly doped silicon as electrode material. A simple and low priced fabrication of microstructured silicon electrodes with locally isolated areas is demonstrated by using well-established silicon processing technologies. Prototypes based on this new tool electrode technology are fabricated. Therewith electrochemical machining of microstructures in stainless steel is successfully demonstrated. Machining gaps down to 10 µm and average surface roughness of 60 nm are achieved. Typical rates of removal between 60 - 240 µm/min are reached. The local isolation of electrode areas advances the machining accuracy.

Keywords: electrochemical machining, micromachining, microstructure

1. Introduction

Microtechnological applications in medical devices, tooling for micro injection moulding or aviation as well as automotive industry demand the usage of construction materials such as steel or titanium. Electrochemical machining offers an interesting way for machining these materials [1].

ECM enables the machining of metals independent of their mechanical properties and achieves high material removal rates. Compared to electro-discharge-machining (EDM) tool wear is extremely low and the surface layer is not damaged. Additional advantages of ECM are:

- no contact between tool electrode and workpiece
- no burr formation
- no finishing process necessary
- smooth surface
- batch processing possible

Size and shape of machined microstructures are affected by tool size and shape as well as the machining gap. For the fabrication of microstructures by ECM it is essential to have a small working gap. This is achieved on the machine side by using the techniques of oscillating tool-electrodes [1] or ultra short voltage pulses [2].

The accuracy of machining is improved in addition by providing an electrically isolating layer on the electrode at the area where current passage is undesirable. If the machining operations should meet high accuracy requirements, this insulating layer must be as thin as possible, for example 10 µm or less.

Different types of insulating materials have already been proposed. The metal core of the electrode is covered by isolating polymer layers or compounds of inorganic compound networks [3]. These approaches suffer from water or hydrogen absorption of organic layers with subsequent decomposition and detachment. The drawback of inorganic compounds is that

owing to the high coating process temperatures only refractory metals can be used as the electrode material. As well, the use of different materials promotes the detachment of the insulating layer due to mechanical stress and hydrogen formation during the electrochemical machining process. A further disadvantage is the thickness of the isolating layer of considerable more than 1 µm. This leads to a larger machining gap and therefore less reproduction accuracy.

Vargas Llona et al. [4] demonstrated uniform electroplating on highly doped silicon wafers. These substrates conduct sufficiently well without the use of an additional seed layer. Gianchandani et al. [5] described a process of utilizing silicon array electrodes for micro-electro-discharge machining.

Our approach now uses micromachined, highly doped silicon as tool electrode material. Silicon is a well-established material in the MEMS-field that can be formed by conventional silicon processing techniques with high accuracy and reproducibility at low cost. By these techniques a simple integration of electrically conductive silicon and isolating silicon oxide areas is possible.

2. Experimental

2.1. Electrode design

The utilization of silicon as electrode material requires a low resistivity ρ that is achieved by highly doped silicon wafers ($\rho = 1 \cdot 10^{-3}$ $\Omega \cdot$cm [6]). This corresponds to an electrical conductivity σ (Eq. 1):

$$\sigma = 1/\rho = 1 \cdot 10^{3} \text{ S} \cdot \text{cm}^{-1} \qquad (1)$$

The typical conductivity of metals at room temperature is larger than $1 \cdot 10^{4}$ S·cm^{-1}. As a consequence, the applicable electrode area has to be kept smaller compared to metal electrodes due to the tenfold higher resistivity of the silicon.

In addition, it's advantageous in terms of a low electrode resistivity to fabricate only the microstructured part of the electrode in silicon and attach it for instance to a metal carrier.

Typical working voltages for electrochemical machining are between 5 and 50 V. The dielectric strength E_{bd} of thermally grown silicon dioxide is larger than $1 \cdot 10^7$ V·cm^{-1}.

For a maximum voltage U_{max} of 50 V a silicon oxide layer thickness d of:

$$d = U_{max}/E_{bd} = 50 \text{ V}/1 \cdot 10^7 \text{ V·cm}^{-1} = 50 \text{ nm} \qquad (2)$$

is sufficient (Eq. 2).

2.2. Fabrication process

The fabrication process of the electrodes is based on standard silicon MEMS processing techniques. Fig. 1 illustrates the process flow.

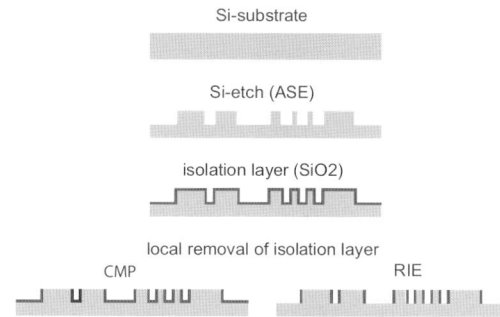

Fig. 1. Process flow of electrode fabrication process.

Highly doped ($\rho = 1$-$5 \cdot 10^{-3}$ Ω·cm) 4"-silicon wafers are used as substrate. The wafers are structured with 100 µm deep pits with vertical sidewalls by advanced silicon etching (ASE®, Surface Technology Systems plc, UK). Afterwards the structured silicon wafer is completely coated with 200 nm silicon dioxide (SiO_2) isolation layer by thermal oxidation.

Following, the silicon oxide layer is locally removed on the front face side by chemical mechanical polishing (CMP) or anisotropic reactive ion etching (RIE) process. These kinds of removal leave the side walls covered with an isolating SiO_2-layer. The silicon oxide on the backside is completely removed by RIE.

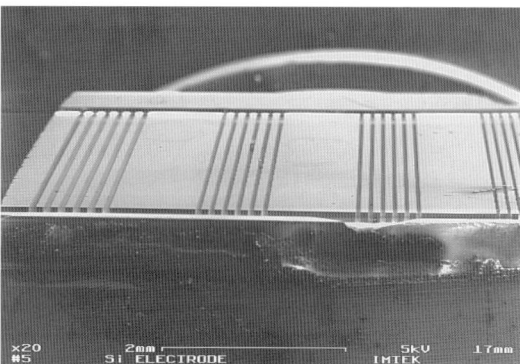

Fig. 2: SEM-picture of a microstructured silicon electrode die with parallel, 100 µm deep channels.

Finally, the electrode dies of size 6 mm·6 mm are cut out of the 4"-wafer by a dicing saw. Fig. 2 shows such a microstructured silicon electrode die.

The silicon dies are fixed to the tip of a square-cut (6 mm·6 mm) brass rod with a length of 100 mm by soldering or electrically conductive epoxy adhesive (see Fig. 3).

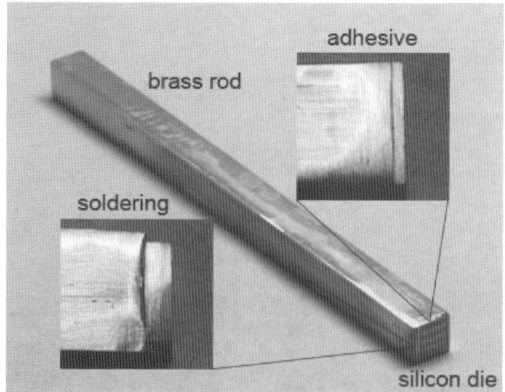

Fig. 3: Different electrode assemblies, the cross section dimension of the brass rod is 6 mm·6 mm.

2.3. ECM set-up

The electrochemical machining die-sinking experiments are performed on a PEM 1360 (PEMTec SNC, F). This machine works with a vibrating tool electrode with constant frequency of 50 Hz and amplitude of 200 µm. Details of the process control are found in [1]. Fig. 4 gives an overview of the set-up of the ECM experiments.

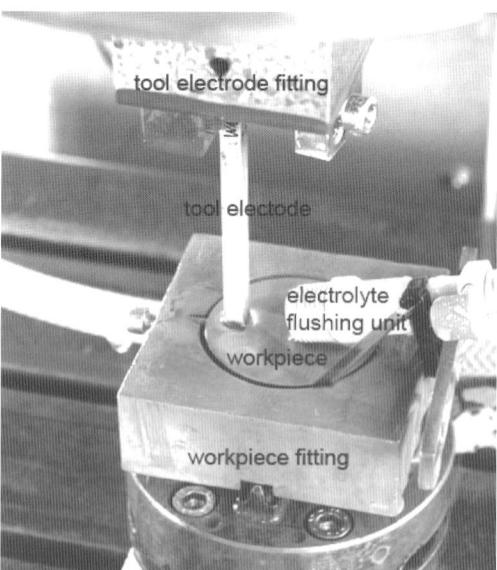

Fig. 4: ECM set-up.

Aqueous $NaNO_3$ solution (10 %$_{wt}$) is used as electrolyte. Workpieces of DIN 1.4108 (CRONIDUR 30, Firth AG, CH) and DIN 1.4441 (316 LVM, Früchtl-Kronos GmbH & Co. KG, D) stainless steel were used for the experiments.

3. Results and discussion

3.1. Surface roughness

Fig. 5 displays a microstructure that was electrochemically machined in stainless steel by a silicon electrode with isolated sidewall. Minimal feature dimensions were 200 μm wide bars. Tab. 1 shows important machining parameters.

Tab. 1: Parameter settings for surface roughness experiments.

Parameter	Setting
Isolation	200nmSiO$_2$
Workpieces material	DIN 1.4108
Working voltage	11 V
Pulse width	7 ms
Rinsing pressure	3,7 bar
Depth machined cavity	60 μm
Machining time	27 s

Average surface roughness Ra has been considerably smoothed by the ECM-process compared to the non machined surface (Ra = 700 nm). At the bottom of the cavity Ra was about 60 nm. The silicon electrode had a mirror finish at the front face side. That has been largely transferred during the ECM-process.

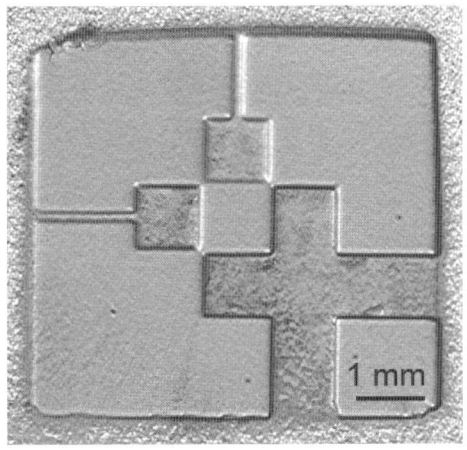

Fig. 5: Micromachined structure in stainless steel machined by silicon electrode with sidewall isolation.

3.2. Sidewall shape

One of the advantages of using silicon as electrode material is the simple integration of isolating areas. Fig. 5 and Fig. 6 compare a cuboid structure machined by an electrode with 200 nm SiO$_2$ sidewall isolation fabricated by CMP versus an electrode without isolation.

The images were taken by an optical profiler based on scanning white-light technology (NewView 5000, ZygoLOT GmbH, D). Tab. 2 overviews important experimental parameters.

The surface area of the cuboid machined with an isolated electrode (see Fig. 6) is larger and the top surface has still the original, rough surface structure.

The sidewall angles of the cuboid in Fig. 6 are steeper compared to these in Fig. 7.

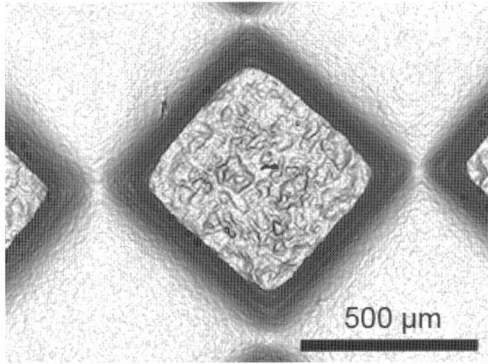

Fig. 6: Cuboid structure machined by electrode with sidewall isolation.

The cuboid machined without isolated electrode (see Fig. 7) features considerably rounded edges.

All sides including the top surface have been machined, which is indicated by a smoother surface compared to the original one.

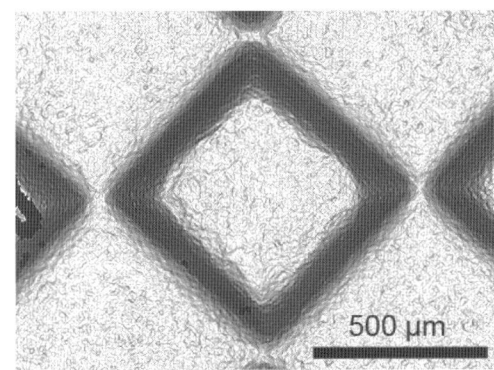

Fig. 7: Cuboid structure machined by electrode without sidewall isolation.

Tab. 2: Parameter settings for sidewall shape experiments.

Parameter	Setting	
Isolation	200nmSiO$_2$	Without
Workpiece material	DIN 1.4108	DIN 1.4108
Working voltage	14 V	13 V
Pulse width	6 ms	5 ms
Rinsing pressure	3,7 bar	3,7 bar
Depth machined cavity	71 μm	80 μm
Machining time	19 s	18 s

3.3. Machining accuracy

The accuracy of ECM with silicon electrodes was tested by machining a parallel bar structure (see Fig. 8). On tool side the channels were 50 μm wide and 100 μm deep. The distance between the channels was 80 μm. The SiO$_2$-isolation on the electrodes front face side has been removed by CMP.

Fig. 8: SEM-picture of machined parallel bar structure.

Fig. 9 displays a cross-sectional profile of one of the machined bars measured non-destructively by an optical profiler based on scanning white-light technology. This experiment was performed with DIN 1.4441 steel. Machining parameter settings are shown in Tab. 3.

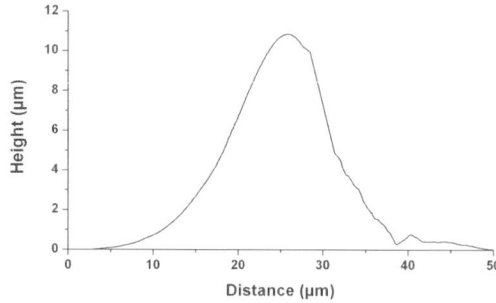

Fig. 9: Cross-sectional profile of machined bar.

The width at the base of the bar is about 30 µm. This means that the machining gap at the bottom is about 10 µm for the isolated electrode. The waviness at the right shoulder of the cross-sectional profile reproduces the surface roughness of the machined material.

Tab. 3: Parameter settings for experiments testing the machining accuracy of ECM with silicon electrodes.

Parameter	Setting
Isolation	200nmSiO$_2$
Workpiece material	DIN 1.4441
Working voltage	15 V
Pulse width	7 ms
Rinsing pressure	3,7 bar
Depth machined cavity	65 µm
Machining time	62 s

Despite the isolation of the electrode the profile shows still a rounded off peak profile typical for ECM. Mechanical erosion by the electrolyte flushing contributes to the profile rounding as well.

Moreover the removal rates of the DIN 1.4441 steel are about four times lower compared to that of DIN 1.4108. Therefore a continuative process optimisation concerning flushing conditions and machining parameters is necessary.

3.4. Electrode assembly

During machining of the samples the current was measured by an inductively coupled current probe connected to an oscilloscope. Typical current densities of soldered silicon electrode dies were between 0,5 and 0,8 A/mm^2, of fixed with adhesive dies between 0,3 and 0,5 A/mm^2.

The differences originate from the varying non-isolated silicon surface as well as the contact resistance of the assembly technique. The contact resistance of fixing the silicon dies by epoxy adhesive is about twice as much as by soldering.

4. Conclusion

The application of a new electrode concept using highly doped silicon with locally isolated areas was successfully demonstrated in the case of micromachining stainless steel.

ECM with such electrodes achieves working gaps down to 10 µm and average surface roughness up to 60 nm at typical rates of removal for the machined stainless steel materials between 60 - 240 µm/min.

Machining accuracy is advanced by local sidewall isolation with silicon dioxide.

Acknowledgements

The authors acknowledge the Hahn-Schickard-Gesellschaft e.V. for funding and supporting this project within the "Talentwettbewerb 2006" program.

References

[1] Foerster R., Schoth A. and Menz W. Micro ECM for production of Microsystems with a high aspect ratio. Microsystem Technologies, 11, 2005, pp. 246-249
[2] Staemmler L., Hofmann K. and Kueck H. ECF- An innovative Technique for Micro Mould Fabrication. Proceedings of 1.st 4M Conference, 2005, pp.375-377
[3] Van Kessel R.P., Rensing P.A., Sanders F.H.M. and Visser C.G. Electrode for electrochemical machining. Patent US 5,759,362, 1998
[4] Vargas Llona L.D., Jansen H.V. and Elwenspoek M.C. Seedless electroplating on patterned silicon. J.Micromech.Mircroeng., 16, 2006, pp. S1-S6
[5] Gianchandani Y.B. and Takahata K. Micro-electrode-discharge machining utilizing semicon-ductor electrodes. Patent US 6,586,699, 2003
[6] Sze S.M. Semiconductor devices, physics and technology. John Wiley & Sons, New York, 1985

Dielectric properties of hydroxyapatite based ceramics

J.P. Gittings[1], C.R.Bowen[1], I.G.Turner[1], A.C.E.Dent, F.R.Baxter[1,2] and J.B. Chaudhuri[2]

[1] Department of Mechanical Engineering, University of Bath,BATH, BA2 7AY.
[2] Department of Chemical Engineering, University of Bath,BATH, BA2 7AY.

Abstract

This paper studies the ac conductivity and permittivity of hydroxyapatite based ceramics (HA) at temperatures from room temperature to 1000°C. HA ceramics were prepared either as dense ceramics or in porous form with interconnected porosity and were sintered in either air or water vapour. Samples were thermally cycled to examine the influence of surface adsorbed water on conductivity and permittivity. Surface bound water was thought to contribute to conductivity for both dense and porous materials at temperature below 200°C. At temperatures below 700°C the permittivity and ac conductivity of HA was also influenced by the degree of dehydration and thermal history. At higher temperatures (700-1000°C), bulk ionic conduction was dominant and activation energies are in the range of ~2eV, indicating that hydroxyl ions are responsible for conductivity.

Keywords: hydroxyapatite, dielectric, sintering, polarisation, bioceramics.

1. Introduction

It has been found since the 1970's that certain ceramics exhibit unique biological properties when placed in an osseous environment [1]. Calcium phosphates (CaP) are a class of ceramics that can be termed 'bioactive'. These ceramics exhibit osteo-conductivity and have been shown to possess a strong CaP-bone interface [2]. CaP materials belong to a family of minerals crystallising with a hexagonal symmetry, which are found extensively in nature [1, 2]. The chemistry of CaP is extremely complex with a series of related inter-substitutions possible, giving rise to various "impure" and/or calcium deficient apatites e.g. natural bone can have up to 8% carbonate [3], resulting in numerous biphasic calcium phosphates being formed.

The general chemical formula of the apatites is $M_5(YO_4)_3X$

where
    M = divalent cation (such as Ca^{2+})
    X = univalent anion (such as OH^-)
    Y = trivalent species usually phosphorous, which forms an oxyanion.

As the name hydroxyapatite (HA) is used for a wide range of compositions beyond the ideal formula – $Ca_{10}(PO_4)_6(OH)_2$ (where the Ca:P ratio is 1.67), caution must be exercised when comparing different results. The exact composition seems to be an important parameter, since the biodegradation rate, for example, depends on it [4, 5]. Tricalcium phosphate – $Ca_3(PO_4)_2$ (TCP) has a Ca:P ratio of 1.5 and is termed a calcium deficient apatite.

The most widely studied CaP ceramics are tricalciumphosphate (TCP), hydroxyapatite (HA) and the newest tetracalciumphosphate (which has lower strengths) [1-6]. The chemical and mineralogical similarity of calcium phosphates with the calcified tissue that they replace may provide a template for osseous growth (regeneration of local environment) and/or subsequent remineralisation [6].

1.1 Structure and dielectric properties of HA

In the HA structure, there is a lattice of hydroxide ions located at the centre of Ca^{2+} triangles along the c-axis of a hexagonal unit cell [7]. The OH^- ions are aligned in columns parallel to the c-axis, along with Ca^{2+} and (PO_4^{3-}) ions [8]. Since the hydroxyl ions within the c-axis columns are thought to have an important role in ionic conduction [7], HA has been regarded as a one dimensional anionic conductor [9].

The study of the dc and ac electrical properties of HA has been of interest for a number of potential applications, including chemical sensors and bone substitutes [10-14]. Nagai et al. [10] examined the surface ionic conduction of HA for humidity sensor applications, since the room temperature conductivity was influenced by relative humidity. Valdes et. al [11] examined the dielectric properties of hydroxyaptite to understand the decomposition of HA to tri-calcium phosphate (TCP; $Ca_3(PO_4)_2$) as a result of the dehydration of hydroxyl ions at elevated temperatures [12]. This was of interest since TCP is thought to have higher bioactivity than HA, but is also more biodegradable. Hoepfner et al. [13] reported the influence of porosity on the room temperature permittivity of HA to understand its interaction with electrical fields applied to improve fracture healing or enhance bone growth. The dielectric properties of polarised HA materials has been examined by Takeda et al. [14].

1.2 Aim of paper

This paper examines the ac electrical properties of HA based ceramics from room temperature to 1000°C and presents relative permittivity, ac conductivity and determination of activation energies. Changes in sintering atmosphere (air or water vapour) were considered to examine the loss of surface adsorbed water (dehydration) or hydroxyl ions (dehydroxylation) during sintering [12]. Samples were thermally cycled to examine the influence of surface adsorbed water on conductivity and permittivity. Since CaP based materials for bone substitute applications often contain tailored porosity to facilitate bone growth, a comparison was made between dense and porous materials.

2. Materials and Methods

2.1 Samples and nomenclature

Dense (>90% theoretical density) HA materials where produced which were pressureless sintered either in air or water vapour to examine the influence of sintering atmosphere on phase evolution, microstructure and electrical properties. The porous material (60-70vol.% porosity) was sintered in water vapour for comparison with the dense material sintered under the same conditions. Details of the fabrication methods can be found in work by Gittings et al. [15]. The following notation was used to describe the three types of HA material examined, *dense(water), dense(air)*, and *porous(water)*.

2.2 Dielectric Measurements

Permittivity and a.c. conductivity were calculated from complex impedance measured in a frequency (f) range of 0.1Hz - 1MHz using a Solartron 1260 Impedance Analyser and a 1296 Dielectric Interface. Based on preliminary testing of the material types, samples were tested from room temperature to 1000°C at 50°C intervals using a voltage of $0.1V_{rms}$ (Cycle 1). In some cases the temperature interval was reduced to 25°C to provide additional data points to determine activation energy. After testing at 1000°C (Cycle 1) the sample was subsequently cooled to 125°C and measurements repeated during reheating of the sample to 1000°C (Cycle 2).

The ac conductivity (admittance) was calculated using Equation 1,

$$\sigma = \frac{Z'}{Z'^2 + Z''^2} \cdot \frac{t}{A}$$ Equation 1

where Z' and Z'' are the real and imaginary parts of the impedance, A is the area of the sample and t is the sample thickness.

The real part of the permittivity was calculated using Equation 2,

$$\varepsilon = -\frac{Z''}{Z'^2 + Z''^2} \cdot \frac{t}{\omega.A}$$ Equation 2

where ω is the angular frequency ($2\pi f$).

3. Results and Discussion

3.1 General features of ac conductivity and permittivity

Figure 1a,b shows the variation in real part of a.c. conductivity and relative permittivity as a function of frequency for the HA porous sample sintered in water vapour during Cycle 1. Data for only 100 °C intervals are shown for clarity.

At low frequencies (<10Hz) there is little or no frequency dependency of ac conductivity for many of the temperatures tested, particularly at elevated temperatures (>400°C). At relatively low temperatures (room temperature to 300°C) there is an initial decrease

in conductivity with increasing temperature, particularly for the porous HA (see Figure 1a). Nagai et al. [10] observed that the HA conductivity decreased when heating from room temperature to 650°C which was thought to be related to the loss of bound water. Nagai et al. [10] examined released water from HA whereby the liberation of water up to 300°C was related to weakly and strongly physisorbed water. Continued water loss at 500-1000°C was thought to be due to dehydroxylation of the OH⁻ lattice.

The relative permittivity exhibits a frequency dispersion in which the magnitude decreases with increasing frequency (see Figure 1b). This is thought to be due to the presence of conductivity in the material, either due to adsorbed water at low temperature or due to ionic conduction at higher temperatures.

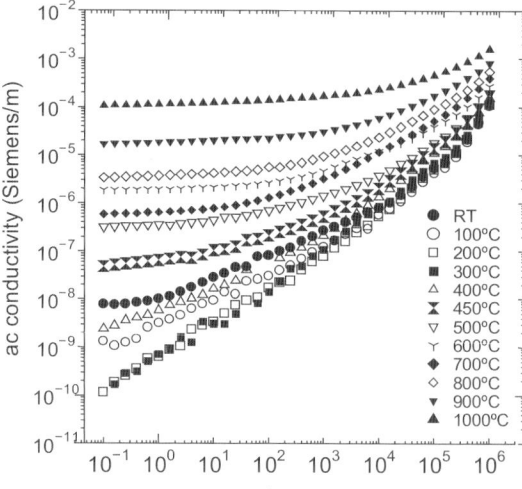

Figure 1(a). Real part of a.c. conductivity for porous(water) for Cycle 1.

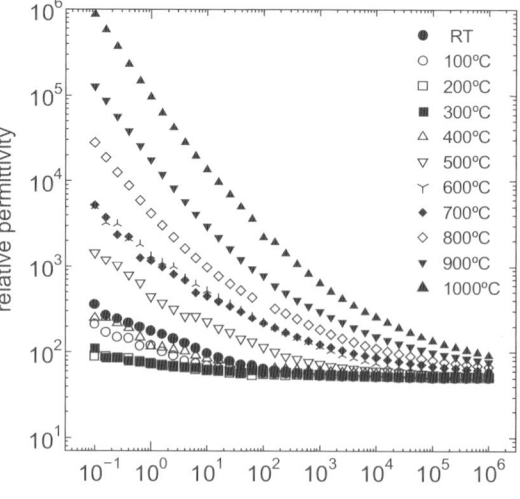

Figure 1b. Relative permittivity for porous(water) for Cycle 1

3.2 Thermal cycling

After testing at 1000°C (Cycle 1) the sample was cooled to 125°C and measurements repeated during reheating of the sample to 1000°C (Cycle 2). Figure 2 shows the low frequency (1Hz) conductivity of the material during Cycle 2 and a comparison with data from Cycle 1. For Cycle 2 the conductivity merely increases with increasing temperature with no initial decrease in conductivity or a small peak in conductivity at ~500-600°C, as was observed during Cycle 1. Similar behaviour was also observed for dense material (not shown). While the initial conductivity drop for Cycle 1 at temperatures less than 200°C can be attributed to the loss of surface bound water, the reason for the peak in conductivity at ~500-600°C is less clear. The conductivity was thought to fall as the material becomes substantially dehydroxylated. The same porous sample was held at room temperature for 1 week and retested (Cycle 3) and conductivity data are also shown in Figure 2. For Cycle 3 the initial decrease in conductivity when heating from room temperature to 200°C has returned, along with a slightly smaller conductivity peak at ~500-600°C. This could indicate that both features are possibly due to surface bound adsorbed water.

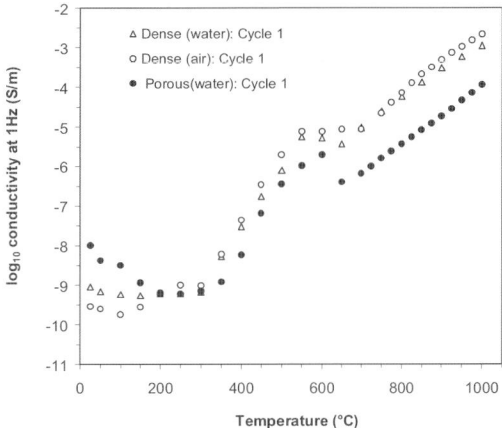

Figure 3. Log$_{10}$ conductivity versus temperature for all samples for Cycle 1

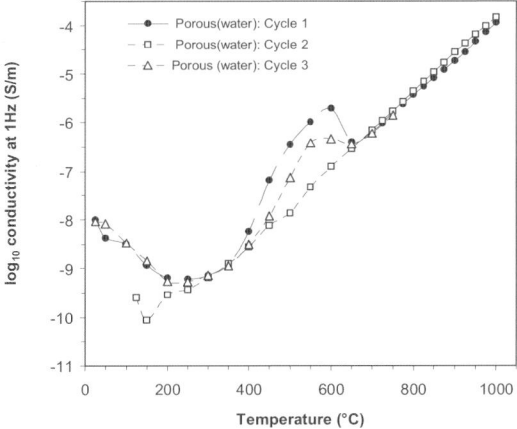

Figure 2. Log conductivity versus temperature for HA porous(water) for Cycles 1-3

3.3 Comparison of dense and porous material

Figure 3 shows that at 500-600°C there is a small peak of conductivity for all three materials followed by a continued rise in conductivity at higher temperatures (700-1000°C). The porous material has the highest conductivity at low temperatures (due to the higher surface area). The dense samples exhibit higher conductivities at the higher temperatures (700-1000°C) since bulk ionic conduction (rather than surface conduction) is probably dominant in this temperature range.

Figure 4 shows the relative permittivity at 1 kHz as a function of temperature which shows that below 250°C the porous HA sample with a high surface to volume ratio and high conductivity exhibits a high permittivity compared to the dense material. The permittivity begins to increase for all materials above 300 °C, as observed for the ac conductivity. For the *porous(water)* material the permittivity initially decreases with increasing temperature (see Figure 1b) between room temperature to 300°C (as was observed for the ac conductivity measurements) and indicates that the frequency dispersion is probably influenced by the presence of conductivity in the material. Above 300 °C the permittivity then begins to rise with increasing temperature for all three materials.

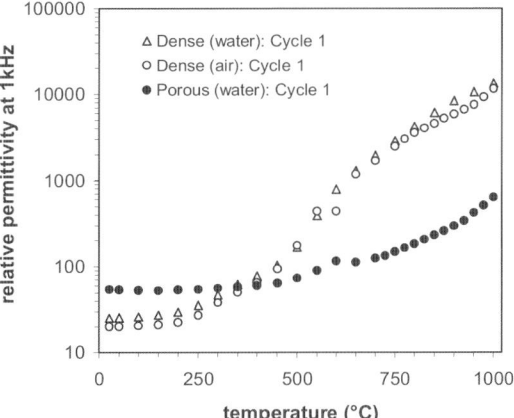

Figure 4. Relative permittivity at 1kHz as a function of temperature for Cycle 1

3.3 Determination of activation energies

The low frequency (1Hz) plateau conductivities where used to produce an Arrhenius plot of the natural log of bulk conductivity against temperature in order to calculate the activation energy for Cycles 1 and 2 for all three materials (see Figures 5 and 6). Activation energies could only be determined at high temperatures (700 – 1000°C) since only in this region was the temperature dependency of logarithm of conductivity versus temperature sufficiently linear (see Figure 3). The determined activation energies vary between 1.86 – 2.23eV. Theoretical consideration shows that activation energy for hydroxyl ion jumps along the c-axis is 2eV [16] – 2.1eV [17] and that for O^{2-} is 1.5eV [15]. The activation for proton conduction is much lower (0.5eV [7]). This indicates that hydroxyl ions are likely to be responsible for ionic conduction in this high temperature range. The higher temperature range of 700-1000 °C in unaffected by the thermal cycling and activation energies are in the range of ~2eV indicating that hydroxyl ions are responsible for conductivity at higher temperatures.

4. Conclusions

This paper has examined the ac conductivity and permittivity of HA based ceramics from room temperature to 1000 °C. At temperatures below 700°C the permittivity and ac conductivity of HA is influenced

160

by processing conditions, structure and degree of dehydration (thermal history). The ac conductivity exhibited a low frequency dependent conductivity followed by a high frequency dependent conductivity.

Thermally cycling of the materials indicated that the initial decrease in conductivity with increasing temperatures from room temperature to 200°C was thought to be due to dehydration of the HA and loss of surface bound water. The surface bound water is thought to contribute to conductivity for both dense and porous materials in this temperature range. The highest conductivity at temperatures <200°C was exhibited by porous HA sintered in water vapour and was attributed to the large surface area of the materials.

Both dense and porous material exhibited a peak in low frequency conductivity in the range of 500-600°C which disappeared during Cycle 2 and reappeared during Cycle 3. This indicates that thermal history and degree surface hydration has an important influence on conductivity in this range. The higher temperature range of 700-1000 °C in unaffected by the thermal cycling and activation energies are in the range of ~2eV indicating that hydroxyl ions are responsible for conductivity at higher temperatures.

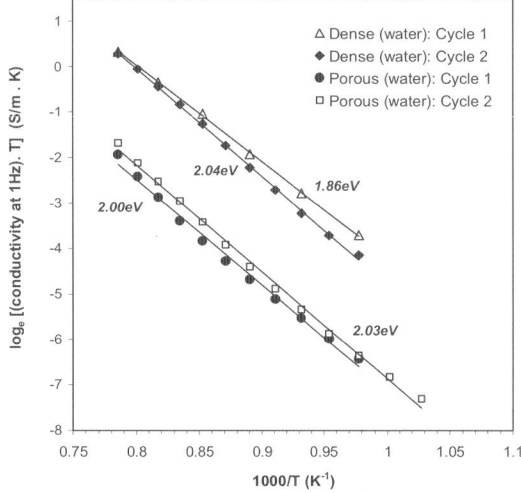

Figure 5. Arrhenius plot of natural log of bulk conductivity against 1000/T for dense (water)

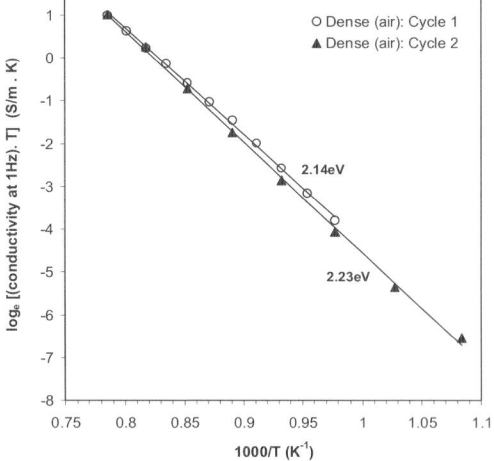

Figure 6. Arrhenius plot of natural log of bulk conductivity against 1000/T (700 – 1000°C) for dense(air)

Acknowledgments

The authors wish to show their appreciation and gratitude to the EPSRC (Grant No. EP/D013798/1) and Great Western Research (GWR) for their funding of this research.

References

[1]   Jarcho, M. (1981). Calcium Phosphate Ceramics as Hard Tissue Prosthesis. *Clinical Orthopaedics and Related Research,* **157**, pp. 259-278.

[2]   Lee, K. Y., Park, M., Kim, H. M., Lim, Y. J., Chun, H. J., Kim, H., Moon, S. H. (2006). Ceramic Bioactivity: Progresses, Challenges and Perspectives. *Annals of Biomedical Materials,* **1**, pp. R31 – R37.

[3]   Hench, L. L. (1998). Bioceramics. *Journal of American Ceramic Society,* **81** (7), pp. 1705-1728.

[4]   Lee, E.J, Koh, Y.H., Yoon, B.H. and Kim. H.E. and Kim, H.W. Materials Letters 61, 2270-2273 (2007)

[5]   Sopyan, I, Mel M., Ramesh S., Khalid K.A., Science and Technology of Advanced Materials, 8 116-123 (2007)

[6]   Tzvetanov, L., Nikolaeva, S., Michailov, I., Tivchev, P. (2002). Bone and Ceramic Interaction in the Bone Union Process. *Ultrastructural Pathology,* **26**, pp. 171 – 175.

[7]   K.Yamashita, K.Kitagaki and T.Umegaki, J.Am.Ceran. Soc. 78[5] 1191-97

[8]   M.P.Mahabole, R.C.Aiyer, C.V.Ramakrishna, B.Sreedhar and R.S.Khairnar, Bull. Mat. Sci., 28 (2005) 535-545

[9]   K. Yamashita, H.Owada, T.Umegaki, T.Kanazawa and T.Futagamu, Solid State Ionics 28-30 (1988) 660-663

[10]   M.Nagai and T.Nishino, Solid State Ionics 28-30 (1988) 1456-1461

[11]   J.J.Prieto Valdes, A.Victorero Rodriguez and J.Guevara Carrio, J.Mat. Res. 10 (1995) 2174-2177

[12]   K.Yamashita, K.Kitagaki, T.Umegaki, T.Kanazawa, J.Mat.Sci.Lett 9 (1990) 4-6

[13]   Timothy P. Hoepfner, Eldon D. Case , Journal of Biomedical Materials Research, Volume 60, Issue 4 , Pages 643 – 650 (2002)

[14]   H.Takeda, S.Nakamura, K.Yamada, T.Tsuchiya and K.Yamashita, Key Eng. Mat 181-182 (2000) 35-38

[15]   J.P.Gittings, I.G.Turner and A.W.Miles, Key Eng. Mat. 284-286 (2005) 349-352

[16]   B.S.H.Royce, Annals of New York Academy of Science, 1974 No238 p131

[17]   M.Sh Kalil, H.H.Beheri, W.I.A.Fattah, Ceramics International 28 (2002) 451-458

Multi-Material Micro Manufacture
S. Dimov and W. Menz (Eds.)

161

Micro Electrical Discharge Machining of Si$_3$N$_4$-based Ceramic Composites

K. Liu, J. Peirs, E. Ferraris, B. Lauwers, D. Reynaerts

Afd. PMA, Department of Mechanical Engineering, Katholieke Universiteit Leuven, Leuven, BE-3001, Belgium

Abstract

The Electrical Discharge Machining (EDM) behaviour and machining properties of advanced engineering Si$_3$N$_4$-based ceramic composites Si$_3$N$_4$-TiN are investigated and discussed in this paper. Two types of EDM machining configurations, micro-EDM milling and die-sinking EDM, are employed in the investigation. Relaxation type of pulse is used, and the performances of EDM process in the form of material removal rate, tool wear and surface quality are studied. These tests result in a performance comparison and a discussion on the ceramic composites material removal mechanism. The feature of material removal mechanism is characterised as chemical decomposition of Si$_3$N$_4$ and TiN at elevated temperature rather than melting/evaporation. The generation of nitrogen gas bubbles leads to a porous and foamy top surface structure. Due to the ideal mechanical and physical property of Si$_3$N$_4$-TiN ceramic composites, an application example - a turbine impeller - as a crucial component in a micro power generation system is manufactured with obtained knowledge in both machining configurations.

Keywords: Micro Electrical Discharge Machining, Ceramic composites, Si$_3$N$_4$-TiN

1   Introduction

Advanced engineering ceramics are attracting more and more attentions in the last decades and employed as critical components in the modern mechanical systems because of their excellent mechanical, physical and chemical properties. Among those various ceramic materials (Al$_2$O$_3$, ZrO$_2$, B$_4$C…), Si$_3$N$_4$ has being regarded as one of the best ceramic material for structural application, owing to its low density, high hardness, high strength at high temperature, oxidation and thermal shock resistance. However, these promising properties also bring difficulties in machining and structuring this material, the wider applications in the industry thus are limited. In recent years, a successful approach by incorporating electrically conductive reinforcements such as TiN, TiC, TiCN, and TiB$_2$ etc. into the silicon nitride matrix is developed [1]-[3]. Among all these mentioned secondary phases materials, the introduction of 30-40 vol. % TiN not just dramatically increase the electrical conductivity of the composite, but also enhance the mechanical properties for instance fracture toughness, strength and wear resistance [4]. On the other hand, the composite is difficult to be machined efficiently by conventional manufacturing methods. Nevertheless, the significantly lowered electrical resistivity provides the possibility of structuring the composite by using other methods such as Electrical Discharge Machining (EDM). It attracts more and more attention in the last few years, major ceramic suppliers and merchants have already commercialised this composite especially for the EDM application.

Researches on electrical discharge machining of this composite have already been performed in the past on the machining behaviours. Liu and Huang [5] studied the effects of wire-EDM conditions on properties, reliability and microstructure of hot pressed Si$_3$N$_4$-TiN composites; furthermore, Liu also examined the electrode wear during the sinking EDM of this material [6]. In [7], Lauwers *et. al.* investigated the machining properties of various EDM configurations and with different dielectric, and proposed the variable material removal mechanisms. However, it still lacks information on the micro/meso-scopic EDM of the Si$_3$N$_4$-TiN ceramic composites.

In this paper, the machining properties of Si$_3$N$_4$-TiN composite are evaluated on both micro milling- and sinking-EDM. The miniaturized input pulse energy is the main difference comparing to macro-scale machining. Surface quality, microstructure, material removal behaviour regarding to the various condition are investigated and discussed. As an application example, the micro-manufacturing of the Si$_3$N$_4$-TiN composite into a miniature turbine impeller, which is a crucial component in a micro power generation system, with obtained the knowledge are demonstrated.

2   Experimental Investigation

2.1   Material properties and microstructure

The Si$_3$N$_4$-TiN ceramic composite employed in this research is obtained from a commercial ceramic supplier Saint-Gobain. Manufacturer's grade is Kersit 601. The measured mechanical and physical properties of the composite are listed in Table 1.

2.2   Experimental Set-up

One of the tools applied in this research is a SARIX SX-100-HPM micro-EDM milling machine. The equipped relaxation type generator and high precision positioning system makes it dedicate for the micro-scale machining [8]. The short discharge duration time (350-400 ns) and low discharge current (< 0.5 A) guarantees the small energy input for each pulse. The usage of hydrocarbon oil as dielectric also assures the smaller sparking gap to attain the machining accuracy. The special designed spindle and clamping head allow the use of thin solid or tubular electrodes with diameters ranging from

40 μm to 1.0 mm combined with external or internal flushing. The automatic electrode feeding system can compensate the tool length due to the wear during machining. However, the rotational speed is limited to 600 rpm.

Another tool is a die-sinking machine, Roboform 350γ, from Charmilles. The advanced generator can generate various pulses from iso-energetic static to relaxation type. It also uses hydrocarbon oil as the dielectric.

Table 1

Mechanical and Physical properties of Si_3N_4-TiN ceramic composite

Supplier		Saint-Gobain
Grade		Kersit 601
Chemical composition	Si_3N_4	64 vol%
	TiN	36 vol%
Binder		Al_2O_3
Grain size (μm)		See Fig. 1
Density (ISO 3369) (g/cm^3)		3.97
Hardness (ISO 3878), (kg/mm^2)	HV_{10}	1508 ± 33
	HV_{30}	1465 ± 6
3-point bending strength (MPa)		979 ± 120
Young's modulus (GPa)		333 ± 3
Fracture toughness (MPa.m$^{1/2}$)	10 kg	8.7 ± 0.7
	30 kg	5.5 ± 0.4
Resistivity (10^{-7}Ω.m)		160
Thermal conductivity (Wm^{-1}K^{-1})	20°C	28
	800°C	19

2.3 Experimental results

It is known that the contradictory effects of the tool electrode wear with the material removal rate and the surface quality in the EDM process; and all are very much related to the given technology parameters. Thus it is impossible to study all the combination of parameters on the outcomes of machining abilities. This research is more focusing on achieving a good surface integrity of the ceramic composites in various machining configurations. More process investigation can refer to [9].

2.3.1 Micro-EDM milling

In the micro-EDM milling of Si_3N_4-TiN composite, tungsten carbide solid rod is used in the experiments. The relationship of the input actual machining parameters with the material removal rate, tool wear and the surface quality are shown in Table 2.

For the finishing regime, the material removal rate and tool wear ratio are difficult to be concluded because of the extremely slow process and unstable machining condition. As can be seen, the micro-EDM milling properties of Si_3N_4-TiN have the same

variation trend as the machining of steel: lower pulse energy is necessary for achieving better surface quality, but not in favour of the machining speed and tool wear.

To further optimizing the surface quality after the micro-EDM milling process, a series of experiments are conducted with gradually lowed energy input of the pulse. In Fig. 2 the relationship between the actual discharge parameters and the surface roughness R_a is plotted. Apparently the surface quality can only be optimized to a certain value. The smoothest surface quality obtained during the tests is 0.74 μm R_a. Further minimized pulse energy cannot provide any change on the R_a, which is unlike the micro-EDM process of steel.

Accordingly the surface topography is examined. In Fig. 3 a), the topography after the semi-finishing process is shown. The surface roughness R_a is 1.65 μm. No regular formed craters like normal EDMed steel surface are appeared; in contrary, the porous, sponge-like surface is revealed. A same phenomenon is also observed on the finest micro-EDMed surface. The cross-section (Fig. 3 b) exhibits this effected surface layer. The thickness is comparably equals to the 10 times of R_a.

2.3.2 Die-sinking EDM

Though the die-sinking EDM machine is not delicately designed for micro manufacturing purpose as aforementioned, it still has modules or technology settings which are specialized for micro-machining. Parameters investigated in the following experiments are set in this range. Copper infiltrated graphite (POCO$^{®}$ EDM-C3) is applied as the tool electrode material. For obtaining low energy input, relaxation type pulse is used. The machining parameters and the performances are listed in Table 3.

Comparing to the EDM milling process, the machining speed of roughing regime for die-sinking EDM is higher, as well as the tool wear. However, the surface quality is relatively worse. As for the finishing regime, even with very low energy input of the pulse, the roughness is somewhat higher than the result from milling process semi-finishing regime; furthermore, it also suffers from the little material removal and elevated tool wear.

The SEM micrographs of Si_3N_4-TiN ceramic composite materials obtained form the die-sinking EDM process at different machining regimes are illustrated in Fig. 4. Similarly, a porous and foamy topography is revealed. Apparently there are no dramatic changes on the microstructure with the varied pulse energy. Moreover, no visible subsurface micro-crack is examined even at high magnification SEM views.

Table 2

Machining properties of micro-EDM milling with WC tool electrode

Machining Regime	Actual open gap voltage (V)	Actual discharge current (A)	Material Removal rate (mm^3/min)	Tool wear ratio (%)	Roughness R_a (μm)	Sparking gap (μm)
Rough	-100	10	0.305	0.05	2.37	15
Semi-finishing	-100	5	0.173	0.92	1.26	10
Finishing	-70	0.5	-	-	0.75	4

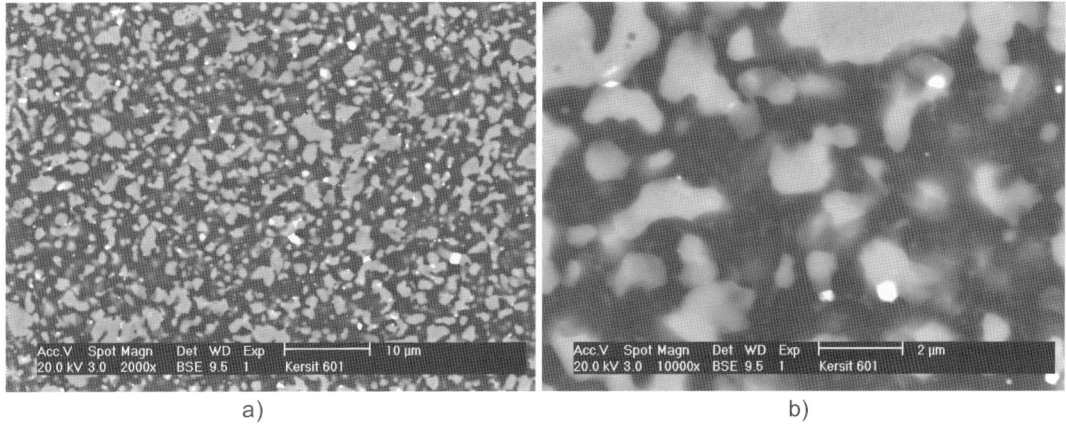

a)   b)

Fig. 1. Scanning microscopic pictures of the Si_3N_4-TiN composites: a) magnification 2000x; b) magnification 10000x. Phases: Grey = TiN; Black = Si_3N_4; White = WC milling ball contamination.

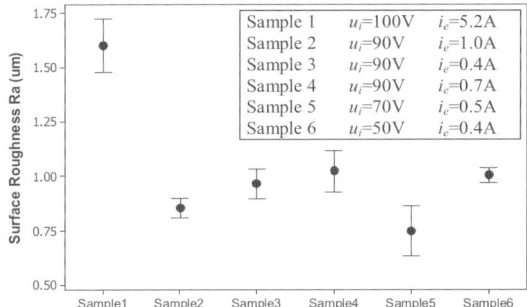

Sample 1	u_i=100V	i_e=5.2A
Sample 2	u_i=90V	i_e=1.0A
Sample 3	u_i=90V	i_e=0.4A
Sample 4	u_i=90V	i_e=0.7A
Sample 5	u_i=70V	i_e=0.5A
Sample 6	u_i=50V	i_e=0.4A

Fig. 2. Surface quality optimization in relation with the actual discharge parameters

2.4 Discussion

As shown in the experimental results, the continuously decreased pulse energy is not giving any benefit for improving the surface quality of Si_3N_4-TiN. These improvements are limited due to its intrinsic material removal mechanism. For the reasons of high material removal and much less tool wear comparing to the machining of steel, the material removal mechanism is not just the melting and evaporation, but also the involvement of chemical reaction, as the proposed in [7]: the decomposition of Si_3N_4 and TiN at elevated high temperature above 1700 °C.

$$Si_3N_4 \rightarrow 3Si + 2N_2 \uparrow$$

$$2TiN \rightarrow 2Ti + N_2 \uparrow$$

This reaction generates enormous amount of gas nitrogen gas bubbles which prohibits the formation of inerratic craters and leads to the creation of the voids, resulting in a foamy and porous top surface structure as shown in the SEM figures aforementioned. Furthermore, it also can be seen that the energy differences in the pulses cannot change the primary microstructure of the surface.

As a proof of the recognized material removal mechanisms, elemental analysis EDAX (Energy Dispersive X-ray Spectroscopy) of the ceramic matrix, specimens after micro-EDM milling and die-sinking EDM are conducted. The spectrums of the collected elemental emissions for each specimen are demonstrated in Fig. 5.

Table 4 lists the quantification of the detected elements. As expected, the content of N is dramatically decreased comparing to the ceramic matrix because of the decomposition. There is also a trace of carbon on both EDMed surface, indicates that there might be a transfer of the tool electrode material on the workpiece surface. However it cannot be confirmed because the machining environment is in the hydrocarbon oil.

Table 3
Machining properties of die-sinking EDM with copper infiltrated graphite tool electrode (Poco EDM-C3)

Machining Regime	Open voltage (V)	Charging Current (A)	Charging time (µs)	Capacitance (nF)	Material Removal rate (mm³/min)	Tool wear ratio (%)	Roughness R_a (µm)	Sparking gap (µm)
Roughing	-200	6	6.4	67	13.78	3.29	2.91	57
Finishing	-120	1	25	1.0	0.032	25.5	1.54	23

Table 4
EDAX quantification (wt%) of each element at Si_3N_4-TiN matrix and the surface after EDM

Elements	C	N	O	Al	Si	Ti
Matrix	–	24.19	5.00	1.69	39.79	29.33
EDM Milling	24.00	13.06	4.06	1.21	30.37	27.29
EDM Sinking	11.13	16.92	10.79	1.40	27.86	31.90

164

Fig. 3. SEM pictures of topography and cross-section of micro-EDM milled Si_3N_4-TiN surface. a) topography with R_a 1.65 µm; b) cross-section with R_a 0.75 µm

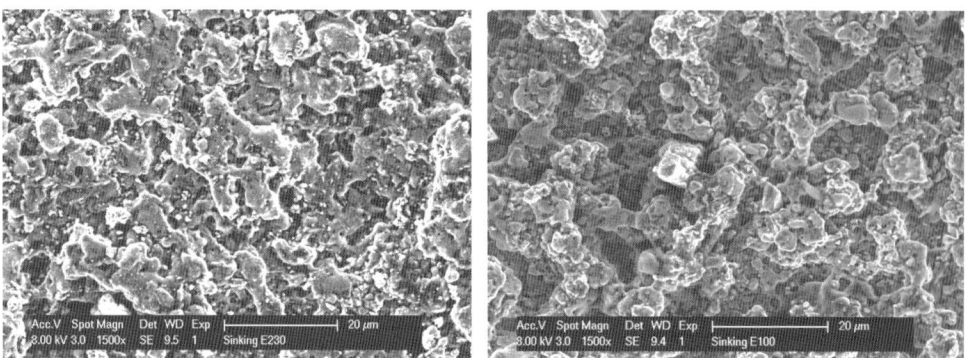

Fig. 4. SEM pictures of topography die-sinking EDM Si_3N_4-TiN surface. a) roughing regime with R_a 2.91 µm; b) finishing regime with R_a 1.54 µm

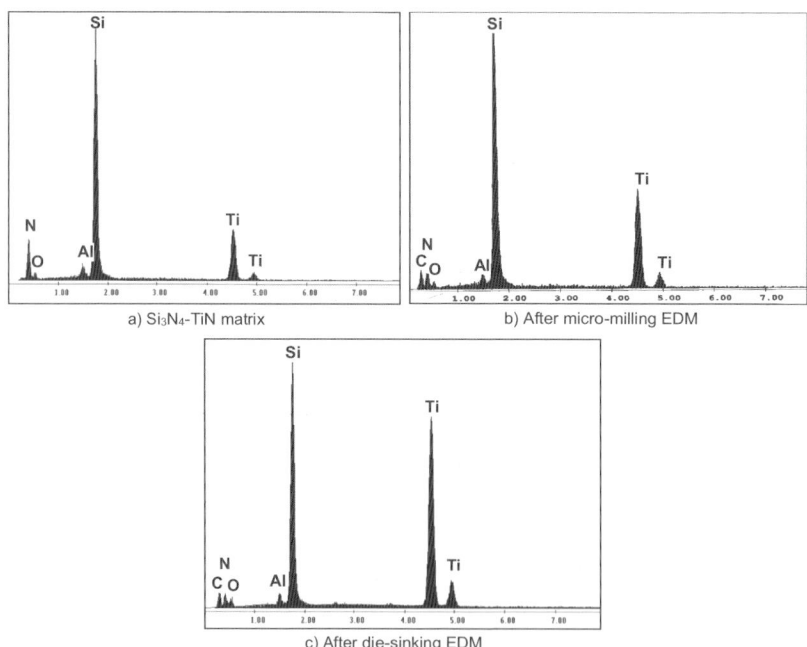

a) Si_3N_4-TiN matrix

b) After micro-milling EDM

c) After die-sinking EDM

Fig. 5. EDAX spectrum of the Si_3N_4-TiN matrix and the machined surface after the EDM process

3   Application

One of the most important application of the Si_3N_4-TiN ceramic composite is as a core component - a gas turbine impeller - of a portable fuel-based power units:. The miniaturised unit system has an overall size less than 1 dm^3 and electrical power output is intended to achieve 1kw with more than 20% efficiency. Thus the turbine impeller is expected to endure inlet temperature up to 1200 K and rotational speed of more than 500,000 rpm. The Si_3N_4-TiN is thus to be an ideal material. The turbine has a mixed axial-radial design and 8 blades with three-dimensional geometry (Fig. 6).

Both micro-EDM milling and die-sinking are employed in the manufacturing of the prototype turbine impellers. The schematic views of the die-sinking process and the milling process are illustrated in Fig. 7 and Fig. 9, respectively.

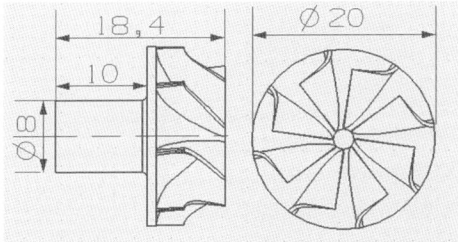

Fig. 6. Dimension of the miniature turbine impeller

In die-sinking process, the graphite electrode, which has a negative shape of a cavity, is produced on a 5-axis micro-milling machine (Kern MMP) with 3-axis machining configuration [10]. In total 10 electrodes are used for manufacturing a turbine. One of the electrodes and the finished product are shown in Fig. 7. Since the electrode milling and die-sinking process can run parallel, the total production time is approximately 15 hours. The dimensional control of the turbine is conducted on a Mitutoyo FN 905 CMM by using a Ø 0.7 mm styli. 600 points on each cavity are measured and results are compared with the CAD model. The deviation map is illustrated in Fig. 8. All the cavities are machined very consistently. However, the tip at the outlet has an overcut about 50 μm, and an unexpected single point undercut (around 36 μm) is observed at all side surfaces.

A close view of the micro-EDM milling process for the turbine manufacturing is presented in Fig. 9. The milling of each cavity starts with Ø1.0 mm WC tool for pocketing and Ø 0.7 μm tool for wall finishing, with layer-by-layer milling process. And the layer thickness is 8 μm and 3 μm, respectively. Due to the low machining speed of finishing regime for this method, only roughing setting is applied. Even with relatively high material removal rate and no need for electrode preparation, it still takes 20 hours for machining one cavity. Thus total machining time for manufacturing a turbine impeller is about 160 hours.

Furthermore, due to the lack of an accurate clamping system, the dimensional control on this finished part is unable to conduct. The improvement is under investigation.

Though it seems there is no benefit for micro-EDM milling on the machining hours, the appearance of the surface quality on the final product are better than the one machined by die-sinking EDM. The reason might be the more open flushing condition for milling process.

4 Conclusion

Micro-scale electrical discharge machining of commercially available Si_3N_4-TiN conductive ceramic composites has been performed and the results reveal the attractive machinability of EDMing this hard, brittle material. The influences of pulse energy on the material removal rate, tool wear, the surface integrity and sparking gap are investigated. With the elevated pulse energy, the machining speed increases, tool wear reduces as well as the surface quality. However the improvement of the surface roughness is limited due to the chemical decomposition of the Si_3N_4 and TiN as a material removal mechanism. The surface also reveals porous, foamy structure.

Precision manufacturing of a miniature turbine with Si_3N_4-TiN ceramic composites by either die-sinking or micro-milling process has also been conducted. Though the micro-milling EDM seems relatively low in machining efficiency, the better flushing condition provides more homogeneous surface quality comparing to die-sinking EDM. To achieve more accurate machining, a clamping system and the machining strategies are still under investigation.

Acknowledgements

This research is sponsored by the Institute for the Promotion of Innovation by Science and Technology in Flanders, Belgium, project SBO 030288, and by the Belgian programme on Interuniversity Poles of Attraction (IAP5/06: AMS). This work is also carried out within the framework of the EC Network of Excellence "Multi-Material Micro Manufacture: Technologies and Applications (4M)".

Fig. 7. A schematic view of the machining process and an enlarged view of the tool electrode, as well as the finished turbine impeller

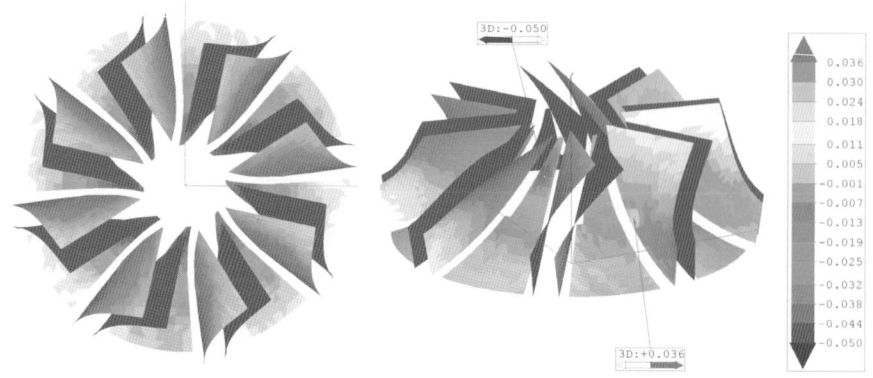

Fig. 8. Comparison results of the sinking turbine with CAD model in top and isometric view

Fig. 9. A close view of the Micro-EDM milling process and a finished turbine impeller

References

[1] Chu CY, Singh JP, Routbort JL, High-temperature failure mechanisms of hot-pressed Si_3N_4 and Si_3N_4/Si_3N_4-whisker-reinforced composites, Journal of the American Ceramic Society, vol. 76 (1993) pp 1349-1353.

[2] Xu HHK, Ostertag CP, Braun LM, Effects of fiber volume fraction on mechanical properties of SiC-fiber/Si_3N_4-matrix composites, Journal of the American Ceramic Society, vol. 77 (1994) pp 1897-1900.

[3] Shin DW, Tanaka H, Low-temperature processing of ceramic woven fabric/ceramic matrix composites, Journal of the American Ceramic Society, vol. 77 (1994) pp 97-104

[4] Herrmann M, Balzer B, Schuberrt C, Hermel W, Densification, microstructure and properties of Si_3N_4-Ti(C,N) composites, Journal of the European Ceramic Society. Vol.12 (1993) pp 287-296

[5] Liu CC, Huang JL, Effect of the electrical discharge machining on strength and reliability of TiN/ Si_3N_4 composites, Ceramics International, vol 29 (2003) pp 679-687

[6] Liu CC, Microstructure and tool electrode erosion in EDMed of TiN/ Si_3N_4 composites, Materials Science and engineering A363 (2003) pp 221-227

[7] Lauwers B, Kruth JP, Liu W, Schacht B, Bleys P, Investigation of the material removal mechanisms in EDM of composite ceramic materials, Journal of Materials Processing Technology, Vol 49 (2004) pp 347-352

[8] Liu K, Ferraris F, Peirs J, Lauwers B, Reynaerts D, Process capabilities of micro-EDM and its application, Proceedings of the 3rd International Conference on Multi-Material Micro Manufacture, 3-5 October 2007, Borovets, Bulgaria, pp.267-270

[9] Liu K, Ferraris E, Peirs J, Lauwers B, Reynaerts D, Process investigation of precision micro-machining of Si_3N_4-TiN ceramic composites by electrical discharge machining (EDM), Proceedings of the 15th International Symposium on Electromachining (ISEM), 23-27 April 2007, Pittsburgh, PA, USA, pp 221-226

[10] Ferraris E, Liu K, Peirs J, Bleys B, Reynaerts D, Production of a miniature Si_3N_4-TiN ceramic turbine impeller by die-sinking EDM, Technical Digest, The 7th International Workshop on Micro and Nanotechnology for Power Generation and Energy Conversion Applications, 28-29th November, 2007, Freiburg, Germany, pp 229-232

Machining of polystyrene by UV laser radiation for patch clamping device fab

S. Wilson[a,b], W.Pfleging[c], A. Welle[d], P.Kirby[b], M.Przylbyski[e]

[a] *Institute for Microstructure Technology, Forschungszentrum Karlsruhe, 76344 Eggenstein-L, DE*
[b] *School of Applied Sciences., Cranfield University, Cranfield, Beds. MK43 0AL, UK*
[c] *Institute for Materials Research 1, Forschungszentrum Karlsruhe, 76344 Eggenstein-L, DE*
[d] *Institute for Biological Interfaces, Forschungszentrum Karlsruhe, 76344 Eggenstein-L, DE*
[e] *ATL Lasertechnik GmbH, Burger Str. 48, 42929 Wermelskirchen, Germany*

Abstract

Laser patterning is of interest for MST applications; direct ablation of polymer material for generating 2D and 3D shapes such as microfluidic channels, curved shapes or micro-holes and alternatively photo-induced change of chemical or physical surface properties. Correct laser choice and process parameters enables new approaches for the fabrication of lab-on-chip devices with integrated functionalities. Laser-assisted ablation and modification of polystyrene (PS) is introduced with respect to the fabrication of polymer devices for high throughput planar patch clamping - a method of measuring the electrical activity of a cell currently a focus for high throughput systems (HTS). There are currently no marketed systems using novel materials that have surface modifications for either individual cell placement, or for dealing with cell networks, a physiologically important consideration for tissue engineering and understanding cell to cell interactions.
Within 4M, a design jointly proposed by FZK and Cranfield University for the fabrication of a polymer patch clamping system, laser micro-drilling of PS and subsequent surface functionalisation for cell adhesion has been investigated as a function of laser and process parameters. High power ArF laser with a pulse of 20 ns as well as high repetition ArF excimer laser sources with pulse lengths of 4-6 ns were used in order to study the influence of laser pulse length on laser drilling and laser induced surface modification. Micro-drilling of PS with diameters down to 1.5 µm have been demonstrated. Furthermore, localized formation of chemical structures suitable for improved single cell and cell network adhesion has been achieved on PS surfaces.

Keywords: laser, ablation, modification, polymer microsystem, cell adhesion, patch clamping, high throughput screening

1. Introduction

Patch clamping is considered the gold standard in cell analysis and gives highly sensitive information about the bioelectric (ion channel) activity of a cell membrane. In comparison with other assay techniques such as fluorescence and binding assays, patch clamping provides the highest sensitivity and information content, good selectivity, flexibility and physiologically relevant measurements. Traditional patch clamping has one major disadvantage; it has very low throughput and high labour costs – an experienced electrophysiologist may be able to make up to 10 measurements per day. Considerable resources have already been put into developing high throughput systems (HTS) for automating patch clamping. To date, commercially available systems are based on the traditional materials of glass or silicon.

1.1. Laser process

In microsystems technology (MST), UV-laser assisted processes are of particular interest for applications in microfluidics, bio-analytics, bio-reactors and micro-optics [1,2,3,4,5]. For packaging microstructured polymers, laser transmission welding has been successfully developed [6,7]. UV-photon induced surface modification of polymers for functionalisation of polymer based micro-devices is a relatively new research field. For this purpose, laser radiation sources or UV-lamp systems may be applied [8]. Excimer laser processing enables high local resolution via direct writing or direct optical imaging of complex structures. The process is in general initiated by direct bond breaking (e.g. separation of side chains or homolytic fissions) leading to the formation of new bonds or radicals. Following the formation or grafting of functional groups, e.g. amino acid or carboxyl groups are possible, which may lead to a change in biocompatibility. This type of functionalisation has been studied in detail for polystyrene (PS) for producing a polymer based automated patch clamping system.

1.2 The design

This paper introduces polystyrene as a possible new material for a patch clamping system. Potential new design approaches and laser-based processes for patterning and surface functionalisation are presented. UV-laser radiation sources emitting a wavelength of 193 nm were used for this purpose. Micro-drilling and surface functionalisation has been established for laser pulse lengths of 20 ns and 4 ns, and thermal and laser bonding techniques [6] are being developed for creating a viable device with integrated microfluidics.

Important for the analysis and synthesis of a design concept are the materials chosen, the concept layout and a cell placement method. Materials selection must result in a biocompatible material with a high dielectric constant and low dielectric loss. The materials must be formable to a high surface finish (average surface roughness, R_a < 0.5μm), and be capable of having holes of 1 – 4μm diameter being formed in a reproducible manner similar to a glass pipette.

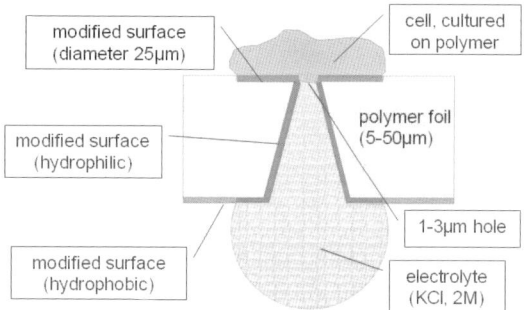

modified surface (diameter 25μm)

cell, cultured on polymer

modified surface (hydrophilic)

polymer foil (5-50μm)

modified surface (hydrophobic)

1-3μm hole

electrolyte (KCl, 2M)

Fig 1: Patch clamping design cross section

Materials should preferable be optically translucent for simultaneous measurement so that other complimentary techniques such as fluorescence can be incorporated. Holes with diameters of 1-3 μm should be produced into a thin polymer foil (thickness 5-50 μm, figure 1). The sidewall angles should be adjusted to provide good feed-through for measuring electrolyte (typically high [K^+]) from the back. A 25 μm diameter surface area modification should be made around the holes for cell placement. Between these holes small tracks with widths of 2 μm are modified for producing networks, as cell neurites grow along the exposed tracks to the next cell. A hydrophilic surface within the holes should support the capillary force which enables electrolyte flow into them. The holes are pitched at 450μm allowing the drilled foils to be bonded to a carrier frame for physical isolation of electrolytes below the cells, thereby preventing electrical cross-talk and signal leakage. Laser micro-material processing is well suited for producing of micro holes in polymers with high aspect ratios and is presented in the next section

2. Experimental Approach

Lasers are used to produce the 1-3μm holes in PS foils, and also to produce the surface modification suitable for the accurate placement of cells over the drilled holes. In order to optimise the surface modification parameters, cell culture test results and surface analysis (XPS and contact angle measurements) were used to produce an array of parameters where cells successfully adhere to PS in a selective way. Antibody staining was performed to visualise guided proteins on modified surfaces.

2.1. Laser drilling and surface modification

Laser-induced micro-drilling and modifications were performed with excimer laser radiation (Lambda LPX 210i, pulse length 20ns and ATLEX-M 300, pulse length 4 ns). Shorter laser pulses significantly reduce thermal contributions to processing. For the high power excimer laser radiation multi-lens arrays were used in order to homogenize the beam to a 'flat top' profile and to have an intensity fluctuation to better than 5%. The short pulse lasers did not require additional homogenisation.

2.2. Cell culture

Cell culture tests were used to assess the practical effectiveness of the surface modifications, and the ability of cells to grow over the laser produced holes. Culturing of L929 cells was performed as described in detail elsewhere [9]. PC12-GFP cells were cultivated in RPMI 1640 media supplemented with 2 mM L-glutamin, 1.5 mg/mL sodium bicarbonate, 4.5 mg/mL glucose, 10 mM HEPES, 1 % non-essential amino acids, 100 units/ml penicillin, 100 µg/ml streptomycin, 10 vol% heat inactivated horse serum (PAA, CatNo. B15-023), and 5 vol% fetal calf serum (PAA, CatNo. A15-649). In both cases, prior to cell inoculation and during cell culture on modified polymer substrates, cell culture medium supplemented with 1 mg/ml Pluronic F-68 (P-5556, Sigma) [9] was used. One day after PC12-GFP plating onto patterned substrates, the culture medium was exchanged and cells were exposed to nerve growth factor from mouse submaxillary glands (NGF-7S, Sigma, N0513) at a final concentration of 50 ng/ml.

3. Results and Discussion

3.1. Laser drilling

For laser micro-drilling laser fluences ε, significantly above the ablation threshold ε_t are necessary. For PS foils, the laser ablation thresholds at 193nm were determined to be 76mJ/cm² at 20 ns pulse length, and 82mJ/cm² for a 4ns pulse length. These correspond well with the documented value of 80mJ/cm² [10]

This laser energy is absorbed to a depth of 60 nm. Debris formation during drilling of micro-holes is significantly reduced by using shorter laser pulses, debris has a significant influence on cell adhesion as described elsewhere [10] and should be avoided. Furthermore, the edges of laser drilled holes show thermally-induced surface defects at the entrance side using 20ns laser pulses.
The diameter of holes at the laser exit side depend on laser fluence, laser pulse length, mask dimensions and the demagnification factor of the objective system. At 0.5 J/cm², a chromium-quartz mask with hole diameters of 50 µm and a demagnification factor of 10; the obtained hole diameter at the laser entrance side is 5-6 µm and at the laser exit side the hole diameter could be adjusted within a range of 1.5 µm up to 2.5 µm. These results have been obtained for a polystyrene foil with a thickness of 50 µm. For fabrication of planar patch clamping system laser drilling of micro-holes is suitable performed by short pulse laser radiation (4 ns).

3.2. Cell Adhesion

The required laser exposure dose for cell adhesion depends on laser fluence and laser pulse. Cell cultivation experiments were performed with both laser pulse lengths (20ns and 4ns) and no significant difference regarding cell adhesion was observed.

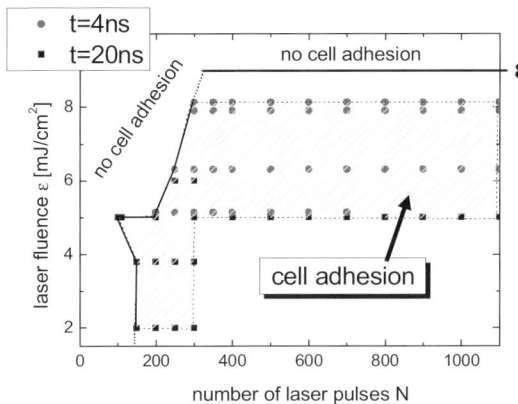

Fig 2: Cell adhesion as a function of laser fluence and pulse number

The adhesion of L929 cells was investigated on PS surfaces as function laser fluences and laser pulse number. Significant cell adhesion on PS surfaces can be obtained only for laser fluences smaller than 10 mJ/cm^2. Here it is necessary to consider that the threshold for the dry etch process ε_c is in the same range. Above this threshold a significant increase of surface roughness and a subsequent removal of functional polar groups is mainly responsible for improved L929 cell adhesion. Not only has the laser fluence shown a threshold for cell adhesion, but also the laser pulse number. For a laser pulse number smaller than 100 pulses no cell adhesion was observed. The threshold of the laser pulse number was determined within an accuracy of 50 pulses. The threshold of laser pulse number varies with laser fluence as shown in figure 2.

For a laser fluence of 6 mJ/cm^2, 250 laser pulses are necessary for subsequent cell adhesion while for laser fluences of 4 mJ/cm^2 150 pulses are required. Cell adhesion is observed within a range of 100 to 1100 laser pulses. It appears that an increasing irradiation dose represented by an increasing laser pulse number has no negative effect on the efficiency of cell adhesion if laser fluences smaller than 10mJ/cm^2 are selected. Modification of PS with ArF excimer laser radiation at low laser fluences enables a high lateral resolution control of cell adhesion.

3.4. Adsorption of albumin and laminin on laser modified surfaces

The adhesion of cell lines on unmodified PS surfaces is generally poor. This is to be attributed to the strong adsorption of albumin from serum containing cell culture media. Since albumin does not present cell attractive peptide sequences the albumin covered surface is passivated with respect to cell adhesion. Due to the altered physico-chemical properties (wettability) of the UV exposed surfaces the competitive adsorption of plasma proteins is influenced. Albumin adsorption onto UV irradiated PS is hindered as it does not adhere well to the COOH bonds created by short wavelength UV exposure, whereas the adsorption of cell attractive proteins increases. As albumin adheres loosely to surface modified areas, it is displaced by the cell attractive protein laminin that strongly adsorbs to surface modified areas. To visualise these phenomena, laminin is stained using an immunofluorescence

protocol (primary antibody: chicken polyclonal anti-laminin (ab14055, abcam), secondary antibody: rabbit anti-chicken IgY (H&L) FITC labeled (Anaspec)). The result is a green fluorescing laminin pattern (adhering showing the modified surface areas (figure 3). Figure 4 shows PC-12 GFP cells growing along the modified laminin coated areas forming a network useful for measuring.

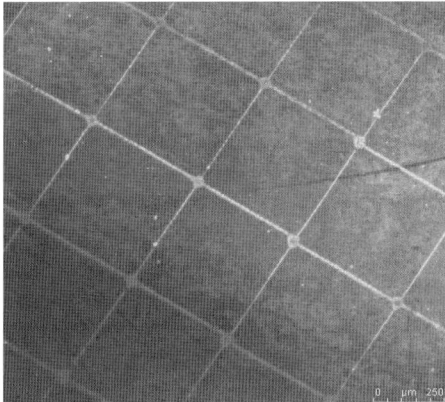

Fig 3: Fluorescence photo: laminin (1.5mm²)

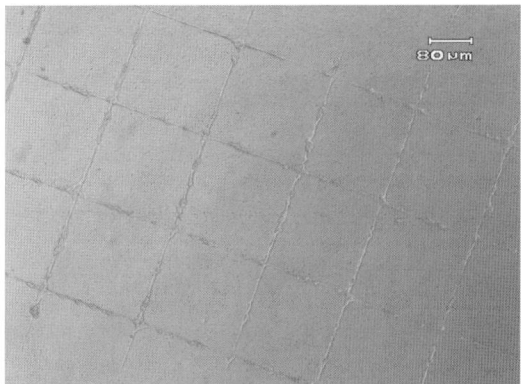

Fig 4: PC12-GFP cells above 2µm holes in PS (6 days)

4 Summary

Laser processing at short wavelengths is an appropriate tool for a selective patterning of polymer surfaces and is suitable for the selective protein adsorption process used for the specific placement of cells for patch clamping measurements. It was demonstrated that these kinds of "patternings" could be combined with high lateral resolution, e.g. ablation and modification of polymer surfaces with respect to of L929 and PC-12 cell adhesion (figure 2). The adhesion of cell clusters, single-cells as well as the formation of cell networks on polystyrene surfaces can be controlled with high accuracy. Ongoing work is being pursued into assessing the quality of adhesion of cells onto modified surfaces as well as the integration of microfluidics into the patch clamping device.

5. Acknowledgements

We are grateful to our colleagues M. Beiser for his technical assistance in SEM. We are indebted to Mrs. M. Marin for her excellent contributions to this research project. We also thank H. Besser and A. de Oliveira for laser material processing. We gratefully acknowledge the financial support by the program NANOMIKRO of the Helmholtz association and the EU within the Sixth Framework Programme ("Network of Excellence in Multi-Material Micro Manufacture (4M)" and "Network of Excellence in Microoptic (NEMO)").

6. References

[1] S. Sinzinger, J. Jahns: Microoptics, Wiley-VCH, Weinheim, FRG, 1999.

[2] W. Ehrfeld, V. Hessel, H. Löwe: Microreactors, Wiley-VCH, Weinheim, FRG, 2000

[3] E. Gottwald, S. Giselbrecht, C. Augspurger, N. Dambrowsky, R. Truckenmüller, V. Piotter, T. Gietzelt, O. Wendt, W. Pfleging, A. Welle, A. Rolletschek, A.M. Wobus, K.-F. Weibezahn, "*A Chip-Based Platform For The In Vitro Generation Of Tissues In Three-Dimensional Organization*", Lab on a Chip 7(2007) p 777-85

[4] A. Brandenburg, R. Edelhäuser, F. Hutter, "Integrated Optical Gas Sensors Using Organically Modified Silicates As Sensitive Films", *Sensors and Actuators* B 11, pp. 361-374, 1993

[5] D.A. Chang-Yen, B.C. Gale, "An Integrated Optical Biochemical Sensor Fabricated Using Rapid Prototyping Techniques", *SPIE* 4982, pp. 185-195, 2003

[6] W. Pfleging, P. Schierjott, C. Khan Malek, "**Rapid fabrication of functional pmma microfluidic devices by CO_2-laser patterning and HPD-laser transmission welding**", Laser Assisted Net Shape Engineering 5, Vol. 2, Editors M. Geiger, A. Otto, M. Schmidt, Meisenbach-Verlag Bamberg 2007, Seite 1207-1220, ISBN: 978-3-87525-261-3

[7] W. Pfleging, O. Baldus
Laser Patterning and Welding of Transparten Polymers for Microfluidic Device Fabrication"
Proc. of SPIE Vol. 6107 (2006) 61075-1 – 61075-12

[8] W. Pfleging, A. Welle, M. Bruns; S. Wilson, "Laser-assisted modification of polystyrene surfaces for cell culture application", Applied Surface Science Vol 253/23 (2007) pp 9177-9184]

[9] A. Welle, S. Horn, J. Schimmelpfeng, D. Kalka, "Photo-chemically patterned polymer surfaces for controlled PC-12 adhesion and neurite guidance" *Journal of Neuroscience Methods* 142, pp. 243-50, 2005.

[10] J.F. Ready, LIA handbook of laser materials processing, Laser Institute of America, 2001

[10] E. Detrait, J.B. Lhoest, B. Knoops, P. Bertrand, "Orientation of cell adhesion and growth on patterned heterogeneous polystyrene surface", *J. Neurosci. Methods* 84, pp. 193-204, 1998

DRIE of non-conventional materials: first results

Samuel Queste, Gwenn Ulliac, Jean-Claude Jeannot and Chantal Khan Malek

Institute FEMTO-ST/Dpt. MN2S, CNRS UMR 6174,
32 Av. de l'Observatoire, 25044 Besançon, FRANCE

Abstract

High speed directional etching of non conventional materials is still insufficiently developed for producing high aspect ratio microstructures. Compared to deep silicon etching, the plasma etching of these materials has suffered from limitations in achievable depth, aspect ratio, verticality and smoothness of surfaces. Inductively coupled plasma (ICP) reactive ion etching (RIE) of quartz crystal, lithium niobate and glass was conducted using fluorine and fluorocarbon based plasma-chemical etching processes. Optimization of etched depth, verticality of the walls, etch rate, etch selectivity towards the etch mask, and surface smoothness was investigated and compared to results of the literature. Deep etching with nearly vertical walls was successfully demonstrated for all three materials.

Keywords: DRIE, quartz, lithium niobate, glass

1. Introduction

A number of materials are of particular interest to microsystem technology. They can be passive structural materials like glass and fused silica for applications like microfluidics and optical applications or active materials like piezoelectric quartz crystal as well as piezoelectric and electro-optic/acoustic lithium niobate ($LiNBO_3$) for actuators and sensors. Effective etching techniques and processes are required to produce vertical and smooth surfaces with high etching rates.

Fabrication of precise features in a controlled fashion made out of non conventional materials is often challenging. The difficulty is reflected in the wide variety of non-conventional techniques, both serial and parallel, used for producing the desired microstructures. However, geometrical variety, aspect ratio and surface quality are often not sufficient, particularly with the smaller structures.

Plasma-based reactive ion etching (RIE) is a very controllable dry process exploiting both chemical and physical processes to remove solid material locally. It is used for highly directional and precise micro-scale etching of materials. However, it is difficult to achieve fine high aspect ratio structures. Reactive ion etching (RIE) using inductively coupled plasma (ICP) RIE system has demonstrated its potential in batch fabrication of deep anisotropic microstructures in silicon and enabled high-aspect-ratio deep silicon etching, which produced great impact on MEMS device fabrication. More recently, high speed directional etching using ICP plasma has being extended to materials other than silicon. Non conventional materials like quartz, lithium niobate and glass can be etched using various fluorine- and fluorocarbon-based chemistries (precursors such as C_4F_8, CHF_3, CF_4, SF_6). However, in contrast to well-studied silicon or silicon dioxide etching, many problems exist in RIE etching of non conventional materials related to the chemical composition of materials which leads to low etch rate and low etch selectivity (to etch mask) and limits the formation of high-aspect-ratio microstructures.

1.1 Plasma etching of quartz

Current manufacturing technology for quartz resonators does not provide a straightforward path for reducing the size of the devices for a large number of sensing applications and advanced dry etching processes have been reported [1,-2].

1.2 Plasma etching of $LiNbO_3$

Ferroelectric lithium niobate is a crystalline material widely used in electronics and communication because of its favourable optical, piezoelectric, electro-optic, elastic, photo-elastic and photo-refractive properties. It is of particular interest as active material for integrated and guided-wave optics due to its large electro-optical, acousto-optical and non-linear optical effects. Exploitation of electro-optical effect in bandgap materials such as $LiNbO_3$ for photonics (non-linear optics and guided optics) and phononics applications is another speciality of the FEMTO-ST institute and requires anisotropic etching of $LiNbO_3$.

Because crystalline metal oxide films tend to be chemically very inert, $LiNbO_3$ is difficult to structure, in particular with high resolution and reasonably fast etch rate. It has been dry etched using chemically reactive gases such as CHF_3, CF_4 and C_3F_8 [3-5].

It has been reported that introducing defect sites in the crystal lattice may enhance the etching rate of any crystal. In particular, this has been carried out using proton-exchange (PE) for surface modification (depth of a few micrometers). PE can be combined with etching of $LiNbO_3$ for example using wet etching or DRIE [6]. At the institute, $LiNbO_3$ waveguides are routinely made by annealed proton exchange for photonic applications (waveguides, photonic crystals).

1.3 Plasma etching of glass

ICP reactive ion etching of various types of glass including pyrex, fused silica a.k.a. amorphous "quartz", and other borosilicates has been reported by several authors [7-9] using various fluorine-based plasmas. It exhibited generally low etch rates and a low selectivity to etch mask and rough surfaces, which limit the formation of high-aspect-ratio microstructures. The etching of glass is complicated by the fact that there is a variety of silica-based oxide glass materials (soda-lime glass, borosilicate glass, etc.) with different compositions.

To date, efforts in silicon dioxide etching have been primarily directed towards realizing features for microelectronics applications (interconnect vias, waveguides, etc.) [10, 11] with limited improvements in etch rate, etch selectivity to masking material, anisotropy, or uniformity of etch across wafer.

In the work reported here, deep reactive ion etching of quartz crystal, lithium niobate, and glass has been investigated as a function of gas chemistry and processing parameters, the influence of which on etched depth, verticality of the walls, etch rate and etch selectivity towards the etch mask was examined.

2. Experimental

2.1 Materials

The samples used in these experiments were wafers of commercial Pyrex 7740, X-, Y- and Z-cuts of $LiNbO_3$ single crystal, and AT-cut single quartz crystal. All the wafers were 4 inches size and polished to mirror finish with the thickness of 0.5mm. The X and Z cuts of $LiNbO_3$ (suitable for proton exchange to tune the optical index) are dedicated to photonic applications and whereas the Y and Y-128 cuts are reserved for phononic applications (that is involving phonons instead of photons).

All the samples underwent a sequential multiple solvent cleaning step before further processing.

2.2 Etch mask

The etch mask used was either evaporated material (Ni, Cr, Al, SiO_2) or electroplated Ni. Prior to the electroplating of the Ni layer, a 100nm of Cr/Ni adhesion/seed layer was deposited by sputtering onto the clean substrate. A positive 2µm thick AZ9260 photoresist was patterned by contact photolithography using a standard process. For the electroplated mask, a 1µm thick layer was electroplated in the photoresist stencil using a typical Watts bath based on Ni sulfamate (lower stress than Ni sulfate) at a current density of 1.5 A/dm² and a bath temperature of 50°C. At the end of the process the photoresist was removed by acetone and the plating layer by RIE. The lift off technique was used for evaporated masks.

All the four-inch substrates were cut into smaller pieces of 1 cm square.

2.3 Deep plasma etching

The etching is carried out using a multiplex AOE reactor of STS with an ICP RF generator of 3 kW and a RIE generator of 1.5 kW. The maximum size of the wafer is 6 inches, and the gas available are SF_6, CF_4, C_4F_8, He, O_2 and Ar. The temperature of the substrate holder was varied between 20 and 80°C. The gas mixture used was C_4F_8/O_2 for quartz and glass etch, SF_6/He and CF_4/He for $LiNbO_3$ etch. The samples were bonded to Si carriers with a thermal grease to cool the sample. Step heights were obtained with a Tencor Alpha-step profilometer. Selectivity and etch rate were estimated from SEM images normalized against profile measurements of the same feature. A profile angle, α, was calculated geometrically from the difference between the top and bottom widths over the etch depth, for narrow trenches and also large open features. This angle reflects the verticality of the wall, with α equal to 90° for a perfectly vertical wall. Uniformity was not measured as the samples were mounted pieces.

3. Results and discussion

3.1 Quartz crystal

Two studies were made, the first one characterizing the etching dependence of quartz on the bias power (120, 180 et 250 W), while other parameters, like antenna power, C_4F_8/O_2 ratio, chuck temperature, chamber pressure, were fixed at 1600 W, 80/5, 60°C and 8 mTorr, respectively. In the second study the etch mask selectivity was investigated and the performance of electroplated Ni and evaporated Ni mask was compared. Etching time was adapted to suit the parameters of the plasma.

Plasma etching of quartz was investigated by ICP RIE etching using a C_4F_8/O_2 plasma. Fig.1 shows SEM photographs of quartz pieces etched at different bias power (120, 180 and 250W). For bias power of 120 and 250 W, the etch profile was relatively vertical (α = 83° and 85° respectively for etch depths of 3.26 and 12.5µm). However, the deviation from verticality at the top of the microstructures is due to erosion of the etch mask during the etching process. A straighter profile with an angle α of 85° is obtained for a 180 W bias power (Fig. 2). For all the samples, the sidewalls of the etch structures were clean and sharp but the edges were roughened. The roughness of the sidewalls reproduces the defects which already existed on the Ni etch mask (faceting effect). The etch rate decreased with the drop of bias power, from 0.735µm/min for 250W to 0.415µm/min for 120 W and conversely the selectivity raised from 27:1 to 100:1. For bias power above 120 W, some micro-trenches at the bottom edge of the sidewall appeared. This phenomenon was already observed for $LiNbO_3$ etch [4] and was attributed to a higher electron density at the edges during etching, leading to higher etching speed.

The etch mask selectivity was studied for electroplated and evaporated Ni masks. It was shown that evaporated Ni gives better selectivity (57:1 compared to 27:1). But the limiting factor of a Ni evaporated mask was the difficulty to deposit a layer thicker than 0.5µm using the lift-off process.

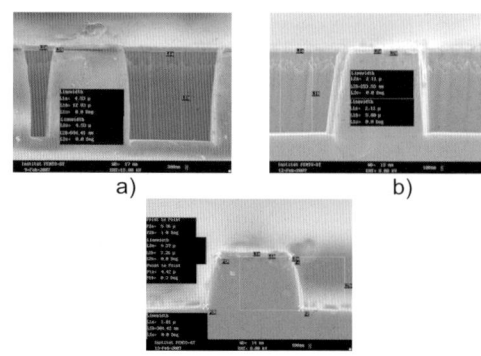

a)   b)

Fig 1: SEM images of quartz microstructure etched in a C_4F_8/O_2 plasma at different bias powers: a) 10 µm *wide and 12.8 µm deep feature etched at 250W; b) 5 µm wide and 5.8 µm deep feature etched at 180W; c) 5 µm wide and 3.3 µm deep feature etched at 120W.*

When etching a quartz AT-cut, our etching results fare well when compared with other studies using fluorine chemistry and they are either better or equivalent. The etching speed of 0.735µm/min is faster than the 0.5µm/min and 0.4µm/min reported by Abe et al. [1] and by Hung et al [12] respectively. No mention of etch profile was reported.

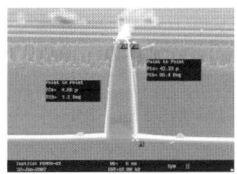

Fig 2: SEM pictures of a 4.7µm wide and 42µm high feature etched with an α angle of 85° in quartz crystal using a C_4F_8/O_2 plasma and 180W bias power.

3.2 Lithium niobate

Standard plasma etching of $LiNbO_3$ crystals was first carried out using Reactive Ion Etching (RIE) in a SF_6 gas on Y-cut crystals with a Ni electroplated mask. It resulted in microstructures with 10 µm depth (aspect ratio 1), an angle of 78° and an etch selectivity of 20 (Fig. 3-a) [13]. More recently, the combination of proton-exchange in X and Z-cuts of $LiNbO_3$ and RIE in a CHF_3 plasma using a sputtered silicon etch mask produced shallow sub-micrometric structures of 1.2 µm depth with a diameter of 400 nm (aspect ratio of 3), an angle of 85° and etch selectivity of 6 (Fig. 3-b) [14].

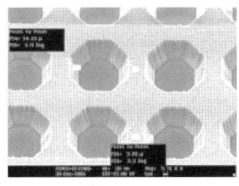 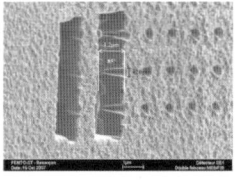

a)   b)

Fig. 3: Structures in $LiNbO_3$ produced by RIE; a) micrometric structures of 10µm width and 10µm depth [24];b) Nanostructures with a diameter of 400 nm and a depth of 1.2 µm in proton-exchanged $LiNbO_3$.

For deep microstructures in $LiNbO_3$, a different chemistry was used as CHF_3 was not available on the DRIE machine. Etching using SF_6- and CF_4-based plasmas was therefore investigated. The first experiments of pattern transfer were performed on the Y-128 cut of $LiNbO_3$ using an SF_6/He plasma to increase the selectivity and varying the usual parameters [coil power, platen power (bias), pressure, temperature, flow rate]. However sidewall verticality was limited to an angle of 60° (Fig. 4-a).

Switching to a different etching chemistry (CF_4/He) an angle on the order of 75° could be obtained on the Y-128 cut of $LiNbO_3$ (Fig. 4-b) and between 70° and 80° on the X-cut (Fig. 5). The influence of plasma chemistry and processing parameters on the etched profile, the selectivity of the etch mask as well as the etching speed using both SF_6/He and CF_4/He plasmas were studied. Fig. 4 shows SEM pictures of etched profiles on the Y-128 cut of $LiNbO_3$ as a function of the

utilized plasma. The bottom of the etched structures is relatively smooth.

The best results using a SF_6/He plasma were obtained by etching a 5.6µm depth with a selectivity of 6 using a 1 µm thick Ni electroplated mask (Fig. 4). A Cr etch mask enabled even better etch profiles.

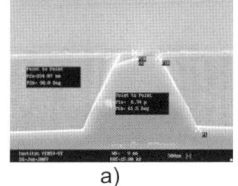 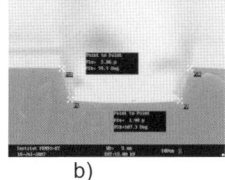

a)   b)

Fig.4: SEM pictures of 2µm deep etched profiles on the Y-128 cut of $LiNbO_3$ as a function of the utilized plasma: a) angle of 60° obtained in an SF_6/He plasma; b) angle of 75° obtained in a CF_4/He plasma.

The trials showed that using a CF_4/He plasma instead of an SF_6/He plasma allowed the etching speed to be multiplied by a factor 5 (0.5µm/mn instead of 0.1µm/mn), the etched profile to be improved by over 13° (75° instead of 62°) while keeping a clean bottom for an etched depth of 2 µm. In the X-cut, using a plasma of CF_4/He and an evaporated Cr mask, an etch profile of 80° could be obtained while keeping an etching speed on the order of 0.5µm/mn and a smooth bottom (Fig. 5).

Several materials were studied as etch mask in a CF_4/He plasma: evaporated Al, SiO_2, Cr and Ni and Ni and Si deposited by sputtering. A selectivity of 4.5:1 was obtained with a Cr evaporated mask.

Fig. 5: SEM picture of etched profile in X-cut $LiNbO_3$ (height 2µm, angle 80°) etched in a CF_4/He plasma.

$LiNbO_3$ is a particularly difficult material to be plasma-etched and a verticality close to 90° has not been reported yet despite the large number of studies. One challenge is that non volatile compounds of LiF type are formed in fluorine chemistry and redeposit, resulting in slower etch rate and less vertical etch profile (larger taper angle) [15, 16]. Exchanging lithium ions with protons (proton exchange) decreases this effect.

In term of etched profile, the best results (85°) so far were obtained with RIE on proton-exchanged lithium niobate similar to the work on DRIE also proton-exchanged crystals (82°) [6]. The highest etch selectivity for etching $LiNbO_3$ was obtained by RIE in an SF_6 plasma using an electroplated Ni mask. The etching was predominantly "physical" or ionic, that is leading to ion etching, using high energy ions produced by a maximum bias power at the limits of our reactor performance. The etch rate was very slow (on the order of 50nm/mn) and it took several hours for etching a 10 µm depth.

One advantage of using ICP etching is the speed of the process. Using DRIE with a CF_4/He, the etch rate was on the order of 0.5µm/mn, which is 10 times higher than with RIE in an SF_6 plasma. For photonic and

174

phononic applications, the target depth is around 2 μm and 10 μm respectively; however, deeper etching of a few tens of micrometers is needed for other applications, which also justifies the use of DRIE. For this purpose, the etch selectivity needs to be optimized, e.g. using a Ni mask which can be deposited in thick layer by electroplating or using a thick silicon layer.

3.3 Glass

Glass was etched using a C_4F_8/O_2 plasma and a 6 μm thick electrodeposited Ni etch mask. A depth of up to 200 μm with an aspect ratio of 20 was achieved with an etch rate comprised between 0.7 and 1μm/min depending on the size of the structures, and an etch selectivity of 18:1, which is comparable to the best result (200 μm) obtained with the same chemistry and a silicon wafer as etch mask [17]. We report a high etch rate when compared with other results obtained with the C_4F_8, SF_6, and C_4F_8/O_2. Concerning the etch profiles, the angles of 82°-88° are on the same order of magnitude as those reported by Li et al. [18, 19] (79°-88° using an SF_6 plasma). Fig. 6 shows an example of such etching.

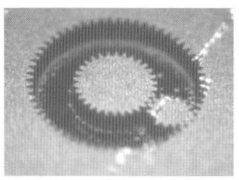 

a)   b)

Fig 6: Gear-wheel etched in pyrex 7740 using ICP RIE in C_4F_8/O_2 plasma: a) 3-D image; b) SEM picture of gear teeth (120 μm height; aspect ratio: 6).

4. Conclusion

Deep etching with nearly vertical walls was successfully demonstrated for three materials using ICP RIE and appropriate etch mask:
- For quartz crystal with the AT-cut, 42 μm was etched at 0.735μm/min using a C_4F_8/O_2 plasma and a Ni evaporated mask.
- For LiNbO$_3$ with the X-cut, 2 μm deep microstructures were etched with an etch profile of 80°, an etch rate on the order of 0.5μm/mn and a smooth bottom using a plasma of CF$_4$/He and an evaporated Cr mask (etch selectivity of 4.5 to 1). DRIE enabled a much faster etching process than RIE.
- For glass, a 120 μm depth was achieved with an etch rate comprised between 0.7 and 1μm/mn in a C_4F_8/O_2 plasma using an electrodeposited Ni mask.

Those results are very competitive with those found in the literature.

Acknowledgements

This work was carried out within the framework of the EC Network Of Excellence "Multi-Material Micro Manufacture: Technology and Applications (4M)" (EC funding FP6-500274-1; www.4m-net.org).

References

[1] Abe T. and Esashi M. "One-chip multichannel quartz crystal microbalance (QCM) fabricated by Deep RIE" J. Sensors and Actuators 82 (2000) pp. 139-143.
[2] Chang, D.T.; Stratton, F.P., Kubena, R.L., and Joyce, R.J "Optimized DRIE etching of ultra-small quartz resonators" Frequency control symposium and 17th European Frequency and Time forum, 4-8 May 2003 IEEE Proc. (2003) pp. 829 – 832.
[3] Tamura M. and Yoshikado S. "Etching characteristics of LiNbO$_3$ crystal by fluorine gas plasma reactive ion etching", Surface and Coatings Technology 169-170 (2003) pp. 203-207.
[4] Yang W. S., Lee H.-Y., Kim W. K., and Yoon D. H. "Asymmetry ridge structure fabrication and reactive ion etching" Optical Materials 27 (2005) pp. 1642 – 1646.
[5] Park W. J., Yang W. S., Kim W. K., et al. "Ridge structure etching of LiNbO$_3$ crystal for optical waveguide applications", Optical Materials 28 (2006) pp. 216 – 220.
[6] Hu H., Millenin A. P., Wehrspohn, Hermann H., and Sohler W. "Plasma etching of proton-exchanged lithium niobate", J. Vac. Sci. Technol. A 24(4) (2006) pp. 1012-1015.
[7] Li X; Abe T., Liu Y., and Esashi M. "Fabrication of high-density electrical feed-throughs by deep-reactive-ion-etching of pyrex glass", J. Microelectromechanical Systems 11(6) (2002), pp. 625-630.
[8] Park J. H., Lee N.-E., Lee J., Park J. S. and Park H. D. "Deep dry etching of borosilicate glass using SF6 and SF6/Ar inductively coupled plasma" Microelectronic Eng. 82 (2005) pp. 119-128.
[9] Jung H. C., Wang S., Hu X., Lee L. and J., Lu W. "Etching of Pyrex Glass Substrates by ICP-RIE for Micro/Nanofluidic applications", J. Vac. Sci. Technology. B. 24(6) (2006) pp. 3162-3164.
[10] Li X., Xing L., Hua X., Fukasawa M., Gottlieb S., Oehrlein S., Barela M., and Anderson H. M. "Effects of Ar and O$_2$ additives on SiO$_2$ etching in C$_4$F$_8$ based plasmas", J. Vac. Sci. Technol. 21(1) (2003) pp. 284–293.
[11] Li X., Xing L., Hua X., Oehrlein G. S., Wang Y., and Anderson H. M. "Characteristics of C$_4$F$_8$ plasmas with Ar, Ne, and He additives for SiO$_2$ etching in an inductively coupled plasma (ICP) reactor", J. Vac. Sci. Technol. 21(6) (2003) pp. 1955 – 1963.
[12] Hung V. N., Abe T., Minh P. N., and Esashi M. "High-frequency one-chip multichannel quartz crystal microbalance fabricated by deep RIE", J. Sensors and Actuators A 108 (2003) pp. 91-96.
[13] Benchabane S., Khelif A., Rauch J-Y., Robert L. and Laude V. "Evidence for a complete surface wave band gap in a piezoelectric phononic crystal", Phys. Rev. E, 73 (2006) 065601(R).
[14] Ulliac G., Courjal N., Chong H. M.H. and De La Rue R.M., to be published.
[15] Shima K., Mitsugi N., and Nagata H. "Surface precipitates on single crystal LiNbO$_3$ after dry-etching by CHF$_3$ plasma" J. of Materials Research 13(3) (1998) pp.527-529.
[16] Nagata H., Mitsugi N., Shima K., Tamai M., and Haga E. M. "Growth of crystalline LiF on CF$_4$ plasma etched LiNbO$_3$ substrates", J. Crystal Growth 187 (1998) pp. 573 – 576.
[17] Kolari K. "Deep plasma etching of glass with a silicon shadow mask", J. Sensors and Actuators A 141 (2008) pp. 677-684.
[18] Li X., Abe T. and Esashi M. "Deep reactive ion etching of Pyrex glass using SF6 plasma" J. Sensors and Actuators A 87 (2001) pp. 139-145.
[19] Li X., Abe T. and Esashi M. "Fabrication of high-density electrical feed-throughs by deep-reactive-ion-ion etching of Pyrex glass" J. Microelectromech. Syst. 11 (6) (2002) pp. 625 - 630.

Carbon nanotubes grown directly on printed electrode of electrochemical sensor

J. Prasek[a], J. Hubalek[a], M. Adamek[a], O. Jasek[b]

[a] Department of Microelectronics, Brno University of Technology, Brno 60200, Czech Republic
[b] Department of Physical Electronics, Masaryk University, Brno 60200, Czech Republic

Abstract

This paper is devoted to the area of electrochemical sensors. In this work several screen-printed thick-film electrodes are prepared. These electrodes are commonly used as the working electrodes of electrochemical sensors. The surface of the electrode has been modified with nanopatterned nanostructures. The nanostructures have been formed as vertically aligned carbon nanotubes that were grown directly on the screen-printed working electrode using plasma enhanced chemical vapour deposition method. The aim was to improve electrochemical properties of the electrode by creating homogeneous and high density carbon nanotubes directly on the thick-film layer. The created structures have been investigated by scanning electron microscopy. The electrochemical properties have been investigated by electrochemical detection of cadmium ions in aqueous solutions. The concentration of cadmium ions in units of µmol/L can be determined with the modified electrode.

Keywords: carbon nanotubes, nanopatterned nanostructures, thick-film electrode

1. Introduction

Electrochemical three-electrode sensors for trace determination of elements in liquids can be utilized in many areas. Their sensing properties are usually determined by the properties of the working electrodes. The current response of the sensor is given by the Levic equation for solid electrodes [1]. It indicates that the current response of the electrode is directly proportional to active electrode area. Therefore it is necessary to create sensor with a high working electrode area. This is in direct contradiction with the need to produce small sensor systems. One possibility of creating high active electrode area on the miniaturized sensor is to create high surface area nanostructures. The nanostructures can be created by nanopatterned structures formed by carbon nanotubes.

Nowadays carbon nanotubes (CNTs) are in the centre of investigation as very perspective material for solid electrodes. For example oxidized multiwall carbon nanotubes (MWNTs) were used as a novel adsorbent for removing Ni(II) from aqueous solution [2]. The metal adsorbance of the carbon nanotubes makes them very promising in heavy metal determination. The glassy carbon electrode modified by a nanoporous composite film was used successfully for the simultaneous voltammetric determination of a trace level of Cd(II) and Cu(II) [3].

One way of fabricating a low-cost small sensor with solid electrodes is use of the thick-film technology (TFT). The advantages of TFT sensors is their low dimensions, good reproducibility, mechanical, electrical properties of electrodes, low cost of the electrodes (e.g. gold electrode containing 1 mg of au), and a well accessible and ecological fabrication process [4, 5]. Thick films fabricated from cermets pastes also have a high mechanical and temperature resistance. It allows them to be used for several deposition methods of sensing materials for the working electrode formation. It predestinates the CNTs exploitation as sensing material on the electrode [6]. The CNTs belong to the most promising nano-materials; they show unique electronic, mechanical and chemical properties [7] that lead to many applications. They can be prepared by arc discharge [6], laser ablation [8] and chemical vapour deposition [9] methods. For industrial applications it is desirable to produce vertically aligned CNT films with uniform properties. Most of the techniques used for industrial fabrication work at low pressure, requiring vacuum systems. However, for industrial application it would be desirable to work at the atmospheric pressure. In electrochemical applications the CNTs are usually used as a material for polymer or epoxy based composites [10]. The aim of this work is to prepare the sensors with carbon nanotubes grown directly on the working electrode using a synthesis of the vertically aligned CNTs in atmospheric microwave plasma torch discharge using plasma enhanced chemical vapour deposition method (PECVD). The electrochemical properties of the electrode will be tested on voltammetric detection of heavy metal ions. For this purpose a screen-printed TFT electrode substrate was designed as a suitable construction for easy material varying of the working electrode and with a high possibility of using in several electrochemical systems.

2. Electrode substrate preparation

The three-electrode sensor for electrochemical measurement was designed and published [11]. Generally for a reference electrode Ag based TFT paste is used that can be electrochemically covered with AgCl layer [12]. It is suggested that an auxiliary electrode is made of Pt paste, but it is possible to use other types of TFT pastes. The same situation is with working electrode (WE), which is usually made of pure Au, Pt or Ag paste.

This sensor is not suitable for new material properties measurement and evaluation because the auxiliary and reference electrodes already made could be damaged during the high temperature CNTs deposition process. Therefore a new substrate (Fig. 1) consisting of the working electrode only was created for the material properties measurement and evaluation.

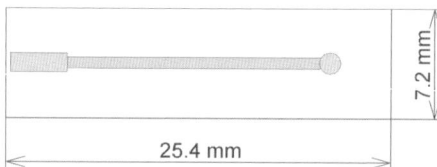

7.2 mm

25.4 mm

Fig. 1. Design of the substrate with WE only.

The substrate was fabricated using screen-printing techniques. The TFT material used for conductive layer was ESL 9912-k paste and ESL 4917 paste for the dielectric layer. For the working electrode the ESL 9912-K paste (Ag) and ESL 5545 paste (Pt) were used (all pastes were ESL ElectroScience, UK).

3. Woking electrode preparation

The nanopatterned working electrode is created in several steps. The bottom layer of the working electrode is formed from the Ag or Pt paste. The carbon nanotubes grow vertically aligned on the working electrode surface forming a nanopatterned structure. This electrode modification by nanotubes has to be homogeneous and the electrode surface has to be completely covered with a high density amount of the nanotubes. To deposit the carbon nanotubes on the working electrode its surface was modified by a thin nickel layer (10 - 20 nm) created by magnetron sputtering or by thermal vacuum evaporation. This layer served as a catalyst in the CNTs growth and argon, hydrogen and methane were used as working gases. The gas flow rates were controlled by electronic flow controllers. Microwave power of 400 W (2.45 GHz, 2 kW max. power) is supplied by a microwave generator and transmitted by a waveguide through a coaxial line to a hollow nozzle electrode. A ferrite circulator protects the generator against the reflected power by rerouting it to the water load. The coaxial line and the electrode accommodated a dual gas flow. Argon (1500 sccm) flowed through a central opening (1 mm) and the deposition mixture, H_2/CH_4 (42/430 sccm), was added by a set of holes in the outer housing. The plasma expands from the central nozzle forming a torch discharge. A quartz tube, 40mm in outer diameter, separates the discharge from the surrounding atmosphere. The base is sealed by a Teflon piece to the flange of the outer coaxial conductor. At the top it is closed by an upper flange with an exhaust tube and a sealed feed through for a substrate holder. The substrate holder is another quartz tube, 18 millimetres outer diameter, fixed at the upper flange. This tube is closed at its top by a quartz window. On the opposite side, i.e. close to the discharge nozzle, two slits are cut through the tube. The electrode was facing the torch during the deposition. Its temperature was measured by Raytek Thermalert TX pyrometer from the back side. After the plasma torch was ignited by an auxiliary rod electrode in flowing argon, the electrode was placed at the desired deposition distance (30 - 50 mm) from the nozzle and the deposition mixture H_2/CH_4 was added. The deposition temperature (Td) was regulated by the deposition distance and it was varied from 950 to 1100 K. The deposition time (td) was 1 to 15 minutes. A more detailed description of the experimental setup and deposition procedure can be found in [13].

4. Experimental

Surface morphology of the deposits was studied by scanning electron microscopy (SEM) with JEOL 6700F microscope equipped with an EDX analyzer. The acceleration voltage was usually 5 kV and the working distance was in the range of 8-9 mm. The pictures were taken at several points from the surface and at cross section of the working electrode.

Electrochemical properties have been tested by cadmium ions detection in 1 mol/L KCl buffer solution using prepared electrode connected against common Ag/AgCl reference electrode. The measurement method was differential pulse voltammetry (DPV) carried out by the Polarographic analyzer PA4 device (Laboratorni pristroje Prague, Czech Republic).

5. Results and discussion

The carbon nanotubes were successfully deposited on the Ag and Pt paste based working electrode with a Ni catalyst. The catalyst layer (10 nm of Ni) was vacuum evaporated on the top of the working electrode. The electrode material selection would influence the electrode surface morphology and the nanotubes growth. This issue was confirmed by SEM analysis of the carbon nanotubes deposits. The cross section of the Pt based working electrode with MWNTs deposited over the working electrode substrate is shown in the figure 2. In this figure it is shown that the Pt layer screen-printed over the alumina substrate is porous and its thickness is approximately 4 µm. The thickness of deposited MWNTs is 1.5 - 2 µm. In the case shown in figure 2 the MWNTs are not vertically aligned.

Fig. 2. Cross section SEM micrograph of the working electrode

The surface roughness and interactions between the printed layer and nickel catalyst are also responsible for different growth results. The surface of the Ag printed layer is rough because of large crystals. Therefore the CNTs deposited have apparent inhomogeneities in their growth (Fig. 3a,b). Large clumps of nanotubes are visible with interspaces not covered with nanostructures. Interspaces are present due to the absence of the catalyst whose active properties are lost in interaction with the electrode surface during high temperature growth and large crystals causing significant surface roughness. In addition, the binder of the metal particles covers them with a thick film producing large intercrystal spaces.

Homogeneously covered parts of the electrode form up a large area of nanoelectrodes that lead huge increasing of the active surface area (see Fig. 3c).

The fabricated electrode was tested for determination of cadmium(II) ions concentration because cadmium was shown as the trace of a heavy metal ion that can be detected [3,14]. The measurements were performed using differential pulse voltammetry in the 1 mol/L of KCl buffer at the scan rate of the 10 mV/s. The output current DPV response was measured under several additions of the

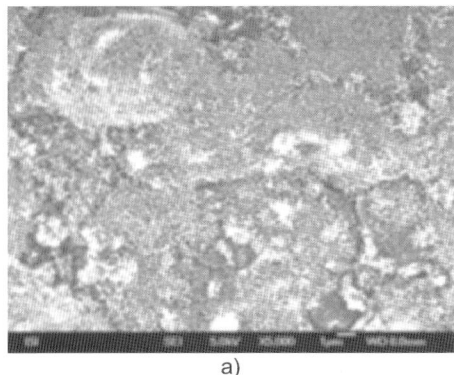

a)

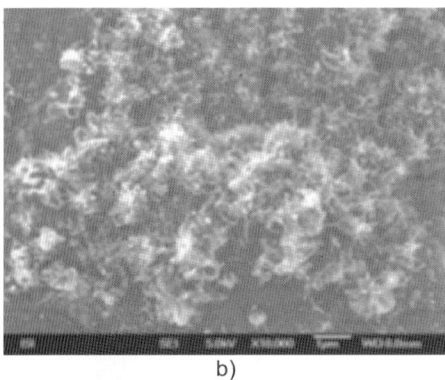

b)

c)

Fig. 3. SEM micrographs of the working electrode with the carbon nanotubes deposited on the Ag layer.

10 mmol/L $CdCl_2$ into the buffer. The responses to the concentrations of the 110 and 220 µmol/L in oxidation process are shown in Fig. 4. The peaks determine cadmium ions oxidation which was monitored on the working electrode. Doubling the concentration from 110 to 220 µmol/L caused more than a double change of the current response. The responses are high in comparison with base line and noise level. The light

coloured ends of the nanostructures, as seen in Fig. 3, are the remainder of the catalyst layer. It is difficult to remove them non-destructively but fortunately the catalyst has not influenced the electrode response.

The detection limit was determined at the level of the 5×10^{-6} mol/L of the cadmium ions. The slope of the calibration curve expressing the sensitivity of the electrode has been determined at the level of 0.016 µA.L/µmol.mm^2 in reduction and 0.135 µA.L/µmol.mm^2 in oxidation process. These results are comparable with the other carbon based material deposited on the working electrode or the standard commercial carbon electrode. The comparison of the fabricated electrode with standard carbon electrode SESV11 from Elektrochemicke detektory (ED, Turnov, Czech Republic) is shown in Fig. 5. It is clear, that the response of the fabricated electrode is three times better than the ones achieved with standard carbon electrode SESV11 as the results in Fig. 5 show. High level of the current response and high sensitivity of the working electrode show and confirm CNTs suitability for the cadmium detection. If inhomogeneities in the CNTs layer are considered, very good results have been obtained with CNTs vertically aligned on the working electrode.

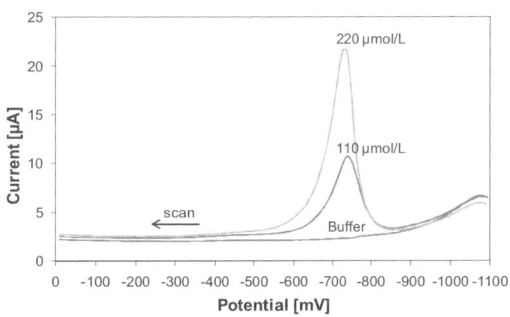

Fig. 4. The current response to two different concentrations of cadmium ions.

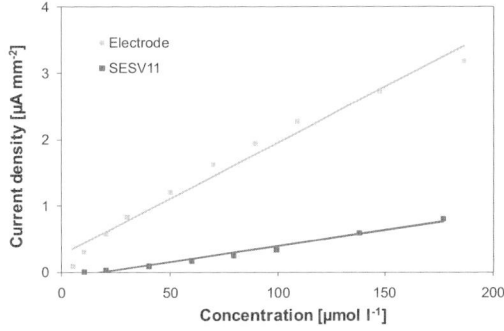

Fig. 5. Calibration curves comparison of fabricated electrode with classical commercial carbon electrode SESV11 obtained in process of reduction.

6. Conclusions

The first sample of the screen-printed electrode with vertically aligned carbon nanotubes formatted on the working electrode surface was prepared and tested in this work. The nanotubes grow homogeneously on the surface but inhomogeneities as open areas without the nanotubes were found. The issue is caused by roughness of the electrode surface and insufficient elimination of the intermetallic reactions of the catalyst layer with the electrode surface. In addition, the results show that residues of nickel catalyst from CNTs growing process do not influence the electrode response because of missing any other peaks in the responses. The PECVD method using Ni catalyst does not confirm growth of vertically aligned CNTs. The impact of the electrode surface quality on the vertical aligning of CNTs has not been found. The results have shown us that the electrode is able to detect the concentrations of 5×10^{-6} mol/L of the cadmium ions. This outcome is significant for the CNTs and method utilization on the nanopatterned working electrode creation if the quality of the nanotubes homogeneity on the first sample is considered. CNTs vertically aligned on the working electrode by plasma deposition are a perspective material and method for the nanopatterned electrode formation.

Acknowledgements

Funding for this work was provided by the Czech academy grant agency under the contract GAAV 1QS201710508, Czech grant agency under the contract GACR 202/05/0607 and Czech Ministry of Education in the frame of Research Plan MSM 0021630503 MIKROSYN.

References

[1] Rieger, P. H. Electrochemistry, Prentice-Hall, USA 1987, ISBN 0- 132-48907-4.

[2] Chen, C. and Wang, X., Industrial & Engineering Chemistry Research 45 (26), 9144-9149, 2006.

[3] Gao, X., Wei, W., Yang, L. and Guo, M., Electroanalysis 18 (5), 485-492, 2006.

[4] Harsányi, G.: Polymer Films in Sensor Applications, Technomic Publishing Co., Lancaster (USA), Basel, 1995. p. 435.

[5] Harsányi, G.: Sensors in Biomedical Applications, Technomic Publishing Co., Lancaster (USA) R, Basel, (Switzerland), 2000, p. 350.

[6] Iijima, S. Nature 354 – 1991, 56.

[7] Ajayan, P. M. Nanotubes from carbon, Chem. Rev. 99 – (1999) 1787.

[8] Thess, A., Lee, R., Nikolaev, et al. Science 273 (1996) 483.

[9] Lee, Ch.J., Kim, D.W., Lee, T.J., et al. Chem. Phys. Lett. 312 (1999) 61.

[10] Pumera, M., Merkoci, A. and Alegret, S. Carbon nanotube-epoxy composites for electrochemical sensing, Sens. and Act. B 113 (2006) 617.

[11] Prasek, J., and Adamek, M., IEEE Sensors 2004 proc. IEEE Sensors 2004. Vienna, Austria: IEEE, TU Wien, 2004, pp. 749 - 752.

[12] Lanz, M., Schürch, D., and Calzaferri, G. Journal of Photochemistry and Photobiology A: Chemistry 120 (1999) 105-117, Elsevier.

[13] Zajickova, L., Elias, M., Jasek, et al. M. Plasma Phys. Control. Fusion 47 (2005) B655.

[14] Liu, G., Lin, Y., Tu, Y. and Ren Z., Analyst 130 (7), 1098-1101, 2005.

Multi-Material Micro Manufacture
S. Dimov and W. Menz (Eds.)

179

An analysis of the effects of nanolayered nitride coatings on the lifetimes and wear of tungsten carbide micromilling tools

D. Zdebski[a], D.M. Allen[a], D.J.Stephenson[a], J. Hedge[a], C. Ducros[b] and F. Sanchette[b]

[a] *Precision Engineering Centre, Cranfield University, Bedford MK43 0AL, UK*
[b] *CEA Grenoble, Labatoire des Technologies des Surfaces, 17 rue des Martyrs 38054 Grenoble CEDEX, France*

Abstract

Micromilling is becoming increasingly important for a wide range of manufacturing tasks in the general field of microengineering, such as milling small channels in micromoulds designed for the fabrication of microfluidic devices by microinjection moulding of polymers. However, micromilling tools, often <1mm in diameter, are rather delicate, fracturing when forces become excessive and, consequently, micromilling can become an expensive process. In an attempt to increase tool lifetimes and reduce costs, micromilling forces have been measured with a microdynamometer and the effects of chromium nitride/titanium nitride and titanium aluminium nitride/titanium nitride coatings have been evaluated as an aid to decreasing tool wear and extending the lifetime of tungsten carbide micromilling tools. The surface finish of the milled workpiece has also been measured to monitor how tool wear affects the resultant milled surface.

Keywords: coatings, micromills, wear

1. Introduction

The cost of micromilling can be reduced if the lifetime of the delicate microtools used in cutting processes can be extended. This research aims at improving tool-lifetime by coating micro endmills with wear-resistant materials. Two coatings have been investigated and compared with an uncoated tool and a tool with a commercially-available coating.

2. Experimental

Two types of 1mm and 0.2mm diameter Unimax tungsten carbide (WC) endmills were purchased from Union Tools, Japan. The first type (Carbide Endmill) was uncoated and the second type (UT Dry) had been coated with a commercial material of unknown composition. Energy dispersive X-ray analysis (EDX) was used to analyse the UT Dry tool coating that was determined to contain aluminium, titanium and nitrogen.

A number of the uncoated tools were coated at CEA by cathodic arc evaporation to produce two other coatings for comparison. The first comprised nanolayers of CrN and TiN and the second comprised nanolayers of AlTiN and TiN. Both coatings have a total thickness of approximately 2µm.

It was noted that the CrN/TiN coatings were gold-coloured, whilst the AlTiN/TiN coatings were grey-coloured, thus making visual identification of the coatings easy for the machine operator and researcher.

Micromilling was carried out on an Evo Kern CNC precision machining centre, installed in the temperature-controlled (20.0 ± 1.0°C), humidity-controlled (50 ± 5% RH) Hexagon Loxham Precision Laboratory at Cranfield University.

Measurement of the micromilling forces was carried out with a Kistler microdynamometer (model 9256C2) positioned as shown in Figure 1.

Examinations of the machined steel (Toolox 33, a prehardened tool steel) surfaces were carried out by white light interferometery (WLI) and scanning electron microscopy (SEM). The tool was also examined by SEM and, in addition, EDX was also used to monitor the chemical changes on the surfaces of the tools during the cutting process.

Figure 2 shows an SEM image of a 200µm diameter uncoated tungsten carbide tool prior to cutting. Figures 3 and 4 show higher magnifications of the flute areas where cutting mainly occurs. Comparison of these images also illustrates the rougher tool surface produced by the CEA coating process. Figures 5 to 8 show the end-on views of the tools after machining for a certain number of 100µm wide edge steps, 500µm deep and 80mm long cuts.

Figure 1. Kistler microdynamometer in the Evo Kern CNC precision machinining centre

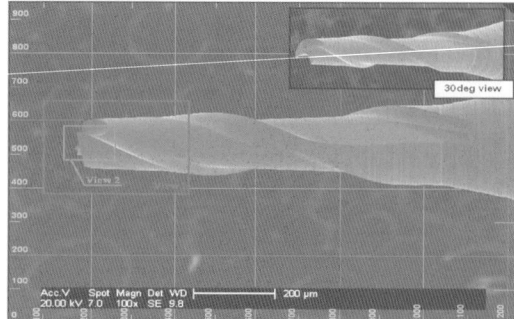

Figure 2. SEM of the uncoated 200µm diameter WC endmill prior to maching

Figure 3. High magnification SEM of the cutting edges of the uncoated 200µm diameter WC endmill prior to machining

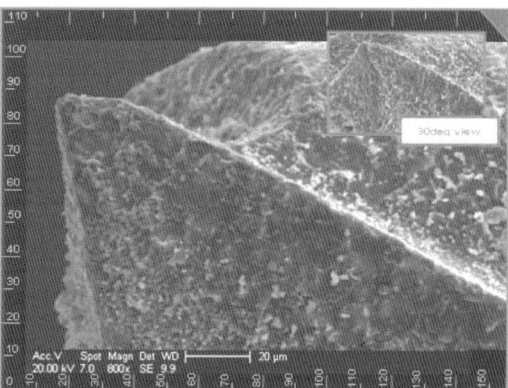

Figure 4. High magnification SEM of the cutting edges of a 200µm diameter WC endmill coated with 2µm of nanolayered AlTiN/TiN prior to machining

The new uncoated tools (Figures 2 and 3) had very sharp flute edges with no visible damage. However, the CEA coatings tended to reduce the sharpness of the cutting edges. A good indicator of tool sharpness is the ratio of feed force to normal force. The ratio is usually higher when the tool is sharper. In Table 1 are shown cutting forces and ratios for cutting steps 100µm wide and 500µm deep in Toolox 33 blocks by 1mm diameter WC tools with different coatings. The cutting speed was 10,000 rpm and the feed rate was 65mm/min. No coolant was applied to minimise external influences for future simulations. Tests with different cooling and lubrication conditions are planned to be undertaken in future work. The cutting forces are roughly the same. However, the ratio of F_{feed} to F_{normal} is slightly lower for the coated tools, indicating a reduction in sharpness.

	Feed force	Normal force	Ratio F_f/F_n
Uncoated tool	6.2 N	6.6 N	0.94
UT Dry	5.7 N	6.9 N	0.83
CEA CrN/TiN	6.2 N	8.3 N	0.75
CEA AlTiN/TiN	6.1 N	7.1 N	0.87

Table 1. Cutting forces and F_f / F_n ratio of 1mm diameter WC tools with different coatings

Wear tests of 1mm diameter tools were carried out under the conditions described before.

Figure 5 shows the uncoated 1mm diameter tool after 15 cuts (having removed approximately 60mm^3 from the workpiece material) and Figures 6-8 show coated tools after 30 cuts (having removed approximately 120mm^3 from the workpiece material). It is apparent that, even on visual inspection, the uncoated tool wears much faster at the cutting edges than the coated tools.

The tool radius was measured at 250µm and 500µm from the tool tip. Analyses of both measurements show similar trends. The measurements at 500µm from the tool tip for one particular 1mm diameter endmilling tool, measured prior to machining and measured at intervals by sequential interruption of the machining process, have been plotted in Figure 9.

All the coated tools tend to show a linear trend in tool radius decrease with the number of cuts. The uncoated tool wears almost twice as fast as the commercially-coated tool and the AlTiN/TiN tool coated by CEA. The AlTiN/TiN-coated tool tends to have a similar trend to wear as the commercially-coated tool that has a very similar type of coating composition.

The best coating material for reducing wear is CrN/TiN, as the change in tool radius is about 20% slower than tools coated with AlTiN/TiN.

The main tool wear mechanism for the specified conditions was abrasion [1].

Figures 10 and 11 show areas where the coating was worn away after 10 cuts. False colour EDX shows graphically specific areas of both wear and material deposition. The green colour [representing tungsten (W) and designated letter A in the figures] signifies areas where the coating has been removed. The red colour [representing titanium (Ti) and designated letter B] signifies still-coated areas and the blue colour [representing iron (Fe) and designated letter C] signifies areas covered by adherent workpiece (Toolox 33) steel.

It is clear that the wear of the tool, appearing in the areas of green colour (A), is mainly on the tip of the flutes and along the sides of the cutting edges [2]. The green colour also shows indirectly the areas which are in contact with the workpiece during the cutting process. The contact areas grow as the tool wears.

Again, it was also observed that the CrN/TiN-coated tool is more resistant against wear than the two coated tools containing aluminum, titanium and nitrogen.

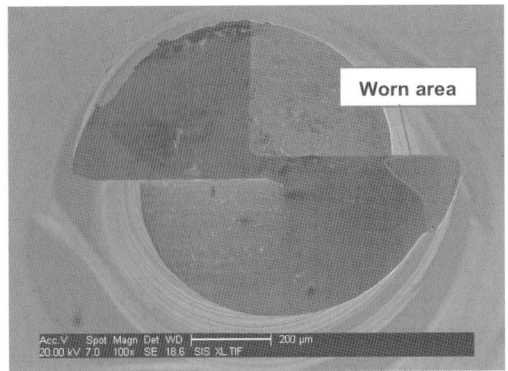

Figure 5. Uncoated tool after 15 cuts

Figure 7. CrN/TiN-coated tool after 30 cuts

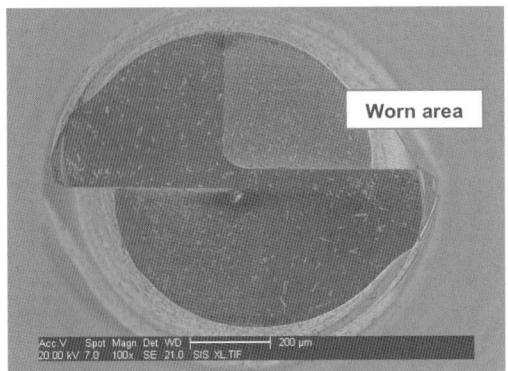

Figure 6. Commercially-coated tool (UT Dry)
after 30 cuts

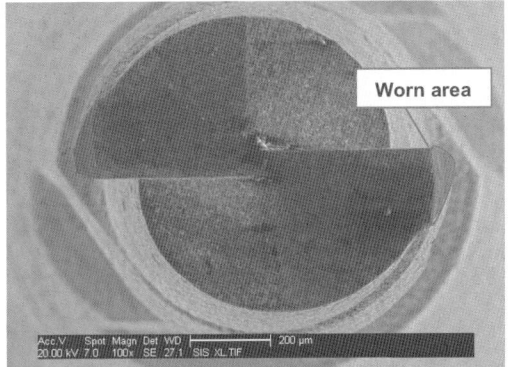

Figure 8. AlTiN/TiN-coated tool after 30 cuts

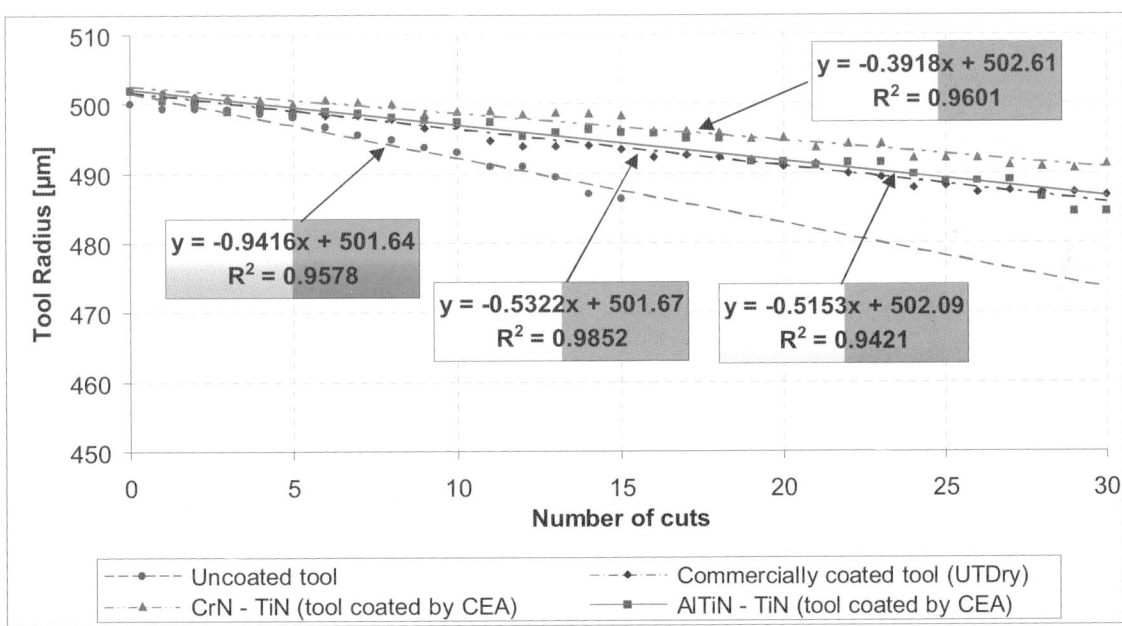

Figure 9. Dependence of tool radius reduction vs. number of cuts for 1mm diameter tools with different coatings.
One cut is 80mm in length

182

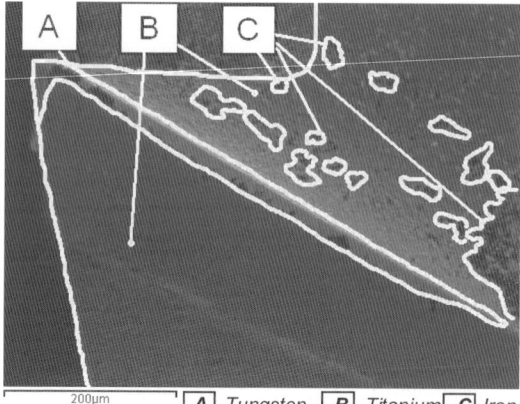

200µm   **A** *Tungsten* **B** *Titanium* **C** *Iron*

Figure 10. High magnification, false colour EDX of the cutting edges of a 1mm diameter WC endmill coated with 2µm of CrN/TiN after ten cuts (side view).

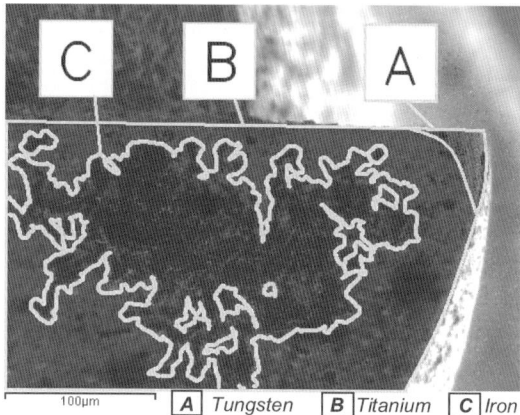

100µm   **A** *Tungsten* **B** *Titanium* **C** *Iron*

Figure 11. High magnification, false colour EDX of the cutting edges of a 1mm diameter WC endmill coated with 2µm of CrN/TiN after ten cuts (front view).

3. Conclusions

The choice of coating method and materials seems to have a significant influence on tool wear. Our preliminary investigations have indicated that all the coatings examined have a beneficial effect on reducing endmill wear. The most efficient coating tested to date is the nanolayered CrN/TiN coating but other coatings and a detailed investigation of the more fragile 200µm diameter endmills will be investigated in future research.

Acknowledgements

DMA and DJS wish to acknowledge the financial support of EPSRC for DZ and JH employed on Grant EP/C534212/1 ("3D-Mintegration"), Cranfield University IMRC (Grant 110) and the invaluable contribution of CEA in coating the tungsten carbide endmills through the 4M Consortium programme.

References

[1] Zhang J.-H., Theory and Technique of Precision Cutting, Pergamon Press, 1991, Chapter 4, The wear characteristics of precision cutting tools, ISBN 0 08 035891 8

[2] Woon K.S. et al, Investigations of tool edge radius effect in micromachining: a FEM simulation approach, Journal of Materials Processing Technology (in the press)

Micromachining of amorphous and crystalline Ni$_{78}$B$_{14}$Si$_8$ alloys using micro-second and pico-second lasers

I. Quintana[1], T. Dobrev[3], A. Aranzabe[2], G. Lalev[3], S. Dimov[3]

[1]CIC marGUNE. Pol. Ibaitarte 5, 20870; Elgoibar; Guipúzcoa, Spain
[2]Manufacturing Processes Department, Fundación Tekniker, Av. Otaola 20, 2060, Eibar, Guipúzcoa, Spain
[3]Manufacturing Engineering Centre, Cardiff University, Cardiff, CF24 3AA

Abstract

The machining response of amorphous and polycrystalline Ni-based alloys (Ni$_{78}$B$_{14}$Si$_8$) to micro-second and pico-second laser processing was investigated. The shape and topography of craters created with single pulses as a function of laser energy together with holes drilled in both materials were studied. The carried out FIB analysis of craters in amorphous and polycrystalline samples revealed that processing both with micro-second and pico-second lasers does not lead to materials crystallization and the short-range atomic ordering of metallic glasses can be retained. When processing the amorphous sample the material laser interactions resulted in a significant ejection of molten material from the bulk that was then followed by its partial re-deposition around the craters. Additionally, there were no signs of crack formation that indicate a higher surface integrity after laser machining. A conclusion is made that laser processing both with short and long pulses is a promising technique for micromachining metallic glasses because does not lead to material crystallisation.

Keywords: amourphous Ni, laser ablation, micro-second laser, pico-second laser, focused ion beam

1. Introduction

Amorphous alloys have attracted considerable interest for many years due to their unique properties for several application areas [1, 2]. Properties like high hardness, fracture toughness and fatigue strength make these materials attractive for manufacturing micro-electro-mechanical systems (MEMS), and die and tool components where minimum size effects and easy microformability are desired [3, 4, 5]. Additionally, amorphous alloys are considered promising tooling materials for micro and nano - structuring and replication [6, 7]. All these applications benefit from their disordered atomic-scale structure. Especially, the amorphous alloys do not have a long-range atomic ordering that can explain the differences in their technological properties in comparison to the crystalline materials [8, 9]. Therefore, it is important for a range of current and potential new application areas to maintain their non-crystalline microstructure during processing. Particularly, if the amorphous alloys are used for producing components incorporating micro and nano features, it is necessary their machining to be performed without any crystallisation, and thus to preserve their only short-range atomic ordering.

During conventional machining of metallic glasses, defects such as crystallisation, burr, and spatter have been observed [10]. In the case of non-conventional machining processes, e. g. ultra-short pulsed laser ablation (femto-second range) and focused ion beam (FIB) milling, it was reported that the carried out micro and nano structuring did not lead to any crystallization, and also the heat affected zone (HAZ) was considered negligible [6, 11, 12, 13]. By performing FIB patterning of Ni - based and Zr - based alloys, it was established that the typical short range atomic ordering of metallic glasses is preserved. Additionally, FIB milling of an amorphous Ni-based alloy showed much better machining response due to lack of any crystalline structural features [6]. Hence, in comparison to their polycrystalline counterparts, a higher surface integrity could be achieved under identical processing conditions. A common characteristic of both processes, FIB milling and femto-second laser ablation, is their high specific processing energy that allows atomic cluster and atom processing to be carried out with almost negligible thermal effects. This suggests that any machining with high energy short pulses should not lead to any changes in material microstructure, and thus the irradiated area should remain amorphous. Thus, a study of laser material interactions when performing structuring with longer pulses, e. g. in the micro-second (μs) range as opposed to femto- and pico-second (ps) laser ablation when the pulse energy is dissipated into the sample in much shorter time intervals, is imperative in order to broaden the usage of laser milling for micro structuring metallic glasses. In particular, this will allow many applications to benefit both from the material enhanced mechanical properties together with high material removal rates achievable with micro- and nano-second pulsed lasers.

In this context, the focus of the research reported in this paper is on investigating the micromachining response of a Ni-based amorphous alloy when performing processing with μs and ps pulsed lasers. In parallel, the machining response of a polycrystalline analogue of the same alloy was used as a reference to assess the effects of the pulse duration on the resulting material microstructure. It was expected a higher surface integrity to be achieved on the amorphous sample because no crystalline structural features were present to facilitate the formation of micro-cracks and other defects.

2. Experimental set-up

The material used in the experiment is a 40 μm - thick foil of amorphous Ni$_{78}$B$_{14}$Si$_8$. The mechanical properties of the material are extremely high, e.g. tensile strength: 1500 - 2000 MPa, and Hardness -

Vickers: 850 Kgf/mm^2. A polycrystalline analogue of the amorphous sample was produced by heat treatment. This was done by placing the foil in vacuum at 550°C for 5 hours, and then leaving it to cool down naturally.

In this study two different laser systems were utilised to perform both pico- and micro- second pulsed ablation. The µs laser processing were performed on a system with a Nd:YAG laser (FOBA Laser 94S) with a wavelength of 1064 nm and a pulse duration set to 10 µs. The ps laser ablation system incorporates a mode-locked Nd:YVO$_4$ laser source with a wavelength of 532 nm and a pulse duration set to 10 ps. The average power delivered by both lasers to the workpiece was measured using a power meter equipped with a high-power laser sensor [14], and thus it was possible to perform measurement when varying laser flashlamp current and pulse frequencies. Both systems employ a set of galvo mirrors for laser beam positioning with high accuracy.

The material response of the laser-machined regions was assessed using scanning electron microscopy (SEM), and also FIB milling and imaging. Both operations were conducted on a cross-beam SEM/FIB system (Carl Zeiss XB1540), which combines a low dose Gemini® Field Emission SEM with a FIB column, Canion 31, and has a gas injection system for gas-assisted etching and deposition and a lithography hardware and software, Elphy Quantum. Benefiting from the carried out research on FIB processing of amorphous and polycrystalline Ni-based samples [6], it was assumed that any evidence of material microstructure changes after laser ablation would be detected by performing spattering tests on the sample by FIB.

The experiments were focused on investigating the shape and topography of the craters created with single pulses as a function of laser energy. Additionally, holes were drilled in the amorphous and crystalline foils in order to study the effects of the exercised heat load on the pre-existing short and long range atomic ordering of both samples.

3. Results and Discussion

3.1 Single pulse craters

Figures 1 (a) and (b) show the SEM images of single pulse craters produced in the amorphous sample by varying the average power of the µs laser. The diameters of the craters were measured, and also the images depict the material that was partially ejected by the vapour and plasma pressure and then re-deposited on the substrate. Part of the ejected material remains around the craters because of the surface tension forces.

Figures1 (c) and (d) show the craters formed in the polycrystalline sample with two different power settings of the µs laser. For comparison, the selected average laser power was identical to that used for the tests conducted on the amorphous foil shown in Figures 1 (a) and (b).

As shown in Fig 1, µs laser ablation causes thermal expansion and material ejection from the crater that is then re-deposited around it. This effect is more pronounced in the amorphous sample. This can be explained with the disordered atomic -scale structure and cohesive energy characteristics of amorphous alloys that affect their thermal properties, in particular their melting and boiling points are not well defined. Also, due to their low thermal conductivity, less heat generated by the absorbed pulse energy is dissipated into the bulk. Then, due to this more concentrated heating more molten material is partially ejected from the craters by the vapour and plasma pressure. This is evident by the bigger diameter of the craters created in the amorphous sample, and also the amount of molten material that remains around them.

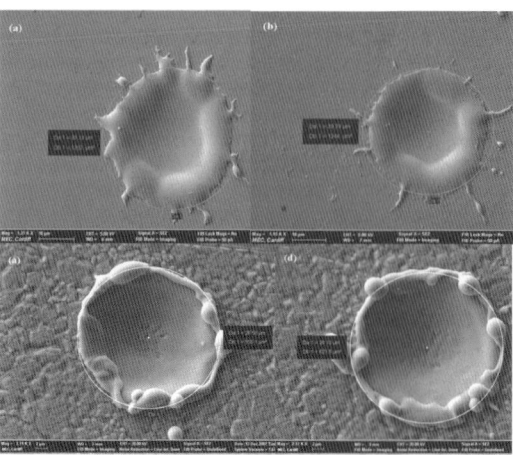

Fig. 1: SEM images of single pulse craters produced by µs laser ablation of amorphous, (a) and (b), and polycrystalline, (c) and (d), Ni$_{78}$B$_{14}$Si$_8$ with an average power: 3 W, (a) & (c), and 3.10 W, (b) & (d).

Similar behaviour has been observed in laser processing of other amorphous materials, including polymers [16, 17] and glasses [18, 19]. However, in the case of laser ablation of polymers with ultra-short pulses, the amount of molten material deposited is lower than that resulting from longer pulses. This can be explained with the fact that the ablation mechanism is not only photothermal, but also photochemical effects occur during laser ablation of polymers, leading to a decrease in re-deposited material around the craters [16, 20].

Figure 2 shows SEM images that depict the evolution of single pulse craters created in the amorphous sample with two different power settings of the ps laser. The craters are not well pronounced and the absorbed energy causes mostly surface expansion. Similar behaviour has been observed in processing metals and polymers with ultra-short laser pulses [16, 21, 22, 23]. Generally, if the laser fluence is lower than the threshold energy for a given material, the photothermal reaction causes surface expansion and the irradiated region is expanded and swollen.

3.2 Drilling holes

Figure 3 shows the drilling holes in amorphous, (a) and (b), and crystalline, (c) and (d), Ni$_{78}$B$_{14}$Si$_8$ created by the µs laser with two different setting of the average power. It is worth mentioning that this µs laser ablation leads to a significant increase in the heat load exercised on the substrate, and as a result of this the material ejected from the holes together with HAZ are higher than those observed in single pulse laser-material interactions. This increase of the thermal load has a more pronounced effect in laser micro-drilling at high power settings as shown in Figures 3 (b) and (d). Additionally, crack formation can be observed on the polycrystalline sample indicating the importance of material microstructure in laser processing. These effects are studied further by investigating the integrity of the resulting surface.

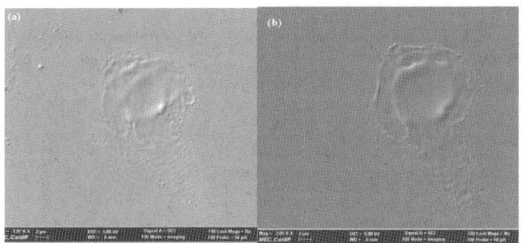

Fig. 2: SEM images of ps laser ablation of amorphous foil with an average power settings: 70mW (a), and 150mW (b).

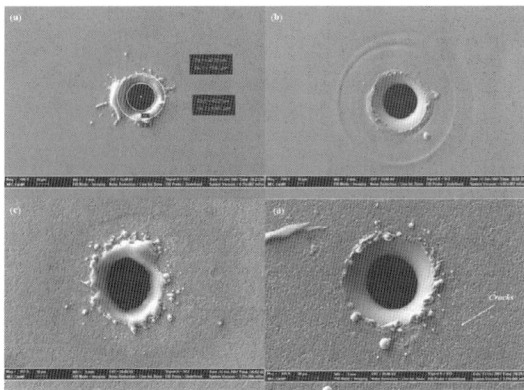

Fig. 3: SEM images of holes machined by μs laser in amorphous, (a) and (b), and polycrystalline, (c) and (d), $Ni_{78}B_{14}Si_8$ foil with an average power outputs: 3.3W, (a) & (c), and 4.7W, (b) & (d).

3.3 Structural characterisation

As it was stressed earlier, machining without crystallisation is necessary in order to preserve the disordered atomic-scale structure of metallic glasses. Therefore, the surfaces of the craters created in the amorphous and crystalline samples by μs and ps laser ablation was investigated by testing their FIB spattering behaviour. Especially, the significant difference in the FIB response of amorphous and crystalline samples [6], was used to detect if there are any changes in the short-range atomic ordering of the amorphous $Ni_{78}B_{14}Si_8$ foil after ps and μs laser processing. In the case of polycrystalline $Ni_{78}B_{14}Si_8$, the crystals' orientation affects the machining response of the material, and results in anisotropic milling-rates that leads to a significant increase of the surface roughness. Figure 4 shows 1 μm deep trenches with dimensions 10 x 4 μm produced by FIB milling in the amorphous, (a) and (b), and polycrystalline, (c) and (d), samples. The spattering behaviour of the surface after μs laser processing was analysed by machining

trenches inside the craters and on non-machined area of the samples as shown in Figures 4 (a) and (c). Figure 4 (b) shows the trench created on the edge of the crater formed by a single ps pulse in the amorphous sample, while in Figure 4 (d) a cross section of the crater created with a single μs pulse in the polycrystalline sample.

By comparing the FIB sputtering results presented in Figure 4 (a) and (b), it is not difficult to see that there is no any evidence of crystallisation in the amorphous sample after μs and ps laser processing whereas the grain structure of the polycrystalline sample is clearly visible in Figure 4 (c), and especially in Figure 4 (d) by using FIB imaging mode. Thus, it can be concluded that the short-range atomic ordering of the amorphous $Ni_{78}B_{14}Si_8$ is not affected by both μs and ps laser processing. To our knowledge, there was only one investigation of the machining response of metallic glasses to laser processing with longer pulses (pulse width higher than 10 ps) [24]. In this research the focus was on the magnetic and structural characteristics of $Fe_{81}B_{13.5}$ $Si_{3.5}$ C_2 metallic glass irradiated with laser pulses in the nanosecond range (λ=532nm, τ=8ns, f=10Hz). The surface characteristics were analyzed by transmission and conversion electron Mössbauer spectroscopy and the results showed an onset of surface crystallisation in the laser-irradiated area. However, our research suggest that the machining response of amorphous alloys can differ, and in the case of Ni-based alloys laser milling with long pulses can be a suitable technique for micro structuring of metallic glasses. Further investigations on different materials have to be performed in order to see whether the disordered atomic-scale structure of other metallic glasses remains unmodified after μs or ns laser micromachining.

Studying the surface integrity of the polycrystalline sample shown in Figure 4 (d), particularly the crack formation resulting from the μs laser irradiation indicates a low energy dissipation capacity in this material compared with its amorphous counterpart. This behaviour might be related to the ordered atomic-scale structure of the crystalline materials. Also, it is important to note that in contrast to crystalline systems, the packing heterogeneity characteristic of metallic glasses would lead to a broader distribution of the activation energy associated with the molecular dynamics of the system, as it is the case with other amorphous alloys and polymers [25, 26]. Additionally, the cooperative nature of atomic motions leads to viscous flow behaviour of the system at high temperatures. These characteristics of disordered atomic-scale structure of metallic glasses provide a better stress accommodation, and laser energy dissipation. Consequently, as we can appreciate in Figures 4 (a) and (d), the probability of crack formation during laser processing would be higher for the crystalline $Ni_{78}B_{14}Si_8$ than for the amorphous one.

4. Conclusion

The FIB analysis of single pulse craters in amorphous and polycrystalline $Ni_{78}B_{14}Si_8$ revealed that processing both with μs and ps lasers does not lead to materials crystallisation and the short-range atomic ordering of metallic glasses can me retained.

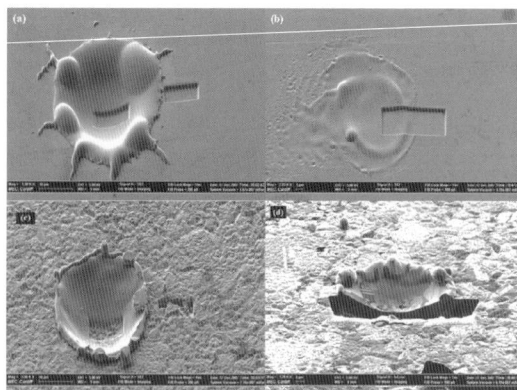

Fig. 4: SEM images of trenches produced in amorphous, (a) and (b), and polycrystalline, (c), $Ni_{78}B_{14}Si_8$; (d) a cross-sectional FIB image of the crater produced by a single μs pulse.

The effect of material microstructure on laser machining was also studied. Different ablation mechanisms were observed to take effect during the laser processing of amorphous and crystalline $Ni_{78}B_{14}Si_8$. When processing the amorphous sample the material laser interaction resulted in a significant ejection of molten material from the bulk that was then followed by its partial re-deposition around the craters due to the surface tension forces. Additionally, there was no signs crack formation that indicates a higher resulting surface integrity after laser machining. The quick cooling down of the sample at the end of each pulse together with the packing heterogeneity and viscous flow behaviour of disordered atomic-scale structures can explain the better stress accommodation and laser energy dissipation observed in metallic glasses.

In summary, the investigation shows that laser processing both with short and long pulses is a promising technique for micromachining metallic glasses because does not lead to material crystallisation. The machining of micro-scale features and micro-structures with high precision and minimal thermal damage can be achieved and the attractive mechanical properties of these materials preserved.

Acknowledgements

The research reported in this paper is funded under the MicroBridge programme supported by Welsh Assembly Government and the UK Technology Strategy Board, and the EPSRC Programme "The Cardiff Innovative Manufacturing Research Centre". Also, it was carried out within the framework of the EC FP6 Networks of Excellence, "Multi-Material Micro Manufacture (4M): Technologies and Applications" and "Innovative Production Machines and Systems (I*PROMS)". The authors gratefully acknowledge the support given to the Networks by the European Commission.

The authors would like also to acknowledge also the Department of Industry, Commerce and Tourism of the Basque Government for supporting financially the CIC marGUNE.

References

[1] A. Inoue *Acta Mater.***48** (2000) *279-306*.

[2] W. H. Wang, C. Dong, C. H. Shek *Materials Science and Engineering* **R 44** (2004) *45-89*.

[3] X. Wang, P. Lu, N. Dai, Y. Li, C. Liao, Q. Zheng, L. Liu *Materials Letter* **61** (2007) *4290-4293*.

[4] T. Fukushige, S. Hata, A. Shimokohbe *J. Microelectromech. Syst.* **14** (2005) *243-253*.

[5] G. P. Zhang, Y. Liu, B. Zhang ¡*Scripta Materialia* **54** (2006) *897-901*.

[6] W. Li, R. Minev, S. Dimos, G. Lalev *Applied Surface Science* **253** (2007) *5404-5410*.

[7] N. Kawasegi et al. *Applied Physics Letters* **89** (2006) *143115*.

[8] J. Das, M. B. Tang, K. B. Kim, R. Theissmann, F. Baier, W. H. Wang, J. Eckert *Phys. Rev. Lett* **94** (2005) *205501*.

[9] B. Zhang and D. Q. Zhano, M. X. Pau, W. H. Wang, A. L. Greer *Phys. Rev. Lett* **94** (2005) *205502*.

[10] M. Bakkal, A. H. Shih, S. B. McSpadden, C. T. Liu, R. O. Scattergood *Int. J. Mach. Tools Manuf.* **45** (2005) *741-752*.

[11] X. Wang, P. Lu, N. Dai, Y. Li, C. Liao, Q. Zheng, L. Liu *Materials Letters* **61** (2007) *4290-4293*.

[12] P. Sharma, N. Kaushik, H. Kimura, Y. Saotome, A. Inoue *Nanotechnology* **18** (2007) *035302*.

[13] W. Jia, Z. N. Peng, Z. J. Wang, X. C. Ni, C. Y. Wang *Appl. Surf Sci.* **253** (2006) *1299-1303*.

[14] COHERENT [WWW]: GS; Power sensor: Coherent FieldMaster LM45 HTD-http://www.coherentic.com/.

[15] J. Hohlfeld, S. S. Wellershoff, V. Conrad, V. Jähnke, E. Matthias *Chemical Physics* **251** (2000) *237-258*.

[16] B. S. Shin, J. Y. Oh, H. Sohn *Journal of Materials Processing Technology* **187-188** (2007) *260-263*.

[17] S. Baudach, J. Bonse, J. Krüger, W. Kautek *Applied Surface Science* **154-155** (2000) *555-560*.

[18] P. Rudolph, J. Bonse, J. Krüger, W. Kautek *Appl. Phys. A* **69** (1999) *763-766*.

[19] M. T. Kasaai, V. Kacham, F. Theberge, S. L. Chin *Journal of Non-Crystalline Solids* **319** (2003) *129-135*

[20] S. Y. Bang *J. KSPE* **22** *38-46*.

[21] A. A. Serafetinides, C. D. Shordoulis, M. I. Makropoulu, A.K. Kar *Applied Surface Science* **135** (1998) *276-284*.

[22] M. Himmelbauer, E. Arenholz, D. Baverle, K. Schilcher *Applied Physics A* **63** (1996) *337*.

[23] S. Baudach, J. Bonse, J. Krüger, W. Kautek *Applied Surface Science* **154-155** (2000) *555-560*.

[24] M. Sorescu *Journal of Alloys and Compounds* **284** (1999) *232-236*.

[25] S. Daewoong, R. H. Dauskardt *J. Mater. Res.* **17**, (2003) *1254*

[26] G. Radons, W. Just, P. Häussler *Collective Dynamics of Nonlinear and disordered systems*, (2005) Springer.

FT-IR study of nanosurface phenomena

I. Markova – Deneva

University of Chemical Technology and Metallurgy –Sofia, 8 St. Kl. Ohridski blvd., 1756 Sofia, e-mail:
vania@uctm.edu

Abstract

IR study of metal nanoparticles in amorphous or crystalline state obtained via water solution of metal salts by borohydride reduction with $NaBH_4$, as well as of nanowires prepared using mesopore ceramic supports has been carried out. FT-IR spectra in mid-infrared region of visible spectrum (4000-400 cm^{-1}) of these nanoscaled materials have been undertaken. IR spectroscopy possibilities have allowed to investigate the nanosurface phenomena and to prove the creation of different chemical bonds such as B-O, B-H, Si-O, O-H in surface atom groups. FT-IR spectra have provided information about the technological and hydrodynamic conditions such as a kind of the initial salt, a type of the reactor used (T, Y, A methods), different ceramic supports used (SiO_2, SiMCM and ALMCM), different variants of the support introducing (surface wetting), as well as about the nanosclaed materials composition, their structure state and nucleation.

Keywords: FT-IR spectroscopy, FT-IR spectra, nanoscaled materials, nanoparticles, nanowires, nanosurface phenomena, chemical reduction, supports.

1. Introduction

IR spectroscopy is widely used in both research and industry as a simple and reliable technique for measurement and has been highly successful for applications in both organic and inorganic chemistry. IR spectroscopy has also been utilized in the field of semiconductor microelectronics. Less data is available for IR spectroscopic studies in the field of inorganic chemistry, as well as in the field of nanoscaled materials.

Our team considers IR spectroscopy, especially FT-IR spectroscopy as a sensitive method to investigate nanoscaled materials surface atom state [1-3]. We have used IR spectroscopy in mid - IR region of the visible spectrum approximately 4000 – 400 cm^{-1} to study the fundamental vibrations and vibrational nanostructures. We have successfully investigated by FT-IR spectroscopy different atom groups formed on the nanoscaled materials surface on the basis of their natural vibrations at a fixed frequencies.

The aim of this work is commonly to study by IR spectroscopy with Fourier transformation the nanosurface phenomena occurring on the interface of nanoparticles and nanowirwes synthesized by chemical reduction via aqueous solutions of metal salts with $NaBH_4$ [4], respectively to investigate on the basis of the FT- IR spectra the surface structure of the nanoparticles and nanowirwes.

2. Experimental details

The investigated metal nanoparticles have been synthesized by a chemical reduction method via water solutions of metal salts with different anion content with $NaBH_4$ at a room temperature and an atmospheric pressure. Three types of reactors providing different hydrodynamic conditions of solutions mixing (this one of the metal salt and the other one of reducing agent) have been used [5].

Amorphous Co nanoparticles have been synthesized via water solutions of $CoSO_4.7H_2O$ using a reactor with an ideal mixing conditions (the so called T- method), while *Co nanoparticles with crystalline structure* have been synthesized using an "antigravity" reactor

working in an ideal displacement regime (the so called A-method) [9].

Cu nanoparticles have been obtained via water solutions of three kinds of copper salts ($CuSO_4.5H_2O$, $CuCl_2.2H_2O$ and $CuCl_2$) using three kinds of reactors working at different hydrodynamic conditions of mixing of both water solutions (T and A-methods mentioned above and the so called Y-method of a full mixing of the solutions).

Ni nanoparticles have been prepared via water solutions of Ni salts with a different anion content ($NiSO_4.6H_2O$, $NiCl_2.6H_2O$) using T - reactor of an ideal mixing of both solutions.

Ni nanowires have been obtained using different ceramic supports (SiO_2, SiMCM and AlMCM) with a different chemical nature (element content) and a different specific surface area [m^2/g]. Two variants of the support wetting procedure have been used: the so-called A-variant (the support is introduced with Ni salt solution to the reducing agent solution) and B-variant (the support is introduced with the reducing agent solution to Ni salt solution) [6]. The weight Ni/support ratio is determined in the range of 10 to 1.

IR spectroscopy study has been carried out with IR spectrophotometer EQUINOX (Bruker) with Fourier transformation in 4000 to 400 cm^{-1} frequency region using KBr matrix, respectively IR spectra of nanoparticles and nanowires sintesized have been undertaken. The samples for IR spectroscopy investigations have been prepared mixing nanoscaled materials amounting to 1 mg and KBr (Spectral purity) amounting to 200 mg and pressed in a tablet form with a diameter 13 mm under a press loading of 8 t/cm^2.

3. Experimental results

3.1. IR study of Co nanoscaled materials (Co nanoparticles)

Co nanoparicles structure has been investigated in previous our works [3,5,6] by means of electron microscopy (TEM, SEM) and X-ray diffraction (XRD). IR spectroscopy used in this work as a method for structure investigations has proved its advantages over the other investigation structural methods in regards to the surface structure and surface phenomena. IR spectra reveal differences as regards a nanoscaled materials structure state.

Fig. 1 presents FT-IR spectrum of heat treated at 200degC amorphous Co nanoparticles. This is a typical infrared spectrum of materials in amorphous state (clear absorption bands are missing). Fig. 2 presents FT-IR spectrum of Co nanoparticles with the same chemical composition, but in a crystalline state also treated at 200 degC. The observed infrared absorptions at characteristic frequencies can be related to the formation of chemical BO_3 and BO_4 groups. The atoms in these groups vibrate in different ways: symmetrical stretching ν_1 (BO_4) at 1030 cm^{-1} and antisymmetric stretching ν_3 (BO_3) at 1310 cm^{-1}, as well as scissoring (deformation vibrations) ν_4 (BO_4) at 520cm^1.

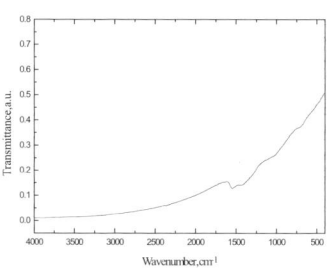

Fig. 1. FT-IR spectrum of amorphous nanoparticles at 200degC

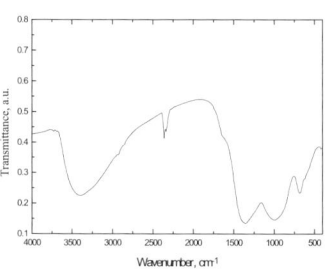

Fig. 2. FT-IR spectrum of crystalline nanoparticles at 200degC

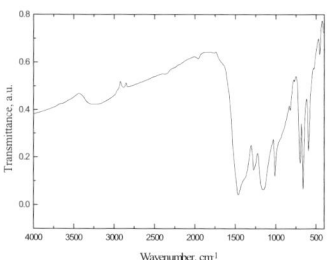

Fig. 3. FT-IR spectrum of crystalline nanoparticles at 750degC

In Fig. 3 is shown FT-IR spectrum of crystalline nanoparticles treated at 750 degC. FT-IR spectra of the crystalline nanoparticles treated over 550 degC are characterized by intensive sharp bands observed in 1500 – 1100 cm^{-1} region. These absorption bands are very sensitive to B-O-B bond angle and to B-O bond length in BO_3 groups.

The different shape of nanoscaled materials FT-IR spectra demonstrates the state of these nanomaterials (amorphous or crystalline). The absorption in this mid - IR регион applies to B-O bonds in BO_3 and BO_4 groups. Our IR investigations demonstrate IR spectroscopy as a useful method when it comes to study nanoscaled interface.

3.2. IR investigations of nanoparticles with different composition

Nanoparticles with different composition have been prepared via water solutions of metal salts with different anion contents (SO^{2-}_4, Cl$^-$) using reactors working at different hydrodynamic conditions of mixing for both solutions: T-method ensuring ideal mixing conditions, Y-method of full mixing conditions and A-method of ideal displacement conditions.

3.2.1. Co nanoparticles

Figs. 4-5 present FT-IR spectra of Co nanoparticles obtained via water solution of $CoSO_4.7H_2O$ (A-method) heat treated at different temperature.

With increasing temperature from 200 to 850 degC the absorption band grows in intensity and shifts to lower frequencies. The quantity of BO_4 groups increases relatively at the expense of a spending of BO_3 groups from the boroxol ring with three non-bridging oxygen. At lower temperatures (200 degC) the bands are broad (Fig. 2), while at higher temperatures (550-850 degC) they are intensive and sharp (Figs. 3, 4, 5).

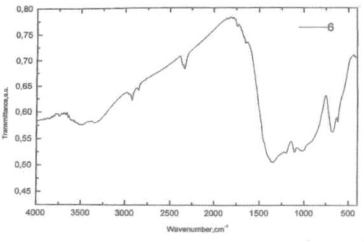

Fig. 4. FT-IR spectrum of Co nanoparicles at 550degC

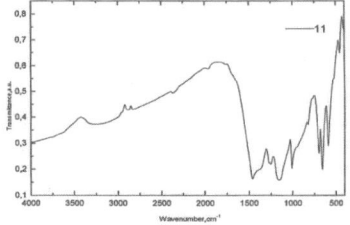

Fig. 5. FT-IR spectrum of Co nanoparticles at 850degC

The synthesis conditions of ideal displacement for both solutions with a mechanic mixer stirring, as well as the anion content of the initial salts used (SO^{2-}_4) determine the formation of isotropic by shape nanoscaled materials.

IR spectra undertaken show the temperature influence to form crystalline nanoparticles resulting from solid phase recrystallization processes, as well as the influence of the method and the reactor used.

3.2.2. Cu nanoparticles

In Figs. 6- 8 have been shown IR spectra of Cu nanoparticles obtained from $CuSO_4.5H_2O$ salt but at different hydrodynamic conditions, respectively A, T and Y methods. In FT-IR spectra of Cu nanoparticles obtained from the sulfate slats nevertheless the reactors used it can be seen clearly expressed bands of absorption in 1500 - 1300 cm^{-1} frequency region and not clearly expressed absorption bands in 1000 - 850 cm^{-1} frequency region. They are assigned to vibrations characteristic for B-O bonds in BO_3 and BO_4 groups forming different structures: antisymmetric stretching $\nu_3(BO_3)$ mode at 1450 cm^{-1} and 1345 cm^{-1}, symmetric stretching $\nu_1(BH_3)$ mode at 1115 cm^{-1}, $\nu_1(BO_4)$ mode and $\nu_1(BO_3)$ mode in 1000 cm^{-1} - 850 cm^{-1} region and deformation $\nu_4(BO_3)$, $\nu_4(BO_4)$ and $\nu_4(BO_2)$ modes in 600 - 430 cm^{-1} region.

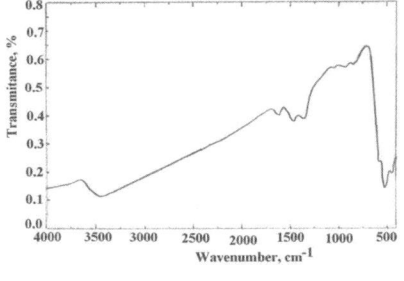

Fig.6.FT-IR spectrum of Cu nanoparticles, T-method, CuSO$_4$.5H$_2$O

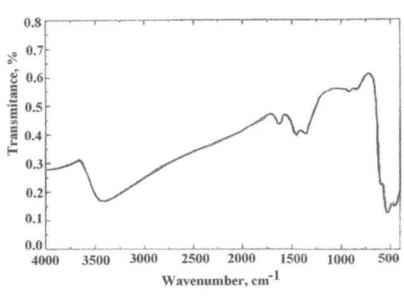

Fig. 7. FT-IR spectrum of Cu nanoparticles, Y- method, CuSO$_4$.5H$_2$O

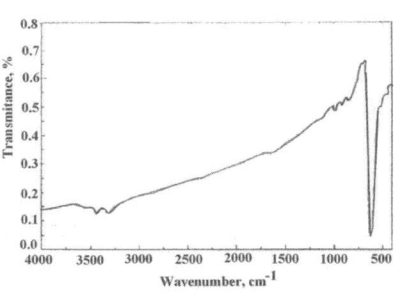

Fig.8.FT-IR spectrum of Cu nanoparticles, A- method, CuSO$_4$.5H$_2$O,

In Fig. 6 presenting FT-IR spectrum of Cu nanoparticles the band at 1115 cm^{-1} which is missed in all other FT-IR spectra can be related to the presence of B-H bonds in BH$_3$ groups. This band is due to stretching vibrations of B-H bonds in the BH$_3$ group.

The characteristic bands at around 3420 cm^{-1} and at 1620 cm^{-1} referring to H-O-H vibration can also be detected in figs. 6-8. An intensive sharp band (at 1620 cm^{-1}) and a slight broad band (at 3420 cm^{-1}) are characteristic for symmetric stretching vibrations $\nu_1(OH)$ of O-H bonds in OH groups and the adsorbed H$_2$O molecules on the nanoparticles surface.

The stretching vibrations at frequencies higher 1000 cm^{-1} characterize short and strong B-O chemical bonds in BO$_3$ group, while the stretching vibrations at frequencies lower 1000 cm^{-1} are typical for weak and longer B-O chemical bonds in BO$_4$ group. The shift of the bands to the higher frequencies is due to an increase of the quantity of BO$_3$ groups in the system probably because of a formation of tetraborate units (a structure unit with less BO$_4$ groups than triborate unit). The shift of the absorption bands to the lower frequencies is due to an arise of BO$_4$ groups in the system with diborate and triborate units which changes the boron coordination from 3(in BO$_3$ groups) to 4(in BO$_4$ groups). IR study of Cu nanoparticles using different initial Cu salts, respectively different reactors working at different hydrodynamic conditions prove IR spectroscopy as a suitable investigation method for the nanoparticles surface

phenomena occurring, as well as a sensitive method as regards to establish the atom groups created on the nanoparticles surface.FT-IR spectra prove the chemical bonds created in surface atom groups.

3.2.3. Ni nanoparticles

Ni nanoparticles obtained via water solutions of Ni salts with different anion content (SO$^{2-}_4$, Cl$^-$) using a reactor working at ideal nixing conditions (T-method) also have been investigated by IR spectroscopy and their FT-IR spectra have been shown in figs. 9 and 10.

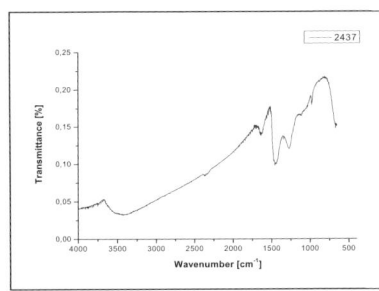

Fig. 9. FT-IR spectrum of Ni nano-particles, NiSO$_4$.6H$_2$O

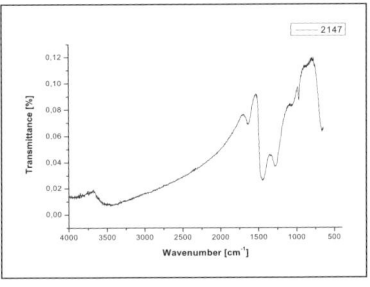

Fig. 10. FT-IR spectrum of Ni nano-particles, NCl$_2$.6H$_2$O

In the case of NiSO$_4$.6H$_2$O salt FT-IR spectrum in Fig. 9 has shown that on the nanoparticles surface have been not formed BO$_4$ atom groups characterizing by weak B-O chemical bonds. But at 1450 cm^{-1} and 1270 cm^{-1} stretching vibrations of shot B-O bonds in BO$_3$ atom groups have been observed, as well as B-H$_2$ bonds have been created vibrating at 970 cm^{-1}. In the case of NiCl$_2$.6H$_2$O salt (Fig. 10) the creation of B-H$_2$ bonds on the nanoparticles surface vibrating at 970 cm^{-1} has been observed too, as well as BO$_4$ groups have been formed.

In both cases the bands at 1630 cm^{-1} due to vibrations of O-H bonds in OH groups and the bands at 3430 cm^{-1} referred to H-O-H bonds vibrations in adsorbed H$_2$O molecules can also be detected.

On FT-IR spectra in figs. 9 and 10 it can be seen the differences in surface groups formation due to different initial salts used for nanoparticles synthesis.

3.2.4. Ni nanowirers

Figs. 11-12 show FT-IR spectra of nanowires obtained via water solutions of NiSO$_4$.6H$_2$O using SiO$_2$ support and A-variant of a wetting procedure, while Fig. 13 presents FT-IR spectrum of nanowires obtained at the same conditions but using NiCl$_2$.6H$_2$O salt. Generally, on FT-IR spectra the following absorption bands have been observed: at 1430 cm^{-1} (BO$_3$ groups), at 800-810 cm^{-1} (BO$_4$ groups), at 650-670 cm^{-1} (B$_3$O$_6$ groups), at 1105-1110 cm^{-1} (Si-O bonds) and at 470-480 cm^{-1} (Si-O bands).

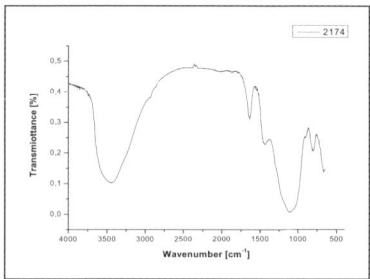

Fig. 11.FT-IR spectrum of Ni nanowires, NiSO$_4$.6H$_2$O, A-variant, SiO$_2$ support

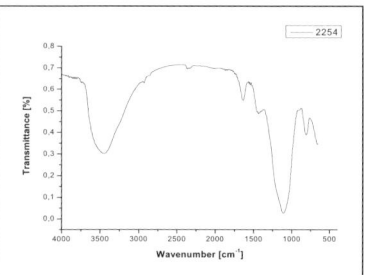

Fig. 12. FT-IR spectrum of Ni anowires, NiSO$_4$.6H$_2$O, B-variant, SiO$_2$ support

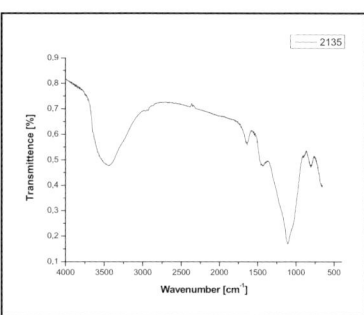

Fig. 13. FT-IR spectrum of Ni nanowires, NiCl$_2$.6H$_2$O, A-variant, SiO$_2$ support

IR spectroscopy method has established the difference between FT-IR spectra due to the different away of the introduction and the surface wetting of the supports.

FT-IR spectroscopy analysis gives an account to initial salt used for the nanostructures synthesis (using NiSO$_4$.H$_2$O salt BO$_3$ groups predominate over BO$_4$ groups, while using NiCl$_2$.6H$_2$O salt BO$_4$ groups predominate), as well as gives an account to the support used.

In Figs. 14 and 15 are shown SEM micrograph of nanoscaled materials investigated in this work by IR spectroscopy [6].

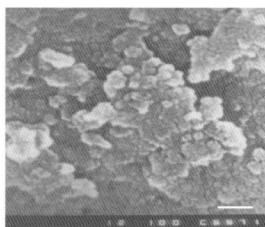 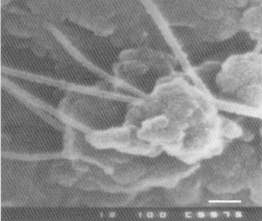

Fig. 14. SEM micrograph of nanoparticles, NiCl$_2$.6H$_2$O [6]

Fig. 15. SEM micrograph of nanowires, NiSO$_4$.6H$_2$O [6]

Conclusion

IR study in mid - IR spectrum of 4000 – 400 cm^{-1} of the nanoscaled materials surface has demonstrated IR spectroscopy as a sensitive method for investigation of nanosurface phenomena on the nanostructured materials

interface. FT-IR spectra undertaken have provided information about the technological and hydrodynamic conditions of the reduction process as regards the initial salt (Ni, Cu, Co sulfates or chlorides), the type of the reactor used (T, Y, A methods), the ceramic support used (SiO$_2$, SiMCM and ALMCM), the different variants of a support introduction and surface wetting.

FT-IR investigations have shown the creation of different bonds and the formation of different groups on the nanomaterials surface. The characterization using FT-IR has been focused on vibrational groups of attached hydrogen, oxygen, boron, respectively on processes occurred on the interface such as atoms reduction and oxidation and molecules adsorption. FT-IR spectra mode is different due to the stretching and deformation vibrations of chemical bonds such as B-O, B-H, Si-O, and O-H in4000 cm^{-1} to 400 cm^{-1} frequency region forming different surface atoms groups.

Using FT-IR for studying surface phenomena, respectively surface attached atoms and their bonds and attached molecules has demonstrated IR spectroscopy as an investigation structural methods for nanoscaled materials in regards to surface phenomena and surface atom groups formation.

Acknowledgement

The author gratefully acknowledges the financial support from the National Science Fund at the Ministry of Education and Science-Bulgaria on a Project NT-5-03, as well as expresses its appreciation to the Scientific Center of UCTM - Sofia under a Contract No 814.

The author is thankful to prof. DSci. I. Dragieva and a scientific researcher Dipl. Eng.K. Alexabdrova both from the IEES - BAS, Sofia for the samples given to be studied in this work and for their scientific help.

References

[1]. I.M.-Deneva, IR spectrosvopy as a method for investigation of nanostructures surface state, Nanosceince and Nanotechnology, 7, eds. E. Balabanova, I. Dragueva, Heron Press, Sofia, (2007), 269-273.
[2]. I.M.-Deneva, K. Alexandrova, I. Dragieva, IR investigations of nanoparticles and nanowires obtained by borohydride reduction method, Nanosceince and Nanotechnology, 7, eds. E. Balabanova, I. Dragueva, Heron Press, Sofia, (2007), 283-286.
[3]. I.M.-Deneva, K. Alexandrova, I. Dragieva, Synthesis and characterization of cobalt nanoparticles, nanowires and their composites, Third International Conference 4M on Multi Materials Micro Manufacture, 3-5 October 2007, Borovec, Bulgaria, Book of Proceedings, p.211.
[4]. I. M.-Deneva, K. Alexandrova, G. Ivanova, I. Dragieva, "IR Spectroscopy Investigation of Metal Amorphous Nanoparticles", *Journal of the University of Chemical Technology and Metallurgy*, XXXVII, 4 (2002) 19-26.
[5]. K. Alexandrova, I. M.- Deneva, at al., "Mechanism of nucleation and Characterization of NiB/SiO$_2$ nanoparticles and nanowires", Journal of the University of Chemical Technology and Metallurgy", Sofia, XXXVIII, 4 (2004) 1031-1038.
[6] K. Alexandrova, I. M.- Deneva, I. Dragieva, "Metallic nanoparticles and nanowires from various aqua solutions", Powder Metallurgy Progress, 6, **2** (2006) 51-58, ISSN 1335-8987.

Multi-Material Micro Manufacture
S. Dimov and W. Menz (Eds.)

191

Micro-extrusion of an ultrafine grained copper can

S. Geißdörfer[a], A. Rosochowski[b], L. Olejnik[c], U. Engel[a]

[a] *Chair of Manufacturing Technology, University of Erlangen-Nuremberg, Egerlandstrasse 11, 91058 Erlangen, Germany*
[b] *Department of Design, Manufacture and Engineering Management, University of Strathclyde, 75 Montrose Street, Glasgow, United Kingdom, G1 1XJ*
[c] *Institute of Materials Processing, Warsaw University of Technology, 85 Narbutta Street, 02-524 Warsaw, Poland*

Abstract

Because of the well known virtues of low cost and high productivity, metal forming technology is well suited for mass production of metal micro-components. However, scaling down traditional metal forming processes proves to be problematic because, among other factors, the relatively coarse grain (CG) structure of micro-billets leads to non-uniform material flow and lack of repeatability during microforming. The aim of the presented study is to investigate a possibility of using an ultrafine grained (UFG) copper for micro-extrusion. The UFG version of Cu is produced by severe plastic deformation at room temperature using 4 and 8 passes of equal channel angular pressing (ECAP). The microstructure and compression properties of the UFG copper are investigated. For visualisation purposes, the microforming process of backward extrusion is carried out at room temperature using half cylindrical billets and transparent tools. The extrusion results, for billets subjected to 4 and 8 passes of ECAP, are compared in terms of the extrusion force, grain flow, shape representation and surface quality and show clearly that applying ultrafine grained material to microforming processes reduces scaling effects.

Keywords: ultrafine grained metal, ECAP, micro-extrusion, in-situ process observation

1. Introduction

Micro technology is gaining an increasing interest due to mobile phones, digital cameras and other consumer electronics products, which become smaller every day. Following this trend, an increasing market for small parts must be satisfied by the production industry [1]. Depending on the required functionality of these parts and the production volume, different manufacturing technologies are available like machining, moulding and forming. In the case of the smallest metallic parts, most of them are produced using machining processes like turning, grinding or milling. For small batch production, machining may be justified. If large quantities of micro-parts are requested, the forming technology is more appropriate due to its high production rate and remarkable accuracy. However, investigations on microforming processes have shown significant differences in the forming behaviour compared to the conventional scale forming, which prevents microforming from being used on a wider scale. Research activities in microforming during the last decade have identified and analyzed the two main size-effects, one with relation to the material flow and another one due to friction. The former can be explained by a dependency of the material flow on the grain size (Hall-Petch effect) and the ratio of the grain size and part's dimensions (contribution of surface grains) [2,3]. The size-effect related to friction has been first investigated using ring upsetting tests scaled-down according to the similarity theorem [4]. A more detailed study has been done using a double cup extrusion test [5,6] which confirmed the previous findings and was used as a basis for theoretical approach [7] describing scale dependency of the friction factor m.

In this paper, the size-effect on friction is assumed to be constant for all of the experiments and thus not analyzed in detail. Since the microstructure has been identified to be one of the main reasons for process-scatter, as well as the uneven shape evolution in can backward extrusion, a novel approach is proposed here to use the ultrafine grained (UFG) material to reduce the scaling effects.

2. Preparation of the ultrafine grained material

2.1 Technology and samples

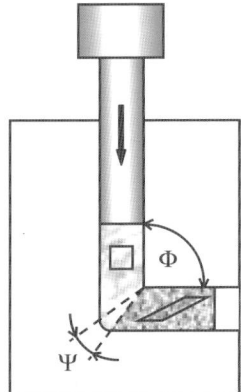

Fig. 1. Schematics of ECAP.

First, a copper (Cu 99.9%) bar was cold forward extruded to reduce its diameter from 20 mm to 13 mm, which was equivalent to deforming it plastically to a strain of 0.86. Next, samples of about 8x8x46 mm were cut from the extruded bar and processed by equal channel angular pressing (ECAP [8]) 4 and 8 times; Fig. 1 explains the principle of the ECAP process used. ECAP is based on forcing the billet material through a constant profile, L-shape channel. Simple shear, taking place along the diagonal plane across the channel turn, deforms the material to a strain of 1.15. To accumulate a larger strain, several passes of the billet through the channel are necessary. The billet rotation between consecutive passes of ECAP was 90° (so called route Bc). The samples were lubricated with dry MoS_2 and fat. Due to high ductility of copper, ECAP was performed at room temperature. Fig.2 displays a round bar produced by forward extrusion, which was subsequently machined into square samples subjected to 4 and 8 passes of ECAP.

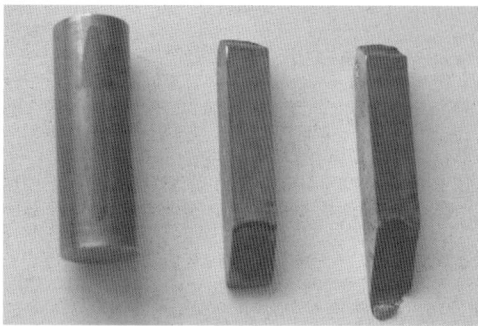

Fig. 2. Forward extruded copper bar and square samples processed by 4 and 8 passes of ECAP.

2.2 Microstructure

The microstructure of the copper samples was investigated using optical microscopy. Fig. 3 presents these results for the samples after forward extrusion and after 4 and 8 passes of ECAP. The estimated grain size of the extruded copper was 30 μm (Fig 3a). Figures 3b and 3c do not allow the grain size to be measured, however, it is obvious that the grain structure was greatly refined after 4 and especially after 8 passes of ECAP.

2.3 Properties

Uniaxial compression was used to characterise the behaviour of copper during micro-extrusion. Since most strengthening due to ECAP is known to occur in the first 1-3 passes, it has been assumed that there would be no much difference between the material subjected to 4 and 8 passes. Consequently, compression testing has been performed for the forward extruded copper and for the UFG copper after it has been extruded and ECAPed 8 times (samples after 4 passes were not compression tested). The cylindrical specimens were prepared with shallow flat lubricant reservoirs on both ends. The specimens were 7 mm in diameter and 9 mm high. The lubricant used was paraffin. The tests were performed using the load control mode of a servo-press with the load rate of 2.5 kN/s. The initial strain rate was about 8×10-3 s^{-1}.

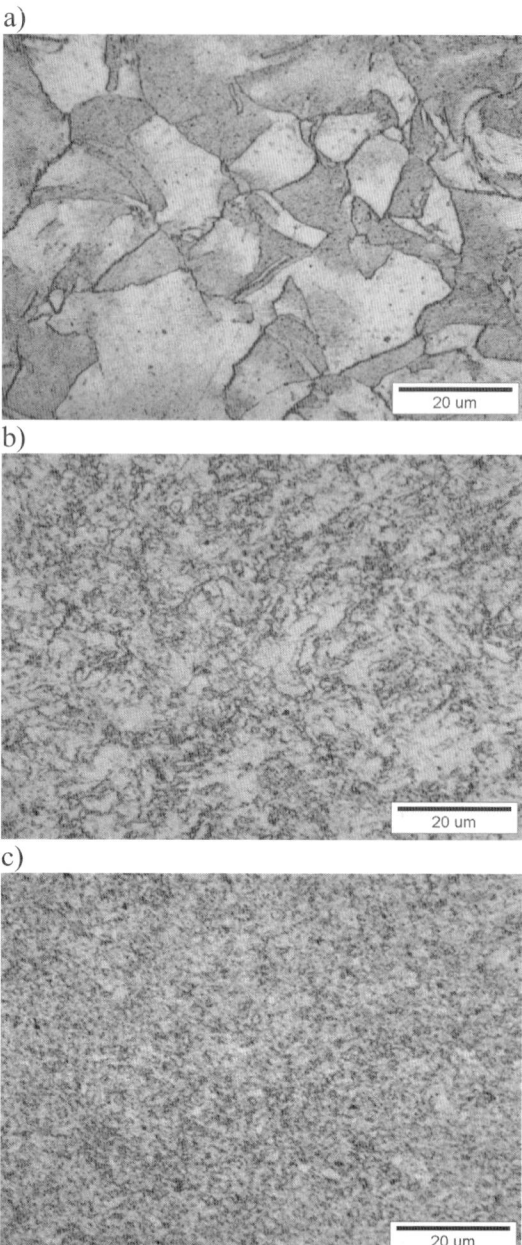

Fig. 3. Optical micrographs of copper structure a) subjected to forward extrusion b) after 4 and c) 8 passes of ECAP at room temperature.

Fig. 4 displays two strain hardening curves obtained in this way. After forward extrusion, the initial yield strength of the tested copper was 280 MPa but very quickly a relatively flat hardening curve was reached at about 400 MPa. The UFG material produced by extrusion and 8 passes of ECAP had a similar characteristic, just at a higher level of 320 MPa and 460 MPa respectively. It can be concluded that, for the material tested, the yield stress increase due to the UFG structure is rather modest compared to a traditionally cold formed copper. From the microforming point of view it is good news because the beneficial UFG structure can be used without much penalty in terms of stresses and forces required for microforming.

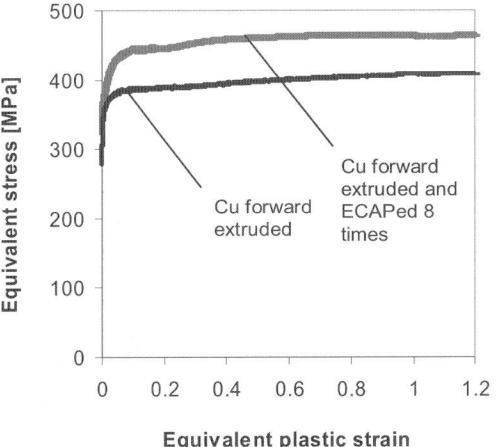

Fig. 4. Strain hardening curves for copper after forward extrusion and after forward extrusion and 8 passes of ECAP.

3. Experiment

3.1 Experimental setup

In order to investigate both, the overall deformation behaviour in terms of the forming force and its scatter and the local forming behaviour in terms of local deformation and shape evolution, a novel experimental setup has been developed [9]. By using a translucent tool, it enables the in-situ observation of the material flow during process and also a post-process analysis of the stress-strain state in the forming area. The process chosen for these investigations is backward extrusion of a can where the die and punch are cut in the centre to obtain a half cylindrical geometry. The die is closed by a side cover made from sapphire. Through this tool, using a CCD-camera system and a telecentric objective, the deformation process can be investigated. The tool consists of a die with an inner diameter of 1 mm and a punch of 0.7 mm in diameter. Thus the resulting wall thickness is about 150 micron. The setup of the tool system used is shown in Fig. 5.

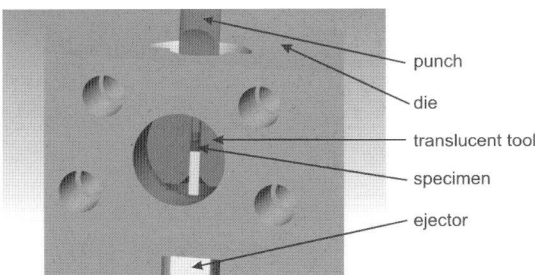

Fig. 5. Setup of the novel tool concept.

3.2 Experimental results

The results of the experiments have been analyzed in

terms of the forming force, process scatter, shape evolution and local deformation behaviour. With respect to the forming force and its scatter, it can be shown (Fig. 6) that UFG copper produced by 4 (NC_4) and 8 (NC_8) ECAP passes shows some noticeable differences. This can be explained by the differences in the material properties and microstructure: the forming force for NC_8 is slightly higher, and, since the material structure of NC_8 is more homogeneous, scatter of the forming force is reduced which can be interpreted as an increased process stability as well as shaping accuracy.

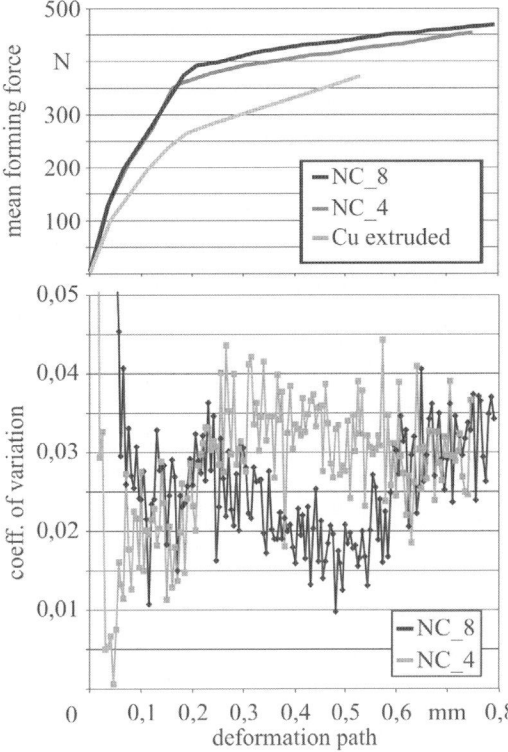

Fig. 6. Forming force and process scatter during backward extrusion of a UFG copper can.

The local deformation recorded by a CCD camera, with resolution of about 2 microns, shows nearly no differences between NC_4 and NC_8.

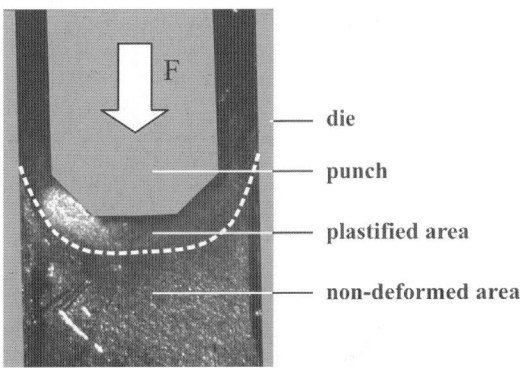

Fig. 7. Local deformation during backward extrusion of an NC_8 UFG copper can.

In both cases the deformation area is tightly and homogeneously distributed around the punch (Fig. 7). In terms of shape evolution, the optical measurements of the top surface of the extruded wall, which are depicted in Fig. 8, clearly show that the surface of NC_8 is less rough than that of NC_4.

a) NC_8

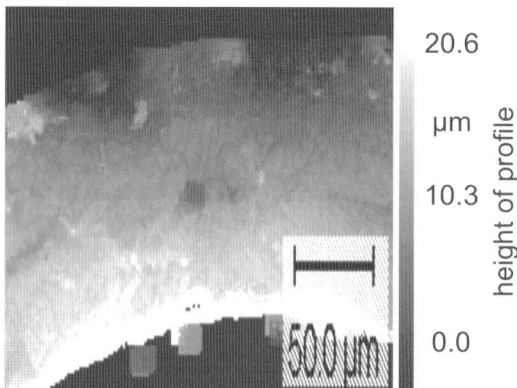

b) NC_4

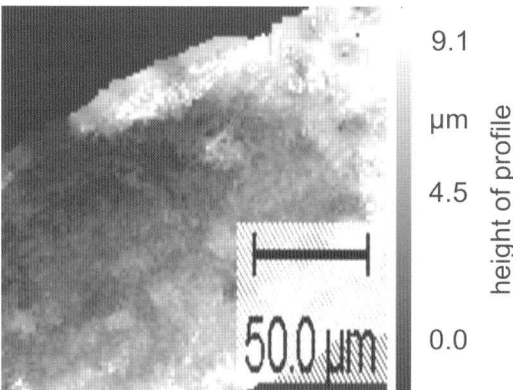

Fig. 8. Optical measurements of the can wall top surface a) profile height of NC_8 material b) profile height of NC_4 material

4. Conclusions

The results show that microforming processes benefit from the UFG structure of the metals formed. Scaling effects, like process scatter and uneven shape evolution, can be significantly reduced compared to the conventional coarse grained (CG) metals. Therefore, it is expected that further research activities, necessary to understand the micro and macro aspects of deformation of CG and UFG metals, will lead to even better results and help introduce microforming into the industrial practice.

Acknowledgements

The authors would like to thank the German Research Foundation for the financial support within the framework of SPP 1138 "process scaling". Also, this work was carried out within the EC Network of Excellence 4M "Multi-Material Micro Manufacturing".

References

[1] R. Wechsun,. et al., Nexus Market Analysis for microsystems, 2000-2005. Nexus Task Force Market Analysis. Wicht Technologie Consulting. München, 2002.

[2] Engel, U.; Tiesler, N.; Eckstein, R.: Microparts - A challenge for forming technology. In: Kuzman, K. (edtr): 3rd Int. Conf. on Industrial Tools, April 22-26, 2001, Celje Slovenia: Tecos 2001, 31-39.

[3] Engel, U.; Eckstein, R.: Microforming - from basic research to its realization. J. Mater. Process. Technolog. 125-126 (2002) 35-44.

[4] Engel, U.; Messner, A.; Tiesler, N.: Cold forging of microparts - effect of miniaturization on friction. In: Chenot, J.L. et al (Edtrs.): Proceedings of the 1st ESAFORM Conf. on Materials Forming.1998; 77-80.

[5] Tiesler, N.: Grundlegende Untersuchungen zum Fließpressen metallischer Kleinstteile. In: Geiger, M.; Feldmann, K. (Edtsr.): Reihe Fertigungstechnik Erlangen Nr. 120, Meisenbach, Bamberg, Germany, 2002.

[6] Tiesler, N.: Microforming – size effects in friction and their influence on extrusion processes. Wire 52 (2002) 34-38.

[7] Engel, U.: Tribology in Microforming. Wear 260(2006), 265-273.

[8] Olejnik, L.; Rosochowski, A.: Methods of fabricating metals for nano-technology, Bulletin of the Polish Academy of Sciences, Technical Sciences, 53/4 (2005) 413-423.

[9] Geißdörfer, S.; Engel, U.: Advanced Approach to Evaluate Local Deformation Behaviour at Microscale. In: Vollertsen, F.; Yuan, S. (Hrsg.): Proceedings of the 2nd ICNFT, (20.-21.09.), Bremen: BIAS-Verlag, 639-64.

Investigation of the mechanical behaviour of thin metal sheets using the hydraulic bulge test

A. Diehl, D. Staud, U. Engel

Chair of Manufacturing Technology, University of Erlangen-Nuremberg
Egerlandstr. 11, D-91058 Erlangen / Germany

Abstract

Ongoing miniaturisation leads to increasing complexity of micro parts linked with continuously decreasing development time. Hence, the demand for reliable material data and means to collect these data in a most efficient way is rising. Since the mechanical properties and thus material forming behaviour are dependent on the stress and strain conditions, the test methods have to be as close as possible to real conditions. Further, due to the so called size effects, data gathered from conventional length scale experiments cannot be used for the description of material used for parts with feature sizes in the micrometer range. In the present paper, the hydraulic bulge test as a means for the mechanical characterisation of thin metal sheets with thicknesses in the range of 25 µm to 500 µm is discussed and compared to data obtained by conventional tensile testing. Challenges due to the small sheet thickness are emphasized and the effect of strain rate on the flow curve is shown. The influence of geometric dimensions on the evaluation of the experiments is investigated by downscaling of the hydraulic bulge test. The material flow curves, as well as the forming limits are discussed in dependence of the sheet thickness.

Keywords: metal foils, mechanical properties, hydraulic bulge test

1. Introduction

Material flow curves, obtained by different means, differ due to various reasons. Amongst influences of temperature and strain rate, the main decisive parameters are the material anisotropy and the stress condition. Hence, the conditions for the determination of material properties should be accommodated to the process conditions, for which the obtained properties are used as an input parameter (e.g. for the process design or FE-Simulation). In case of forming processes, the stress and strain conditions are generally not uniaxial, making the commonly used tensile test a rather unadequate means for the determination of mechanical properties. Additionally, regarding the mechanical characterisation of metal foils, only very small strains can be reached using tensile testing [1]. This fact is impeding the determination of quantities like the material hardening exponent n which are derived from the material flow behaviour.

Resulting from these difficulties, the biaxial hydraulic bulge test as a means for the mechanical characterisation of sheet metals gains increasing importance. While this test is well established in conventional sheet metal characterisation, where significantly larger strains compared to tensile testing were achieved, there is only little knowledge about the applicability in case of metal sheets with thicknesses smaller 0.4 mm. Due to this lack of experience the European norm draft for the determination of forming limit curves of metallic materials is limited to minimum sheet thicknesses of 0.4 mm. In [2] hydraulic bulge tests have been performed with brass foils (thicknesses ranging from 150 µm to 600 µm) using different bulge diameters. A decrease in flow stress was observed, when using the small diameter. A correlation between the dome height resulting from different bulge diameters using the hydraulic bulge test and the channel height evolving during micro channel

hydroforming has been drawn by [3]. In [4] an aero-bulge test for copper foils with thicknesses ranging from 10 µm to 100 µm is described. A comparison with results obtained by conventional testing methods, however, has not been performed.

In this paper, the hydraulic bulge test is adopted for the mechanical characterisation of very thin aluminium (Al 99.5) sheets with thicknesses ranging from 25 µm to 500 µm. The influence of geometric dimensions on the material behaviour is investigated by performing scaled experiments. Results obtained from these experiments are compared with material data obtained by tensile tests. Further, the specifics resulting from the small sheet thicknesses are emphasized.

2. Hydraulic Bulge Test

Using the hydraulic bulge test the material flow behaviour as well as the forming limits can be determined for biaxial stress conditions. A conventional set-up for the experiments was used as schematically shown in Fig. 1. The tool consists of a movable upper part including the drawing die and a fixed hollow cylinder as lower tool with a fluid inlet. The specimen is placed in the cylinder, clamped by the drawing die and bulged by oil pressure. The standard circular die opening has a diameter of 115 mm. For the scaled experiments the radius of the die opening was scaled according to the foil thickness using a scaling factor λ, as shown in Table 1. The sheet thickness t = 200 µm was defined to represent a scaling factor λ = 1.

In order to achieve sufficient formability and to ensure comparability the spring hard Al 99.5 sheet material was annealed. The recrystallised microstructure was characterised by metallographic means. For all sheet thicknesses similar mean grain sizes L_G of approximately $L_G \approx 25$ µm were achieved.

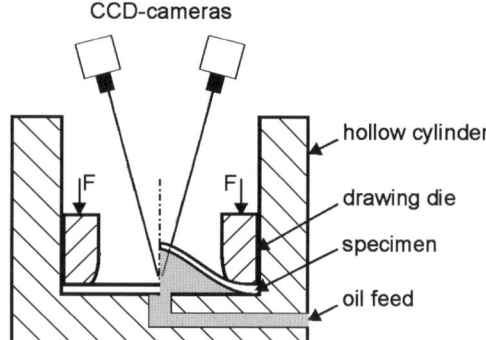

Fig. 1: Schematic of the experimental set-up for the hydraulic bulge test

During the hydraulic bulge test, the specimen is bulged into the die opening by continuously increasing fluid pressure, causing material flow in the whole bulged area. The rate of the pressure increase is controlled by the velocity of the pressure relay valve´s cylinder piston. For the determination of the flow curves the current pressure as well as the major and minor strains in plane direction have to be measured. The pressure is recorded using sensors at the oil feed, while the strains were determined using an optical measurement system (ARAMIS). This system recognises a stochastical pattern applied to the specimens´ surface by recording the bulging process via two CCD cameras. The optical strain measurement system allocates coordinates to every pixel in the image and by referring all data of the recorded pictures to the initial state picture it provides the requested information on the pole height and strain distribution.

If the die diameter is sufficient large in relation to the sheet thickness, the bulged material can be treated as a membrane shell assuming uniform stress distribution over the sheet thickness and neglecting bending stresses. According to the membrane theory [5], the flow stress σ_f at the pole can be determined by equation 1:

$$\sigma_f = \frac{p \cdot R}{2t} \qquad (1)$$

where p is the hydraulic pressure, R the curvature radius at the pole and t the actual thickness of the sheet at the pole. The curvature radius was calculated using a parabolic approximation considering all measured points. The sheet thickness at the pole is determined by the values of major and minor strain assuming volume constancy.

Table 1
Scaled dimensions of the hydraulic bulge tests

Al 99.5, annealed, T = 550°C, 1,5 h, $L_G \approx$ 25 µm		
sheet thickness t	die diameter d	scaling factor λ
200 µm	115 mm	1
100 µm	57.5 mm	0.5
50 µm	28.8 mm	0.25
25 µm	14.4 mm	0.125

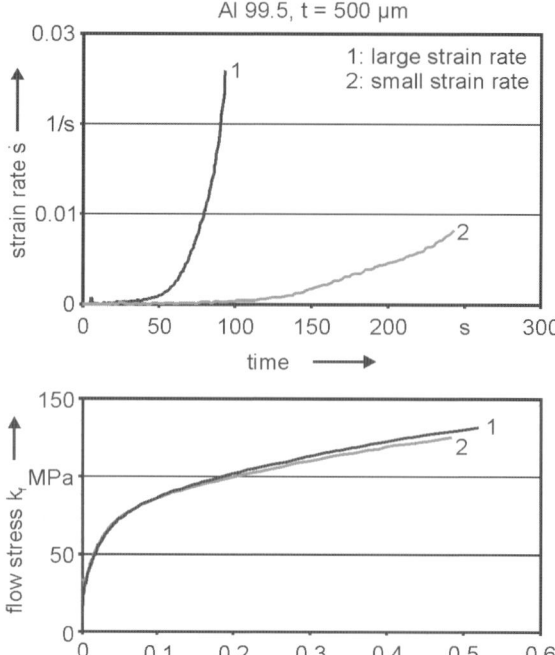

Fig. 2: Influence of the strain rate on the flow curve obtained from the hydraulic bulge test

3. Mechanical Properties

3.1 Flow Curve

In order to qualify the hydraulic bulge test as a means for the determination of flow curves and forming limits of metal sheet materials with thicknesses smaller than 0.4 mm, it is necessary to compare results gained from the hydraulic bulge test with results from conventional experiments (e.g. tensile test) for both macro and micro scale. As a reference for the macro case annealed (see Table 1) Al 99.5 sheet material with thickness of t = 0.5 mm was used.

An important parameter regarding the hydraulic bulge test is the rate of pressure increase determining the forming velocity of the bulged material. By variation of the velocity of the cylinder piston, defining the rate of pressure increase, the strain rate of the material can be varied. It must be noted, that the strain rate is not constant during the test, due to the thinning of the material at the pole and the resulting decrease of

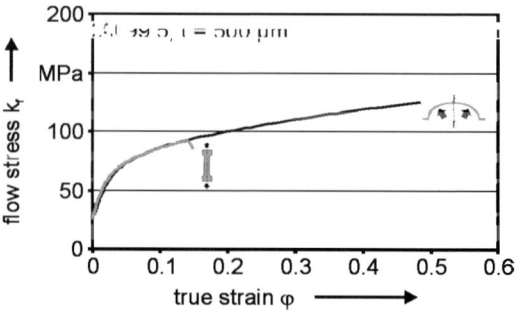

Fig. 3: Comparison of flow curves obtained by hydraulic bulge test and tensile test for the reference sheet thickness of t = 500 µm

material volume contributing to material flow. The effect of different slopes of the strain rate is shown in Fig. 2 for annealed Al 99.5 sheet material with t = 500 µm. A significant increase in the flow stress can be observed for higher strain rates, hence, a comparison of two bulge test is only possible, for experiments with similar slopes of the strain rate over time.

As can be seen in Fig. 3 there is good agreement between flow curves obtained by hydraulic bulge test and tensile test for the reference sheet thickness verifying the test set up. Further, a significant increase in the strain to fracture can be observed when using the hydraulic bulge test compared to tensile testing.

For smaller sheet thicknesses differences between the flow curves determined by the two different methods can be observed. While for t = 25 µm, different slopes of the flow curves are present (Fig. 4), for 50 µm ≤ t ≤ 200 µm similar slopes of the flow curves are measured (exemplarily shown in Fig. 4 for t = 200 µm). However, for the latter thicknesses the absolute value of the flow stress obtained by the hydraulic bulge test exceeds the flow stress obtained by tensile testing for the same true strain.

These differences could either be put down to the different stress conditions or to anisotropic behaviour of the material which both gain influence with decreasing sheet thickness. Uniaxial tensile test were performed with rolling direction parallel to the applied force, while in bulge testing the material behaviour is averaged over the whole plane. Here further investigations will be performed including determination of material anisotropy and evaluation of flow curves obtained by bulge tests in specific directions.

A further characteristic of the flow curves calculated from the data of the hydraulic bulge test is the large scatter of the flow stress for small true strains as can be seen in Fig. 4 for t = 25 µm, where the flow stress is clearly overestimated at the beginning of the flow curve. Due to the large curvature radii at the beginning of the hydraulic bulge test, the approximation of the curvature radius R for the calculation of the flow stress is rather inaccurate. Hence, for small true strains a significant scatter of the calculated flow stress is

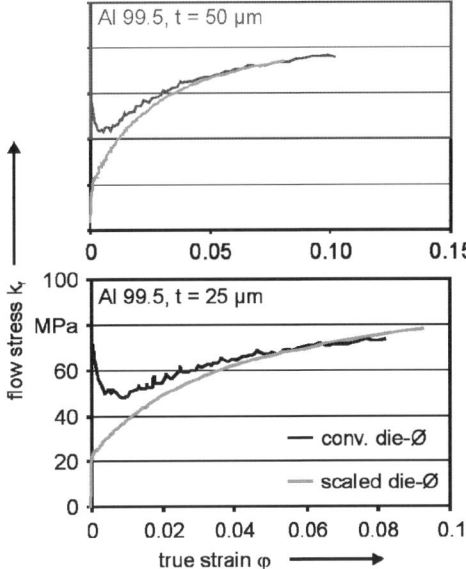

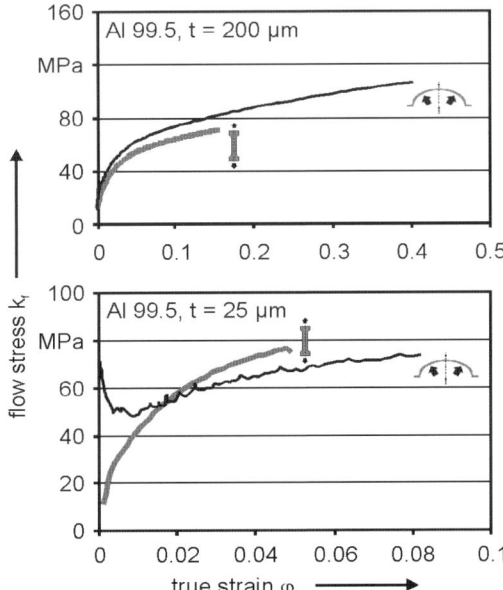

Fig. 4: Flow curves obtained by the hydraulic bulge test and tensile test for t = 200 µm and t = 25 µm

Fig. 6: Comparison of flow curves obtained with conventional die diameter (Ø=115 mm) and scaled diameter (see Table 1)

present and the actual beginning of material flow cannot be determined.

The influence of geometric dimensions on the material behaviour was investigated by scaling down the die diameter of the drawing die (see Table 1). As can be seen exemplarily for t = 50 µm (λ = 0.25) in Fig. 6, using smaller die diameter leads to better resolution of the flow curves for smaller strains. For larger strains both curves converge. For the case of t = 25 µm a significant improvement of the measured flow curve could be achieved. Due to the smaller die opening, the initial phase of the experiment exhibiting large curvature radii and hence evaluation inaccuracies is minimised. Thus, for the mechanical characterisation of thin sheet material with small maximum strains, smaller die openings seem to be more promising for achieving reliable results with high resolution.

Despite the necessity for further investigations to understand the relevant parameters of the hydraulic bulge test adopted to thin metal sheets in detail, it must be noted that for all sheet thicknesses a significant increase in the maximum strain could be achieved by the hydraulic bulge test compared to tensile testing (Fig. 5). Except for a sheet thickness of t = 50 µm the relative difference of the maximum strain is increasing with decreasing foil thickness underlining the need for a

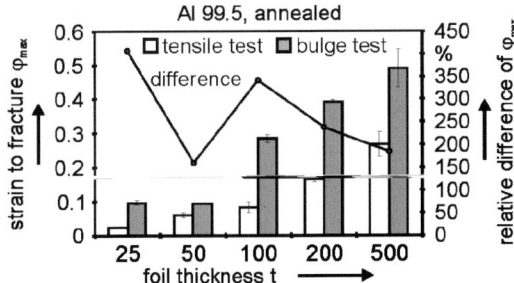

Fig. 5: Strain to fracture of tensile and bulge test and its relative difference in dependence of foil thickness for annealed Al 99.5

qualification of the hydraulic bulge test as a means for the determination of flow curves of thin sheet material. Reaching strains of about $\varphi_{max} \approx 0.1$ for the smallest sheet thickness t = 25 µm it is possible to determine the strain hardening exponent n which is the slope of the logarithmic stress-strain curve in the range $0.05 < \varepsilon < 0.2$. In Table 2 the n-values are listed in dependency of the sheet thickness.

Table 2
Dependence of the hardening exponent n on the sheet thickness in bulge tests

Hydraulic Bulge Test, Al 99.5, annealed					
sheet thickness t [µm]	25	50	100	200	500
hardening exponent n	0,19	0,23	0,24	0,26	0,27

As can be seen from Table 2, the strain hardening behaviour is depending on the sheet thickness. The continuous decrease of the strain hardening exponent with decreasing sheet thickness can be put down to the increasing share of surface grains with decreasing foil thickness. Strain hardening mainly occurs due to dislocation pile up at grain boundaries to adjacent grains. For thin sheets and a large share of surface grains the ratio of free surface to material volume is increasing, hence more grain boundaries with contact to the surface are present which do not contribute to the strain hardening mechanism.

3.2 Forming Limits

Defining the maximum major and minor strain values in the pole area as the forming limit condition, it is possible to determine supporting points of the forming limit diagram. In the present study only circular specimens are being used resulting in one point of the forming limit diagram for each foil thickness with the minor strain being in the range of the major strain (Fig. 7). In future work the minor strain will be varied by using different elliptical dies, which enables the determination of further supporting points of the right side of the forming limit diagram.

The formability of thin metal sheets is strongly dependent on the sheet thickness, as can be seen from the forming limits of circular specimens (Fig. 7). Due to larger microstructural inhomogeneity and the lower ability of material flow in thickness direction the formability is decreasing with decreasing foil thickness. The application of smaller die diameters has no significant influence on the forming limits, since the forming limits of experiments with conventional die diameter are in the same range as those of the scaled experiments.

4. Conclusions

In the present paper the hydraulic bulge test has been investigated as a means for the mechanical characterisation of thin metal sheets. It has been shown that is possible to determine flow curves of sheet material with thicknesses smaller than 400 µm in a significantly larger strain range compared to tensile test. However, there are still differences regarding the flow curves obtained by tensile test and hydraulic bulge test that will be addressed in future investigations. A higher resolution of the flow curves for larger strains could be achieved by scaling down the drawing die opening according to the sheet thickness. The material forming limits were not affected by the reduction of the formed area and showed a decreasing formability of the sheet material with decreasing sheet thickness.

Acknowledgements

The authors gratefully acknowledge the support from the German Research Foundation (DFG). Also, this work was carried out within the framework of the EC Network of Excellence "Multi-Material Micro Manufacture: Technologies and Applications (4M)".

References

[1] Diehl, A.; Engel, U.; Geiger, M.: *Mechanical Properties and bending behaviour of metal foils.* In: Menz, Fillon, Dimov (Ed.) : 2nd International Conference on Multi-Material Micro Manufacture, Grenoble, France (2006), Oxford, UK: Elsevier, S. 297-300

[2] Picart, P., Michel, J.F.: *Effects of Size and Texture on the Constitutive Behaviour for Very Small Components in Sheet Metal Forming*, Adv. Technol. of Plast. Vol. II, Proc. of the 6th ICTP, Sept. 19-24, 1999, 895-900.

[3] Mahabunphachai, S., Koç, M.: *Investigation of Size Effects on Material Behaviour of Thin Sheet Metals in Hydraulic Bulge Testing and Micro-channel Hydroforming.* In: Clemson University (Ed.): Proceedings of the 2nd International Conference on Micromanufacturing, 10.-13. September 2007, Greenville, USA (2007), S. 145-149

[4] Hoffmann, H., Hong, S.: *Tensile Test of very thin Sheet Metal and Determination of Flow Stress Considering the Scaling Effect*, Annals of the CIRP Vol. 55/1, 2006

[5] Gologranc, F.: *Beitrag zur Ermittlung von Fließkurven im kontinuierlichen hydraulischen Tiefungsversuch.* Dissertation, Institute for Metal Forming Technology, University of Stuttgart, Germany, 1975.

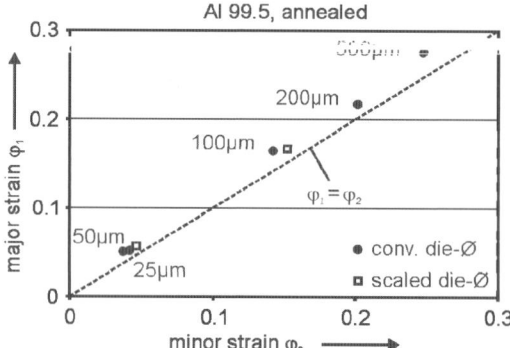

Fig. 7: Forming limits for circular specimens using conventional die-diameter (Ø=115 mm) and scaled diameter (see Table 1)

3D micro and nanostructuring of an epoxy based resist by electron beam lithography

G. Rius, J. Bausells, C. Martín, A. Llobera and F. Pérez-Murano

Institut de Microelectrònica de Barcelona, IMB-CNM-CSIC, Barcelona, SPAIN

Abstract

We present the results of producing three dimensional micro- and nanostructures on an epoxy based resist. Epoxy based resists are very interesting in microsystems technology for their good mechanical properties, that allow to produce high aspect ratio microstructures. We have optimized the definition of free standing structures by either using electron beam lithography alone or combining electron beam lithography and UV optical lithography. To tune the energy and dose of the electron beam exposure properly, Monte Carlo simulations are used.

Keywords: negative resist, nanofabrication, electron beam lithography

1. Introduction

Fabrication of polymeric micro- and nanostructures is of great interest for Microsystems Technology. Polymers are materials which are simple to process, compatible with many technologies, easy to modify in order to include specific functions (like, for example, sensing of specific bioentities). Polymers are also widely used in the area of microfludics, and in consequence they find large applications for developing lab-on-a-chip systems.

On the other hand, polymers are the base material for many resists used in lithography. Photoresists are widely used in microelectronics and microsystems to transfer patterns by UV lithography. Smaller patterns can be obtained using electron beam lithography (EBL), where an electron beam sensitive resist is locally exposed to high energy beams that either induce the cross-linking of the polymer chains (negative resists) or its breaking in smaller molecules (positive resists).

Although lithography is usually employed to define two dimensional patterns on surfaces, here we present results on the direct structuring of polymers using electron beam lithography. We show how free standing structures can be realized by tuning the dose and energy of the electrons. Then, specific zones of the polymers are cross-linked as a consequence of the spatial distribution of absorbed energy in the resist. The result is a three-dimensional polymeric structure after development.

2. Experimental

Experiments are performed using an epoxy based resist [1]. Similar to SU8, it is a negative-tone chemically amplified photoresist negative resist based on EPON SU8 resin. It is photosensitized by the addition of a triaryl sulfonium salt, e.g. Cyracure UVI from Union Carbide, which is also called Photo Acid Generator (PAG). SU8 is widely used for the fabrication of microstructures [2]

The resist is deposited at 2500 rpm spin speed for 45 s, which provides a resist layer of ~5 µm. The pre- and post- exposure processing conditions are identical and fixed. They consist of baking from 65 to 95 °C for 20 minutes, on air. Synthesized to be exposed under UV light, the performance for electron irradiation is established experimentally, but the assessment of electron scattering simulations helps to determine the trajectories of incoming irradiation. Indeed, scattering processes studied by Monte Carlo simulations for the SU8 resist suggest that beam energy strongly determines the electron penetration depth in the polymer.

3. Results

Based on simulations in 5 µm thick layer of SU8 to determine the electron penetration depth as a function of beam energy, it is found that beam energies >18 keV can be used to fully expose the thick layer (Figure 1).

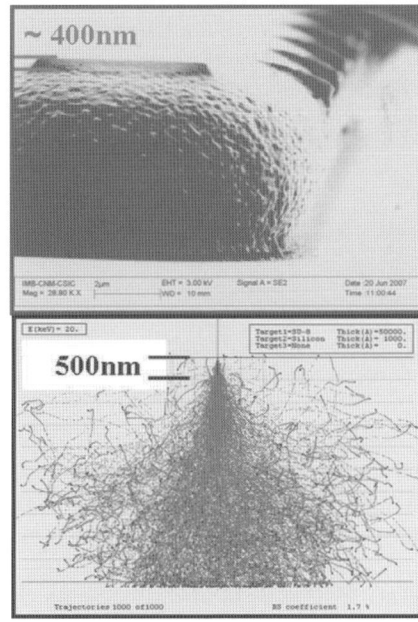

Fig. 1 Tilted SEM image of a line produced by exposing a 5 µm width line at 20 keV (left). The resulting line profile in depth is in agreement with the beam spread within the resist shown in electron trajectories simulations (right). Low forward scattering is achieved just for 400 nm in depth.

The simulations also indicate how the required dose as a function of energy has to be tuned. Beam energies of 18 and 20 keV are tested and both achieve to fully expose the whole resist thickness if an appropriate dose is used. But 20 keV is more suitable since it guarantees a stronger adhesion.

The feature shape distortion is remarkable. Squared patterns become nearly round due to the deviations that electrons suffer along the thick resist. Figure 1, left, shows the result of exposing a line of a 5 μm in width. In the top, electrons are little deviated from the incoming axis only for a thickness of about 400 nm in agreement with the simulations (Figure 1, right).

From these results, it is clear that design and exposure conditions are crucial. They determine and limit the pattern that can be obtained. However, it is possible to take advantage of the strong energy dependence of electron penetration depth in order to fabricate free standing structures.

The evaluation of shape dependence on energy is determined by exposing narrow lines (that is, lines defined as Single Pixel Lines, SPL, in the design, i.e. zero linewidth) at different beam energies (3, 5, 10, 15 and 20 keV) with wide transversal lines exposed at 20 keV to hold them (defined as 5μm linewidth in the design). As an example, anchored lines performed at 5, 10 and 15 keV are shown in Figure 2 (top). Not only penetration depth is proportional to the beam energy, but also a strong influence on linewidth is found. At low energies, 3-5 keV, the results of line exposure are more similar to strips than to lines, whereas beyond 15 keV, lines are narrower, but also deeper.

The dependence of line depth with dose is shown more in detail in Figure 2 (bottom). It evidences wider and deeper free standing structures as a function of increasing exposure dose, which correlates with the predictions of Monte Carlo simulations of punctual electron irradiation in conditions equivalent to the experimental ones.

Regarding the result of the exposures performed at 20 keV, the depth-dose dependence is clear (Figure 2, bottom and Figure 3, top). Increasing the internal dose factor by 1 in each line from left to right side leads to correspondingly deeper lines, in particular, lines from 0.6 μm to more than 2.5 μm in depth are found.

Line exposure results are summarized in the plot of Figure 3. For 15 and 20 keV, the two curves represent the measured linewidth on top and bottom, while for 3 and 5 keV uniform linewidth is found for all the depth of the strip.

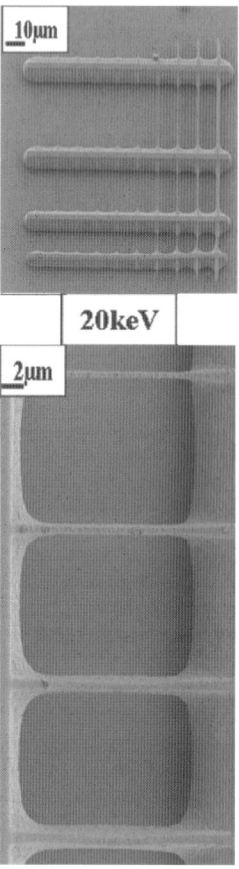

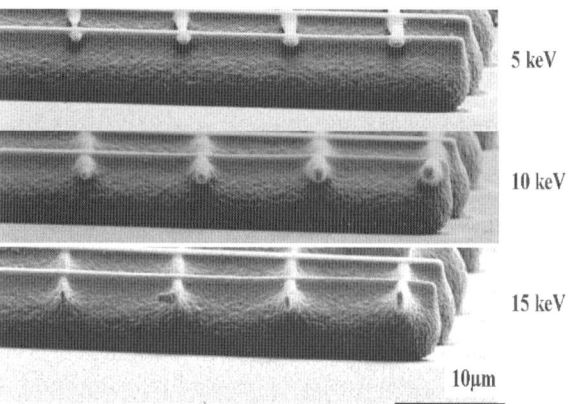

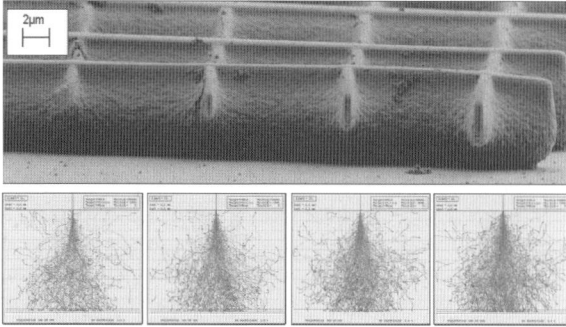

Fig. 2 (Top) Profile SEM images of lines defined at different beam energies. At higher beam energies the penetration range is higher and linewidth increases with dose (from left to right). (Bottom) Lines produced at 20keV show good correlation with electron trajectories simulations, where line depth is proportional to exposure dose.

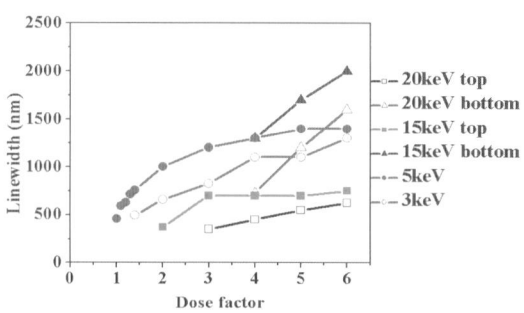

Fig. 3 (Top) Top view of lines produced at 20keV with increasing dose from left to right. Linewidth is proportional to exposure dose. (Bottom) Dependence of linewidth with dose factor shows that at higher energies forward scattering gets relevant (red and blue lines).

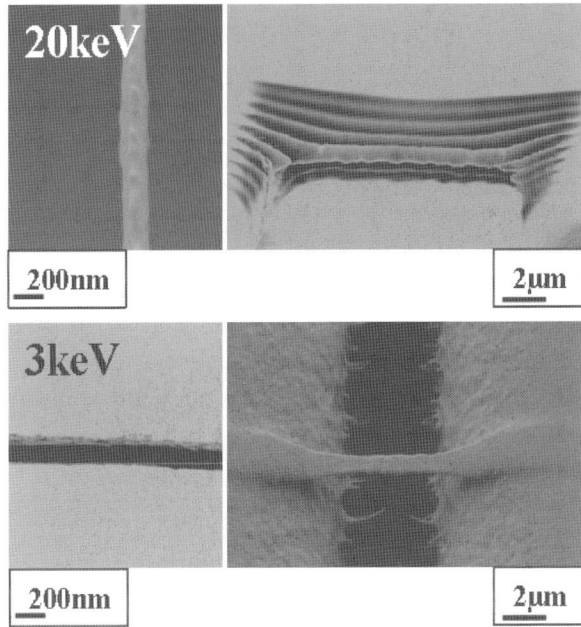

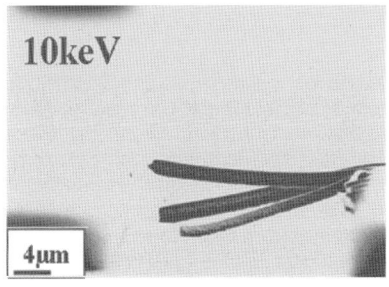

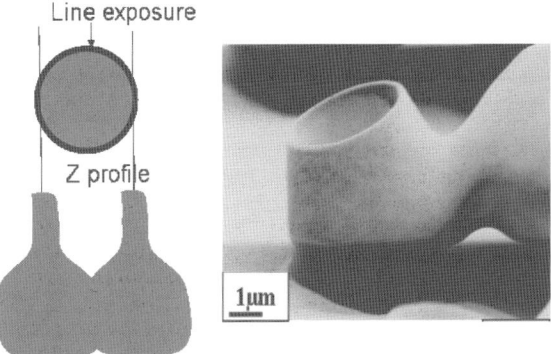

Fig.4 Experimental results for lines defined at 20 and 3 keV. For 20 keV, line is 200 nm in width and 750 nm in depth, whereas at 3 keV linewidth is 0.5 µm and depth is 140 nm.

Fig. 5 Free standing cantilevers or 3D structures can be defined combining the proper beam energies and design.

As can be seen, top values show approximately the expected trend of higher resolution at higher energies, but the line widening with increasing dose has a significantly different increasing rate. Also, electrons at 3 keV beam energy may be so easily stopped that the whole exposure process differs from the rest of tendencies.

Concerning to linewidth, resolution is also determined by beam energy and dose. At 20 keV, linewidth of less than 200 nm are achieved with ~ 650 nm remaining thickness (55 nm on the top of the line), whereas at 3 keV, minimum linewidth is around 365 nm, but remaining thickness does not reach 200 nm (Figure 4). Somehow, resolution is higher at low energies.

The combination of different beam energies (grey scale lithography) suggests that a great diversity of structures can be fabricated, from single or double clamped beams (Figure 5, top, and Figure 3), to channels or many other fancy configurations (Figure 5, right) that could be used for many applications, for example, microfluidic devices [3]. Microfluidic devices are obtained when combining high beam energy to pattern the channel walls and very low beam energy to define the top cover of the channel.

Taking advantage of forward scattering events and the characteristic electron stopping power in the resist, specific shapes are obtained. As an example, in Figure 5, right, bowl-shaped structures are obtained when irradiating a 0.1µm circunference at 20keV. The structure is produced by taking advantage of the profile shown in figure 2 bottom, where the top of the resist consist on a sharp definition and the bottom is widen, overlapping with the opposite spart of the circumference. Such kind of objets could be used as containers for particles and represent an additional way to prepare molds, for example, to be used in soft lithography.

A final approach consists on the combination of EBL with UV lithography. One of the interesting feature of the resist is that it is both electron and photon sensitive. These properties allow to produce very complex structures without a penalty on the fabrication time [4]. EBL is optimal to produce high resolution features and three dimensional structures, but it presents a very low throughput. On the other hand, UV lithography presents very high throughput.

Figure 6 shows examples of free-standing structures fabricated by the combination of UV and electron beam lithography. After depositing a 5 µm thick resist layer on a silicon surface, the sample is exposed to UV radiation through an optical mask to define the holding structures. The optical dose is selected so that the full thickness of the resist is exposed. After this step, the sample is transferred to the chamber of the scanning electron microscope where it is exposed to the electron beam. The exposure consists of an array of lines that will define nanometer scale free-standing beams between the holding structures. In this case, the electron dose is chosen so that only the very top layer of the resist becomes cross-linked, ensuring a good spatial resolution. The electron beam exposure is performed blindly (i.e., without alignment with respect to the optical pre-defined features), because at this stage of the processing, these latter features do not present any contrast. After this step, the sample is developed to remove all the non exposed areas and post-baked. As it can be observed in figure 6, the resulting beams are clearly defined, their width is of few hundreds of nanometers and they are pretty straight, indicating good mechanical properties like a low internal stress. This example represents a new approach to the fabrication of nanomechanical polymeric devices.

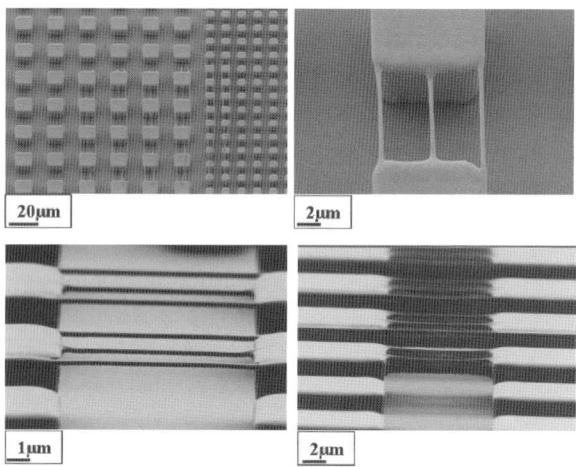

Fig.6 Combination of EBL with UV shows that high resolution structures can be obtained. As an example, EBL is used to define high resolution features anchored in UVL defined features. Lines are defined at 5 keV (left) and 20 keV (right).

4. Conclusions

We have shown that by tuning the energy and dose of the electrons, three dimensional structures can be directly fabricated on an electron sensitive epoxy based resist. In addition, since the resist is also sensitive to UV radiation, combination of optical and electron beam lithography allows to fabricate complex structures with high flexibility in the design. Application of the resulting structures in the areas of microfluidics, soft lithography and biochemical sensing are foreseen.

Acknowledgements

This work was partially supported by European project Novopoly, NMP3-CT-2005-013619. We acknowledge MicroResist GmbH for supplying of the polymeric material.

References

[1]   mr-L 5005 XP, under development by MicroResist GmbH. http://www.microresist.de

[2]   K.Y. Lee, N. LaBianca, S.A. Rishton, S. Zolgharnain, J.D. Gelorme, J. Shaw and T.H.P. Chang. Micromachining applications of a high resolution ultrathick photoresist, Journal of Vacuum Science & Technology B, 13 (6), 3012-3016 (1995).

[3]   P. Mali, A. Sarkar and R. Lal, Facile fabrication of microfluidic systems using electron beam lithography, Lab on a Chip 6 (2)310-315 (2006).

[4]   V. Kudriashov, X.C. Yuan, W.C.Cheong and K. Radhakrishnan, Greyscale structures formation in SU8 with e-beam and UV, Microelectronic Engineering 67-8, 306-311 (2003).

Explosive welding of Ni- based amorphous foils for micro-tooling applications

R.M. Minev [a], S.S. Dimov [b], S.R. Koev [c], G.Lalev [b], N.H. Festchiev [a]

[a] Department of Materials and Manufacturing Engineering, Rousse University, 8 Studentska, 7017 Rousse, Bulgaria
[b] Manufacturing Engineering Center, Cardiff University, Cardiff, CF24 3AA, UK
[c] BOM Ltd, Basarbovo, 7071 Rousse, Bulgaria

Abstract

In spite of the commercial advantages the available engineering materials for IC and MEMS processes are not able to meet the manufacturing demands for 3D high-aspect-ratio nano/micro structures and high precision. There is a group of energy assisted processes, such as laser ablation, e-beam and ion beam machining that could provide the needed high specific processing energy to create 3D microstructures. However, the required surface integrity of the manufactured nano/micro structures cannot be achieved without developing appropriate materials with adequate processing response. Thus, to broaden the range of micro-engineering products and multiply their capabilities the introduction of "novel" compatible amorphous or composite materials is required.

The study presents the capability of the explosive welding technology to create a bimetallic sandwich with amorphous Ni-based alloys foils (40 μm thick) without affecting the structure of the materials. Direct patterning by Focused Ion Beam (FIB) was used to produce masters from these materials for injection moulding and hot embossing tools. It was demonstrated that high feature resolution and surface quality of the manufactured nano/micro structures can be easily achieved by employing this technological chain.

Keywords: metallic glasses, micro tooling, explosive welding

1. Introduction

Employing the conventional engineering materials for processing of MEMS has its commercial advantages but these materials will not be able to meet the manufacturing demands for 3D high-aspect-ratio nano/micro structures and high precision. Thus, to broaden the range of micro-engineering products and multiply their capabilities the introduction of specifically tailored nano-structured or amorphous materials that are reasonably compatible with IC/MEMS batch-fabrication is required.

The creation of such capabilities necessitates the development and characterisation of materials which microstructure is optimised for performing surface processing at nano scale. In particular, to produce micrometer scale features and surfaces with sub-50 nm roughness, atom cluster processing is necessary with a group of mechanical (e.g. diamond machining), and energy assisted processes: ultra-short pulse laser ablation and excitation; electro discharge machining; micro milling; e-beam and ion beam (FIB) machining.

This study is aiming to produce specific amorphous/crystalline material composite for further processing and final commercialisation in new products by attaining a "step change" in the performance of complementary micro manufacturing processes, e.g. achievable surface quality, dimensional accuracy and functional properties of micro-engineering products.

The materials traditionally processed in the IC manufacturing (silicon, range of polymers and copolymers, ceramics etc.) are normally used in monocrystal, nanocrystalline or amorphous states and do not introduce significant surface integrity problems. However these materials do not possess the required mechanical and functional properties required by a range of micro manufacturing and micro tooling applications. Different lithographic techniques have been utilised to overcome this problems and to fabricate complex 3D structures. Subsequent pattern transfer from the structured polymeric resist to a wafer is performed via etching processes. An alternative technique is to use the patterned resist as a sacrificial template for producing metallic 3D masters through electroforming for several replications by thermal imprinting, embossing or moulding. And so, different non metallic negative replication materials are used to overcome the problems associated with the direct fictionalization of the metallic parts. This study is focused on up-scaling tasks associated with the production of metallic materials for micro and nano manufacturing applications that possess not only superior mechanical and physical properties but also can be processed cost-effectively.

Amorphous alloys (metallic glasses) have been attracting the attention of researchers for many years as materials with a wide range of useful properties, e.g. the surface hardness and wear resistance [1-2]. Compared with the crystalline counterparts, the abilities to sustain larger reversible, elastic deformations particularly under compressive stress conditions, make the metallic glasses promising tool materials for hot embossing and micro/nano replication since the features sizes can be reduced without risking functional failures.

2. Conventional metallic glasses

At California Institute of Technology (Caltech) in 1960, Pol Duwez et al. discovered [3] that if a molten metal (specifically, the binary metallic alloy $Au_{80}Si_{20}$) is under cooled uniformly and rapidly enough, e.g. at $1 \times 10^6\,°C\,s^{-1}$ an amorphous glassy structure could be obtained. With no crystal defects, mechanical properties of the metallic glasses are superior, in particular: (i) strength (twice that of stainless steel, but lighter); (ii) hardness (for surface coatings); (iii) toughness (more fracture resistant than ceramics); (iv) elasticity (high yield strength).

The absence of grain boundaries means that the material is resistant to corrosion and wear, as well as possesses soft magnetic properties, specifically in the alloys of glass formers (B, Si, P) and ferrous transition metals (Fe, Co, Ni).

3. Bulk metallic glass (BMG) scales up

In 1969, Chen and Turnbull formed amorphous spheres of ternary Pd-M-Si (M = Ag, Cu, Au) alloys at critical cooling rates of 100 to 1000°C s^{-1}, specifically $Pd_{77.5}Cu_6Si_{16.5}$ with a diameter of 0.5 mm [3].

Compared to crystalline steel and Ti alloys (Table 1 and Fig. 1) the commercialized Zr-based glasses have similar densities but high Young's modulus (96 GPa) and elastic strain-to-failure limit (ε_{el} = 2%). The glasses have high tensile yield strength (σ_y = 1.9 GPa), i.e. a high strength-to-weight ratio, making them a possible replacement for Al [3].

Both conventional metallic glasses and BMG are good candidates for micro manufacturing application. Although BMG has the advantage of the 3D macro exterior the conventional materials like Ni-based amorphous alloys has some obvious advantages:

- Higher strength (>2700 MPa),hardness (>8.2 GPa);
- Lower cost;
- Traditional usage of similar crystalline alloys (e.g. Ni-based) for replication masters;
- Maximum service temperature in air >500 K;
- Higher rigidity (150 GPa) than Zr based alloys.

In addition, the conventional amorphous Ni based alloys possess lower thermal diffusivity and conductivity than the crystalline counterpart in the range from room temperature to crystallization temperature, and it is possible to directly de-sublimate the material by short pulse laser or sputter it by FIB without significant thermal effects [4-5].

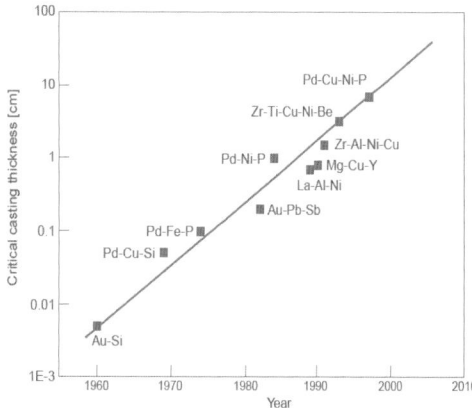

Fig. 1. Development of the critical casting thickness [3].

Table 1 Properties of Vitreloy (Zr-Be-Ti-Cu-Ni) compared to metal alloys [3].

Properties	Vit1	Al	Ti	Steel
Density g cm^{-3}	6.1	2.9	4.3	7.8
Yield strength, GPa	1.9	0.63	1.32	1.60
Elastic strain limit	2%	0.5%	0.5%	1.0
Fracture toughness, K_{1c} MPa.m$^{-3/2}$	140	45	115	154
Specific strength GPa.g^{-1}cm^{-3}	<0.32	<0.24	<0.31	<0.21

Explosive fabrication and more specifically welding and powder compacting is the main if not the single existing method to produce composite materials by cladding amorphous metals (metallic glasses) on polycrystalline metallic substrates. The speed of energy dissipation in this process allows the amorphous or ultrafine grain structure (UFG) to be preserved and the resulting layered composite to benefit from the unique properties of metallic glasses. The development of such composites would result in improvements of micro replication masters in regards to their surface quality, wear resistance and feature resolution. Currently, the most widely used technology for fabricating such masters is electroforming, and by applying it embossing tools and injection moulding inserts incorporating sub-micron features can be produced. The main applications include refractive and diffractive micro lens arrays, optical components and micro fluidic structures. The development of amorphous/crystalline composite materials could enhance the performance of replication tools for serial manufacture of components for these application areas. The proposed technological chain would look as follows:

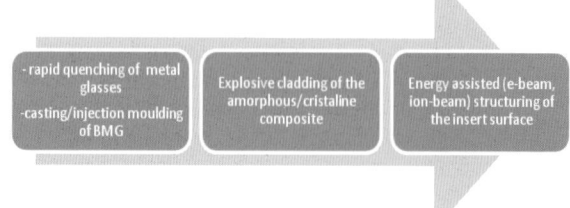

Fig.2. Technological chain for micro tooling with amorphous/crystalline composite inserts.

4. Explosive welding of the amorphous alloys

Our results and the achievements of other researchers [6-12] show that for satisfactory explosive cladding of laminated amorphous/crystalline composites it is necessary to achieve the following processing conditions:

1. Strong and continuous fusion between the amorphous foil and the substrate;
2. Sustain the amorphous structure of the coating;
3. Avoid brittle fracture of the amorphous material;
4. Prevent intensive wave formation that could violate the surface integrity (Fig. 3).

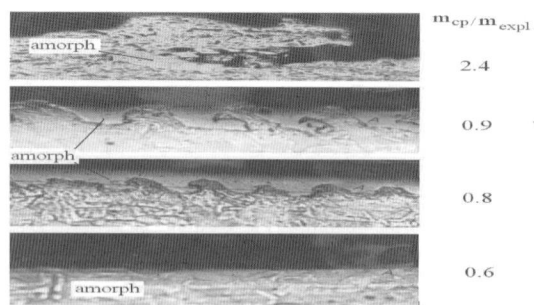

Fig.3 Wave formation on amorphous $Ni_{78}B_{14}P_8$/Cu interface (optimal value for m_{CP}/m_{Expl} ~0.8)

It is also necessary to define and abide appropriate welding parameters that ensure the requirements listed above. To achieve such processing conditions the relationship between the welding parameters (e.g. the mass proportion of the cladding plate and explosive - m_{CP}/m_{Expl} and impact velocity - V_p) and the quality of the joints presented in Fig. 4 is used [6].

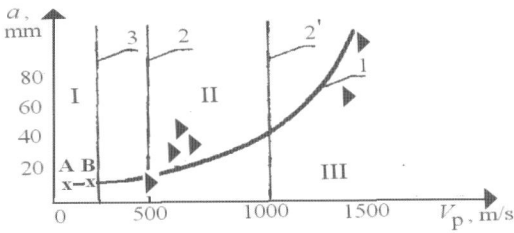

Fig. 4 Diagram of amorphous $Ni_{78}B_{14}P_8$ foil/Cu-substrate welding conditions.

Since the wave formation at the amorphous/crystalline interface is essential for the mechanical soundness of the composite it is appropriate to use the wave amplitude (a) as a criterion for the quality of the joint. It was shown that (a) have to be of optimal value, 5-15μm, in order to achieve a good adhesion without significant trespassing of the substrate material into the coating [6].

Line (1) represents the parabolic correlation between V_p and (a) [13-14]. Point B is associated with the welding conditions where no interface wave formation is observed. Point A – boundary conditions for satisfactory bonding. The processing conditions in the AB section result in some weak bonding without wave formation. The lines 2 and 2' show the critical theoretical (following Wittman equations [15]) and experimental technological parameters under which the energy dissipation is enough to bring the foil into crystalline condition. By analogy it is possible to estimate V_p for which a brittle behavior of the amorphous foil could be observed. Subsequently we can distinguish the following areas on the diagram: (I) welding the foil to the substrate without any material changes; (II) brittle behavior; (III) structural crystallisation of the foil.

This study explores the possibility to weld the amorphous $Ni_{78}B_{14}P_8$ foils with 40 μm thickness onto different substrate materials (Stainless Steel, Mild steel, Cu, Al). The technological conditions for producing such laminated amorphous/crystalline composites are provided in Table 2.

Table 2 Parameters for explosive welding

Welding design	Inter-stice mm	Set angle α^o	Impact velocity V_p	m_{CP}/m_{Expl}
	1	8	1400	2.4
	1	0	1500	3.0
	1	0	720	0.8
	1	0	590	1.2
	0.8	0	-	1.0
	0	0	780	0.9

The quality of the welding was studied by metallographic analyses and the structural changes in the amorphous foil were examined by x-ray diffractometry. The results show that if the process is carried out according to the parameters in area (I) of the diagram the amorphous state of the material is maintained and the corresponding complex of foil properties remained unchanged. If the joining conditions lead to intensive heat dissipation or interface wave formation crystalline phases in the deposited material were immediately detected.

Table 3 Properties of the substrate and related cracking of the coating

Substrate material	Elongation (A%)	Crack density (mm^{-2})
Stainless Steel	5	0
Mild steel	25	1-2
Copper	40	10
Aluminium	45	30

One of the most pronounced problems of the explosive cladding of the amorphous foils was the formation of 'net'-cracking in the coating. However, in the carried out experimental study it was possible to overcome this problem by applying the materials with reduced dynamic plasticity. It was wrongly suggested in the previous research [6] that increasing of the plasticity by heating the substrate would reduce the crack formation in the amorphous coating. The most important conclusion that can be derived from the experiments is that increasing the yield strength and reducing the ductility (A%) (Table 3) of the substrate are important for the successful foil cladding. To overcome the crack formation, the substrate and the coating must have similar A%. The ultimate conditions for cladding the $Ni_{78}B_{14}P_8$ foil (A%<1.5) onto Al substrate (A%>20) (Fig.5,a) demonstrate that when the substrate material does not have sufficient resistivity to dynamic upsetting the amorphous material runs out of deformation resources and dynamically cracks occur. If the plasticity of both materials does not differ from one another by more than 90% ($Ni_{78}B_{14}P_8$ foil – Stainless Steel) the amorphous material does not crack. In this case it forms specific surface wave lines of deformation (Fig.5,b). The wave lines could be interpreted as local luxation of the material. These lines can be easily removed during the planarization of the samples by polishing.

Fig.5 Crack formation in Al/ $Ni_{78}B_{14}P_8$ 70x50 mm sample (a) and (b) - deformation lines on the surface of Steel/$Ni_{78}B_{14}P_8$ composite

5. FIB 3D structuring of the amorphous surface

FIB has attracted the attention of researchers worldwide due to its high patterning flexibility, high resolution, and relatively simple processing steps [16-18]. By applying FIB technology, it is possible to eliminate many of the lithography steps (including an e-beam, photon or x-ray exposure, development of resist, oxygen plasma etching, etching of metal substrate, resist stripping, dry etching and/or wet etching) and pattern Ni substrate directly. What's more, FIB can be used to optimize the resulting surface properties by varying the ion beam parameters and by utilizing various gas-assisted-etching techniques.

Here we present the structuring capabilities of FIB technology in patterning of amorphous $Ni_{78}B_{14}Si_8$ substrates [16]. A polycrystalline analogue of the alloy was also processed as a reference material to indicate the influence of the amorphous structure on the material behavior under ion bombardment. Ion beam fluence related cross section profile evolution of trenches was examined by and SEM. The ion fluence in $nC/\mu m^2$ was calculated according to equation: $f_i = I_{ion} \cdot t_s / A$, where I_{ion} is the Ga^+ ion beam in nA, A is the target area size, μm^2 and t_s - the exposure time in sec.

Both amorphous and crystalline 40 μm thick $Ni_{78}B_{14}Si_8$ laminates were used in the experiment. The material has extremely high mechanical properties. It has been reported that for Ni based metallic glasses with metalloid concentration higher than 20% [19], electrons can be arguably described as free charges and the electronic transport is believed to be dominated by s-like nearly free electrons, resulting in dramatically change in the transportation properties of the alloy compared to crystalline Ni. Therefore, in the amorphous structure, the energy dissipation during the ion exposure is minimized and the sputtering yield is increased. On the other hand, the non-crystalline structure predetermines better surface integrity since no crystalline defects, point defect agglomerates and grain boundaries present in the material.

The FIB structuring was conducted on a XB1540 Carl Zeiss system. Rectangular trenches were produced to study the effects of the scan speed and ion fluence on the sputtering yield and evolution of the cross-section profile. The target area of the trenches was 10×4 μm and the ion beam probe was 0.5 nA.

The results for the FIB response of amorphous and crystalline substrates are shown in Fig. 6. The experiments revealed that for amorphous $Ni_{78}B_{14}Si_8$, with the increase of the ion beam scan speed, the sputtering yield first decreased and then kept constant. The milled depth was proportional to the ion fluence. Patterning of polycrystalline $Ni_{78}B_{14}Si_8$ demonstrated the anisotropy-milling rate for grains with different orientation. Trenches with significant improvement and perfect surface finish were produced in amorphous $Ni_{78}B_{14}Si_8$.

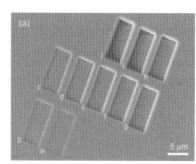 

Fig.6. Images of trenches produced by FIB in amorphous (a) and crystalline (b) $Ni_{78}B_{14}Si_8$ with ion fluence f_i, in the range 0.38- 9.92 $nC/\mu m^2$.

The profile evolution analysis revealed that the surface finish of the FIB milled trench is dependent on scan speed, ion fluence and applied scan strategies. The achieved results confirmed that the re-deposition effects could be easily overcome by increasing the ion beam scan speed. It has been proven that the amorphous metallic material due to the lack of crystalline structural features (grains, lattice defects, etc.) gives advantages in obtaining good surface finish. Surface integrity of the amorphous alloy was better than the polycrystalline one where identical milling conditions were applied.

6. Conclusions

1. It was shown that the technological parameters for obtaining successful laminated composites of crystalline/amorphous metals through explosive cladding are within the range of $m_{CP}/m_{Expl} = 1$, and $V_p = 800m/s$. An appropriate combination of materials would be achieved if the ductility of the substrate and amorphous foil are similar, e.g. Stainless steel/Ni-P-B.

In this experimental study an area of 4-5 cm^2 free of defects was produced.

2. The experiments show that FIB direct pattering could be a potential alternative technology to lithography based pattern transfer technique for producing Ni-alloy mould/insert for micro/nano tooling and hot embossing applications. Because of the superior mechanical properties and improved surface finish the amorphous Ni based alloys could offer better quality and increased life cycle of the tooling components.

Acknowledgements

The research reported in this paper was carried out within the framework of the EC FP6 Networks of Excellence "Multi-Material Micro Manufacture (4M): "Technologies and Applications" and the FP6 Integrated Project "Charged Particle Nanotech" (CHARPAN).

References

[1] A.A.Kundig, M.Cucielli, P.Uggowitzer, A.Dommann, Microelectronic Engineering, v.67-68, Issue I, pp 405-407, 2003

[2] B. Prakash, K. Hiratsuka, Tribology letters, v.8, No2-3, pp.153-160, 2000

[3] M.Telford, Bulk Metallic Glass, Materials Today, March, pp.36-43, 2004.

[4] M. Ektessabl, T. Sano, Review of Sc. Instruments, v. 71, Iss.2, 1012-1015, 2000

[5] D.Schubart, F.Vollertsen, M.Kauf, Modeling Simul.Mater.Sci.Eng. 5, pp.79-92, 1997

[6] Minev R., Festchiev N., Expl. Plating of Amorphous Met.Foils, Proc. Int. Seminar on High Energy Working of Rapidly Solidif. Mat. (HERAPS'88), Novosibirsk, pp.161-166, 1988

[7] Wood N., Machine and Tool Blue Book, No1, p.78, 1980

[8] Cline C.F., Scripta Metalurgica, No 11, p.1137, 1977

[9] Hammerschmidt M., Metallwissenschaft und Technik, N034, 1980

[10] Prummer R., Explosive welding of Metallic Glasses onto Metall, 2-nd Meeting on Explosive Working of Materials, Novosibirsk, 1982

[11] D.J. Vigueras, O.T. Inal, A. Szecket , Explosive Welding of an Amorphous Ribbon to a Mild Steel Substrate, Explomet 85, Marcel Dekker Inc., N. York, Basel, pp.927-942, 1986

[12] K. Zhang, X. Li, Study on the Multilayer Explosive Clading of Thin Amorphous Foils, 3rd Int. Conf. on Mech. and Phys. Behaviour of Mat. under Dynamic Loading, J. Phys. IV France 01, pp.229-234, 1991

[13] Deribas A., Physics of Expl. Work-hardening and Welding, Novosibirsk, Nauka, 1980

[14] Kudinov V.M., Explosive Welding in the Metalurgy, Moskva, Metallurgia, 1978

[15] Wittman R., Proc. 2-nd Symp. Expl. Working of Met., Marianske Lasne, p.153, 1973

[16] Wuxia Li, Roussi Minev, Stefan Dimov, and Georgi Lalev, Patterning of Amorphous and Polycrystalline $Ni_{78}B_{14}Si_8$ with a Focused Ion Beam, Appl.Surf.Sc., v.253, issue 12, pp.5404-5410, 2007

[17] C. Ochiai et al., J.Vac.Sci.Tech. B 19 p.933, 2001

[18] W. Li, T. Shen, Appl.Phys.Lett., v. 87. pp.113-123, 2005

[19] J.Ivkov, E.Babic, H.Lieberman, J.Phys.: Condens. Matter 1, 551, 1989

Process characterisation including process chains

Hot embossing of high aspect ratio sub-µm structured surfaces for micro fluidic applications

M. Heckele[a], M. Worgull[a], T. Mappes[b], G. Tosello[c], T. Metz[d],
J. Gavillet[e], P. Koltay[d], H. N. Hansen[c]

[a] Forschungszentrum Karlsruhe (FZK), Institute for Microstructure Technology (IMT),
D-76344 Eggenstein-Leopoldshafen, Germany
[b] University of Karlsruhe (TH), Institute for Microstructure Technology (IMT),
D-76344 Eggenstein-Leopoldshafen, Germany
[c] Technical University of Denmark (DTU), Department of Mechanical Engineering,
DK-2800 Kgs. Lyngby, Denmark
[d] Laboratory for MEMS Applications, Department of Microsystems Engineering (IMTEK),
University of Freiburg, George-Koehler-Allee 103,79110 Freiburg, Germany
[e] French Atomic Energy Commission (CEA), Laboratory of Innovation for New Energy Technologies and
Nanomaterials (LITEN), 38054 Grenoble, France

Abstract

Sub-micro structured surfaces allow modifying the behaviour of polymer films or components. Especially in micro fluidics a lotus-like characteristic is requested for many applications. Structure details with a high aspect ratio are necessary to decouple the bottom and the top of the functional layer. Unlike to stochastic methods patterning with a LIGA-mould insert it is possible to structure surfaces very uniformly or even with controlled variations (e.g. with gradients). In this paper the process chain to realize polymer sub-micro structures with minimum lateral feature size of 400 nm and up to 4 µm height is presented.

Keywords: Hot embossing, micro fluidics, micro-nano structured surfaces

1. Introduction

A new generation of passive, capillary driven micro-fluidic systems is expected to enable advanced management of liquid and gas. In such enhanced flow structures, fluids are guided through channels with geometrically modified surfaces. The realization of lotus–like patterns on surfaces requires precise control of the structuring method. The manufacturing method of such structured surfaces should also enable mass-fabrication capability, for example by polymer replication. To demonstrate the moulding of sub-µm structures, an indirect tooling technology for the manufacturing of the mould insert and the hot embossing process for polymer replication have been selected.

1.1. Wetting behaviours of structured surfaces

The liquid repellence of a surface is principally governed by a combination of its chemical nature (i.e. surface energy) and, in case of stochastic surfaces, by its topography at micro-scale dimensional range (i.e. surface roughness). Although unstructured low surface energy materials often exhibit high water contact angle values [1] this characteristic is not yet sufficient to yield super-hydrophobic behavior. In order to obtain this behavior, the difference between the advancing and the receding contact angles (contact angle hysteresis) must be minimal. Effectively, contact angle hysteresis can be regarded as the force required to move a liquid droplet across the surface; i.e. in the case of little or no hysteresis, very little force is required to move a droplet, hence it rolls off easily [2][3].
Theoretical study for idealized rough hydrophobic surfaces predict that contact angle hysteresis initially increases with surface roughness [4] until eventually a maximum value is reached. Greater roughness scales beyond this lead to a fall of the contact angle hysteresis due to the formation of a composite interface (where the liquid is unable to completely penetrate the surface).

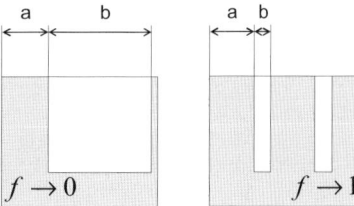

Fig. 1. Surface fraction f for a columnar structure.

The phenomena can be described by the Cassie-Baxter equation (Eq. 1), where porous surface topography, superimposed to the roughness of a solid surface, causes air to become trapped in voids (i.e. prevents liquid from wicking) [5]:

$$\cos\theta* = f \cdot \cos\theta + f - 1$$

Where: $\theta*$ is the apparent contact angle, f is the surface fraction (i.e. the total area of solid-liquid interface in a unity of plane geometrical area parallel to the surface), and θ is the contact angle of the rough surface.
The surface solid fraction for a simple porous surface structure such as a columnar structure with square section is defined as follows:

$$f = a^2 / (a+b)^2$$

Where: a is the pillar width and b is the spacing

between two consecutive pillars (see Fig. 1). By increasing the void between two micro structures (i.e. decreasing the width of the structures), a smaller surface fraction can be obtained (see Eq. 2) and therefore a larger contact angle (see Eq. 1) is achieved. A hydrophobic substrate can become super-hydrophobic; under certain conditions a hydrophilic substrate can act as a hydrophobic one.

1.2. Latest developments on fluid management at micro scale

Recently several approaches where made to use superhydrophobic gradients to passively move droplets on horizontal surfaces. Beside several approaches using chemical gradients (e.g. [6]), systematic investigations of a structural gradient using pillars with varying size and a constant pitch were recently performed [7]. However, the fabricated devices could achieve the movement of droplets only by adding vibration energy to overcome the contact angle hysteresis. A major role in [7] played the relatively high depinning force when a droplet leaves one of the pillars. This effect could be probably avoided by introducing structures in the sub-μm range where the depinning force is small due to the downscaled size of each single structure.

2. Design

A lotus effect can be achieved by honeycomb structures (Fig. 2) and can be attained over a broad range of geometric parameters. To compare different wetting behaviours depending on the ratio between void and bulk characteristic lengths, two different designs were investigated and manufactured. In particular, the pitch of the micro-structured pattern was maintained constant (pitch = 4 μm), whereas the diameter of the combs and the thickness of the separating walls varied. The two patterns realized have the following geometrical characteristics:

1. Wall thickness = 1000 nm, comb diameter = 3000 nm (surface fraction f = 0.25).
2. Wall thickness = 400 nm, comb diameter = 3600 nm (surface fraction f = 0.10).

The height of the structures is 4 μm resulting in an aspect ratio of the sub-micro walls between 4 and 10.

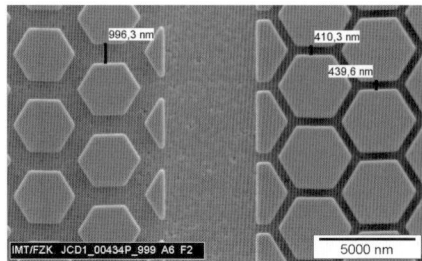

Fig. 2. Honeycomb structures after E-beam lithography. The two patterns correspond to two different surface fractions on the final surface: f=0.25 (left) and f=0.10 (right).

The height of the structures (and therefore the aspect ratio) was chosen in order to obtain the conditions for hydrophobic surfaces (i.e. formation of a liquid/air composite interface due to the air volume trapped into the voids) and also to allow the de-

moulding during the replication by hot embossing without damaging the separating walls.

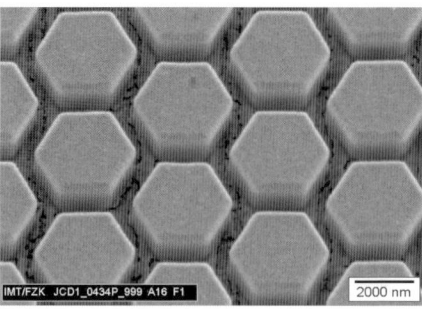

Fig. 3. Detailed view of honeycomb structures after E-beam lithography.

The structured surface was of an area of 20x10mm² split into two different areas of 10x10mm² with different diameter and pitch of the honeycomb structures as described. The two designs were placed as close as possible to each other to analyze an effect at the interface between both designs (see Fig. 2).

3. Tooling technology

To replicate this design via hot embossing into a thermoplastic material it is necessary to manufacture a tool presenting the negative pattern on its surface. In the following, the manufacturing steps to transform the two-dimensional pattern into the three-dimensional embossing stamp are described.

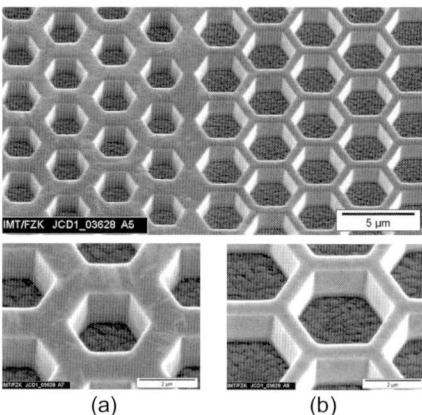

(a)   (b)

Fig. 4. Electroplated gold mask for X-ray lithography. The mask refers to the two selected designs with 1 μm thick walls (a) and 400 nm walls (b).

3.1. Electron beam lithography

In the first step the design was written via electron beam (E-beam) lithography into a 3.2 μm thick photo resist (PMMA). The E-beam machine was operated at maximum acceleration voltage of 100 kV. In order to avoid damage of structures during the de-moulding step of the hot embossing process, vertical sidewalls on the mould insert have to be manufactured. Hence, to avoid any undercuts in the mold inserts already at the beginning of the process chain, the dose during E-beam lithography had to be optimized. Therefore variations in dose had to be performed to calibrate the E-beam process and to select the beam parameters for the writing procedure. Finally, writing was performed with a current of 10 nA and a dose of 800 µC/cm^2 (see Fig. 2 and Fig. 3).

3.2. X-Ray lithography

In order to manufacture a mask to be used for X-ray lithography, the grooves generated by the development of the polymer were filled with gold by electroplating (see Fig. 4). To obtain a gold layer with a homogeneous absorption a gold sulphidic bath was employed. The thickness of this gold mask was in the range of 2.20 ± 0.05 µm.

This mask was used for the transfer of the design to a 4 µm and also to a 10 µm thick resist (PMMA). Different aspect ratios of the structures up to 20 (referring to the sidewalls) were achieved. With these two expositions two different moulds for replication were fabricated. As shown in Fig. 5, a deformation of the structures was observed due to a shift of the mask during the X-ray lithography process.

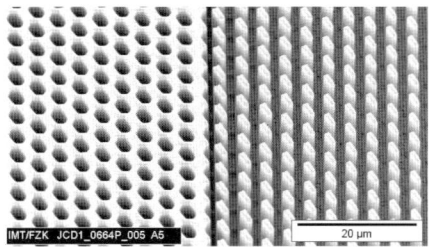

Fig. 5. Resist pattern (10 µm high) achieved by X-ray lithography. Distortion of the pattern is due to a mask shift during the process.

3.3. Electroplating - Nickel shim fabrication

Based on the X-ray lithography two nickel mould inserts were fabricated by electroplating. These mould inserts used for replication by hot embossing were characterized by a diameter of 4 inch and a thickness of approximately 300 µm. Only the sample with structures of 4 µm height has been successfully electroplated up to now. Representative for these mould inserts, a nickel shim with optical structures in the similar dimension of structures, is shown in Fig. 6. Honeycomb structures of the microstructured nickel mould insert obtained from the 4 µm thick resist are shown in Fig. 7.

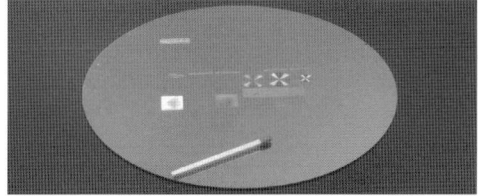

Fig. 6. Representative for the fabricated mold inserts: a 4 inch nickel shim with µ-structured areas.

4. Replication by hot embossing

The replication of the mould inserts was carried out by hot embossing (Fig. 8). Hot embossing is well suited for the replication of large structured areas with high aspect ratio sub-µm structures on a thin residual layer. Especially the short flow distances and the moderate flow velocities will produce low stress in the moulded

part [9]. Furthermore, the precise vertical de-moulding, guaranteed by the precise vertical guidance of the crossbars of the machine, is essential for successful de-moulding of the filigree lotus structures.

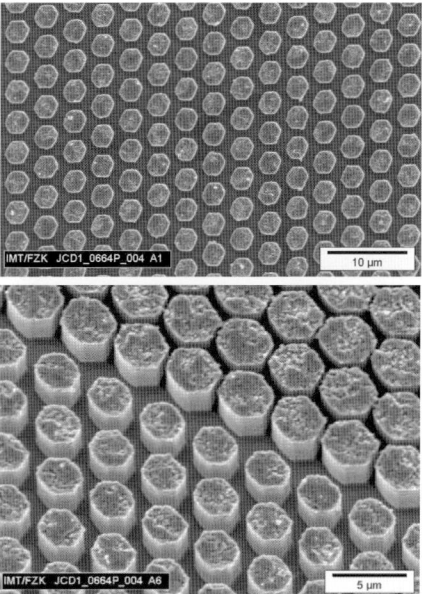

Fig. 7. Honeycomb structures of the microstructured nickel mould insert.

Because of the influence of shrinkage of the polymer, the challenge of this task is the de-moulding without deformation or damage of the structures, in particular the honeycomb structures with sidewalls in the range of 400 nm. For the actual 4 µm high structures a good quality was achieved. The next generation with 10 µm height will require a precise control of the de-moulding parameters.

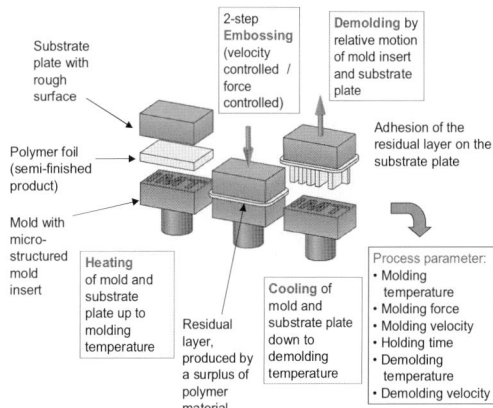

Fig. 8. Schematic view of the hot embossing process.

The replication of high aspect ratio micro-nano structures into polymer with a very low defect rate will ensure the function of the lotus effect. The lotus structures can be replicated in a wide range of polymer materials, beginning from low temperature materials like PS or PMMA up to semi-crystalline high temperature materials like PEEK. For the first approach PMMA was

212

employed. Example of sub-micron honeycomb structures moulded in PMMA by hot embossing is shown in Fig. 9. The structure size of these honeycombs is in the same range like the structures used for the lotus effect.

IMT/FZK JCD1_0063P_001_5 A8 A6/F1   2 µm

Fig. 9. Sub-µm honeycomb structures replicated in PMMA by hot embossing (pattern pitch is 2 µm and wall thickness is 500 nm.

5. Functional test

The polymer (PMMA) substrate structured via hot embossing was tested by means of contact angle measurements. Droplets of distilled water were deposited on the three different areas of the sample (droplet diameter = 1300 ± 20 µm, volume 1.1±0.1 µl, see Fig. 10). Measured contact angles were:

- Reference area (without structuring) = 81±4°
- Area 1 (1000 nm thick walls) = 87±2°
- Area 2 (400 nm thick walls) = 107±6°

By introducing the sub-µm honeycomb structures on the surface, the hydrophilic property of the PMMA substrate (contact angle lower than 90°) has been turned apparently into a hydrophobic one.

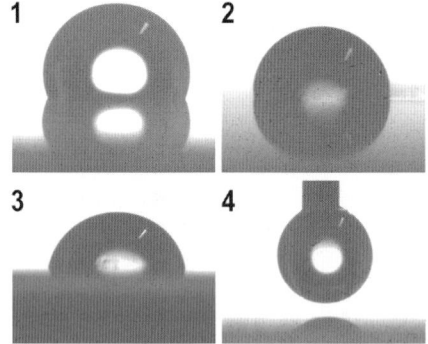

Fig. 10. Different wetting behaviours due to the surface sub-micro structuring of the polymer substrate:honeycomb with wall thickness of 400 nm (1), honeycomb with wall thickness of 1000 nm (2), substrate without structuring (3). Water droplet before the deposition (4).

6. Conclusion

A new process chain for the manufacturing of polymer for sub-µm structured surfaces was established. The presented tooling process steps included E - Beam writing, mask fabrication and X-ray exposure with subsequent nickel electroplating. Hot embossing was used to replicate the micro-nano

honeycomb structures on a polymer substrate.

Surface sub-µm structuring can be employed to provide particular properties to the polymer surface for enhanced flow functionality suitable for microfluidic applications. In this research work, in particular, sub-µm honeycomb structures were used to provide hydrophobic property to hydrophilic polymer surface. Alternatively, hydrophobic substrates can be turned into super- hydrophobic by using the presented surface structuring.

Advanced design solutions employing this new technology include micro-nano fluidic systems where the surface of micro-sized channels present sub-µm-structures, as well as micro fluidics systems whose surface incorporates structural geometrical gradients generating a driving force to move liquid samples along channel structures. Research work is being carried out by the authors and results will be present in future publications.

Acknowledgements

The present research was performed within the framework of the European Network of Excellence "Multi Material Micro Manufacture: Technology and Applications" (4M) (European Community funding FP6-500274-1; www.4m-net.org) and within the activity of the Cross Divisional Project "MINAFLOT" (Micro- and Nano-structured Surfaces for Liquid and Gas Management in Microstructured Flowfields) supported by 4M and the European Community (Project no. FP6-500274-2). Furthermore the collaboration of the Polymer Technology Division (4M Work package 4) and of the Microfluidics Application Division (4M Work package 10) is gratefully acknowledged.

References

[1] Tsibouklis J et al. Poly(perfluoroalkyl methacrylate) Film Structures: Surface Organization Phenomena, Surface Energy Determinations, and Force of Adhesion Measurements. Macromolecules. 33:22 (2000) 8460-8465.

[2] Furmidge C. Studies at phase interfaces. I. The sliding of liquid drops on solid surfaces and a theory for spray retention. Journal of Colloid Science. 17:4 (1962) 309-324.

[3] Miwa M et al. Effects of the surface roughness on sliding angles of water droplets on superhydrophobic surfaces. Langmuir. 16:13 (2000) 5754-5760.

[4] Johnson RE and Dettre RH, Contact angle hysteresis. I. Study of an idealized rough surface. Adv. Chem. Ser. 43 (1964) 112.

[5] Cassie ABD and Baxter S. Wettability of porous surfaces. Transactions of the Faraday Society. 40 (1944) 546-551.

[6] Moumen N, Subramanian RS and McLaughlin JB. Experiments on the motion of drops on a horizontal solid surface due to a wettability gradient. Langmuir. Vol. 22, no. 6 (2006) 2682-2690.

[7] Shastry A, Case MJ and Bohringer KF. Directing droplets using microstructured surfaces. Langmuir. Vol. 22, no. 14 (2006) 6161-6167.

[9] Heckele M and Schomburg WK. Review on micro molding of thermoplastic polymers. Journal of Micromechanics and Microengineering. 14 (2004) R1-R14.

Multi-Material Micro Manufacture
S. Dimov and W. Menz (Eds.)

Influence of process parameter variation on ceramic feedstock flow behaviour

T. Hanemann[a,b], J. Aroni[a]

[a]Forschungszentrum Karlsruhe, Institut f. Materialforschung III, D-76021 Karlsruhe, Germany
[b]Albert-Ludwigs-Universität Freiburg, Institut f. Mikrosystemtechnik (IMTEK), D-79110 Freiburg, Germany

Abstract

With respect to feedstock development for different ceramic injection molding techniques the influence of various process parameters during feedstock development was investigated systematically. First the dispersant concentration at the fillers surface was changed in a wide range. The impact on the particle size distribution was measured. Second the size and the geometry of the used stirrers for compounding in an unsaturated polyester resin as polymer matrix were varied. The resulting composite flow properties at a fixed solid load and different temperatures were determined experimentally using a cone and plate rheometer. Increasing dispersant amounts at the alumina surface lead to a change of the particle size distribution and to a significant composite viscosity drop. The use of different stirrers affects directly the composite viscosity as well as the flow behaviour to a certain extent.

Keywords: Ceramic feedstock development, dispersants, rheological properties

1. Introduction

In microsystem technologies ceramic parts carrying a microstructured surface with details below 100 µm and aspect ratios larger than 1 become more and more important due to their outstanding thermomechanical properties and chemical stability in harsh environments. For a successful commercialization a low cost fabrication has to be established. This means a realization of a process chain covering the single steps
1. Filler conditioning (e.g. surface treatment)
2. Compounding
3. Molding
4. Thermal postprocessing (debinding, sintering)

In the last years different variants of injection molding techniques using reactive resins, wax or thermoplastics as binder in the investigated feedstock systems have been developed. These techniques cover the whole replication field starting from rapid prototyping up to mass fabrication [1-5].

For the realisation of dense ceramic parts using polymer binder-ceramic filler-composites a powder load of at least 50 vol% is necessary, which causes a significant increase of the composite's viscosity [6]. In literature a large number of publications deals with the realisation of feedstock systems with large filler loads, mostly related to wax, paraffin or thermoplastic binders applying different kinds of dispersing agents for a reduction of the feedstock viscosity [7,8]. In 2005 Zürcher and Graule gave a comprehensive overview of the influence of the dispersant structure on the flow behaviour of zirconia/organic solvent/dispersions [9]. Detailed rheological investigations of composites consisting of a polymer reactive resin as matrix and surface treated ceramic fillers like alumina and zirconia using different types of organic dispersants have been published in 2006 and 2007 [10-12].

In feedstock development the use of dispersing agents is strictly recommended. The addition of suitable dispersants allows for a viscosity reduction at constant load or, consequently, a load increase at constant viscosity. Furthermore the homogeneity of the feedstock and the greenbody strength, important for demolding, is improved.

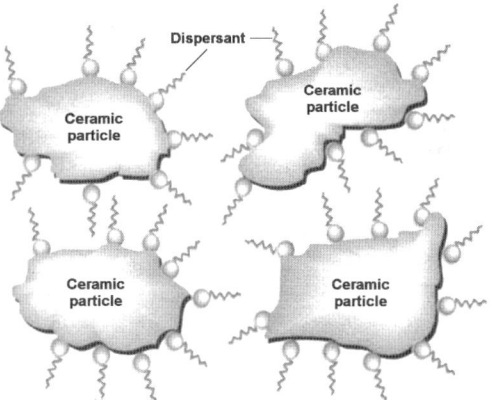

Fig. 1. Dispersants attached at particle surface.

On a molecular level dispersants, designed for the use in polymer matrix materials, are small amphiphilic molecules or oligomers as well as polymers. These molecules adsorb via van-der-Waals-forces or hydrogen-bridge-linkage on the filler's surface and reduce the interactive forces between the individual filler particles preventing reagglomeration and supporting wetting by the binder. Figure 1 shows exemplarily the surface attachment of a mono-functional dispersant molecule like a polyethyleneglycolalkylether to ceramic particles, dispersed in a polymer matrix.

The addition of fillers to a liquid changes the temperature influence on the composite's viscosity, which can be described with an Arrhenius-type approach [6] (1), η_1 and η_2 are the apparent viscosities at the two different temperatures T_1 and T_2, R the gas constant and ΔE_a is the flow activation energy, which depends mainly on the composition of the investigated system. In general an increasing solid load causes a raise of the flow activation energy which is equivalent with an improved sensitivity to temperature changes.

$$\ln \frac{\eta_1(T_1)}{\eta_2(T_2)} = \frac{\Delta E_a}{R}\left(\frac{1}{T_1} - \frac{1}{T_2}\right)$$

(1)

With respect to compounding even a slight temperature increase yields a noticeable viscosity drop, especially at high loads close to the maximum accessible filler load. Low viscous polymer-based reactive resins like polymethylmethacrylate solved in methylmethacrylate or unsaturated polyester solved in styrene can be used as model systems for high viscous polymer melts and for a rapid prototyping of microstructured parts made of plastic, ceramic and metals [3-5]. The apparent viscosity of the pure reactive resin is under ambient conditions below 5 Pa s and enables a rapid composite processing using simple laboratory equipment e.g. for dispersant screening experiments [10-12].

In this work the influence of the used dissolver stirrer size and geometry as well as the dispersant amount on the composite flow behaviour will be discussed.

2. Experimental

2.1. Ceramic powder surface coating

The commercial Almatis alumina CT3000 SG was selected as ceramic test material and treated with a dispersant of the polyethyleneglycolalkylether-type (Brij72, see Fig. 2, SigmaAldrich) always doubling the dispersant amount starting with 3.3 g per kg ceramic. The molecule consists of a small hydrophilic moiety, which can adsorb at the alumina surface, and an extended aliphatic tail, which can interact with the nonpolar polymer matrix. To ensure a homogenous surface hydrophobization a large sized rotary evaporator equipped with a 5 l evaporation flask was used (Buechi AG). 1 kg of dry alumina was coated with the dispersant in 1 l of ethanol at 50°C. After two hours the ethanol was removed, the remaining coated alumina was dried at elevated temperature. The resulting particle size distribution was measured using a Beckman Coulter LS230. The relative large batch amount enables a feedstock preparation in bigger volumes. Table 1 lists the different compositions realized by surface coating.

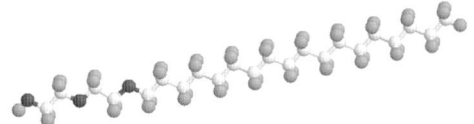

Fig. 2. Molecular structure of Brij72.

Table 1
Investigated alumina/Brij72 compositions.

Name	Alumina content (g)	Brij content (g)	d_{50}-value (µm)
Sample 1	1000	0.0	1.9
Sample 2	1000	3.3	2.0
Sample 3	1000	6.7	2.0
Sample 4	1000	13.3	2.3
Sample 5	1000	26.6	5.0

2.2 Reactive resin based feedstocks

An unsaturated polyester resin (Roth GmbH) with a polymer content around 65 wt% and styrene as reactive thinner was used as polymer binder. Composites containing 50 wt% coated alumina (22.4 vol%) have been prepared using two different sized dissolver stirrers (diameters 42 and 29 mm, see Fig. 3).

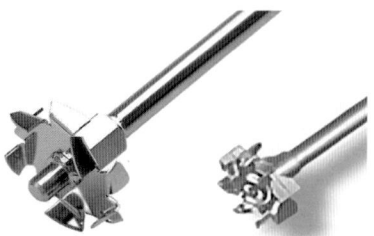

Fig. 3. Applied different sized stirrers.

Despite the fact, that both stirrers are claimed to be dissolvers stirrers with a donut-like flow profile allowing large shear forces during compounding, the shape and the rotor blades are totally different; the smaller stirrer shows more similarity to a turbine or blade mixer. The latter one generates smaller shear forces and a reduced deagglomeration potential in comparison to the larger dissolver stirrer. This one needs at least a sample volume around 40 ml, the smaller one only a sample volume around 20 ml for successful mixing in a suitable sized glass beaker. An untreated alumina, dispersed in the resin (sample 1), was used as reference material. The low filler content enables the use of a cone and plate rheometer (CVO50, Bohlin) avoiding experimental complications like sticking or gap emptying at larger shear rates. All viscosity measurements were done at 20, 40 and 60°C in the shear rate range between 1 and 200 1/s. The experimental uncertainty of the obtained data is in the range of ± 5%.

3. Results and Discussion

3.1. Impact of surface coating on the particle size distribution

Figure 4 shows the influence of the surface coating on the differential particle size distribution. The uncoated alumina (sample 1) possesses two main fractions: a small one between 100 and 400 nm and a large fraction around 2 µm. Increasing dispersant amounts cause a reduction of the fine fraction (sample 3) which disappears at the largest Brij concentration (sample 5) completely. In the latter case only the main fraction remains which is shifted to larger values between 5 and 6 µm. As a consequence the average particle size value d_{50} increases especially at very large dispersant amounts (see Table 1).

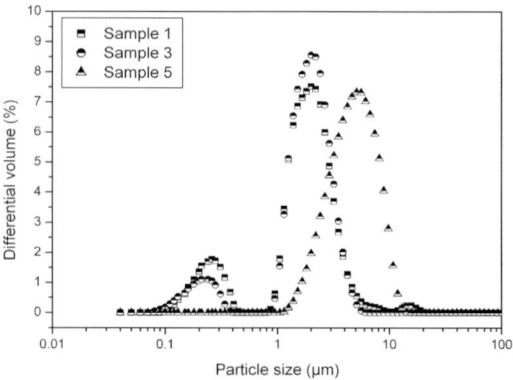

Fig. 4. Influence of surface coating on particle size distribution.

3.2. Impact of stirrer size and dispersant concentration on the flow behaviour

It has been shown earlier, that the used stirrer type and dispersant concentration have a strong influence on the resulting unsaturated polyester-alumina-composite viscosity [10]. The stirrer blade geometry affects directly the deagglomeration capability, with respect to particle deagglomeration sharp blade edges as in case of the large dissolver stirrer are favourable in comparison to smooth blades as in turbine stirrers.

Increasing dispersant amount can influence the composite viscosity in different manners [9-11]:
   a)   viscosity drop
   b)   viscosity drop with local viscosity minimum for a certain dispersant concentration
   c)   viscosity increase

Figure 5 shows for the samples 1 and 5 the flow curves at 20, 40, and 60°C using the 42 mm dissolver stirrer for dispersion. As expected with increasing temperature the viscosity drops significantly. Sample 5 shows at all investigated temperatures a reduced viscosity, especially at larger shear rates. In a rough approximation all flow curves show Newtonian flow behaviour. Table 2 lists for all investigated composites using the large stirrer for compounding the viscosities at a shear rate of 100 1/s. Increasing dispersant amounts cause a pronounced viscosity drop at all investigated temperatures (percentage reduction @20°C: 41%; @40°C: 32%; @60°C: 35%).

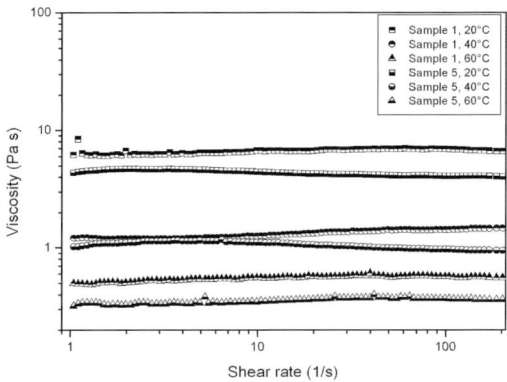

Fig. 5. Flow curves of the samples 1 & 5, using the 42 mm stirrer for compounding.

Table 2
Composite viscosity (large stirrer).

Sample	Composite viscosity(@100 1/s)		
	@20°C	@40°C	@60°C
Sample 1	6.89	1.43	0.57
Sample 2	5.08	1.16	0.50
Sample 3	4.63	1.11	0.48
Sample 4	4.97	1.14	0.45
Sample 5	4.07	0.97	0.37

In case of the composites using the 29 mm stirrer the flow curves are different (Fig. 6). Whilst the addition of uncoated alumina to the unsaturated polyester matrix (sample 1) yield a Newtonian flow in the investigated shear rate range the surface coating introduces a slight pseudoplastic flow at 40°C and 60°C. As in the previous mentioned composites increasing temperatures result in a viscosity reduction.

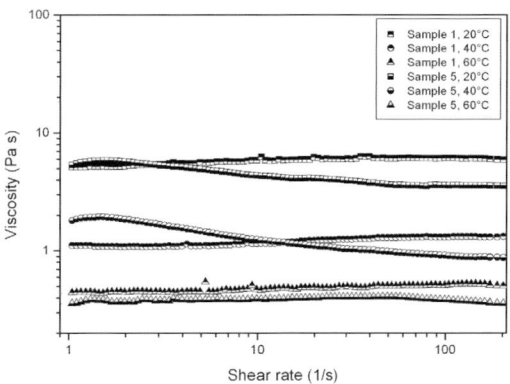

Fig. 6. Flow curves of the samples 1 & 5, using the 29 mm stirrer for compounding.

Table 3 summarizes for all mixtures fabricated by the 29 mm stirrer the viscosity data. Increasing dispersant amounts lead again to a significant viscosity reduction (percentage reduction @20°C: 42%; @40°C: 31%; @60°C: 27%).

The stirrer size and the shape cause two basic differences in the resulting flow curves and viscosity values: Firstly, almost all viscosity values listed in Table 3 are smaller than the ones presented in Table 2, especially at the lowest measuring temperature. The differences vanish at larger temperatures. Secondly the mixture (sample 5) prepared with the 29 mm stirrer show a pseudoplastic flow at 40°C and 60°C, whilst the same sample prepared with the 42 mm stirrer show a Newtonian flow.

Table 3
Composite viscosity (small stirrer).

Sample	Composite viscosity(@100 1/s)		
	@20°C	@40°C	@60°C
Sample 1	6.12	1.31	0.52
Sample 2	4.70	1.14	0.52
Sample 3	4.46	1.06	0.42
Sample 4	4.66	1.10	0.43
Sample 5	3.58	0.91	0.38

Hence for screening purposes the small stirrer may be applied exploiting the small sample volume necessary despite the slight differences in the resulting viscosity data. These differences are expected to be significantly larger in case of highly filled composites or feedstocks suitable for the different variants of micro powder injection molding.

3.3. Impact on the flow activation energy

As shown earlier using micro- and nano-sized ceramics [11,13] surface coating and the resulting surface hydrophobization has a strong influence on the flow activation energy, which is a reliable measure for the viscosities sensitivity to temperature changes. In general increasing solid filler content (uncoated filler) cause a significant flow activation energy increase especially close to the critical filler load [14,15]. A surface hydrophobization cause a better coupling of the filler to the polymer chains, the sensitivity to temperature changes drops, especially in case of nanosized fillers with their large specific surface areas.

216

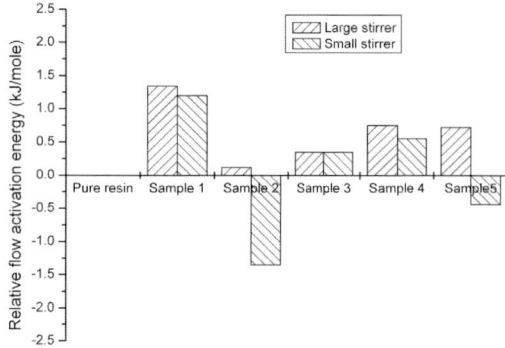

Fig. 7. Relative flow activation energy change.

At a small filler load, as investigated here, only a small impact on the flow activation energy can be expected as published earlier [11,13-15]. Figure 7 shows the flow activation energy change relative to the pure polyester resin. The addition of untreated alumina to the resin cause a slight flow activation energy increase independent of the used stirrer type. Within the experimental error the values are almost identical. Increasing Brij72 concentrations cause in all cases a reduction of the flow activation energy, but the numerical values scatter, especially the ones for sample 2 and 5. In both cases the use of the small stirrer results in a small reduction of flow activation energy change relative to the pure resin. An unequivocal trend cannot be observed, which is due to the relative low solid load of 22.4 vol% in the composite far away from the critical filler load. The scattering of the numerical value is within the range observed in earlier investigations using microsized fillers [11,13-15].

4. Conclusion

The influences of the dispersant concentration and the dispersing conditions on the composite viscosity have been investigated systematically. Increasing dispersant amounts cause a shift to larger average particle sizes. A reduction of the dissolver stirrer size and especially the blade geometry cause changes in resulting flow properties, which can be enhanced in case of larger solid loads in the composite. At constant solid load increasing dispersant amounts cause a pronounced viscosity reduction at all measured temperatures. As in earlier investigations a clear impact on the flow activation energy cannot be observed. With respect to feedstock development the Brij72 concentration should be as large as possible to ensure a low viscosity value. As a drawback very large dispersant amounts can reduce the greenbody strength due to a reduced coupling between the coated filler and the polymer matrix resulting in molding defects as well as a reduced sinter density. Hence in all cases a compromise between the different criteria has to be found experimentally for each investigated binder-filler-system individually.

Acknowledgements

The authors gratefully acknowledge the financial support by the European Commission within the 4M-Network of Excellence and the Deutsche Forschungsgemeinschaft DFG (SFB 499). TH thanks R. Schillinger for sample processing at IMTEK.

References

[1] Ruprecht R, Finnah, G, Piotter, V. Microinjection Molding – Principles and Challenges. In: Löhe D and Haußelt J (Eds) Microengineering of Metals and Ceramics, Wiley-VCH, 253-287, Weinheim, 2005.

[2] Bauer, W, Haußelt, J, Merz, L, Müller, M, Örlygsson, G, Rath. Micro Ceramic Injection Molding. In: Löhe D and Haußelt J (Eds) Microengineering of Metals and Ceramics, Wiley-VCH, 325-356, Weinheim, 2005.

[3] Hanemann T, Honnef K, Hausselt J. Rapid prototyping of microstructured ceramic and metal parts using reaction molding techniques. Proc. 1st Intern. Conf. on Multi-Material-Micro-Manufacture (4M), 29.06.-01.07.2005, Karlsruhe, FRG.

[4] Hanemann T, Honnef K, Hausselt, J. Process chain development for the rapid prototyping of microstructured polymer, ceramic and metal parts: Composite flow behaviour optimization, replication via reaction molding and thermal postprocessing. Intern. J. of Adv. Manuf. Techn. 33 (2007) 167-175.

[5] Hanemann T, Bauer W, Knitter R, Woias P. Rapid prototyping and rapid tooling techniques for the manufacturing of silicon, polymer, metal and ceramic microdevices. In: Leondes CT (Ed) MEMS/NEMS Handbook: Techniques and Applications. Springer Publisher, 187-255, Berlin 2006, Vol. 3.

[6] German RG. Powder Injection Molding. Princeton: Metal Powder Industries Federation, 1990.

[7] Xie Z-P, Luo J-S, Wang X, Li J-B, Huang Y. The effect of organic vehicle on the injection molding of ultra-fine zirconia powders. Materials and Design 26 (2005) 79-82.

[8] Trunec M, Dobsak P, Cihlar J. Effect of powder treatment on injection moulded zirconia ceramics. J. Europ. Ceram. Soc. 20 (2000) 859-866.

[9] Zuercher S, Graule T. Influence of dispersant structure on the rheological properties of highly-concentrated zirconia dispersions. J. Europ. Ceram. Soc. 25 (2005) 863-873.

[10] Hanemann T. Influence of dispersants on the flow behaviour of unsaturated polyester-alumina-composites. Composites A 37 (2006) 735-741.

[11] Hanemann T. Viscosity change of unsaturated polyester-alumina-composites using polyethylene glycol alkyl ether based dispersants. Composites A 37 (2006) 2155-2163.

[12] Hanemann T, Heldele R, Haußelt, J. Structure-property relationship of dispersants used in ceramic feedstock development, Proc. 4M 2007 – 3rd Intern. Conference on Multi-Material-Micro-Manufacture (4M), 03.-05.10.2007, Borovets, Bulgarien, 73-76.

[13] Hanemann T, Heldele R, Haußelt J. Particle size dependent viscosity of polymer-silica-composites, Proc. 4M 2006 - 2nd Intern. Conference on Multi-Material-Micro-Manufacture (4M), 20.-22.09.2006, Grenoble, Frankreich, 191-194.

[14] Hanemann T. Influence of particle properties on the viscosity of polymer-alumina-composites, Ceramics International, 2008, online available, doi: 10.1016/j.ceramint.2007.08.007.

[15] Hanemann T, Honnef K. Process chain development for the realization of zirconia microparts using composite reaction molding, Ceramics International, 2008, online available, doi: 10.1016/j.ceramint.2007.10.005.

Micro injection moulding: an experimental study on the relationship between the filling of micro parts and runner designs

C.A. Griffiths, S.S. Dimov, E.B. Brousseau

Manufacturing Engineering Centre, Cardiff University, CF24 3AA, UK

Abstract

To increase productivity and thus reduce the unit cost, often micro moulding tools incorporate multiple cavities. For this a runner design must be selected, the main function of the runner system is to facilitate the flow of molten material from the injection nozzle into the mould cavity. Therefore, the micro injection filling process depends on the optimum design of runner systems. In this context, the paper reports an experimental study that investigates the flow behaviour of the polymer melts in micro cavities with a particular focus on the relationship between the filling of micro parts and the size of the runner system. In particular, the runner size effects on the micro injection moulding process were investigated. The filling performance of spiral-like micro cavities was studied as a function of runner size in combination with melt temperature, mould temperature, injection speed and holding pressure time employing the design of experiment approach. In addition, the results were analysed further to identify the effects of the runner size together with flow properties of polymers, PP and ABS, on the behaviour of the micro injection moulding process.

Keywords: micro fabrication, injection moulding, runner system, polymer processing

1. Introduction

Micro Injection Moulding (IM) of polymer materials is one of the key technologies for manufacturing micro devices as it provides reliable and cost effective means of producing micro components in large quantities.

Consistent replication is a key issue for serial manufacture and in the last 30-40 years, a rich repository of polymer processing knowledge was created in macro moulding. Unfortunately, such know-how cannot be employed directly in micro IM due to scale effects [1]. For example, for any given geometric shape, smaller objects have a higher surface to volume ratio (SV_R) than larger ones with the same shape. Such high SV_R increases the heat dissipation which affects the moulding performance of polymers. Thus, existing tool and part designs and polymer processing methods should be applied cautiously in micro IM and most likely some proven designs and processing strategies at macro scale should be re-considered taking into account these scale effects [2].

This observation also applies when considering existing methodologies for designing runner systems [3]. In particular, the runner system is one of the most important basic elements of thermoplastic injection moulds [4]. Its main function is to facilitate the flow of molten material from the injection nozzle into the mould cavity. To increase productivity, often moulding tools incorporate multiple cavities and runner systems that are designed for producing many components from a single shot volume.

One of the most important conditions for consistent replication is the runner system to deliver a polymer melt to all cavities at same time and with as small as possible variations of pressure and temperature [5]. Therefore, for micro components the increased heat dissipation due to high SV_R should be a dominant factor in selecting the most appropriate runner system design. In addition, during the filling stage, a frozen layer is formed along the walls of the mould that affects the flow behaviour. In particular, a thicker frozen layer results in a lower flow of polymer melt, and as the flow reduces, the heat loss increases and thus the frozen volume, too. The resulting flow resistance can then lead to excessive pressure in order to fill the multiple cavities [6]. To avoid this, it is necessary to analyse the effects of different runner designs on the thickness of the formed frozen layer and employ monitoring techniques such as the measurement of maximum cavity pressure (P_{max}) during the filling stage [7].

Thus, the optimum design of runner systems is an important pre-requisite for the production of quality parts with the micro injection filling process. Therefore, this paper investigates the flow behaviour of the polymer melt in micro cavities with a particular focus on the relationship between the filling of micro parts and runner designs. The paper is organised as follows. The next section discusses the important factors in designing runner systems. Then, the experimental set-up and the test tool used to investigate the effects of runner sizes on the flow behaviour are described. Next, the design of experiments is discussed together with the approach adopted for analysing the results. Finally, the experimental results are presented and the relationship between runner sizes and the melt fill of multiple micro cavities is analysed.

2. Runner System

2.1. Design considerations

Several issues should be taken into account when designing runner systems. These include:
• *Polymer material.* Heat loss during the melt fill can prevent flow, so for high and low viscosity polymers an appropriate runner size is necessary. The heat loss in the material occurs firstly at the runner walls, where a vitrified layer of polymer acts as an insulation for the higher melt temperature (T_b) at the core of the flow. The selected T_b must be maintained long enough for the cavity to be filled completely. Once filled with the volume required, the temperature in the core should be high enough to apply the holding pressure. During the holding pressure time (t_h), the material is packed out in

the cavities long enough for it to solidify and counteracts any contraction during cooling.

• *Injection moulding machine*. The pressure, temperature the runner size, notably its cross section, results in T_b that is less affected by wall temperature. However, there are two economic implications that are associated with large runners and speed capabilities together with its minimum and maximum shot weights should be considered. The ratio of runner to part weights is important because micro part volumes with large or small runner systems can be outside the machine shot weight range.

• *Mould design*. This includes part size, number of cavities and the selected layout. The choice of the runner type must be based on the available tool space and include adequate distance between the part cavities. Available technologies/methods for machining the cavities can also influence the runner design, especially the runner size in order to minimise the tool manufacture cost.

• *Part design*. The cooling time of the runner and the part depends on their dimensions. In particular, an increase in. The first is that the runner cooling time can exceed that of the parts, and thus lead to an increase of the cycle time. Secondly, as the runner is not part of the final product this represents an extra material cost. An optimum runner should provide flow control within a reduced working area, and ideally should be as small as possible with a cooling time equal to that of the parts.

2.2. Runner cross section

The cross section has an impact on the thermal losses in the runner system, and thus on ensuring that an optimum viscosity is maintained for each specific material. Three main types of runner cross sections are typically used: round, trapezoidal and parabolic. Trapezoid and parabolic, in this investigation only the runner type with a circular geometry is studied as it is considered optimum in regard to temperature losses [8].

In addition, in micro IM it is difficult to estimate the optimum size of the runner, e.g. its diameter (D), based on the empirical knowledge that exists at the macro scale [3]. Therefore, the effects of the runner diameter on the behaviour of the micro IM process is investigated by taking into account material and process related factors.

3. Experimental setup

3.1. Part design and tool manufacture

The part used to analyse the runner size influence in the filling of micro cavities is a spiral (see Fig. 1(a)) that incorporates eight unequal sections with a total length of 29 mm and a cross-section of 500 x 250 μm (Table 1).

Table 1. Spiral lengths

Section	1	2	3	4	5	6	7	8
Length [mm]	1	3.5	2.5	7.5	1.5	6.5	0.75	5.75
Total [mm]		4.5	7	14.5	16	22.5	23.25	29

Three tools were manufactured with four identical

and symmetrically positioned micro cavities for replicating the spiral (Fig. 1(b)). Due to the symmetrical design, the branches of the runner to each part are balanced and its cross section is round with an overflow for the melt front. The diameter D of the runner cross section varies in the range from 1 to 3 mm for these three tools.

All three tools were made from brass and the cavities were machined using micro milling with surface finishes of 0.12Ra, 0.1Ra and 0.27Ra for the 1mm, 2mm and 3mm runners respectively.

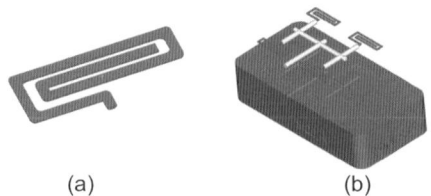

Fig.1. (a) The spiral (b) Tool configuration.

3.2. Condition monitoring

In this study, pressure and temperature variations in the runner area were investigated using a piezoelectric force transducer and thermocouples, respectively. Each of the three tools had been modified to accommodate the condition monitoring sensors as it is shown in Fig. 2(a).

In order to measure the pressure, a 1mm measuring pin (MP) and a force transducer were positioned in the centre of the runner system in the moving half of the tool as it is shown in Fig. 2(b).

Temperature readings were taken directly from the runner area of each tool. Two holes were drilled in the fixed half of the tool to accommodate 500μm diameter K type thermocouples as shown in Fig. 2(a). In particular, temperature readings were taken at the entry and at the end of the runners, and the difference between them was used as an indication of the thermal efficiency of the runner.

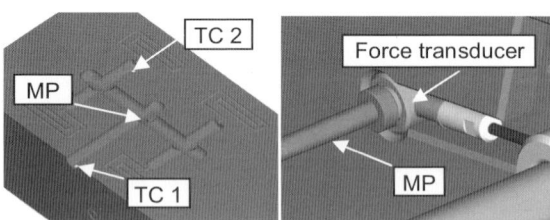

Fig.2. (a) The positions of thermocouples, TC1 & TC2, and measuring pin (MP) (b) The force transducer behind MP.

3.3. Design of experiments

The filling performance of micro cavities relies heavily on the speed and the temperature control during injection. Therefore, in addition to the runner diameter (D), the effects of T_b, mould temperature (T_m), injection speed (V_i) and t_h were investigated in this study. Taguchi orthogonal arrays (OA) were employed to ensure that the experimental results were representative of a broad processing window. In addition, by employing OAs the experimental results could be used further to optimise the process by identifying the best combination of processing parameters, and also the most significant of them in regard to the runner performance.

Two commonly used materials in IM, PP and ABS, were selected to conduct the planned experiments. For each combination of runner size and material used, given that four factors at three levels were considered, a Taguchi L9 OA was selected. The three levels of control for V_i and t_h were the same for all materials, while the levels for T_b and T_m were different (Table 2).

Table 2. L9 orthogonal array for PP and ABS

	t_h (s)		T_b (°C)		T_m (°C)		V_i (mm/s)	
	PP	ABS	PP	ABS	PP	ABS	PP	ABS
1	0		220	220	20	40	200	
2			250	250	40	60	500	
3			270	280	60	80	800	
4	2		220	220	40	60	800	
5			250	250	60	80	200	
6			270	280	20	40	500	
7	4		220	220	60	80	500	
8			250	250	20	40	800	
9			270	280	40	60	200	

The output of each experiment was assessed by measuring the flow length of the mouldings, the temperature and P_{max} in the runner cavity. Given that three runner sizes and two different materials are considered, six L9 OAs were defined. In addition, ten trials were performed for each combination of controlled parameters in these six OAs. Thus, in total 10 x 9 x 6 = 540 experimental trials were carried out.

3.4 Simulation

When a plastic is sheared, heat is generated. The amount of the released heat is determined by the product of viscosity η and shear rate γ. In particular, the higher the η and higher the γ the higher the heat generation. To determine the runner effects on τ and γ a finite element analysis (FEA) simulation using Moldflow software and a dual domain flow model was conducted. The simulation used experiments 1-3 from table 2, the t_h factor was omitted and a injection time factor (t_i) was used to replace the V_i factor. A total of three simulations are conducted for each runner and material.

4. Analysis of the results

4.1. Flow length

Fig. 3 presents a summary of the flow length results obtained from all 540 trials. Given that there are four cavities, the maximum and minimum average flow lengths achieved during the experiments are provided.

For the 3mm runner, the maximum average flow length show that both PP and ABS only achieved 90% filling of the cavities. Both materials had unequal filling for the four cavities while a higher variation between the maximum and minimum lengths was observed in the case of ABS.

For the 2mm runner, PP filled completely the cavities in all 9 experiments. This shows that this runner size was more efficient than the 3mm one. For ABS, the maximum filling achieved was 90%, which was similar to that observed with the 3mm diameter

runner while the minimum length was higher, 77%. Thus, for both materials the 2mm runner can be considered more efficient.

For the 1mm runner, PP filled completely the cavities in all 9 experiments. Thus, it is difficult to judge whether this runner size is more or less efficient than the 2mm one. However, it is evident that it is more efficient than the 3mm diameter runner. The maximum filling achieved in the ABS experiments was 79.5% while the minimum length was 72.4%. Although, the difference between high and low flow lengths is relatively small compared to the 2 and 3 mm runners, by looking at the maximum flow length results achieved with the three different runner sizes it is considered that for ABS the 1mm runner is the least efficient one.

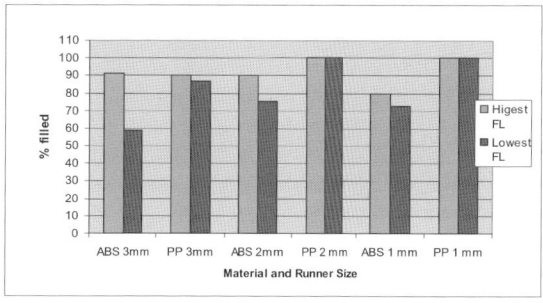

Fig.3. Maximum and minimum average flow lengths.

4.2. Temperature

For each trial, the temperature changes between TC1 and TC2 in the runner cavities was measured to judge about the size effects. Fig. 4 presents the average temperature variations for each combination of runner diameter and material.

For the 2mm and 3mm runners, a temperature increase between the beginning and the end of the runner system was observed for both materials, with a higher increase for PP in both cases.

For the 1mm runner, PP exhibited a marginal temperature increase while a decrease was observed in case of ABS. These results show clearly that the 1mm runner causes the lowest deviation from the set T_b of the three sizes investigated in this study.

If a temperature increase within the runner system is required in order to improve the filling, the 2mm runner can be regarded the best choice of the three sizes considered in this research.

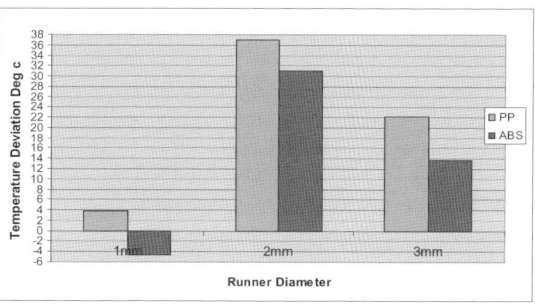

Fig.4. Temperature changes in the runner system.

4.3. Pressure

The runner size effects on P_{max} in the runner cavities were also analysed. Fig. 5 presents the average P_{max} for each combination of runner size and material.

Both materials were subjected to a higher P_{max} with the decrease of the runner diameter. In particular, in case of PP the average P_{max} is doubled with the decrease of the runner size from 3 to 2 mm.

From the carried out experiments, it can be concluded that to extend the tool life it will be desirable to use a bigger size runner because of the P_{max} reduction with the increase of the runner diameter.

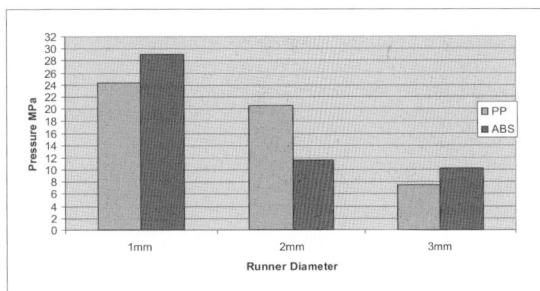

Fig.5. Runner cavity pressures.

4.4 Simulation

High shear rates tend to occur in the feed system due to the high volumetric flow, the average shear results taken from the three simulations for each runner and material identify that the runner size decrease resulted in an increase in both τ and γ (Table 3). The associated heat generation from such an increase could account for the temperature results for the 2mm runner (fig 3), though the heat generation appears to be counteracted by the SV_R for the 1mm runner.

Table 3. Simulation shear results

Runner	3mm	2mm	1mm
PP τ[MPa]	.035	.061	.094
PP γ [1/s]	144.9	318.0	590.6
ABS τ[MPa]	.090	.143	.273
ABS γ [1/s]	90.3	107	195.3

5. Conclusions

The paper reports an experimental study that investigates the flow behaviour of the polymer melts in micro cavities with a particular focus on the relationship between the filling of micro parts and the size of the runner system. The following conclusions can be made based on the reported research:

1. The flow length results for both PP and ABS showed that the 2mm size runner had the optimum surface to volume ratio and shear heating balance in regard to the filling performance. It is important to note that an increase of the runner dimensions did not have a positive effect because both materials failed to fill the micro cavities with the larger 3mm runner.

2. A temperature increase from the set T_b was measured for both materials and all three runner sizes, except for ABS with the 1mm runner system. The use of the 2mm runner resulted in the highest increase of the average temperature while the 1mm runner was the least subjected to temperature variations. For PP, the temperature variations in the runner system do not seem to affect the filling performance. In particular, the micro cavities were completely filled when using the 1mm and 2mm runners while not with the 3mm. This suggests that PP is not sensitive to temperature losses. On the contrary, the results for ABS suggest that the flow temperature affects the filling performance. In particular, the highest flow length was obtained when the highest temperature increase was recorded using the 2mm runner system. In contrast, for the 1mm runner the decrease in temperature led to the lowest flow length.

3. For both materials, an increase in pressure with the reduction of the runner size was observed. The use of the 1mm runner resulted in the highest pressure, with P_{max} doubled and trebled in comparison to the results obtained with the 3mm runner system for PP and ABS, respectively.

Finally, from the simulation shear results it is important to stress that in micro IM the polymer properties become an even more important factor in selecting the runner design. Experimental studies and simulations of runner melt flow behaviour should precede the tool manufacture.

Acknowledgements

The research reported in this paper is funded by the EPSRC Programme "The Cardiff University Innovative Manufacturing Research Centre" and the EC FP6 Project "Surface Enhanced Micro Optical Fluidic Systems (SEMOFS)". Also, it was carried out within the framework of the EC FP6 Networks of Excellence, "Multi-Material Micro Manufacture (4M): Technologies and Applications".

References

[1] Fleischer J and Kotschenreuther J. Manufacturing of micro molds by conventional and energy assisted processes. 4M2005 Conference on Multi-Material Micro Manufacture, Karslruhe, Germany, June 29 – July 1, (2005) 9-17.
[2] Yao D and Kim B. Scaling issues in miniaturization of injection molded parts. J. Manuf. Science & Engineering. 126(4) (2004) 733-739.
[3] Griffiths CA, Dimov SS and Brousseau EB. Micro injection moulding: the influence of runner systems on flow behaviour and melt fill of multiple micro cavities. Proc. ImechE (B): J Eng. Manufacture, submitted for publication.
[4] Javierre C, Fernandez A, Aisa J and Claveria I. Criteria on feeding system design: conventional and sequential injection moulding. J. Mat. Process. Tech. 171(3) (2006) 373-384.
[5] Yen C, Lin JC, Li W and Huang MF. An abductive neural network approach to the design of runner dimensions for the minimization of warpage in injection mouldings. J. Mat. Process. Tech. 174(1-3) (2006) 22-28.
[6] Spina R. Injection moulding of automotive components: comparison between hot runner systems for a case study. J. Mat. Process. Tech. 155-156 (2004) 1497-1504.
[7] Min BH. A study on quality monitoring of injection molded parts. J. Mat. Process. Tech. 136(1-3) (2003) 1-6.
[8] Tang SH, Kong YM, Sapuan SM, Samin R and Sulaiman S. Design and thermal analysis of plastic injection mould. J. Mat. Process. Tech. 171(2) (2006) 259-267.

Multi-Material Micro Manufacture
S. Dimov and W. Menz (Eds.)

221

A study of factors affecting the performance of micro square endmills in milling of hardened tool steels

P. Li[a], P. Aristimuno[b], P. Arrazola[b], A.M. Hoogstrate[c], J.A.J. Oosterling[c], H.H. Langen[a], R.H. Munnig Schmidt[a]

[a] Department of Precision and Microsystems Engineering, Delft University of Technology, Delft, the Netherlands
[b] Manufacturing Department, Faculty of Engineering, Mondragon University, Spain
[c] APPE Precision Manufacturing, TNO Science and Industry, Eindhoven, the Netherlands

Abstract

Proper setting of cutting conditions is critical for the performance of micro endmills in micro milling of hardened tool steels. In this paper, the influence of the cutting parameters on the wear behaviour of micro square endmills is presented. The selected parameters are cutting speed, depth of cut, and feed per tooth; Central Composite experimental Design (CCD) was used for a statistical analysis of the influence of these parameters. A quadratic model was fitted to describe the performance of the tool wear; the ANOVA analysis shows that the quadratic model gives a good prediction of the experimental results. On considering the magnitudes of the coefficients it is seen that the feed per tooth has a greater influence on the tool wear than cutting speed and depth of cut within the tested process window. By applying this method, the micromilling process can be planned to achieve an optimum tool wear performance for a tool-workpiece combination.

Keywords: micromilling, tool wear, design of experiments

1. Introduction

Direct machining of hardened tool steel by micro endmills has great advantages over EDM process in throughput time and manufacturing cost for the dies and moulds industry to produce micro moulds or macro moulds with micro features. However, as found in former experiments [1], the severe tool wear and premature breakage of micro endmills become the bottleneck for the development of this technology and make it difficult to apply micromilling in industry. Due to the tool problems, the workpiece quality is not satisfying requirements in terms of burr formation, surface quality and form accuracy.

There are several reasons for the poor performance of micro endmills, such as unsuitable tool geometry, unknown cutting conditions, lack of knowledge in machine tools, machinability of workpiece material, and wrong milling strategies. Among all these aspects, the cutting conditions play an important role for the tool wear/breakage. In [2] it was reported that the error in depth of cut led to tool breakage. On the one hand, because of the scaling effect, the micro cutting tools become vulnerable to excessive cutting conditions; big cutting force will break the tool directly from first contact. On the other hand, the process parameters for micromilling of hardened tool steel are unknown. There is no handbook available for micromilling as a reference. At this moment the selection of cutting parameters are mainly from the recommend values by tool suppliers, which are mostly based on trial and error practice.

Statistical design of experiments (DOE) is an efficient method for planning experiments so that the data obtained can be analyzed to yield valid and objective conclusions. [3] Well chosen experimental designs maximize the amount of "information" that can be obtained for a given amount of experimental effort. Moreover, the significant input variables can be identified through a proper design; therefore the process can be optimized in terms of the response.

Experimental design has been used in research to study conventional hard milling process [4] and surface roughness in micromilling [5]. No tool wear model for micromilling was seen from literatures.

In this paper, the DOE method was applied to study tool wear behaviour of micro endmills in milling of hardened tool steels on the purpose of a better understanding of the process and planning of machining parameters.

2. Experimental design and setup

2.1. Experimental setup and instruments

The experiments were done on a commercial micromilling machine, KERN EVO, which locates at the Mondragon University in Spain, shown in Fig. 1. The hybrid bearing spindle with speed up to 50000rpm was used in this study. The position accuracy on workpiece of the machine is about 2μm. The tool clamping interface is HSK 25; collet type tool holders are used. Nano type Blum laser tool setting system is equipped and Renishaw infrared touch probe is used for

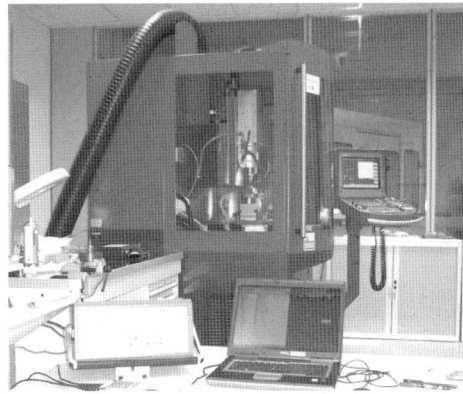

Fig. 1. Experimental setup.

workpiece coordinate setting.

The used cutting tools were 2-flute TiAlN coated Ø 0.5mm square carbide endmill, as shown in Fig. 2. The cutting length is 0.8mm; the helix angle is 30°. The cutting edge radius was checked by making a cross section of the cutting tool; the result is about 2µm. For comparison, these endmills were produced from same batch to avoid quality variation. Minimum Quantity Lubrication was used. The workpiece material is Böhler W300 (SAE H11) with 54 HRC; its chemical composition (%) is: C 0.37, SI 1.18, MN 0.35, P < 0.005, S 0.004, CR 5.01, MO 1.29, V 0.32. The dimension of the workpiece is 20×20×10mm (L×W×H). The surfaces of the workpiece were ground at forehand to achieve a good flatness.

To exclude the influence of complex geometries, simple slot milling was chosen to test the tool wear behaviour of the micro endmill. For a fair comparison, same amount of workpiece material ($20mm^3$) was removed at each different combination of cutting conditions; tool wear was measured after machining.

The dynamic runout at the cutting tool tip was measured by drilling a hole in resin under same speed as the test and checking the diameter of the drilled holes afterward. Stability of the process was checked by examining the milled surface, which was not a problem. The tool wear was measured by Keyence VHX-100 microscope and FEI Quanta 600 Scanning Electron Microscope; the resolution of the pictures is about 0.3µm/pixel, which depends on the magnification of the lens. Cutting force was monitored by Kistler MiniDyn 9256C2. The workpiece quality was checked by Mitutoyo Surftest 500 and WYKO NT 3300 white light interferometer.

2.2. Experimental design

In this research, Central Composite Design (CCD) is adopted to study the relation between cutting conditions (cutting speed v_c, depth of cut a_p, and feed per tooth f_z) and the wear of micro endmills, and to optimize the cutting conditions in order to plan the cutting conditions to achieve an optimum tool performance. Because it was known that micro tools show different dominant wear type/mechanism from conventional endmills [1], a quadratic relation is first assumed, and the interaction between input variables cannot be excluded at forehand.

CCD is an efficient design method for fitting second-order response surface equation. By proper choice of the number of centre points, the design will have some beneficial properties, such as orthogonality, rotatability, and uniformity of precision [6]. In this study, 6 centre points are used. Each input variable has 2

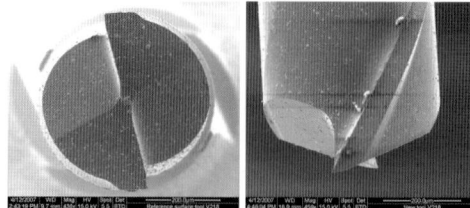

Fig. 2. Picture of a new Ø 0.5mm endmill.

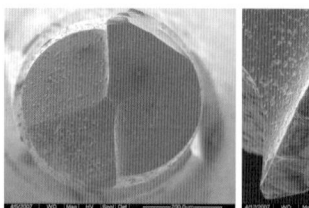

Fig. 3. Tool wear on end face and flank face.

levels, so there are altogether 8 (2^3) factorial points. Besides the 6 augmented points, there are 20 (8+6+6) experiments in total. The selection of levels for the input variables was based on former experiments; the maximum v_c is limited by the achievable spindle speed. The coded design variables are shown in Table 1.

3. Results and discussion

3.1. Tool wear type and measurement

During tests, no tool breakage happened under the tested conditions. Fig. 3 gives an example of the wear of the tested Ø0.5mm square endmill, both top view and side view. From this figure, it is seen that the main wear type of the micro endmill is the fracture of the two cutting edge corners, which changes the geometry of the endmill largely. For conventional endmills, the flank wear is normally used as the tool wear criteria; however from the side view of the worn micro endmill, it is clear that the flank wear is not dominant, and difficult to measure, because of which it cannot be used here to evaluate the wear of the micro endmill. Therefore, it is decided in this study that the wear of the micro endmill is defined as the reduction of the tool cutting diameter at the end face of the cutting tool.

The result of the tool wear measurement is shown in Fig. 4. The quality of the tool wear data was checked by 4-plot, namely Run sequence plot, Histogram plot, Lag plot, and Normal probability plot. The 4-plot showed that the process is in statistical control.

Table 1
Coded design variables

Coded variables x_i	Input variables		
	v_c (m/min)	a_p (µm)	f_z (µm)
-1.68	20.73	19.6	1.96
-1.00	31.42	40.0	4.00
0.00	47.12	70.0	7.00
1.00	62.83	100.0	10.00
1.68	73.51	120.4	12.04

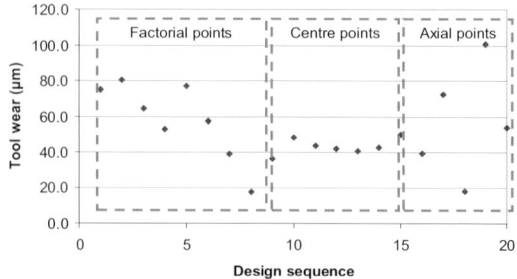

Fig. 4. Tool wear measurement result.

Table 2
Analysis of variance for tool wear

Source	d.f.		SS	MS	F
Total	8222.25	19			
Regression	8031.04	9		892.34	59.18**
Lack of fit	115.81	5		23.16	1.54
Pure error	75.39	5		15.08	

**Significant at the 5% level.

Therefore, these data can be used for further analysis.

3.2. Analysis of the results

A quadratic model is proposed to describe the relation between input variables and the response. The coefficients of the quadratic model were fitted by least square method. After regression, the model is:

$$\hat{y} = 131.65 + 0.25v_c - 88.09a_p - 15.32f_z + 0.01v_c^2 + \\ 1444.64a_p^2 + 1.41f_z^2 - 4.91v_c a_p - 0.09v_c f_z - 55.78a_p f_z \quad (1)$$

where $\hat{y}$ is the estimated yield, v_c, a_p, and f_z are input variables. The R^2 measures how well the regression fits the observed data. In this case, the R^2 is 0.9767, which shows that the model explains 97.67% of the variability in tool wear behaviour.

The result of ANOVA analysis is shown in Table 2, which gives the statistical significance of the regression by comparing the mean square of regression against the estimated value of the pure error. The F regression is 59.18, which is much bigger than the critical value of significance value at 5% level F (9/5)=4.77. This means that the yield of the experiments can be well described by a quadratic function. The F lack of fit is 1.54, which is smaller than the critical value of F test at 5% level F (5/5)=5.05; therefore no important terms are missing or misspecified in the functional part of the model.

In Eq. 1, the sign of the coefficients show the positive or negative influence of the input variable on tool wear. For example, for the main effects, the coefficient of v_c is positive, which means that the tool wear will increase with the increase of v_c. The coefficients of a_p and f_z are negative, which means that the tool wear magnitude will decrease with the increase of these variables. However the influence of the input variables is complicated by their interaction and the quadratic terms. Besides not all of these terms have same significant effect on the response.

The significance of all the coefficients in Eq. 1 was tested by observed t-test; it was done by dividing the absolute value of the coefficient value by the standard value of the coefficient. The observed t-result was compared with the critical t-value to decide if this coefficient is statistically different from zero. If the absolute value of the observed t-result is much greater than the critical t-value, it can be concluded that this coefficient is statistically significant. Otherwise the coefficient is not statistically different from zero; it can be omitted from the equation and its effect is pooled into the error term. The result of the t-test is that the f_z and its quadratic term, and the interaction $v_c f_z$ and $a_p f_z$ have significant effect on the tool wear; all other terms can be omitted. The adjusted tool wear model is shown in Eq. 2. The R^2 value of the new model is 0.96.

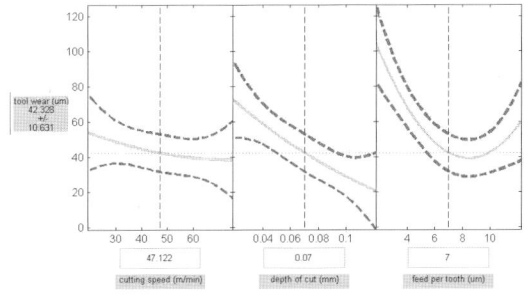

Fig. 5. Interaction plot of the response surface model.

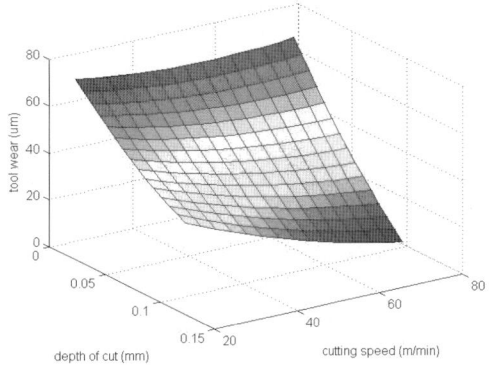

(a) f_z is fixed at 7μm

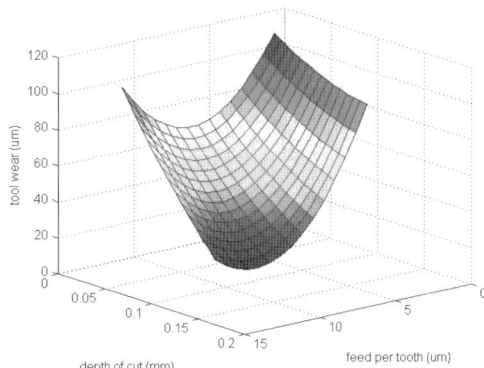

(b) v_c is fixed at 47.12m/min

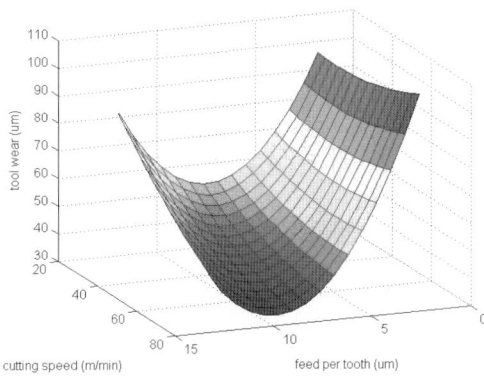

(c) a_p is fixed at 0.07mm

Fig. 6. Effect of input variables on tool wear.

$$\hat{y} = 139.25 - 16.03f_z + 1.38f_z^2 - 0.05v_cf_z - 70.90a_pf_z \quad (2)$$

Fig. 5 is the interaction fit and plot of the response surface model. Each plot shows the fitted relationship of the tool wear to the independent variable at a fixed value of the other two independent variables. The first plot has v_c as the independent variable. The second and third plots have a_p and f_z respectively. The 95% confidence intervals are also plotted on this figure.

From Fig. 5, it can be seen that within the tested process window, the tool wear decreases with the increase of the v_c and a_p. As described in the section 2, the material removal was kept same for each cutting condition set, the higher the cutting speed, the shorter the machining time; therefore it is reasonable to see the trend of decreasing tool wear with increasing v_c. Another reason for this observation could be that the tested speed range is limited by the spindle speed; the critical v_c for severe tool wear is not within the range of the test. When increasing a_p, the unit force on the cutting edge is kept same, but the machining time is short, so it is beneficial for the tool wear.

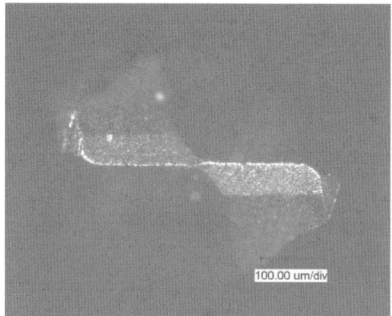

(a) v_c=47.12m/min, a_p=0.07mm, f_z=1.96μm

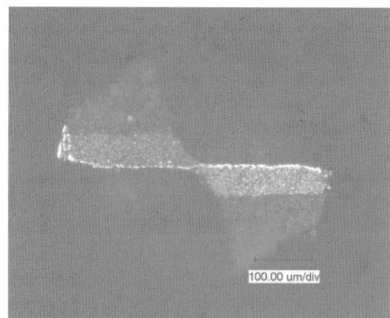

(b) v_c=47.12m/min, a_p=0.07mm, f_z=7μm

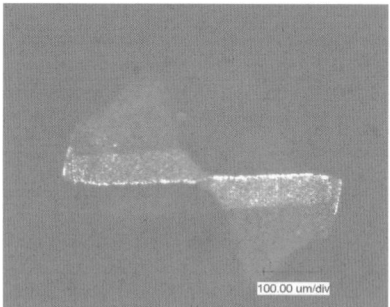

(c) v_c=47.12m/min, a_p=0.07mm, f_z=12.04μm

Fig. 7. Tool wear at different feed per tooth, (a) 100.6μm, (b) 42.2μm, (c) 54.0μm.

The effect of f_z has different effect on the tool wear from v_c and a_p. From Fig. 5, it can be seen that the tool wear magnitude decreases with the increase of f_z until a certain level, then further increase of f_z will lead to increase of tool wear. This is because that the increase of f_z will increase the unit force on the cutting edge. When f_z is above a critical value, the stress on the cutting edge corner will be too big and the cutting edge will be broken.

In Fig. 6 the relation between the input variables and the tool wear can be further studied. Fig 7 shows the tool wear at different cutting conditions.

From Fig. 5-6, it is seen that there exists an optimum value for f_z to achieve a minimum tool wear value when v_c and a_p are fixed. For example, when v_c is 62.83m/min (coded 1), a_p is 0.10mm (coded 1), the tool wear achieves a minimum value of 13.67μm at f_z of 9.51μm. By applying this method, the cutting conditions can be planned to achieve minimum tool wear for a tool-workpiece combination.

4. Conclusions

Design of experiment is used to study the effect of cutting conditions on tool wear in micromilling of hardened tool steel. It was observed that micro endmills show different wear type from macro endmills. From the ANOVA analysis of the magnitudes of the coefficients, f_z has a greater effect on the tool wear than that of cutting speed and depth of cut within the tested range. From the interactive response surface model plot, it is seen that there exists an optimum f_z value that achieves a minimum tool wear when v_c and a_p are fixed. By applying this method the cutting conditions can be planned to achieve an optimum tool wear under a tool-workpiece combination.

Acknowledgements

This work is being supported by the Innovation Research Program (IOP) of the Dutch Government and Launch Micro Project (FP6-NMP) of the European Committee, reference Nr. 11795.

References

[1] Li P, Oosterling JAJ and Hoogstrate AM. Performance evaluation of micromilling of hardened tool steel. Proceedings of the 2nd International Conference on Micromanufacturing, USA, 2007, pp. 219-224.

[2] Bissacco G, Hansen HN and Chiffre LDe. Improving axial depth of cut accuracy in micromilling. Proceedings of the Fourth Euspen International Conference, UK, 2004, pp. 386-387.

[3] NIST/SEMATECH e-Handbook of Statistical Methods, www.itl.nist.gov/div898/handbook, 2006

[4] Vivancos J, Luis CJ, Ortiz JA and Gonzalez HA. Analysis of factors affecting the high-speed side milling of hardened die steels. Journal of Materials Processing Technology, 2005, vol. 162-163, pp. 696-701.

[5] Wang W, Kweon SH and Yang SH. A study on roughness of the micro-end-milled surface produced by a miniatured machine tool. Journal of Materials Processing Technology, 2005, vol. 162-163, pp. 702-708.

[6] Petersen RG. Design and analysis of experiments. Marcel Dekker INC., New York, 1985.

Multi-Material Micro Manufacture
S. Dimov and W. Menz (Eds.)

225

Manufacturing and verification of tools for ECF

K. Hofmann[a], L. Staemmler[b], H. Kück[a,c]

[a] Institute of Micro- and Precision Engineering (IZFM), University of Stuttgart, 70569 Stuttgart, Germany
[b] now: Greiner Bio-One GmbH, 72636 Frickenhausen, Germany
[c] Hahn-Schickard-Institute for Micro Assembly Technology (HSG-IMAT), 70569 Stuttgart, Germany

Abstract

The electrochemical milling with ultra short voltage pulses (ECF) displays an important progress in micromachining of hard materials. Machining a workpiece with conventional milling the removal takes place by shape cutting. Therefore mechanical forces are applied to tool and workpiece. In contrast, using electrochemical milling, the material removal occurs by an electrochemical reaction. Therefore the workpiece as well as the tool are submerged into an electrolyte and the surface of the workpiece is etched by a galvanic current. Hereby the so called working distance is formed between tool and workpiece, which goes linear with the pulse amplitude and pulse on time in a first approximation. As a result, there are no mechanical forces applied to the tool. This allows the use of very thin tools.

To achieve the highest precision with this technique, it is necessary to manufacture very precise tools and to verify their shape and dimensions. In addition the use of rotating tools could be a promising strategy to speed up the ECF process and reduce the roughness. Therefore we introduce a method to produce very thin rotation-symmetric tools with high precision using the ECF technique. While the tool rotates the diameter is reduced by a one sided removal of material similar to machining with a turning lathe.

To verify the shape and the dimensions of these tools a commercial laser measuring system for tool setting and breakage control was integrated into the ECF machine. Algorithms to determine the tool diameter and the toolshape are installed. Further algorithms have to be developed to characterize more details of the tool like tilt and run-out error.

Keywords: ECF, rotational symmetric tool, manufacturing, measuring

1. Introduction

Miniaturization proceeds not only in the semiconductor industry but also in other branches like medical engineering and automotive industry. The allocation of micro devices is getting more and more important. With electrochemical milling with ultra short voltage pulses (ECF), invented by Schuster et al. [1,2], a new technique is available to manufacture structures with micrometer feature size in conductive materials. Tool and workpiece are submerged into an electrolyte and the workpiece is electrochemically etched by a galvanic current. The so called working distance is formed between tool and workpiece. In a first approximation it varies linearly with pulse width and pulse amplitude. As a result, there is no contact between tool and workpiece and therefore no mechanical forces are applied to the tool and hence no tool wear occurs. This allows the use of very thin tools within the ECF process.

For smallest structures both the tool diameter and the working distance can be reduced. Nevertheless for the latter the exchange of electrolyte within the working gap becomes more and more difficult. A promising strategy could therefore be the use of rotating tools to generate a flow of electrolyte around the tool. This will enhance the exchange of the electrolyte in order to speed up the process and reduce the roughness. The rotation of the tool demands an extremely precise positioning of the tool on the rotational axis to avoid a run-out error. Thus the tool is manufactured in the ECF machine itself using the ECF-process. Therefore the electrochemical parameters have to be adapted to machine the tool instead of the workpiece.

To achieve the highest precision with this tech-

nique, it is not only necessary to manufacture very precise tools but also to verify their shape and dimensions. This allows to adapt the NC-data to the measured tool diameter.

The aim of the experiments shown here is to manufacture and verify rotation-symmetric tools for the ECF process.

2. ECF Technique

ECF is a technique to produce structures with micrometer feature size into electrochemically active materials, even in very hard materials like stainless steel [3]. Therefore tool and workpiece are immersed into an electrolyte where they form a double layer capacitance on their surfaces. To prevent corrosion both tool and workpiece are hold at a constant cathodic potential. For the dissolution process in addition to the constant potentials short voltage pulses in the range of 10 ns to 200 ns are applied between tool and workpiece. During these pulses the double layer capacitances are charged over the resistance of the electrolyte.

Because the value of this resistance depends on the length of the path of the electric current in the electrolyte a large distance between tool and workpiece leads to a slow charging of the double layer capacitance whereas a small separation leads to a fast charging (fig. 1). Due to the fact that charging takes place only during the short pulses the double layer capacitance is charged to a potential high enough for an electrochemical reaction only in close proximity around the tool, the so called working distance. Areas of the surface that are farther away are not affected by the ECF process.

3D-forming of the workpiece is achieved by moving the tool similar to conventional milling machines. If the feed rate is higher than the dissolution the tool comes in contact with the workpiece forming a short circuit. That is detected by the ECF machine. The motion of the tool stops and an evasion strategy starts to release the contact.

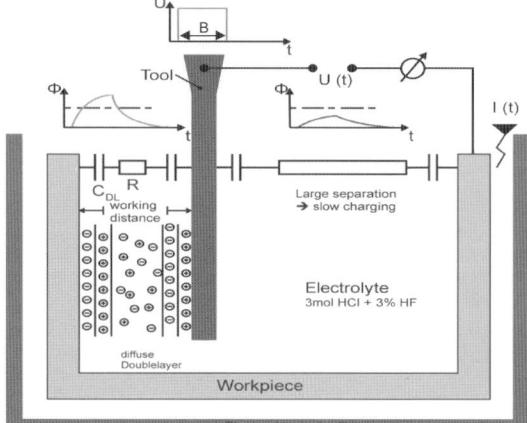

Fig. 1. Sketch of the electrochemical cell in the ECF process

To achieve the aim of the work presented here it is necessary to machine the tool instead of the workpiece. For that only the electrochemical parameters have to be adapted. This is mainly done by inverting the pulse and shifting the constant tool potential.

3. Experiments

All ECF experiments shown here were carried out on an ECF machine developed by ECMTEC GmbH (Holzgerlingen, Germany). It consists of a xyz-stage with a traversing range of 100 mm in each direction. The stage is driven according to CAD/CAM-data files that are processed by a PC-software. The potentiostat and the pulse generator were especially developed for the ECF process. The pulse generator fits into the processing head of the ECF machine. This is necessary to keep the distance between pulse generation and tool as short as possible. The adjustment range for the pulse width is between 10 ns and 1600 ns and different duty cycles can be set. The pulse generator also detects the contacts between tool and workpiece by measuring the current through the tool. The potentiostat is equipped with an input that stops the potential control while the tool is in contact with the workpiece.

The electrochemical cell is a basin made of PTFE where a platinum-wire (Pt-wire) is horizontally spanned. The wire acts as counter electrode for the machining of the tool and has a diameter of 100 µm.

A high precision spindle was added to the ECF machine with a concentricity smaller than 2 µm. Figure 2 shows the schematic setup of a tool. It consist of a straight W-wire (purity: 99.9+%) with a diameter of 200 µm. The wire is held by a plastic cone that is tightened via a modified Allan screw. Although the use of the plastic cone is harmful for accuracy it is necessary for the insulation of the wire and therefore it limits the surface areas charged by the pulses. Nevertheless the plastic cone increases the run-out errors of the W-wire.

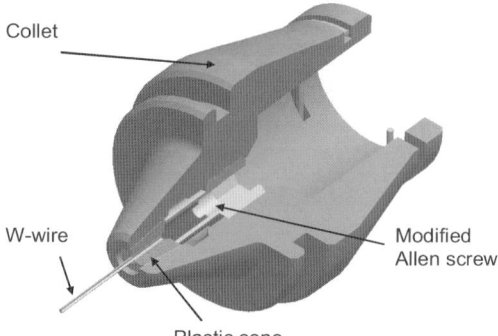

Fig. 2. Schematic setup of the tool

To lessen these errors the diameter of the W-wire is reduced in situ in the ECF machine. Here the tool rotates and the diameter of the W-wire is reduced by a one sided removal of material similar to the machining with a turning lathe. Therefore the tool is placed over the Pt-wire laterally shifted to the side, which is the start position for the tooling process. The machining setup is shown in fig. 3. The relative position between W-wire and Pt-wire can be determined by using the tool as a sensing device and touching the Pt-wire with the tip and the side of the W-wire while the tool is rotating. The contact is detected by measuring the short circuit.

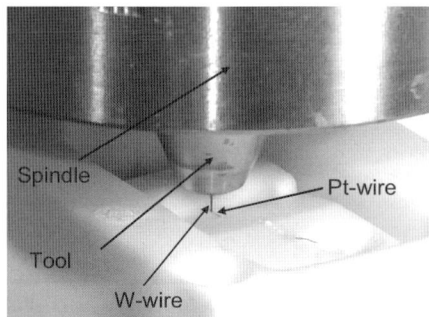

Fig. 3. Machining setup

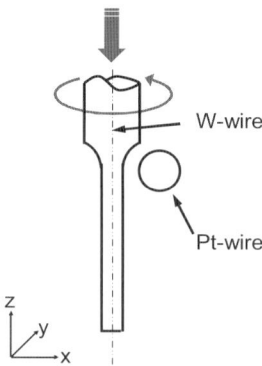

Fig. 4. Scheme of the strategy to reduce the tool diameter

While it is rotating, the tool is moving from the start position into negative z direction until it gets in contact with the Pt-wire. Then the contact is released by moving the tool backwards. As soon as the contact is released the movement stops and short voltage pulses are applied to the tool for a period of 30 s. The constant tool potential is set to 100 mV, the pulse on-time is 100 ns and the duty cycle is 1:8. During this time the electrochemical reaction takes place and the tungsten is dissolved until the distance between W-wire and Pt-wire is equal to the working distance. After this time the tool moves again into negative z direction until it gets in contact with the Pt-wire, releases the contact and the W-wire is machined for another 30 s and so on (fig.4).

Due to this lathe-like technique used to reduce tool diameter the achieved tool stay precisely on the rotational axis of the spindle. Furthermore, since the trimming of the wire is done in situ, the tool can directly be used without any respanning, keeping the tool in its precice position.

All the results shown in this paper were made in aqueous electrolyte containing 2 M NaOH.

To verify the shape of the tools a commercial laser measuring system (Blum-Novotest GmbH, Germany) is used to measure the diameter of the tool. It starts by the tip and moves along to tool axis measuring the tool diameter every 20 µm.

4. Results

4.1 Manufacturing of rotation-symmetric tools

To investigate the manufacturing of a rotation-symmetric tool a W-wire with a diameter of 200 µm was reduced down to 46 µm using a constant tool potential of 100 mV and 100 ns pulses with an amplitude of 2.4 V. The movement of the tool corresponds to the strategy described above. The counter electrode had a diameter of 100 µm. A SEM image of the resulting tool is shown in fig. 5 and fig. 6. Fig. 5 shows the tool in a side view. It can be clearly seen that the material removal does not take place symmetrically around the axis of the 200 µm W-wire but around the rotational axis of the spindle. Hence there is a shift between the centre axis of the W-wire and the thinner cylindrical part of the tool. This shift is the run-out error of the initial tool and in this case it is in the range of 50 µm. Mainly this large initial run-out error is caused by the use of the plastic cone to clamp the W-wire in the collet. It is difficult to manufacture this plastic part with an accuracy high enough for this application. Nevertheless it is necessary to manufacture this part out of a non-conductive material to insulate the W-wire from the other metallic parts of the tool to keep the surface areas charged by the pulses as small as possible.

The thinned part of the tool does not show a run-out error concerning the spindle axis anymore.

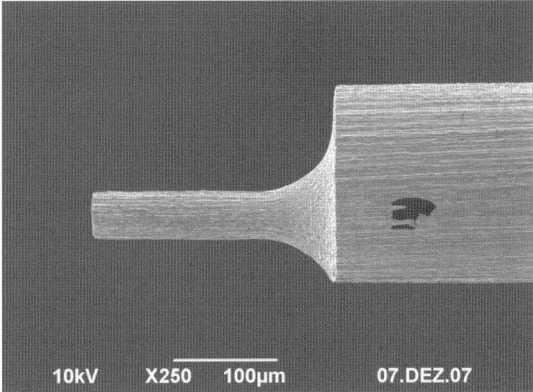

Fig. 5. SEM image of an rotation symmetric tool made with ECF

Fig. 6. SEM image of a rotation symmetric tool made with ECF

The surface of the machined part of the tool in fig. 5 and fig. 6 shows a high roughness, which is the consequence of the ECF process [4]. The roughness may be reduced be changing the process parameters. Using shorter pulses or pulses with lower amplitude the roughness of the tool should be decreased. In consequence this would reduce the material removal rate. On the other hand a rougher surface could have a positive influence on the exchange of electrolyte in the working gap using rotating tools.

4.2 Verification of tool shape and diameter

Fig. 7 shows the measured result of the tool diameter with the laser measuring system. The graph shows an xz-plot of the positions, where the tool interrupted the laser beam. The triangle data points represent the right side wall of the tool, the circles the left wall. Both give an impression of the tool shape. The squares show the tool diameter, i. e., the difference between the coordinates measured on the left and on the right side of the tool. Mark that to illustrate the tool's shape the scale in fig. 7 starts at 40 µm (see abscissa at the top). In all three cases the ordinate shows the distance between tool tip and measuring point. Measurement of the diameter results in a repeatability of 0.27 µm which was derived from a set of 100 measurements (data not shown).

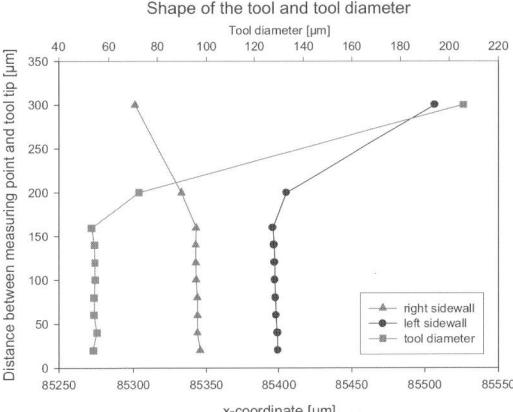

Fig. 7. Measurement of tool shape and diameter with conventional laser measuring system for tool setting and breakage control

5. Conclusion

To manufacture very thin tools with low run-out error for the ECF process a spindle was installed to the ECF machine with a true running accuracy of the spindle of less than 2 µm. Using this spindle it could be shown that it is possible to manufacture rotation-symmetric tools using the ECF technique although the tool has an initial run-out error. A tungsten wire with a diameter of 200 µm could be processed so that a cylindrical tool with 46 µm diameter and no run-out error results.

Further it could be shown that these tools can be verified using a commercial laser measuring system. The repeatability for this system is below 0.5 µm.

In further experiments the influence of tool rotation on the ECF process including the influence of the roughness of the tool surface will be investigated.

Acknowledgement

This research project (16IN0372) was funded by the Bundesministeriums für Wirtschaft und Technologie (InnoNet) and VDI/VDE-IT.

We also like to thank R. Schuster (Universtität Karlsruhe, Germany), T. Gmelin (ECMTEC, Germany) W. Reiser (Blum-Novotec GmbH, Germany) for the helpful discussions, comments and support.

References

[1] Schuster R, Kirchner V, Allongue P, Ertl G. Electrochemical Micromachining, Science 289 (2000) 98-101.

[2] Schuster R. Electrochemical Microstucturing with Short Voltage Pulses. ChemPhysChem 8 (2007) 34-39.

[3] Cagnon L, Kirchner V, Kock M, Schuster R, Ertl G, Gmelin WT, Kück H. Electrochemical Micromachining of Stainless Steel by Ultrashort Voltage Pulses. Z. Phys. Chem. 217 (2003) 299-313.

[4] Staemmler L. Mikrobearbeitung verschleißfester industrietauglicher Stahlwerkstoffe durch elektrochemisches Fräsen mit ultrakurzen Spannungsimpulsen. Abschlussbericht AiF FV-Nr. 180ZN (2007)

Multi-Material Micro Manufacture
S. Dimov and W. Menz (Eds.)

229

Electrochemical finishing of nickel microstructures

S. Kissling, K. Bade

Forschungszentrum Karlsruhe, Institut für Mikrostrukturtechnik,
Hermann-von-Helmholtz-Platz 1, 76344 Eggenstein-Leopoldshafen, Germany

Abstract

One method to manufacture high aspect ratio metallic microstructures is the LIGA technique. The acronym LIGA stands for the German words for lithography, electroforming and moulding. A resist layer (e.g. PMMA) is structured using deep X-ray lithography. The resist is developed and the resulting mould is filled with metal by electroplating. Though electroplating is an essential part of the LIGA process there are still challenges concerning the deposit surfaces. Nevertheless, extremely precise metal structures can be manufactured. In particular, the sidewall surface quality can be in the sub-micrometer range. But due to irregularities during the deposition process, resulting in rough or wavy surfaces, the emerging surface does not meet required tolerances. For this reason, a finishing process is necessary. Electrochemical techniques such as electro- or plasmapolishing have been evaluated. Electropolishing, a common anodic dissolution technique widely used in industry to obtain smooth, bright and burr-free surfaces, as well as plasmapolishing, also a technique based on the anodic dissolution are presented. First results of both an electropolished and a plasmapolished nickel microstructure are reported.

Keywords: nickel microstructures, electropolishing, plasmapolishing, LIGA

1. Introduction

The manufacturing of precise metallic micro components can be achieved by the LIG(A) process chain [1, 2], e.g. the combination of x-ray lithography and electroplating (Fig. 1). In many cases high precision could be achieved, but the growth front of the metal being deposited tends to roughen with increasing deposit thickness. Reduction of the roughness during electroplating is in some cases possible by use of complex additive chemistries or by more elaborate deposition techniques such as reverse pulse plating. Furthermore, the current density distribution is influenced by the pattern layout which may lead to an inhomogeneous thickness distribution on the substrate scale and also on the pattern scale itself. The inhomogeneous current distribution increases the waviness within the thickness profile.

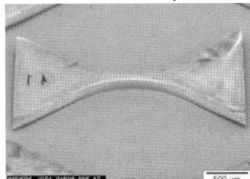

Fig. 1: SEM of metallic micro tensile specimen for fatigue investigations [3]

For characterization of the surface we chose the R_a roughness value which is often applied and used here for trend analysis. Ra is defined by Eq. 1, where L is the evaluation length and z(x) the height at a point x. For deeper discussion on surface roughness refer to [4].

$$R_a = \frac{1}{L}\int_0^L |z(x)|dx \qquad (1)$$

An additional finishing step improves the surface quality. Electrochemical finishing processes are well known for good surface finishing with reasonable results.

A very common technique in achieving burr-, stress-free and shiny surfaces is electropolishing. Another not so well established process for the electrochemical surface finishing is plasmapolishing. Both techniques are characterised in short and first results presented in this paper.

2. Experimental details

Finishing experiments were carried out with micropatterns. The PMMA pattern (Fig. 2), 320 µm thick, is filled with 120 µm nickel, deposited from a nickel sulphamate electrolyte at a current density of 10 mA/cm² [5]. For further LIGA processing details see [2]. The electropolishing was performed with remaining PMMA, plasmapolishing was carried out in the stripped state due to processing constraints.

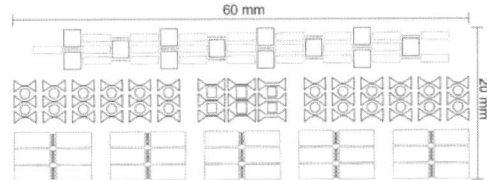

Fig. 2 Schematic of the layout with different micro tensile specimen

Roughness measurements were carried out with a profilometer (Tencor P2, tip radius 10 µm) with a scan range of 1000 µm (unless otherwise indicated).

2.1 Principle of Electropolishing

The electropolishing process, anodic dissolution, is often strongly simplified as reverse electroplating. An essential prerequisite to achieve a surface brightening is the presence of a passive layer and a high viscosity electrolyte boundary layer at the workpiece. Anodic dissolution in the active potential region would result in inhomogeneous removal.

Fig. 3 displays the principle. When direct current is applied under specific conditions to this electrolytic cell with an appropriate electrolyte at an appropriate temperature, the high frequency components of the height distribution of the workpiece are removed. This process, depending on time, will result in a rounding of the rough edges. Furthermore, brightening and levelling is only possible if all the necessary conditions (temperature, time, current density, electrolyte movement) are met. [6]

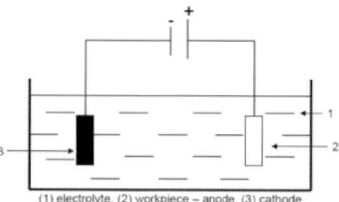

(1) electrolyte, (2) workpiece – anode, (3) cathode

Fig. 3: Schematic of electropolishing process

Electropolishing was performed on micro tensile specimens (see Fig 2). See Table 1 for further details on the parameters.

Table 1
Electropolishing conditions [6, 7]

Electrolyte composition	working conditions
Sulphuric acid* 60 vol %	60 °C
Phosphoric acid* 20 vol %	800 mA/cm²
Water 20 vol %	4 min

* concentrated acids were used

2.2 Plasmapolishing

Comparable to electropolishing plasmapolishing means passivation (oxide layer formation) and transpassive metal dissolution at the workpiece. It differs clearly in some parameters, e.g. high voltage (200 - 400 V) and lower concentration of the electrolyte components. When high voltage is applied at the workpiece a plasma discharge is visible at the electrode-electrolyte interface. Fig. 4 shows a schematic of the plasmapolishing equipment. To date, processible materials are ferrous materials, copper alloys as well as titanium and its alloys [9, 10, 11].

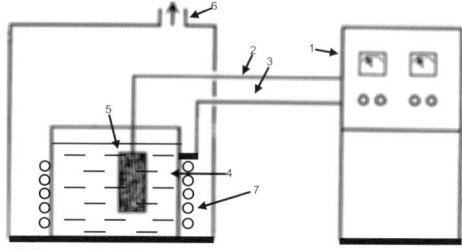

(1) Power supply unit, (2) connecting wire – anode (workpiece), (3) connecting wire – cathode (tank), (4) electrolyte, (5) workpiece, (6) exhaust system, (7) cooling

Fig. 4: Schematic of plasmapolishing equipment after [12]

Plasmapolished surfaces offer a bright and shiny surface with R_a < 0.1 µm (depending on the initial surface condition). A comparison between electropolishing and plasmapolishing is presented in Table 2.

Table 2
General comparison of

	Electropolishing	vs. Plasmapolishing
Electrolyte	concentrated	depleted
Viscosity	medium-high	low
Temperature	RT-60°C	80-85°C
(electrolyte/workpiece surface)		< 100°C
Voltage	up to 60 V	200-400 V
Removal rate*	30 µm/min	2.5 µm/min

* [10]

3. Results and discussion

3.1 Initial state

Fig. 5 displays the initial state of the microstructures after the electroforming process (120 µm). The surface appears dull, the average roughness R_a ranges between 0.6 µm and 2.1 µm. The uneven, wavy and grainy surface of the deposited nickel is visible in the SEM pictures.

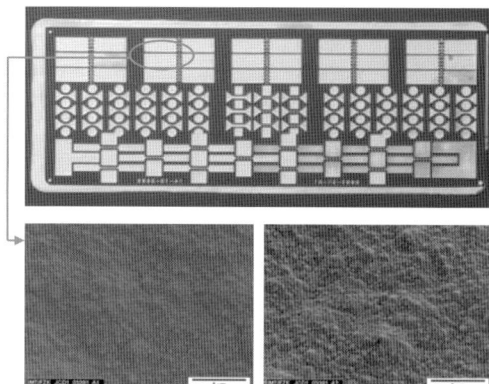

Fig. 5: Initial state of nickel microstructures The left picture indicates the top view, the right picture shows the surface under an angle of 30°

3.2 Electropolishing

After electropolishing the surface has a shiny and bright appearance though on closer inspection milky cords are visible over the entire microstructure (Fig. 6). A very fine-grained surface has developed. Furthermore the waviness was reduced from micro to macro scale.

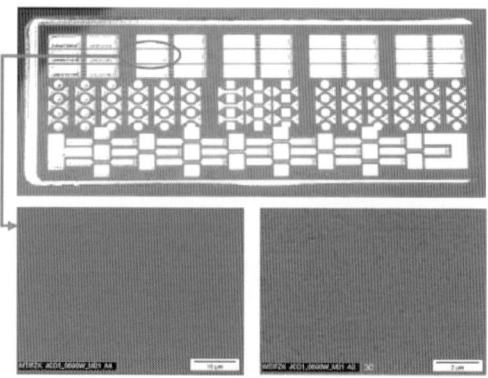

Fig. 6: After electropolishing

As expected two regions have formed during the electropolishing: the boundary area (higher material removal) and the inner area of the single micro specimen.

The average roughness R_a ranges between 0.2 µm and 1.2 µm. Fig. 7 is a schematic diagram of the R_a change before and after the electropolishing process. The greyscale indicates the increasing R_a. It can be seen that the R_a value diminished in the majority of the measurements. Nevertheless some R_a values increased during the electropolishing process see for example no. 16, 17, 28 or 32.

A reason for this might be the specimen size in correlation with the electropolishing effect (higher material removal rate at rough edges). Larger structures in the layout, no. 1 to 15 and 31 to 39, exhibit a layout mensural area of about 4.7 mm to 2 mm and 3.5 mm to 1 mm respectively compared to the 1.2 mm to 1.2 mm dimension of the smaller structures (no. 16 to 30). The boundary area forms a frame of approximately 500 µm at each single structure. With a measuring length of 1 mm at the larger structures the measurement took place in the inner area. The smaller structures were measured with a measuring length of 500 µm. But only the boundary area is present.

It can be assumed after all that the inner area is smoother. All the larger structures beside some outliers (no. 2, 32, 37) have a R_a value of 0.6 µm or better. The smaller structures have an average R_a value up to 1.2 µm.

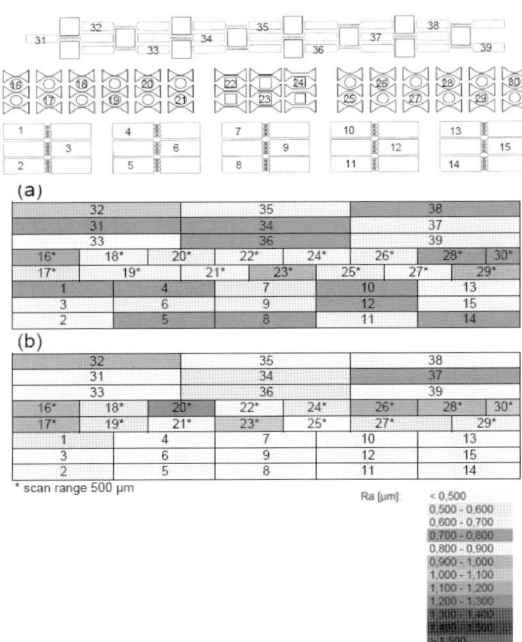

Fig. 7: R_a before (a) and after (b) electropolishing. The numbers indicate the measuring points on the layout and on the schematic

3.3 Plasmapolishing

After plasmapolishing the surface still has a dull but slightly brighter appearance. The average roughness R_a ranges between 0.5 µm and 1.0 µm. An additional patchy surface pattern is detected (Fig. 8). At the patch edge a step is visible in the SEM.

The grainy structure has disappeared, the surface is smoother. But some unevenness is still visible. The rough edges are not as sharp as in the initial state. A reason for this might be the stripped PMMA. An interesting point to notice is the step height up to 8 µm which can be seen in the SEM pictures (Fig. 8). Interestingly, the R_a values at the top and the bottom of the step are relatively close. Tab. 3 details some top and bottom values.

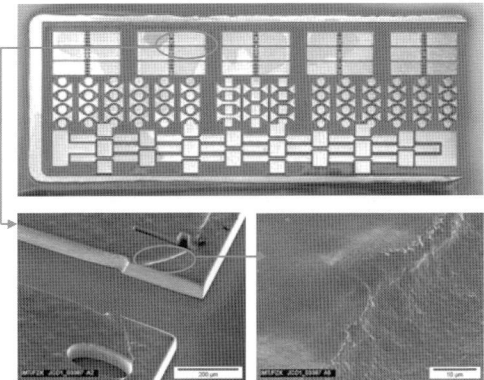

Fig. 8: After plasmapolishing

Tab. 3
R_a values of top and bottom at the step

R_a top [µm]	R_a bottom [µm]
0,5875	0,5975
0,5450	0,5950
0,3950	0,7900
0,6125	0,6725

Fig. 9 is a schematic diagram of the R_a change before and after the plasmapolishing process. The greyscale indicate the increasing R_a. Obviously the R_a values are lowered during the plasmapolishing process. The average roughness varies with position in the layout area. This variation is still visible after plasmapolishing, but the R_a is moderately reduced

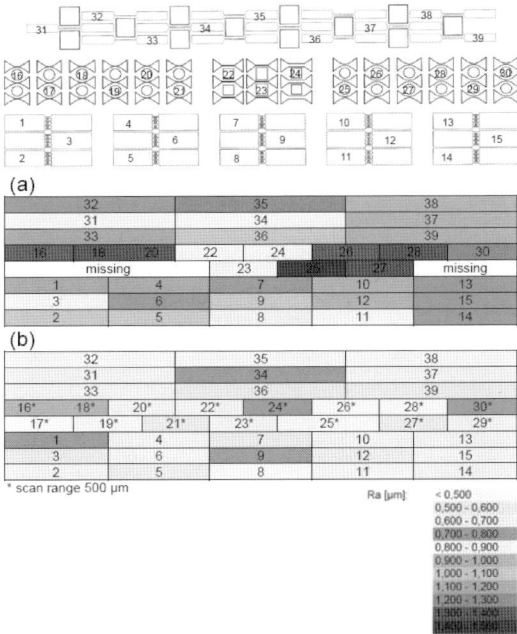

Fig. 9: R_a before (a) and after (b) plasmapolishing. The numbers indicate the measuring points on the layout and on the schematic

(range of R_a 1.4 µm → 0.5 µm) in the whole area. It is not completely understood why there are some very high R_a values (> 1.5 µm) at the small tensile specimens in the centre of the pattern, see for example no. 16, 18 and 20. A plausible explanation could be the roughness increase with increasing thickness. Indeed, at the above mentioned positions are showing large thicknesses. But the plasmapolishing process is also yet not clearly understood. Further investigation is necessary.

Nevertheless, the surface does not appear bright and shiny as it would be expected. The R_a has improved about 50% of the initial roughness value.

4. Conclusions

This paper describes two electrochemical techniques, electro- and plasmapolishing, to obtain a better surface finishing for LIGA micro patterns. Both electropolishing and plasmapolishing showed a surface modification. In general a reduction of R_a is achieved. Some effects were recognized:
- boundary layer and inner layer: this could be due to current density distribution or convection
- edge rounding: the different initial state - stripped or non-stripped PMMA – seems to be an explanation. This could be applied for remaining sharp edges during electropolishing
- increased R_a: a reason for this case might be the thicker deposition thickness due to current density distribution in the initial state.
Nevertheless both electrochemical finishing processes are a promising application yielding highly defined top surfaces of LIGA parts but the documented effects have to be studied in more detail in the future.

Also more work is necessary for a better understanding of measuring and characterising the surface of micro patterns. Difficulties in using the R_a value could arise due to short scan length resulting in falsified roughness or poor differentiation of waviness and roughness.

Acknowledgements

The authors would like to thank the Beckmann-Institut für Technologieentwicklung e.V. Oelsnitz, Germany for processing the plasmapolishing of the micro patterns.

References

[1] E. W. Becker et al., Microelectron. Eng. 4 (1986) 35-56.
[2] C. Khan Malek, V. Saile, Microelectr. J. 35 (2004) 131-143
[3] J. Aktaa et al., Scripta Mater. 52 (2005) 1217-1221
[4] J. M. Bennett, L. Mattsson, Introduction to Surface Roughness and Scattering, Optical Society of America, 1999
[5] J. Th. Reszat et al., Proc. Micro System Technology 2005, Franzis Verlag, Poing 2005, pp 200
[6] D. Landolt, Electrochim. Acta 32 (1987) 1-11
[7] H. Dettner, J. Elze: Handbuch der Galvanotechnik Band 2/2, Carl Hanser Verlag München, 1964, pp. 876 – 881, pp. 926 – 928
[8] P. V. Shigolev: Electrolytic and Chemical Polishing of Metals (second edition), Tel-Aviv: Freund, 1974
[9] Patent DE 10207632
[10] www.dtva-ev.de/Vortrag-Workshop-Plasmapolieren-210307.pdf (Jan. 2008)
[11] Beckmann-Institut für Technologieentwicklung e.V., www.beckmann-institut.de (Jan. 2008)
[12] A. L. Yerokhin et al., Surface & Coatings Technology 122 (1999) 73 – 93

Wire electro discharge grinding: surface finish optimisation

A. Rees[a], E. Brousseau[a], S.S. Dimov[a], H. Gruber[b], I. Paganetti[b]

[a] *Manufacturing Engineering Centre, Cardiff University, CF24 3AA, UK*
[b] *AGIE AG für Industrielle Elektronik, Losone, Switzerland*

Abstract

This paper investigates the technological capabilities of a micro machining process for performing Wire Electro Discharge Grinding (WEDG). In particular, micro Wire Electrical Discharge Machining (µWEDM) is employed in combination with a rotating submergible spindle to perform WEDG. In this paper, the effects of different factors on the achievable surface finish after WEDG are investigated. In particular, an experimental study employing the Taguchi parameter design method is conducted to identify the most important main cut machining parameters that affect the surface quality of the machined parts. Then, the obtained results are used to analyse the effects of the investigated parameters on the achievable surface roughness, and ultimately to select the optimum technological parameters for performing WEDG. The process parameters that statistically have a significant influence on the surface finish are presented. The study shows that by optimising the main cut machining parameters of WEDG a level of surface finish comparable to that of µWEDM can be achieved.

Keywords: Micro EDM, micro machining, WEDG

1. Introduction

Micro wire electrical discharge machining (µWEDM) is a widely employed material removal process used to manufacture micro components requiring intricate shapes and profiles with high levels of surface finish in the nanometre range. Unlike traditional cutting and grinding processes, which rely on the force generated by a harder tool or abrasive materials to remove the softer workpiece material, the EDM process utilises electrical sparks or thermal energy to erode the unwanted material and generate the desired shape.

By applying µWEDM it is possible to machine complex, ruled surfaces, and precision components employing wire electrodes with diameters down to 0.02 mm and achieve a surface finish down to Ra 0.07 µm. The electrode is usually a plain brass or coated wire, such as zinc coated brass or coated steel wires. Varying wire orientations can be achieved during machining by controlling the position of both the upper and lower guiding heads in the horizontal planes. In this way a variety of ruled surfaces can be generated. To perform the operation, both the workpiece and the wire electrode are submerged in either de-ionised, de-mineralised water or hydrocarbon oil [1]. The dielectric insulating properties aids in avoiding the electrolysis effect on the wire electrode during the EDM process [1]. In addition, the removal of material through µWEDM when the workpiece is submerged leads to temperature stabilisation in the processing area and efficient flushing especially in cases where the workpiece has varying thickness [2].

To broaden the application area of µWEDM and the range of parts manufactured applying this technology, a rotary submergible spindle can be added to allow the machining of cylindrical components. This type of machining is termed as WEDG [3]. Traditionally however, the WEDG process is applied in combination with EDM die-sinking [4] as illustrated in Figure 1. Such an implementation of WEDG has demonstrated extremely high surface finishes when used in conjunction with processes like lapping as reported by Masuzawa et al [5]. Generally though, surface finishes

for WEDG are in the region of Ra 0.8 µm [6]. Although the process of WEDG is capable of producing electrodes with a diameter of 5 µm when used in combination with EDM die-sinking [4], the main application for this technology is restricted to the manufacture of on-the-machine electrodes or pins. In particular, such traditional implementation of WEDG shows limitations for producing cylindrical components with high aspect ratios features.

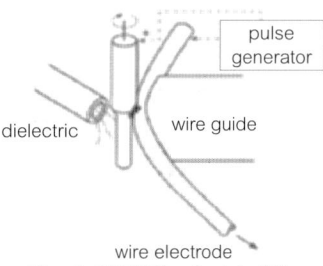
Fig. 1. WEDG principle [7]

In this research, the process of WEDG is investigated when implemented in combination with µWEDM. In this case, no wire guide is required at the point of contact between the electrode wire and the rotating workpiece. This characteristic provides a higher degree of design flexibility allowing the process to lend itself well to the manufacture of cylindrical components. Such WEDG implementation however is limited to the machining of diameters in excess of 60 microns [3].

Figure 2 depicts the difference in machining "footprints" between conventional µWEDM and WEDG after performing one main and three trim cuts operations if the process is not optimised for the specific process conditions that arise as a result of the workpiece rotation. In particular, in these trial cuts both tungsten carbide workpieces were produced using the same technological parameters that were developed for conventional µWEDM. The comparison clearly shows that a better surface finish is obtainable by performing µWEDM, approximately Ra 0.20 µm, compared to Ra

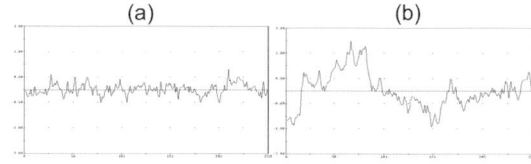

(a)   (b)

Fig. 2. Surface profiles resulting after (a) conventional µWEDM and (b) WEDG

0.51 µm achieved by WEDG. It is not difficult to conclude that the selected processing parameters do not take into account the changes in the EDM conditions due to the workpiece rotation and thus are not suitable for WEDG. Therefore, it is important to study the factors that have such a detrimental effect on the resulting surface integrity.

For this reason, an experimental study that investigates the parameters affecting the process behaviour in WEDG when used in combination with µWEDM is reported in this paper. In particular, the objective is to demonstrate that by optimising its process parameters it is possible to produce axis-symmetric components with surface finish that matches that achievable when performing a conventional µWEDM.

2. Implementations of WEDG

As illustrated in Figure 1, the WEDG process relies on an electrical discharge between a travelling wire and a rotating electrode. Extensive research has been carried out on WEDG when implemented with µEDM. In particular, it is a proven method for on-the-machine manufacture of micro electrodes for EDM drilling and milling [4, 8].

In recent years, a number of researchers have investigated the implementation of WEDG with WEDM. For example, Piltz et al [3] studied the effect of three different approaches for producing cylindrical components through the process of EDM. In particular the process behaviour in terms of pulse stability, hydrodynamic behaviour of dielectrics and machine dependent gap and feed control were investigated. Attempts to optimise surface finish were not covered during this research.

A similar approach for machining of free form cylindrical parts was investigated by Qu et al. [9-10] that extended the capabilities of the conventional WEDM technology by introducing an additional rotary axis to the machine set-up. In particular, the effects of pulse on-time, part rotational speed and wire feed rate on the surface integrity and roundness of produced parts were analysed. However, the process was studied in the context of machining macro-components and thus, its findings are not directly applicable at the micro scale.

In the study conducted by Juhr et al. [11], the importance of the correct selection of process parameter for performing the main cut during WEDM was highlighted. The research concluded that the material properties and surface finish resulting after the main cut could be improved only marginally by performing follow up cuts. Therefore, when machining micro components employing the WEDG process, the surface quality obtained after the main cut is very important and determines to a larger extent the achievable final surface finish.

Therefore, this research investigates the effects of spindle speed, flushing pressure, pulse OFF time, open circuit voltage and pulse ON time on the resulting surface finish after performing the main cut in WEDG. An experimental study was carried out employing the Taguchi parameter design method in order to identify the most important factors affecting the surface quality. The obtained results were used to analyse the individual contributions of these factors on the achievable surface roughness and also to select the optimum technological parameter for performing WEDG.

3. Experiments

3.1. Machining set up

To study the effects of the main cut technological parameters of WEDG on surface finish a rotating submergible spindle is added to a conventional machine for µWEDM. Figure 3 shows the system configuration and its working principle. A 50 µm brass coated steel wire was used as an electrode. A test piece 3 mm in diameter composed of 94% tungsten carbide and 6% cobalt was used in this experimental study. The workpiece material was produced through sintering with an average grain size of 0.3 µm. In the proposed experimental set-up, the test piece was fixed on the machine employing the collet of the rotating spindle.

3.2. Experimental Design

As it was already stated the main objective of this experimental study is to investigate the influence of a set of process parameters on the resulting surface roughness after WEDG. In particular, the parameters considered in this experimental study are:

- *Spindle speed.* The speed at which the workpiece is rotated.
- *Flushing.* Although the workpiece during the experiments was completely submerged, extra flushing was introduced through the upper and lower heads of the machine. The extra flushing does not only provide additional evacuation of the debris from the erosion area, it also helps the wire electrode to remain stable during the machining.
- *Pulse OFF time.* This is the time duration between the pulses. It is set by the power-supply controller of the machine.
- *Open circuit voltage.* The voltage between the electrode and the workpiece when the distance between them is too great to allow ionisation of the dielectric fluid.
- *Pulse ON time.* The time when the spark's electric current may flow as set by the machine power-supply controller.

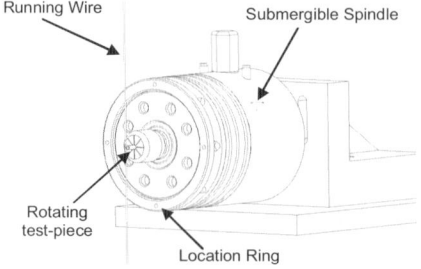

Fig. 3. Experimental design set up

These particular process parameters were selected because only their values can be modified by the machine operator when performing WEDG. In Table 1, the parameter values used to carry out the machining trials are provided. The range of each parameter ensures that consistent results are obtained from the process. Three values for each parameter were selected in order to study their effects in an experimental "window" as broad as possible.

Given that five parameters at three levels each are considered, it is possible to design 243 different experiments. However, by employing an orthogonal array (OA) based on the basic Taguchi L_{27} OA [12], the number of experiments was reduced to 27. These 27 experimental runs were carried out in a random order to minimise the influence of possible stochastic factors on the resulting surface finish after WEDG.

The surface topography was measured once at the same position on each machined cylindrical component by employing a white light interferometric profiling microscope. Then, in order to obtain the Ra measurements, the scanned profiles were analysed along a sampling length of 250 μm and a high-pass filter was applied to remove their waviness components.

Table 2 shows the 27 combinations of process parameters used in the experimental runs together with the achieved surface roughness with each of them. Due to time and cost constraints, the experiments were not repeated and thus, only one run was performed for each parameter combination.

4. Analysis of the results

4.1. Optimum parameters levels

For each parameter level, the average Ra value achieved was calculated to determine its effects on the resulting surface roughness. In particular, the value of a given parameter is considered to be optimum, the best of the selected three levels, if its corresponding average Ra roughness is the lowest [12]. Figure 4 shows the results obtained for the five analysed parameters in this experimental study.

By applying this method, it is possible to identify the theoretical best set of machining parameters within the investigated processing window in regards to achievable surface roughness, in particular:

- A3: spindle speed = 1500 rpm
- B3: flushing pressure = 2 bar
- C1: pulse OFF time = 42.5 μsec
- D1: open circuit voltage = 100 volts
- E1: pulse ON time = 4.5 μsec

4.2. Confirmation experiment

Given that the identified optimum combination of parameters did not correspond to any of the 27

Table 1. Machining parameters and their levels

Parameter	Code	Level 1	Level 2	Level 3
Spindle speed (rpm)	A	500	1000	1500
Flushing pressure (bar)	B	0	1	2
Pulse OFF time (μsec)	C	42.5	27.5	12.5
Open circuit voltage (V)	D	100	150	200
Pulse ON time (μsec)	E	4.5	28.55	52.4

Table 2. L27 orthogonal array

	A	B	C	D	E	Ra (μm)
1	1	1	1	1	1	1.32
2	1	1	1	1	2	1.09
3	1	1	1	1	3	0.64
4	1	2	2	2	1	1.66
5	1	2	2	2	2	1.70
6	1	2	2	2	3	2.12
7	1	3	3	3	1	1.27
8	1	3	3	3	2	1.58
9	1	3	3	3	3	1.70
10	2	1	2	3	1	1.09
11	2	1	2	3	2	3.20
12	2	1	2	3	3	2.09
13	2	2	3	1	1	1.34
14	2	2	3	1	2	0.89
15	2	2	3	1	3	1.14
16	2	3	1	2	1	1.02
17	2	3	1	2	2	0.80
18	2	3	1	2	3	1.54
19	3	1	3	2	1	1.05
20	3	1	3	2	2	1.17
21	3	1	3	2	3	1.04
22	3	2	1	3	1	0.87
23	3	2	1	3	2	0.96
24	3	2	1	3	3	1.18
25	3	3	2	1	1	1.51
26	3	3	2	1	2	1.05
27	3	3	2	1	3	0.93

experimental runs, one more trial was carried out in order to confirm its validity. The surface roughness achieved with this combination of parameters' values was Ra 0.57 μm. If this result is compared with the other measurements in Table 2, it is clear that the achieved surface finish is better than the lowest roughness, Ra 0.64 μm, obtained in the 27 experimental runs. Additional machining tests and surface measurements would be beneficial to confirm this initial result. However, this suggests that the identified combination of parameters' values within the considered processing window results in the best surface roughness.

In addition, a subsequent trim cut machining operation was performed on the workpiece and the surface roughness obtained was Ra 0.21 μm. This is comparable with the value of Ra 0.20 μm achieved with the conventional μWEDM by applying an optimised technology (see Section 1). This result demonstrates that after a thorough process optimisation the surface finish of components produced by WEDG can match the one achievable with the conventional μWEDM.

4.3. Significant parameters

Based on the results in Table 2, an analysis of variance (ANOVA) was carried out in order to assess the significance of each machining parameters on the achievable surface roughness. The results of ANOVA are given in Table 3.

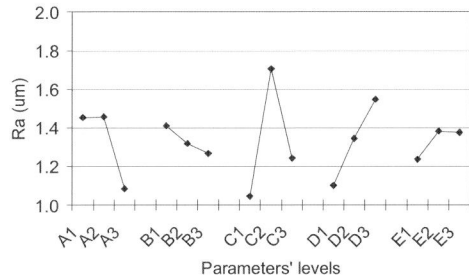

Fig. 4. Main effects

236

Table 3. ANOVA

Parameter	DoF	SS	V	F
A (spindle speed)	2	0.824	0.412	2.40
B (flushing)	2	0.095	pooled	-
C (Pulse OFF time)	2	2.061	1.031	6.01
D (Open circuit voltage)	2	0.905	0.452	2.64
E (pulse ON time)	2	0.122	pooled	-
Error	20	3.430	0.171	
Total	26	7.220		

Note: DoF denotes the degree of freedom, SS – the sum of squares, V – variance and F - variance ratio.

For each parameter, the variance ratio value, F, was compared with the values from standard F-tables for given statistical levels of significance. In this way, it was observed that:

- The parameters "*flushing*", "*pulse ON time*" and "*spindle speed*" had no statistical significance on the achievable surface roughness.
- The parameter "*open circuit voltage*" is statistically significant at 90% confidence level.
- The parameter "*pulse OFF time*" has the highest statistical significance in regards to the achievable surface roughness, a confidence level of 99%.

4.4 Comparison with µWEDM

The optimum results obtained from the investigation show that the process of WEDG requires the discharge energy to be reduced to allow the process to achieve a surface roughness that is comparable to the µWEDM process. To evaluate whether a reduction in discharge energy to the level used in WEDG will improve the surface roughness of conventional µWEDM, a further experiment was carried out. In particular, the surface roughness of two test pieces machined with µWEDM was compared. The first test piece was produced using set-up parameters optimised for WEDG and the second one using the conventional parameters optimised for µWEDM. The resultant surface finishes from the WEDG and µWEDM parameters were Ra 0.19 µm and Ra 0.20 µm respectively. The result demonstrates that for conventional µWEDM, reducing the discharge energy to the level required in WEDG is specific to that process.

5. Conclusions

In this study, the effects of five process parameters on the achievable surface finish after WEDG were investigated. The conducted experimental study showed that only the process parameters, pulse OFF time and open circuit voltage, have statistically significant effects on the obtainable surface roughness.

The results also demonstrated that by optimising the main cut machining parameters of WEDG it is possible to obtain a surface finish comparable to that achievable with a conventional µWEDM. In particular, by applying the identified processing parameters it was possible to achieve Ra 0.57 µm after the main cut and then to bring the roughness down to Ra 0.21 µm through a subsequent trim cut.

For the process of WEDG to obtain high degrees of surface finish, the technological parameters are significantly different from the process of conventional µWEDM. In particular, lower discharge energy is required to provide a lower value of surface roughness in WEDG. However, applying the same value of discharge energy to conventional µWEDM does not improve the surface finish.

Acknowledgements

The research reported in this paper is funded under the MicroBridge programme supported by Welsh Assembly Government and the UK Department of Trade and Industry, the EPSRC Programme "The Cardiff Innovative Manufacturing Research Centre" and the ERDF Programme "Supporting Innovative Product Engineering and Responsive Manufacture". Also, it was carried out within the framework of the EC FP6 Networks of Excellence, "Multi-Material Micro Manufacture (4M): Technologies and Applications" and "Innovative Production Machines and Systems (I*PROMS)". The authors gratefully acknowledge the support given to the Networks by the European Commission.

References

[1] Kunieda M, Lauwers B, Rajurkar KP, Schumacher BM. Advancing EDM through fundamental insight into the process. CIRP Annals, 54(2) (2005) 599 – 622.

[2] Ho KH, Newman ST, Rahimifard S, Allen RD. State of the art in wire electrical discharge machining (WEDM). Int. J. Machine Tools & Manufacture. 44 (2004) 1247-1259.

[3] Piltz S, Roehner M, Uhlmann E. Manufacturing of micro cylindrical parts by electrical discharge machining processes. Proceedings of the 1st Int. Conf. Micromanuf., ICOMM, (2006), 265 – 269.

[4] Masuzawa T, Fujino M and Kobayashi K. Wire electro-discharge grinding for micro-machining. CIRP Annals, 34(1) (1985) 431-434.

[5] Masuzawa T, Yamaguchi M, Fujino M. Surface finishing of micropins produced by WEDG. CIRP Annals, 54(1) (2005) 171 – 174.

[6] Rajurkar KP, Levy G, Malshe A, Sundaram MM, McGeough J, Hu X, Resnick R and De Silva A. Micro and nano machining by electro-physical and chemical processes. CIRP Annals, 55(2) (2006) 643 – 666.

[7] Fleischer J, Masuzawa T, Schmidt J, Knoll M. New applications for micro-EDM. J. Mat. Processing Tech. 149 (2004) 246 – 249.

[8] Zuyuan Yu, Masuzawa T, Fujino M. 3-D Micro-EDM with simple shaped electrode. CIRP Annals, 47(1) (1998) 169 – 172.

[9] Qu J, Shih AJ, Scattergood RO. Development of the cylindrical wire discharge machining process, Part 1: concept, design and material removal rate. Trans. of the ASME. 124 (2002) 702-707.

[10] Qu J, Shih AJ, Scattergood RO. Development of the cylindrical wire discharge machining process, Part 2: concept, surface integrity and roundness. Trans. of the ASME. 124 (2002) 708-714.

[11] Juhr H, Schulze HP, Wollenberg G, Kunanz K. Improved cemented carbide properties after wire-EDM by pulse shaping. J. Mat. Processing Tech. 149 (1-3) (2004) 178-183.

[12] Roy RK. A primer on the Taguchi method. Van Nostrand Reinhold, New York, 1990.

Multi-Material Micro Manufacture
S. Dimov and W. Menz (Eds.)

237

Improved bonding strength in hybrid micro parts by using plasma

W. Michaeli, T. Kamps

Institute of Plastics Processing at RWTH Aachen University, 52056 Aachen, Germany

Abstract

Micro injection moulding is established as one of the most common manufacturing processes for thermoplastic polymers due to its high degree of automation and the short cycle times. With micro assembly injection moulding, offline joining process steps can be avoided by overmoulding components of the micro system directly in the injection mould. Overmoulding can be used to generate movable or fixed combinations of different materials. One of the materials combined in a hybrid micro system is thermoplastic polymers, whereas the other one can be selected from a wide range of materials, e.g. technical ceramics, glass, or metals. Micro assembly injection moulding provides several advantages compared to other joining processes. However, functional integrity of the micro system is an end requirement. In the case of a joint micro structure, the bonding strength between two components affects the stability of the whole micro system and is thus important for the part's quality. As tests conducted at IKV show, a plasma treatment (plasma activation) of the insert parts significantly increases the bonding strength. Inserts of metal and glass have been overmoulded with several polymers, and the influence of different plasma gases and duration of the treatment on the feasible bonding strength is shown.

Keywords: micro assembly injection moulding, plasma, surface treatment, overmoulding

1. Introduction

Products in micro system technology distinguish high functional integrity. In many cases, this aspect can only be achieved by combining different materials to provide a new spectrum of advantageous properties. Joining processes which meet the special requirements of micro system technology are needed for this task.

Micro injection moulding is established as one of the most common manufacturing processes for polymeric materials in micro system technology. Thanks to a high automation level and short cycle times, injection moulding is established also in micro system technology to produce large numbers of micro components at low cost.

1.1. Micro assembly injection moulding

Due to more and more complex joining operations in micro system technology, the approach to use the injection moulding process for both forming and joining is pursued quite often. The benefit of this technology is a short process chain whose steps are run predominantly directly in the injection mould, as is shown in Fig. 1. Also shown are two example geometries: composite parts made from two polymers or one

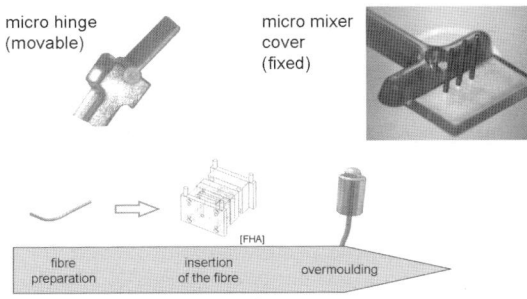

micro hinge (movable)

micro mixer cover (fixed)

[FHA]

fibre preparation — insertion of the fibre — overmoulding

micro assembly injection moulding

Fig. 1. The process chain of micro assembly injection moulding and example parts.

polymer and another (non-plastic) material like metal, ceramics, or silicon, can be manufactured in one process step. The joint can be movable or rigid, dependent on the material's compatibility and the joining forces in the joint area. Thus, the various options of micro assembly injection moulding differ in the materials being used, the operation sequence and the resulting functionality [1,2,3].

Based on the broad spectrum of properties that can be realised by material combination, possible areas of application are medical and communication technologies, and micro fluidics and automotive sectors [2].

Joint surfaces with functional properties are typical for hybrid micro systems as a result of the composite design. Depending on the design of the joining area, different bond mechanisms take effect where the materials meet. A criterion to estimate the bond quality of rigid hybrid parts is the tensile strength of the joint. When inserts are overmoulded by a polymer the bonding strength is mostly dominated by a force-fit as long as no positive joint is given. It has also been found that in such cases material compatibility can considerably enhance the effective forces [4,5].

Surface treatment techniques are well known to augment the adhesive forces between materials. This approach is helpful if the optimisation of the process alone no longer meets to the requirements. A compareison of different pre-treatment techniques in respect of micro injection moulding showed so far, that the plasma process leads to the most promising results [4].

1.2. The use of plasma in micro assembly injection moulding

Plasma is a gas in an excited state which consists among others of highly reactive particles. The plasma atmosphere is technically generated by charging a gas with energy, e.g. by means of a laser beam or an electrical field. In order to decrease the temperature of plasma, low pressure plasma processes are applied in a sealed off process chamber into which the process is

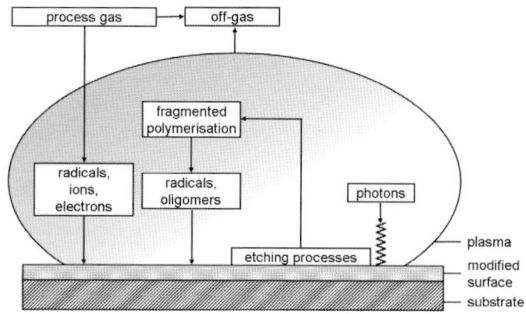

Fig. 2. The plasma process [6].

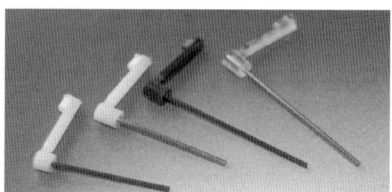

Fig. 3. Test specimen.

introduced: a vacuum pump keeps the pressure constant by simultaneously purging the chamber (Fig. 2). The most common process gases are oxygen, nitrogen, or inert gases like argon. Dependent on the substrate's properties which are exposed to the plasma, chemical reactive combinations are created on the surface, and a cleaning and removal of surface residues or even a polymerisation takes place. These effects overlay during the treatment and are of different intensity, so that an over-treatment is possible.

In multi-component injection moulding, the use of plasma is well known to improve or even enable the bonding of two (non-compatible) polymers. But also when a plastic and a non-plastic are combined, the effects of plasma are useful [2,4,6].

2. Objective and experimental setup

2.1. Objective

The objective of the trials is to optimise the encapsulation of non-plastic inserts. Criteria for a good joint between the insert parts and the overmoulded polymer are bonding strength, leakproof tightness, temperature resistance, and flash free surfaces. For investigations on the bonding strength, a cylindrical test specimen made of plastics and different insert materials is used, similar to Fig. 3.

2.2. Plasma treatment of capillaries

The capillaries used for the plasma trials measure a diameter of 0.4 mm and a length of 30 mm. After injection moulding, the overlay with plastics can be

chosen from 0.5 mm to 5.0 mm.

The capillaries were cleaned with acetone in an ultrasonic bath and dried overnight in an oven prior to the plasma treatment. The plasma was generated in PICO laboratory equipment from Diener Electronic GmbH, Nagold, Germany. As summarised in Table 1, the inserts were exposed to the plasma of different process gases over various lengths of time. The generator power of the equipment is 100 W, the process pressure was kept at about 0.3 mbar.

2.3. Injection moulding trials

The pre-treated capillaries were placed into the cavity of the injection mould, whose design is illustrated in Fig. 4. The mould is equipped with a hot runner which was run at the nozzle temperature of the plasticising unit of the injection moulding machine (Ferromatik Milacron ELEKTRA 30). Due the time-limited effect of plasma on the surface (see Chapter 3.1.2), the overmoulding had to be performed within two hours after the treatment. When overmoulding the glass capillaries, the melt tempera-ture was kept as low as possible in order to quickly solidify the polymer around the insert and thus avoid a frontal loading which would cause the capillaries to break. The inserts made from steel could be overmoulded with standard temperatures.

2.4. Determination of the bonding strength

The joined specimens were tested on a standard tensile testing machine. The pull-out velocity was set to 10 mm/min. During the test, the insert part is pulled out of the plastics part at a constant speed of 10 mm/min. The displacement of the insert and the resulting force are recorded. The result of the testing is a force-deformation diagram, in which the maximum force indicates the failure of the structure. This value was used as a benchmark for all investigations.

movable half　　　　　　　　　　　　　**fixed half**

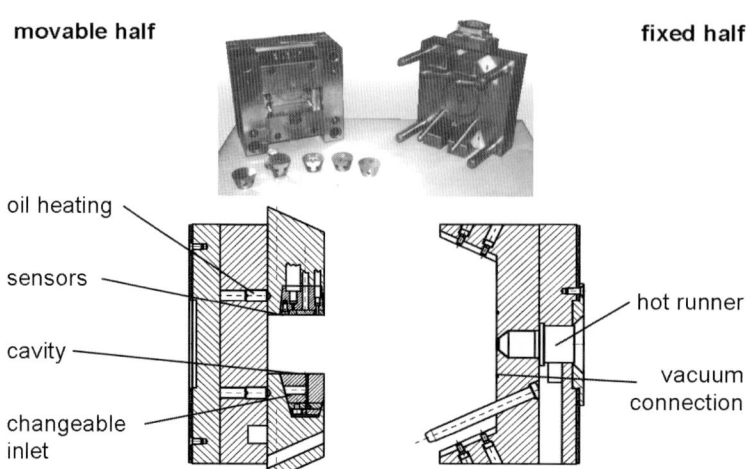

oil heating

sensors

cavity

changeable inlet

hot runner

vacuum connection

Fig. 4. Design of the injection mould.

Table 1
Overview of the experiments with plasma.

treatment duration	insert material / plasma process gas and polymer			
	Steel		glass	
	gas	polymer	gas	polymer
–	Ar, O_2	POM Group I[a]	O_2	POM, PA Group II
3 min	Ar, O_2			
5 min	Ar, O_2		O_2	POM, PA Gr. II
7 min	Ar, O_2			
10 min			O_2	POM, PA Gr. II
15 min	Ar, O_2	POM Gr. I		
20 min			O_2	POM, PA Gr. II
30 min	Ar, O_2			

[a] Polymers of Gr. I: ABS, PE-LD, PC; Gr. II: PP, PMMA

3. Results

3.1. Influence of the plasma process

3.1.1 Surface characterisation

For the comparison of the steel and glass insert surfaces prior to and after the treatment, the roughness, wettability, and the chemical constitution by means of electron spectroscopy for chemical analysis (ESCA) was examined.

To summarise the result towards a possible mechanical manipulation of the surface (roughness), when measured the etching effect of plasma on the pins, it could be seen that the process had little effect. Also optically, the surfaces did not change. Their appearance in REM is shown in Fig. 5.

Capable of determining the chemical constitution of the outer atomic layers to less than 10 nm the ESCA was used to analyse the plasma effect on the metal capillaries. With oxygen as a plasma process gas, the percentage of oxygen augments from 35 % to over 40 %, whereas the percentage of carbon drops from about 47 % down to 20–30 %. If argon is used as a process gas, both effects diminish. A one-digit percentage of argon dissolves in the substrate.

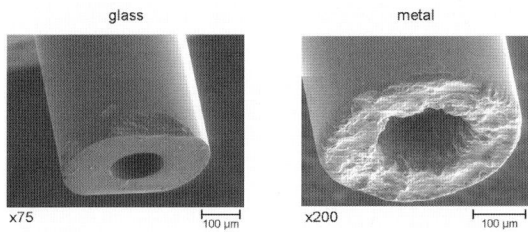

Fig. 5. Surface of the glass and metal inserts.

3.1.2 Influence of plasma process times

The time effect of the plasma process was further observed in order to identify suitable time-frames for the plasma process and the following joining with plastics using micro assembly injection moulding. The contact angle of water on treated and untreated metal plates was used as a criterion to measure the effect of the plasma activation (OCA). A good wettability is reflected in a small contact angle. Fig. 6 points out, how on the one hand the plasma process augments the wettability of the substrate and on the other hand the effect is time-dependent. The activation survives independently from the duration of the plasma treatment by at least one day, but during the first hours, the activation diminishes the most. The duration of the treatment is responsible for how fast the course of decay is reached. In order to assure constant starting conditions prior to the overmoulding of the capillaries, the joining had to be performed within one to three hours of the treatment.

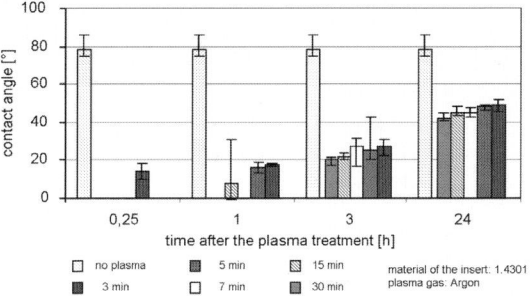

Fig. 6. Influence of the duration of plasma treatment and fading effect.

3.2. Bond quality of the overmoulded capillaries

As mentioned above, the bonding strength was used as a benchmark to quantify the quality of the joint of polymer and insert. The tensile forces for the composite parts with an insert made from stainless steel (1.4301) are displayed in Fig. 7. Here the overlay of polymer and insert was set to 4.0 mm. The values of untreated capillaries are opposed to ones which were treated over 15 min with oxygen plasma in combination with different polymers according to Table 1.

With all material combinations having been investigated, the tensile force after plasma treatment augments substantially. Apart from the parts with POM, the bonding strengths nearly double. Nonetheless, best absolute values are reached with POM. Additionally new tests with PA 6 polymer materials provide even better results.

Fig. 8 shows the complimentary results from the tests with glass inserts. The borosilicate glass

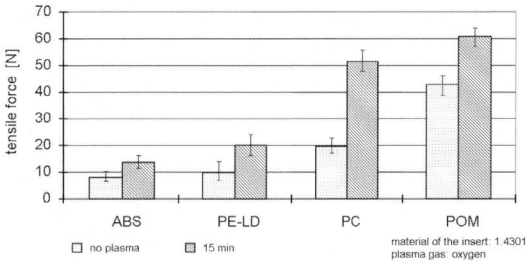

Fig. 7. Tensile forces of hybrid micro parts made of stainless steels and different polymers.

capillaries were overmoulded with an overlay of 2.0 mm. This diagram is extended with data concerning the time effect of the plasma treatment. Contrary to the results with overmoulded steel inserts, the effects in overmoulding glass inserts appear to be more complex.

If PA is used, the most significant outcome can be observed. In this case the bonding strength does not increase, similar to when steel is used, but drops down to a third of the original level. With PMMA as polymer, the course of the tensile forces is comparable to PA. However, the minimum value of the force is reached independently from the duration of plasma treatment in this case. If PP is used for overmoulding, the values are more or less not affected by the plasma treatment. The strength of the given hybrid structure of glass and polymer could only be enhanced by means of oxygen plasma when POM was used for the polymer component. However, the variations of the measurements are throughout very high.

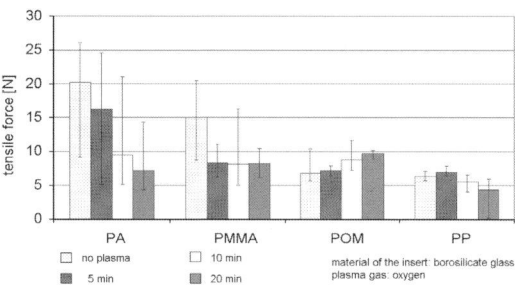

Fig. 8. Tensile forces of hybrid micro parts made of stainless steels and different polymers.

4. Discussion, summary and prospects

The results presented in this paper deal with the potential of the plasma technology to increase feasible bonding strengths of rigid hybrid micro parts (different polymers with glass and stainless steel) being processed in micro assembly injection moulding. As a criterion to quantify the bond quality, the tensile force of the structure was used.

Firstly the time effect of the plasma process was investigated. The contact angle representing the activation effect after the plasma treatment rapidly changes during the first few minutes after the treatment. The duration of treatment determines how fast the activation drops down to a level from which the effect decreases more or less constantly. The course of the data is typical for plasma activation and is mainly dominated by reactions of the (at the beginning highly active) substrate surface with the surrounding atmosphere, hence the activation effect decays.

The overmoulded steel capillaries throughout showed a better adhesive bonding with plasma treatment than without. Effects occurring here are a cleaning of the surface which correlates with the dropping percentage of carbon in the substrate's surface, and probably also a chemical change (activation by reaction) which supports the adhesive strength. The trend to augmented adhesive interaction involves a better bond quality. The results concerning the parts with overmoulded glass inserts are inhomogeneous, as explained above. While bonding strengths of parts with untreated capillaries are fine, especially with PA and POM, the tensile forces are smaller and more or less levelled after the plasma

treatment. A reason for this decrease may be the inert surface of glass; the cleaning effect of the plasma removes hydrocarbon residues on the substrate which develops adhesive bonding towards the polymer during overmoulding when omitting the plasma process. Closer information will be available with ESCA data of the glass inserts.

The trials regarding the capability of the plasma process for micro assembly injection moulding are presently completed with argon, hydrogen and nitrogen as plasma process gases. Further investigations on the chemical constitution of the treated substrate surfaces are presently run with the aim to gain more detailed information about the physical-chemical interaction between plastics and insert, and thus help to answer the question, which is the dominating reason for the improvement of the bond quality.

The effect of injection moulding process parameters on the feasible bonding strength was investigated in a former test series. As these results showed, POM grades are best suitable for the use in micro assembly injection moulding as the best tensile strength performance was reached with this polymer [2]. The capability of POM for the encapsulation of glass and steel inserts could be replicated in the works presented in this paper.

As a conclusion it can be said that using plasma is a highly practical approach to enhance bonding strengths in hybrid micro systems if process parameter variations alone do not lead to the desired strength performance. Additionally, the moulding process not only requires to be configured towards a good bonding strength of the hybrid parts but mainly towards a good quality of the polymer mouldings. Which plasma process settings are best suitable for which material combinations, is the main objective of further trials concerning potential industrial applications.

Acknowledgements

The research presented in this paper is conducted as part of the Collaborative Research Centre SFB 440 "Assembly of Hybrid Microsystems" and is supported financially by the DFG (Deutsche Forschungsgemeinschaft), to whom we extend our thanks.

References

[1] Ziegmann C.: Kunststofftechnische Prozesse für die Mikromontage. Dissertation, RWTH Aachen University, Germany, 2001
[2] Opfermann D.: Untersuchungen zur Verbundfestigkeit beim Mikro-Montagespritzgießen. Dissertation, RWTH Aachen University, Germany, 2007
[3] Schmachtenberg E., Johannaber F.: Assembly injection moulding – processing principles and definition. In: Handbook of the symposium Montagespritzgießen, Formschluss – Kraftschluss – Stoffschluss. Lehrstuhl für Kunststofftechnik, 2007.
[4] Michaeli W., Opfermann D.: Increasing the feasible bonding strength in micro assembly injection molding using surface modifications. Proceedings of the 64th ANTEC, Charlotte, May 7-11. SPE, 2006
[5] Amesöder A., Kopczynska A., Ehrenstein G.-W: Multi-component Processing: Plasma Makes for a Strong Bond. Kunststoffe International 2003/09
[6] Schmachtenberg E., Hegenbart A.: Plasma Processes: The Functionalisation of Plastic Surfaces, Kunststoffe International 2006/01

Multi-Material Micro Manufacture
S. Dimov and W. Menz (Eds.)

241

Force analysis in micro milling Al 6082 T6 in various engagement conditions

G. Bissacco[a], T. Gietzelt[b], H.N. Hansen[c]

[a] *Department of Mechanics and Innovation (DIMEG), University of Padova, via Venezia 1, 35131 Padova, Italy*
[b] *Forschungszentrum Karlsruhe Institut für Mikroverfahrenstechnik, 76021 Karlsruhe, Germany*
[c]*Department of Mechanical Engineering (MEK), Technical University of Denmark (DTU), Produktionstorvet 2800 Kgs. Lyngby, Denmark*

Abstract

This paper discusses the issues related to force measurement in micro milling and presents the results of the experimental investigation performed in an on going Cross Divisional Project within the 4M network of Excellence, aiming at force analysis and process characterization in micro milling. Reliable force measurement in micro milling is shown to be a challenging task. Measured forces are affected by contributions coming from the machining system. Based on the performed measurements, tool engagement has been demonstrated to occur at each tooth passing, even at feeds per tooth as low as 2 μm.

Keywords: Micro Milling, Cutting Force measurement, Micro Tools

1. Introduction

Micro milling is becoming an established process for manufacturing of micro mechanical components and components containing micro features in a wide range of materials. Main advantages of this process over competing processes used for micro fabrication are the capability to generate complex 3D geometries, the high material removal rate, the relatively low surface roughness and the wide range of machinable materials [1], [2]. Recent progress in the field of micro cutting is summarized in [3]. On one hand micro milling is a process widely available in precision mechanics workshops benefiting from the established knowledge for the corresponding macro scale process. On the other hand the mechanical interaction between the tool cutting edges and the workpiece material, which is the basis for material removal in micro milling, is responsible for tool deflections, tool wear and tool breakage which are difficult to handle in an ordinary production workshop. In fact as the size of the tool and the material removal unit are downscaled, a number of size effects arise and the whole material removal geometry changes as compared to conventional size cutting processes [4]. A review of the mechanics of machining at micro scale is presented in [5]. Cutting forces represent the fingerprint of the process as they are directly related to the process parameters, average cutting edge radius and integrity and wear of the tool. Thus cutting forces can be used for selecting process parameters, estimating the accuracy of the machined parts and in process monitoring of tool wear and process stability. It is apparent then that there is a strong need for documentation concerning micro milling forces, made widely available to machinists, process planners and tool designers. However cutting force measurement in micro milling is a difficult task and requires expertise and dedicated equipment. It is therefore a task better suited for research laboratories, while workshop engineers are the likely users of such information. In order to contribute to answer this need, a Cross Divisional Project [6] has been funded by the 4M NoE, with the objective to establish a basis for a knowledge database for cutting forces in micro milling

operations. The data collected is used within the project for identification of functional correlations between cutting parameters at micro scale, cutting forces, and machining accuracy, taking into account the different machining systems. Furthermore, the collected data will be made available for the development and validation of predictive analytical models.

This paper discusses the issues related to force measurement in micro milling and presents the initial results of the experimental investigation performed at the Technical University of Denmark within the framework of the abovementioned 4M Cross Divisional Project.

2. Characteristics of micro milling forces

The forces exchanged between the tool and the workpiece in micro milling processes are discontinuous and occur at each tooth engagement. The intensity of the force varies at each instant as the engaged tooth proceeds along the engagement angle, as a consequence of the variation of the uncut chip thickness. However cutting forces are not linearly proportional to the instantaneous uncut chip thickness due to the variation of the ratio between the uncut chip thickness and the cutting edge radius which generates a variation of the pattern of the plastic deformation within the workpiece material.

Cutting force measurement is important in order to generate the data necessary for proper process set up and optimization. However reliable measurement of forces in micro milling is a difficult task and requires a number of precautions. Indeed, with tool diameters as low as 100 μm, depths of cuts below 5 μm, and rotational speeds above 30000 rpm, cutting forces have amplitudes in the order of 10^{-2} N and frequencies in the order of few KHz. Therefore force measurement in micro milling is a dynamic measurement. Dynamometric platforms for such a task must have high sensitivity and very high signal-to-noise ratio, high resonance frequency in all measurement directions and be provided of a data acquisition system that allows sampling at frequencies of several tens of KHz simultaneously on each sampled channel. Furthermore a close correspondence between nominal and actual engagement conditions must be ensured, particularly with respect to the axial depth of cut, for which variations of few µm constitute large percentage variations. As a consequence, machine tools characterized by high positioning accuracy and repeatability must be used for the tests and special care must be taken to control or compensate machine tool thermal deformations.

3. Experimental set up

The micro milling experiments performed at DTU were carried out on a 3 axis vertical milling machine with a positioning repeatability of 1 µm along all axis, provided with a high speed attached spindle with a maximum rotational speed of 50000 rpm. A three components measuring platform Kistler Minidyn 9256C2, provided by FZK, was used for force measurement. Such dynamometer is characterized by a maximum measuring range of 250 N in all three directions and natural frequencies of 4.0 KHz, 4.8 KHz and 4.6 KHz in X, Y and Z respectively. The experimental set up is shown in Fig. 1. Milling tools selected for the CDP were two fluted, TiAlN coated flat end mills by Hitachi, with diameters of 0.5 mm, 0.2 mm and 0.1 mm. Tool neck lengths were 3 mm, 1 mm and 0.5 mm respectively, selected in order to be able to realize an aspect ratio of 5 for each tool diameter. For each tool, the diameter was measured at FZK by means of laser profilometer and visual inspection was carried out by means of SEM in order to ensure the absence of critical defects that could affect the force level during milling as for instance damages of the coating (Fig. 2). Tool run out was measured for each tool after chucking using an inductive displacement sensor with a resolution of 0.1 µm. Workpiece materials were an aluminium alloy Al 6082 T6, a maraging steel 1.2709 and pre-sintered zirconia. Although the complete experimental plan was carried out at DTU, involving all materials, tool diameters and process parameters combinations, in this paper only a sub-set of the acquired data is presented and discussed.

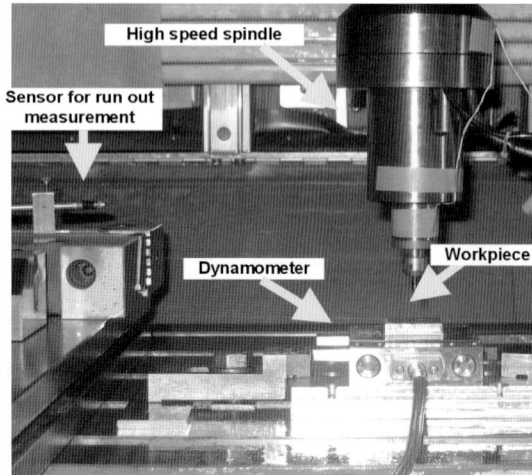

Fig. 1. Experimental set up used at DTU.

4. Experimental plan and measurement method

Force measurement was performed during machining in several engagement conditions obtained by varying selected machining parameters. Tool engagement in flat end milling is geometrically defined

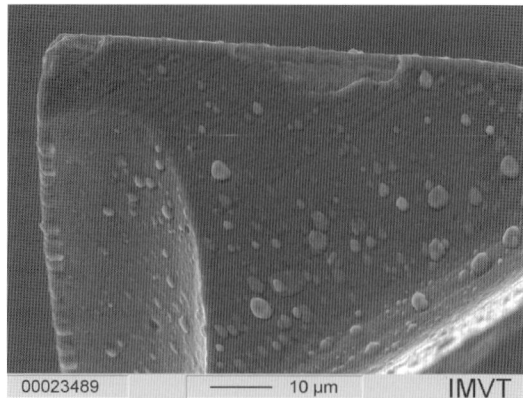

Fig. 2. SEM image of a 200 µm end mill showing coating damages at the cutting edges.

by the combination of axial dept of cut (a_p), radial depth of cut (a_e) and feed per tooth (f_z). In order to allow a comparison across different process sizes corresponding to different tool diameters, the engagement parameters were defined in a modular way, as a fraction of the tool diameter (a_p/D, a_e/D, f_z/D). According to the overall experimental plan, the varied parameters were the radial depth of cut, the feed per tooth, the tool diameter and the workpiece material, while the rotational speed and the axial depth of cut were maintained at constant value. The results presented in this paper refer to the sub-set of the experiments performed on the aluminium alloy 6082 T6 with end mills of 0.2 mm in diameter. The explicit values of the process parameters tested are reported in Tab. 1.

Tab. 1 Process parameters settings.

a_e/D	0.1	0.3	0.5	0.7	1
f_z/D	0.01	0.02			
a_p/D	0.05				
n [rpm]	32000				
D [µm]	200				
Milling mode	Down milling				

The machining layout consisted of a series of parallel passes, performed in down milling mode, at constant axial depth of cut, with different radial depths according to the values in Tab. 1. The machining passes were performed in the direction of the machine tool X axis, while the step over direction was opposite to the machine Y axis. The cutting fluid used throughout the tests was a commercial oil emulsion with oil concentration of 5% in volume. The direct application of the cutting fluid on the workpiece using conventional lubricant supply systems generated forces on the dynamometer far larger than those produced by the cutting action of the tool. Therefore it was necessary to apply the cutting fluid in small amounts on the workpiece surface before each machining pass. Furthermore the temperature difference between the fluid and the dynamometer top plate generated a variation of the preload of the dynamometer, inducing an apparent large signal drift. Therefore a suitable time was allowed for thermal stabilization between the application of the cutting fluid and the beginning of data acquisition. For each process parameter combination, a minimum of 3 test repetitions were performed and for

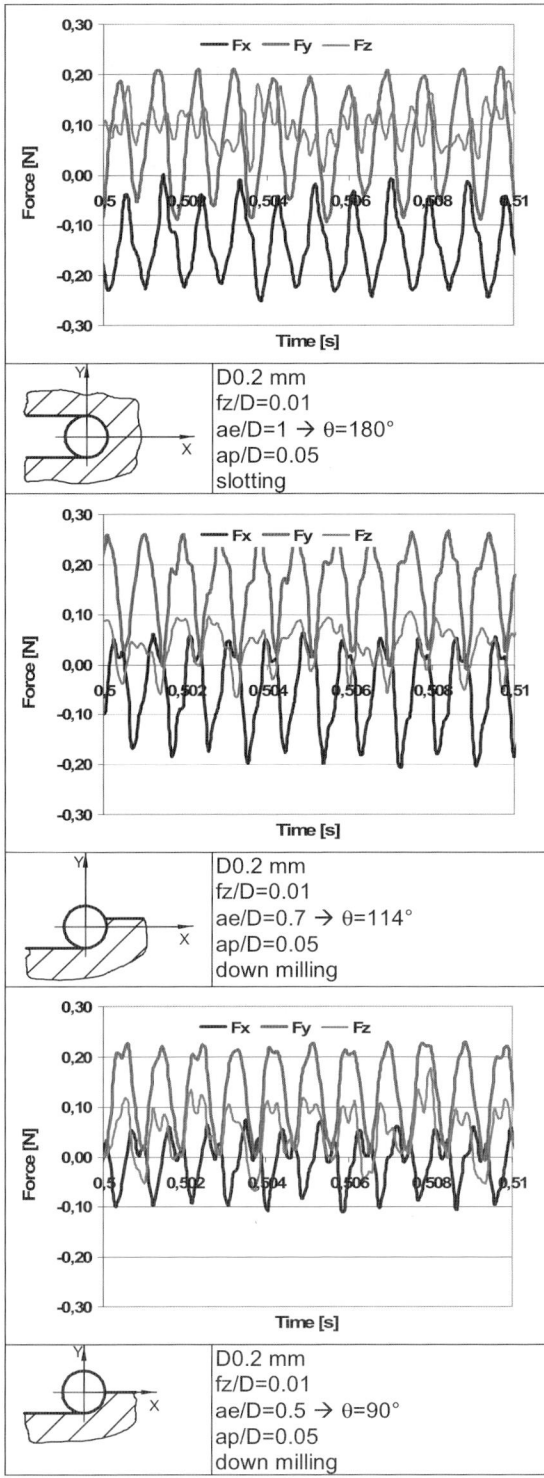

Fig. 3. Force profiles for a_e/D 1, 0.7 and 0.5.

each force acquisition approximately 2500 peaks were evaluated, in order to have a significant representation of the repeatability of the phenomenon. A dedicated CNC program was written, which allowed tool resting time in order to reset the system between each acquisition.

5. Results and discussion

The acquired data were corrected for drift and

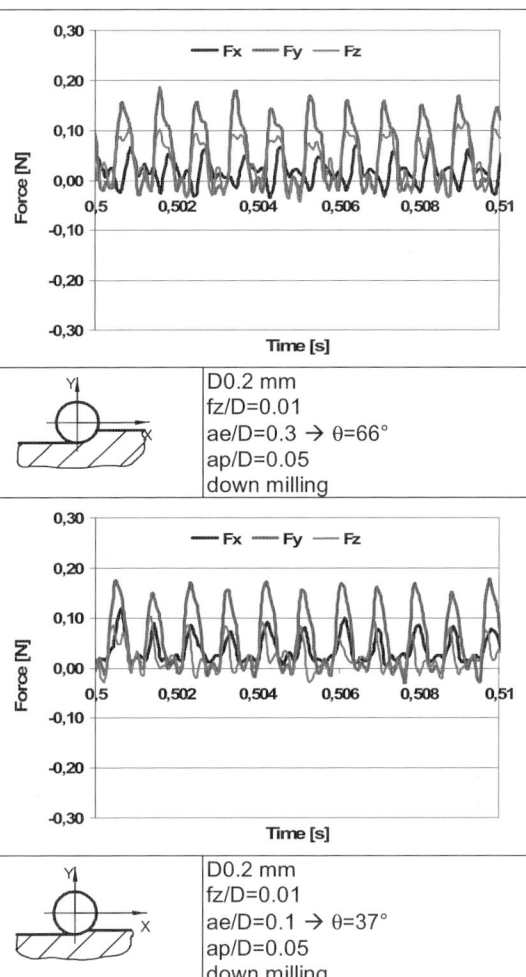

Fig. 4. Force profiles for a_e/D 0.3 and 0.1.

offset. Due to the limited dynamic response of the dynamometric platform, high frequency harmonics corresponding to the resonance frequencies of the dynamometer were superimposed to the force signals. The acquired data was then filtered with a low pass filter with a cut-off of 4 KHz. This is expected to leave the force signal unaffected with regard to the harmonics generated by the cutting action. In fact the tooth passing frequency corresponding to a rotational speed of 32000 rpm is 1.067 KHz, thus approximately one fourth of the lowest resonance frequency of the dynamometer.

It is important to notice that the measured signal is the result of a complex action that does not involve only the interaction between the tool and the workpiece, but also the machining system. Indeed what is measured is the force acting on the dynamometer. Since the dynamometer is mounted on a moving table, inertia forces, due for instance to straightness errors of the guideways, will add to the cutting forces. The decision concerning what belongs to the process and what belongs to the system is a critical one in the analysis of cutting forces in micro milling.

A series of profiles of the forces acting on the tool for f_z/D 0.01 and a_e/D from 0.1 to 1 (constant feed per tooth and different engagement angles) is shown in Fig. 3 and Fig. 4. For all engagement conditions displayed the force peaks corresponding to each tooth engagement are clearly visible. From the diagrams it can be noticed that the time between homologous points on two succeeding peaks corresponds well to the tooth passing period calculated on the basis of the nominal rotational speed. Furthermore the peaks have a rather uniform shape and height within the displayed time window. This demonstrates that the tool engages the material and removes a chip at each rotation, although the feed per tooth is only 2 μm and therefore of the same order of magnitude of the cutting edge radius. This has been a debated issue in literature, with some authors [7] suggesting that, at low feed per tooth, the tool slides over the workpiece surface and only every several rotations it actually engages the material, removing a chip much thicker than the nominal one. If this had been the case, higher peaks would have been clearly distinguished.

From Fig. 3 and 4 it is seen that peak forces for all components and all engagement conditions are always lower than 0.25 N. The peak value for the force component along the Y axis is almost unaffected for radial depths from 1 to 0.5. This is consistent with the fact that the maximum uncut chip thickness is the same in all three cases, while for lower radial depths the uncut chip thickness decreases, with a consequent decrease of the peak values, as seen in Fig. 4.

Fig. 5 and Fig. 6 show the average peak force as a function of the radial depth respectively for $f_z/D=0.01$ and $f_z/D=0.02$. The average forces and their standard deviation are calculated on the basis of three repetitions, thus taking into account the peak force variability observed within a single acquisition as well as that due to subsequent test runs. All the test runs have been performed with the same tool. As can be seen from the diagrams, the measured peak forces are characterized by a large spread. This is indeed a typical problem when performing tests at this length scale, where the transient condition of the machining system

and stochastic perturbation can affect the measurement results dramatically. The spread of the results is expected to increase further when more than one tool is used for the tests. The large standard deviation observed does not allow to draw sound conclusions, however, on the basis of the average peak values, a tendency towards an increase of the force with increasing engagement angle can be observed for the force component parallel to the feed direction Fx.

6. Conclusions

A subset of the results of an on going Cross Divisional Project within the 4M network of Excellence, aiming at force analysis and process characterization in micro milling, has been presented and briefly discussed.

Reliable force measurement in micro milling, with tool diameters as low as 100 μm and rotational speeds of more than 30000 rpm is a challenging task. Highly specialized equipment and a number of precautions are necessary. Besides the contribution due to the tool workpiece interaction, measured signals are affected by contributions coming from the machining system. Furthermore the variability of the micro geometry of nominally identical tools strongly affects the measurements. Based on the performed measurements, tool engagement has been demonstrated to occur at each tooth passing, even at feeds per tooth as low as 2 μm.

Acknowledgements

The authors would like to acknowledge the EU Network of Excellence 4M - Multi-Material Micro Manufacture, for the support given to this work.

References

[1] Alting L., Kimura F., Hansen H. N., Bissacco G., 2003, Micro Engineering, Annals of the CIRP, 52/2:635-657.

[2] Masuzawa, T., 2000, State of the art of micromachining, Annals of the CIRP, 49/2:473-488.

[3] D. Dornfeld, S. Min, Y. Takeuchi, 2006, Recent Advances in Mechanical Micromachining, Annals of the CIRP, 55/2:745-768.

[4] G. Bissacco, H. N. Hansen, L. De Chiffre, 2005, Micromilling of hardened tool steel for micro mould making applications, Int. Journal of Materials Processing Technology, Vol. 167, Issues 2-3, p. 201-207.

[5]] X. Liu, R.E. Devor, S.G. Kapoor, 2004, The mechanics of machining at the microscale: assessment of the current state of the science, Trans. ASME J. Manuf. Sci. Eng. 126, p. 666–678.

[6] G. Bissacco, T. Gietzelt, 2007, Setup of a data base for micro milling, for optimization of manufacturing of high aspect ratio features in multi materials using micro milling, electroforming and polymer replication, 4M Cross Divisional Project.

[7] X. Lai, H.Li, C. Li., Z. Lin, J. Ni, 2008, Modelling and analysis of micro scale milling considering size effect, micro cutter edge radius and minimum chip thickness, Int. J. Machine Tool & Manufacture, Vol. 48, p. 1–14.

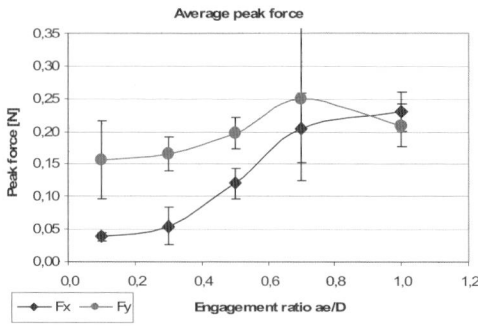

Fig. 5. Average peak force values for $f_z/D= 0.01$ at various radial depths.

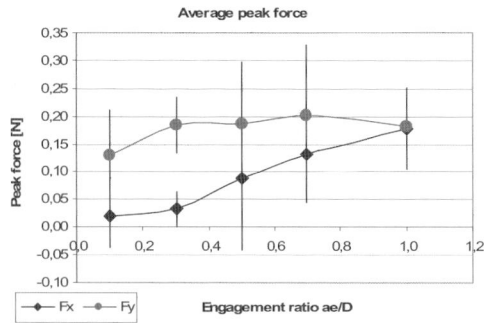

Fig. 6. Average peak force values for $f_z/D= 0.02$ at various radial depths.

Micro-Injection moulding: surface treatment effects on part demoulding

C.A. Griffiths[1], S. S. Dimov[1], E.B. Brousseau[1], C. Chouquet[2], J. Gavillet[2], S. Bigot[1]

[1]*Manufacturing Engineering Centre, Cardiff University, Cardiff CF24 3AA, UK*
[2] *French Atomic Energy Commission (CEA), Laboratory of Innovation for New Energy Technologies and Nanomaterials (LITEN), 38054 Grenoble, France*

Abstract

Micro injection moulding as a replication method is one of the key technologies for micro manufacture. The understanding of process constraints for a selected production route is essential at both the design stage and during mass production. In this research a tool surface treatment is used to study the effects of demoulding a part with micro features. In particular a tool coated with diamond like carbon (DLC) will be compared to an identical tool without coating. Through a range of experimental trials the effects of four process parameters, namely melt and mould temperature, and cooling and ejection time will be used to evaluate the demoulding process. Using two polymer materials PP and ABS, a special attention is paid to the forces present in demoulding and conclusions are made about the influence of DLC surface treatments and the factors affecting demoulding.

Keywords: Micro Injection Moulding, Surface treatment, Demoulding, Micro-fluidics,

1. Introduction

Microfluidic technologies have found a large variety of new applications in fields such as biotechnology, cytometry, medical diagnostics and micro chemistry. The successful development of such new micro devices is highly dependent on manufacturing systems that can reliably and economically produce large quantities of micro components. In this context, micro injection moulding (IM) of polymer materials is one of the key technologies for micro manufacturing.

In order to achieve an economical and reliable production of microfluidic parts it is important to study systematically the factors that affect the micro IM process. An important step in micro injection moulding which can affect the mechanical properties of the produced components is part demoulding. During the solidification process of the moulding cycle, the polymer melt shrinks onto the mould cavity walls and features. The part-mould forces that develop at this stage have to be overcome for subsequent part removal. To avoid yielding when breaking the bond between the polymer and the tool cavity, the maximum equivalent stress applied for part removal should not exceed the tensile yield stress of the material [1]. Thus, the factors that influence the demoulding process have to be studied carefully to avoid destroying parts and features and/or introducing further internal stress to a component through plastic deformation.

In this paper, the effect of different surface treatment on the demoulding behaviour of parts with micro features is reported. The paper is organised as follows. The next section discusses the important factors affecting ejection behaviour, especially part-mould forces and surface treatment methods. Then, the experimental set-up and test tool used to investigate the effects of cavity coatings on the demoulding forces are described together with the design of the carried out experiments. Finally, the experimental results are presented and the interdependence between tool surface treatment and demoulding forces in micro IM is analysed.

2.0 Demoulding factors

2.1 Part-mould forces

In polymer IM, the predicting the adhesion forces between the part and the tool is a complex task due to its dependence on part geometry and on process parameters such as the temperature and the pressure used during the process. Ejection forces (F_E), also called release forces (F_R), have been identified as a total friction between the tool and the polymer interface.

Previous research studies on IM forces and demoulding behaviour found that there are instances in which the friction effects can be difficult to explain. In particular, [2] showed that injection pressure did not affect F_E noticeably and that during processing the coefficient of friction (μ) is different to published data. [3] observed that the number of ejector pins affects the part-mould forces. More specifically, an increase in the number of ejector pins resulted in a reduced stress distribution in the moulded part. In another study, F_E was found to increase with the increase of the tool surface roughness [4]. In [5, 6], the holding pressure and surface temperature of the cavity were found to substantially influence F_E.

Together with high surface to volume ratio (SV_R) and high aspect ratio micro features, present challenges in micro IM call for the decrease of part-mould forces and tool wear, and thus to maintain optimum mechanical and structural stability for replicating quality parts and increasing tool life.

2.2 Tool Coatings

One method that can be used for improving the wear resistance of tool surfaces is to apply surface treatments. In particular, the wear of a surface can be reduced with traditional methods such as heat treatment and nitriding. In addition, previous research found that techniques like physical vapour deposition (PVD) and chemical vapour deposition (CVD) resulted in moulds with significantly better wear resistance [7-9]. At the same time, the surface quality of the moulded parts was improved due to reduction of the part-mould forces.

The surface treatment of tools using Pulsed Laser Deposition (PLD) of diamond like carbon (DLC) coatings results in tools with hard surfaces of up to 70Gpa. Optimisation of deposition can lead to DLC surfaces with μ in the range of 0.05-0.2μ, an order of magnitude lower than that of ceramic coatings [10]. Another role that tool coating can fulfill is to protect against undesirable polymer and tool interactions. In particular, metal tools employed to produce micro parts for medical products run the risk of releasing metal ions [11]. For example, nickel is a common contact allergen and at the same time it is a material that is commonly used for the manufacture of micro tools. By coating the cavities, a barrier between the tool and the polymer can be created. Furthermore, due to the amorphous nature of DLC coatings it is possible to introduce tunable antibacterial elements and thus to counteract contamination [12]. Based on the findings of previous studies, it is clear that surface treatments can reduce part-mould forces and tool wear. This research investigates the effects that the tool coating can have on part demoulding in micro IM.

3.0 Experimental set-up

3.1 Test materials

Two commonly used materials in injection moulding, Acrylonitrile Butadiene Styrene (ABS), and Polycarbonate (PC) were selected to conduct the planned experiments. The machine used to perform the micro injection moulding tests was a Battenfeld Microsystem 50.

3.2 Part design and tool manufacture

The part design used in this study is a 15mm x 20mm x 1mm micro fluidics platform (Figure 1). The design includes features commonly found in micro fluidics components such as reservoirs and channels. The pin dimensions are 500 µm in diameter and 600 µm in height, and the cross section of the main channels is 200 x 200 µm. Two identical tools were manufactured in brass. They were produced using micro milling. The moving and fixed halves of the mould were assembled to a primary mould and then inspected for parallelism and shut off of the mating faces.

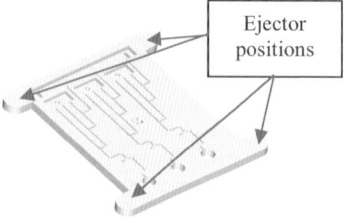

Figure 1. Micro fluidics platform and Ejector positions

3.3 DLC Surface treatment

The DLC thin film was elaborated in a Low Frequency Plasma Enhanced Chemical Vapor Deposition reactor. Prior to deposition, the substrate were cleaned first in acetone and ethanol by an ultrasonic washer and then in a Ar + H$_2$ etching plasma. In order to improve adhesion a Si-C:H intermediate layer (0.5µm) was deposited on the substrate using a plasma of tetramethylsilane (TMS) and argon. Then, 2µm DLC film was deposited onto the Si-C:H interlayer. Mechanical characterisations of this DLC coating were also performed and values are summarised in Table 1.

Table 1. Mechanical properties of DLC film

Properties	Values
Hardness [GPa]	22 ±2
Young Modulus [GPa]	160 ±10
Friction coefficient	0.05
Wear rate [mm^3.N^{-1}.m^{-1}]	5 10^{-7}

3.4 Force measurements

In this study, variations in force during the ejection stage of the IM process were assessed using a Dynisco PCI piezoelectric force transducer with a measuring range from 0 to 10,000 N. The sensor output signals were downloaded onto a PC using a National Instruments cDAQ-9172 USB data acquisition unit and the measured values were accessed through the National Instruments Labview 8 software. Each tool had to be modified to accommodate the force transducer. An ejector sub assembly was manufactured to house the four pins used for part removal. To carry out the force measurements, the transducer was positioned in the middle of the ejector plate sub assembly (Figure 2). When the ejector assembly moves the transducer is subjected to a mechanical load and proportional voltage that is monitored with a National Instruments NI 9205 16-bit module.

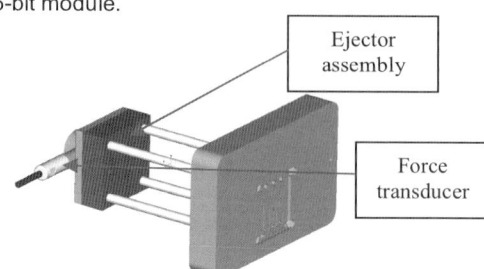

Figure 2. Force transducer and ejector assembly

Table 2. L9 fractional orthogonal array for ABS

Trial	T$_b$ [°C]		T$_m$ [°C]		t$_c$ [s]		t$_e$ [s]	
	Value		Value		Value		Value	
1	A1	220	B1	40	C1	1	D1	0
2	A1	220	B2	60	C2	5	D2	5
3	A1	220	B3	80	C3	10	D3	10
4	A2	250	B1	40	C2	5	D3	10
5	A2	250	B2	60	C3	10	D1	0
6	A2	250	B3	80	C1	1	D2	5
7	A3	280	B1	40	C3	10	D2	5
8	A3	280	B2	60	C1	1	D3	10
9	A3	280	B3	80	C2	5	D1	0

Table 3. L9 fractional orthogonal array for PC

Trial	T$_b$ [°C]		T$_m$ [°C]		t$_c$ [s]		t$_e$ [s]	
	Value		Value		Value		Value	
1	A1	280	B1	80	C1	1	D1	0
2	A1	280	B2	100	C2	5	D2	5
3	A1	280	B3	120	C3	10	D3	10
4	A2	300	B1	80	C2	5	D3	10
5	A2	300	B2	100	C3	10	D1	0
6	A2	300	B3	120	C1	1	D2	5
7	A3	320	B1	80	C3	10	D2	5
8	A3	320	B2	100	C1	1	D3	10
9	A3	320	B3	120	C2	5	D1	0

3.5 Design of experiments

Due to the fact that the filling performance of micro moulds relies heavily on the temperature control during injection, the effects of barrel temperature (T_b) and mould temperature (T_m) were investigated. After filling it is necessary for the part temperature to be sufficiently low to facilitate the demoulding without part deformation. Therefore the cooling time after filling (t_c) and the use of a delay for the ejection time (t_e) were also taken into account. Thus, given that four factors at three levels each were considered, a Taguchi L9 orthogonal array (OA) was selected. (see Table 2 and 3).The response of each tool surface treatment to each set of control parameters was analysed by measuring F_E during the part ejection. Given that two tool surfaces, untreated, and DLC and two materials, PC and ABS, are investigated, four L9 OAs were defined.

4. Analysis of the results

4.1 Average Force results

In this study, L9 OAs were employed to ensure that the experimental results were representative of the considered processing window. For each trial, the effects of the applied surface treatments on F_E were investigated and then based on the conducted trials the F_E mean values were calculated for each of the OAs as shown in Figure 3. For the untreated tool on average both ABS and PC results were subjected to the highest demoulding forces of all the experiments. ABS had a higher average than PC. For the tool with the DLC coating, both materials experienced a reduced demoulding force compared to the untreated tool. The average ABS results were the lowest of all experiments, and compared to the untreated tool results there were a F_E reduction of 41.6% for the DLC treated tool. For PC, the reduction of F_E was 10.68%.

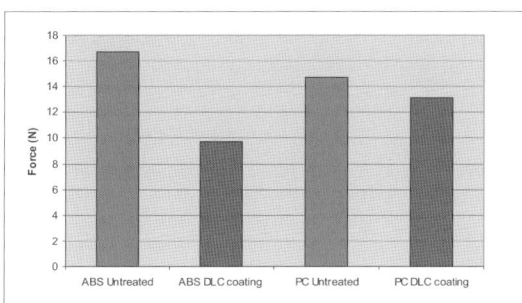

Figure 3. The average demoulding force for the OAs

4.2 Optimum parameters levels

The average demoulding force based on trials conducted for each combination of control parameters was calculated in order to determine the optimum parameter levels for the investigated surface treatment and polymers employing the Taguchi method. The value of a given parameter is considered to be optimum, the best of the selected three levels, if its corresponding average F_E is the lowest. Figure 4 shows the results of the experiments. By applying this method, it is possible to identify theoretically the best set of parameters within the investigated processing window with respect to F_E. The theoretical best set of

processing parameters are provided in Table 4.

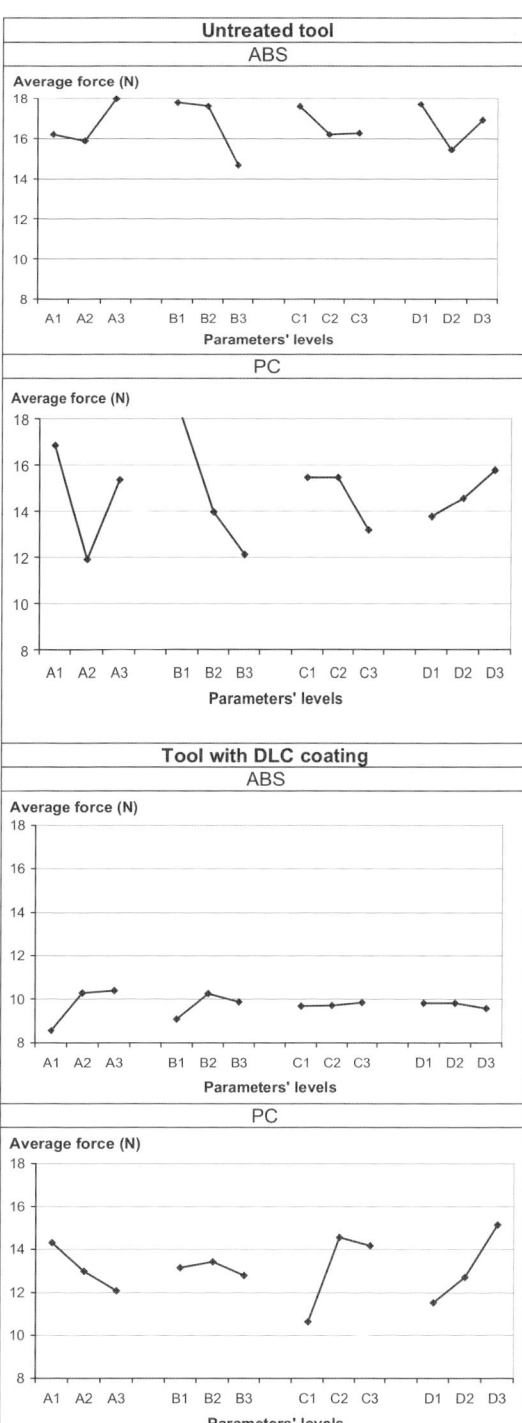

Figure 4. Main effects for each combination of surface treatments and polymers

Table 4. The theoretical best set of processing parameters

	T_b [°C]	T_m [°C]	t_c [s]	t_e [s]
Untreated & ABS	250	80	5	5
Untreated & PC	300	120	10	0
DLC & ABS	220	40	1	10
DLC & PC	320	120	1	0

4.3 Parameters contribution to optimum performance

Based on the experimental results, an analysis of variance (ANOVA) was performed in order to assess the contribution of each processing parameter to the resulting demoulding behaviour. Table 5 shows the percentage contribution of each parameters. Based on this analysis and the selection of the best parameters levels (Table 4), it is possible to compute the lowest theoretical demoulding force for each combination of surface treatment and polymer as shown in Table 6.

Table 5. Percentage contribution of each parameter

	Untreated tool		DLC coating	
	ABS	PC	ABS	PC
T_b	10.3	27.7	72.9	12.4
T_m	38.8	42.3	23.3	-
t_c	-	-	-	48.2
t_e	11.1	-	-	35.0

Table 6. The lowest theoretical demoulding force

	Untreated tool		DLC coating	
	ABS	PC	ABS	PC
F_E [N]	12.62	9.33	7.90	7.99

5. Conclusions

The paper reports an experimental study that investigates part demoulding behaviour in micro IM, with a particular focus on the effects of surface treatments on the demoulding forces. In particular, the demoulding performance of a representative microfluidics part was studied as a function of tool surface treatment in combination with four process parameters, T_b, T_m, t_c and t_e, employing the design of experiment approach. The following conclusions can be made based on the reported research:

- The average demoulding forces measured for both PC and ABS showed clearly that surface treatments reduce significantly F_E in comparison with untreated tools.

- From the conducted experiments, it is immediately apparent that there is not a unique selection of parameter levels as far as the demoulding behaviour is concerned that can be considered optimum for each type of surface treatment or polymer investigated in this research.

- By conducting an ANOVA analysis it was possible to assess process parameters' contribution to optimum performance. The lowest theoretical demoulding forces computed for each combination of tool treatment and polymer showed again that DLC coatings reduce significantly the demoulding forces for the polymers considered.

Finally, it is important to stress that in micro IM the polymer properties become an even more important factor in selecting surface treatments. Experimental studies and simulations of demoulding behaviour should precede the tool manufacture.

Acknowledgement

The research reported in this paper is funded by the EPSRC Programme "The Cardiff Innovative Manufacturing Research Centre" and the EC FP6 Project ""Surface Enhanced Micro Optical Fluidic Systems (SEMOFS)". Also, it was carried out within the framework of the EC FP6 Networks of Excellence, "Multi-Material Micro Manufacture (4M): Technologies and Applications".

References

[1].Navabpour, P., et al., *Evaluation of non-stick properties of magnetron-sputtered coatings for moulds used for the processing of polymers.* Surface and Coatings Technology, 2006. **201**(6): p. 3802-3809.

[2].Sasaki, T., et al., *An experimental study on ejection forces of injection molding.* Precision Engineering, 2000. **24**(3): p. 270-273.

[3].Bataineh, O.M. and B.E. Klamecki, *Prediction of local part-mold and ejection force in injection molding.* Journal of Manufacturing Science and Engineering-Transactions of the Asme, 2005. **127**(3): p. 598-604.

[4].Pouzada, A.S., E.C. Ferreira, and A.J. Pontes, *Friction properties of moulding thermoplastics.* Polymer Testing, 2006. **25**(8): p. 1017-1023.

[5].Pontes, A.J. and A.S. Pouzada, *Ejection force in tubular injection moldings. Part I: Effect of processing conditions.* Polymer Engineering and Science, 2004. **44**(5): p. 891-897.

[6].Pontes, A.J., et al., *Ejection force of tubular injection moldings. Part II: A prediction model.* Polymer Engineering & Science, 2005. **45**(3): p. 325-332.

[7].Mitterer, C., et al., *Industrial applications of PACVD hard coatings.* Surface and Coatings Technology, 2003. **163-164**: p. 716-722.

[8].Heinze, M., *Wear resistance of hard coatings in plastics processing.* Surface and Coatings Technology, 1998. **105**(1-2): p. 38-44.

[9].Cunha, L., et al., *Performance of chromium nitride and titanium nitride coatings during plastic injection moulding.* Surface and Coatings Technology, 2002. **153**(2-3): p. 160-165.

[10].Voevodin, A.A., M.S. Donley, and J.S. Zabinski, *Pulsed laser deposition of diamond-like carbon wear protective coatings: a review.* Surface and Coatings Technology, 1997. **92**(1-2): p. 42-49.

[11].Grill, A., *Diamond-like carbon coatings as biocompatible materials-an overview.* Diamond and Related Materials, 2003. **12**(2): p. 166-170.

[12].Hauert, R., *A review of modified DLC coatings for biological applications.* Diamond and Related Materials, 2003. **12**(3-7): p. 583-589.

Strategies for material removal in laser milling

P V Petkov, S Scholz and S Dimov

*Manufacturing Engineering Centre, Cardiff University, Queen's Buildings,
The Parade, Newport Road, Cardiff, CF24 3AA, UK*

Abstract

Laser milling with microsecond pulses is a thermal material removal process usually associated with detrimental effects such as heat affected zones (HAZ), a recast layer and debris. Process optimisation can lead to considerable reduction of the above mentioned negative effects. In this context, the research investigates the effects of tool path optimisation and material removal strategies on the resultant surface quality and edge definition. The conducted experimental study shows clearly that the applied milling strategies have a significant effect on the resulting surface topography and the edge definition. Also, the research demonstrates that by optimising the laser path and material removal strategies it is possible to reduce significantly the thermal load when milling micro features, and thus to minimise HAZ and other secondary effects.

Keywords: laser ablation, laser milling, machining strategies, tool path optimisation

1. Introduction

For pulsed laser machining the actual process of removing material from the substrate takes place during the laser pulse. Several mechanisms exist for material removal, depending on the laser pulse duration and some material specific time parameters [1, 2]. The following material dependent time parameters are considered important:

- τ_e - electron cooling time;
- τ_i - lattice heating time;
- τ_L - laser pulse duration.

As a rule $\tau_e << \tau_i$ and, for most materials, τ_i is in the picosecond range. Three different ablation regimes exist according to the laser pulse length:

- femtosecond pulses - $\tau_L < \tau_e < \tau_i$;
- picosecond pulses - $\tau_e < \tau_L < \tau_i$;
- nanosecond and longer pulses - $\tau_e < \tau_i < \tau_L$.

During the laser pulse, the energy is absorbed by the top surface layer of the substrate where heat is generated and the material starts melting, raising to the vaporisation temperature. The long pulse duration allows a thermal wave to propagate into the material through conduction and evaporation occurs from the liquid material. The molten material is partially ejected from the cavity by the vapour and plasma pressure, but a part of it remains near the surface, held by surface tension forces. After the pulse, the heat quickly dissipates into the bulk of the material and a recast layer is formed [3].

Secondary effects from machining with nanosecond and longer pulses are heat affected zones, recast layer, micro cracks, shock wave, surface damage and debris from ejected material (Fig. 1). Additionally, the vaporised material forms plasma from the leading edge of the pulse and this is sustained during the rest of the pulse. Due to the plasma shielding effect (the plasma absorbs and defocuses the pulse energy) a higher irradiance (fluence) is required for deeper penetration [4].

2. Factors influencing the removal process

Overall accuracy of the laser milling process depends on the accuracy of the hardware/software of the machine and the removal strategies employed [5].

Some of the most important factors are: laser source; optical components; NC stages; software accuracy (model and beam path); control system and strategies for material removal.

In this paper special attention will be drawn to the strategies for material removal. Very often this factor is ignored and as a result the outcome of the removal process is not satisfactory.

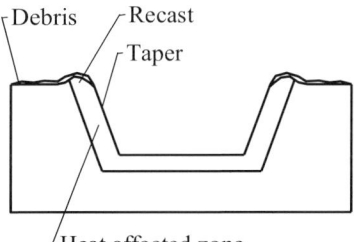

Figure 1 Secondary effects of laser machining

3. Design of experiment

A test structure as shown in Fig. 2 was designed to investigate the influence of the material removal strategy on the final part quality. The structure comprises of radially repeating 8.7 mm long segments, with a variable width between 40µm up to 800µm. The depth is kept constant at 150µm. Four different removal strategies as shown on Figure 3 were applied using the same processing parameters as they are given in Table 1. In particular, the strategies investigated in this research are:

(1) Random "hatching" inside together with a cut along the border as shown in Figure 3(a);
(2) Profile cuts only, Figure 3(b);
(3) Hatching only, Figure 3(c)
(4) The cuts have the same configuration as those for strategy (a), and the only difference is in the profile cut along the border where the beam is tilted in order to produce vertical walls, Figure 3 (d).

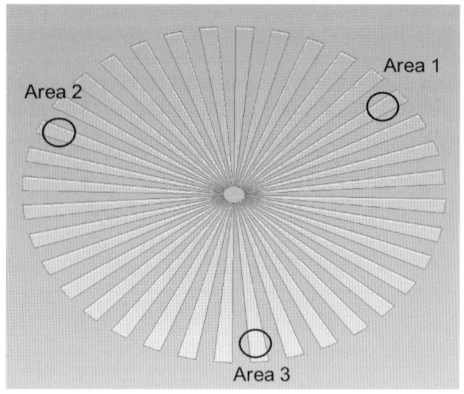

Figure 2 Test structure and measurement areas

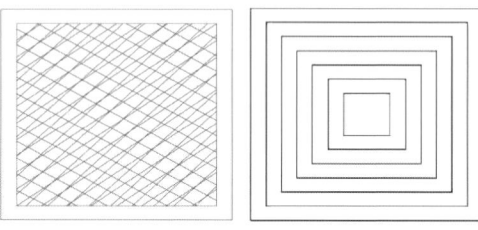

(a) random hatching and   (b) profile cuts only
a profile cut along
The border

 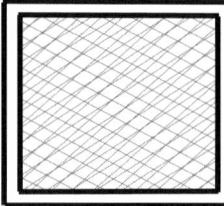

(c) hatch cuts only   (d) vertical walls

Figure 3 Material removal strategies

Two test pieces were produced (stainless steel BS316 and brass BS2874 CZ121, later referred to as stainless steel or steel and brass respectively), each containing four structures as shown on Figure 4. These two materials were selected because they have different ablation thresholds and thermal conductivity.

Based on previous investigations better results were expected for the stainless steel sample. The active control of the removal rate per layer available on the microsecond laser milling system used in this research was disabled in order to compare solely the effects of the four removal strategies.

After the milling operation, the samples were cleaned in ultrasonic bath with a light degreaser and thus to remove the debris from the ablation process without affecting the resulting surface roughness. Then, the roughness of different machined areas was measured by employing a white light interferometer, MicroXAM-100-HR, and scanning a patch 250µmX330µm. The depth results were obtained on an optical system, MITUTOYO Quick Vision Accel, by performing 3

measurements in each area, and then the average was used in the experimental study.

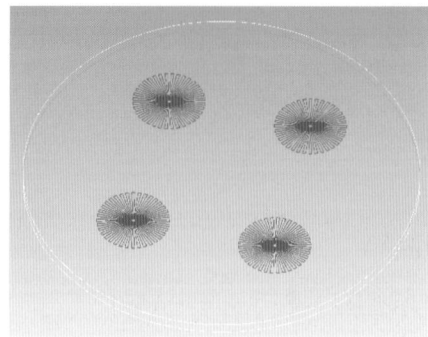

Figure 4 Test piece

Table 1 Laser parameters

	SS BS316	BrassBS2874 CZ121
Power	71.04%(9.5W)	56%(1.2W)
Frequency	40 kHz	15 kHz
Scanning speed	400 mm/s	150 mm/s
Wavelength	1064 nm	1064 nm
Pulse duration	10µs	10µs

4. Results and discussion

The roughness and the depth of the milled structures, four on the SS316 and brass substrates, were measured in three places as shown in Figure 2, and the results are provided in Table 2.

Table 2 Roughness and depth results

		SS316		Brass	
		Ra, µm	Depth, µm	Ra, µm	Depth, µm
Str.1	1	1.84	202	3.5	164
	2	3.46	207	3.3	175
	3	3.32	209	3.5	167
Str.2	1	13.37	204	3.5	201
	2	15.45	212	3.9	213
	3	12.47	209	na	210
Str.3	1	3.83	190	3.8	127
	2	3.04	207	3.9	139
	3	3.61	202	3.4	131
Str.4	1	2.16	195	na	156
	2	2.63	206	6.36	164
	3	2.29	204	6.7	161

Additionally, scanning electron microscope (SEM) images in the milled areas were taken in order to compare the resulting surface topography and border profiles after applying each of the four investigated machining strategies.

Figures 5-8 show SEM pictures taken from the steel sample near the centre, middle sector and at the end of the test segments. As it can be seen from these figures the selected milling strategy has a significant effect on the resulting surface topography and border profile.

Since the CAD model and tolerances used to generate the laser path programmes and all laser parameters are exactly the same any differences in the machining results have to be attributed solely to the employed material removal strategies. The best average and the most consistent surface roughness (see Table 2) and edge definition was achieved when Strategy 4 was applied, Figure 3(d). Only in Area 1 the resulting roughness is better on the sample machined with Strategy 1.

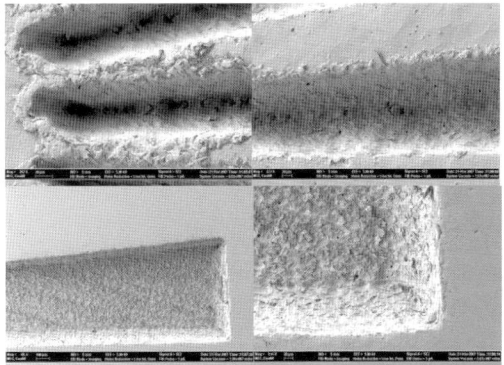

Figure 5 Strategy 1 results on stainless steel

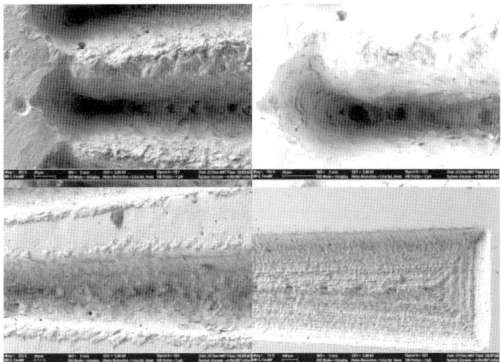

Figure 6 Strategy 2 results on stainless steel

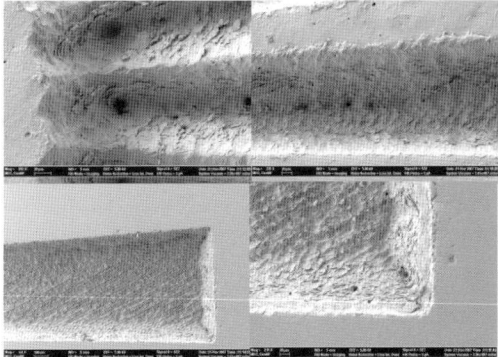

Figure 7 Strategy 3 results on stainless steel

The results for the brass substrate are provided in Figures 9 to 12. The effects of milling strategies on the resulting surface roughness and segments' depth are different from those observed on the SS316 sample. In addition, there was a very high variation of machined depth when comparing the results for the segments milled with different strategies. And, this is in spite of the fact that the laser parameters during the machining

of the brass sample were kept the same as it was the case with the four stainless steel structures.

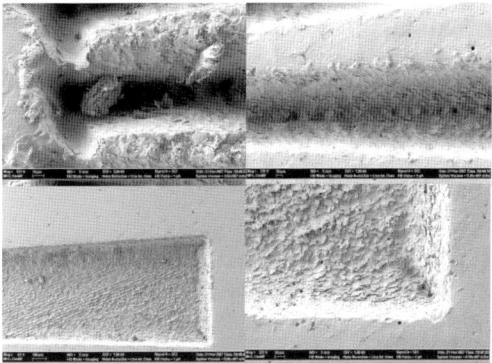

Figure 8 Strategy 4 results on stainless steel

This can be explained with the dissimilar melting temperatures and thermal conductivity of the investigated two materials that result in different heat built up after the machining. While for the brass thermal conductivity is 115 W/mK for the steel is only 13.4 W/mK. However, in spite of the better heat dissipation in brass surface defects and especially the re-cast layer are noticeably higher in comparison with the stainless steel sample due to its relatively low melting temperature, 885 °C and 1371 °C, respectively.

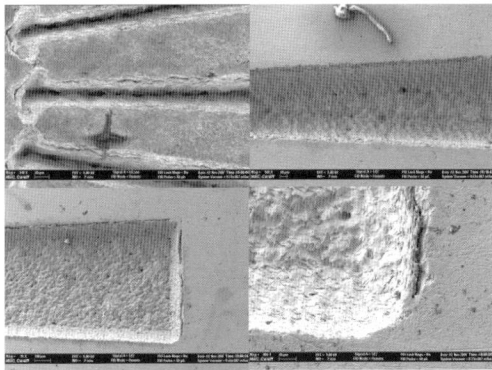

Figure 9 Strategy 1 results on brass

The results from roughness and depth measurements are provided again in Table 2. Please note that it was not possible to measure the surface roughness in some areas of Structures 2 and 4. The best result in terms of roughness and surface consistency was achieved by applying Strategy 1. Although, in the inner sector the machining results for Strategy 3 looks the same as those for Strategy 1, there is a significant difference in the way these two strategies are applied. In order to produce vertical walls a synchronised movements of both the stage and the galvo mirrors are necessary. Hence, additional error is introduced.

The best results would be achieved by applying strategies that do not require any stage movement, e.g. Strategies 1 and 3 compared with Strategy 4, and ensure the most consistent coverage of the ablated surface with single craters that exhibit some irregularities in their shape (Figure 13). At the start of the each segment where their width is only 40μm due to the close proximity of the neighbouring laser paths, there is not sufficient time for the heat to dissipate into the bulk of the material, and therefore heat builds up in the milled area. As a result of this more material is removed and the edge definition is relatively poor. The other negative effects in these narrow areas are the increase of HAZ and in general more pronounced secondary effects as cracks and material re-deposition. These effects can be easily seen when comparing the SEM images in the top left corner in Figures 5 to 12.

Figure 10 Strategy 2 results on brass

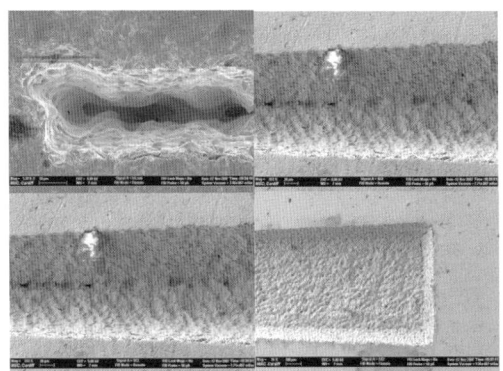

Figure 11 Strategy 3 results on brass

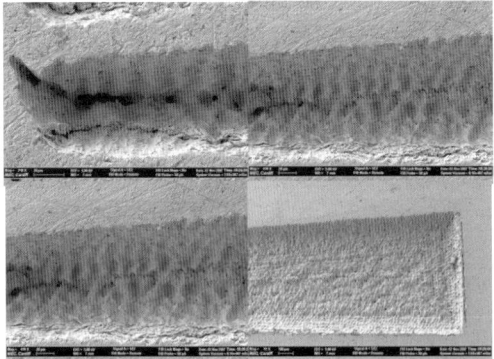

Figure 12 Strategy 4 results on brass

Looking at the depth results in Table 2 it is easy to see the interdependence between measurement results and applied milling strategies. In both materials, SS316 and brass, the mean depth achieved exhibits the same trend – in descending order - strategy 2, strategy 1, strategy 4, strategy 3. Variations in depth measurements could be attributed to material inconsistencies.

In regards to edge definition, as it was expected the machined contours are sharper when the milling strategies include border cuts instead of laser passes normal to the side walls.

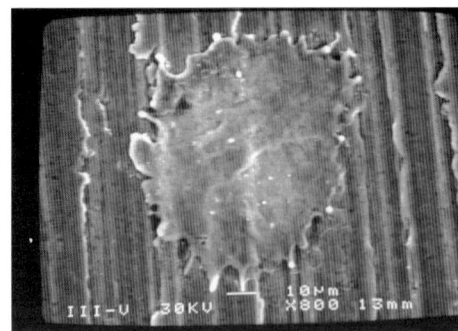

Figure 13 Single pulse crater

5. Conclusions

A very important, but commonly ignored aspect of laser milling with long pulses is discussed in this paper. In particular, the effects of tool path optimisation and material removal strategies on the resultant surface quality and edge definition were investigated. The conducted experimental study shows clearly that the applied milling strategies have a significant effect on the resulting surface topography and the edge definition.

Also, the research demonstrate that by optimising the laser path and material removal strategies it is possible to reduce significantly the thermal load when milling micro features, and thus to minimise HAZ and other secondary effects.

Acknowledgements

The authors would like to thank the Technology Strategy Board, the Welsh Assembly Government and the UK Engineering and Physical Sciences Research Council (EPSRC) for funding this research under the MicroBridge Programme and the EPSRC Programme "The Cardiff Innovative Manufacturing Research Centre". Also, this work was carried out within the framework of the EC Networks of Excellence "Innovative Production Machines and Systems (I*PROMS)" and "Multi-Material Micro Manufacture: Technologies and Applications (4M)".

References

[1] Momma C, Nolte S, Chichkov B N, von Alvensleben F and Tünnermann A, 1997, "Precise laser ablation with ultrashort pulses", Applied Surface Science, Vol. 109-110, pp 15-19

[2] Kautek W and Krüger J, 1994, "Femtosecond pulse laser ablation of metallic, semiconducting, ceramic and biological materials", Proc. SPIE, Vol. 2207, pp 600-610

[3] Dörbecker C, Lubatschowski H, Lohmann S, Ruff C, Kermani O and Ertmer W, 1996, "Influence of the ablation plume on the removal process during the ArF-excimer laser photoablation", Proceedings SPIE, Vol. 2632, pp 2-9

[4] Chichkov B N, Momma C, Nolte S, von Alvensleben F and Tuennermann A, 1996, "Femtosecond, picosecond and nanosecond laser ablation of solids", Applied Physics, A63, pp 109-115

[5] Pham, D. T., Dimov, S. S., Petkov, P. V., Petkov, S. P., (2002) Laser Milling. In Proc Instn Mech Engrs Vol 216 Part B: J Engineering Manufacture (2002)

Multi-Material Micro Manufacture
S. Dimov and W. Menz (Eds.)

253

Investigations in Variothermal Injection Moulding of Microstructures and Microstructured Surfaces

W. Michaeli[a], F. Klaiber[a], S. Scholz[b]

[a]Institute of Plastics Processing, RWTH Aachen University, Aachen, Germany
[b]The Manufacturing Engineering Centre, Cardiff University, Cardiff, CF24 3AA, UK

Abstract

Telecommunication, information and medical industries have a high growth potential. A key technology for those industries is the replication of microstructures. Precise microstructured parts with functional surfaces can be produced economically by injection moulding. The whole process chain (thermal mould condition, moulding, demoulding, measurement and analysis) must be analysed carefully to ensure the highest precision and reliability. To enable the precise production of such structures fundamental studies were conducted at the Institute of Plastics Processing (IKV). The studies considered several polymers (PMMA, POM) on the one side and various test structures on the other side. In addition an innovative external inductive heating unit was analysed and implemented into the process to heat the cavity surface efficiently. Using this technique cavity surface temperature increase rates of up to 60 K/s have been achieved. A pyrometer was implemented for contact less instant temperature measurement, and controller was used to realise preset cavity temperatures by regulating the inductor power. With the dynamic inductive heating system the moulding accuracy of the microstructures could be increased drastically. The final step of the process chain comprises of the measurement and analysis of the microstructured moulded parts. To analyse the microscopic deviation between the mould cavity and the surface of the moulded part scanning electron microscopy (SEM) and white light interferometry (WLI) was used.

Keywords: Injection Moulding, microstructures, dynamic heating

1. Introduction

Microstructured surfaces find applications in many areas. By integrating microstructures, surface properties like self cleansing effects, lens coating or the flow behaviour in tubing can be optimised [1, 2, 3, 4]. To achieve good moulding results of microstructures through the process of injection moulding, it is necessary to heat the mould to temperatures close to the melting point of the polymer. In order to attain such high temperatures in the mould and to guarantee demoulding without damage at the same time, the process of variothermal temperature control has been proposed. This study aims to advance microstructure replication by using injection moulding and hybrid injection compression moulding and investigates the advantage of external heating in the injection (compression) moulding process. An induction system that can be positioned in the open mould by robot control is used as external heating to heat the mould cavity. Additionally, the mould is held at a constant temperature using water temperature control in order to guarantee a demoulding free of damage. Inductive heating as an external heat source has been proven successful and it has been shown that surface structures with high power densities and therefore short heat-up times can be heated selectively [5, 6]. For process control, IKV developed a fully automated system composed of an injection moulding machine, robot and inductor heating system which is extended by a pyrometer (touchless temperature measurement device). A control algorithm allows to prescribe defined mould temperatures and to regulate the induction power. The process control is simulated in Matlab/Simulink. The thermal homogeneity of the induction heating is analysed in experiments runs and heat-up rates using induction heating are investigated.

The experiments show that heat-up rates of 60 K/s can be realised using external induction heating. The IKV system is used for experiments on the replication of microstructures using injection moulding and injection compression moulding with and without induction heating. In order to use both process variations a mould is used with a variable contour core and the capability of hybrid injection compression moulding. The microstructure that was used for the experiment is a LiGA honeycomb structure with a structural height of 100 µm. All replication investigations were done with different materials (Roehm: PMMA DQ501, PMMA 6N, Ticona: POM C 52021). The respective process parameters for each combination of structure, material and process are held constant during the experiment. Different techniques were used to qualify and quantify the replications. SE-, light- and macro-microscopy were used to identify the quality of the replicated structures. A white light interferometer microscope (WLI) was used to measure the replication depth.

2. Machine set-up and layout of the control system

In Figure 1 the inductive heating unit including the pyrometer can be seen. The inductor is placed by a Wittmann W721 robot in front of the cavity. The robot places the inductor at any given place in front the cavity with high precision and reproducibility. The inductive heating unit for injection moulds can be operated by setting the inductor power and the heating time (mode 1) [7, 8]. Preliminary experiments with operation mode 1 have shown that the heat-up process is uncontrolled and not reproducible [9, 10]. It can furthermore lead to an uncontrolled overheating of the cavity surface, that can result in destruction of the fragile microstructures. In addition too high mould temperatures can have a negative influence onto the mould and the part. In particular the mould steel can loose its stiffness and a colour change can arise on the mould surface.

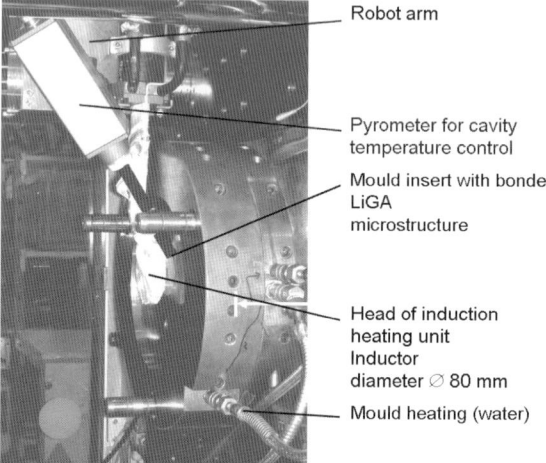

Figure 1: Inductive heating unit

A more innovative way of operating an inductive heating unit is the direct presetting of cavity temperature and heating time. In this case the inductor power is controlled in dependence of the set temperature and the measured temperature (mode 2). Therefore a control algorithm is designed, in which desired cavity surface temperatures can be pre-set. It is necessary to instantly measure the cavity temperature. In our experiments a Maurer KTR 2300-1 Pyrometer was integrated into the inductive heating unit. For an exact temperature measurement the pyrometer was calibrated on the radiation properties (emission and reflection) of the cavity surface. After calibration and adjustment of the pyrometer the cavity temperature can be preset at the inductive generator. Inside the generator a PI-controller regulates the temperature. The control system was designed to achieve high heat-up rates and at the same time it avoids a temperature overshooting. The process was simulated in Matlab/Simulink and validated with inductive heating experiments. In Figure 2 a comparison of measured and calculated temperatures can be seen. "T Q" is the temperature measured by tactile thermocouple which is placed 2 mm behind the cavity. "T P" describes the temperature measured by the pyrometer. The simulation is based on a simplified unsteady heat conduction equation.

$$\frac{\mathrm{d}}{\mathrm{d}t}E_{th}(t) = P_{in}(t) - P_{out}(t) \qquad (1)$$

For the based control-volume the change of the energy content is equal to the difference between power in- and output. The power input into the control-volume is described as

$$P_{in}(t) = \eta \cdot P_{eff} \cdot L(t) \qquad (2)$$

Where η describes the degree of efficiency of the inductor, P_{eff} is the effective power of the medium

frequency generator and L(t) is the power factor. The power factor can either be preset at the medium frequency generator (mode 1) or it can be controlled by the PI controller in dependence of set and actual temperature (mode 2). The power outflow through the mould can be written as

$$P_{out}(t) = \lambda_{steel} \cdot \frac{\pi \cdot d_{inductor}^2}{4 \cdot s} \cdot \left(T(t) - T_{base}\right) \qquad (3)$$

where λ_{steel} describes the heat conduction coefficient of steel, d is the diameter of the inductor and s is the distance that has to be taken into account for heat conduction (distance from directly heated volume till approximately the mould base temperature T_{base} is reached). The parameter s is empirically determined. The time dependent change of the inner thermal energy $E_{th}(t)$ can be calculated through

$$\frac{d}{dt}E_{th}(t) = c_{steel} \cdot m_{control-volume} \frac{d}{dt}T(t) \qquad (4)$$

where c_{steel} is the heat capacity of steel and $m_{control-volume}$ is the mass of the control-volume.

$$m_{control-volume} = \rho_{steel} \cdot \pi \cdot d_{inductor}^2 \cdot \delta \qquad (5)$$

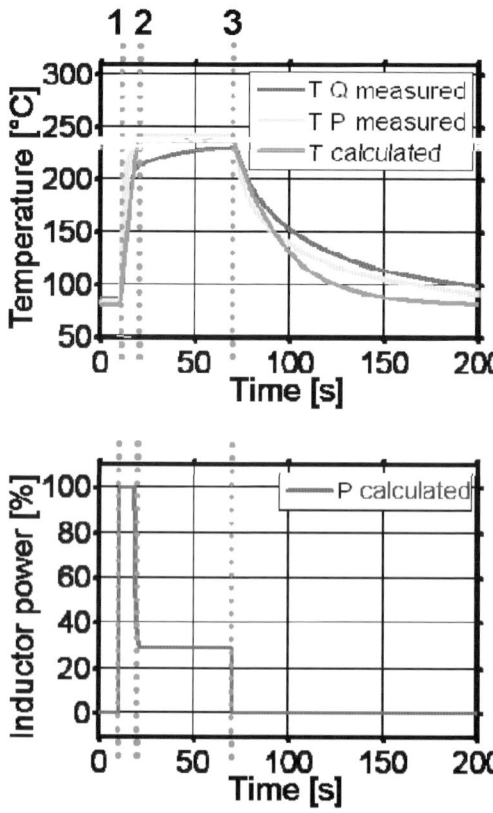

Figure 2: Temperature and inductor power over time

where ρ_{steel} is the density of steel and δ is the depth of the directly induction heated volume (Eddy currents are induced up to this depth). With that the fundamental differential equation for the simulation can be written as shown in equation (6).

$$P_{in}(t) - \lambda_{steel} \frac{\pi \cdot d_{inductor}^2}{4 \cdot s} \cdot (T(t) - T_{base}) = $$
$$c_{steel} \cdot m_{control-volume} \cdot \dot{T}(t) \tag{6}$$

This differential equation is used for the simulation of the heat-up process in Matlab/Simulink. With the apparent temperature differences the heat loss through the cavity surface is small compared to the processes inside the mould plates and is therefore neglected. For the simulation of the heat-up process the process variables shown in Table 1 were used. At point of time 1 in Figure 2 the heat-up starts and the power increases to the maximum. When the set temperature is reached the power is decreased by the controller to an amount with which the temperature can be held constant at the set point (point of time 2). At point of time 3 the end of the heating time is reached and the induction heating unit is switched off.

With the simulation the controller adjustment of the medium frequency generator could be supported. The controller adjustment was verified and optimised in practical investigations. The integrated PI controller has two parameters that can be changed. The proportional factor K_P and the reset time T_N. The controller was adjusted so that a power factor of 100 % could be used with no temperature overshoot. With these adjustments cavity surface heat up rates of 60 K/s can be achieved.

3. Moulding experiments

Process variable	Unit	Value
Effective inductor power P_{Eff}	W	15000
Inductor efficiency h	-	0,2
Heat conduction coefficient l_{steel}	W/(mK)	55
Distance till base temperature s	m	0,2
Inductor head diameter $d_{inductor}$	m	0,08
Induction penetration depth δ	m	0,04
Density r_{steel}	Kg/m³	7800
Heat capacity c_{steel}	J/(kg K)	467

Table 1: Process variables for simulation

The variothermal moulding process using controlled external inductive heating was verified with moulding experiments. For the moulding trials different mould inserts were used. The results of the moulding experiments with the LiGA honeycomb structure with a structural height of 100 µm and a width across flats of 40 µm are shown. Two different moulding techniques, injection moulding and injection compression moulding, were used to replicate the microstructures. Both

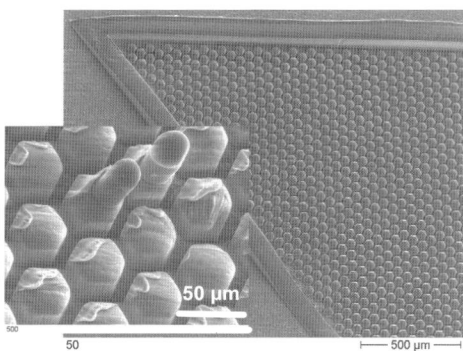

Injection moulding

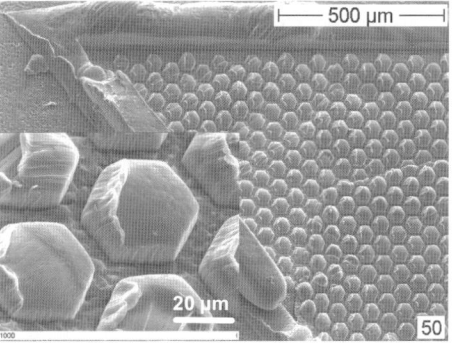

Injection compression moulding

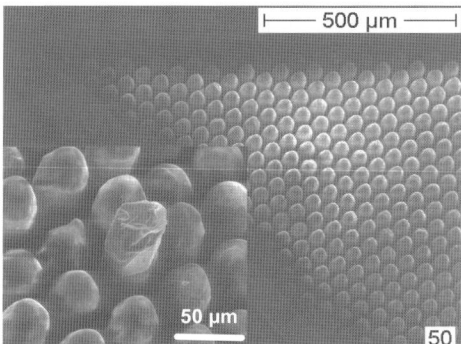

Injection moulding with induction

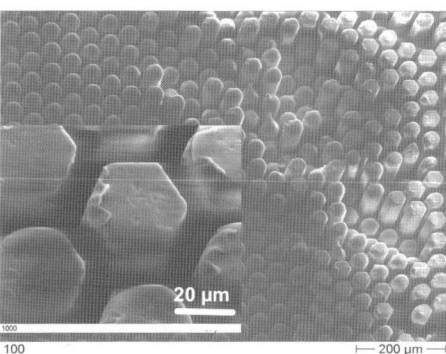

Injection compression moulding with induction

Figure 3: Moulded LiGA microstructures

techniques were used in combination with isothermal and variothermal mould heating. In injection moulding the moulding process is started by closing the mould, injecting the molten polymer and then holding pressure is applied until the gate freezes. In injection compression moulding the mould is first closed to the compression gap and then simultaneously the injection and the compression phase take place. By injection compression moulding a much more homogeneous pressure distribution within the cavity is achieved. A backflow of the melt into the plastification cylinder is prevented by applying holding pressure. Figure 3 shows the moulding results. The material used is POM. The LiGA microstructure could be well replicated with POM. The structure depth was measured using white light interferometry (WIM).

WIM measurements	Structure depth
Injection moulding (IM)	23 µm
Injection-compression moulding (ICM)	20 µm
Variothermal IM	87 µm
Variothermal ICM	100 µm

Table 2: Microstructure replication depths

With increasing cycle numbers a clogging of the honeycomb structures took place. This can be explained due to missing draft angles of the LiGA microstructures. The fact that some honeycomb pillars rip off can be explained with demoulding forces that increase with the moulding depth. Best moulding results could be achieved with the POM C 52051 material. Complete replication of the microstructures was achieved using variothermal ICM. This was confirmed by the white light interferometry measurements. If the inductive heating unit is used the structure height of the moulded parts could be increased from 23 µm to over 87 µm (Table 2). This is a result of the increased cavity temperature during melt injection. But it has to be kept in mind that the structure is not fully replicated because of the clogging of the honeycomb structures. Table 2 shows the results of partial area of the microstructured surface.

The experiments show that the replication results are generally improved when external induction heating is applied. Best results are achieved when using injection compression moulding with induction heating (Figure 3). However, the process window in this case is relatively narrow and problems arise in the cooling so that already perfectly moulded structures can be damaged during the process of demoulding and the mould be blocked. Comparing the materials, POM is in clear favour to PMMA since the demoulding characteristics of POM are better due to the partly crystalline structure of the material. Additionally, POM has a lower viscosity in the mould. Comparing the two PMMA grades, better replication is achieved using PMMA 6N OQ due to the added demoulding additives that are lacking in the PMMA DQ 501 material.

4. Outlook

This research work shows that the process parameters have a significant influence on the processing results. The use of an inductive heating system for the moulding of micro structures and LiGA-generated cavities shows very promising results. With this device cavity surface heat-up rates of up to 60 K/seconds can be realised to accurately replicate microstructures. Future research projects will investigate the application of a special coating on the microstructures to reduce the demoulding problems.

Acknowledgement

The investigations set out in this report received financial support from the German Research Foundation (DFG) as part of SFB/TR 4 "Process Chains for the Replication of Complex Optical Elements", to whom we extend our thanks.

References

1. N.N. (2007) NEXUS Market Analysis for MEMS and Microsystems III, 2005-2009. WTC Wicht Technologie Consulting
2. Barthlott, W.; Neinhuis C., Purity of the sacred lotus, or escape from contamination in biological surfaces. Planta (1997) 202, pp. 1-8
3. Bechert, D.W.; Buse, M.; Hage, W., Experiments with three-dimensional riblets as an idealized model of shark skin. Experiments in Fluids 28 (2000), pp. 403-412
4. Baker, K. M., Highly corrected close-packed microlens arrays and moth-eye structuring on curved surfaces. Applied Optics 38 (1999) No.2 , pp. 352-356
5. Weber, A.; Schinköthe, W.; Completely integrated induction heating and pulsed cooling for injection moulding, *Proceedings of the 19th Stuttgarter Kunststoff-Kolloquium*, 9.-10.3.2005, Stuttgart
6. Tewald, A., Entwicklung und Untersuchung eines schnellen Verfahrens zur variothermen Werkzeugtemperierung mittels induktiver Erwärmung. Dissertation, University of Stuttgart, Stuttgart, 1997
7. Schaumburg, C.: Mikrospritzgießen mit induktiver Werkzeugtemperierung. Dissertation, University of Stuttgart, 2001
8. Gärtner, R., Dynamic mould temperature control with an inductive heating system, Proceedings of the 3rd European Conference "Injection Moulding 2002", 24-25 October 2002, Copenhagen, Denmark
9. Michaeli, W.; Schmachtenberg, E.; Gärtner, R.; Schönfeld, M.; Seul, M.; Schröder, T.; Thornagel, M., Spritzgießwerkzeugtechnik, Proceedings of the 22nd International IKV Colloquium, 10.-12.3.2004, Aachen
10. Gärtner, R., Analysis of the process chain for the production of micro-structured parts by injection moulding. Dissertation, RWTH Aachen University, 2005

TEM/SEM and FT-IR characterization of biocompatible magnetic nanoparticles

K.Alexandrova[a], I. Markova – Deneva[b], A. Gigova[a], I. Dragieva[a]

[a]Institute of Electrochemistry and Energy Systems, Bulgarian Academy of Sciences, Sofia, Bulgaria, kremena@bas.bg
[b]University of Chemical Technology and Metallurgy, 1756 Sofia, 8 Kl. Ohridski Blvd., Bulgaria, vania@uctm.edu

Abstract

Fe-Co-Cr-B(N,C,O,H) nanoparticles were synthesized by chemical reduction in aqueous solutions of cobalt precursor complexes such as (ethylenediamine)dichloro cobalt chloride $[Co(en)_2.Cl_2]Cl$ and aqua solutions of $FeCl_2.4H_2O$ and $CrCl_3.6H_2O$ with sodium borohydride as reducing agent. During the synthesis a reactor insuring hydrodynamic conditions of ideal mixing for solutions at a room temperature and atmospheric pressure was used. Transmission electron microscopy (TEM), scanning electron microscopy (SEM) and Fourier transformation infra red spectroscopy (FT-IR) investigations of nanoparticles obtained were carried out. The influence of the applied d. c. magnetic field during the synthesis on their properties were established. It is visible that a d. c. magnetic field induces a chain arrangement of nanoparticles with higher hydroxide content, possessing higher coercive force and lower transmittance determined by FT-IR.

Keywords: TEM/SEM, FT-IR investigations, nanoparticles, borohydride reduction

1. Introduction

Our work team has a long time experience working on synthesis of nanomaterials using reduction of aqueous salt solutions or complexes with water solutions of sodium borohydride or potassium borohydride [1]. During the synthesis there are possibilities to obtain different in size, structure or shape nanoparticles with specific conductivity/resistivity or ferromagnetic properties.

The obtained from the noble metals gold, silver and platinum nanoparticles contain less than 0.2 wt. % boron included in them by the reducing agent [2]. Nanoparticles are also obtained from non-noble metals such as iron, cobalt or nickel and in these particles, chains or wires the boron content is much higher [3, 4]. The content can be controlled and this element is often used to absorb harmful radiation during cancer treatment.

Recently the biocompatibility of such nanoparticles has been thoroughly investigated for different medical applications [5]. The elements - hydrogen, oxygen, nitrogen and carbon comprise 96.6 wt. % of the human body content [6]. The same atoms bind and form chemical bonds with the surface of the nanomaterials synthesized by the borohydride method (BH). As this case proves too, the formation of cobalt-nitrogen bond and its transfer to the nanoscale ferromagnetic particles is possible [6].

The aim of this work is to investigate, visualize and present with the help of TEM/SEM and FT-IR techniques the processes of nucleation and organization of the chemical bonds between the attached surface atoms. A subsequent task was to synthesize and investigate nanoparticles consisting of iron, cobalt, chromium proportionally to their ratio in the human body.

2. Experimental details

Metallic nanoparticles from $FeCl_2.4H_2O$ 7.10^{-2} M salt solution mixing with $Co[(en)_2.Cl_2]Cl$ and $CrCl_3.6H_2O$ 3.10^{-3} M solutions were produced by (BH) method in a reactor of ideal mixing [4]. A d.c. magnetic field (intensity 700 Oe) was applied during the synthesis for some of the prepared samples. After completion of the full reduction for period of two minutes the black precipitate is filtered and washed out many times with distilled water, and acetone. It is dried in a vacuum drier $(5.10^3$ Pa) for about 4 to 24 hours, and the yield is weighted.

The boron weight percentage content of the samples is analytical determined by titrimetric method. Iron, cobalt and chrome weight percentage content are determined by *JEOL Superprobe-733* scanning electron microscope/microanalyser and mounted System 5000 *(HNU SYSTEMS) EDS X-ray system with Si detector*.

The specific surface area (SSA) of the obtained nanoparticles is determined by BET method (*AREA meter, Strohlein*, nitrogen flow, at temperature 78K).

The obtained nanoparticles morphology is characterized by electron microscope (*TEM/SEM JEM 200 CX JEOL – Japan*), the image is obtained in a transmission and scanning mode in vacuum 10^{-7} torr and accelerating voltage 100 kV.

Powder X-ray diffraction (XRD) patterns were collected using a TUR-M62 apparatus (Germany) with Co-Kα radiation. Data interpretation was carried out using the JCPDS database. Average crystallite sizes were determined from the XRD peaks using Scherrer's equation. Moessbauer spectra of the samples were recorded at 295 K on a electromechanical type spectrometer (Wissenschaftliche Elektronik GmbH, Germany) working in a constant acceleration mode. A [57]Co/Cr (activity $\cong 5\ 0$ mCi) source and an α-Fe standard were used. The experimental spectra were treated using the least squares method for sextets Sx1, Sx2, Sx3,Sx4 and doublet (Db).

Quantum Design, Vibrating Sample Magnetometer (VSM) is used for all magnetic measurements.

FT-IR spectroscopy study has been carried out with IR spectrophotometer EQUINOX (Bruker) with Fourier transformation in 4000 to 400 cm^{-1} frequency region using KBr matrix and FT-IR spectra of ferromagnetic nanoparticles synthesized have been undertaken. The samples for FT-IR spectroscopy investigations have been prepared mixing nanoscaled materials amounting to 1 mg and KBr (*spectral purity*) amounting to 200 mg and pressed in a tablet form with a diameter of 13 mm under a press loading of 8 tons/cm^2.

3. Results and discussion

Experimental data about the properties of nanoparticles investigated in this article are given in Table 1.

Table 1
Experimental data for the chemical content, the specific surface area (SSA), size and magnetic parameters of the samples

Parameters	# of samples	
	2458 (with m. f.)	2459 (without)
Iron [wt. %]	90.75	90.67
Cobalt [wt. %]	4.06	3.86
Chromium [wt. %]	2.11	2.10
Boron [wt. %]	3.13	3.86
SSA [m^2/g]	42.32	30.95
Nanoparticle diameter [nm]		
i) according BET	18	24.6
ii) acording TEM	34.2	41÷44
iii) for XRD phases:		
- Fe$_3$O$_4$/γFe$_2$O$_3$	11.2	10.6
- Fe, Fe-Co, Fe-B	5.8	5.6
Moessbauer spectra:		
i) Sx1,x 2,x 3, x4	67	72
ii) Db	33	28
Coercive force, H$_c$ [Oe]	207	32
Magnetization, M [emu/g]		
i) M$_{saturation}$	65.1	77.3
ii) M$_{remanence}$	13.5	3.7
Aspect ratio M$_r$/M$_S$	0.20	0.04
Weight of samples [g]	1.31	0.70

The analysis of the two types of samples, obtained with (#2458) and without (#2459) applying d. c. magnetic field displayed a similarity in the chemical content in terms of the elements iron, cobalt and chromium.

Boron content in the samples obtained with applying of magnetic field is smaller, which corresponds to the higher hydrogen content in concordance to some of our previous investigations. Under the influence of the applied magnetic field hydrogen organizes the nanoparticles in chain-like formations.

The higher values of the specific surface area confirm that nanoparticles obtained in magnetic field are smaller in size. The comparative estimations of grain size diameters based on BET and TEM measurements are similar.

3. 1. TEM/SEM investigations of Fe-Co-Cr-B(N,C,O,H) nanoparticles

Figures 1 and 2 present TEM micrographs of nanoparticles obtained with or without applying a d. c. magnetic field during the synthesis.

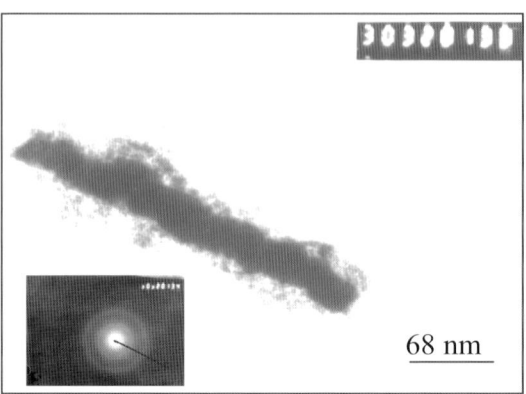

Fig. 1. TEM micrograph of Fe-Co-Cr-B(N,C,O,H) nanoparticles obtained with applying of a d. c. magnetic field

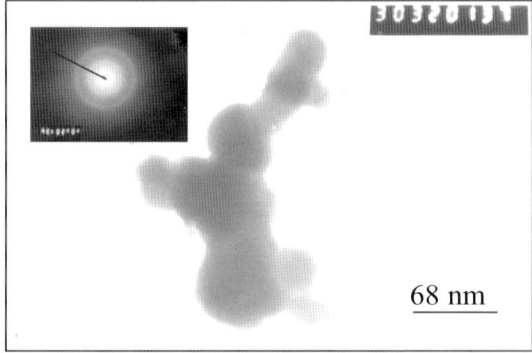

Fig. 2. TEM micrograph of Fe-Co-Cr-B(N,C,O,H) nanoparticles obtained without applying of a d. c. magnetic field

Figure 1 visualizes the accumulation of iron hydroxides and oxides upon the ferromagnetic chain. The Moessbauer data for the same sample show increase in the doublet part responsible for the oxide phases.

On Figures 3 and 4 SEM micrographs of the same nanoparticles are presented. The coercive force of this type of nanoparticles (chain) is higher and reaches up to 207 Oe, while in nanoparticles obtained without magnetic field hydroxide shells lack (Figure 2). They have a smaller doublet and much lower coercive force – 32 Oe. The values of magnetization and aspect – ratio follow the general rules for ferromagnetic particles self-organization.

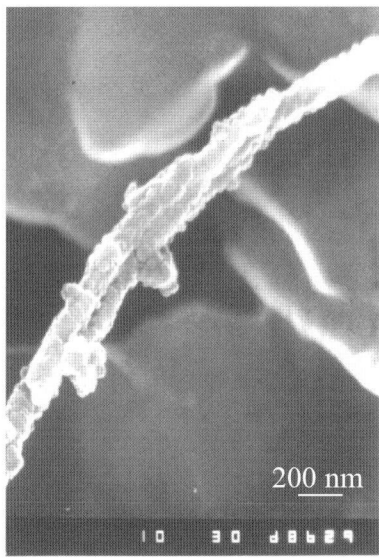

Fig. 3. SEM micrograph of Fe-Co-Cr-B(N,C,O,H) nanoparticles obtained with applying of a d. c. magnetic field

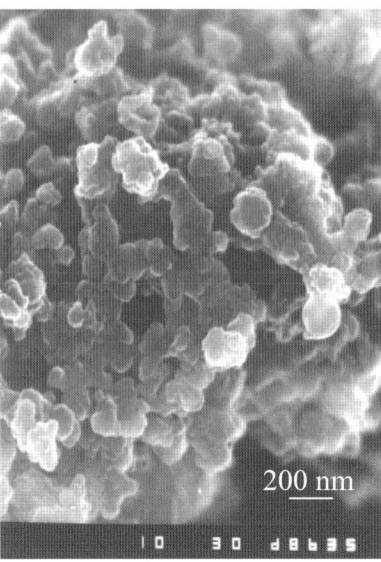

Fig. 4. SEM micrograph of Fe-Co-Cr-B(N,C,O,H) nanoparticles obtained without applying of a d. c. magnetic field

The d. c. magnetic field applied during the synthesis induces a chain arrangement of nanoparticles. This phenomenon was observed earlier at the synthesis of nanoparticles obtained by the borohydride reduction method [7, 8].

These nanoparticle chains hold on the surface quantities of OH groups. The OH groups have been observed in TEM micrographs and in FT-IR spectra [9].

The yeild of obtained samples as a weight by using identical concentrations of initial solutions undoubtedly shows the inclusion of oxygen and the

formation of greater quantity of oxides/hydroxides upon the formed nanoparticle chains. The smaller particles forming nanoparticle chains allow obtaining a higher yield (1.31g) after filtration, while the loss of bigger nanoparticles obtained without applying of d. c. magnetic field during the synthesis is higher (up to 30%).

The determined via XRD data diameters of the crystallites of the $Fe_3O_4/\gamma Fe_2O_3$ phases and of Fe/Fe-Co/Fe-B - phases are close in values. They not allow precise evaluation of the influence of the applied magnetic field on grain size.

3. 2. FT-IR study Fe-Co-Cr-B(N,C,O,H) nanoparticles

Figure 5 present FT-IR spectra of nanoparticles obtained by chemical reduction with $NaBH_4$ in aqueous solutions of Co (en)$_2$ complexes [Co(en)$_2$.Cl$_2$]Cl and aqua solutions of $FeCl_2.4H_2O$ and $CrCl_3.6H_2O$ respectively applying and not applying a magnetic field during the synthesis.

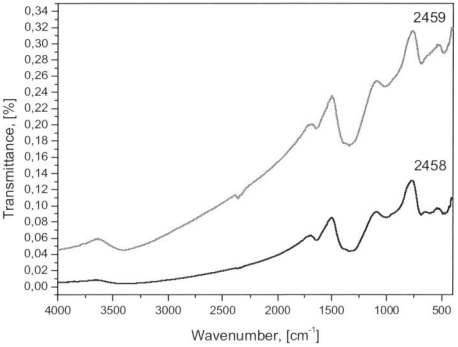

Fig. 5. FTIR spectrum of Fe-Co-Cr-B(N,C,O,H) nanoparticles #2458 and #2459

FT-IR spectra of Fe-Co-Cr-B(N,C,O,H) nanoparticles in Figure 5 prove the creation of chemical bonds between N-H, C-N, B-O, H-O atoms in different attached to surface atom groups and molecules and give information about the synthesis conditions. FT-IR spectrum of nanoparticles obtained in the presence of magnetic field during the synthesis differ from FT-IR spectrum of nanoparticles obtained without the presence of a magnetic field during the synthesis.

The absorption bands observed in 3500-400 cm^{-1} frequency range characterize stretching vibrations ν (NH$_2$) mode of N-H bonds in NH$_2$ groups (at 1339.2 ÷ 1338.0 cm^{-1}) and stretching vibrations ν(CN) mode of C-N bonds (at 1020.7÷1006.8 cm^{-1}), as well as stretching symmetric ν^s (BO$_4$) mode of vibrations of B-O bonds in BO$_4$ groups (at 686.5 ÷ 685,5 cm^{-1}), respectively deformation δ (BO$_4$) mode vibrations of B-O-B bonds in the same BO$_4$ groups (489.6 ÷ 485.5 cm^{-1}).

Absorption bands characteristic for stretching and deformation vibrations of O-H bonds in OH groups and H$_2$O molecules have appear at

$3398.7 \div 3387.4$ cm^{-1}, and $1648.9 \div 1634.9$ cm^{-1} respectively.

FT-IR spectra undertaken catch the change in the synthesis technological conditions such as a magnetic field applied or not during the synthesis and show the influence of the magnetic field on the nanoparicles absorption. They prove the formation of different atom groups such as NH_2, CN, BO_4, respectively the adsorption of OH free groups and H_2O molecules on the nanomaterials surface.

4. Conclusion

The d. c. magnetic field applied during the chemical reduction synthesis of Fe-Co-Cr-B(N,C,O,H) nanoparticles influences on the magnetic parameters, as well as on the chain formation phenomenon from the smaller particles holding OH groups on the their surface.

FT-IR spectroscopy study of the obtained nanoparticles proves the influence of the synthesis conditions and the d. c. magnetic field application on the creation of chemical bonds in different atom groups attached to the surface, respectively on the nanoparticles absorption/transmittance in the mid-IR range. The synthesized nanoparticles in the presence of a d. c. magnetic field decreasing significantly their transmittance.

Acknowledgments:

This investigation was carried out with financial support of the National Science Fund at the Ministry of Education and Science under Project NT-5-01/2006, NT-5-03/2006. Our team wants to thank to Prof. Bekir Aktas from the Gebze Institute of Technology, Turkey for the magnetic measurements of the samples and to Prof. Ivan Mitov and Dr. D. Paneva for Moessbauer investigations.

References

[1] I. Dragieva, Z. Stoynov, K. Klabunde, Synthesis of Nanoparticles by Borohydride Reduction and their Applications, *Scripta Mater.* 44 (2001) 2187.

[2] S. Stoeva, K. Klabunde, C. Sorensen, I. Dragieva, Gram-Scale Synthesis of Monodisperse Gold Colloids by the Solvated Metal Atom Dispersion Method and Digestive Ripening and Their Organization into Two- and Three-Dimensional Structures, *Journal of the American Chemical Soc.,* 124, 10 (2002), 2305.

[3] G. Glavee, K. J. Klabunde, C. Sorensen, G. Hadjipanayis, Sodium Borohydride Reduction of Cobalt Ions in Nonaqueous Media. Formation of Ultrafine Particles (Nanoscale) of Cobalt Metal, *Inorganic Chemistry* 32, 4 (1993), 474.

[4] I. Markova-Deneva, K. Alexandrova, I. Dragieva, Synthesis and Characterization of Cobalt Nanoparticles, Nanowires, and Their Composites, *Proceedings of the 3rd International Conference on Multi-Material Micro Manifacture,* 4M 2007, 211, Eds. Stefan Dimov, Wolfgang Menz and Yuli Toshev.

[5] C. Popov , S. Bliznakov, S. Boycheva, N. Milinovik, M. D. Apostolova, N. Anspach, C. Hammann, W. Nellen, J.P. Reithmaier, W. Kulisch, Nanocrystalline Diamond/Amorphous Carbon Composite Coatings for Biomedical Applications., *Diamond & Related Materials*. 2008 (in press).

[6] I. Dragieva, S. Stoeva, P. Stoimenov, E. Pavlikianov, K. Klabunde, Complex Formation in Solutions for Chemical Synthesis of Nanoscaled Particles Prepared by Borohydride Reduction Process, *Nanostructured Materials*, 12 (1999) 267.

[7] D. Mehandjiev, I. Dragieva, M. Slavcheva, Study of the Interglobular Structure of Metal Powder Obtained by the Borohydride Reduction Process, *J. Magn. Magn. Mat.* 50 (1985) 2525.

[8] I. Dragieva, D. Buchkov, D. Mehandjiev, M. Slavcheva, Amorphous Magnetic Powders - Some Special Peculiarities, *J. Magn. Magn. Mat.,* 72 (1988) 109.

[9] I. Markova-Deneva, K. Alexandrova, I. Dragieva, FT-IR spectroscopy investigations of nanostructured materials, *9th National Workshop Nanoscience & Nanotechnology, Sofia,* November 28-30, 2007, Topics B, B4 (in press).

Template fabrication incorporating different length scale features

G. Lalev[1], P. Petkov[1], N. Sykes[2], V. Velkova[1], S. Dimov[1], D. Barrow[2]

[1] *Manufacturing Engineering Centre, Cardiff University, Newport Road,Cardiff, CF24 3AA, UK*
[2] *metaFAB, Cardiff University, Newport Road, Cardiff, CF24 3AA, UK*

Abstract

A cost effective methodology for pattering of Nano Imprint Lithography (NIL) templates with different length scale features is proposed. The approach relies on selecting the optimum processing window of different technologies for cost effective micro and nano patterning. Very promising results were obtained when first fused silica templates were structured by F2 laser ablation at 157 nm without inducing phase transformation of the material. It was demonstrated that nanoscale features and complex 3D microscale features could be machined with a Focused Ion Beam (FIB) over the existing topography produced by laser ablation. Thus, a large area (up to several square centimetres) of the NIL templates is easily patterned with micro- and even meso-scale features by laser ablation while nano- and micro-scale features could be introduced by FIB machining.

Keywords: Focus ion beam (FIB), laser ablation, micromachining, nanostructuring, 3D machining, Nano Imprint Lithography

1. Introduction

Nano Imprint Lithography (NIL) is considered as the next generation lithography (NGL) technology with potential application in integrated photonic devices, nanoelectronics, life sciences, patterned media and next generation memory devices. Today the industry looks to NIL as an enabling technology for novel devices and also as a means for replacing high-end substrate materials with low cost polymers. According to the International Technology Roadmap for Semi-conductors (ITRS) NIL is a as a very promising solution considered for the 32 nm technology node, and beyond. The technology is also envisaged to fulfil the technical requirements of other non-mainstream applications outside of semiconductors, such as in biotechnology.

Recently, Step and Flash Imprint Lithography (S-FIL™) was proposed as one of the most promising nanoimprint techniques for nanopatterning large surface areas [1]. It was demonstrated that in the sub-50 nm regime, the resolution of the S-FIL™ process is only limited by the resolution of the template fabrication processes. Particularly, the resolution of the imprinting technology is strictly dependent on the ability to create a 1x master template, and improvements in feature resolution can be achieved without new light sources, optical systems or resist materials. In this sense, imprint lithography is a multi-generational technique that can be used to facilitate device and process prototyping at several upcoming lithography nodes. Since the patterning process is carried out at room temperature and ambient pressure using UV curable resists it is possible to achieve accurate overlay and reduce significantly process defects [2]. The main advantage of nanoimprinting compared to other high throughput techniques is the ability to replicate accurately 2.5D and 3D structures incorporating features at different length scales, and varying forms from simple shapes to complex diffractive optical elements.

It is worth stressing that thermal NIL is now more widely employed for nano- and micro-pattering, largely because it is cheaper and works with a broad range of polymer materials. However with this imprinting process the replication of features at different length scale (e.g. micro- and nano-scale features in one go) is an issue due to the difference in polymer flow, different effective pressure between nano- and micrometric patterns and varying residual thickness. As a result, uniformity over the entire wafer is difficult to achieve because the applied pressure must be uniform over the wafer surface. Using thermal imprinting a good printing uniformity was demonstrated over an 8-in. wafer for pattern dimensions from 250 nm to 100 μm [3]. Currently, there is demand for fabrication of monolithic devices requiring much smaller features (e.g. sub 50 nm gratings for photonic application combined with micro fluidic channels, etc.).

On the other hand, UV-NIL has advantages including (i) absence of thermal expansion that impedes precise alignment, (ii) low imprint pressure and low viscosity of the uncured resist, which allows uniform pattering over large areas, (iii) almost negligible difference in the residual layer thickness and very importantly (iv) to imprint nanometric and micrometric patterns, simultaneously. The only limitation for the process is the cost effective fabrication of 2.5D and 3D masters/templates. Different pattering techniques can by utilized for nano- and micro-scale structuring in the range from sub 50 nm to hundreds of micrometers. However, all of them have some limitations in terms of either limited resolution or processing speed

Currently, the templates' pattering in mask making facilities is mostly undertaken using e-beam lithography (EBL) employing shaped and Gaussian beam tools. The latter offers finer resolution, but with significantly longer writing time. For direct writing of 3D nanoscale features, single beam FIB technology could be successfully applied but the slow milling speed makes it unacceptable for industrial applications. To achieve cost-effective 3D nanopatterning of large surface areas, any potential solution should integrate the capabilities of high-resolution massively-parallel beam-based technologies with those of high-throughput replication techniques. Projection maskless pattering (PMLP) technology can fulfill all the above mentioned requirements and is, therefore, ideally suited for the fabrication of complex 3D nanoimprint templates. Such a system is currently under development within the European Framework Program Six Integrated project CHARPAN [4], where, through a programmable mask, a massive structured beam (incorporating hundreds of thousands highly parallel beams) patterns the wafer at relatively high speed compared to the conventional single beam FIB systems.

In this paper, a feasibility study on an alternative approach for cost effective structuring of UV-NIL templates is reported. The proposed approach combines the capabilities of two complementary technologies with different cost effective processing windows, F2 laser ablation and FIB machining, for 2.5D and 3D structuring of NIL templates incorporating different length scale features.

2 Nano- and micro-structuring of UV-NIL templates by FIB

2.1 Fabrication of templates with 2.5D nano- and micro-scale features

It has already been reported that the FIB technology could be applied for producing complex 2D and 3D micro and nano structures using a layer-by-layer fabrication method [5]. Li *et al* [6] demonstrated the use of FIB milling for structuring fused silica as an alternative solution to EBL for the fabrication of S-FIL™ templates with simple 2D structures. Template fabrication employing FIB is much simpler compared to the processing chains associated with the use of EBL because several processing steps can be omitted. In particular, the processing steps involved in the fabrication of templates when FIB milling is applied are depicted schematically in Fig. 1 .

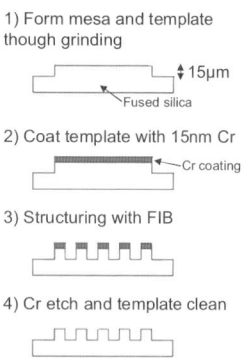

1) Form mesa and template though grinding

2) Coat template with 15nm Cr

3) Structuring with FIB

4) Cr etch and template clean

Fig. 1 Main steps for FIB patterning of UV-NIL templates.

Prior to FIB processing, a 15 nm layer of Chrome (Cr) is deposited on the top surface by thermal evaporation. This is required to limit the surface charging effect and pattern drift during ion-beam exposure. Cr was preferred to other evaporation sources such as gold because Cr appears to adhere better to the fused silica surface. Once the evaporation is completed, the template is solvent cleaned again before FIB exposure.

The results of the experimental investigation into the existing functional dependence between the milling depth and ion fluence for fused silica reported in [6] were used to set-up the processing parameters. In this feasibility study a Carl-Zeiss XB 1540 FIB/SEM cross-beam system was used to structure the active area of the template. The FIB patterning was externally controlled by the Raith Eliphy Quantum lithography

hardware and software. To speed up the patterning process, especially when 2.5D relief structures have to be produced, FIB milling can be applied only for pattering of a Cr mask on a quartz template followed up by ICP/RIE etching to fabricate them. Gratings fabricated in this way are shown in Fig 2.

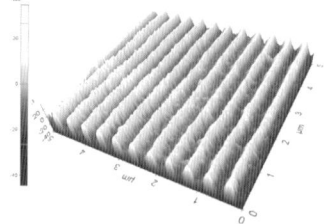

Fig. 2 AFM image of 150 nm grating in fused silica with 40 nm depth realized through dry etching.

Initially, a 15 nm Cr layer was patterned through FIB sputtering. Next, ICP/RIE etching was carried out for 15 min followed up by striping of the remaining Cr layer with Microposit 18 for 300 seconds and then rinsing with deionised water for a further 300 seconds. The sample was further cleaned with solvents to eliminate any other residues from the etching. The carried out AFM measurements showed that the resulting trenches were 40 nm deep, which is more that twice deeper than the Cr pattern.

To prepare the template for S-FIL™ imprinting the process chain includes further processing steps including NMP cleaning and Piranha etch. Finally, the template active area has to be coated with the anti-sticking agent, Relmat, supplied by Molecular Imprints Inc., to facilitate the release of the template from the imprinted polymer.

It is worth stressing that because a FIB/SEM cross-beam system is employed for patterning the template it is possible to carry out a simultaneous inspection of the patterned areas with the integrated SEM. This offers a viable solution for inspecting some critical structures of templates prior to imprinting, and thus to address metrology issues associated with the fabrication of S-FIL™ templates, highlighted by several research groups [7]. In this way, the inspection is performed on the machine and it is not required to use alternative inspection solutions that add complexity to the fabrication process [8].

2.2 Fabrication of NIL templates with complex 3D micro- and nano-structures

Conventional FIB systems operate using bitmap data which define 2D cross-sections. However, to fabricate complex 3D surfaces, it is required to design stacks of layers in order to define the structure accurately. Thus it is impractical to design and import manually the bitmap files required for producing successive layers. One possible way to overcome this problem is to employ GDSII stream format for converting the geometrical design data available in 3D CAD models into a format that can be used to control the FIB milling process through a proper lithography hardware and software, e.g. Elphy Quantum. The method is described in detail elsewhere [9]. The opportunities offered by this CAD-CAM approach were demonstrated by fabricating an array of square micro lenses.

In this research, the same approach was applied

for producing complex 3D geometries incorporating different length scale features, nano structures over a micro topography. In particular, the test structure used in this feasibility study was Moth's eye lens. For this a 10x10 µm lens array, was surface patterned with nano-scale lenses (each 150 nm in diameter) in hexagonal arrangement. First, the 3D CAD model of the lens was generated. Then, this 3D model was sliced into layers and exported into GDSII file format. The template fabrication procedure was the same as that described in the previous paragraph. By optimizing the FIB exposure parameters, a 2x2 array of Moth's eye lenses was successfully milled into fused silica, as shown in Fig 3.

Thus, it was demonstrated that by adopting this CAD-CAM approach, complex topographies incorporating micro- and nano-scale features can be fabricated on NIL template with the necessary shape accuracy.

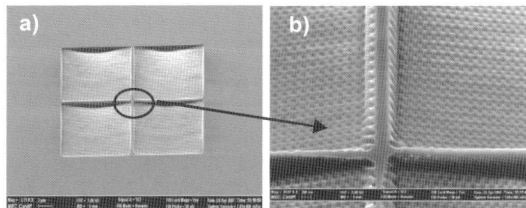

Fig. 3 a) SEM image of a FIB fabricated 2x2 array of Moth's eye lenses and b) SEM image detail of the patterned bottom of the microlenses with nanolenses in hexagonal arrangement.

3 Nano and micro structuring of UV-NIL templates by FIB/laser ablation

Since FIB machining is a very slow process it would be advantageous to apply a faster process for fabrication of large microscale features on relatively big surface areas that do not require nanometer resolution and precision. In this research, laser ablation, which has been used successfully for machining diffractive and refractive microstructures on polymer substrates [10,11] was investigated as a complementary technology to FIB.

3.1 Laser sources and experimental set-up

Dielectric materials, e.g. fused silica and glass, require the use of UV laser sources due to their poor approbation of longer wavelengths. In particular, synthetic quartz that is usually utilised as a UV-NIL template material exhibits over 90% transmission of wavelengths in the range from 180 to 600nm.

An alternative solution to UV laser sources is the use of lasers with ultrashort pulse durations, and thus to achieve energies per photon exceeding the band-gap of fused silica. Therefore, initially in this research the feasibility of structuring directly fused silica employing a pico-second laser source was investigated. The characteristics of the laser source that was used are provided in Table 1.

Table 1 PS Laser source characteristics

Beam quality	M2 <1.5
Pulse duration	~ 8 ps
Wavelength	355 nm
Power	15 mW
Pulse frequency	10 kHz
Spot size	15 µm

Due to relatively long wavelength, 355nm, the absorption coefficient was very low, ~ 92% light transmission. Although it was possible to structure the substrate by increasing the power, the resulting roughness was quite high, in the region of Ra 0.6 µm. In spite of the short pulse duration, a further increase of the power was not possible due the high thermal load on the substrate that triggered a material phase transformation from amorphous to polycrystalline. Polycrystalline materials are not ideal for subsequent dry etching due to the increased roughness of the sidewalls. Thus, this manufacturing route was ruled out as suitable for template fabrication.

To overcome this problem two other possibilities were considered, either a further reduction of the pulse duration, going to the femto-second (fs) range, or the use of deep ultraviolet (UV) laser sources. In this study, the latter was selected due to its relatively high material removal rates in comparison to fs-lasers. In particular, a vacuum UV laser source at 157 nm was chosen because it can provide photon energies exceeding the band-gap of fused silica [12] and material removal rates of about 100 nm per pulse can be achieved.

The set-up used in this research incorporated a LPF220 excimer laser source, summarized in table 2

Table 1 F2 Laser source characteristics

Pulse duration	26 ns
Wavelength	157 nm
pulse energy	40mJ
Frequency	2 kHz
fluence	in excess of 2J/cm2

This was coupled through a 2.5m N2 perfused (<20ppm O$_2$) enclosed beamline to an x-y-z-ø workpiece holder (50nm lateral resolution) with the beam imaged and focused through a 31x Schwartzfeld reduction lens. The beamline incorporated twin, fly-eye homogenisers for the creation of a uniform beam profiles at the mask plane. A white light through-the-lens viewing system allowed focusing and substrate-beam registration. The 157nm laser beam was shaped using a metal mask, which was fabricated using a Thales Bright 130fs laser with 2µm spot.

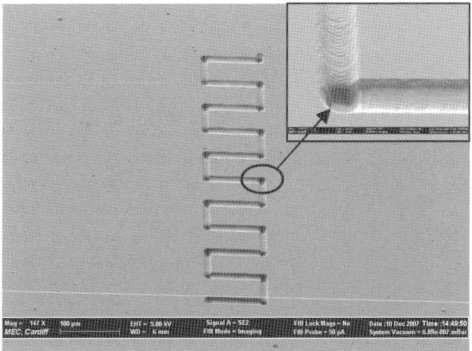

Fig 4. SEM image of grooves with 10 µm width and total length of 1.5 mm machined by F2 laser ablation in fused silica.

3.2 Experiments and results

A fused silica template was successfully patterned using F2 laser at 157 nm wavelength.. Since the set-up described in the previous section uses 31x demagnification projection optics, to ablate the required micro structures, masks were fabricated using the integrated fs laser. In Fig. 4, an example of 10 µm width grooves with a total length of 1.5 mm machined in fused silica is given.

After carrying out some initial test, the capabilities of FIB and F2 laser ablation were combined in order to produce structures that incorporate different length scale features over a relatively large surface area. Fig 5 gives an example of micro and nano features machined by FIB over a pre-existing topography created by F2 laser. In particular, first a 10x6 array of holes were patterned in the fused silica template with pitch of 20 µm with the F2 laser. The holes were 10 µm in diameters and had different dept varying from 150 nm to 3µm. Next, inside the holes 2.5µm and 1.5µm diameter concave microlenses were fabricated by FIB machining utilising the CAD-CAM approach described in Section 2.2. The achieved resolution and shape accuracy of the lenses show clearly that the fused substrate has not undergone phase transformation as a result of the F2 laser ablation. In addition, 100 nm line gratings were machined inside the holes by FIB to test whether the microstructured material is suitable for nanostructuring (bottom right insert in Fig 5.)

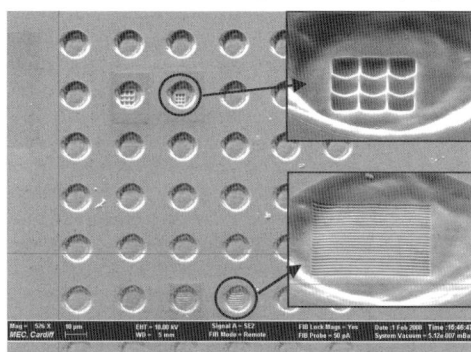

Fig 5. SEM image of a 10x6 array of micro-holes machined by F2 laser in fused silica. The top and bottom right inserts show the FIB machined concave micro lenses and 100 nm line grating, respectively.

4. Conclusions

The paper describes a cost effective method for fabrication of NIL templates incorporating different length scale features by combining the capabilities of laser ablation and FIB machining. It was demonstrated that F2 laser ablation at 157 nm does not trigger any phase transformations in the fused silica substrate. This is a very important point, since this enables the subsequent overlay FIB machining. Different features including both concave micro lenses and nanogratings were FIB machined over the micro features fabricated by laser ablation. Thus, a large area, up to several square centimetres of the NIL templates, can be cost effectively patterned with micro- and even meso-scale features by laser ablation while nano- and micro-features can be added by follow-up FIB machining.

The proposed fabrication route for producing UV-NIL templates incorporating a wide range of micro and nano features simultaneously can find diverse applications. For example, the fabrication of nanogratings onto micro machined features can be used to achieve function integration in novel lab-on-a-

chip or point of care devices, where the micofluidic channels can be monolithically integrated within the optical sensing system.

Acknowledgements

The process development reported in this paper is funded under the European FP6 Project "Charged Particle Nanotech" (CHARPAN), the metaFab and MicroBridge programmes supported by the Welsh Assembly Government and the UK Technology Strategy Board, and the EPSRC Program "The Cardiff Innovative Manufacturing Research Centre". Also, it was carried out within the framework of the EC FP6 Networks of Excellence, "Multi-Material Micro Manufacture (4M): Technologies and Applications".

References

[1] D. Resnick, S.V. Sreenivasan, and C.G. Willson, "Step & flash imprint lithography", Materials Today 8 (2005) 34.

[2] J. Choi, et al., "Distortion and Overlay Performance of UV Step and Repeat Imprint Lithography", Microelectronic Engineering Vol. 78-79,(2005) 633-640.

[3] C. Perret, C. Gourgon et al., "Characterization of 8-in. wafers printed by nanoimprint lithography" Microelectron. Eng. 73–74 (2004) 172.

[4] (www.CHARPAN.com)

[5] Fu Y., Bryan N.K.A, Fabrication of three-dimensional microstructures by two-dimensional slice by slice approaching via focused ion beam milling, J. Vac. Sci. Technol. B 22, (2004) 1672.

[6] Li W., Lalev G.M., Dimov S., Zhao H., Pham D.T., A study of fused silica micro/nano patterning by focused-ion-beam, Applied Surface Science, Volume 253, Issue 7, 30 January (2007), 3608-3614.

[7] Myron L.J., Gershtein L., Gottlieb G., Burkhardt B., Griffiths A., Mellenthin D., Rentzsch K., MacDonald S., Hughes G., "Advanced Mask Metrology Enabling Characterization of Imprint Lithography Templates", SPIE Microlithography Conference, Vol 5752 February (2005) 384-391.

[8] Resnick D. J., Myron L. J., Thompson E., Hasebe T., Tokumoto T., Yan C., Yamamoto M., Wakamori H., Inoue M., Ainley E., Nordquist K.J., Dauksher W.J., "Direct die-to-database electron beam inspection of fused silica imprint templates", J. Vac. Sci. Technol. B Vol 24, No6, Nov/Dec (2006) 2979-2983.

[9] G Lalev, S Dimov, J Kettle, F van Delft and R Minev, "Data preparation for FIB machining of complex 3D structures", Proceedings of the Institution of Mechanical Engineers, Part B, Journal of Engineering Manufacture, Vol 222 (2008) 67-76.

[10] G.P. Behrmann, M.T. Duignan, "Excimer laser micromachining for rapid fabrication of diffractive optical elements" Appl. Optics 36 (1997) 4666-4674.

[11] K. Zimmer, D. Hirsch, F. Bigl, "Excimer laser machining for the fabrication of analogous microstructures", Appl. Surf. Sci. 96–98 (1996) 425-429.

[12] P.R. Herman, R.S. Marjoribanks, A. Oettl, K. Chen, I. Konovalov, S. Ness, "Laser shaping of photonic materials: deep ultraviolet and ultrafast lasers" Appl. Surf. Sci. 154-155 (2000) 577-586.

Process modelling and simulation

Micro injection moulding: simulation of melt flow behaviour

C.A. Griffiths, S.S. Dimov, E. B. Brousseau and M. S. Packianather

Manufacturing Engineering Centre, Cardiff University, Cardiff CF24 3AA, UK

Abstract

Micro injection moulding as a replication method is one of the key technologies for micro manufacture. The understanding of process constraints for a selected production route is essential at both the design stage and during mass production. In this research, an existing Finite Element Analysis (FEA) system is used to study the effects of four process parameters, namely melt and mould temperature, injection speed and part thickness. A special attention is paid to the melt flow sensitivity when filling micro channels, particularly the factors affecting shear rate and flow front temperature. The results obtained from two different simulation models are presented for two polymer materials, PP and ABS and conclusions are made about the important factors affecting part quality.

Keywords: Micro Injection Moulding, Finite Element Analysis, Viscosity, Process Parameters, Response Surface Methodology

1. Introduction

With the rapid development of micro engineering technologies there is an increasing trend towards product miniaturisation. The development of new micro devices is highly dependent on manufacturing systems that can reliably and economically produce micro components in large quantities. In this context micro injection moulding of polymer materials is one of the key technologies for micro manufacturing. Consequently, it is important to study the factors that affect the replication capabilities of the micro injection moulding process.

For effective processing of micro parts, polymer viscosity is a factor of significant importance. The balance between keeping the polymer temperature sufficiently high to fill the cavity and ensuring at the same time that it does not fluctuate above a critical point is essential for achieving a stable process. Existing research in micro injection moulding has found that high melt and mould temperatures, and high injection speeds facilitate the filling of micro cavities with high aspect ratios [1]. However, high process settings can lead to a temperature related melt fracture of polymers resulting from the increase of the shear stress. In this context, the behaviour of pressurised polymer materials in contact with tool surfaces, particularly micro channels, could be an important factor affecting the part quality [2]. The effects of these various factors can be simulated using existing Finite Element Analysis (FEA) systems. In particular, to simulate polymer flow of generalised Newtonian fluids these FEA systems employ viscosity models such as the Cross Williams-Landel-Ferry (WLF) together with the Hele-Shaw flow model.

In this context, the objective of this study is to investigate the simulated flow behaviour of the melt in a micro cavity with a particular focus on the polymer shear stress (τ), the flow front temperature (T_{ff}) and the flow length. To achieve this, a series of simulation runs is carried out and the effects of a range of process variables on the filling behaviour are analysed. The paper is organized as follows. First, the factors affecting the viscosity of polymer melt flows are presented. Then,

two different modelling approaches developed within the Moldflow Plastics Insight 5.1 software are described together with the design of experiments method employed for planning the simulations. Finally, the results are reported and conclusions are made about the important factors affecting part quality.

2. Viscosity model

Viscosity models should represent the behaviour of polymer melts, fundamentally their viscosity and temperature [3]. The Cross WLF viscosity model, with the temperature and pressure dependence factor, is a mathematical expression that describes the shear thinning behaviour of polymers and is widely used for numerical simulations. In particular, the modified Cross WLF equation describes the viscosity η as [4]:

$$\eta = \frac{\eta_0}{1 + \left(\dfrac{\eta_0 \gamma}{Tau}\right)^{(1-n)}} \tag{1}$$

where: η_0 represents the zero shear viscosity at the centre of the polymer flow cross section, γ is the shear rate and n and Tau are data fitted coefficients. More specifically, n represents the shear rate sensitivity and $(1 - n)$ characterises the slope of the line over the pseudo plastic region in the logarithmic plot of η and γ [5].

The Cross WLF model uses η_0 as a function of temperature and pressure and is considered to be more effective at high and low γ in comparison to the Power law model [6].

3. Simulation models

In this study FEA is used to simulate polymer flow in micro cavities. It is well known that FEA models are widely used to simulate replication processes at macro scale and in the last two decades significant advances in this field were reported. Also, some attempts were made to apply such models for simulating injection moulding of micro parts and features [1, 7] and a good correlation between actual process behaviour and simulation results was reported [8].

The Finite Element Method (FEM) was employed in this research to create a model for simulating the polymer filling behaviour in micro cavities. By applying Finite Difference Method (FDM) the initial mesh was generated for the geometrical model that was analysed. Each element was examined using a mesh statistic tool in order to verify its accuracy within the model domain. Then, the accuracy of the imported geometric model was adjusted through node and element modifications to achieve a good reproduction of the model surface boundaries.

The CAD model once imported and meshed by applying such a hybrid FEM-FDM approach is used for dual domain analysis of laminar flow in generalised Newtonian fluids utilising the Hele-Shaw flow model. In addition, a second simulation approach was adopted by applying a tetrahedral element mesh to the CAD model in order to perform a complementary 3D analysis based on the Navier-Stokes flow model. By implementing the Cross WLF viscosity model together with the orthogonal FDM modified dual domain and 3D meshes [9], it was possible to carry out the required simulation analyses.

For the simulation study reported in this research it was required to specify injection time (t_i), melt temperature (T_b), and mould temperature (T_m) for each simulation run. In addition, to take into account the high surface to volume ratio in micro injection moulding, a size factor was introduced. This size factor is a global thickness multiplier (GTM) that allows any increase or decrease of the component dimensions to be taken into account when analysing the flow behaviour. In our case, the GTM variation factor was set at 5% thus varying the FEA model with an increases or decreases of part thickness by 25 µm.

The simulation models created in this way were used to investigate flow length and possible flow instabilities that can lead to surface defects [10].

4. Experimental setup

4.1 Test part

To validate the proposed approach for simulating the melt flow behaviour at micro scale a test part was used to create a FEA model. This part is shown in Fig. 1.

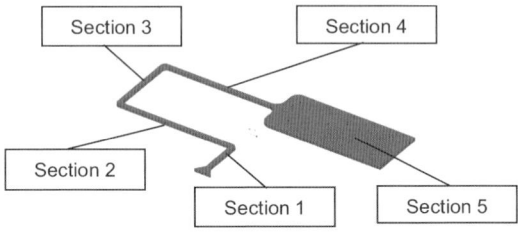

Fig. 1. Test part.

It consists of four runners with unequal lengths that lead to a rectangular section. The dimensions of these five sections are given in Table 1. All corners have a radius of 0.5 mm. The runner system has a square cross section with dimensions 0.5 x 0.5 mm and a surface to volume (S/V) ratio about 30% higher than that of a circular cross-section (Table1).

Table 1. Test part and tool design

Section	1	2	3	4	5
Length [mm]	5	14.5	7.3	14	15
Total length [mm]	5	19.5	26.8	40.8	55.8
Volume [mm³]	55.80				
Area [mm²]	285.5				
S/V ratio	5.11				

4.2 Planning of simulation experiments

The filling performance of micro moulds is highly dependent on temperature control and V_i. To simulate their effects on τ, T_{ff} and the flow length, two commonly used materials in injection moulding, Polypropylene (PP) and Acrylonitrile Butadiene Styrene (ABS) were selected.

In this simulation study, the effects of T_b, T_m and t_i on the filling behaviour of the test part were investigated using Taguchi Design of Experiments. In particular, maximum, minimum and medium values of these process parameters within their recommended processing windows were utilised in the carried out simulation runs for the selected two materials (Table 2). Also, as it was already mentioned, to take into account the high surface to volume ratio of micro parts the GTM factor was introduced. Through it, an increase and decrease in the overall thickness of the parts can be modelled and thus to account for some of the scale effects in micro injection moulding.

To assess the effects of T_b, T_m, t_i and GTM on the melt flow behaviour, the Moldflow software tools for conducting Design of Experiments (DoE) and factorial analysis were applied. Finally, the simulation results were analysed applying the Response Surface Methodology (RSM) [11]. In particular, RSM creates mathematical models representing the interrelation between one or more responses of the set input parameters. The results are an evaluation of the main effects of the control factors presented as criteria weightings and RSM graphs of the three most influential factors.

Table 2. Design of experiments factors and levels

Inputs	PP	ABS
t_i [s]	0.1	
	0.3	
	0.5	
Melt temperature T_b [°C]	220	
	250	
	280	
Mould temperature T_m [°C]	20	40
	40	60
	60	80
Global thickness multiplier [µm]	475	
	500	
	525	

5. Experimental results

5.1 Analysis of shear stress and flow front temperature results

The analysis of τ and T_{ff} is based on the dual domain simulation model only. The results in Table 3 show the importance of the considered four factors in order of their rank weightings when their effects on resulting T_{ff}, τ and overall quality are analysed. For both, PP and ABS, t_i was the most important factor based on the overall rank weightings, followed by T_b and T_m. However, PP showed a higher dependence on t_i while the influence of T_b and T_m was more pronounced for ABS. The GTM factor had a negligible effect when all criteria weightings were considered. In addition, t_i had the highest influence on τ for both PP and ABS. This is also observed for the flow front temperature when PP is considered. In the case of ABS however, T_b was the most influential factor on T_{ff}.

Table 3. Moldflow DoE Results

Criterion weightings	Control Factors	PP	ABS
τ rank weighting (%)	t_i	83.41%	64.28%
	T_b	13.81%	17.87%
	T_m	1.79%	17.64%
	GTM	0.89%	0.17%
T_{ff} rank weighting (%)	t_i	81.59%	25.05%
	T_b	11.54%	43.65%
	T_m	6.17%	31.17%
	GTM	0.32%	0.00%
Overall quality rank weighting (%)	t_i	76.06%	46.31%
	T_b	21.05%	27.47%
	T_m	1.77%	26.05%
	GTM	1.04%	0.01%

In addition, the graphs in Fig. 2 to Fig. 5 show the interactions between the factors that affect τ and T_{ff} during the injection process. In particular, these RSM graphs depict the results of the carried out factorial comparisons of the multi-dimensional patterns of responses to varying control parameters.

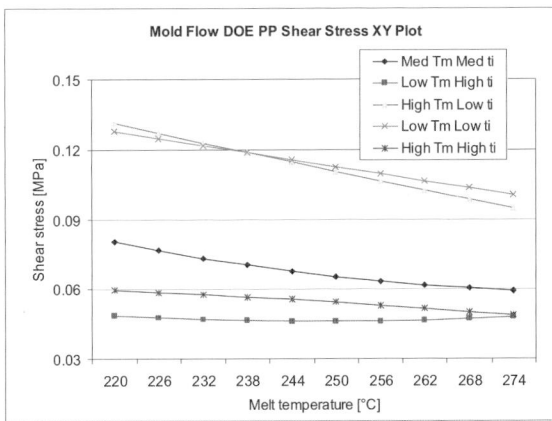

Fig. 2. PP shear stress.

Fig. 2 shows that low t_i settings result in a higher τ. The medium and high t_i led to significantly lower τ than that at low t_i. For all settings, an increase of T_b resulted in a decrease of τ, except for the combination of low T_m and high t_i. In addition, this low T_m and high t_i setting resulted in the lowest τ achieved in all simulation runs.

For ABS, the graph in Fig. 3 shows that again low t_i results in a higher τ. It is interesting to note that for the combination of low t_i and T_b below 244°C, τ exceeded its critical value. Similarly to PP, the medium and high t_i settings led to much lower τ than that at low

t_i. In addition, at low and medium t_i a continuous decrease of τ was observed with the increase of T_b. The high t_i setting led just to a small reduction of τ in the T_b range from 220°C to 250°C, while outside it, τ levelled out and showed only a small increase at higher end temperatures.

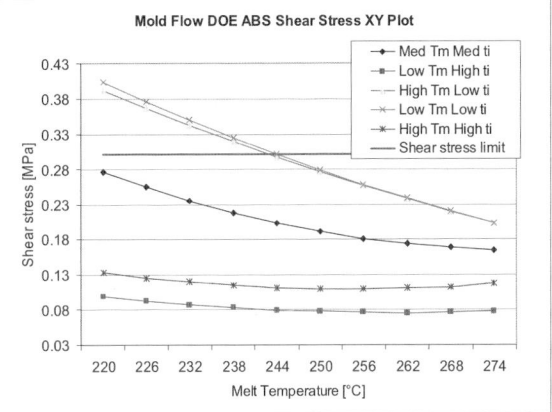

Fig. 3. ABS shear stress.

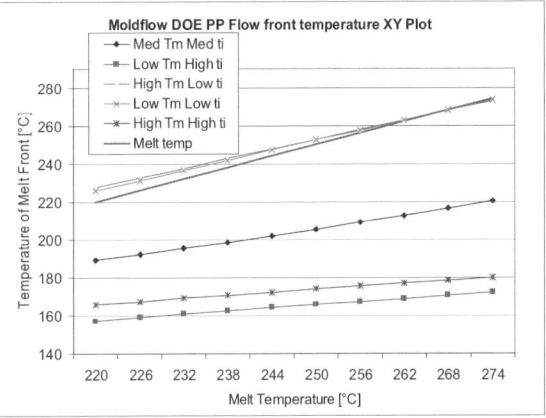

Fig. 4. PP Flow front temperature.

For PP, the graph in Fig. 4 shows that for all settings T_{ff} increases with the increase of T_b and this dependence is less distinct at high t_i. Low t_i settings led to a significant increase of T_{ff}, and at T_b below 260°C, T_{ff} exceeded the PP melt temperature. This indicates shear heating at low t_i combined with T_b below 260°C.

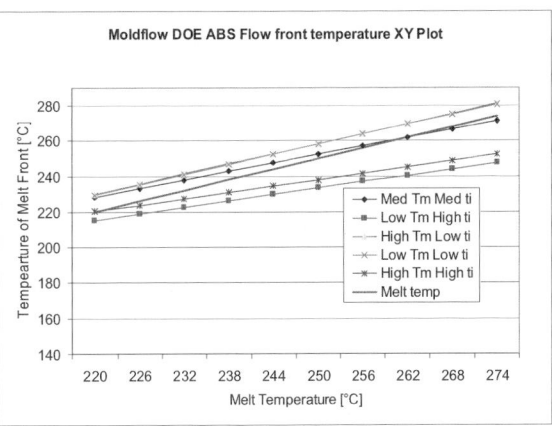

Fig. 5. ABS Flow front temperature.

For ABS, Fig. 5 shows that similarly to PP, for all settings T_{ff} increases with the increase of T_b. In addition, T_{ff} exceeded the ABS melt temperature at low and medium t_i and T_b below 260°C. This indicates again the existence of shear heating at these settings.

5.2. Analysis of flow lengths

In Table 4, the experimentally obtained maximum and minimum flow lengths for both materials, PP and ABS, reported in [2], are compared with the results attained using the dual domain and 3D simulation analyses. Overall, the dual domain simulation model resulted in 100% filling of the cavity in all tests. The 3D simulation model gave a lower estimation in comparison to the dual domain simulation results. In particular, on average the filling was between 70-80% for the 3D simulation model.

Compared to the actual results the dual domain flow analysis overestimated the polymer flow length in all simulation runs. The 3D flow analysis underestimated the polymer flow length in PP simulation runs. However, for ABS there was both an overestimation and underestimation of the flow lengths

Table 4. The results as a percentage of the maximum and minimum flow length

Flow length simulations	Actual % filled	Dual domain simulation % filled	3D simulation % filled
PP Min	84.4	100	73
PP Max	97.9	100	79.3
ABS Min	52.4	100	73.9
ABS Max	80.55	100	72.1
Average %	78.8	100	74.6

6. Conclusions

In this paper important factors affecting the flow behaviour in micro injection moulding were discussed and two FEA models were proposed to simulate the process. The following conclusions can be drawn from the study.

- Regarding the flow length, the simulation models were not in agreement. The dual domain flow analysis showed a 100% filling of the polymers in all simulation runs. The 3D flow analysis showed a 70-80% filling. Compared to the actual results Overall, 3D simulations for both materials were closer to the actual results.
- The T_{ff} and τ analysis results can be utilised to identify process conditions leading to surface defects. For example, high τ is the cause of unstable flow fronts and excessive shear heating that leads to material degradation. In addition, inconsistent T_{ff} across a micro part indicates potential problems in filling the cavity.
- For both PP and ABS, the simulation study showed that overall t_i is the most important factor affecting part quality. In particular, low t_i results in high τ and in the case of ABS the critical τ limit is reached at low t_i and low to medium T_b.
- For both PP and ABS, T_{ff} at low to medium t_i can result in an increase of T_b above its set-up level.

Hence, at this processing window shear heating occurs at the melt front and the frozen layer along the cavity walls.

Acknowledgement

The research reported in this paper is funded by the EPSRC Programme "The Cardiff Innovative Manufacturing Research Centre" and the EC FP6 Project ""Surface Enhanced Micro Optical Fluidic Systems (SEMOFS)". Also, it was carried out within the framework of the EC FP6 Networks of Excellence, "Multi-Material Micro Manufacture (4M): Technologies and Applications".

References

[1] Su YC, Shah J and Lin L. Implementation and analysis of polymetric microstructure replication by micro injection moulding. J. Micromechanics and Microengineering. 14 (2004) 415-422.

[2] Griffiths CA, Dimov SS, Brousseau EB and Hoyle RT. The effects of tool surface quality in micro-injection moulding. J. Mat. Processing Tech. 189 (2007) 418-427.

[3] Greene JP. Numerical Analysis of injection moulding of glass fiber reinforced thermoplastics Part 1 Injection flow and pressure. J. Polymer Engineering and Science. 37 (1997) 590-602.

[4] Theilade URO. Surface micro topography replication in injection moulding. PhD Thesis, DTU Department of Manufacturing Engineering and Management, Technical University of Denmark.

[5] Helleloid,GT. On the computation of viscosity shear rate temperature master curves for polymetric liquids. J. Applicable Mathematics. 1 (2001) 1-11.

[6] Young W. Effect of process parameters on injection compression molding of pickup lens. J. Applied Mathematical Modelling. 29 (2005) 955-971.

[7] Shen YK, Shie YJ and Wu WY. Extension method and numerical simulation of micro injection moulding. Int. Comm. Heat and Mass Transfer. 31 (2004) 795-804.

[8] Yuan S, Hung NP, Ngoi, BKA and Ali MY. Development of micro replication process. J. Materials and Manufacturing Processes. 18 (2003) 731-751.

[9] Galantucci LM and Spina R. Evaluation of filling conditions of injection moulding by integrating numerical simulations and experimental tests. J. Mat. Processing Tech. 141 (2003) 2066-275.

[10] Grillet AM, Bogaerds ACB, Peters GWM and Baaijens FPT. Numerical analysis of flow mark surface defects in injection moulding flow. The Journal of Rheology. 46 (2002) 651-669.

[11] Antony J. Multi-Response optimization in industrial experiments using Taguchi's quality loss function and principal components analysis. J. Quality and Reliability Engineering. 16 (2000) 3-8.

Implementation strategies for the optimization of micro injection moulding simulations

G. Tosello[a], A. Schoth[b], H.N. Hansen[a]

[a] Technical University of Denmark (DTU), Department of Mechanical Engineering,
Produktionstorvet, Building 427S, DK-2800 Kgs. Lyngby, Denmark
[b] Laboratory for Process Technology, Department of Microsystems Engineering (IMTEK),
University of Freiburg, George-Koehler-Allee 103,79110 Freiburg, Germany

Abstract

In polymer micro manufacturing technology, software simulation tools adapted from conventional injection moulding can provide useful assistance for the optimization of moulding tools, mould inserts, micro component design, and process parameters. Conventional implementation methods of simulation are not suitable for micro injection (µIM) application and are limiting the possibility to extend the use of existing packages for the modelling and the simulation of polymer micro parts. Different strategies optimized for the set-up the simulation of a miniaturized part with micro features are presented. Model design and mesh issues are discussed, as well as dynamic implementation of the flow constrains for the creation of an effective interface between the machine and the polymer flow in the simulation software. The results of the different methods are evaluated by means of a quantitative study which compares the simulated results and the actual micro injection moulding experiments.

Keywords: micro injection moulding, simulation, cavity injection pressure, cavity injection time

1. Introduction

Simulation programs in polymer micro technology are applied with same purposes as in conventional injection moulding. To avoid the risks of costly re-engineering, the functions of the final products as well as the manufacturing steps are simulated extensively before starting the actual manufacturing process: important economic factors are the optimization of the moulding process and of the tool using different simulation techniques.

Simulation tools can work adequately from a qualitative point of view but numerical values cannot be calculated as precisely as necessary [1]. In addition to that, most programs have difficulties in simulating exactly the filling of microstructures with high aspect ratio. The reason is that commercial software tools developed for macroscopic applications do not consider microscopic aspects properly.

However, a proper implementation strategy employed during the set-up of the simulation can greatly improve the quality (i.e. the accuracy) of the simulated results. During this research work, an extensive experimental data base (based on actual micro injection moulding experiments) has been set in order to carry out a comparative study with simulations results. Simulations were carried out using a commercially available code and their implementation was performed applying different approaches. Results were compared in a quantitative study to indicate the method which is leading to the most accurate results. In particular, performance indicators such as cavity injection pressure, injection cavity time and flow front position have been selected for the analysis.

2. Experimental validation of simulation of µIM

Validation of simulation software is an essential step in order to assess the capability of the implemented mathematical model to predict what is actually taking place during the process under investigation. In order to validate the polymer flow (i.e. filling) simulation in µIM, different approaches can be employed: a micro cavity partially filled in subsequent steps (the so-called short shots method) [2,3,4], flow-melt visualization method using a in-cavity high speed camera [5], the use of flow marker positions to trace the evolution of the flow front [6,7], the comparison of process parameters levels (i.e. injection pressure) sampled during the process. Eventually a comparison can be performed between the simulation and experimental results.

3. Experimental micro injection moulding

In order to establish a consistent database to be used for the simulation software validation, experimental micro injection mouldings were produced on a Battenfeld Microsystem50 injection moulding machine. In-cavity injection pressure samplings were executed using a piezoelectric pressure sensor applied at the injection location (where the melt is pushed into the cavity by the injection piston) of a two-cavity micro structured mould (see Fig. 1). Pressure samplings over time allowed the determination of the cavity injection speed and the punctual value of the cavity pressure during the filling of the cavity.

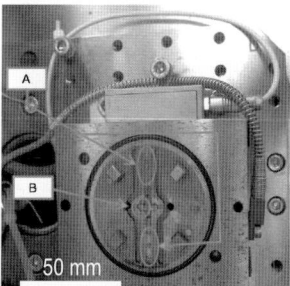

Fig. 1. Two-cavity micro mold (A) equipped with in-cavity pressure sensor at injection location (B).

The moulded component was a tensile bar test part with the shape of a thin plate (15x3x0.3 mm³) including three micro features having semi-circular section (150 μm radius) and lengths from 1500 μm up to 2000 μm (see Fig. 2, 4E, 4F). The part was moulded using a commercially available polystyrene (PS) polymer grade. PS is a relevant material in the field of micro injection moulding for its high flowability and optical properties (i.e. high transparency).

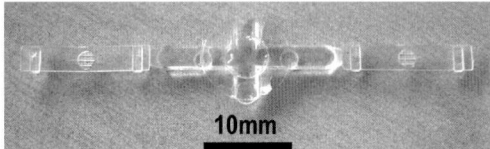

Fig. 2. Two micro injection moulded components and runner system.

The average weight of the complete moulded part was 119 mg and distributed as follows: 17% for each of the two parts and 64% for the miniaturized runner system. It is calculated that a single micro feature had a weight in the range from 0.14 up to 0.18 mg depending on the length.

Micro injection moulding was executed using the following process parameters settings: temperature of the melt = 220°C, temperature of the mould = 70°C, injection speed = 100 mm/s, total injected volume = 130 mm³. Mouldings were executed increasing the injected volume after each production batch. Partially filled moulded parts (i.e. short shots, see Fig. 3) were produced by injecting the following volumes (expressed in mm³): 55, 65, 75, 85, 90, 95, 100, 105, 110, 120, 125, 130 (complete part). For each batch, the cavity injection pressure was recorded with sample rates up to 25kHz (i.e. pressure samplings at 40 ns interval).

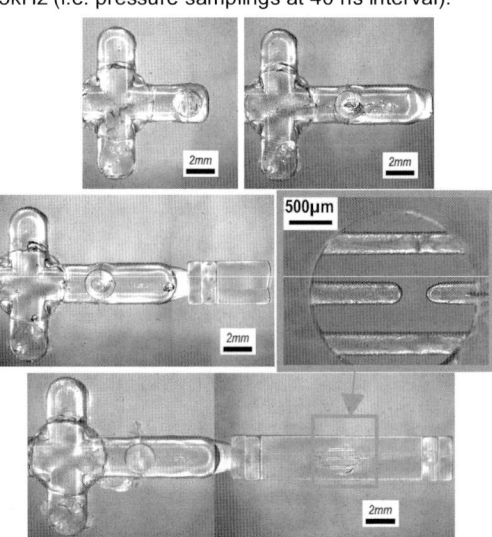

Fig. 3. Examples of short shots produced by μIM (polymer = PS, volume = 55, 75, 95, 125 mm³) and detail of the micro features filling (125 mm³).

During the injection moulding process, an automatic execution of the process (including ejection and handling of the polymer micro parts) was performed for each batch. Firstly, 50 cycles where carried out to stabilize the process in the current process parameter set-up. Subsequently, 10 parts obtained from the following 30 cycles have been randomly collected and analyzed. These randomly selected parts have been weighted and inspected with a calibrated optical microscope. During the all investigation, more than 800 micro moulded parts were produced. Moulded part weight for all batches showed a standard deviation of 0.35 mg.

4. Micro injection moulding simulation

The commercial software program Moldflow Plastics Insight® MPI 6.1 was employed for simulation. The main material functions implemented were Cross-WLF (William Landel Ferry) for viscosity and two-domain Tait for pvT. Details on the mathematical-physical models of the simulation including the finite element formulation are given in [8]. Three dimensional filling simulations were performed setting the melt temperature equal to the barrel temperature and the mould temperature equal to the temperature acquired by the sensor in the mould. Mould temperature was monitored with a closed-loop control system by the μIM machine.

Boundary conditions related to the part (i.e. modelling and meshing) as well as the dynamic management of the interface machine/polymer flow/cavity were investigated and are described in the following sections.

4.1. Part modelling and meshing

When simulating conventional macro-moulded parts, even though the part cavity is modelled and meshed with a three dimensional mesh, the sprue-runner system is usually meshed using simplified one-dimensional elements in order to save computational time. This is due to the fact that the volume of the sprue-runner system is very low when compared to the part volume, i.e. to 1 to 5%, and it is safely assumed it is not going to affect the results of the simulation sensibly.

On micro injection moulding simulation, as mentioned in the previous section, the runner system can easily account for more than 50% of the injection volume. The thermo-mechanical history of the melt flow is heavily influenced by the dynamic evolution through the whole runner, gate and finally the cavity. Therefore, a full three-dimensional meshing of the whole system composed by the runner (see Fig. 4A), the two gates (see Fig. 4B) and the two parts considered as a one complete moulded part was carried out. Furthermore, actual volume left on the surface by the ejectors and the pressure sensor were solid-modelled as part of the moulded product (see Fig. 4C, 4D). Mesh tolerances were also analysed and optimized to fit the need for accuracy of the μIM application. In particular, an average length of the side of a single element (i.e. tetrahedron) of 90 μm was adopted. Moreover, a tolerance between the meshed model and the CAD solid model of 30 μm was used. These settings were chosen as trade-off between the accuracy of the part modelling (especially needed for the micro features modelling, see Fig. 4E, 4F) and the resulting number of elements (i.e. the computational time). The model had a volume of 107 mm³ and was modelled using 1215156 tetrahedrons 3D solid elements.

4.2. Flow-time dependence implementation

When performing the micro injection moulding process, the injection speed (implemented as piston speed) is usually set as parameters to define the evolution over time of the melt to fill the cavity.

Different approaches can be employed to increase the accuracy. Two methods have been selected and are presented in the following.

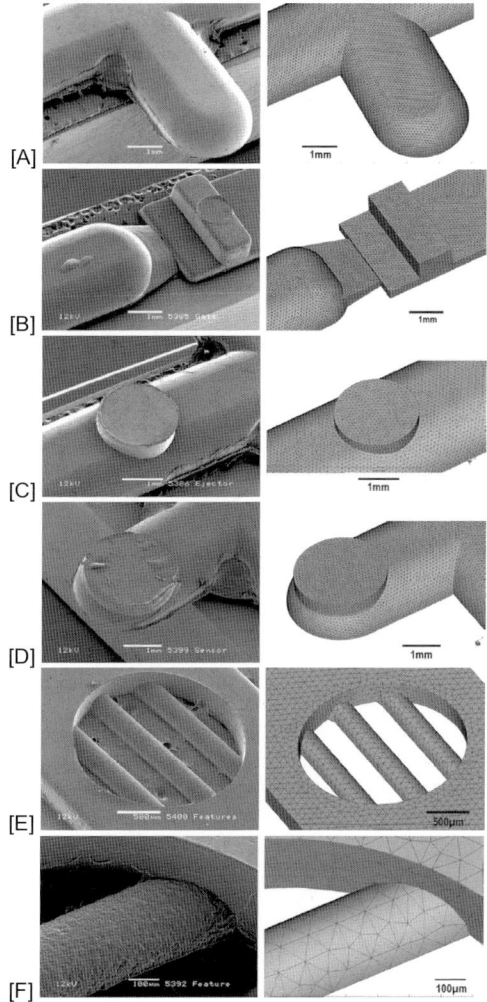

[A]

[B]

[C]

[D]

[E]

[F]

Fig. 4. SEM images of the micro moulded parts (left) and related three-dimensional meshed model for the injection moulding simulation (right).

A.   Injection time – From the injection cavity pressure plot (see Fig. 5) given by the pressure sensor the actual cavity injection time is determined and implemented into the simulation. The software will calculate the flow in order to fit the given time constrain. In particular, an initial transition time will be allowed to the flow rate to reach a stable value of the flow rate, simulating the delay due to the acceleration of the piston. On the other hand, when such value is reached, it is kept constant during the remaining injection time until the complete filling (condition not verify in reality). An experimental cavity injection time of 83±1 ms was determined and implemented.

B.   Cavity injection pressure Vs time – From the injection cavity pressure plot (see Fig. 5) given by the pressure sensor the actual cavity injection pressure over time is determined and implemented into the simulation. To obtain the experimental time-constrained pressure condition at the injection location, a definite flow rate (i.e. piston injection speed) is necessary and

is calculate by the software. Therefore the delay due to the actual piston acceleration is taken into account in the simulation, as well as the actual flow rate during the filling of the whole cavity. Also, the physical condition of the polymer (through a punctual value of the injection pressure) is determined. In the experiment, the pressure repeatability was calculated for each sampled value through the all cavity injection time for three randomly chosen mouldings. A standard deviation of 1.7 MPa was calculated.

The method (A) is carried out using a filling simulation performed in speed control. On the other hand, to implement the method (B) a packing simulation is carried out (i.e. in pressure control).

5. Results and discussion

The cavity pressure profile is a fundamental factor directly correlated to the quality of the part and of the process and it is the critical process parameter for the precision moulding of high accuracy thermoplastics [9] as micro moulded components. It is therefore of great importance that the injection pressure profile is simulated accurately in order to obtain reliable results.

First of all, it is important to observe that the maximum experimental cavity injection pressure (46 MPa) was reached at the cavity injection time of 83 ms, which corresponded to the complete part (injected volume of 130 mm^3). On the other hand, the experimental short shots corresponding to 125 mm^3 (injection time of 51ms) shows that, despite the fact that the main flow front reached the end of the cavity, one micro feature in the middle of the moulded was not filled yet (see detail in Fig. 3). The complete μ-feature filling was obtained during the last 32 ms of the cavity filling time when the polymer could still flow, the piston was applying the needed pressure and the flow rate was very low (160 mm^3/s instead of 2500-3500 mm^3/s calculated during the cavity filling) determining an intermediate flowing condition between a filling and a packing phase.

The comparison of the experimental and simulated pressure plot shows that (see Fig. 5):

•   Simulation A (i.e. implementation of the cavity injection time) appeared inaccurate in terms of pressure Vs time prediction and flow front position. The cavity injection pressure was much lower than the experimental for most of the cavity injection time. On the other hand the maximum cavity injection pressure at the very end of filling was of 55 MPa, i.e. 20% larger than in the experiments. Moreover, simulated short-shots showed that the micro features were completely filled before the main flow front has reached the end of the cavity. Complete micro features filling happened after 75 ms and complete part filling at 88 ms.

•   Simulation B (i.e. implementation of the cavity injection pressure profile) showed a pressure profile following the same trend as the experimental. On the other hand, despite the fact that the entire experimental pressure plot was implemented, the simulation stopped after 76 ms (complete filling of the cavity), the micro features were filled before the flow front reached the end of the cavity (after 50 ms) and the final filling/packing phase could not be predicted.

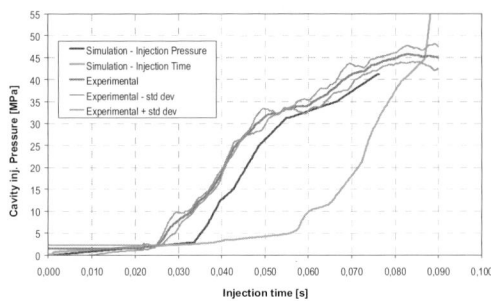

Fig. 5. Comparison of experimental and simulated cavity injection pressure profiles.

In order to assess the accuracy of the two implementation methods, the partial filling time determined with the experimental short-shots analysis was compared with the predicted partial times (Fig. 6).

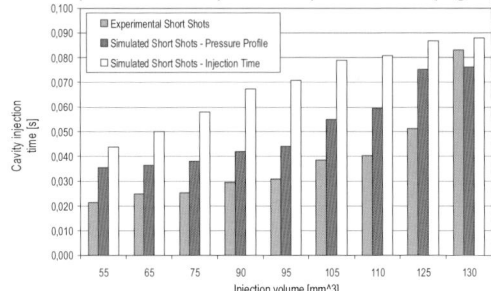

Fig. 6. Comparison of cavity injection times at different injection volumes.

The simulations are clearly predicting a slower filling, even though to a different extent, depending on the adopted method. An accurate implementation (B) resulted on a reduction of the error on the calculation of the partial filling time. In particular, while the injection time implementation A produced an overestimation of the partial filling time of more than 100%, the pressure profile implementation B resulted on an overestimation of less than 50%. This means that the dynamic conditions of the flow could be calculated with higher accuracy by using the more advanced approach. This would bring to the fact that the calculation of the shear rate is more accurate and therefore the viscosity values of the material during the filling are determined closer to the reality.

6. Conclusion and outlook

Optimized simulations of micro injection moulding have the potential to enable more effective design phase of polymer-based micro product. The implementation strategies of simulations can heavily influence the results and they have to be careful selected. In particular, the importance of an accurate modelling and three-dimensional meshing of the whole injection system has been shown. Moreover, the importance of reliable experimental data in terms of cavity injection pressure to be implemented in the simulation software has been demonstrated. In particular, the accuracy of the filling step could be improved of at least 50% in the injected volume/injection time domain. Pressure profile using an optimized implementation strategy could be simulated with higher accuracy when compared with a first approach based on the injection cavity time only.

Results in terms of accuracy of the flow front prediction were not satisfactory, showing dysfunctional behaviour of the predicted flow.

As final result, it can be concluded that the use of experimental data from actual moulding can greatly improved the quality of µIM simulations. Data should not be limited to the actual injection time, but also dynamic characteristics of the flow (i.e. pressure) should be implemented.

Further research will address other dynamic implementation approaches (e.g. injection speed or flow rate over time, etc). Moreover, other aspects to be investigated and evaluated with a quantitative method as the one presented in this paper are: the use of rheological model suitable for micro polymer application (which can effectively take into account, for example, the wall slip effect), heat transfer coefficient value optimized for micro scale polymer flow, the influence of pressure on the polymer melt viscosity, the use of an elongation viscosity model.

Acknowledgements

The present research was performed within a collaborative research activity program carried out between the Micro/Nano and Precision Manufacturing Research Group at the Department of Mechanical Engineering of DTU (Denmark) and the Laboratory for Process Technology at IMTEK (Germany). Dr. A. Schoth and the IMTEK staff are acknowledged for their hospitality to the corresponding author, for providing the experimental set-up and making available the simulation software. PhD students R. Jurischka and T. Hösel (both at IMTEK) are acknowledged for their help in connection with the µIM experiments and the process sampling respectively. DTU is acknowledged for sponsoring the study under its PhD funding programme. The work was carried out under the framework of the 4M NoE "Multi Material Micro Manufacture: Technology and Applications" (European Community founding FP6-500274-1; www.4m-net.org) and in connection with the activities of the Processing of Polymer Technology Division (4M Work Package 4).

References

[1] Piotter V et al. Microsystem Technologies. 8 (2002) 387–390.
[2] Jaworski MJ et al. Proceedings of the 62nd Annual Technical Conference ANTEC (2003) 642–646.
[3] Whiteside BR et al. International Polymer Processing. Vol. XXI-5 (2005) 162–169.
[4] Mehta MN et al. Proceedings of the 62nd Annual Technical Conference ANTEC (2003) 3550–3554.
[5] Han X et al. Polymer Engineering and Science. 46:11 (2006) 1590–1597.
[6] Tosello G et al. Proceedings of the 3rd International Conference on Multi-Material Micro Manufacture (4M) (2007) 259–262.
[7] Gava A, Tosello G et al. Proceedings of the 9th International Conference on Numerical Methods in Industrial Forming Processes (NUMIFORM) (2007) 307–312.
[8] Kennedy P. Flow Analysis Injection Molds, Hanser, 1995.
[9] Kazmer D and Barkan P. Polymer Engineering and Science. 37:11 (1997) 1880–1895.

Geometry Optimization of Micro Milling Tools

J. Fleischer, M. Deuchert, C. Kühlewein, C. Ruhs

*Institute of Production Science (wbk), Universität Karlsruhe (TH), Kaiserstrasse 12,
76131 Karlsruhe, Germany*

Abstract

The geometry of micro milling tools currently in use have been adopted from macro tools, assuming that chip formation and process kinematics are analogical in both types of tools [1]. Experience has proved that micro tools respond to influences in a very different way than macro tools [2]. Oftentimes, structural details such as the rake angle and the twist angle impede further miniaturization and are impossible to achieve with conventional manufacturing techniques. Therefore it is necessary to get a comprehensive understanding of the entire process by taking a structure mechanical and cutting technological approach to micro milling tools in order to be able to optimize them. Another objective consists in the production of these miniaturized milling tools by means of force-free procedures such as laser ablation and electrical discharge machining.

The present state of research already puts the deficits of the currently available tools on display. Insufficient manufacturing tolerances of ±10 μm, constitute a substantial change of cutting conditions for the commonly used lateral infeed or feed per tooth of a few micrometers. Sometimes, only one cutting edge is engaged, which results in increased wear and, therefore, reduced durability, increased cutting forces, minor surface quality and a higher probability of milling cutter breakage. For that reason, a single-edged geometry has been proposed. It guarantees clear adjustment of the process parameters feed per edge and lateral infeed. For that purpose, stability analyses of simple stylus geometries have been conducted by means of FEM simulations. The resulting tool with a diameter down to 30 μm was machined on the EDM-machine at the wbk (Sarix SX 100). First tests have been carried out that prove the ability of these tools to cut steel.

Keywords: micro milling, FEM, EDM, milling tool

1. Motivation

At present the geometries of micro milling cutters are created by scaling down macro tools. Due to the increasing miniaturization of components [3], it is becoming ever more complex to produce the required tools. Furthermore researches have shown that micro tools respond to influences in a very different way than macro tools. Another problem is that the accuracy during the manufacturing process is insufficient [4]. Therefore it is necessary to get a comprehensive understanding of the entire process by taking a structure mechanical and cutting technological approach to micro milling tools in order to be able to optimize them to realize smallest diameters down to 30 μm.

2. Simulation

FEM simulations using the general-purpose code ABAQUS developed by ABAQUS, Inc. have been used to analyse the stability of simple stylus geometries. In order to verify the analysis, several output values were identified: besides the stress distributions and the maximum principle on the tool also the maximum deflection of the tool was investigated. For defining an optimized geometry shape, simulations were performed for three different geometries, all with a diameter of 300 μm. Geometry 1 and 2 are trapezium-shaped geometries whereas geometry 3 has a semi-circular shape (see Fig. 1). Several simulation runs were conducted for each geometry shape to reflect the reality:

- effects of the centrifugal load due to the eccentric mass and the rotation speed of the cutting tool
- effects of the cutting force due to the machining, subdivided in the force acting on the major flank face, on the minor flank face and on the rake face
- superposition of the forces acting on the major flank face, on the minor flank face and on the rake face to the cutting force
- superposition of the centrifugal load and the cutting force

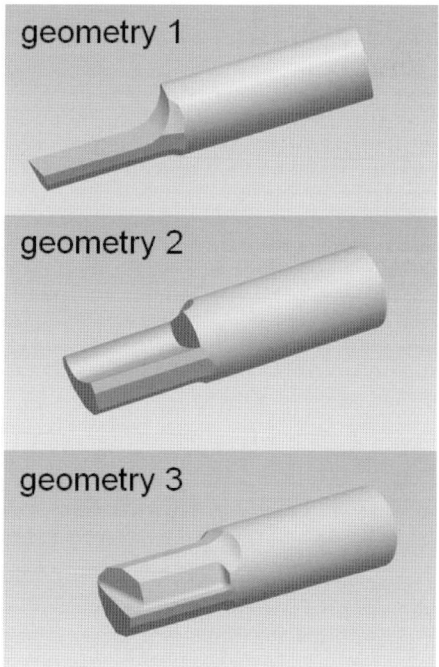

Fig. 1. Micro cutting tool geometries.

A fine mesh has been adopted at the major and minor flank face as well as at the rake face to allow proper representation of the stress distribution in the cutting area due to the cutting forces. Fig. 2 shows the mesh configuration of the micro cutting tool.

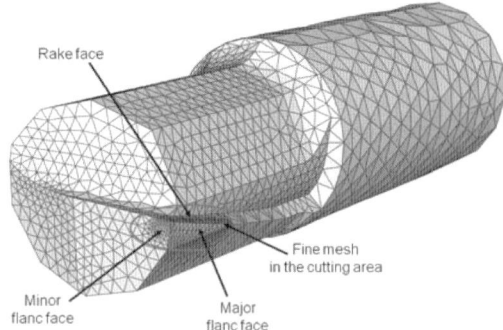

Fig. 2. Meshing of the micro cutting tool (geometry 3).

The forces acting on the different faces were determined experimentally by using a force measurement platform and modeled as a specified distributed surface load (*DSLOAD), the centrifugal load acting on the entire cutting tool was modeled as a specified distributed load (*DLOAD) modifying the data lines to define the centrifugal loads. In order to define the variation of the load magnitude during simulation, the *AMPLITUDE option was used in ABAQUS to represent the run-up of the cutting tool from standstill up to a maximum rotation speed of 160.000 min^{-1}. Hereby, the acceleration is subdivided into 17 steps increasing the rotation speed by 10.000 min^{-1} for each step.

In comparison with the trapezium-shaped geometries (geometry 1 and 2), the results showed a 30% higher stability of the semi-circular geometry (geometry 3) and therefore a smaller deflection enabling a more accurate machining of the work piece. Fig. 3 illustrates the stress distribution of the three different geometries as well as the deflection. For improving the representation of the deflection the scale factor has been increased by several orders of magnitude. Based on the obtained simulation results the analyses of trapezium-shaped structures were stopped and further investigations were performed for the semi-circular geometry.

The optimized geometry was scaled down from 300 µm to different sizes, namely 150 µm, 125 µm, 100 µm, 75 µm, 50 µm and 30 µm to re-investigate the influence of the cutting force and centrifugal load for the different diameters in the next simulation runs. The simulation results allowed a comparison of the behaviour of the different tool sizes and the relocation of the front end showed, that the smaller the tool is the less the influence of the centrifugal load gets. Additionally, in order to make a statement about the concentricity tolerance, the effects of an eccentric chucking of the micro cutting tool in the machine tool are currently addressed. Therefore, the axis of rotation of the cutting tools was translated by 3 µm in 1 µm-steps in different directions (along the positive and negative x- and y-axis and along the imaginary 45° and 135° axis). First results indicate that the influence of the eccentric chucking needs to be considered. A quantitative evaluation is currently performed.

As a result of the simulations conducted so far FEM is a useful simulation software for designing micro milling tools. If all necessary boundary conditions are taken into account in the simulation model, a statement about the optimized geometry due to the obtained stress distribution is feasible. Furthermore, the FE simulation enables the output of some variables that cannot be obtained in the experiments or only with very high efforts, for example the deflection of the micro milling cutter due to an eccentric chucking.

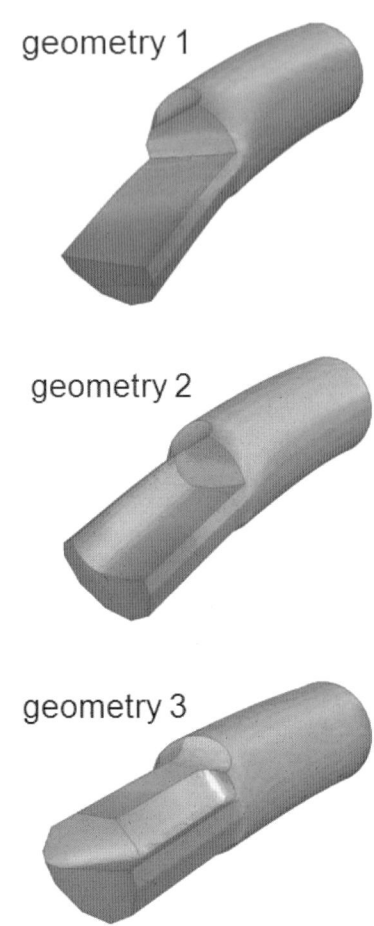

Fig. 3. Structure mechanical FE simulation of different micro cutting tool geometries.

3. Manufacturing of these single-edged micro milling tools

After the optimization by simulation via the FEM-software tool ABAQUS the geometry was manufactured to verify the results of the researches. Therefore the manufacturing method WEDG (wire electro-discharge grinding) was chosen [5, 6]. The main advantage of this method is that there are no forces between the work piece and the electrode during the discharge process. So it is adapted for producing very small and filigree structures [7].

The micro milling tools were manufactured by an EDM-machine (Sarix SX100), which was especially developed for high precise standards. In addition to the common equipment, the facility was upgraded by a wire unit, which were needed for the manufacturing process mentioned above. A wire diameter of 100 μm was chosen to assure a stable erosion process. Furthermore the whole facility was situated in an air-conditioned area to avoid any negative effects caused by the environment.

For the production of such micro milling cutters a blank made of carbide has to be chucked in the spindle. It is very important to use a clamping system with a high accuracy to ensure a defined position of the blank in relation to the wire. First of all the helical geometry of the blade has to be machined by die sinking. This procedure includes a simultaneous rotational and translational movement. The form used for this production step was machined by micro milling. After that small parts are cut of the blank by WEDG, like it is illustrated in the Fig. 4. For each cut the work piece needs to be positioned over the wire and sunk to start the erosion process. The first spiral-shaped milling cutters, which were manufactured by this routine, had a diameter of 300 μm. The clearance angle on the front side was 10°. The flanks behind the secondary cutter were reduced by 5 μm and the remaining cutter circumference is reduced by 10 μm. A total of ten flanks are processed along the circumference. There was an angle of 36° between one flank and the next.

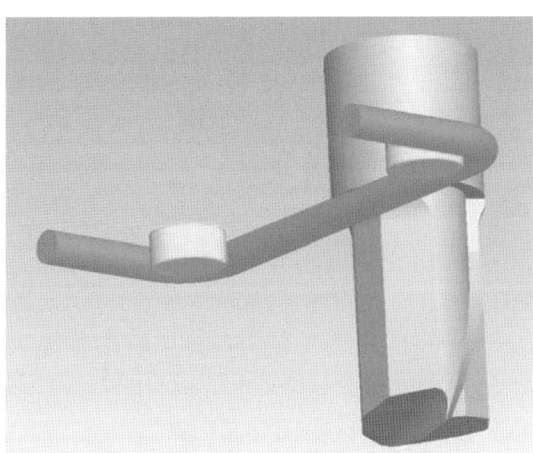

Fig. 4. Milling cutter production procedure.

To ensure a stable erosion process the right parameters are needed. The following parameters have been optimized and set during several preliminary series of tests:

Initially a frequency of 190 kHz was used, because the higher the frequency and the smaller the working pulse is, the smaller are the resulting craters, which are typical for surfaces manufactured by EDM. The working gap had to be as small as possible, because a wide spark gap needs a higher voltage to overcome the dielectric in the working gap. So the gap was about 5 μm wide to achieve a high quality of the surface. If the working gap gets too small, it cannot be sufficiently cleared of the erosion products, which causes process-disturbing short circuits. It was very important to minimize the electric current, because this leads to better surfaces of the work piece. That is why the electric current was set to 0 A, so no internal capacities will be added to energy storage, because the discharges are controlled by capacity immanent in the system only. System capacity is less than 10 pF here.

That is why the electric current of 0 A was chosen. A voltage of 80 V was adjusted. The erosion process became instable at a voltage of less than 80 V, so this was the smallest value possible. Fig. 5 shows a micro milling cutter with a diameter of 100 μm manufactured by WEDG.

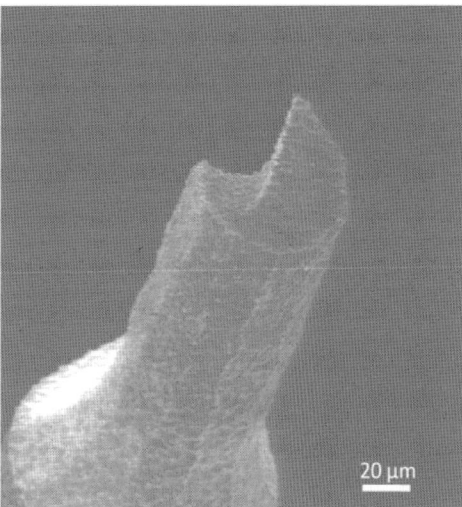

Fig. 5. Cutter Ø 100 μm, helical shape (geometry 3).

4. Results

In order to verify the functionality and performance of these cutters several milling experiments were conducted. The research showed that the WEDG-machined helical geometry (geometry 3) as simulation proved was the most stable option. Milling cutters with a diameter of 300 μm (feed per tooth 5 μm, infeed 10 μm, total depth 50 μm) and 45 μm (feed per tooth 1 μm, infeed 1 μm, total depth 10 μm) were used for manufacturing the grooves in brass displayed in Fig. 6.

Fig. 6. Groove, machined by a 45 μm milling tool in brass.

What becomes apparent is increased burr formation. The potential influencing factors include high cutting edge rounding and notchiness resulting from crater formation in the manufacturing process and the fact that spiralization is yet to be optimized. Geometry adaptation has been established as one possible solution to allow for chip removal, the minimization of burr formation and the reduction of cutting forces. Another possibility to improve the milling process is to perfect the EDM manufacturing process and to achieve better surfaces, less erosion craters and sharper cutting edges.

The comparison of the force measurement of the one edged cutter and a commercially available two edged cutter shows clearly the advantage of the one edged cutting tool (Fig. 7). The force peaks of the feeding force of the one edged milling cutter are at a similar level. The failure in the concentricity of two-edged cutters causes different equivalent chipping thicknesses and on this account different chipping forces which can lead to tool breakage. The forces were detected by a force measurement platform beneath the work piece (which is made of brass MS58).

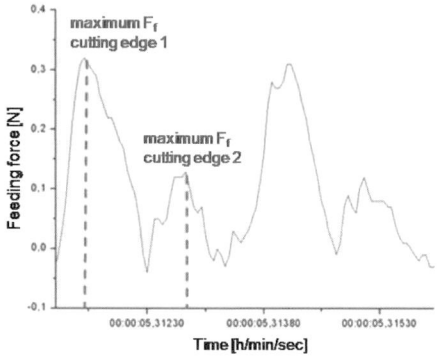

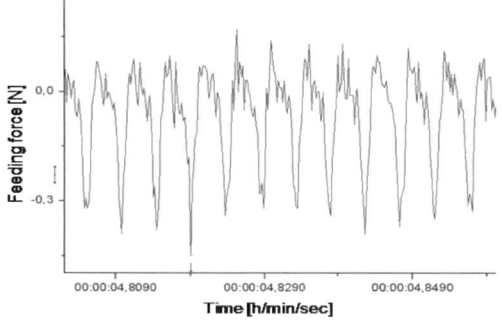

Fig. 7. Force measurement of a two-edged cutter (top) and a one-edged cutter (bottom)

5. Summary

At present the tolerance of the radial misalign-ments of micro milling cutters are larger in size than the feed per tooth during the manufacturing process. This leads to higher wear at the cutting edge and a lower durability of the micro milling tool. To avoid this disadvantage a single-edged cutter was created and manufactured. The optimization process was supported by FEM-simulation to reach maximum stability. The research started with a simulation of a simple stylus geometry and ended with an optimized single-edged structure. To verify the simulation results some milling tests took place and showed the potential of well defined micro milling tools.

Acknowledgements

The authors wish to thank the German Research Foundation (DFG) for their support. "Structuring guidelines and machining procedures for micro milling tools".

References

[1] E. Uhlmann, M. Füting, K. Schauer: Optimierung von Mikrofräswerkzeugen in der Werkzeug-planungsphase, wt Werkstattstechnik, Jahrgang 94 (2004), H. 11/12
[2] D. Oberschmidt: 16IN 0121 – MiCuTool Innovative Herstellungsverfahren für Mikrozerspanwerkzeuge, InnoNet-Kongress, 6. November 2006
[3] J. Hesselbach, A. Raatz, J. Wrege, H. Herrmann, H. Weule, C. Buchholz, H. Tritschler, M. Knoll, J. Elsner, F. Klocke, M. Weck, J. Bodenhausen, A. Klitzing: „mikroPRO - Untersuchung zum internationalen Stand der Mikroproduktionstechnik, wt Werkstattstechnik, Jahrgang 93 (2003), H. 3
[4] E. Uhlmann, S. Piltz, K. Schauer: Dynamische Werkzeuganalysen in der Mikrozerspanung wt Werkstattstechnik, Jahrgang 93 (2003), H. 3
[5] J. Fleischer, T. Masuzawa, J. Schmidt, M. Knoll: New Applications for Micro-EDM, Euspen 2004
[6] J. Schmidt, M. Simon, H. Tritschler, R. Ebner: μ-Fräsen und μ-Erodieren für den Formenbau, wt Werkstattstechnik, Jahrgang 91 (2001), H. 12
[7] J. Schmidt, J. Fleischer, M. Knoll: Electrodes for Micro-EDM, EUSPEN International Topical Conference on Precision Engineering, Micro Technology, Measurement Techniques and Equipment (2003), Proceedings - Volume 1, pp: 177-179, ISBN: 3-926832-30-4

On the force between two metallic plates of a gripper immersed in a nonpolar fluid

D. Dantchev, K. Kostadinov

Institute of Mechanics, Bulgarian Academy of Sciences, Acad. G. Bonchev St. Bl. 4, 1113 Sofia, Bulgaria

Abstract

We analyse, as a function on the temperature T and the chemical potential μ, the total force $F_{tot}(T,\mu,L)$ between two metallic plates of a gripper separated at a distance L from each other and immersed in a nonpolar fluid which can be liquid, or gas. In our approach we take into account the direct substrate-substrate van der Waals interaction, the van der Waals interactions between the molecules of the fluid with the other molecules of the fluid as well as with the constituent elements of the substrate, and the interaction between the plates generated by the fluctuations of the density of the fluid (i.e., the Casimir force). We suppose that both plates are equal and strongly prefer the liquid phase of the fluid. Under such boundary conditions both the direct plate-plate van der Waals interaction, as well as the Casimir force, are forces of attraction of the plates toward each other. In the phase space (temperature, chemical potential), we identify the regions where the net interaction force is the strongest. It turns out that these regions are close to the bulk critical point of the fluid ($T=T_c, \mu=\mu_c$), and near the so-called capillary condensation regime $T<T_c,(L/a)(\Delta\mu/k_BT)=O(1)$, with $\Delta\mu=\mu-\mu_c<0$ and a the characteristic distance between the molecules of the fluid. These regions shall be avoided in order to prevent sticking of the plates of the gripper on each other.

Keywords: grippers, van der Waals forces, Casimir effect, thin films

1. Introduction

Handling and fixing of micro parts reliably and precisely is the main bottleneck in micro assembly and is far from being solved today. Further handling and fixing strategies must be developed taking into account the forces appearing in small distances between the gripper's plates and the micro object or between the object and a surface to taken off or to be placed on. Also the environmental influences are usually not mastered yet and they have to be studied.

Tools for assisting the robot in micro assembly tasks are not available today. Components such as micro part feeders and miniature grippers for micro objects also put some problems of interaction in small distances with micro objects that should be studied in order to find ways to control these operations.

In this article we study the force between two metallic plates of a gripper immersed in nonpolar fluid.

If a fluid is confined by parallel plates at a distance L (see Fig. 1) and is in contact with a particle reservoir with a chemical potential μ and temperature T, the grand canonical potential $\Omega_{ex}(T,\mu,L)$ of the fluid in excess to its bulk value $AL\omega_{bulk}(T,\mu)$ depends on L. Then the effective force F_{tot} between the plates normalized per cross sectional area A is

$$F_{tot}(T,\mu,L) = -k_BT\frac{\partial\omega_{ex}(T,\mu,L)}{\partial L} \, , \qquad (1)$$

where k_B is the Boltzmann's constant,

$$\omega_{ex}(T,\mu,L) = \omega(T,\mu,L) - L\omega_{bulk}(T,\mu) \\ = \Omega_{ex}(T,\mu,L)/A \qquad (2)$$

is the excess grand canonical potential per cross sectional area A, $\Omega(T,\mu,L) = A\omega(T,\mu,L)$ is the total grand canonical potential, and $\omega_{bulk}(T,\mu)$ is the density of the bulk grand canonical potential. Besides

temperature T, chemical potential μ, and film thickness L, the force also depends on which boundary conditions the surfaces impose on the system. The order near the surfaces can be either reduced or -- which is the generic case for liquids confined by solid substrates – increased due to effective surface fields generated by the confinement. The latter case is known as (+,+) boundary conditions (for a more precise definition see below). For this case, the schematic phase diagram of a fluid film with thickness L is shown in Fig. 2.

In order to calculate the effective force between the

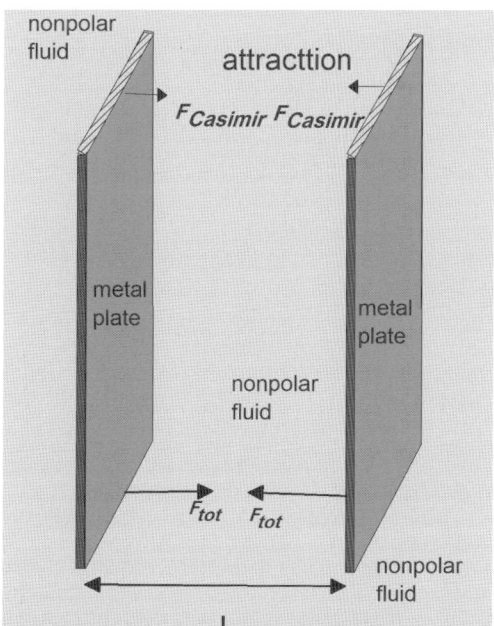

Fig. 1. The total and the Casimir force acting on the plates of a gripper. Note that they both are forces of attraction, i.e. the Casimir force causes stronger attraction between the plates.

plates we consider long-ranged pair interactions between the fluid particles decaying with distances r between each other asymptotically $\sim J^l r^{-6}$ and long-ranged substrate potentials $\sim J^{l,s} z^{-3}$ acting on fluid particles at a distance z from the flat surface of a semi-infinite substrate. As it has been shown in [1] in the case of d-dimensional systems, in which the interaction decays asymptotically with the distance $\sim r^{-d-\sigma}$ (with $\sigma > 2$), the effective total force $F_{tot}(T, \mu, L)$ between the plates can be written in the form:

$$F_{tot}(T, \mu, L) \simeq (\sigma - 1) H_A(T, \mu) L^{-\sigma} \xi_0^{\sigma - d}$$
$$+ k_B T L^{-d} X \left[\frac{L}{\xi_t}, \frac{L}{\xi_\mu}, l\left(\frac{L}{\xi_0}\right)^{-\omega_l}, s\left(\frac{L}{\xi_0}\right)^{-\omega_s}, g_\omega \left(\frac{L}{\xi_0}\right)^{-\omega} \right]. \quad (3)$$

In the first term of the above equation X is dimensionless, universal scaling function, which takes into account the part of the force that is due to the fluctuations of the density of the fluid between the plates, i.e. the Casimir force [2-7], ξ_t is the bulk correlation length, i.e. $\xi_t = \xi(t \to \pm 0, \Delta\mu = 0) \simeq \xi_0^\pm |t|^{-\nu}$ at bulk coexistence $\mu = \mu_c$ and for $t = (T - T_c)/T_c \to \pm 0$, while $\xi_\mu = \xi(t = 0, \Delta\mu \to 0) \simeq \xi_{0,\mu} |\beta_c \Delta\mu|^{-\nu/\Delta}$ is the bulk correlation length at the critical temperature $T = T_c$ with $\beta_c = (k_B T_c)^{-1}$. For $T > T_c$ and $\Delta\mu = 0$ one has $\xi_0 = \xi_0^+$, while for $T < T_c$ and $\Delta\mu = 0$ one has $\xi_0 = \xi_0^-$ with the ratio ξ_0^+/ξ_0^- being universal; $\xi_{0,\mu}$ is the same for $\Delta\mu \to +0$ and $\Delta\mu \to -0$.

The second term in Eq. (3) stems from the free energy contribution $H_A L^{-(\sigma - 1)}$, where

$$H_A(T, \mu) = A_{l,s}(T, \mu) + A_l(T, \mu) + A_s(T) \quad (4)$$

with the so-called Hamaker constants $A_{l,s}$, A_l and A_s. Here A_l represents that part of the Hamaker constant which is generated by the long-ranged part of the fluid-fluid interactions, i.e., the dispersion interaction, A_s is the part due to the direct long-ranged interactions between the two substrates, and $A_{l,s}$ is the corresponding term generated by the long-ranged tails of the substrate potentials acting on the fluid particles (the explicit dependence of $A_{l,s}$, A_l and A_s on the parameters of the model will be presented below). In Eq. (3) $\omega \simeq 0.81$ (with $d=\sigma=3$) is the standard Wegner's correction-to-scaling exponent for short-ranged systems, while $\omega_l = \sigma - (2 - \eta)$ and $\omega_s = \sigma - (d + 2 - \eta)/2$ are the correction to scaling exponents due to the long-ranged tails of the fluid-fluid and substrate-fluid interactions, respectively. For realistic systems, i.e., for the "genuine" non-retarded van der Waals interaction, which governs the nonpolar fluids, one has $d=\sigma=3$ and, then,

$$\eta = 0.034, \beta = 0.329, \nu = 0.631, \Delta = 1.567,$$
$$\omega = 0.81, \omega_l = 1.03, \omega_s = 0.52. \quad (5)$$

In Eq. (3) the factor g_ω is the dimensionless scaling field associated with the Wegner-type corrections, while

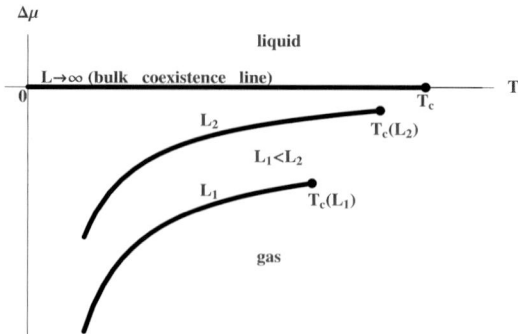

Fig. 2. The phase diagram of a thin fluid film bounded by metal plates, which strongly prefer the liquid phase of the fluid. The bent gas-liquid coexistence curves correspond to capillary condensation transitions for $L=L_1$ and $L=L_2$ with $L_1 < L_2$.

l and s are dimensionless nonuniversal coupling constants with l being proportional to the strength J^l of the long-ranged tail in the fluid-fluid interaction and s being proportional to the contrast between the substrate potential and the fluid-fluid interaction integrated over a half-space (see below).

As it has been demonstrated in [1], when $L \gg L_{crit}$, with

$$L_{crit} = \xi_0 (2^{\sigma+1} |s|)^{\nu/\beta}, \quad (6)$$

one can expand the scaling function X in Eq. (3) which leads to:

$$X \simeq X^{sr}\left[\frac{L}{\xi_t}, \frac{L}{\xi_\mu}\right] + s\left(\frac{L}{\xi_0}\right)^{-\omega_s} X_s^{lr}\left[\frac{L}{\xi_t}, \frac{L}{\xi_\mu}\right]$$
$$+ l\left(\frac{L}{\xi_0}\right)^{-\omega_l} X_l^{lr}\left[\frac{L}{\xi_t}, \frac{L}{\xi_\mu}\right] + g_\omega \left(\frac{L}{\xi_0}\right)^{-\omega} X_\omega^{sr}\left[\frac{L}{\xi_t}, \frac{L}{\xi_\mu}\right]. \quad (7)$$

One concludes that the leading behaviour of the scaling function X very near the bulk critical point $L/\xi_t = 0$, $L/\xi_\mu = 0$, i.e. when $L/\xi_t \ll 1$ and $L/\xi_\mu \ll 1$, can then be determined by only taking into account the short ranged part of the fluid-fluid and the substrate-fluid interactions, i.e. will be given by the function $X^{sr}\left[L/\xi_t, L/\xi_\mu\right]$. This normally greatly simplifies the analytical and the numerical difficulties of the calculation. If $L \leq L_{crit}$, however, one has to take the long-ranged tails of the interactions into account in the whole "critical region of the finite system", i.e. when $L/\xi_t = O(1)$ and $L/\xi_\mu = O(1)$. Having in mind that for the most fluids ξ_0 typically is of the order of 3 Å one has

$$L_{crit} \simeq 60 |s|^{1.918} \quad \text{nm}. \quad (8)$$

As discussed in [1] for a variety of systems $s \in [1, 2]$. However, for some systems such as [3]He or [4]He films near their bulk liquid-gas critical point and confined by Au substrate [8] s can be as large as 4. Thus, one has that L_{crit}, depending on the system, is between 60 nm and 1 µm. In the current article, on the example of a gripper immersed in a typical nonpolar fluid system with $L=100-150$ Å, $s=1$, and $l=0.3$, we will calculate the total force of attraction between the gripper plates, taking into account the long-ranged tails of the interactions involved and the fluctuations of the density of the fluid.

The structure of the article is as follows. In the next section, we define the model in which the concrete analytical and numerical calculations are performed. Section **3** will present the obtained results for the T and μ dependence of the force. The article closes with a discussion presented in Section **4**.

2. The model

Within the density functional approach for inhomogeneous fluids confined between two parallel flat plates at a distance L the grand canonical functional $\Omega[\rho(\mathbf{r})]$ of the fluid:

$$
\Omega[\rho(\mathbf{r})] = \int f_{HS}[\rho(\mathbf{r})]d^3\mathbf{r} + \int V(z)\rho(\mathbf{r})d^3\mathbf{r} \\
+ \frac{1}{2}\iint \rho(\mathbf{r})w(\mathbf{r}-\mathbf{r}')\rho(\mathbf{r}')d^3\mathbf{r}\,d^3\mathbf{r}' - \mu\int\rho(\mathbf{r})d^3\mathbf{r}
\tag{9}
$$

has to be minimized with respect to the local number density $\rho(r)$ [9]. The plates exert a substrate potential $V(z)$ with z as the normal distance from one wall. For an individual wall $V(z\to\infty) \sim z^{-\sigma}$ where $\sigma=3$ for a genuine van der Waals interaction; μ is the chemical potential and f_{HS} is the bulk free energy density of a hard-sphere system acting as a reference system; $w(r)$ is the fluid-fluid potential. In Eq. (9) the integrals run over the slab volume.

In order to simplify the calculations one can further simplify the continuum functional in Eq. (9) by replacing it by its lattice version. The grand potential functional for this lattice gas system then is

$$
\Omega[\rho(\mathbf{r})] = k_B T \sum_{\mathbf{r}\in\Lambda}\rho(\mathbf{r})\ln[\rho(\mathbf{r})] + [1-\rho(\mathbf{r})]\ln[1-\rho(\mathbf{r})] \\
+ \frac{1}{2}\sum_{\mathbf{r},\mathbf{r}'\in\Lambda}\rho(\mathbf{r})w(\mathbf{r}-\mathbf{r}')\rho(\mathbf{r}') + \sum_{\mathbf{r}\in\Lambda}[V(z)-\mu]\rho(\mathbf{r})
\tag{10}
$$

where Λ is a simple cubic lattice in the region $0 \le z \le L$ occupied by the fluid. Here and in the following all length scales are taken in units of the lattice constant a of the order of a molecular diameter (and thus are dimensionless) so that the particle density $\rho(\mathbf{r})$ is dimensionless and varies within the range [0,1]. In Eq. (10) the terms in curly brackets correspond to the entropic contributions, while the other terms are directly related to the interactions present in the system. We suppose that the fluid potential $w(\mathbf{r}) = -4J(\mathbf{r})$ is given by the expression

$$
J(\mathbf{r}) = J^l_{sr}\left\{\delta(|\mathbf{r}|)+\delta(|\mathbf{r}|-1)\right\} \\
+ J^l/(1+|\mathbf{r}|^{d+\sigma})\theta(|\mathbf{r}|-1),
\tag{11}
$$

where $J(\mathbf{r})$ describes the interaction between the fluid particles, whereas the one between the fluid and the substrate particles is given by

$$
J^{l,s}(\mathbf{r}) = J^{l,s}_{sr}\delta(|\mathbf{r}|-1) + J^l/|\mathbf{r}|^{d+\sigma}\,\theta(|\mathbf{r}|-1).
\tag{12}
$$

Here $\delta(x)$ is the discrete delta function and $\theta(x)$ is the Heaviside step function (with the convention $\theta(0)=0$);

ρ_s is the number density of the substrate particles in units of a^{-d}. Once the interaction $J^{l,s}(\mathbf{r})$ is specified, one can determine the precise form of the substrate potential

$$
V(z) = -\rho_s J^{l,s}_{sr}\left[\delta(z)+\delta(L-z)\right] \\
+ v_s\left[(z+1)^{-\sigma}+(L+1-z)^{-\sigma}\right],
\tag{13}
$$

where

$$
v_s = -4\pi^{(d-1)/2}\frac{\Gamma\left(\dfrac{1+\sigma}{2}\right)}{\sigma\,\Gamma\left(\dfrac{d+\sigma}{2}\right)}\rho_s J^{l,s}.
\tag{14}
$$

We consider the layers closest to the substrate to be completely occupied by the liquid phase of the fluid (which implies that we consider the strong adsorption limit), i.e., $\rho(\mathbf{r}_\parallel, z=0) = \rho(\mathbf{r}_\parallel, z=L) = 1$, which is achieved by taking the limit $J^{l,s}_{sr}\to\infty$ and is known as (+,+) boundary conditions applied to the system under consideration.

The variation of Eq. (10) with respect to $\rho(\mathbf{r})$ leads to an equation of state for the equilibrium density $\rho^*(\mathbf{r})$. For a given geometry and surface potential $V(z)$ the solution of the equation determines the equilibrium order-parameter profile $\rho^*(\mathbf{r})$ in the system. Inserting this profile into Eq. (10) one then derives the grand canonical potential of the system and, applying the definition (1), one obtains the effective force $F_{tot}(T,\mu,L)$ between the plates of the gripper. The results are presented in the next section.

3. Evaluation of the force of attraction F_{tot} between the plates of the gripper

Evaluating the Hamaker terms A_i, A_s and $A_{i,s}$ one obtains:

$$
A_l(T,\mu) = -\frac{4\pi^{(d-1)/2}}{\sigma(\sigma-1)}\frac{\Gamma\left(\dfrac{1+\sigma}{2}\right)}{\Gamma\left(\dfrac{d+\sigma}{2}\right)}J^l\rho_b^2(T,\mu),
\tag{15}
$$

$$
A_s(T,\mu) = -\frac{4\pi^{(d-1)/2}}{\sigma(\sigma-1)}\frac{\Gamma\left(\dfrac{1+\sigma}{2}\right)}{\Gamma\left(\dfrac{d+\sigma}{2}\right)}J^s\rho_s^2(T,\mu),
\tag{16}
$$

$$
A_{l,s}(T,\mu) = \frac{8\pi^{(d-1)/2}}{\sigma(\sigma-1)}\frac{\Gamma\left(\dfrac{1+\sigma}{2}\right)}{\Gamma\left(\dfrac{d+\sigma}{2}\right)}J^{l,s}\rho_b(T,\mu)\rho_s(T).
\tag{17}
$$

where ρ_s is the number density of the substrate, while ρ_b is the bulk limit, for given T and μ, of the number density of the fluid. Note that while $A_i > 0$, $A_s > 0$, one has $A_{i,s} < 0$, i.e. while *the fluid-fluid and the direct substrate-substrate interactions contribute to a force of attraction between the plates, the fluid-substrate interaction diminishes this force acting in the opposite direction.* Adding the right- and the left-hand sides of Eqs. (15), (16) and (17), one obtains the total Hamaker term H_A, see Eq.(4), as:

$$H_A(T,\mu) = -\frac{4\pi^{(d-1)/2}}{\sigma(\sigma-1)} \frac{\Gamma\left(\frac{1+\sigma}{2}\right)}{\Gamma\left(\frac{d+\sigma}{2}\right)} \times$$

$$\left[J^l \rho_b^2(T,\mu) - 2J^{l,s}\rho_b(T,\mu)\rho_s(T) + J^s \rho_s^2(T,\mu) \right] \quad (18)$$

$$\sim -(\alpha_s \rho_s - \alpha_l \rho_b)^2 < 0,$$

where we have taken into account that the parameters of interactions J^l, J^s, and $J^{l,s}$ can be connected [10] to the atomic polarizabilities α_l and α_s : $J^l \sim \alpha_l^2$, $J^s \sim \alpha_s^2$ and $J^{l,s} \sim \alpha_l \alpha_s$. Note that this implies

$$\frac{J^l}{\alpha_l^2} = \frac{J^s}{\alpha_s^2} = \frac{J^{l,s}}{\alpha_l \alpha_s}, \quad (19)$$

wherefrom one derives that, say, $J^s = \left(J^{l,s}\right)^2 / J_l$, and thus Eq. (18) can be rewritten in the form

$$H_A(T,\mu) = -\frac{4\pi^{(d-1)/2}}{\sigma(\sigma-1)} \frac{\Gamma\left(\frac{1+\sigma}{2}\right)}{\Gamma\left(\frac{d+\sigma}{2}\right)} J^l \left[\rho_b - \frac{J^{l,s}}{J^l}\rho_s \right]^2 < 0. \quad (20)$$

The result above implies that the Hamaker term is *always negative* and, therefore, the corresponding contribution to *the force between the plates bounding the fluid will always be a force of attraction, independently on the concrete fluid and the material, from which the plates of the gripper are made, provided the both plates are from the same material.* With $d = \sigma = 3$ for real nonpolar fluids Eq. (20) further reduces to

$$H_A(T,\mu) = -\frac{\pi}{3} J^l \left[\rho_b - \frac{J^{l,s}}{J^l}\rho_s \right]^2 < 0. \quad (21)$$

If one takes, as example, such well investigated fluid as helium bounded by gold plates one can easily evaluate H_A. Using the data given in [8], one obtains that H_A is of the order of 6.8×10^{-21} J, wherefrom one concludes that, for L of the order of 10^{-8} m, the force of attraction (per unit area) between the plates of the gripper will be of the order of 1.4×10^4 N/m^2.

As already stated above, in addition to the attraction of the plates, which is due to the dispersion forces in the system, one also observes strong forces of attraction between them. They are due to the fluctuation of the order parameter, i.e. to the Casimir effect, see Eq. (3), and due to the competition between the liquid-like and the gas-like phases of the fluid in the capillary condensation regime, see Fig. 2 and the discussion in section III.B of Ref. [1]. An estimation of the Casimir effect can be obtained by knowing the so-called Casimir amplitude $X^{sr}(0,0) \simeq -0.9$ [11], see Eq. (7), and by making use of the observation that the maximum of the Casimir force is, roughly, about 10 times deeper than this value when one scan the thermodynamic space [1]. Again, for L of the order of 10^{-8} m, as above, the numerical estimation gives that the fluctuation induced force is a force of *attraction* and is in of the order of 8×10^4 N/m^2 (per unit area) for a fluid with a critical temperature around the room temperatures. The dependence of this estimate on the

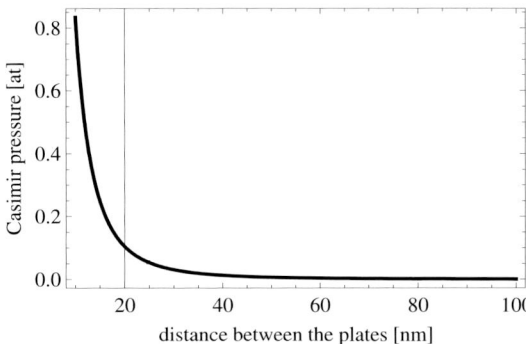

Fig. 3. The Casimir force decays with the distance as L^3 and, thus, it is easy to modify the estimations of the force given in the main text for any distance L between the plates of the gripper. The corresponding result is plotted in the current figure.

distance between the plates is illustrated in Fig. 3.

4. Discussion and concluding remarks

In the current article, we presented a model and on its basis did study the force acting between the two metallic plates of a gripper immersed in a nonpolar fluid. This force shall be taken into account when one designs the control and management of the plates of a gripper that is supposed to perform a given operation in a concrete type of fluid with a micro object. In the phase space (temperature, chemical potential), the regions where the net interaction force is the strongest have been identified. They are close to the bulk critical point of the fluid $(T = T_c, \mu = \mu_c)$, and near the so-called capillary condensation regime $T < T_c$, $(L/a)(\Delta\mu/k_BT)$ $= O(1)$, with $\Delta\mu = \mu - \mu_c < 0$. These regions shall be avoided in order to prevent sticking of the gripper plates on each other.

Acknowledgment:
This work was part funded by the European Commission FP6 Integrated Project HYDROMEL.

References

[1] Dantchev D, Schlesener F and Dietrich S., Physical Review, E **76** (2007) 011121.
[2] Casimir H B G, Proc. K. Ned. Akad. Wet. **B 51** (1948) 793.
[3] Casimir H B G, Physica **19**, (1953) 846.
[4] Fisher M E and de Gennes P-G, C. R. Acad. Sci. Paris Serie **B 287** (1978) 207.
[5] Brankov J G, Dantchev D M, and Tonchev N S, *Theory of Critical Phenomena in Finite-Size Systems Scaling and Quantum Effects* (World Scientific, Singapore, 2000).
[6] K. A. Milton, Journal of Physics, **A 37** (2004) R209.
[7] Krech M., Dietrich S., Physica Review, Lett. **66** (1991) 345; Physical Review, A **46** (1992) 1886; **46** (1992) 1922.
[8] Dantchev D, Rudnick J and Barmatz M., Physical Review, E **75** (2007) 011121.
[9] Evans R., Advanced Physics, **28** (1979)143.
[10] Attard P, Bèrard D R, Ursenbach C P, and Patey G. N., Physical Review A 44 (1991) 8224.
[11] Vasilyev O, Gambassi A, Maciolek A, and Dietrich S., Euro. Phys. Lett. **80** (2007) 60009.

A study of the gate size effects on the process of optical data storage micro-scale replication

D. S. Trifonov[a], Y.E. Toshev[b]

[a.]Institute of information Technology, Bulgarian Academy of Sciences, 1113 Sofia, Bulgaria
[b.]Institute of Mechanics and Biomechanics, Bulgarian Academy of Sciences, 1113 Sofia, Bulgaria

Abstract

The present paper offers 3D CAD models of the gate system of an optical disc mould that are developed. The type of the gate system is known as a "Krauss Maffei" system, in which the central hole of the polymer substrate is formed by breaking off the circle gate from the polymer substrate. The gate depth and the gate position are defined as variable parameters and combined with different variants of the processing conditions in mould filling simulation. A range of improved variants of all the variable parameters is obtained with the help of iterative steps within the frame of the simulation code. A modified gate system is proposed in the process of research, which allows a gate with a lager depth parameter to be used. The modified system allows also alteration of the gate position with respect to the central hole. The best results are achieved, using the proposed modified gate system in the case, when a gate with a larger depth is used and the gate position is located symmetrically towards the central hole.

Keywords: optical data storage replication, 3D CAD model, simulation, nanostructures

1. Introduction

1.1. Optical data storage micro-scale replication and nanotechnology

Nanostructures can be replicated on large surfaces using injection molding of thermoplastic materials. Polymer injection molding is highly suitable for the mass fabrication of high precision micro- and nanostructures [1]. Optical discs have become increasingly used for information storage in recent years. The disc information structures are both in the micro/submicro range and in the nanometer range, and in the form of pits, are moulded into the disc during the injection moulding process. The key reason for the success of Compact Discs (CD) as storage media is the fact that injection molding makes it possible to transfer Gbytes of data points (surface pits) in a few seconds onto a cheap polymer carrier. This task has become even more important for the new formats (DVD and successors) with higher storage densities, where structure sizes of below 100nm have to be replicated [2, 3, 4].

1.2. Definition of the technical problem in optical data storage replication

The production of polymer substrates for optical discs requires high dimension accuracy, exact copying of the information structure and low birefringence. The experience gained up to now in the field of injection moulding of optical disc substrates has shown that the mould design and the gate parameters occupy a central place in the production process [5, 6, 7, 8, 9].

In the process of injection moulding of optical disc substrates, a hot polymer melt is injected in the thin cooled mould cavity under high speed and pressure with the help of a circle gate located in the cavity centre. This means that substrate indicators such as birefringence, residual tensions and copying of the information structure are highly dependent on the behavior of the melt flow, filling the mould cavity. It becomes clear that great internal tensions are available in the final product as a result of improper location, form and dimensions of the circle gate, which result at the end in high level of the main indicator for substrate quality - the birefringence. It follows from this, that the construction solution and the scheme of the circle gate location play a significant role in the quality of the final product. Finally, a "statement of the problem" for optical disc moulding starts with the fundamental difficulty of filling and packing a mould cavity of much greater length than depth, without incurring high levels of moulded-in stress and orientation. For instance, a 120 mm diameter erasable media disc, which is 50 times longer melt flow path than part thickness [10, 11, 12, 13].

In spite of the progress in this area, the approach applied up to now by different manufacturers in solving the problem of cavity filling and formation of the central hole substrate has not declined from the basic method - filling through a thin film circle gate located at the place where the central hole is to be formed. This requires the use of a complex and precise cropping punch or another system, removing the circle gate, in order to form the central hole. On the other hand, the necessity to shape the central hole by separating the circle gate from polymer substrate, calls for minimal depth of the circle gate enabling its easy removal. This approach influences badly the level of the product internal tensions and the orientation phenomena, accompanying the process of filling [14].

Computer simulation is occupying a significant role in the most of the recent papers, which consider the optical disc moulding, but they concern mainly the advantages of the program or the creation of nonlinear viscoelastic models without an analysis of the influence of the gate system geometry and parameters on the

284

mould filling process. Probably, the reason of this is that the gate system is the key for useful optical disc moulding and its design represents Know-How of the companies, working in this area [15].

The purpose of the paper is to develop 3D CAD models of an optical disc mould and the gate system and to provide a primary research to study the effect of the gate size and the gate position towards central hole of the polymer substrate on the process of mould cavity filling. A range of correct gate parameters at appropriate processing parameters is to be obtained, so that easy mould cavity filling to be attained, using the 3D CAD models developed and computer simulation code.

Another purpose of the work is to give a possibility for future investigations of the gate size effects on the process of precise pits replication.

2. 3D CAD model of the gate system

The 3D modelling of an object is the first stage to be considered when computer simulation techniques are used. An author's flow chart of the stages, necessary to study the effect of the gate system design on the process of precise pits replication, using injection moulding is shown (see Fig. 1 and Fig. 2).

The mould design concept is based on a well-known type of an optical disc mould, in which the central hole of the polymer substrate is formed by breaking off the circle gate from the polymer substrate, using a moving bush in the opposite direction of the sprue bush. The general disadvantage of this construction is the necessity to use a minimal depth of the circle gate for its easier removal, which contradicts to the support of the optimal conditions of the process of mould cavity filling. Another disadvantage is the impossibility to change the gate position towards the central hole. The reason for this is the necessity to

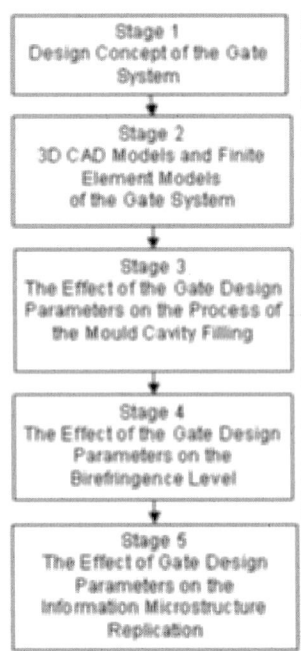

Fig. 1. The basic stages of the entire scheme of the Investigations.

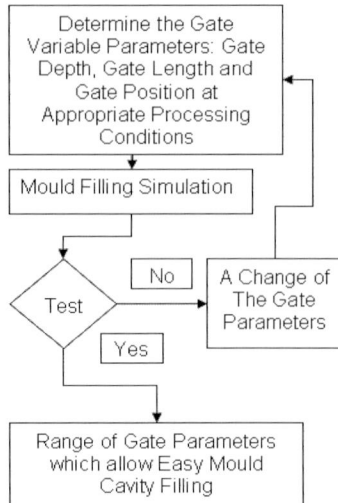

Fig. 2. The stage 3 (see Fig. 1).

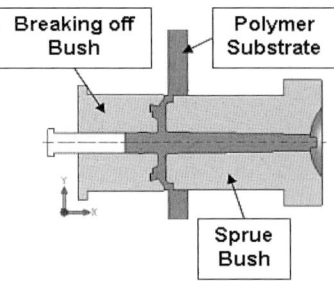

Fig. 3. The "Krauss Maffei" type of the gate system.

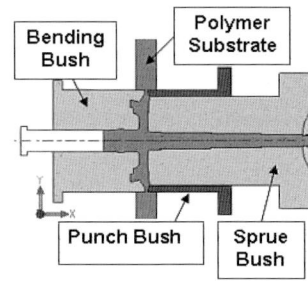

Fig. 4. The modified gate system.

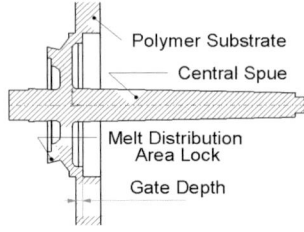

Fig. 5. 3D CAD model of the "Krauss Maffei" type of the gate system.

locate the gate in the edge, located in the opposite direction towards the sprue bush, which is detailed described of the authors of this concept [14].

285

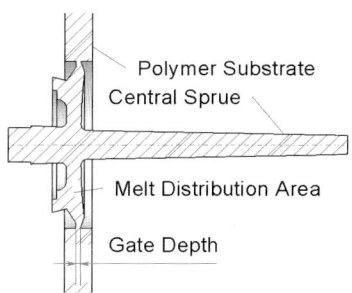

Fig. 6. 3D CAD model of the modified gate system.

In the process of our work, on the bases of a previous author's punch system design, a modified gate system design is realized. The central hole is formed by moving "a bending bush" and a punch bush in a direction, opposite to the sprue bush. As a result of the motion of the bending bush, the central gate is bent and at this moment the punch bush splits the gate from the polymer substrate. This allows the formation of the central hole of the polymer substrate in the case when a gate with a greater depth is used and gives a possibility to change the gate position towards the central hole (see Fig. 4).

The 3D CAD models are built on the bases of an integrated 3D CAD model of the gate system and the polymer substrate, considered as one object. The gate system consists of a central sprue, a melt distribution area and a circle gate (see Fig. 5 and Fig. 6).

3. Mould filling simulation and results

All the experiments are carried out in the case when the polymer substrate outer diameter is equal to 120.20 mm and the polymer substrate thickness is 1.30 mm. The gate depth and the position of the gate towards the central hole thickness are chosen as variable parameters (see Fig. 5 and Fig. 6).

A mould filling simulation program MoldflowXpress (integrated with SolidWorks) is used in the primary investigation to study the effect of the gate size and the gate position towards the central hole on the mould cavity filling process. The program can predict whether the injection moulding will be successful. The algorithm is based on the gate geometry, the plastic injection location, the type of plastic material and the processing conditions (melt and mould temperatures and injection time).

A polycarbonate plastic material Makrolon (PC) of the BAYER AG supplier is used. A PC Intel Pentium processor (2.7 GHz) with 1200 MB DDRAM is used.

A number of simulations of the two types of the gate system at values of the gate depth equal to 0.20 - 0.40 mm at intervals of 0.10 mm and at different positions of the gate towards the central hole thickness are provided. The parameters selected in three variants of the processing conditions (melt and mould temperatures) are combined: (a) Tmelt=280°C; Tmould=80°C (b) Tmelt=300°C; Tmould=100°C (c) Tmelt=320°C; Tmould=120°C). In many cases the results show that the mould cavity filling process is not sufficiently good at the parameters above given.

Experiment 1: "Krauss Maffei"-type of the gate system: A) Processing conditions: a) Tmelt=280°C; Tmould=80°C; b) Tmelt=300°C; c) Tmelt=320°C; Tmould=120°C; B) Gate depth=0.20–0.40 mm at intervals of 0.10 mm.

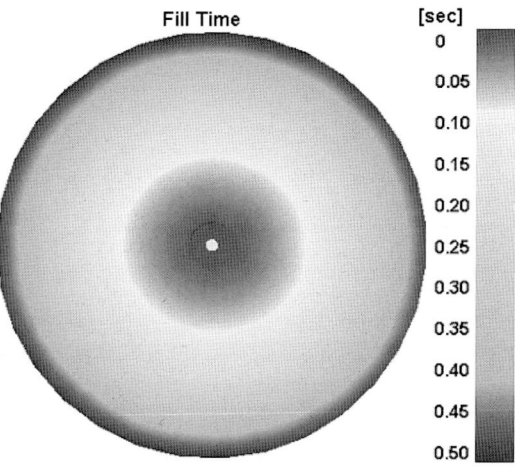

Fig. 7. Mould filling simulation with "Krauss Maffei" gate system type: gate depth=0.40 mm; Tmelt=320°C; Tmould=120°C.

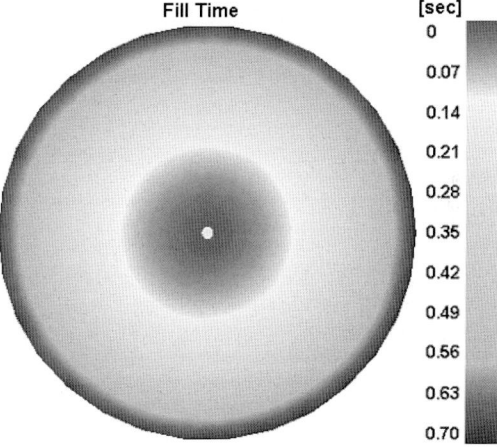

Fig. 8. Mould filling simulation with modified gate system: gate depth=0.30 mm; Tmelt=300°C; Tmould=100°C.

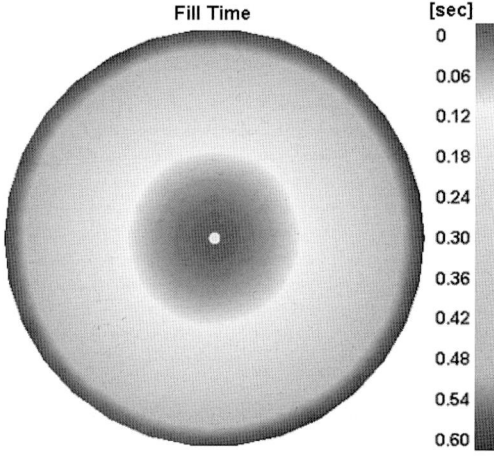

Fig. 9. Mould filling simulation with modified gate system: gate depth=0.30 mm; Tmelt=320°C; Tmould=120°C.

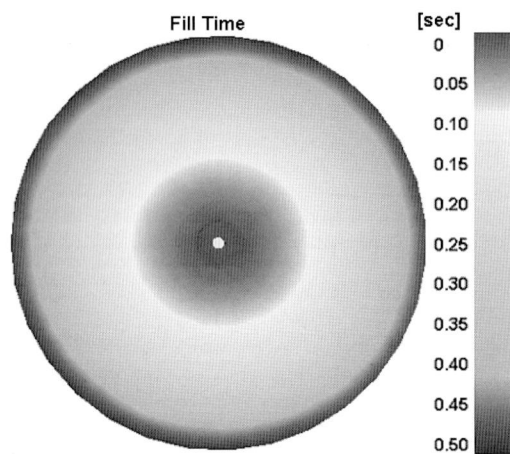

Fill Time [sec]

Fig. 10. Mould filling simulation: Modified gate system: gate depth=0.40 mm; Tmelt=320^{0}C, Tmould=120^{0}C.

The results of the experiment with the "Krauss Maffei" type of the gate system show that it is difficult to fill the mould cavity when the gate size is equal to d=0.20 – 0.28 mm and d=0.32 – 0.36 mm. Good results are obtained when d=0.29 – 0.31 mm and d=0.37 – 0.40 mm, but only in the case, when Tmelt=320^{0}C and Tmould=120^{0}C. Some good results of investigation are shown (see Fig. 7).

Experiment 2: Modified Gate System: A) Processing conditions: a) Tmelt=280^{0}C; Tmould=80^{0}C; b) Tmelt=300^{0}C, Tmould=100^{0}C; c) Tmelt=320^{0}C; Tmould=120^{0}C; B) Gate depth=0.20 – 0.40 mm at intervals of 0.10 mm.

The results of the experiment with the modified gate system show that it is possible to fill the mould cavity at all gate depths chosen, except in the case, when the gate depth is equal to 0.20 - 0.30 mm and Tmelt=280^{0}C and Tmould=80^{0}C. Some good results of the experiments with the modified gate system are shown (see Fig. 8, Fig. 9 and Fig. 10).

The best results with the modified gate system in respect to an easy mould cavity filling are obtained, when the gate depth is equal to 0.31 -0.40 mm at all of the processing parameters.

4. Conclusions

3D CAD models of an optical disc mould and the gate system are developed, in which the central hole of the optical disc is formed by breaking off the circle gate from the polymer substrate. The simulation program MoldflowXpress (integrated with SolidWorks) is used for initial investigation to study the effect of the gate size and the gate location towards the central hole thickness on the mould cavity filling process.

Two ranges of correct gate sizes are determined, in which an easy mould cavity filling is attained. The results of the experiments with this type of the gate system show that it is difficult to fill the mould cavity, except in the case, when greater values of the mould and melt temperatures are used.

A modified gate system is obtained in the working process. The modified system allows the change of the gate position towards the central hole and the use of a gate with a greater depth.

The results of the experiments with the modified

gate system show that it is possible to fill the mould cavity at all variable parameters chosen, including also some cases, when the values of the gate depth and the melt and mould temperatures are minimal.

The correct gate dimensions for the two gate systems at appropriate processing parameters are obtained in the process of computer simulation. The best results are achieved, using the modified gate system at a greater depth of the central gate and when the gate is symmetrically located towards the central hole.

The approach presented and the 3D CAD models of the gate system can be used as a ground for future development of the mould filling simulation, oriented towards more precise determination of the resulting injection pressure, the residual stresses distribution and the temperature distribution.

Acknowledgements

We are grateful to 6FP 4M Network of Excellence.

References

[1] Schift H., David C., Gabriel M, Gobrecht J., Heyderman L.J., Kaiser W. and Köppel S. Micro- and Nanoengineering MNE'99, Rome, Italy, Sept. 22-24 (1999).

[2] Schift H., Glaus F., Gobrecht J., Haas B., Ketterer B., D'Amore A., Simoneta D., Kaiser W. and Gabriel M., Paul Scherrer Institut, Switzerland, PSI annual report (2000).

[3] Schift H., Bächle D., Gobrecht J., Kaiser W., D'Amore A. and Gabriel M. Paul Sherrer Institut, Switzerland, PSI scientific report (2002).

[4] Tzarnorechki O., Borissova D. IIT/WP-169B, (2003).

[5] Hatch D., Kazmer D. The International Journal of Advanced Manufacturing Technology 18, No 5 (2001).

[6] Shyu G.D., Isayev A.I. and Lee Ho-Sang, Korea-Australia Rheology Journal 15, No 4 (2003) 159-166.

[7] Trifonov D., Tzonev N., Tzarnoretchki O. Cybernetics and Information Technologies 2, No 2 (2002) 113-117.

[8] Beich W.S. Photonics Spectra, March (2002) 127.

[9] Tokuhara S., Onisawa Y., Iyoshi S., Toba H. Daicel Chemical Industries Ltd Research Center 1239, boshi - ku, imeji, yogo, Japan (2002) 671-712.

[10] Keyes D. R., Lamonte R., Mc-Nalli D., Birtritte M. Photonics Spectra, October(2001) 31.

[11] Galic G. Managing Partner, Galic/Maus Ventures (1988).

[12] Swiss Moulds AWM targets cycle time. Optical Disc Systems, November-December (1998) 26.

[13] Zhou H., Li D., Advances in Polymer Technology 20, No 2 (2001) 125-131.

[14] Eichsleder M., Eusemann N., Kunststoffe 77 (1987) 150-153.

[15] Gábor J., Ovacs K. Periodica Polytechnica Ser. Mech. Eng. 49, No 2 (2005) 115-122.

Numerical modelling and experimental characterization of short pulse laser microforming of thin metal sheets

J.L. Ocaña, M. Morales, C. Molpeceres, O. García, J.A. Porro, J.J. García-Ballesteros

Centro Láser UPM. Ctra. de Valencia, km. 7,3. 28031 Madrid. Spain

Abstract

Continuous and long-pulse lasers have been extensively used for the forming of metal sheets for macroscopic mechanical applications. However, for the manufacturing of micro-mechanical systems (MMS), the applicability of such type of lasers is limited by the long relaxation time of the thermal fields responsible for the forming phenomena. As a consequence, the final sheet deformation state is attained only after a certain time, what makes the generated internal residual stress fields more dependent on ambient conditions and might difficult the subsequent assembly process. The use of short pulse (ns) lasers provides a suitable parameter matching for the laser forming of an important range of sheet components used in MMS. The short interaction time scale required for the predominantly mechanic (shock) induction of deformation residual stresses allows the successful processing of components in a medium range of miniaturization (particularly important according to its frequent use in such systems). In the present paper, a discussion is presented on the specific features of laser interaction in the timescale and intensity range needed for thin sheet micro-forming with ns-pulse lasers along with relevant modelling and experimental results and a primary delimitation of the parametric space of the considered class of lasers for the referred processes.

Keywords: laser micro-forming, residual stresses, shock waves, numerical modelling, short pulse lasers

1. Introduction

The increasing demands in MMS fabrication are leading to new requirements in production technology. Especially the packaging and assembly require high accuracy in positioning and high reproducibility in combination with low production costs.

Conventional assembly technology and mechanical adjustment methods are time consuming and expensive, so that accurate positioning of smallest components represents a key assignment in micro-manufacturing.

Although initially assembling the components with widened tolerances before precisely micro-adjusting them in a second step has proven to be more time and cost efficient, as mounted micro-components are typically difficult to access and highly sensitive to mechanical forces and impacts.

For this reason, contact-free laser adjustment processes offer a great potential for accurate manipulation of micro-devices.

Laser forming, usually indicating laser thermal forming, is a flexible rapid prototyping and low-volume manufacturing process [1–5], which uses laser-induced thermal distortion to shape sheet metal parts without tooling or external forces.

Laser thermal forming has many technological advantages compared to the conventional forming technologies, including design flexibility, production of complex shapes, forming of thick plates, and possibility of rapid prototyping.

However, it is hard for laser thermal forming to maintain material's properties of shaped metal parts because of thermal effects resulting in undesirable microstructure change, including recrystallization, and phase transformation even with no melting involved during the process [3]. Additionally, thermal forming may melt or burn the surface and even result in small crack on the surface.

Laser shock forming is a non-thermal laser forming method using the shock wave induced by laser irradiation to modify the curvature of the target [6]. It has the advantages of laser thermal forming (non-contact, tool-free and high efficiency and precision) while its non-thermal character makes it possible to maintain material properties or even improve them by inducing compressive stress over the target surface, (a desirable feature in industry for shaped metal parts to resist corrosion and fatigue cracks, see [7]).

In this paper, laser shock micro-forming is studied using both numerical and experimental methods for a thin metallic film in a one-side pinned configuration. The effect of laser parameters on the concrete deformation mechanism is investigated experimentally and data obtained from experiments are then used to validate the corresponding simulation model. Curvatures before and after laser micro-forming characterizing the net bending effect of the process were measured using reflection confocal microscopy.

2. Model Description

The developed calculational model is integrated by two principal modules conceived respectively for the analysis of the problem of laser shock waves generation and propagation under two different but complementary approaches [7–9].

On one side, LSPSIM is a one-dimensional model intended for the estimation of the pressure wave applied to the target material in Laser Shock experiments [8,9].

LSPSIM describes the material-tamper gap assuming an only phase of evolution that can be extended to the end of the processes, i.e., obtains the target-confining medium gap amplitude by solving the coupled system of energy and impulse equations subject to the thermofluiddynamic conditions imposed by the laser energy deposition (see refs. [7,9]).

On the other hand, on the basis of the time-dependent pressure profile calculated by LSPSIM, the HARDSHOCK code (based in the FEM commercial code ABAQUS®) solves the shock propagation problem into the solid material, with specific consideration of the material response to thermal and mechanical alterations induced by the propagating wave itself (i.e. effects as elastic–plastic behaviour, changes in elastic constants, phase changes, etc.).

For the kind of problems considered in this paper, a 3D version of the HARDSHOCK code is used. From the point of view of time differencing, the usual strategy of explicit differencing for the initial fast shock propagation phase followed by standard implicit differencing for the analysis of the final residual stresses equilibrium is used.

Concerning material behaviour, in Laser Shock processes the material is stressed and deformed in a dynamic way, with strain rates exceeding 10^6 s^{-1}, for which static stress-strain relations are essentially invalid to characterize the material response. Instead, a dynamic elastic limit has to be used. In the reported calculations such limit has been fixed according to Johnson-Cook model [10], applicable to the typical behaviour shown by the different materials through their dynamic strain–stress curve. The actual material yield strength is taken according to Von Mises' criterion.

A sample of the 3D simulated geometry is shown in figure 1. It corresponds to the incidence of a uniform laser beam of cylindrical shape on a planar surface with a pinned end as boundary conditions. The FEM element used for the mechanical simulation is C3D8R an 8-node brick reduced integration with hourglass control.

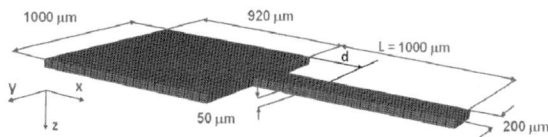

Fig. 1. Geometrical properties of a typical FEM Model used for the numerical simulation of one-end pinned thin metal strip subject to laser shock forming.

3. Numerical Results

The described model has been applied to study the effect of laser pulse energy and laser spot position effects on the net metal sheets bending angle. All the results presented in this paper refer to Stainless Steel AISI 304, whose assumed mechanical properties are shown in Table 1 [11].

Using LSPSIM, the resulting plasma pressure is applied to the thin sheet. Plastic deformation induced by the shock wave generates a residual stress distribution in the beam. Stress in the direction of the beam (S_{11}) in the incident surface is in compression whereas in the rear surface is tensile. This stress distribution produces a local bending in the direction of the laser beam (see figure 2).

3.1. Analysis of the influence of laser pulse energy on the bending angle

As a first step in the analysis, the influence of the laser pulse total energy on the bending angle of typical specimens has been analyzed. In figure 3 the thin

sheet deformations for different laser pulse energies are displayed. One only laser pulse incident on a selected area of the specimen (laser spot centred at $d = L/3$, see figure 1) has been considered for this simulation.

Table 1
Mechanical properties of stainless steel AISI 304

Property	Value
ρ Density [kg.m^{-3}]	7896
ν Poisson's ratio	0,25
E Elastic Modulus [MPa]	193
T_m Melting Temperature [K]	1811
T_0 Test Temperature [K]	300
Johnson-Cook parameters:	
A [MPa]	350
B [MPa]	275
C	0.022
N	0.36
M	1
$\dot{\varepsilon}_0$ [s^{-1}]	1

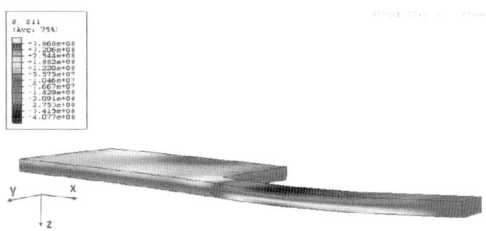

Fig. 2. Sample result of the S_{11} (longitudinal sheet direction) residual stresses field induced in a typified specimen after the application of a short laser pulse at $d = L/3$.

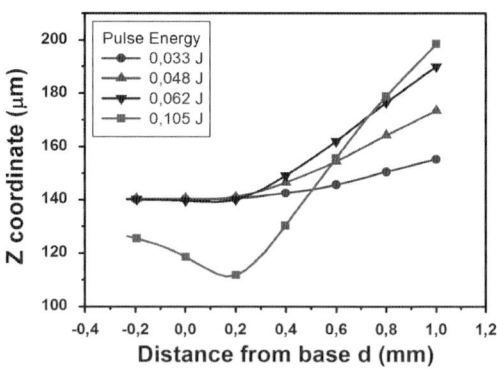

Fig. 3. Numerical deformation curves of reference thin sheet specimens subject to ns laser pulses irradiation at $d = L/3$ for different laser pulse energies

Figure 4 shows the net bending angle of the metal sheet (determined for small angles as the ratio between the apparent bar end displacement to the bar length) as a function of the laser energy pulse. As it can be observed, after a critical value of pulse energy, the bulk material deformation starts to have a deleterious effect on the net bending angle: The higher plastic deformation at the laser incidence region pushes down the bar as a whole then partly compensating the effect of the local bending.

As it can easily understood, this deleterious effect is dependent on the concrete point of application of the laser pulse over the specimen. This is the reason for the analysis performed in Section 3.2.

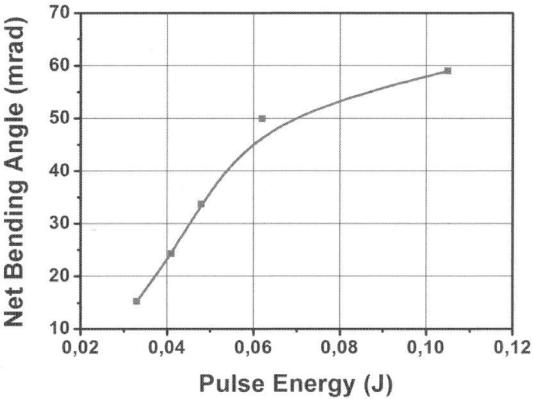

Fig. 4. Net bending angle of reference thin sheet specimens subject to ns laser pulses irradiation at d = L/3 as a function of laser pulse energy

3.2. Analysis of the influence of laser spot position on the bending angle

In this point, the analysis of the influence of the laser incident position (laser spot centre position) on the final specimens bending is performed.

In this case, and as a consequence of the combination of two main coupled mechanisms, namely a local bending due to plastic deformation produced by shock evolution in the sheet and an overall displacement angle due to bending moment produced by applied pressure far from the clamping, a maximum in the net bending angle is obtained when the laser spot is applied at about d = L/3, a result that could also be obtained from a pure theoretical analysis in this geometrically simple case.

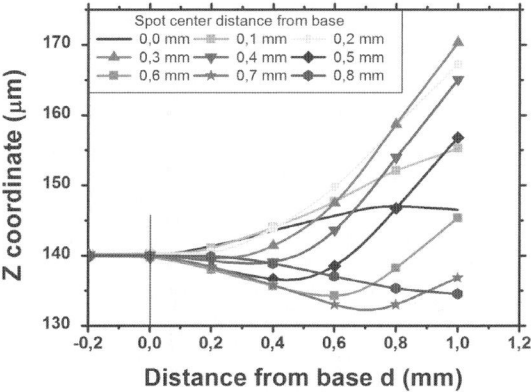

Fig. 5. Numerical deformation curves of reference thin sheet specimens subject to 0.048 J, 10 ns laser pulses irradiation at different distances from the pinned end

In figure 5 the deformed profiles of the clamped sheet subject to an equal-energy laser pulse (0,048 J) applied at different distances from the clamping base are displayed, and in figure 6 the variation of the net

(i.e. apparent) bending angle is analyzed against the distance of the laser point incident to this base. The referred maximum at d = L/3 can be clearly observed in this figure.

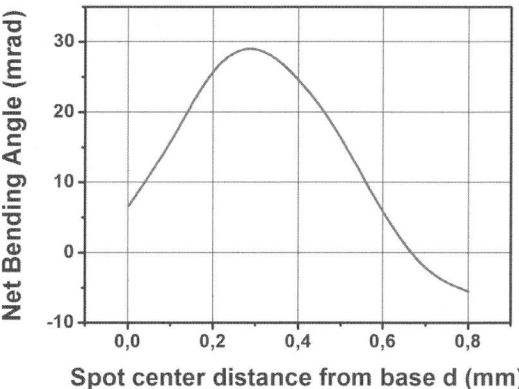

Fig. 6. Net bending angle of reference thin sheet specimens subject to 0.048 J, 10 ns laser pulses irradiation as a function of spot center distance from base, d.

4. Experimental Setup

In order to validate the laser shock micro-forming results obtained on a computational basis, the corresponding test specimens reproducing the required energy and geometry irradiation conditions were systematically treated at the available experimental facility. Table 2 shows the corresponding working conditions.

Table 2
Laser working conditions for specimens irradiation

Irradiation system parameters	Value
Laser energy per pulse [J]	Up to 1.05
Pulse length FWHM [ns]	9.4
Original Beam radius [mm]	15
Mask radius [μm]	750
Energy per pulse (after mask) [mJ]	33-105
Final spot radius [μm]	175
Confining layer	Air

The practical irradiation system used for the experiments is photographically shown in figure 7. Each specimen was fixed on a holder by means of a computer controlled stage. The laser pulse was conducted to the interaction area by means of a reflecting mirror and a focussing lens. In order to obtain a smaller spot size and to reduce the applied pulse energy, a pinhole mask was placed before the focussing lens.

In figure 8 a SEM photograph of the geometry of the treated specimens is shown together with the representative areas of laser interaction at each particular case. As it can be observed the specimens were machined in sets containing several beams from a larger metal sheet by means of the laser micromachining workstation available at UPM Laser Centre.

Fig. 7. Laser shock microforming experimental setup used at UPM Laser Centre

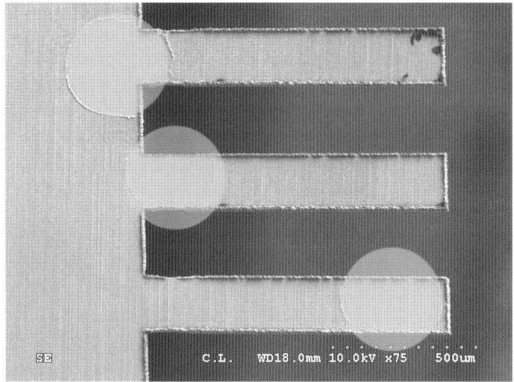

Fig. 8. SEM photographs of test pieces irradiated at different distances to clamping base.

5. Experimental Results and Discussion

The suitability of laser microbending of thin metal strips by means of ns pulsed lasers with average power in the range of several Watt has been experimentally demonstrated.

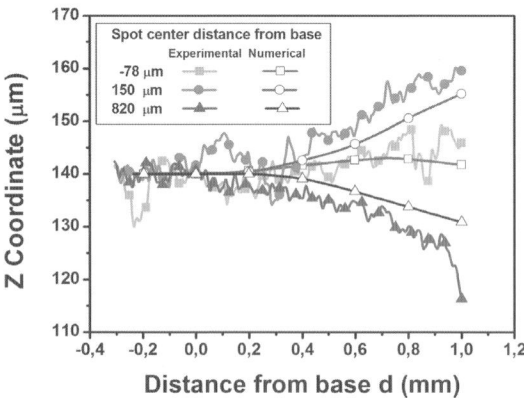

Fig. 9. Comparison of numerical vs. experimental deformation geometry for three different experimental conditions (distance from clamping base). The experimental profiles are typical of confocal microscopy.

Beam deformation was measured for the different specimens using reflection laser confocal microscopy, the observed experimental profiles being in good agreement with the numerical model predictions, as shown in Fig. 9.

From the conceptual point of view, simulations of single-end pinned targets show the presence of two opposite bending components:

1. Local bending at beam incidence position due to local plastic strain. This bending is produced in the direction of the laser beam.
2. Overall angular displacement from beam clamping due to shear stress in the beam. This bending is produced against the direction of the laser beam.

Additionally, the simulations have shown as critical parameters:

1. Pulse energy: there is a minimum (threshold) laser energy density required to produce a permanent (plastic deformation) material bending and, on the other end, for pulse energies exceeding a certain value, the overall angular displacement is the most important effect, the net local bending angle loosing its relative importance and even disappearing.
2. Laser spot centre position: a maximum in global bending has been obtained for laser spot applied at d = L/3. From this point on, overall angular displacement increases but net bending angle is reduced.

As a final observation, and provided that the laser forming possibilities of short pulse lasers have not only proved but also adequately characterized, according to the authors' experience, the use of ns laser pulses is expected to provide a really suitable parameter matching for the laser bending of an important range of MMS sheet components

Acknowledgements

Work partly supported by Spanish MEC Projects PSE020400-2006-1, PSE020400-2007-2 and CIT0205002005-11.

References

[1] Vollertsen, F. et al.: J. Mater. Process Technol. 151 (2004) 70.
[2] Widlaszewski, J.: M. Geiger, A. Otto (Eds.), Laser Assisted Net Shape Engineering, 4, Meisenbach-Verlag, 2004, p. 1083.
[3] Dirscherl, M. et al: J. Laser Micro/Nanoeng. 1 (2006) 50.
[4] Schmidt, M. et al.: J. Laser Appl. 19 (2007) 124.
[5] Geiger, M. et al.: Dubowski (Eds.), Second International Symposium on Laser Precision Microf. Proc. SPIE, vol. 4426, 2002,
[6] Zhang, W. et al.: ASME Trans. 126 (2004) 10.
[7] Ocaña, J.L: et al.: C.Phipps, M. Niino (Eds.), High-Power Laser Ablation II SPIE Proceedings, vol. 3885, 2000, p. 252.
[8] Ocaña, J.L: et al.: M. Geiger, A. Otto (Eds.), Laser Assisted Net Shape Engineering, vol. 3, Meisenbach-Verlag, 2001, p. 199.
[9] Ocaña, J.L. et al.: Appl. Surf. Sci. 238 (2004) 242.
[10] Johnson, G.R. et al: Int. J. Eng. Fract. Mech. 21 (1985) 31.
[11] Akbari-Mousavi, S.A.A. et al.: Materials and Design 29 (2008) 1.

Multi-Material Micro Manufacture
S. Dimov and W. Menz (Eds.)

291

Modelling the Solidification-Structure of Al Micro-Castings as a function of their Aspect Ratio and Mould Pouring Temperature

J-F. Charmeux[a], R. Minev[a], S. Dimov[a], E. Minev[a]

[a]*Manufacturing Engineering Center, Cardiff University, Cardiff, CF24 3AA, UK*

Abstract

Producing micro-castings trough vacuum investment casting is known to be associated with high cooling rates due to small scale of the castings. High cooling rates together with alloy composition might be the main factors affecting the final metallographic structure of castings' alloys during the solidification process. When using Al-Si-Mg casting alloys, the size of the dendritic structure can be used for a non-destructive test to assess the mechanical properties and overall quality of the castings. Also the ability of the alloys to be structured by different mechanical and energy assisted processes is highly dependant on their metallographic structure. Based on earlier experimental results, this paper proposes an empirical model describing the degree of dendrite cell refinement in cast micro-features as a function of their AR and mould pouring temperature. Additionally, the paper reports the strong correlation between the DCS refinement and the changes in the mechanical properties of the castings through MHV measurements following a Hall-Petch equation type.

Keywords: micro-casting, dendrite cell refinement, investment casting.

1. Introduction

The capability of the Vacuum Rapid Investment Casting (VRIC) to produce complex micro and meso-micro components was studied in details in [1-3]. It was shown that features with size less than 150 µm, ribs with aspect ratio (AR) higher than 50, and surface roughness in the range of 5 µm could be reproduced with accuracy of 10% or less. This makes the technology quite promising in Rapid Prototyping (RP) micro features and components especially by applying RP techniques to produce accurate sacrificial patterns or casting clusters through direct shell processes.

In a more recent paper [4], the authors investigated the degree of dendrite refinement when producing micro components using Al-Si-Mg casting alloys. An empirical model (Fig. 1) was proposed to describe the evolution of DCS in micro thin walls with different thicknesses as a function of their AR for a given mould/metal materials and processing parameters.

Following is the form of the proposed model which was chosen to perform the nonlinear regression analysis of the experimental data:

$$DCS = Span \cdot e^{-k \cdot AR} + Plateau \qquad (1)$$

This model named exponential decay with plateau has three parameters, *Span, k* and *Plateau* that could be associated with different physical aspects of the casting process. In particular, the *Span* describes quantitatively the degree of cells' refinement encountered in each specimen in respect to its overall size and complexity. The coefficient *k*, represents the slope of the curve and reflects the changes in the degree of under-cooling achieved during solidification. *Plateau* is a threshold value that defines the refinement limit for the existing functional dependence between DCS and AR of the features. It is dependant on the thermodynamic characteristics and the freezing range of the casting alloy, and also on its ability to form dispersive structures under the specific cooling down conditions for a given investment material.

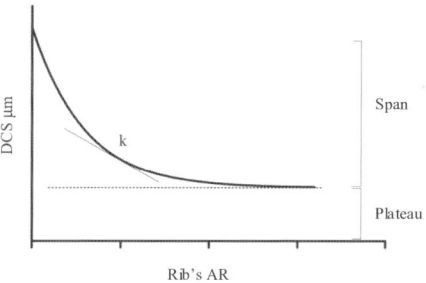

Fig. 1 Model: Exponential decay with plateau.

Following are the main conclusions derived from the experimental study conducted in [4]:

1. The microstructure refinement in castings incorporating meso/micro scale features is significant. In particular, the smallest dendrite size measured in features with a thickness of 250 µm was less than 5 µm. The grain refinement of the casting alloys should be expected to facilitate the replication process and also to reduce the shape instability of the channel type features.

2. The group of ribs having thicknesses in the range of 0.5-0.25 mm underwent similar solidification conditions regardless of their thicknesses.

3. The influence of features' AR on resulting material microstructure is significant when casting meso/micro scale components. For example, the increase of AR from 1 to 50 in the ribs with a thickness in the range from 250 to 500 µm led to more than 3 times reduction of DCS (Fig. 2). This phenomenon could be explained with the significant increase of the cooling rates when reducing the feature sizes that are highly dependent on mould thermal properties and processing temperatures.

4. The porosity distribution in meso/micro features of

the castings was more uniform. This is due to almost simultaneous solidification of casting alloys in such features. In particular, the apparent porosity in meso/micro features was reduced to less than 0.1 % while in the core of the castings it was typically around 1 to 5 %.

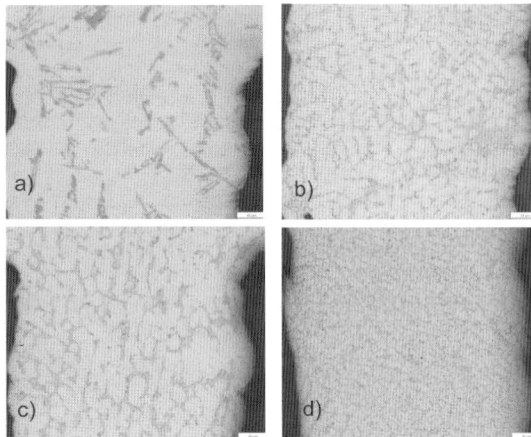

Fig.2 The dendrite refinement of 0.3 mm thick ribs: Mould temperature of 450°C, a) AR 1 and b) AR 40; Mould temperature of 250°C, c) AR 1 and d) AR 40 (x100 magnification)

The aim of this study is to optimize the model's parameters reported in [4] and to introduce a general expression of this model as a function of the mould pouring temperature (T_M) for cast micro features with dimensions 0.5-0.25 mm. Additionally, an attempt was made to find a relationship between the observed structural refinement in these cast micro-ribs and their respective mechanical properties.

2. Experimental designs and settings

The two moulds temperatures used in our experiments were 250°C and 450°C respectively. A detailed description of the manufacturing technology together with the methodology and experimental settings used in this investigation is provided in [4].

Due to the micrometre scale of our rib samples, Micro Hardness Vickers (MHV) measurements were taken on the group of ribs with dimensions 0.5-0.25 mm using a load of 0.05 kg. This approach was chosen in our investigation since it provides a simple and non-destructive method for assessing the resistance of the material to plastic deformation.

3. Results and discussion

3.1. Model's parameters optimization

The purpose of this section is to optimize the best-fit parameters previously obtained from the regression analysis performed in [4] and to calculate their 95% confidence interval (CI) using a Monte Carlo data simulation [5]. Since it was demonstrated in [4] that the group of ribs having thicknesses in the range of 0.5-0.25 mm underwent similar solidification conditions regardless of their thicknesses, we calculated the model's optimized parameters based on the average

DCS refinement exhibited by each of the two 0.5-0.25 mm thick group of ribs. The SDs characterizing the DCS variability of each of the two 0.5-0.25 mm thick group of ribs was also calculated in order to be implemented in the Monte-Carlo data simulation.

The two averaged sets of DCS data were first analysed using a nonlinear regression similar to that used in [4] so that new parameters could be calculated. Using the latter parameters' values, we generated for each processing temperature a group of fifty sets of data with a random scatter equivalent to the SDs calculated for each group of ribs, respectively 1.53 and 2.60 μm for the mould temperatures of 250°C and 450°C respectively. Doing so, the simulated data would have been similar to what one would have observed if the experiments were repeated 50 times.

Finally, each set of the generated fifty sets of data was analysed individually as in [4], providing us with fifty different values for each respective parameter. These values, which followed a Gaussian distribution, were then used to calculate the 95% CI on the parameters.

After calculation of the best-fit parameters, the optimized model was compared to the original DCS data to verify whether or not the goodness of its fitting was improved. By analysing the coefficient of determination R^2 and the SD of the residuals S_{xy}, it appeared that the quality of its fitness to the DCS data slightly improves for the T_M of 450°C. Considering the T_M of 250°C, no significant changes were observed compared with the regression analysis performed using the independent DCS values of all rib thicknesses. Therefore, the parameters calculated in this section were kept for the rest of this study to develop the expression of our general model. These parameters are displayed in table1.

	Mould Temp. T_M	
Model optimized Parameters	250°C	450°C
SPAN	8.65	18.65
K	0.162	0.092
PLATEAU	5.24	8.65

Table 1. Model best-fit parameters.

3.2. Expression of the general model

The aim of this section is to propose a general model characterizing the evolution of the DCS of our cast micro-features as a function of T_M. To do so, each parameter (Span, k and Plateau) was expressed as a function of T_M assuming a linear relationship of each parameter between the two processing temperatures (Fig. 3). Then, by implementing these equations into equation (1), we obtained the general expression of our empirical model:

$$DCS = \left(5.E^{-2} \cdot T_M - 3.85\right) \cdot \exp^{-\left(\left(-3.5.E^{-4} \cdot T_M + 0.25\right)AR\right)} + \left(1.7.E^{-2} \cdot T_M + 0.98\right)$$

where T_M - mould temperature; AR- Aspect Ratio.

This model has been used to generate several theoretical curves with different mould temperatures of 150, 250, 350, 450 and 550°C (Fig. 4).

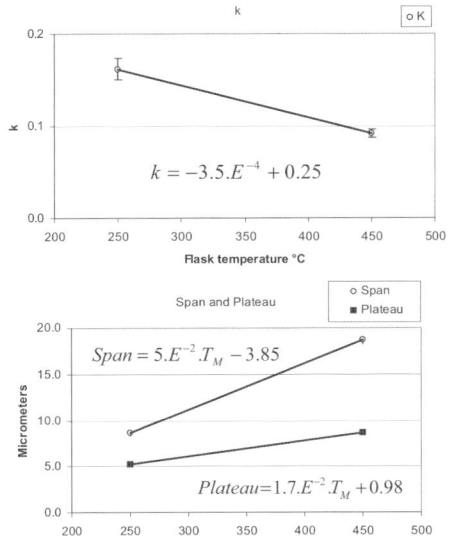

Fig.3 Plot of the model's optimized parameters as a fuction of T_M together with their 95% CI.

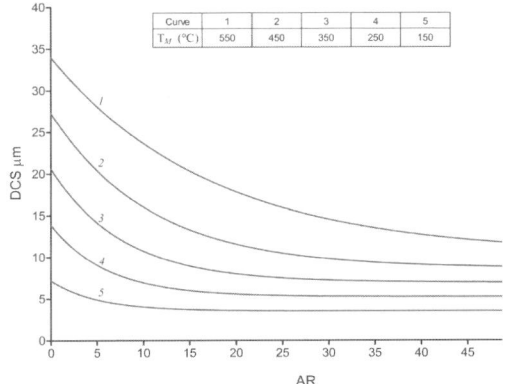

Fig.4 Evolution of the DCS of cast micro-features as a function of their AR for different T_M.

The generated curves of figure 4 highlight the importance in lowering as much as possible the mould pouring temperature in order to reduce the influence of AR upon the resulting dendritic structure and thus produce micro casting exhibiting both finer metallographic structure and higher structural homogeneity.

3.3. Mechanical properties of cast micro-features

The microstructure of the casting micro-features directly affects the mechanical properties (hardness, ductility and strength) of the castings and therefore their machinability (e.g. by micro-milling) if additional processing is required to reach the desired functionality of the micro-part. Also, finer microstructures in Al-Si casting alloys lead to superior thermal conductivity properties of these alloys [6]. Therefore, it is interesting and practically important from an engineering point of view to examine the micro-mechanical properties of the alloy. This data could be of interest for developing

various micro-casting demonstrators such as cooling devices, heat exchangers, micro-fluidics channels, micro-gear boxes, delicate jewellery, polymer replication micro-tools etc…

It appeared from the MHV analysis that the measured MHV was strongly dependent on the ribs' AR. No significant distinction could be made between MHV measurements taken for the whole range of thicknesses produced under similar cooling down conditions (i.e. similar T_M). Then, in order to compare the results obtained under our two experimental conditions, the arithmetic mean between each set of MHV data was calculated and plotted (Fig. 5).

Generally, the lower the mould temperature and consequently the finer the metallographic structure, the higher the measured MHV, as highlighted by the moving average trend lines displayed in figure 5. The total increase in MHV was of approximately 10% and 13% for the processing temperatures of 250°C and 450°C respectively. Thus, one might expect a similar increase in the tensile properties of these ribs while increasing their AR. Also, both curves exhibited a similar trend, which was characterized by a linear increase of the MHV up to an AR of approximately 25 before stabilising. The fact that the MHV did not increase further for AR above 25 correlated well with the Plateau threshold observed for the structural refinements of the rib castings.

Therefore, and to conclude this work, an attempt was made to identify a possible Hall-Petch relationship [7] between the MHV measurements performed on the ribs with thicknesses in the range of 0.5-0.25 mm and the structural refinement described by our model. To do so, we plotted on the graphs of figure 5.20 the evolution of the MHV as a function of the inverse square root of the DCS values provided by our model. This plot was similar to that of the Hall-Petch equation, which describes the dependence of the metals' yield strength on grain size. Through a linear regression, it appeared that the data correlated by more than 70% as highlighted by the values of the coefficient of determination R^2 of figures 6.a and 6.b. Thus we could link the evolution the MHV of an Al-Si-Mg casting alloy to its structural refinement employing a Hall-Petch equation type such as:

$$MHV = MHV_0 + K_H *(DCS)^{-1/2}$$

where, similar to the Hall-Petch equation, MHV_0 and K_H are constant for a given material and processing parameters.

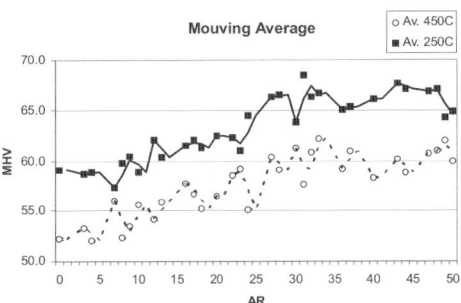

Fig.5 Evolution of MHV in cast micro-ribs as a function of their AR.

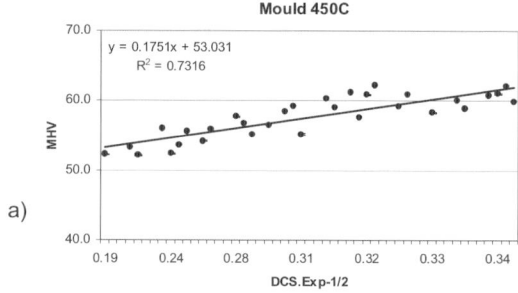

a)

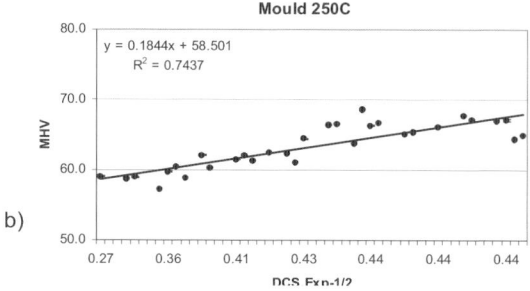

b)

Fig.6 Dependence of DCS upon metal hardness based on a Hall-Petch equation type: a) T_M of 450°C; b) T_M of 250°C.

4. Conclusions

Based on the conducted experimental study the following conclusions can be derived:

o Based on the 0.5-0.25 mm thick group of cast micro-features, an empirical model describing the structural refinements of the as-cast DCS as a function of the mould pouring temperature and features' AR was proposed.

o The model showed that lowering the mould pouring temperature is essential for producing micro casting with high AR exhibiting both fine metallographic structure and high structural homogeneity.

o The DCS refinement observed in our micro-features led to a total increase in hardness of approximately 10% and 13% for the mould processing temperatures of 250°C and 450°C respectively.

o Finally, it was found that the DCS refinement proposed by our model correlated well with the increase of the metal hardness according to a Hall-Petch equation type.

5. Acknowledgments

The research reported in this paper is funded under the MicroBridge programme supported by Welsh Assembly Government and the UK Department of Trade and Industry and the EPSRC Programme "The Cardiff Innovative Manufacturing Research Centre". Also, it was carried out within the framework of the EC FP6 Networks of Excellence, "Multi-Material Micro Manufacture (4M): Technologies and Applications" and "Innovative Production Machines and Systems (I*PROMS)". The authors gratefully acknowledge the support given to the Networks by the European Commission.

6. References

[1] J-F.Charmeux, R.Minev, S.Dimov. Capability Study of the Fcubic Direct Shell Process for Casting Micro-components, 4M Conference, Karlsruhe, 2005, pp.227-230, ISBN 0-080-44879-8.

[2] J-F.Charmeux, R.Minev, S.Dimov. A Comparative Study of Three Technologies for Producing Castings with Micro/Meso-scale Features, 4M Conference, Grenoble, 2006, pp.327-330, ISBN 0-080-45263-9

[3] J-F.Charmeux, R.Minev, S.Dimov. Benchmarking of Three Processes for Producing Castings Incorporating Micro/Meso-scale Features with High Aspect Ratio. IMechE part B, 2006. In press.

[4] J-F. Charmeux, R. Minev, S. Dimov, E. Minev. Metallographic Investigation and Solidification-Structure Modelling of Al Micro Castings. Proceedings of the 3rd International Conference on Multi-Materials Micro Manufacture, Borovets, 2007. pp. 217-220.

[5] Gentle, J. E. 1998. Random number generation and Monte Carlo methods. New York: Springer.

[6] Vázquez-López, C., Calderón, A., Rodriguez, M. E., Velasco, E., Cano, S., Colás, R., Valtierra, S. 2000. Influence of dendrite arm spacing on the thermal conductivity of an aluminum-silicon casting alloy Journal of Materials Research 15(1), pp. 85-91.

[7] Hall, E. O. 1970. Yield point phenomena in metals and alloys. London: Macmillan.

295

Simulation of Microforming Processes by Applying a Mesoscopic Model

S. Geißdörfer[a], U. Engel[a], M. Geiger[a]

[a] *Chair of Manufacturing Technology, University of Erlangen-Nuremberg, Egerlandstrasse 11, 91058 Erlangen*

Abstract

Continued miniaturization in many fields of forming technology implies the need for a better understanding of the effects occurring while scaling down from conventional macroscopic scale to microscale. At microscale, the material can no longer be regarded as a homogeneous continuum because of the presence of only a few grains in the deformation zone. This leads to a change in the material behaviour resulting among others in a large scatter of forming results. A correlation between the integral flow stress of the workpiece and the scatter of the process factors on the one hand and the mean grain size and its standard deviation on the other hand has been observed in experiments. Conventional FE-simulation, is not able to consider the size-effects observed when scaling down processes. Actually the reduction of the flow stress the increasing scatter of the process factors and a local material flow being different to that obtained in the case of macroparts. For that reason, a new simulation model has been developed taking into account the size-effects. The present paper deals with the theoretical background of the new mesoscopic model, its characteristics like synthetic grain structure generation and the calculation of micro material properties - based on conventional material properties. The verification of the simulation model is done by carrying out various experiments with different mean grain sizes and grain structures but the same geometrical dimensions of the workpiece.

Keywords: Simulation, microforming, mesoscopic model

1. Introduction

Microproduction technology is supposed to be one of the key technologies of the next years. As an evidence, the NEXUS market analysis for microproduction technology [1] predicts an increasing market volume in this area of up to 25 billion US$ in 2009. One of the driving forces for this demand of microparts are the increasing sales figures of technical consumer products like digital cameras, camcorders, mobile phones and MP3 players. The metallic parts needed for these electro-mechanical systems, with characteristic geometrical dimensions in the range of few millimetres are up to now mostly manufactured by turning, milling or electrochemical processes. Putting in mind that these processes are predominately appropriate to small quantity production, other technologies like microforming are more likely to be used. If this technology - which is well suited for serial production at industrial scale - is applied, some specifics have to be regarded. When scaling down the geometrical dimensions from conventional scale to micro scale, so-called size-effects appear which are mainly caused by the material forming behaviour and a change in the friction conditions. The influence of the material on the forming process has been investigated by various microforming processes showing a significant dependency of the forming results on the initial state of the material structure. Further investigations have confirmed these influences to be the main reasons for the occurring size-effects [2, 3]. An approach to describe these effects using simulation methods from a global point of view was done by [4]. Based on Ashby's theory, the influence of a free surface on the forming behaviour of microparts was considered in a simulation model by subdividing the material into three volume fractions with different material properties. This enables the description of the

reduction in the flow stress of open die processes when scaling down to microscale.

As the size-effect on the material flow is mainly controlled by the impact of the material structure, in this paper an approach is described to consider local deformation behaviour depending on the local material structure within a finite element simulation.

2. Simulation approach

The first attempt to describe the size-effect in the material forming behaviour has been given by [4]. This approach considers the influence of the different material forming behaviour on the integral flow stress while decreasing the size of the workpiece but it is not able to consider the influence of the material structure on the scatter of the forming results and the shape evolution. Thus, the material definition of the simulation system has to be enhanced by describing the size-dependent local material forming behaviour [5]. This has been realized by the development of a mesoscopic simulation model which generates - based on the Monte-Carlo-Potts algorithm - a synthetic material structure mapping the real material structure quite well. The material properties for each of the synthetic grains have been calculated individually based on its size and position within the specimen (cf. Fig. 1).

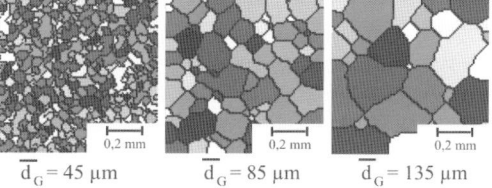

$\overline{d}_G = 45\ \mu m$ $\overline{d}_G = 85\ \mu m$ $\overline{d}_G = 135\ \mu m$

Fig. 1: Synthetic material structures according to the theory of Monte-Carlo Potts

As it is given by the theory of Hall-Petch and Ashby [6] the materials plastic response to external forces is mainly controlled by grain size and the position of the grain within the specimen. The first can be described by the correlation between grain size and flow stress as given by eq. (1) where τ is the shear stress, τ_0 the critical shear stress, K the Hall-Petch slope and d_G the mean grain size.

$$\tau = \tau_0 + \frac{K}{\sqrt{d_G}} \tag{1}$$

If forming behaviour on scale of some few grains is to be considered eq.(1) can be modified (eq.(2)) taking into account the influence of neighbouring grains on the mechanical properties of a single grain.

$$\tau = \tau_0 + \frac{m_{1,2} \cdot \tau_i \cdot \sqrt{\delta}}{\sqrt{d_G}} \tag{2}$$

In eq.(2), $m_{1,2}$ represents a transformation matrix considering the different sliding systems in adjoining grains, τ_0 is the critical shear stress in the considered grain and the factor δ represents the distance between pile-up source and the dislocation. In this case, dislocations can be assumed to pile-up until the stress concentration at the grain boundary exceeds the stress τ_i. From this point on, dislocation sources will be activated in neighbouring grains.

Following the fact that eq.(2) is valid for only a single grain, the Hall-Petch factor K_i (valid for a single grain) is given by eq.(3) and the correlation between the integral Hall-Petch factor K (cf. eq.(1)) and K_i by eq.(4):

$$K_i = m_{1,2} \cdot \tau_i \cdot \sqrt{\delta} \quad \text{and} \quad K_i = K \cdot \xi_i \tag{3)(4}$$

Thus, in case of a single grain, eq.(2) can be rewritten as:

$$\tau = \tau_0 + \frac{K \cdot \xi_i}{\sqrt{d_G}} \tag{5}$$

The newly introduced factor ξ_i describes the amount of dislocation pile-up at the grain boundary region. Considering the influence of the free surface of the forming behaviour of the specimen and thus the non-capability of dislocations to pile up in this region, a more detailed view on the influence of the grain boundary on the properties is given in eq.(6).

$$\xi_i = \frac{\bar{k}_{nG}}{k_G} = \frac{1}{k_G} \frac{1}{\alpha_G} \sum \alpha_c \cdot k_{nG} \tag{6}$$

In eq.(6) $\bar{k}_{nG}$ represents the mean value of yield stresses k_{nG} of the adjoining grains related to the real contact area fraction α_c/α_G of the grains in case of 3D studies. In case of a free surface grain boundary it can easily be seen by eq.(6) that there $k_{nG}*\alpha_c$ is zero and thus decreasing the Hall-Petch slope K_i for the considered grain. Furthermore, this approach even reflects the influence of the free surface on grains being located not directly at the surface due to the consideration of the ratio between k_G and k_{nG}. Applying

this model to macroscopic dimensions there is a large number of smallest grains within the considered specimen and thus the influence of the free surface on the integral material properties is rather low.

A further enhancement of the mesoscopic model is done based on the theory of Meyers and Ashworth [7] considering the dislocation pile-up in the region of a grain boundary in a more detailed way. This is achieved by a subdivision of each single grain into two main regions, actually the grain boundary volume fraction α_{GB} yielding a higher flow stress $k_{f,GB}$ due to the occurring dislocation pile-up compared to the rather low flow stress $k_{f,I}$ within the material volume fraction α_I yielding no pile-up. The integral material behaviour of a single grain is expressed by linear superposition of the yield stresses of these two volume fractions.

$$k_f = \alpha_I \cdot k_{f,I} + \alpha_{GB} \cdot k_{f,GB} \tag{7}$$

In order to simplify the analytical model, the shape of a single grain is considered to be spherical. Thus, the above mentioned two diametrical volume fractions α_I and α_{GB} can be expressed by:

$$\alpha_{GB} = 2\left[3\left(\frac{t}{d_G}\right) - 6\left(\frac{t}{d_G}\right)^2 + 4\left(\frac{t}{d_G}\right)^3 \right] \tag{8}$$

and

$$\alpha_I = 1 - \alpha_{GB} \tag{9}$$

with a thickness of the grain boundary layer t and grain size d_G. Inserting eq.(8) and eq.(9) into eq.(7) leads to

$$k_f = k_{f,I} + 6\left(k_{f,GB} - k_{f,I}\right) \cdot t d_G^{-1} - \\ 12\left(k_{f,GB} - k_{f,I}\right) \cdot t^2 d_G^{-2} + 8\left(k_{f,GB} - k_{f,I}\right) \cdot t^3 d_G^{-3} \tag{10}$$

Obviously, different cross sections of a single grain will produce different area fractions of α_G and α_{GB}. Hence Meyers and Ashworth recommended using mean values of t and d_G, $\bar{t}$ and $\bar{d}_G$ with values according to following approach.

$$\bar{t} = 1.57 \cdot t, \bar{d}_G = \frac{\pi}{4} d_G \tag{11}$$

Considering the variation of the thickness t of the work hardening layer to be

$$t = \left(k_1 k_2 d_G\right)^{1/2} = k_{MA} d_G^{1/2} \tag{12}$$

it has to fulfil two effects: the fluctuation of the stress field varies with d_G if the grain size is decreased, leading to a dependency $t = k_1 d_G$ and the dislocation spacing to be unchanged and the dislocation interactions will dictate a constancy in t, thus $t = k_2 d_{G0}$. Assuming the term $\bar{t} d_G^{-1}$ to be approximately equal to $2t d_G^{-1}$, eq.(10) can be written as

$$k_f = k_{f,I} + 12 k_{MA}\left(k_{f,GB} - k_{f,I}\right) \cdot d_G^{-1/2} - 24 k_{MA}^2 \\ \left(k_{f,GB} - k_{f,I}\right) \cdot d_G^{-1} + 16 k_{MA}^3 \left(k_{f,GB} - k_{f,I}\right) \cdot d_G^{-3/2} \tag{13}$$

In case of grains being in the micrometer range the $d_G^{-1/2}$ term dominates and thus the Hall-Petch relation is obtained where the Hall-Petch slope K is equal to

$$K = 12k_{MA}\left(k_{f,GB} - k_{f,I}\right) \qquad (14)$$

Finally, inserting eq.(3) and eq.(5) into eq.(7), the flow stress of the boundary region $k_{f,GB}$ and the inner region $k_{f,I}$ of the grain can be calculated by

$$k_{f,GB} = k_{f,0} + \frac{K \cdot \xi_i}{\alpha_{GB} \cdot \sqrt{d_G}} \qquad (15)$$

and

$$k_{f,I} = k_{f,0} \qquad (16)$$

The thickness of the grain boundary region t can be calculated by inserting eq.(14) into eq.(12). For the investigations in this paper, t was assumed to be

$$t = 0.133 \cdot d_G^{0.7} \qquad (17)$$

3. Verification of integral deformation behaviour

The flat upsetting test is well suited for the investigations of the influence of the material structure on the microforming results due to its robustness against varying friction conditions and the plain strain condition in the centre of the specimen required for 2D simulation purpose. In this test, the specimen is deformed transversally. The experiments are carried out using a universal testing machine (UTS 5K with a Walter & Bai controller) which is equipped with a HBM 5 kN force measurement system. The specimens are made from CuZn15, a material frequently used for the production of microparts in electronic applications. The dimensions of the cylindrical workpieces are 0.5 mm in diameter and 3 mm in length. Even if the influence of friction on the experimental results is very small due to the chosen test, MoS2 is used as solid lubricant to keep friction reproducible.

The variation of material structures for different specimen is realized by applying different heat treatments on the specimen. This leads to a variation of mean linear grain sizes in a wide range, determined by standard methods (DIN 50601). A number of 25 specimen per batch is chosen to get an adequately reliable statistical basis for the experimental investigations. The evaluation of the flat upsetting tests is done in terms of mean load $\overline{F}$ as well as its scatter, characterized by standard deviation s_F, respectively variation coefficient $s_F / \overline{F}$. In case of fine grains the total grain structure can be regarded as almost homogeneous yielding rather reproducible results. However, for coarse grains, the grain structure might be more different from specimen to specimen even if the nominal mean grain size is kept constant, causing the increasing scatter in material behaviour. In order to verify the mesoscopic simulation model the same number of simulations using different material structures but identical mean grain sizes have been done. By comparing the results of simulation and experiment it can be shown that the mesoscopic model is very well suited to represent the forming process (Fig. 2).

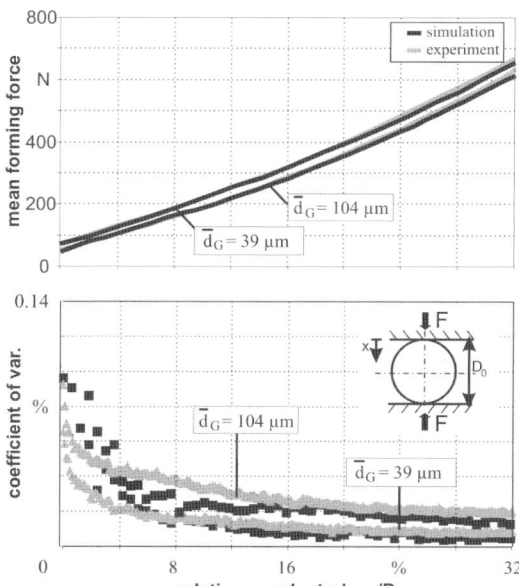

Fig. 2: Comparison between simulation and experiment using flat upsetting test in terms of mean forming force and its coefficient of variation

4. Local deformation behaviour

If different synthetic material structures are applied to the specimen, significant differences in the local deformation behaviour, represented by the scatter of the equivalent plastic strain, can be seen (Fig. 3).

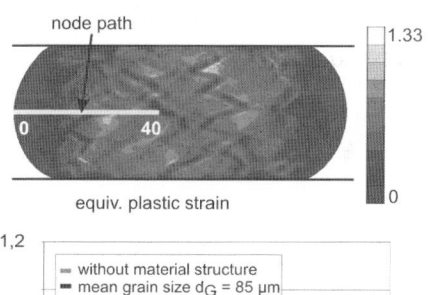

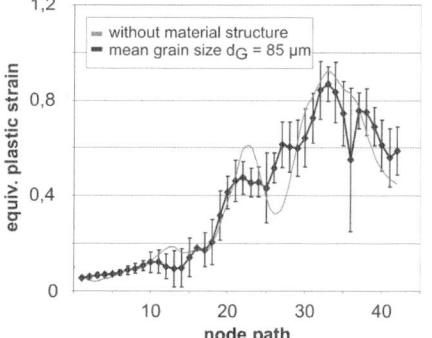

Fig. 3 Local deformation in dependency of grain constellation

If the local deformation behaviour as a result of the mesoscopic simulation model has to be verified by experimental results, a novel experimental setup has been designed where an in-situ analysis of the local deformation is possible. The forming process chosen for these investigations was the can-backward extrusion process which is quite sensitive to the influence of differences in the local deformation behaviour on the process results like the shape evolution of the can. The tool (shown in Fig. 4), which is positioned within a universal testing machine (described above), consists of a punch and a die, both half cylindrical to perform the deformation operation on the half-cylindrical specimen. The die is closed by a translucent window made from sapphire.

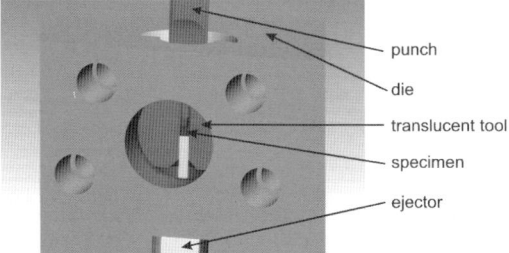

Fig. 4 Setup of the novel tool concept

The experiments have been carried out using the material CuZn15 at two different grain size levels (d_G =32 μm and d_G = 123 μm) and two different punch diameters of d_P = 0.7 mm and d_P = 0.8 mm. The deformation velocity has been chosen to be rather low at v = 0.1 mm/min in order to avoid dynamic effects and early tool failure. The process has been recorded by a CCD-camera and a telecentric objective, thus an in-situ observation of the deformation through the translucent tool has been achieved with a resolution of about 1-2 micron. From the digital images of the material with d_G = 32 μm and d_P = 0.7 mm recorded during the process differences in the local material flow can be seen which now can be attributed to the impact of the material structure on the deformation. This is even more true if the grain size is increased and the wall thickness reduced.

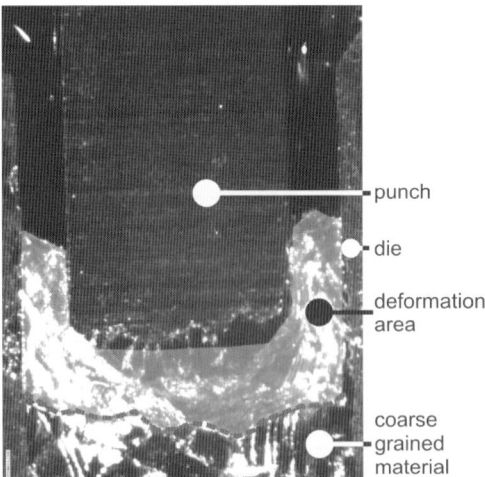

Fig. 5 Analysis of the local displacement, d_G = 123 μm, d_P = 0.7 mm

In order to measure the local displacements, the recorded images have been analyzed by using the software ARAMIS, frequently used in the area of sheet metal forming for the calculation of forming limit diagrams. By using this tool, the deformation paths in dependency of the material structure can be calculated. In Fig. 6, the local displacement at one time step is shown. It must be noted that especially in case of small local deformations there is a difference between fine and coarse grained material (e.g. in the front of the

punch). If the deformation is large, this difference is reduced as most of the grains at this stage are contributing to the deformation.

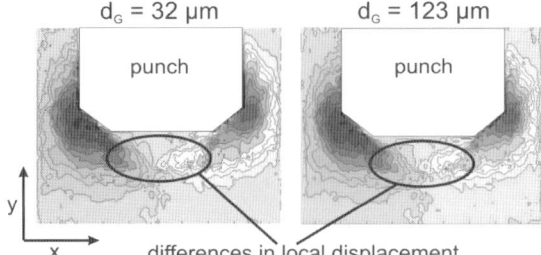

Fig. 6 local displacement in x-direction calculated by image processing

Discussion

By applying the novel tool setup for the analysis of the local deformation behaviour it has been shown, that the material structure is affecting the local deformation and as a result the shape evolution. As it has been shown by the results of the first tests, by using this strategy local deformation in dependency of the material structure can be calculated and in the future used to qualify the mesoscopic simulation model on a local level. In addition if there are differences detected by this new method in comparison with the simulation data, also the mesoscopic model will be enhanced to consider anisotropic deformation behaviour at a local level.

Acknowledgements

The authors would like to thank the German Research Foundation (DFG) for the financial support within the framework of SPP 1138 *Process Scaling*. Also, this work was carried out within the framework of the EC Network of Excellence "Multi-Material Micro Manufacture: Technologies and Applications (4M)"

References

[1] R. Wechsun et al., NEXUS Market Analysis for MEMS and Microsystems III. Nexus Task Force Market Analysis. Wicht Technologie Consulting. München, 2006.

[2] M. Geiger, M. Kleiner, R. Eckstein, N. Tiesler, U. Engel, Microforming, Keynote Paper, Annals of the CIRP, 50 (2) (2001) 445-462.

[3] U. Engel, R. Eckstein, Microforming - from basic research to its realization, J. Mater. Process. Technol. 125-126 (2002) 35-44.

[4] U. Engel, A. Meßner, M. Geiger, Advanced Concept für the FE-Simulation of Metal Forming Processes for the Production of Microparts. In: Altan, T. (Edtr.): Advanced Technology of Plasticity 1996, Vol. II, p. 903 - 907

[5] Geiger, M.; Geißdörfer, S.: Mesoscopic Model - Advanced Simulation Of Microforming Processes.. In: Prod. Eng. Res. Devel. (2007), Nr. 1,Springer, S. 79-84

[6] M.F. Ashby, The Deformation of Plastically Non-homogeneous Materials, Phil. Mag. 21, (1970) 254-255.

[7] M. A. Meyers, E. Ashworth, Phil. Mag. A, 1982, 46, 737.

Influence of Force Components on Thin Wire EDM

A. Herrero[a], S. Azcarate[a], A. Rees[b], A. Gehringer[c], A. Schoth[c], J.A. Sanchez[d]

[a] Micro & Nano Technologies Dep., Fundacion Tekniker, Avda. Otaola 20, 20600 Eibar, Spain
[b] Manufacturing Engineering Centre, Cardiff University, Cardiff, CF24 3AA, UK
[c] IMTEK, University of Freiburg, Georges-Koehler Allee 103, EG-79110, Freiburg, Germany
[d] Dep. of Mechanical Engineering – Faculty of Engineering of Bilbao, Avda. Urquijo s/n, 48013 Bilbao, Spain

Abstract

Apart from the important role that Micromachining and Ultraprecision machining has provided to the development of improved or innovative miniaturised products, these techniques have also attracted the interest of the researchers to obtain the highest accuracy and a thorough analysis of the principles governing the material removing mechanisms. The present article exposes the theoretical analysis of some aspects of the thin WEDM that drop the process accuracy in terms of minimum machinable slot or corner over/undercutting. The scaled electrode dimensions and the reduced power supply with respect to the normal process causes a different influence of the process variables and contributes to obtain complementary information about the WEDM process. The different force components contributing to the wire deformation are discussed and some of them are analyzed from a theoretical point of view presenting analytical calculations to evaluate their expected magnitude and pointing out the difficulties to obtain an experimental characterisation of each phenomena.

Keywords: thin WEDM, process accuracy, process forces

1. Introduction

Wire EDM is a industrially well established machining process that provides reliability and high accuracy. Nevertheless the knowledge about the erosion mechanism, the scale of the different phenomena arising from the sparking process and the influence on the wire electrode is still limited [1].

Some usual errors like wire lag, bowing, corner accuracy, white layer depth, etc. have been partially solved with the last generation machines thanks to the application of last generation controls for path definition, wire tension, etc. and the results of the research performed in the past by several authors [2,3, 4 among others] that identified different algorithms to limit the influence of the aforementioned errors.

Since the beginning of the WEDM analysis, many authors focused their efforts in the identification of the different physical and chemical processes arising from the sparking process. Stressing the influence of the geometrical error in most applications, the research carried out until now has identified the most important force components acting on the wire: electrostatic, electromagnetic, dielectric flushing, spark generation, wire traction and wire feed. Apart from these components, the thermal effects should always been looked after. For normal wires, the values obtained by different authors provide the magnitude of the different effects with some dispersion.

All these aspects are still nowadays an important research field but the inclusion of smaller wires (down to Ø0.02mm) has introduced not only a new range of applications, but also a new field for technology analysis. The research on both scales constitutes an important chance for the better understanding of the wire EDM process.

In a previous job [5], the experimental analysis and characterisation of a Ø0.03mm thin wire performance was addressed, the present paper analyses some of the reasons for such performance, specially the effect of the force components acting on thin wires.

2. Thin Wire EDM limitations

At present wires and machining equipment for thin WEDM are not capable to perform the machining with high productivity and low errors in a reliable way because both electrical (specially current) and mechanical parameters cannot extract the highest material performance avoiding rupture. This is the reason why the operator must apply soft working conditions (low current and traction forces) in aims of reliability, hampering the process capabilities in terms of accuracy and productivity.

As it was already introduced in the past [5], the slot produced by thin wire when cutting 120~130 aspect ratio features in steel or tungsten carbide is circa 60% the wire diameter while the corner undercut/overcut can be as big as 2 times the wire diameter for small angles (10°). The magnitude of such errors can constitute an important handicap in the machining of precision punches/dies or moulds for micro-replication.

Concerning the research aspects, the process does also imply important difficulties for experimental characterisation. The measurement of some critical process variables is harder to do than that in conventional WEDM: the spark frequency is circa 1MHz; (over the acquisition rate and the minimum resolution of most commercial current probes and high speed acquisition cameras, something similar happens with the measurement of currents circa 40 mA); the ~10 times scale of the wire makes difficult to apply many of the displacement measuring techniques (illumination aspects, focal distance, depth of focus, etc. for optical systems; wire alignment for inductive or capacitive sensors, etc.; touch surface magnitude for tactile probes).

The analysis and inspection of the machined components must be carefully addressed [5,6]: specially for micro-features, even with the latest metrology techniques, component clamping or cleaning constitute important challenges.

3. Electrostatic force

The electrostatic force acts during the pause (OFF time) of the spark cycle due to the attraction between charges of opposite signs (wire and part surface prior to spark). The magnitude and direction can be calculated by differential integration if the charge distribution on the wire and the part surface is defined.

For thin wires, it is possible to perform some assumptions that simplify the calculation of the charge in the wire: i) electric charge in the wire is assumed to be concentrated in its axis; ii) the electric field vectors entering the wire present a radial distribution towards its axis; iii) during the OFF time the voltage decays linearly from the wire to the part surface in the radial direction (fig. 1).

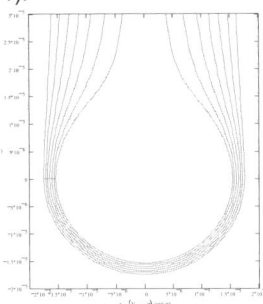

Fig. 1 Assumed distribution for voltage constant lines

The charge distribution in the wire can be calculated using the law of Gauss (1) for a close section around the electrode.

$$\int_O E \, dS = Q/\varepsilon \qquad (1)$$

Being E, the electric field (negative radial gradient of the voltage in the radial direction, which has been supposed to be linear for each radius); $\int_O dS$ the integration on loop around the wire; Q is the charge in the wire; ε is the electrical permeability of the dielectric.

Given the assumed voltage distribution, it is possible to calculate the negative voltage gradient in the orthogonal direction for each point at the surface of the machined path (electric field) and, estimate the charge distribution on the surface applying the law of Coulomb (2).

$$\Delta E(\theta) = -\Delta V/\Delta n = q(\theta)/(4 \cdot \pi \cdot \varepsilon \cdot \Delta n^2) \qquad (2)$$

Being θ the polar coordinate used for integration; $\Delta E(\theta)$ the normal electric field for each differential surface; ΔV the voltage variation; Δn the orthogonal to surface distance; $q(\theta)$ the positive charge on the surface.

Each differential surface will produce a force vector acting on the wire, given by the law of Coulomb ($r(\theta)$ is the distance from each point on the surface to the wire). The electrostatic force acting on the wire can be calculated Integrating all force components. For the exposed example, the components of the force in the feed direction produce an electrostatic force attracting the wire to the part (3), while the components in the orthogonal direction are balanced.

$$F_{electrostatic} = \int_\theta [-q(\theta) \cdot Q/(4 \cdot \pi \cdot \varepsilon \cdot r(\theta)^2)] \cdot \sin(\theta) \cdot d\theta \qquad (3)$$

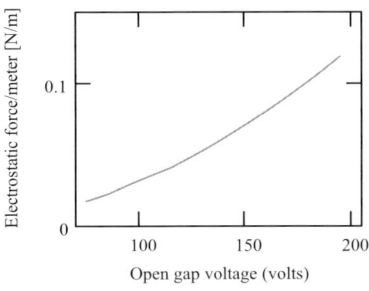

Fig. 2 Calculated Electrostatic force per unit length for a Ø0.03mm wire

The force depends specially on the open gap voltage. Figure 2 shows the analysis of the forces for a Ø0.03mm wire in a straight slot cutting a 3.6 mm height WC part, using oil as dielectric. The obtained values do not differ much to those provided by different authors, the applied voltage, gap and permeability is not very different to the normal WEDM.

4. Electromagnetic force

The electromagnetic force appears due to the creation of a magnetic field around the direction of a moving charge. In the EDM process the current will only appear during the discharge (ON time).

The direction of the current inside the part is difficult to estimate and, for the wire, it will depend on the spark location with respect to the conductivity blocks (fig 3).

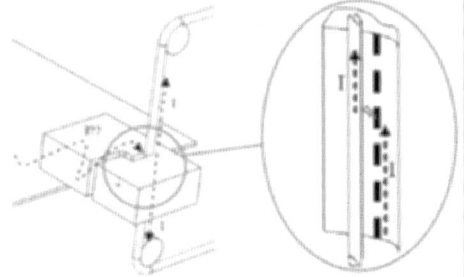

Fig. 3 Assumed path for the current in the part

The discharge will happen from the wire to a given position on the part surface, assuming that current will flow on the part surface tracing a line parallel to the wire, the wire electrode will be subjected to an attracting force between two parallel wires given by the law of Biot-Savart (4).

$$F_{electromagnetic} = \mu_{WC} \cdot \mu_0 \cdot I^2 \cdot L/(2 \cdot \pi \cdot r_w) \qquad (4)$$

Being μ_{WC} and μ_0 the magnetic susceptibility of the part and vacuum; I the discharge current; L the part height and r_w the wire radius.

The force depends specially on the discharge current but the obtained values are negligible due to the low discharge assumed for calculations (fig.4, the adopted values are those measured in real discharges with thin wires).

The values obtained for conventional EDM are similar to those provided by other authors, nevertheless, the experimental validation of this component is difficult to perform. The force values, if compared to other forces arising form the spark, are negligible and it is difficult to design a test to separate the influence of this force.

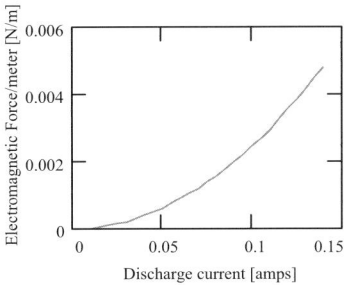

Fig. 4 Calculated Electromagnetic force per unit length for a Ø0.03mm wire

5. Dielectric flushing

The dielectric fluid surrounds all the erosion zone because both part and wire are submerged in thin WEDM. At the same time, the dielectric is injected from the top and bottom guides with a higher pressure (~6 bar for high aspect ratios) in order to clean the debris and cool the part. Oil and deionised water are used as dielectrics in market available systems but, according to the experience of the authors in machines using any of these dielectrics, the differences are minimum in terms of gap size, machining feed or machining parameters.

Trying to estimate the effect that dielectric causes on the wire: from a empirical point of view, it is possible to obtain higher cutting rates and cut wider components with higher pressures but generally this implies larger gaps. Doing an analytical estimation of the pressure executed by the dielectric is not an easy task. The problem of two opposite jets running inside a microchannel (formed by the wire and the part) and submerged in a fluid is a state of the art problem which is being analysed even for simple symmetric cavities.

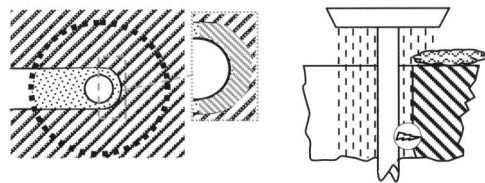

Fig. 5 Dielectric flushing scheme: top view (left); section (right)

Considering the discharge gap as a micro-channel closed by the diameter of the wire and the part (fig. 5, detail) the hydraulic diameter is too reduced, obtaining very low Reynolds numbers (<20). It is also possible to assume the microchannel as the fraction of the flow leaving the upper guide and entering the part (fig.5, left), in this case the flux is also laminar but the obtained Reynolds numbers are bigger (~100).

The Darcy-Weisbach equation can be used to calculate pressure drops in micro-channels [7]. Considering any of both sections, the pressure drop is excessive for the flow to reach the centre of thick workpieces (fig. 6).

This is probably the reason why some authors have reported that it is possible to assume that sparks occur in a dry plasma [7] (the high spark rate creates bubbles around the wire and new sparks appear in the absence of dielectric), neglecting the influence of the dielectric force[8].

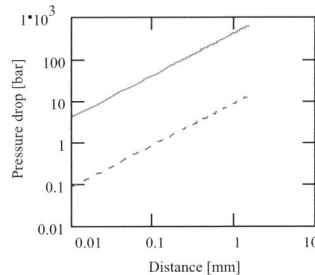

Fig. 6 Pressure drop caused by a constant flow inside a microchannel constrained by the discharge area (red); machined slot area (blue)

At the same time, the higher fluid pressure will make the dielectric jet get deeper into the machined slot to remove debris and cool the part. This effect, together to a second dielectric pumping effect caused by the plasma bubbles pushing the dielectric inwards/outwards the spark area can be the reason why lower wire breakage, higher productivity or wider components machining can be achieved with higher dielectric pressures. In this case the bigger gaps can be due to the dragging force executed by the dielectric on the wire and the turbulences caused by the plasma bubbles.

6. Spark force

The force caused by the spark on the electrode has been considered as an impulse or a travelling pressure wave by several authors in the past. It is not really well understood what happens when the spark appears: the appearance of bubbles, debris, instantaneous thermal sources, etc. at the same time in random locations at a very high repetition rate makes of the process analysis a difficult challenge.

Recent jobs [8,9] have brought some light on the problem. They consider the force as that cause by the generation of a plasma bubble that expands and collapses creating the force on the electrode.

Trying to apply a similar point of view for WEDM from a theoretical point of view, the bubble is a plasma that contains a mixture of gases created during the ablation of electrode, dielectric and part material. The cooling of this material will form the debris and the re-solidified layer of the WEDM process.

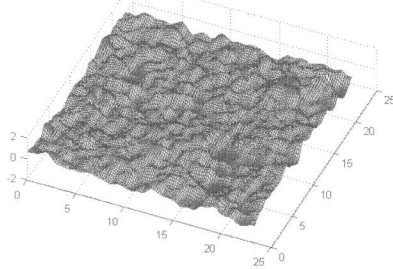

Fig. 7 Surface sparked by thin WEDM (X100 Confocal microscopy)

Identifying the portion of electrode and part material that is ablated during the process can be done by analysing the crater on both of them. For single spark tests this is easier than for real machining conditions, but the obtained values are hardly ever the same even if the same conditions are applied. The real machining process modifies the energy distribution within the dielectric (the spark is ramified by debris, etc.), part and electrode.

Just as some authors [10] have made in the past, a crater mean size can be obtained analysing the sparked surface. Optical microscopy (fig. 7) reveals a high heterogeneity in the dimensions of the craters that makes difficult to achieve the desired value.

Despite the random nature of the EDM process, if FFT filtering techniques are applied to different slides in radial and Cartesian directions, a mean value can be defined for the crater diameter and depth (fig. 8).

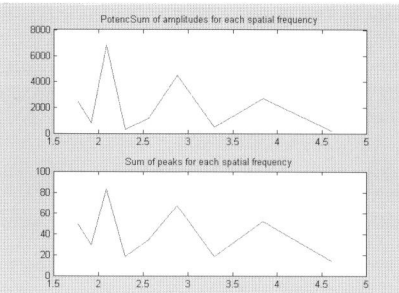

Fig. 8 Sum of all peaks (amplitudes on the top and unitary below) at different spatial frequencies for the different slices in the Cartesian and radial directions

The values obtained for the crater diameter under different conditions agree with those intuitive values guessed by the operator in the microscope. These values can be used to filter the surface and, comparing the filtered surface to the measured surface, define a volume for the sparked volume and the re-solidified volume. Dividing the measured area by the crater area, the number of craters per unitary surface and the re-solidified volume per crater are estimated.

The crater dimensions can be also used to estimate a initial bubble volume (considering the crater as part of a sphere).

If re-solidified volume and crater volume on both wire and part are assumed as ablated, the ablated number of mols for each material (electrode and part) can be calculated (5)

$$N_{mol} = V_{part} \cdot \rho_{WC}/MW_{WC} + V_{wire} \cdot \rho_W/MW_W \qquad (5)$$

Being N_{mol} the number of mols; V_{part} and V_{wire} the calculated crater and resoldified layer volume per aprk in part and wire; MW the molecular weight on the part (assumed to be tungsten carbide) and the wire (assumed to be tungsten). If the plasma is assumed to be a perfect gas, the volume per mol in normal conditions (273.15 K and 1 atm) will be 22.4 litters.

In order to achieve ablation, the temperature should be high enough to change the state of the materials of the electrodes (6203 K for WC). At that temperature the bubble volume should be that calculated out of the crater dimensions.

For a perfect gas, the partial pressure of the bubble at the temperature in which the part material is ablated can be calculated out of the conditions of pressure, volume and temperature at normal conditions. For Ø0.03mm WEDM, the obtained crater dimensions are Ø1.92~2.88μm and 0.12~0.38μm depth for both pat and wire. The calculated pressures range from 30 to 140 atm. Values that fit results obtained by other authors [11].

7. Conclusions

The present job introduces an effort to analyse the scale and the influence of different forces on thin wires in the WEDM process. Electrostatic and spark forces are definitely important in thin wires while the low current applied in the process seems to reduce the effect of electromagnetic force with respect to normal WEDM. The influence of dielectric flushing is difficult to analyse because the large pressure drop in microchannels seems to make the influence different to the conventional process. FFT filtering can be used to analyse scanned surfaces and estimate mean values for crater dimensions in EDM for different calculations.

The proposed analysis solutions are compared to results obtained by other authors in normal WEDM but a thorough experimental validation should be perform in the future job.

Acknowledgements

The authors wish to thank the Spanish Ministry for Education and Science and the National Plan for Research, Development and Innovation 2004-2007. Thanks also to the members of the 4M - Network of Excellence promoted by the European Commission for their inputs, opinions and discussions.

References

[1] Schumacher BA, "After 60 years of EDM the discharge process remains still disputed", J. of Mat. Proc Tech 149, 2004, pp. 376-381

[2] Dekeyser WL and Snoeys R., "Geometrical accuracy of wire-EDM", ISEM 9,1989, pp. 226–232.

[3] Beltrami et al. "A simplified post process for WEDM", J. of Mat. Proc Tech 58, 1996, pp. 385-389

[4] Kinoshita et al. "Study on EDM with wire Electrode; Gap Phenomena", Annals of the CIRP 25, 1976, pp. 141-145

[5] Herrero et al., "Error analysis of thin wire EDM corner machining", ISEM XV, 2007, pp. 121-126

[6] Herrero et al., "Discussion on Thin WEDM Error Analysis and Characterisation", Proc. of the 1st 4M Int. Conf., 2005, pp. 165-168

[7] Tamura T., Kobayashi Y. "Measurement of impulsive forces and crater formation in impulse discharge" J. Mat. Proc. Tech. 149, 2004, pp. 212-221

[8] H. Obara et al. "Simulation of Wire EDM", ISEM12, 1998, pp. 99–108.

[9] Klocke F. Et al. "Force measurements in the Micro Spark Erosion with various electrode materials, polarities and working media" ISEM XV, 2007, pp. 263–268.

[10] Pandit et al. "Data dependent system approach to EDM process modelling from surface roughness profiles", Annals of the CIRP 29, 1980, pp.107-111

[11] Descoeudres A., "Characterization of electrical discharge machining plasmas", Thesis nr. 3542, EPFL, 2006

Systems: novel product and system designs

Multi-Material Micro Manufacture
S. Dimov and W. Menz (Eds.)

305

An integrated all-optical microfluidic particle sorter

S. Valkai, H. I. Kirei, L. Oroszi and P. Ormos

Institute of Biophysics, Biological Research Centre, Hungarian Academy of Sciences,
H-6726 Szeged, Hungary

Abstract

A fully integrated microfluidic sorter is introduced. It is able to count, characterize and sort micrometer sized particles and cells. All functions of the device are performed by light. The objects to be sorted are counted optically, they are characterized by measuring their fluorescence. Even the sorting itself, directing the particles into different channels is performed by the pressure of light. The device is built by photopolymerization, from a light cured optically clear resin upon a glass plate support. The whole structure is created in a single photolithography step. The microfluidic channels and optical waveguides that carry the illuminating, detecting and sorting light form a single integrated structure. The supporting units, like sample reservoirs, pumps, light sources, light detectors are easily connected to the device from the outside. The device is optimized for simplicity. It is a proof-of-concept instrument, it demonstrates that it is possible to build simple optically driven microfluidic systems that perform complicated functions.

Keywords: microfluidics, integrated optics, cell sorter

1. Introduction

Cell sorting is a basic procedure in life sciences and medical diagnostics. Fluorescence activated cell sorting (FACS) is a highly developed technology (1). In the typical process, fluorescent markers are identified by optical measurements and the cells are separated according to the content of these markers. Available instruments can separate cells based on such optical signatures very fast and very efficiently. The device and method is central to a very large variety of biochemical tests. Typically, these highly complex instruments are rather expensive, consequently their use is limited to large, well equipped laboratories.

There is a promising new trend in biochemical instrumentation, where large, expensive instruments are replaced with small, simple devices. The approach relies on scaling down the system to micrometer characteristic sizes, and such microfluidic systems can realize more and more tasks (2). The concept, also called lab-on-a-chip technology, has numerous advantages. In the genomics approach, there is a demand for a large number of possibly straightforward tests on small amounts of samples that can be performed outside sophisticated laboratories (3). Such tests often may not even need the high performance offered by the traditional complex, cumbersome units. There is room for instrumentation where simple biochemical analysis is performed by easy to use uncomplicated devices.

In this new technology functions of biochemical procedures are duplicated in microfluidic systems. Cell sorting based on fluorescent characterization is a typical task that should also be solved within microfluidics. There have been several successful attempts in this direction. Devices with different levels of complexity and integration have been introduced. Miniaturization is not a simple scaling down of size, small sizes also demand different solutions of details. A number of schemes were introduced where the sorting is done by various mechanisms, for example pressure gradients (4,5), or electrokinetic forces (6). The observation, characterization of the particles to be separated is always performed optically, and in microfluidics systems to achieve this function integrated optical components are added (6, 7)

In the micrometer size regime the pressure of light of reasonable power (10mW) is sufficient to move high refractive index particles in aqueous environment. This property can be used to separate particles, and a number of vastly different schemes have been introduced to use light even for the actual separation step. Optical manipulation started as a tool to manipulate individual particles in the form of optical tweezers, where a focused laser light of large numerical aperture is grabbing a high index of refraction particle (8). The force of light was also used in bifurcating flow channels to deflect cells between the branches in solutions of different complexity (9, 10), in systems analogous to FACS machines.

Recently new sorting schemes have also been introduced where complex light patterns can separate more particles simultaneously. Multiple sorting centers are created in optical lattices, and with clever arrangements highly efficient sorting can be realized based on different parameters of the particles to be separated (size, index of refraction, etc) (11, 12).

The devices introduced earlier were built with varying levels of complexity. The channels, optical waveguides for the analyzing light are integrated from several components of various materials into a single system. However, it is characteristic to all systems introduced earlier that the sorting step is performed in a complex optical arrangement: the laser light is introduced into the device in a microscope through a microscope objective that has to be positioned very accurately to achieve proper function, and the optical system gets more complicated with the complication of the optical potential pattern applied.

In the design of the microfluidic devices

robustness is also a central requirement and systems with the highest simplicity in design, production and use would be preferred. If it can be produced in a simple cheap procedure, then an efficient "throw away" instrument would be obtained for use in rough environment that would provide advantages by expanding the envelope where such investigations can be done.

In this work we present a microfluidic FACS device that is built with the lowest possible complexity. The structure: microfluidic and optical parts are prepared in a single step as a single unit. The system is fully integrated, a microscope is not needed for the operation.

2. Materials and methods

A Y-shaped structure was chosen: an incoming stream of particles is divided into two. The channel carrying the stream before the bifurcation is interrogated with light: an illumination and observation optical waveguide is applied to the channel. The separation is achieved by light carried to the channel at the Y junction on both sides enabling redirection into both directions. Fig. 1. shows the outlay of the channel geometry.

The device is built as a three layer structure. Layer one, the base is a microscope slide. The main component is the second layer, the structure containing the channels and the optical waveguides (Fig.1.). This is built on the glass surface by photopolymerization. First, a layer of the polymer is created on the glass support by spin coating of the desired thickness (typically 40 µm) and the structure is produced by illuminating the layer through a mask (MicroT Ltd, Hungary).

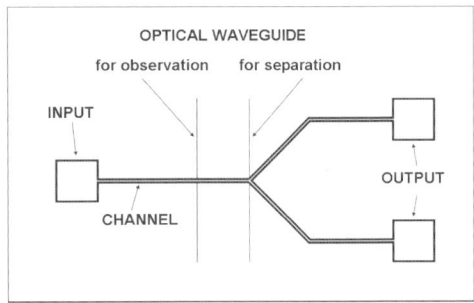

Fig.1. The layout of the separator. The structure is built in a single photolithography step. Height of the structure: 40 µm. Width of the channel: 40 µm. Width of the channel wall: 40 µm. Width of the optical waveguide: 10 µm. The structure is covered with a layer of PDMS.

We used Norland Optical Adhesive 61 photopolymer (Norland Co, USA) and it was polymerized by light from a Hg lamp (HBO100, Zeiss, Germany). The reason to use this polymer was that it has a good mechanical stability, at the same time it is also transparent to light at the wavelengths used in our experiments. Finally, in the third layer the structure was covered by a soft layer of silicone rubber PDMS (Polydimethylsiloxane, Sylgard 184, Dow Corning, USA). The PDMS was covered and pressed with a glass sheet and the fluid connectors are attached to this. Clamping down the cover glass

ensures tight fit. Fig. 2. shows the ready device: in a. the structure before covering, and in b. the finished device with the input connector in place on the microscope before use.

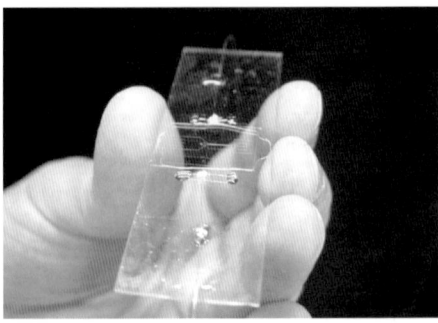

a.

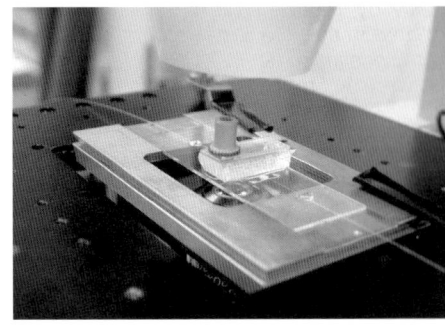

b.

Fig.2. Photograph of the separator. **a.** The structure built by photolithography on a cover slip glass plate glued upon the central part of a slide. The optical fiber is attached to the integrated waveguide. **b.** The finished device on the microscope. The structure is covered with a PDMS block.

The structure is most simple: channel and waveguides are produced as a monolithic unit. Connection to the fluid channels is achieved through standard microfluidics connectors (Upchurch N-125S, USA), glued on the cover glass and fluid is pumped by a SP100IZ syringe pump (WPI, USA).

The optical connections were realized by attaching fibers to the ends of the integrated waveguides, The waveguides extended to the edge of the glass support, the fibers were positioned and attached to the ends by manipulators and they were simply glued to each other by a photoresist (also from Norland, NOA 81).

Both for detection and separation laser light was used. Detecting light came from simple diode laser (in the presented experiments we used an excitation wavelength of 405 nm) coupled into the single mode fiber. The light at the output of the opposite fiber was fed into a photodetector, this simple arrangement was used to characterize the object passing in front of the fibers. Separating light came from a higher intensity fiber laser IPG YLM-2-1070 (wavelength 1070 nm, maximum intensity: 2W), also carried to the device by fibers. The light of the laser was divided into two beams by a beamsplitter to allow deflection into both directions. The efficiency of the coupling was determined in separate experiments: typically 40% of the light was lost in each connection.

In the experiments polystyrene beads were used to characterize the system: fluorescent and non fluorescent beads were distinguished and separated. The fluorescent beads (Fluorescbrite YG Carboxylate Microspheres, Polysciences, USA) contained the dye FITC with absorption maximum at 441 nm and emission at 486 nm. The light passing through the channel was measured at the output of the opposite waveguide behind a colored glass high wavelength pass filter of 450 nm. The light was detected and measured with a photomultiplier Hamamatsu H5783-01 (Japan).

The system was observed on a microscope: an inverted microscope Zeiss Axiovert 135 was used. We monitored the device by a TV camera, and recorded the events in a computer for detailed analysis. The motion of the beads was followed by individually by Particle Tracking Velocimetry (PTV) implemented in the iTrack developed in our institute. All characteristic parameters (position, trajectory, velocity) were determined and stored separately, and this way any desired parameter of the flow or separation could be determined.

3. Results

3.1. Detection and characterization of the beads

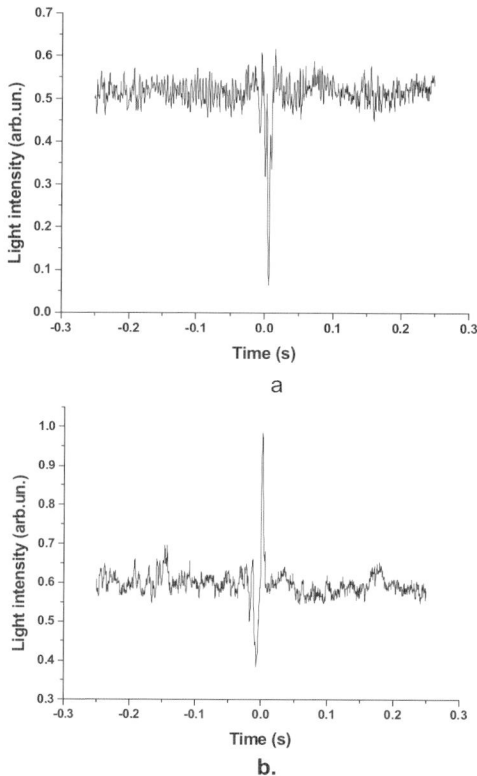

Fig.3. The light in the integrated optical detection component. **a**: A non fluorescent bead is passing between the illuminating and detecting waveguides. **b**: A fluorescent bead is passing between the waveguides

First the particles passing in front of the observation optical waveguides have to be characterized. This is the basis of sorting. We tested and demonstrated this task by identifying fluorescent and nonfluorescent polystyrene beads. The colored glass high pass filter through which the observation light is measured transmits a small portion of the light even at the wavelength of the diode laser that the highly sensitive photomultiplier detector can safely detect. Consequently, if a non fluorescent bead passes in front of the waveguides, it blocks the light, and a decrease of light intensity is detected as a signal of the event. On the other hand, a fluorescent bead causes an increase of the light at the emission wavelength – both cases are shown in Fig.3. The fluorescent and non fluorescent beads are represented by positive and negative light pulses. We note that the specific characteristics of the signals vary to a certain extent, they are not totally identical. This is due to the fact that the beads pass the detecting waveguides in different positions in the channel. However, the general characteristics, i.e. that in the case of the non fluorescent bead only light decrease is observed and in the case of the fluorescent bead the light increase is dominant is always true. Reliable characterization of the beads can be achieved this way.

3.2. Separation of the beads

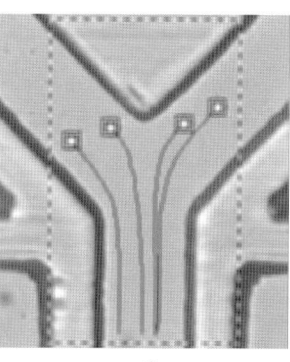

a.

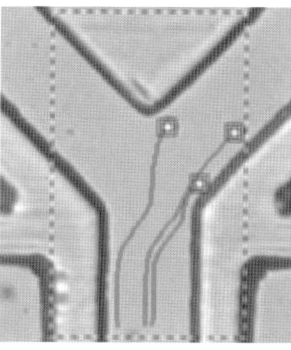

b.

Fig.4. Beads moving in the separator. Traces of several typical followed beads are shown. **a**. There is no separating light: beads are deflected according to their position in the channel before the Y-junction. **b**. The separating light is coming from the left: all beads are deflected to the right.

The beads are separated with light coming from the waveguides at the Y-junction. The images in Fig. 4. show traces of beads in the channel at the bifurcation. Fig.4.a. shows the state where there is no separation: the flow is laminar, and the beads end up in the bifurcating channels depending on their position within the channel before the branching. If the separating light is turned on, the beads are pushed into a single channel, as shown in Fig.4.b., where light comes from left.

We studied the separation in great detail: traced a statistic number of beads and analyzed how effectively it is possible to deflect the selected beads into the target channel. The conditions for perfect separation are depending primarily upon the laser intensity and flow velocity. Such an investigation of the system is shown in Fig. 5. where the separation efficiency is studied as a function of separating laser intensity. The definition of the efficiency is the following:

$$\eta = 2 * (\frac{a}{t} - 0.5)$$

where **a** is the number of beads going into the desired channel, **t** is the total number of beads and if the distribution of the beads before the branching is uniform η changes between 0 and 1. A separation efficiency 0 means that the beads are not affected by light while 1 means perfect separation. The figure shows that a laser intensity of 0.5 Watts is sufficient to achieve practically perfect separation.

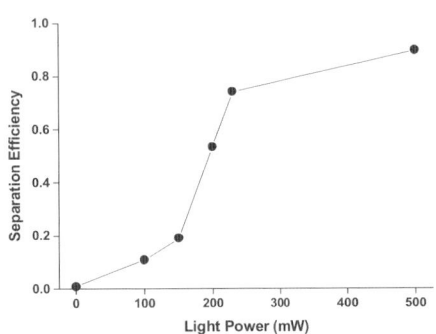

Fig.5. The efficiency of sorting as a function of the sorting light intensity

We note the sigmoidal shape of the intensity dependence. While a separation level of 70-80 % is achieved at an intensity of 200 mW. Further increase of efficiency needs significantly higher intensity. This is due to the fact that some beads move at the side of the channel and they are more difficult to move. Significant improvement could be achieved if the movement of the beads would be localized to the center of the channel by hydrodynamic focusing.

4. Conclusion

We have built a fully functional separator. The key structure of the device is built by photopolymerization in a single photolithography step. All components (channel walls, optical waveguides for both detection and separation) form a monolithic unit. Only the accessory components, like baseplate, connectors, etc. have to be provided separately. Light for both detection and separation is carried to the device by optical fibers, they are easily attached to the planar structure. Thus, the simple device is a stand-alone system: it is able to function independently from any complex optical system like a microscope. This device has to be regarded as a proof-of-concept instrument, it demonstrates that it is possible to build simple optically driven microfluidic systems that perform complicated functions.

Acknowledgement

This work was supported by the grant Országos Tudományos Kutatási Alap NK72375.

References

[1] Givan A. L., (2001) Flow Cytometry—First Principles, 2nd edn., Wiley Liss, New York

[2] Whitesides G.M (2006) The origins and the future of microfluidics Nature **442**, 368-373

[3] Chena C.C., Zappe S., Sahin O., Zhang X.J., Fish M., Scott M., Solgaard O. (2004) Design and operation of a microfluidic sorter for *Drosophila* embryos Sensors and Actuators B 102 59–66

[4] Chena C.C., Zappe S., Sahin O., Zhang X.J., Fish M., Scott M., Solgaard O. (2004) Design and operation of a microfluidic sorter for *Drosophila* embryos Sensors and Actuators B 102 59–66

[5] Kruger J., Singh K., O'Neill A., Jackson C., Morrison A. and O'Brien P., (2002) Development of a microfluidic device for fluorescence activated cell sorting J. Micromech. Microeng., 12, 486-494

[6] Lee G-B. Fu L-M, Yang R Y and Pan Y-J (2003) Micro flow cytometers using electrokinetic forces with integrated optical fibers for on-line cell particle counting and sorting Proc. 7th Internatlonal Conference on Miniaturized Chemical and Biochemical Anaiysts Systems, Squaw Valley, California USA

[7] Wang Z., El-Ali J., Perch- Nielsen I.R., Mogensen K.B., Snakenborg D., Kutter J.P. and Wolff A. (2004) Microchip Flow Cytometer with Integrated Polymer Optical Elements for Measurement of Scattered Light Proc. IEEE MEMS p. 710-713

[8] Ashkin A. and Dziedzic J.M. (1987) Optical trapping and manipulation of viruses and bacteria Science, 235, 1517-1520

[9] Applegate Jr. R.W., Squier J., Vestad T., David J.O., Marr W.M., Bado P,. Dugand M.A. and Saidd A.A. (2006) Microfluidic sorting system based on optical waveguide integration and diode laser bar trapping Lab-on-a-Chip 6, 422–426

[10] Wang M.M, Tu E, Raymond D.E., Yang J.M., Zhang Z, Hagen N., Dees B.,Mercer E.M., Forster A.H., Kariv I., Marchand P.J., Butler W.F. (2005) Microfluidic sorting of mammalian cells by optical force switching Nature Biotechnology, vol 23 number 1 jan 2005 83-87

[11] Cizmar T, Siler M., Sery M., Zemanek P., Garces-Chavez V., Dholakia K., (2006) Optical sorting and detection of submicrometer objects in a motional standing wave. Phys. Rev. B 74, 035105

[12] MacDonald M.P., Splading G.C. and Dholakia K., (2003) Microfluidic Sorting in an Optical Lattice. Nature **426**, 421

Multi-Material Micro Manufacture
S. Dimov and W. Menz (Eds.)

Feasibility of polymers for wafer scale capping of RF MEMS

P.J. Bolt, J.E. Bullema, R. Korbee, R. Kusters

TNO Science and Industry, Eindhoven, The Netherlands

Abstract

This paper concerns the feasibility of polymer capping of RF-MEMS devices, replacing traditional silicon solutions. The advantage would be less costs and potential for both further miniaturisation and integration of electrical functions in the cap. One of the challenges is the resistance against expoxy overmoulding as part of the traditional back-end process chain. This involves temperatures of 175°C and pressures of 10MPa, which the cap has to withstand. Calculations are made and experiments carried out to investigate the feasibility of selected polymers. It is shown that nanofillers will lift the polymers mechanical properties comfortably above the minimum established demands.

Keywords: polymer, capping, RF MEMS, simulation, materials testing

1. Introduction

A growth market is RF-MEMS filters for tele-communication. These are made in silicon fabs on waferscale. The actual RF filter structures often need a clearance (air cavity) between the filter and the package. The latter is normally made by moulding over of the die with an epoxy.

In order to preserve this air cavity in the overmoulding process, a protective cap is used that can withstand overmoulding pressures between 8 and 10MPa and temperatures between 160 and 200°C. Another demand is to resist reflow temperatures for lead free solder (soldering profiles up to 260°C peak temperature). Typical the air cavities have lateral dimensions in the range from a few 100μm to 1000μm. The total height of the cap should preferably not exceed 100μm. Other technical demands are that the capping doesn't interfere with the RF properties of the devices and provides sufficient hermiticity to keep the internal conditions (pressure, composition of internal atmosphere) stable. Often care must be taken that metal coatings on the filter elements are not corroded by gases from the capping material. In such cases halogens should not be present in the capping material.

The capping is a mid-end process, i.e. directly after fabrication of the filter wafer and before (compatible with conventional) back-end processing. State of the art in RF-MEMS capping are flat silicon caps which are positioned with pick and place techniques. For future generation of products, capping technologies are looked for, which reduce costs and allow further miniaturization (footprint reduction).

An alternative is wafer scale capping with a silicon capping wafer, which has to be processed for enabling (wire bonding or flip chip) interconnects. Both from ease of processing and cost perspective, use of polymer wafer level capping would be attractive here. Although the importance of polymers in MEMS is growing [1], their application is not established for this application.

Polymers that are used in semiconductor industry for e.g. structural (gap fill, stress buffer for passivation layers) or electric (dielectric interlayer) purposes are mostly applied as dry or wet films and patterned with lithographic techniques. Photosensitive poly-imides and epoxies are commonly used for this purpose. Very common is the use of thermosetting epoxies for packages, encapsulants and underfills. Thermoplastic liquid crystal polymers receive recently attention because of their hermitic and RF properties [2, 3].

With regard to capping of RF-MEMS, the feasibility of SU-8 epoxy capping was shown in [4], applied by a sacrificial layer method. Because SU-8 contains halogens, it will however not generally be applicable.

TNO carries out a study into the feasibility of polymer caps, with regard to both functional demands as well as the embedding in the process chain. This paper reports on one aspect, the strength to withstand the overmoulding step. Model system was an air cavity of 300μm x 600μm.

Firstly, a numerical analysis was carried out into the effect of thickness and shape on the deflection of a cap when overmoulded. It showed that a material with 2.5GPa Young's modulus should suffice for capping the 300μm x 600μm area.

Of primary importance is the softening of the cap during the overmoulding process. Because of lack of data in literature as well as from material suppliers, the effect of temperature on the stiffness was measured for commercial films, made out two types of candidate materials, LCP and poly-imide. Also, for a poly-imide precursor solution the effect of nanofillers on the high temperature stiffness was measured.

Finally, LCP and poly-imide film were used in experiments in which overmoulding was simulated with a purposely built apparatus.

2. Numerical analysis

FEM simulations of the cap deformation under a 9MPa external pressure were performed, see Fig. 1. Inner dimensions were 300μm x 600μm, the wall width was 20μm and height was 35μm. FEM code was MSC.Marc and 2005r3, linear elastic eight node isoparametric brick elements (type 7), were used. The Young's modulus was set at 2.5 GPa and the Poisson ratio at 0.25.

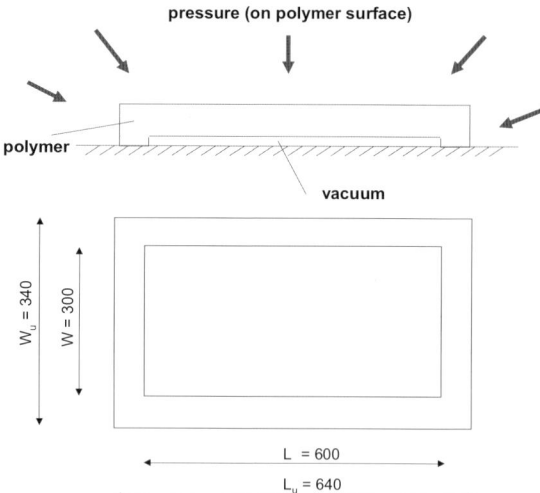

Fig. 1. Geometry of CAP deflection FEM calculations

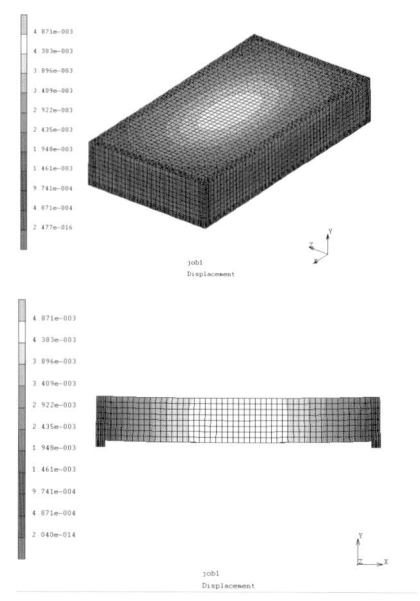

Fig. 2. Simulated displacement [mm]

Table 1
. Calculated deflection (in μm) of the cap centre

cap tickness	cap deflection
50	21
75	9.4
100	5.7

Fig. 2 depicts an example of a deformed cap geometry and Table 1 shows calculated deflections at the cap centre. It should be noticed that the geometry of the wall or rim which are part of the cap structure or on which the cap rests, affects the results. In case of a wall height of 15μm instead of 35μm, the deflection of a 100μm thick cap would be 4.8μm (and 8.2μm for a 75μm thick cap). If a large single 100μm thick cap is considered covering a matrix of cavities (with a 15μm

high rim), the deflection would be reduced to 3.3μm.

Deflections were also calculated analytically for an unclamped thin plate, see Fig. 3. For reference, a 100μm thick plate results in a deflection of 3.2μm.

Further optimization is possible if the cap is given a concave dome shape, which will not be elaborated here.

How thinner the cap, how larger the deflection, which determines the minimum wall height. There is an optimum thickness, for a minimum total height of the cap. According to Fig. 3 this is for a cap thickness of 56μm, resulting in a total height of 74μm. A thicker cap however has the advantage of less strain in the cap and less shear stress at the surface with the die.

A remark has to be made on the effect of thermal expansion when the cap is heated. The cap will deform and deflect outwards due to the CTE mismatch with the silicon substrate. If this is taken into account, FEM calculations showed that the total deformation is reduced by a factor of two.

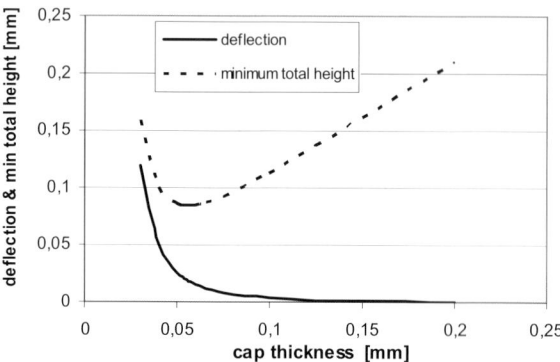

Fig. 3. Deflection and sum of deflection and cap thickness (minimum total height of cap) versus plate thickness.

3. Materials testing

3.1. LCP and poly-imide film

The calculations so far suggest that a Young's modulus of 2.5GPa (at the overmoulding temperature) should be large enough for a 100μm thick cap to withstand the overmoulding pressure without unacceptable deformation.

Two types of materials had emerged as first candidates for the capping application, namely LCP and poly-imides. Because of lack of experimental data, these were subjected to dynamic mechanical analysis (or DMA) for assessing the effect of temperature on their elastic properties. The samples were analyzed at a heating rate of 5 ºC/min.

LCP has good dielectric, barrier and humidity absorption properties and is consequently of large interest for packaging applications. A disadvantage however is their large anisotropy. Tests were carried out on 100 μm gauge (not reinforced) LCP film from Kuraray.

Poly-imides are widely used in semiconductor industry, and are mostly applied on wafers by spincoating followed by curing and lithographic structuring. For these test however, a 130μm gauge Kapton poly-imide film was used.

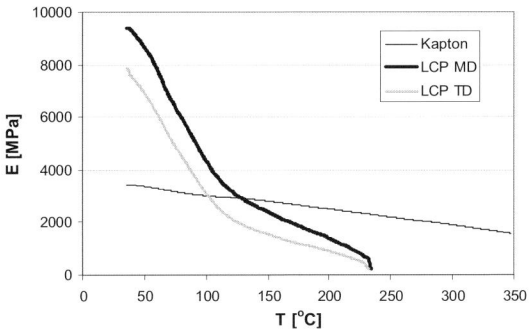

Fig. 4. DMA analysis on poly-imide Kapton and LCP film (latter in two orientations)

Fig. 4 depicts the measured curves. The modulus of Kapton falls in the range that is suitable for plastic capping. Actually, the 2.5GPa used in the calculations in Section 2 coincides well with the 175°C stiffness of Kapton. The stiffness of the tested LCP grade (an extruded film), although starting at a high value at room temperature drops quickly with temperature and will make it difficult to meet the stiffness requirements for overmoulding at 160°C to 200°C.

In general, the properties of LCP are strongly anisotropic (with a factor three stiffness variation for some grades), but for the tested grade at room temperature this was reduced significantly due to its manufacturing process (by film extrusion). At high temperatures the anisotropy is however again strong. The graph shows also that the elastic modulus of LCP decreases continuously as the temperature rises. It does not stay relatively level until the glass transition point in order to decrease sharply afterwards as crystalline or amorphous polymers tend to do.

3.2. Poly-imide with nanofillers

Nanofillers are used to improve the strength or other properties (such as hermiticity) of polymers. As part of the present feasibility study, the effect of mixing nanofillers in a precursor for an already relatively stiff poly-imide (dielectric interlayer) film was investigated. Fig. 5 shows that the nanofillers did increase the high temperature stiffness with 40%.

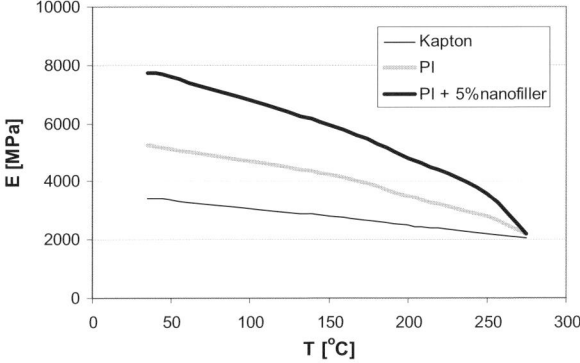

Fig. 5. DMA analysis on poly-imide Kapton film and film from a poly-imide precursor with and without nanofiller.

4. Experimental overmoulding simulations

In order to be able to validate the calculations, and recommendations that results from them, as well as the DMA analyses, a practical test was defined and executed. A tool was made to simulate the overmoulding process, see Fig. 6.

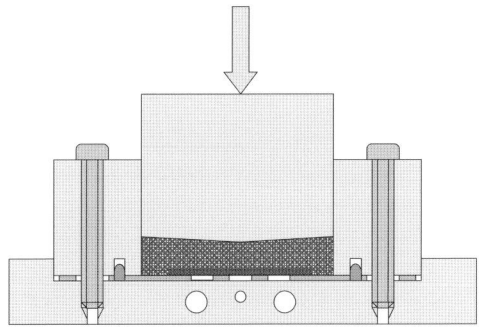

Fig. 6. Overmould simulation apparatus

The polymer capping film is placed on top of a template with a hole pattern, which is fixed on a base plate and under a ring. After heating of the tool (with heat rods in the base plate) the space above the film is filled with an epoxy compound and compressed by a punch. During the curing of the epoxy and subsequent cooling, the punch force and hence internal pressure are kept constant. Afterwards the punch is removed and both the geometry of film and epoxy surface (which has become a negative of the unreleased film surface) are measured, see Fig. 7.

A series of tests was carried out with a 130µm thick stainless steel template with a variety of apertures, see Fig. 8, on 100µm gauge LCP and 80µm gauge Kapton poly-imide film. Overmould pressure was 10MPa and temperature was 165°C.

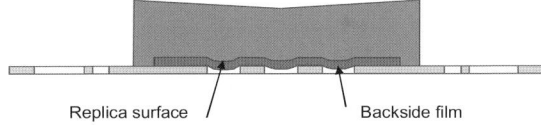

Replica surface   Backside film

Fig. 7 Template with deformed film and expoxy replica.

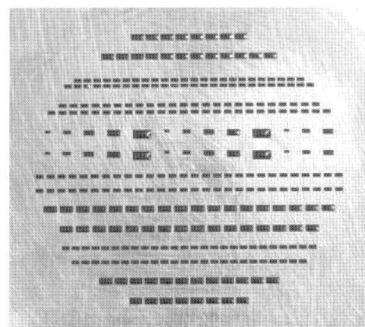

Fig. 8 Template with a pattern of holes ranging from 0.2mm x 0.4mm to 0.7mm x 1.4mm.

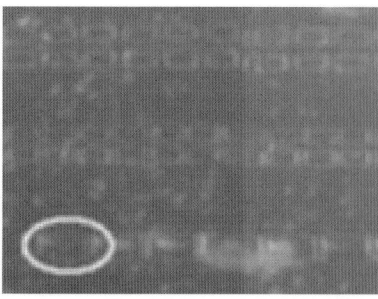

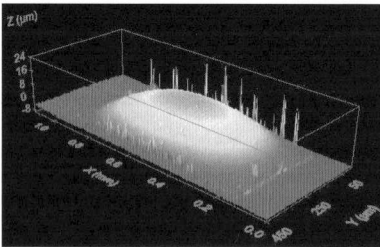

Fig. 9 Top: poly-imide film after overmoulding, with protrusions at site of 0.5mm x 1.0mm holes and 0.3mm x 0.6mm holes.
Bottom: confocal microscope scan of the encircled 0.5mm x 1.0mm protrusion.

Table 2
Measured and calculated deflection of the poly-imide film for various aperture sizes

aperture size [mm]	deflection range [μm]	calculated [μm]
0.2 x 0.4	1 – 1.1	1.4
0.3 x 0.6	2 – 7	7
0.4 x 0.8	n.a.	22
0.5 x 1.0	29 – 49	54
0.7 x 1.4	80 – 104	208 [a]

[a] N.B. experimental value cannot be larger than template gauge of 130μm

In agreement with the measured low stiffness at the test temperature, the LCP film failed. In effect, the film ruptured at the larger holes. The tested poly-imide film remained intact, see Fig. 9. Table 2 shows measured and analytically calculated deflection values, which agree reasonably (though calculated values are too high) except for the largest aperture.

5. Conclusions

Both numerical and analytical analyses of RF-MEMS cavity caps indicate that a modulus of elasticity of 2.5 GPa is large enough to make a cap withstand the overmoulding pressure.

Dynamic mechanical analysis tests show that poly-imides have sufficient high temperature stiffness to fulfil this condition. It was also shown that nanofillers re-inforced poly-imides (added in precursor phase) had a 40% increased high temperature stiffness.

A dedicated test apparatus was built and used for experimental simulation of the overmoulding of polymer film cavity caps. The experiments showed the feasibility of poly-imide as capping material. And last but not least reasonable agreement between measured and calculated deflection of the caps.

As a final conclusion, this study confirmed the feasibility of polymer capping from the mechanical point of view. What follows is the further selection and development of polymers, processing technology and equipment for embedding of wafer level polymer capping technology in the total RF-MEMS production chain.

Of prime importance will be bonding and adhesion technology, the effect of and how to overcome thermal expansion mismatch and barrier properties of the polymers (nanofiller reinforced or with coatings).

Acknowledgements

This work has been carried out in the framework of Point One project MEMSland.

References

[1] Liu, C., Recent developments in Polymer MEMS, Advanced Materials, vol. 19, 2007, pp. 3783-3790.

[2] Guoan Wang; Thompson, D.; Tentzeris, E.M.; Papapolymerou, Low cost RF MEMS switches using LCP substrate, Proc. 34th European Microwave European Conference, Volume 3, ISBN 1-58053-992-0, 2004, pp. 1441 – 1444.

[3] Gilleo, K; Jones, D.; Pham-Van-Diep, G.; Wang; T Thermoplastic Injection Molding: New Packages and 3D Circuits, Proc. Electronic Circuits World Convention 10, Anaheim, California, February 22-24, 2005.

[4] Franosch, M.; Oppermann, K.-G.; Meckes, A.; Nessler, W.; Aigner, R., Wafer-level-package for Bulk Acoustic Wave (BAW) filtersMicrowave Symposium Digest, 2004 IEEE MTT-S International, vol.2, 2004, pp 493 – 496.

Sub-Micron Referencing System for Ultraprecision Machining Processes

C. Brecher[a,c], M. Weinzierl[a], A. Rashid[b], R. Schmitt[c], D. Köllmann[c],

[a] Fraunhofer Institute for Production Technology IPT, Germany
[b] System 3R Intl. AB, Sweden
[c] Werkzeugmaschinenlabor (WZL), RWTH Aachen University, Germany

Abstract

The set-up of ultraprecision machining processes is characterized by manual process steps which require a lot of personal skill and experience to full fill sub-micron requirements in form accuracy. Besides the fact that these manual process steps require a lot of time, they individualize each ultraprecision machined work piece and therefore prevent ultraprecision machining processes from becoming universal and cost efficient machining processes for high precision work pieces. To overcome this deficit, automation solutions are developed within the European Integrated Project (IP) »Production4µ« which enable the realization of efficient ultraprecision process chains with a high level of accuracy. In this paper, a sub-micron referencing system is introduced, which has been developed within this IP to contribute to the high accuracy process chains by enabling the automated and repeatable clamping of work pieces with sub-micrometer deviations from their original position. This does not only enable the efficient combination of different machine-tools and processes but also allows for an increase in product quality.

Keywords: precision machining, work piece clamping, automation

1. Introduction

Clamping and referencing of work pieces in ultraprecision machining is a crucial task concerning the manufacturing quality in terms of form accuracy. Especially, when manufacturing a multiple number of similar work pieces, the clamping accuracy is a major aspect in the production chain. In order to keep the sub-micron accuracy level which is achievable with state of the art ultraprecision machine tools, the only way today is to machine a set of work pieces from the solid as well as an extensive alignment of the diamond tools by test cuts. The more and more complex the designs of optical and mechanical micro components get, the more effort is required for the alignment of the work pieces. By applying novel tool exchange methods which use optical metrology systems the sub-micron form accuracies which are demanded for ultraprecision machined geometries can be fulfilled [1]. Despite the excellent results which are achieved with this method, even a small scale serial production requires a high effort, since the tools have to be exchanged back and forth to machine each individual work piece. The more sophisticated approach is a flow line production as is applied in conventional machining using palletizing and referencing systems which enable a zero-point referencing on each machine tool in the production chain. However, the true repeatability of those systems lies within an accuracy of 2 – 5 µm which is acceptable for conventional machining but exceeds the tolerance limits of ultraprecision machining. Therefore, a passive alignment chuck/pallet system with a sub-micron clamping repeatability has been developed by System 3R within the European Integrated Project (IP) »Production4µ«. The passive alignment is based on the elastic averaging principle. Elastically flexible features in the pallet encounter with highly stiff x, y and z references on the chuck. A high pull force is applied to settle the pallet down on the stiff references and to lock it in a highly stable position on the chuck. An extremely

high precision manufacturing of the chuck's reference points ensures for a sub micron repeatability of the pallet position on the chuck. The system accuracy which describes the combination of different chucks with different pallets has been measured with a maximum deviation < 0.5 microns. At Fraunhofer IPT, this sub-micron clamping device has been tested for ultraprecision machining purposes. The results from ultraprecision machining (fly cutting) of a sample work piece are described in the paper. The sample's geometry has been chosen according to a demonstrator work piece which will be realized within the European IP »Production4µ«.

2. MACRO design and working principle

MACRO is a high precision palletized workholding system manufactured by System 3R international AB. In addition to holding the work piece appropriately, the pallet system ensures high repeatable accuracy in location of the work piece. The system also provides for indexing the work piece at 90 degrees with four indexing positions. The system comprises a receiver chuck that is usually fixed on a machine table and a pallet that carries the work piece throughout the process chain. The repeatable accuracy of this system is characterized through maximum deviation of the pallet from its initial position when it is detached and re-clamped on the chuck. For the MACRO a repeatable accuracy of $\pm 2\,\mu m$ which is valid for both, with changing of chucks or pallets and indexing of a pallet on the same chuck. This accuracy is achieved through a high precision interface between the pallet and the chuck. Design of the system is based on the principle of elastic averaging which assumes that the system is grossly over constrained but each contact element is relatively flexible and when forces are applied to clamp the system, the elements deform elastically and errors average [2].

The repeatable accuracy of the system over the

entire usable life is ensured through intuitive design and state of the art manufacturing technology. All contact surfaces on the chuck are rigid while the contact surfaces on the pallet are rigid in Z coordinate and flexible in the X and Y coordinate. The rigid contact surfaces on the chuck are made from carbide steel and all surfaces, both on the chuck and the pallet are grinded with high precision.

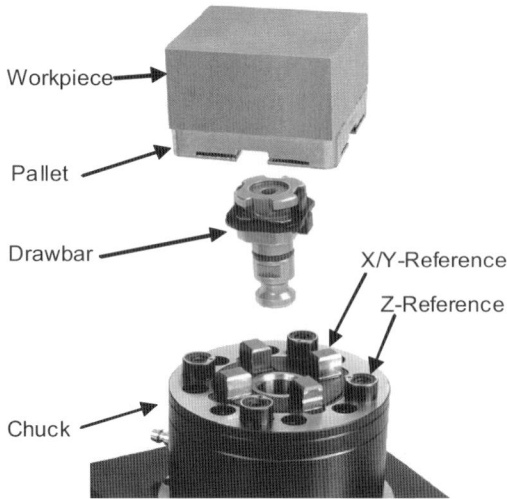

Fig. 1: Elastic Averaging Mechanism in the MACRO System

MACRO-Nano is the next generation system that provides a repeatable accuracy below one micron. This high level of repeatable accuracy in the system is achieved through state of the art manufacturing technology. The rigid contact surfaces both on the chuck and the pallet are prepared through high precision grinding followed by a lapping operation. In this case the design of the pallet has also been improved to avoid sub-micron deflection due to clamping forces. Testing of this system has shown a repeatable clamping accuracy well below one micron. Three chucks have been tested each with ten different pallets for measuring locating precision. The measurements include angular alignment of the pallet along its entire length and breadth, centricity of the pallet and parallelism in Z coordinate. Figure 2 shows the worst of ten precision values for the three chucks in each category of measurement; best observed values are as low as 0.1 µm.

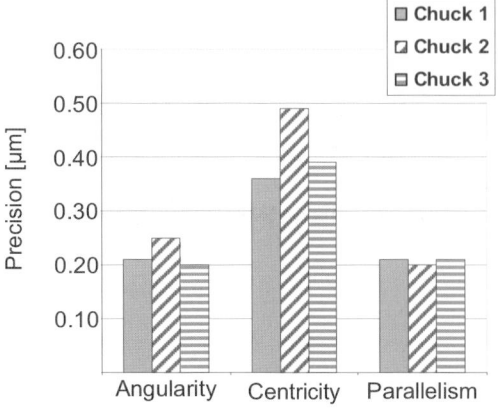

Fig. 2: Precision in pallet location on the Macro-Nano chuck

3. Testing of the macro system in an ultraprecision machining environment

The MACRO-Nano chuck system has been tested for ultrapecision machining purposes at the production site at Fraunhofer IPT. The aim of the machining test was to evaluate the in process behaviour and to measure the deviations of the sample work piece's geometry when using the system as a referencing system.

The machining of the samples as been performed on an ultraprecision machining centre LT Ultra MMC1100-2Z which enables the machining of optical surfaces with a form deviation below 0.1 µm / 100 mm PV. The sample work piece has been machined with a centre feature surrounded by a reference plane as shown in (**Fig. 3**). This sample geometry has been chosen to correspond with mould insert for micro optical components which has been defined as a demonstrator within the European IP »Production4µ«. Another influence on the choice of the sample geometry was the measurability using optical and tactile measurement systems which are described below.

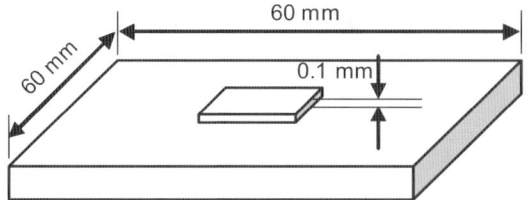

Fig. 3: Test geometry for machining samples

The machining of such a surface geometry requires the subsequent machining of four lower parts towards the edges of the work piece after planing the entire work piece surface. With a conventional ultraprecision machining set up, this is done by indexing the work piece with a highly precise rotary table or by cutting the steps at four different positions within the machine coordinate system. In both cases, the highly precise interaction of at least two machine axes is required. This requirement can be met with state of the at ultraprecision machine tools [3] but still is subject to errors caused by misalignment and inadequate referencing of the work piece in the four different positions. By indexing the work piece about a high precision reference, a symmetric centre artefact can be machined in four identical operations without excessive referencing operations. The previously determined repeatable accuracy of the MACRO-Nano chuck with a maximum deviation < 0.5 µm (**Fig. 2**) is a very promising solution for this task.

3.1 Test Sample Machining

After planing the work piece surface by fly cutting, the machining of the surrounding surface has been started by cutting a ledge at one side of the work piece. The bottom surface of this first ledge has been considered as the reference surface for the latter analysis since it has been machined without detaching the pallet from the chuck and thus represents the accuracy of the ultraprecision machine tool.

The other three ledges have been machined into the work piece surface by indexing the pallet on the MACRO-Nano chuck. In order to do so, the pallet has been released form the chuck, rotated by 90° about its vertical axis and clamped again on the MACRO-Nano chuck. In this way, all four ledges of the surrounding surface have been machined into the work piece without changing the basic set-up the process (Fig. 4).

Fig. 4: Sample machining using the MACRO-Nano chuck/pallet system

3.2 Characterization of the work piece geometry

After machining the surface to the specified geometry, the transition seems between the four ledges were clearly visible as straight lines. This is a typical indicator for form deviations caused by misalignments between the different tool paths resp. the four different work piece positions. Taking a closer look at the alignment between the plane of the tool path and the reference plane of the work piece, defined by the MACRO-Nano chuck, it becomes visible, that a relative tilt between the both planes results in a relative tilt between the bottom surfaces of the four ledges. These relative tilts can be considered as an indicator for the accuracy of the MACRO-Nano chuck – if all tilts between the bottom surfaces are the same, the repeatability of the MACRO-Nano chuck is high. Large deviations, indicate a low repeatability of the MACRO-Nano chuck. In order to qualify the orientation of the four bottom surfaces, optical and tactile measurement methods have been applied.

3.2.1 Work piece Characterization by Deflectrometry Measurement

The deflectometry measurement system is a non-contact, absolute and full surface measurement method for reflective surfaces [4]. In general, light hitting a surface is reflected, refracted or absorbed. For the deflectrometry measurement system only the reflected part of the light is important. The more of the surface is mirrored the merrier. Since a surface area is mirroring, if it is microscopic plane. This means that the structures of the surface roughness must be under the measures of the wave length [5]. The primary measurand by the phase measured deflectometry is the local slope of the surface. This occurs by scanning the surface with a laser by detecting and measuring the reflection angles of the beam. Hence the topography is reconstructed and plotted via mathematical integration [6] (**Fig. 5**).

The deflectometric measuring setup consists of:

- A camera
- The surface under test
- A structured illumination, this could be a diffuse reflective screen on which via projector a structure is displayed

The deflectometry impressed with its defined local resolution and furthermore by its possibility for a full surface inspection. The measurement setup is relative rugged against vibrations and other perturbations and may be used for plane and spherical surfaces but also for free formed surfaces [7].

Within the project Production4µ the deflectometry is intended to measure the surface of a work piece which located on a MACRO-Nano pallet which will be placed in the measuring envelope and subsequently be inspected. The measured data will be collected by a data accommodator and provided for the machine control. Afterwards the measuring object will be placed in the process chain again. On the other side it will be investigated to which extent this method can be used for a machine integration. For that purpose the deflectrometry measurement system has been the preferred system to be uses for the characterization of the MACRO-Nano chuck via characterization of the machined surface.

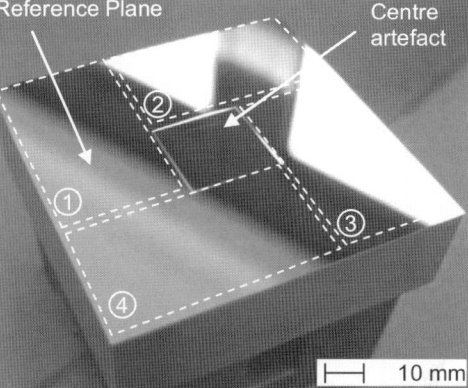

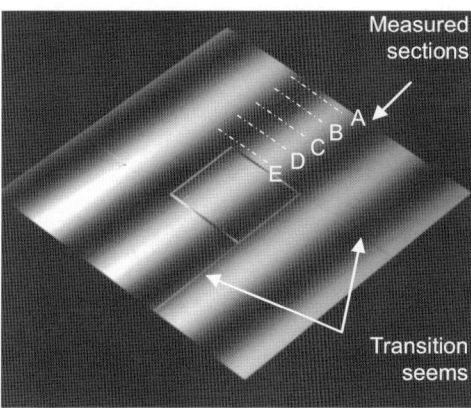

Fig. 5:   Test sample geometry and measured features (surfaces: 1-4; sections: A-E)

The information about the deviation between the relative surface orientations was defied by comparing the surface normal vectors . The mean angle between the surface normal vectors was determined to be 0.259° by spatial vector multiplication. The largest deviation from this mean orientation was 0.633° between surface 1 and surface 3 (**Fig. 5**). In terms of repeatability, this means that a maximum deviation of about 0.6 µm has been detected on the work piece surface.

Since special attention must be paid to the right adjustment of the measurement parameter setup which affect the measurement range and the resolution, a alternative characterization of the work piece surface has been performed with a tactile measurement method.

3.2.2. Characterization of the work piece surface by tactile measurement

The characteristics of the transition seems between the four ledges can be considered as an indicator for the repeatable accuracy of the chuck as well – if all seems are equal, the repeatability of the chuck is high. Large deviations between the transition seems indicate a low repeatability of the chuck. Therefore, step heights have been determined by tactile measurement at five sections on each transition seem (Fig. 5). The variations between the step heights (Δh_i) at each section indicate the repeatable accuracy of the MACRO-Nano chuck system. Table 1 lists the average step height at five sections (A - E) at each transition seem. The deviations ($\Delta_1 - \Delta_4$) from the average step height are caused by misalignments when indexing the pallet on the chuck. that the largest deviation from the average step height has been determined to an absolute value of 0.5 µm at the outer section A at the transition seem between ledge 1 and ledge 4. This result confirm the previous measurements. The tactile measurements have been performed using the Taylor Hobson Talysurf system [8].

Section	Average step height	Δ_1	Δ_2	Δ_3	Δ_4
A	2.2	0.1	0.2	0.2	-0.5
B	1.9	0.1	0.0	0.3	-0.5
C	1.6	0.1	0.0	0.3	-0.4
D	1.2	0.1	0.0	0.3	-0.4
E	0.9	0.0	0.1	0.3	-0.4
Centre	0.5	-0.1	0.1	0.4	-0.4

(All values are in µm)

Table 1: Mean step height and average deviations a the four transitions

4. Applications

Within the European IP »Production4µ«, automated process chains for the high volume production of high quality work pieces are realized. Besides the initial precision referencing of work pieces with sub-micron accuracy [9] iterative quality loops are integrated into the process chains. These quality loops are mainly determined by machining, measuring and re-machining of work pieces according to the measured deviations. In order to do so, a highly reproducible clamping of work pieces is necessary. The MACRO-Nano chuck provides trough its passive alignment principle a fast and convenient solution to realize such an iterative quality loop as an cost efficient alternative to active referencing methods.

5. Summary and outlook

With the System 3R MACRO-Nano chuck, a high accuracy work piece clamping device featuring lapped references has been introduced. After the characterization of the repeatable accuracy of the device itself, the MACRO-Nano chuck has been tested in a diamond machining environment. The form deviations of the machined test sample have been characterized by two independent measurement methods with a maximum deviation of < 0.6 µm. No adverse effects on the work piece shape or surface quality due to elastic deformations or vibrations caused by machining forces have been recognized.

Acknowledgements

The achievements presented in this paper are the results from the EC-funded Integrated Project »Production4µ«.

The authors would like to thank the European Commission for their support which enabled the works done in the field of ultraprecision and micro system technology.

References

[1] Brecher, C., Lange, S., Weinzierl, M., Peschke, C. In: Active and Passive Tool Alignment in Ultraprecision Machining for the Manufacturing of Highly Precise Structures. Annals of the German Society for Production Engineering (WGP). XIII/1, 2006

[2] Alexander H. Slocum, Alexis C. Weber; precision passive mechanical alignment of wafers, Journal of Microelectromechanical Systems, Vol. 12, No. 6, December 2006

[3] Weck, M.; Day, M.: Maschinenkonzepte zur Mikrozerspanung mit Diamantwerkzeugen, IDR 35(2001) 4, pp. 311-322

[4] 3D Shape GmbH: Product data sheet: Optical 3D-Sensor for Specular Surface Measurement.

[5] Kammel, S. In: Deflektometrische Untersuchung spiegelnd reflektierender Freiformflächen. Universitätsverlag Karlsruhe 2005

[6] N.N.:Measurement of Aspheric Surfaces with 3D-Deflectometry. URL: www.dgao-proceedings.de/download/106/106_a18.pdf [25.10.2006]

[7] N.N. In: Vermessung spiegelnder Oberflächen. The learned journal QZ Qualität und Zuverlässigkeit 08/2006

[8] N.N.: Taylor Hobson Ltd. URL: http://www.taylor-hobson.com/talysurfpgi1240.htm

[9] Brecher, C., Weinzierl, M. In: New approaches for an automised production in ultraprecision machining. Proceedings of the 4M 2007 Conference

Active microvalves for micro-fluidic networks in plastics – selecting suitable actuation schemes

A.Boustheen[a], F.G.A. Homburg[a], J.E. Bullema[b], A. Dietzel[a,c]

[a]Micro and Nano Scale Engineering, Eindhoven University of Technology, The Netherlands
[b]TNO Science and Industry, Eindhoven, The Netherlands
[c]Holst Center, Eindhoven, The Netherlands

Abstract

Using active microvalves liquid flow in microsystems can be precisely controlled and timed. Plastic microfluidic networks offer high flexibility in the material selection and potentially also allow for low cost mass fabrication. For selecting a suitable micro-actuation scheme, the different options are compared on the basis of actuation performance parameters. For thermal-expansion, electrostatic, electroactive, piezoelectric and shape memory actuation principles the work density is derived from basic actuator physics and literature material parameters. For the targeted actuator dimensions also frequency, stroke and force characteristics are calculated. These are compared with actuator performance targets typical for micro-fluidic networks: forces between 160µN and 16mN, stroke of 50µm, repetition frequencies ranging from 100Hz to few mHz. As a result, only electroactive polymer and thermal actuation principles remain as viable options and shall in further work be experimentally evaluated using a modular design with interchangeable actuators.

Keywords: Polymer microvalves, actuation principle evaluation

1. Introduction

Polymer microfluidic systems offer a number of advantages: low production costs, flexibility in terms of device area, material choices, manufacturing processes, and biocompatibility. Electrically addressable microvalves are basic functional elements allowing precise control and timing of fluid flow in complex fluidic systems. A detailed review of microvalves and several actuation schemes is given in [1] and [2]. Various factors such as work density, force and stroke of the employed actuation principle, frequency of operation, dimensions, closing pressure, but also life time stability and system- and in some cases bio-compatibility have to be taken into account when the most suitable microvalve actuation scheme is selected. Here we discuss and compare: electrostatic, electroactive, thermo-pneumatic, shape memory actuation, and piezoelectric actuation. Electromagnetic actuation is not considered since design and fabrication is more complex. In this paper, the actuator selection criteria are delivered by requirements typical for micro-valves suitable for micro-fluidic networks but a similar approach can be used for other application fields as well.

2. Problem definition

An array of active microvalves shall be fabricated concurrently with the other lab-on-chip components on planar substrates which limits their out of plane dimensions in the range of a few µm to 0.5mm. Here, an out of plane dimension of about 100µm is targeted. For an active microfluidic valve the channel dimensions are typically about 50µm×50µm. The required actuator stroke is therefore also 50µm. The in plane dimensions of the actuators are assumed not to exceed dimensions an order of magnitude larger than the channel dimensions, resulting in a maximum surface area of the actuators of 0.250mm² enabling dense lateral packing of channels for highly parallel networks. The response time of the actuator should allow for channel closing/opening in milliseconds but for some applications (for instance disposable, actuate once systems) a few seconds may be sufficient. The forces required to close the valve for a flow rate in water between 1µl/s and 100µl/s can easily be calculated assuming laminar flow conditions to derive the pressure drops across the microvalves. This pressure multiplied by the valve surface area results in force targets ranging between 160µN and 16mN. In the following, actuator performance characteristics will be derived from basic considerations and literature values. To calculate the operating frequency the mechanical natural frequency will be taken for all actuation schemes except for thermally controlled ones where actuation speed is limited by ($\sim 1/4\tau$) of their thermal time constants. Finally, these performance characteristics shall be compared to the micro-valve-requirements translated into actuator target values.

3. Thermal expansion actuators

Thermal expansion of materials in a confined cavity results in a pressure increase and can be used for transduction. The work density w is described by

$$w = \rho\Delta h \qquad (1)$$

where ρ is the density of the material and Δh is the change in enthalpy for the given temperature difference. Water, fluorinert-77, and paraffin are the actuating materials with good thermal conductivity and thermal expansion coefficients and are used for the calculations. For micro-fluidic applications a temperature range of 10C to 70C can be allowed. In accordance with the problem definition a cylindrical actuation cavity with surface area of 0.250mm² and height of 100μm is assumed. The increase in volume ΔV (for $\Delta T=60K$) for paraffin is 10% [3]. The work density w can also be derived as

$$w = (Fh)/V \qquad (2)$$

where F is the nominal force on the membrane (P times surface area of the cavity), h is the mean displacement resulting from expansion. Equation (2) can be used to calculate the actuation stroke and force. The obtained values reflect maximum values obtainable from the thermal actuators. Transmitting materials (membrane) can reduce the efficiency of the conversion from thermal to mechanical energy.

Table1: Thermal actuator results

Material	Stroke (μm)	Force (μN)	Work density (J/m³)	Frequency $(1/4\tau)$Hz
Water	1.20	1300	2.5116e8	3.46
Fluorinert-77	8.4	80	1.178e8	0.845
Paraffin	10	191	3.0675e8	2.68

4. Electrostatic actuators

The electrostatic force F between two parallel electrodes separated by a distance d_0 is given by

$$F = (1/2)\varepsilon V^2 (A/d_0^2) \qquad (3)$$

where V is the voltage between the electrodes, ε_0 is the permittivity of free space, A is the surface area of the electrodes. This force can be used for actuation when one of the electrodes is fixed and the other is movable and suspended by a spring. At pull-in voltage V_p the electrodes come in contact resulting in the maximum stroke (d_0) of the actuator [4].

$$V_p = \sqrt{(8Kd_0^3/27\varepsilon A)} \qquad (4)$$

where K is the stiffness of the spring, in our case a membrane. The work density w is then given by

$$w = (1/2)[C_1 V_1^2 - C_2 V_2^2]/(Ad_0) \qquad (5)$$

where the subscripts 1 and 2 indicate the initial and the final position of the actuator and C the capacitance. For a 1μm gold layer on top of a 20μm PDMS layer forming the movable electrode, K can be calculated. Calculations are performed for a d_0 value of 50μm and the results are given in table 2. Higher voltages can be applied if a thin dielectric layer such as PDMS or SiO$_2$ is deposited on the surface of one electrode. Instead of pull-in voltage the breakdown voltage of the dielectric layer will then limit the achievable forces.

Table2: Electrostatic actuator results

V_p(V)	Force (μN)	Stroke (μm)	Work density (J/m³)	Frequency (Hz)
890	350	50	1.4e3	14e3

Other electrostatic actuator schemes (like comb-actuators) will show similar trends but will require more complex integration schemes to be usable as micro-valve actuators.

5. Electroactive polymer actuators

Two types can be distinguished: (i) Ionic Electroactive Polymers (IAP) and (ii) Electronic Electroactive Polymers (EAP). IAP actuators have low operating voltages and frequencies because they rely on ionic diffusion and are considered to be impractical for the desired application [5]. EAP actuators are generally dielectric elastomers or electrostrictive polymers embedded between two electrodes. When an electric field is applied the dielectric compresses in the direction of the electric field and expands in other directions. The effective actuation pressure P for an EAP actuator is given by [6]

$$P = \varepsilon\varepsilon_0 E^2 = \varepsilon\varepsilon_0 (V/d)^2 \qquad (6)$$

Where E is the applied electric field, ε is the dielectric constant of the elastomer. The energy density w is given by:

$$w = (1/2)(S_z)^2/E_y \qquad (7)$$

Table 3: EAP actuator results

Actuator	Field (V/μm)	k	Force (N)	Stroke (μm)	Work density (J/m³)	Frequency (Hz)
Silicone HS3	72	2.8	0.032	47.4	1.4e4	1.6e4
Dow corning 730	80	6.9	0.097	40	4e4	3.35e4
Isoprene rubber latex	67	.03	2.7	10.4	4.6e3	4.3e4

where S_z is the polymer thickness strain in the z direction calculated using equations from [6]. For calculating actuator force, stroke and frequency it is assumed that a dielectric polymer of 100μm is embedded between two circular membranes of surface area 0.250mm². For the selection of actuation voltages, reported electric field strength was used [6]. The results are given in table 3.

6. Piezoelectric actuation

Piezoelectric materials respond to an externally applied field with strain and vice versa. Thereby, electrical input is converted into mechanical transduction. Commonly used piezoelectric materials are polycrystalline ceramic materials such as PZT (lead zirconate titanate), ZNO (zinc oxide). The work density w is given by

$$w = (\tfrac{1}{2})(d_{xx}E)^2 E_y \qquad (8)$$

[7], where d_{xx} is the strain coefficient tensor (typically d_{33} or d_{31}) and E_y is the Young's modulus of the piezoelectric material. It is assumed that the piezoelectric actuator has an out of plane dimension of 500µm. The stroke is calculated by multiplying strain and the thickness and the force is calculated by multiplying stiffness with stroke. The stiffness of the actuator is calculated using the thickness and the stiffness coefficients. In order to get sufficient stroke for the microvalve it is necessary to have an out of plane dimension of 500µm instead of the desired 100µm

Table 4: Piezoelectric actuator results

Actuator	V	d_{33} $(10^{12}V/m)$	Force (N)	Stroke (µm)	Work density (J/m^3)	Frequency (MHz)
PZT402	100	285	0.941	0.03	5.4e2	3.6
ZNO	100	11.7	0.084	0.001	2	6.2
BaTiO$_3$	100	127	0.784	0.013	2e2	5.7
PTI	100	68	0.39	0.007	5.4	5

7. Shape-memory-Actuators

By phase transformations shape memory alloys return to their previous shape at a higher temperature after subjected to a deformation at a lower temperature. The lower temperature phase is called martensitic and the higher temperature phase is called austenitic. Changing between these two phases will result in strain and thus actuation. The most common SMA materials are NiTi alloys. A NiTi membrane of the targeted dimensions can actuate only in one direction and the work density can be described by equation (1). By using a preload spring, mechanical reversible cycling between the two phases is enabled. A simplified more practical approach for description of this two-way actuation system is taken. Force displacement lines can be plotted independently for the two phase states of a membrane with previously defined dimensions and for the preload spring. In a balanced two-way system a spring and the SMA deliver comparable forces and strokes. Therefore, it is reasonable to assume spring stiffness as the average of the stiffness of the two phases. The equilibrium points of the force/displacement lines of the two phases and the spring give actuation force and stroke values. Strains up to 2% can be achieved in NiTi for several thousand actuation

cycles. [7] The work density can now be calculated using equation(2). Because the spring volume was neglected the work density is overestimated by approx. a factor of 2. The resulting data are given below.

Table 5: Shape memory actuator results

Force (N)	Stroke (µm)	Work density (J/m^3)	Frequency $1 / 4\tau$ (Hz)
17N	1.5µm	1.e6	125

8. Comparison of actuator characteristics with micro-valve target assumptions

The previously discussed micro-valve target values are plotted along with the just derived actuator characteristics in figures 3 and 4.

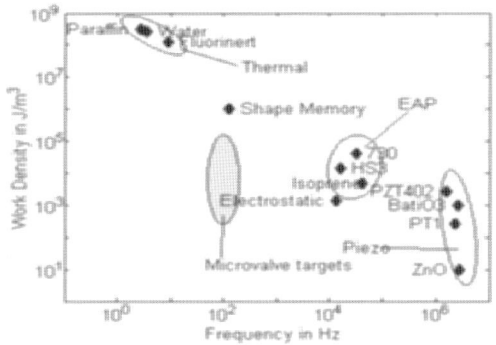

Figure 3: Work density and frequency for various actuation principles. Calculated values from all the actuation principles and the target values for micro-valves are displayed.

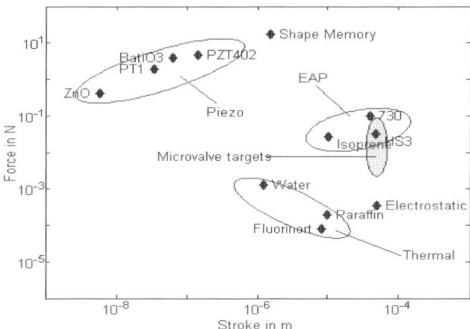

Figure 4: Force and stroke for various actuation principles. Calculated values from all the actuation principles and the target values for micro-valves are displayed.

In terms of work density piezo actuators show lower values than other actuators but still are within or close to the targeted regime. Their operating frequency is more than sufficient, nevertheless they would need an additional transmitter to achieve the targeted stroke which would make the valve design and fabrication too complex. Thermal-expansion actuators show the highest work density among the considered actuation schemes, though their actuation frequency overlaps only with the slow end of the target specification. Therefore they may be considered for non time-critical applications like use-once disposables. Thermal actuators show strokes close to the required values but less force than the all other schemes with the exemption of electrostatic actuation. Electrostatic actuation and electroactive polymers show work densities, strokes and forces comparable to the targeted values and frequencies higher than required.

But achieving a high stroke with electrostatic actuation requires higher voltage which could lead to the electrical breakdown of the system when operated in liquid environment. In the presence of impurities, the fluid could also become ionic. EAP actuators overlap with the initial specifications but require very high voltages. However, it is worth considering whether and how much targeted specifications can be softened in order to allow for alternatives. The main advantage of a thermal actuator is that there is no need for a secondary transmitter to operate a valve. Reducing the micro-channel aspect ratio would require less stroke for valving but would result in different flow characteristics. The frequency of the thermal driven actuators can be increased by allowing more areal space for the actuation system (for better thermal dissipation). Similar to that of piezoelectric actuation, thinner EAP layers require less voltage but also give less stroke and energy density.

9. Modular microvalve

From above considerations we can expect that EAP and thermal actuators show a good potential for scaling and tuning in order to approach the targeted specifications but this has also to be experimentally verified.

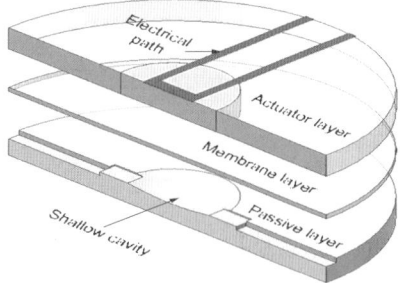

Figure 5: Schematic of the modular microvalve.

Hence it is essential that the fabricational platform is to be modular. To achieve this modularity and to implement various actuator design within a single platform a layer based approach is proposed. The functional components of the fluidic system are fabricated in different layers, hence can be designed independently of each other for specific applications. Therefore the microvalve has to fulfill both the design and the fabricational specifications. The microvalve design consists of three distinct layers. (1) Passive layer containing fluidic reservoirs and cavities, (2) Actuator layer with actuators, (3) a membrane as an actuating element. A layer based thermal actuator is to be fabricated as a first step. The thermal actuator will be based on the paraffin expansion which will act on the membrane to achieve actuation. The passive layer will consist of channels and cavities. The membrane will rest on the shallow cavity to control the flow of the fluid. The actuator part can be replaced by another actuation scheme without changing the passive layer design.

9. Outlook

The main goal of the project is to develop suitable fabricational platform for microfluidic devices. To allow for direct comparison a modular valve design (see figure 5) shall in further work be used. Thereby, also challenges of fabrication and bio-compatibility can be investigated. In this way thermal and EAP actuators can be evaluated and directly compared for the same passive micro-channel sub-system.

Acknowledgement

This research is funded by the Dutch government through IOP project IPT06210B.

References

[1] K.W. Oh and C.H. Ahn, "A review of microvalves," *Journal of Micromechanics and Microengineering*, vol. 16, 2006, pp. R13-R39.
[2] R.G. Gilbertson and J.D. Busch, "A survey of micro-actuator technologies for future spacecraft missions," *The british Interplanatery soceity*, vol. 49, 1996, pp. 129-138.
[3] E. Carlen, E. Carlen, and C. Mastrangelo, "Surface micromachined paraffin-actuated microvalve," *Microelectromechanical Systems, Journal of*, vol. 11, 2002, pp. 408-420.
[4] R. Gupta, "Electrostatic Pull in Test Structure Design for in Situ Mechanical Property Measurements of Microelectromechanical Systems MEMS," 1997.
[5] S.A. Wilson et al., "New materials for micro-scale sensors and actuators: An engineering review," *Materials Science and Engineering: R: Reports*, vol. 56, Jun. 2007, pp. 1-129.
[6] R. Pelrine et al., "High-field deformation of elastomeric dielectrics for actuators," *Materials Science and Engineering: C*, vol. 11, Nov. 2000, pp. 89-100.
[7] P. Krulevitch et al., "Thin film shape memory alloymicroactuators," *microelectromechanical Systems, Journal of*, vol. 5, 1996, pp. 270-282.

Low-power humidity sensor for RFID applications

L. Löfgren[a], B. Löfving[a], T. Pettersson[a], B. Ottosson[a], C. Rusu[a], S. Haasl[a], K. Persson[a],
O. Vermesan[b], N. Pesonen[c], P. Enoksson[d]

[a] *The Imego Institute, Arvid Hedvalls Backe 4, SE-400 14 Göteborg, Sweden*
[b] *SINTEF ICT, N-0314 Oslo, Norway*
[c] *VTT, Wireless sensors, FIN-02044, Espoo, Finland*
[d] *Chalmers University of Technology, Micro and Nanosystems group, SE-41296 Göteborg, Sweden*

Abstract

Wireless sensors incorporated in RFID systems are important in several industrial, consumer and logistics applications. By extending RFID tags to sensing applications, the products become smarter. Application areas for these smart tags include; health care (verification of the environmental conditions during transport or in storage of e.g. diapers, bandages, etc.), food monitoring (food quality during transport, storage and sales) and construction industry (e.g. building material).

In this paper, a small, very low power and low cost humidity sensor tailor made for passive RFID applications is presented. The sensor consists of a glass chip substrate with a sub-micron interdigitated gold electrode structure covered with a humidity sensitive polyimide layer. The humidity absorbed by the sensing layer is measured capacitively. Finite element modeling and analytic calculations were used to determine the design of the interdigitated electrodes and the optimal thickness of the polyimide layer. A read-out electronics circuit was designed and used to evaluate the sensor. Sensors were fabricated and calibrations have been made to verify their function. The sensor response was close to linear from below 20 to above 90 %RH and its response time was proven to be at least as short as that of the climate chamber, namely 0.1 %RH/s. The concept can easily be adapted to measure a range of other parameters such as temperature or the presence of certain substances.

Keywords: humidity sensor, passive RFID tag, e-beam lithography, nanoimprint lithography (NIL)

1. Introduction

RFID sensor network applications are emerging and moving towards commercialization. As they do so, a new level of intelligence and information will start to spread through society: the possibility for a wide number of different actors to accurately, remotely and quickly access information about their environment or the quality of a certain product. A requirement for this to happen is the development of suitable sensors with key requirements being low power consumption and a low price.

RFID (Radio Frequency Identification) systems consist of RFID tags (i.e., transponders) and readers (i.e., interrogators). There are three types of RFID transponders: passive, semi passive and active. Passive RFID transponders do not contain a power source and they operate only when powered by a nearby RFID reader. Active RFID transponders contain a battery and are constantly powered. Semi-passive RFID tags are hybrid implementations of active and passive transponders. RFID tags can be thought of as "electrical bar codes", which contain identification data, or remote memories.

One functional area of great relevance to many supply chain applications is the ability to monitor environmental parameters using an RFID transponder with built-in sensor capabilities. Parameters of interest may include temperature, humidity, shock, security and tamper detection. Wireless passive systems based on Micro System Technology (MST) for implantable pressure monitoring, and for neural recording have been reported [1, 2]. Wireless sensors are generally important in several industrial and consumer applications and a low-cost humidity sensing RFID tag has the potential of being an important tool in the food and medical distribution chain and storage as well as in the monitoring of indoor and outdoor environment.

For maximum wireless range and remote energy supply, low-power micro sensor systems are essential, making capacitive detection one of the most suitable sensing methods. Interdigitated electrodes with sensing material (e.g. polymer film on top of electrodes) were chosen since it allows for simple batch processing, miniaturization and low-cost. This sensor configuration with polyimide film [3-6] is predominantly used in both university research and commercial applications but is not developed for passive RFID tags.

In this paper, the design, micro fabrication and characterization of a capacitive humidity sensor for very low power applications, in the µW range, is presented.

2. Design and simulations

Within the IntelliSense project [7], whose framework this sensor was developed in, the specifications were chosen such that the sensor could be used with a passive RFID tag. The resulting requirement was a low-power capacitive sensor with a working point of 10 pF at 2 V, 10 kHz and an active sensor area no bigger than 1 mm^2.

The solution is a set of interdigitated electrodes covered with a sensitive polymer, a generic measurement method suitable for measuring a range of different parameters (humidity, pH, temperature, etc) depending on the chosen polymer (Fig. 1, 2).

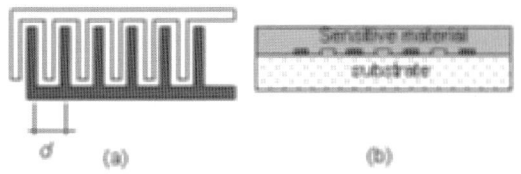

Fig 1. Sensor based on interdigitated electrodes: (a) top view of electrodes indicating the electrode periodicity *d*, (b) cross-section.

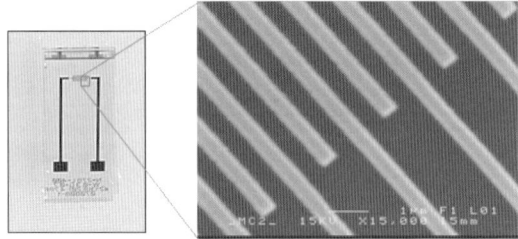

Fig. 2. Sample sensor with SEM picture of interdigitated electrodes.

Interdigitated electrodes are used in all kinds of applications, in areas like non-destructive testing, telecommunications, chemical sensing and biotechnology [3,5,8,9]. In case of the humidity sensor, the polymer, which functions as a dielectric between the electrodes, absorbs or releases moisture and its dielectric properties (permittivity) change as a function of the relative ambient humidity, thus the capacitance changes.

Polyimide is well suited for the application due to a high water uptake and a high diffusion rate resulting in high sensitivity and short response time. Polyimide experiences a change in dielectric constant from $\epsilon = 3.0$ at 0% RH to $\epsilon = 4.2$ at 100% RH [4].

Analytical model and Finite Element Simulations (FEM) in ANSYS and Comsol MultiPhysics have been used for the calculation and simulation of the sensor capacitance. The results from ANSYS, Comsol and the analytic calculations all lie within 10% of each other.

There are several parameters involved in designing the sensor, like electrode dimensions (spacing, length, number, etc) and polymer layer properties (thickness, processing, etc.). The most important of these is the ratio between the thickness of the polymer layer and the electrode periodicity, characterized by the distance *d* from one electrode to the next (Fig. 1). For the best possible sensitivity, the dimensions of the finger electrodes need to be correlated with the thickness of the polymer so that the main part of the electric field lines lie within the polymer (Fig. 3). If too thin a layer is used, the permittivity change of the polymer will not affect the total capacitance to the desired extent. Too thick a layer will on the other hand react more slowly to a changing environment i.e. in case of the humidity sensor the time for water to diffuse through the polymer will be increased. Literature and simulations show that 95% of the electrical field lines are within the polymer when the thickness of the polymer is at least half the electrode periodicity (Fig. 4) [3,5,9].

The viscosity of the polyimide placed certain limits on the thickness that could be evenly spin coated onto

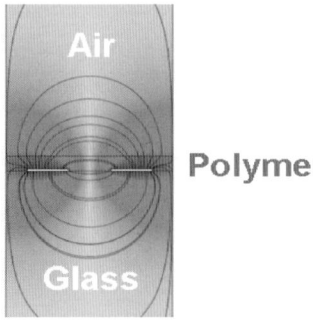

Fig. 3. FEM simulation of the field lines between a pair of interdigitated electrodes, seen through a cross section.

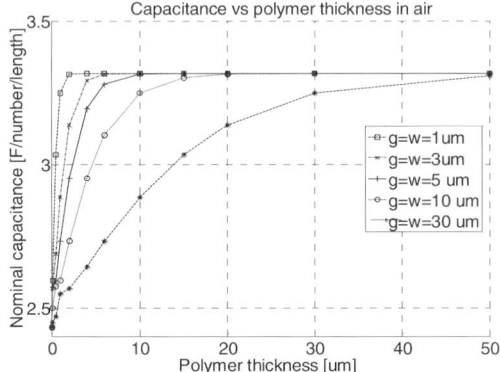

Fig. 4. Simulation of sensor capacitance dependency on polymer thickness for different electrode configurations. *g* is the gap between electrodes, *w* is the finger width and the periodicity $d=2(g+w)$. Sub-strate permittivity $\epsilon = 4.5$, polymer permittivity $\epsilon = 3.0$.

the wafer. Initial tests indicated that a 3 µm thickness could be achieved and suitable dimensions of the fingers were found to be 300 nm in width and 700 nm gap in between fingers making the total periodicity $d = 2$ µm.

The finger length was fixed at 400 µm and the total number of fingers was the parameter used to adjust the capacitance of the sensor to the required 10 pF. From an initial design with 1000 fingers, tests and further simulation iterations led up to the final design with 558 fingers.

3. Fabrication

The sensors were fabricated on 4-inch Pyrex wafers, initially coated with 200 nm Au using a 15 nm thick Ti adhesion layer. E-beam lithography (JEOL JBX-9300FS) was used to pattern the designed interdigitated structure, with 300 nm wide fingers and 700 nm gap in between fingers, onto the wafers. The gold and titanium layers were dry etched using ion milling (Oxford Ionfab 300).

A 50 nm plasma enhanced chemical vapor deposited (PECVD) silicon nitride layer was deposited on the wafers using an STS PECVD, decreasing the

Fig. 5. Sensor mounted directly on a pair of contacts.

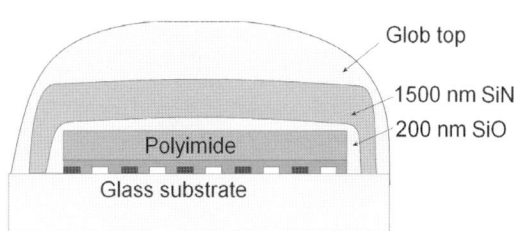

Fig. 6. Sketch of multi layer sealing of the reference chip. Layer thicknesses are not drawn to scale. Glob top thickness is in the range of mm.

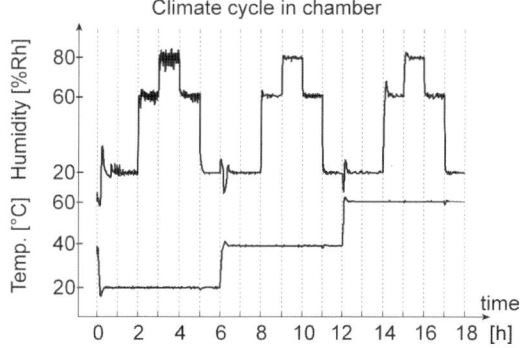

Fig. 7. Temperature and humidity cycling scheme for sensor testing in climate chamber.

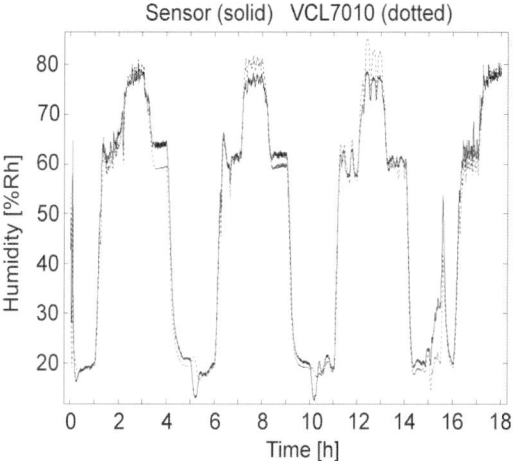

Fig. 8. Comparison between the developed sensor (solid line) and the humidity measurement of the climate chamber (dotted line).

risk of voltage breakthrough and improving the mechanical stability of the fingers. During the process development stage, the wafers were protected with photoresist and diced prior to cleaning and spin coating of the polyimide. Polyimide PI2723 (Dupont Co.) was spin coated onto the wafers, to form a 3 μm thick layer. The polyimide was annealed in a nitrogen atmosphere at 450 °C.

The sensor was initially glued, wire bonded and tested on FR4 and aluminum oxide substrates. However the humidity and temperature sensitivity of the substrates turned out to be highly influential on the sensor, adding signals in the order of 0.1 pF. The solution for independent testing of the sensor was to mount contacts directly onto the sensor pads using conductive glue thus removing the substrate altogether (Fig 5).

The influence of the substrate on the sensor read out shows the need for a reference sensor. This reference sensor is to be used together with the sensor on the final RFID tag to subtract signals from temperature fluctuations and other sources of noise from the substrate. The requirements on the reference sensor are that it is insensitive to humidity yet in every other aspect identical to the actual sensor. This was accomplished by sealing off a humidity sensor using layers consisting of 200 nm silicon oxide and 1500 nm silicon nitride on top of the polyimide layer and cover these with a glob top (Fig 6). The silicon dioxide layer relieved stress between the polyimide and the thick and hard but brittle silicon nitride layer. The silicon oxide and silicon nitride were deposited using PECVD.

4. Measurements

The final sensor was tested in a climate chamber (VCL7010) using a test sequence for temperature and humidity (Fig. 7). Measurements showed that any attempt to take out the analog sensor signals from the climate chamber through cables to an external A/D-converter suffered severely from the humidity and temperature sensitivity of the cables themselves and thus a hermitically sealed box with measurement electronics (Smartec UTI sensor-to-time signal converter with Atmel ATmega48 microcontroller used for data handling) was designed and fabricated for use inside the climate chamber. Sockets through the box wall made it possible to plug in the sensor.

The sensor read out was compared with the humidity sensor of the chamber for one climate cycle (Fig. 8). The noisy character of the graphs is due to the humidity tune-in behavior of the climate chamber and is not due to sensor accuracy. When the two curves are compared one can see that our sensor follows the VCL7010 sensor closely in almost every point. It can also be seen that the sensor is at least as fast as the climate chamber at 0.1 %RH/s. Furthermore, measurements showed that the sensor has a wide measurement range, being virtually linear from 20 to 90% RH (Fig 9).

The reference sensor's insensitivity to humidity

324

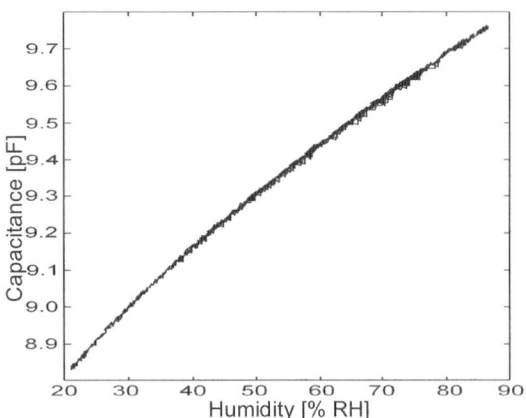

Fig. 9. Measurement of sensor capacitance vs. humidity.

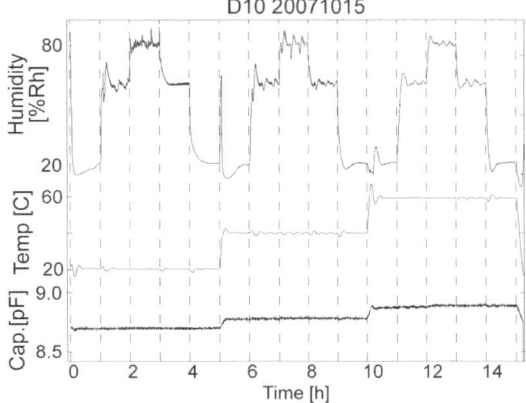

Fig. 10. Climate chamber test of reference chip. The humidity and temperature test cycle is included on top with the sensor read out at the bottom.

was tested and confirmed (Fig 10). A slight temperature dependence was noted, underlining the need for the reference sensor in applications. The operating point of the sensor was around 9 pF with a 1 pF measuring range.

5. Conclusions

A small, low-power and potentially low-cost humidity sensor for passive RFID applications has been constructed using a sub-micron interdigitated electrode structure with spin-coated polyimide.

The next step in the sensor development is to adjust the wafer process for a scaled-up, low-cost batch production. The focus is on replacing the slow and costly e-beam lithography with nano-imprint lithography, but a new design will also reduce the total chip size to obtain an increased number of chips per wafer. Smaller size is also an asset in RFID tag applications.

Acknowledgements

The humidity sensors were developed in the IntelliSense project as part of the Nordite program sponsored in Sweden by Vinnova.

References

[1] DeHennis A. and Wise K. A wireless microsystem for remote sensing of pressure, temperature and relative humidity. JMEMS (2005) Vol.14, 12- 22.

[2] Najafi K. et al. A wireless batch sealed absolute pressure sensor. Proc. 14th Transducers / Eurosensors (2000) 585.

[3] Laconte J. et al. High-sensitivity capacitive humidity sensor using 3-layer patterned polyimide sensing film. Proceedings of IEEE, Sensors (2003) Vol.1, 372-377.

[4] Johari H. Development of MEMS sensors for measurements of pressure, relative humidity and temperature. M. Sc. Thesis, Worcester Polytechnic Institute, USA, 2003.

[5] Patel K.S., Kohl P.A. and Bidstrup-Allen S.A. Novel technique for measuring through-plane modulus in thin polymer films. IEEE Trans. Components, Packaging, and Manufacturing Technology, Part B (1998) Vol.21, 199-202.

[6] Laville C. and Pellet C. Interdigitated humidity sensors for a portable clinical microsystem. IEEE Trans. Biomedical Engineering (2002) Vol.49, 1162-1167.

[7] IntelliSense project, www.intellisenserfid.com

[8] Mamishev A. V. et al. Interdigital sensors and transducers. Proceedings of the IEEE (2004) Vol.92, 808-845.

[9] Kummer A. M. and Hierlemann A. Configurable electrodes for capacitive-type sensors and chemical sensors. IEEE Sensors Journal (2006) Vol.6, 3-10.

Author Index

Subject Index

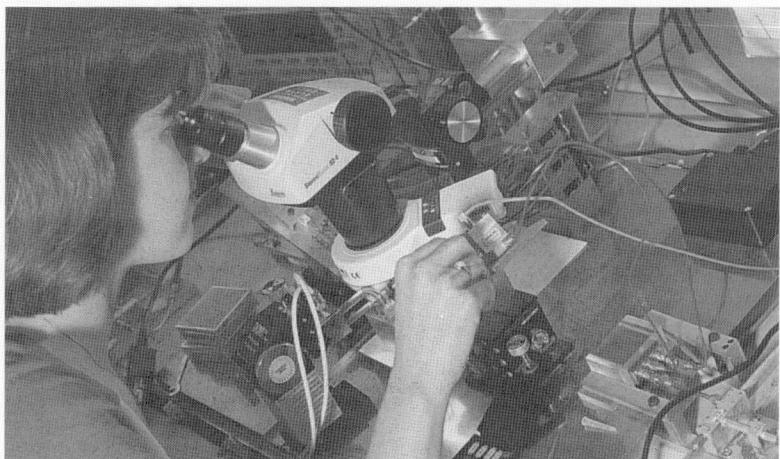
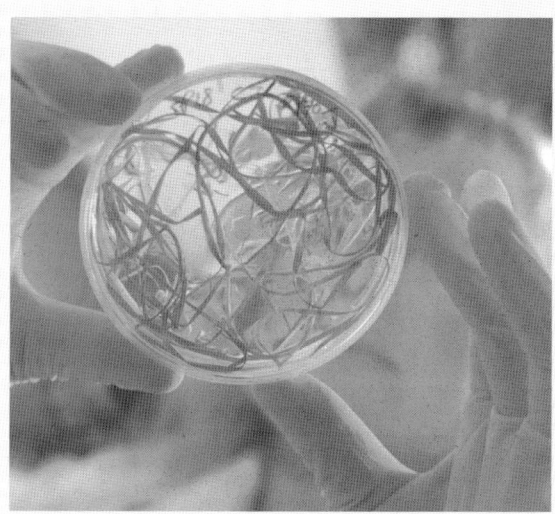

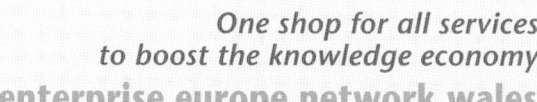

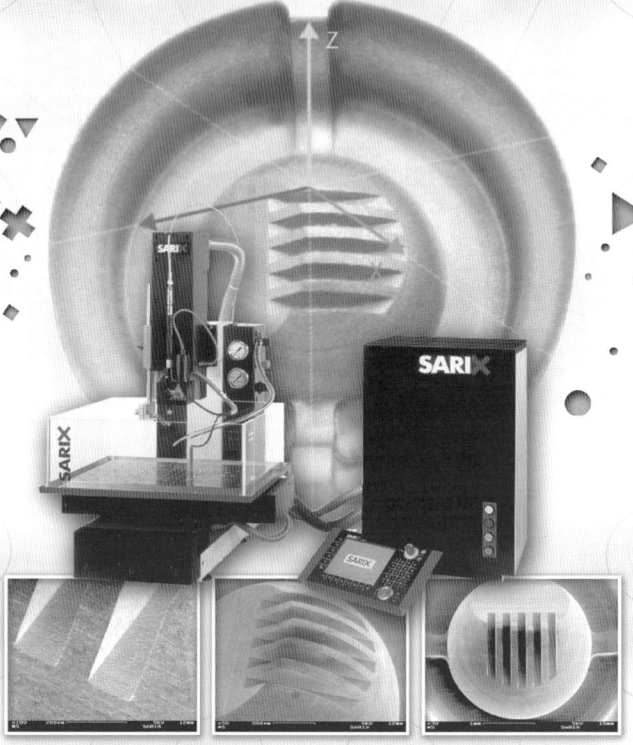

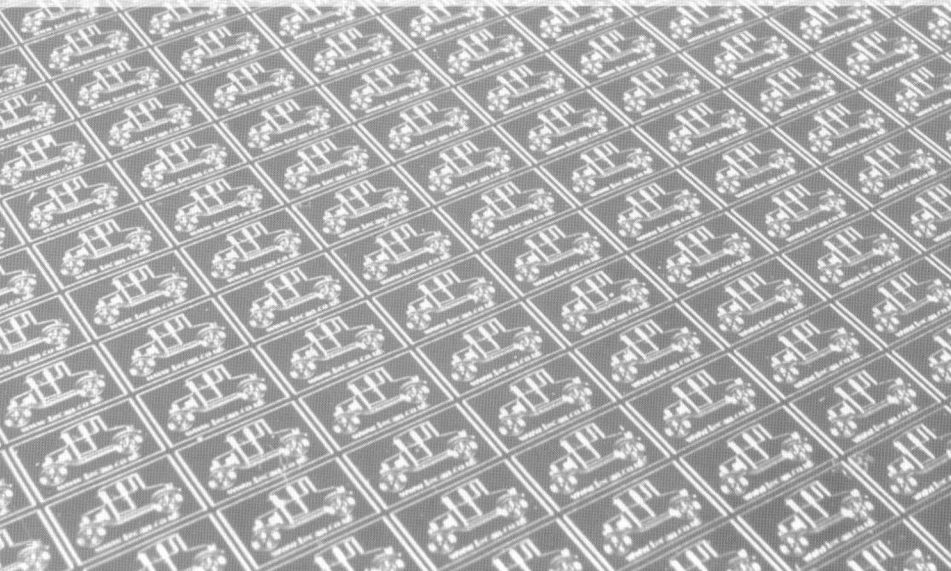

What is the Nanotechnology Knowledge Transfer Network?

The Nanotechnology Knowledge Transfer Network (KTN) was established in 2007 by the Technology Strategy Board as a successor to the MNT Network.

The Nanotechnology KTN facilitates the transfer of knowledge and experience between industry and research, offering companies dealing in small-scale technology access to information on new processes, patents and funding as well as keeping up-to-date with industry regulation. The four broad areas that the KTN focuses on are:

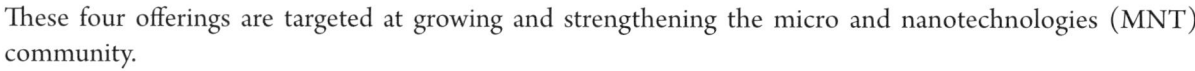

- Promoting and facilitating knowledge exchange
- Supporting the growth of UK capabilities
- Raising awareness of Nanotechnology
- Providing thought leadership and input to UK policy and strategy

These four offerings are targeted at growing and strengthening the micro and nanotechnologies (MNT) community.

What can the Nanotechnology KTN offer you?

By joining the Nanotechnology KTN, you can get access to a broad range of benefits and services, designed to support the exploitation and commercialisation of MNT.

- **Brokerage & Partnering**

Facilitating the finding of a specific technology/capability or a long-term supplier in the MNT sector, and the finding of partners for research and business, and to link collaborators for developing research consortia.

- **Networking**

Increased market opportunities and sales potential through a series of flagship events, exhibitions, sector focus groups, international trade missions and half-day facility visits at your organisation.

- **Promotion**

Free listing on the online UK MNT Directory, enabling you to promote your organisation and its capabilities, facilities and products and services. The UK MNT Directory is accessible to organisations worldwide and has an average of over 15,000 user sessions a month.

- **Intelligence**

Maintain or increase market position with fast access to new opportunities and technologies, the latest developments in standards and legislation, and updates on health and safety and quality measures.

- **Finance**

Save time and effort by signposting and assisting with appropriate investment sources including the EU 7th Framework Programme, Grants for R&D and the Technology Programme.

Join the Nanotechnology KTN

To become a member, register today at: **www.nanotechnologyktn.com**

Are you ready to connect with the 4M community?

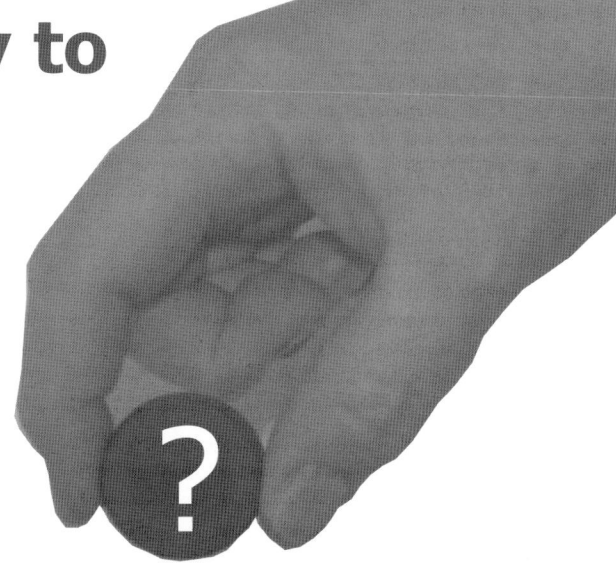

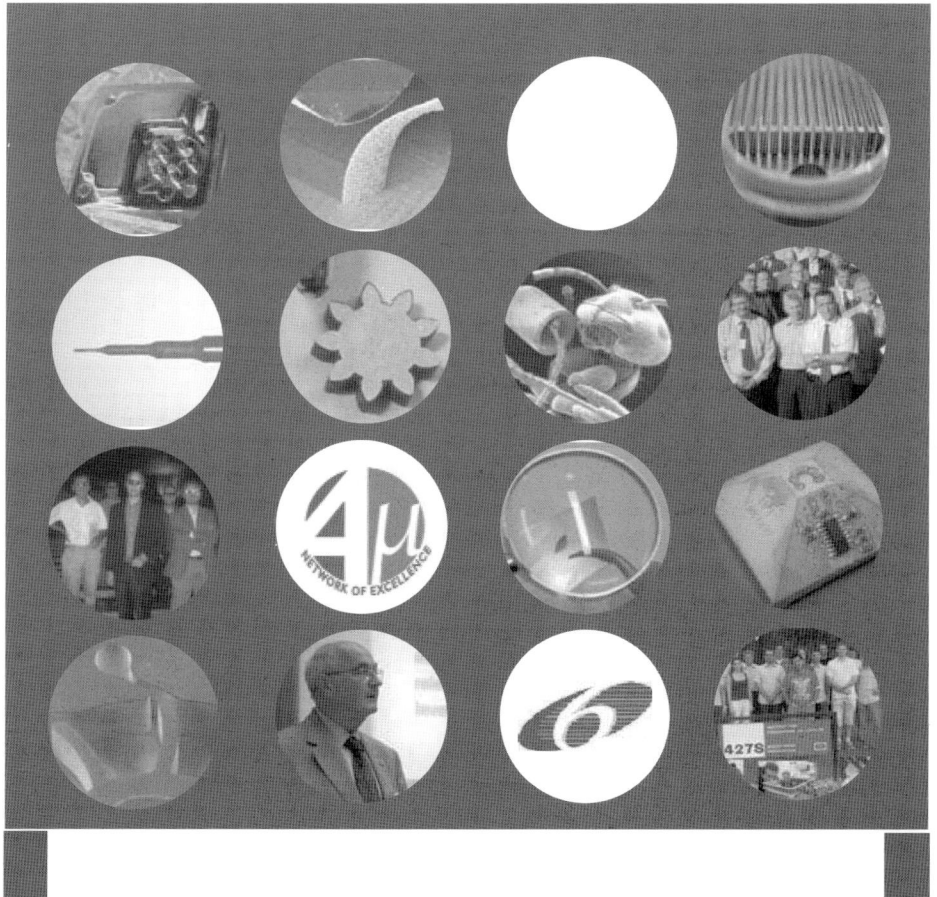

The 4M Network of Excellence in Multi Material Micro Manufacture, is to become the 4M Association. For more information on how you can join visit:
www.4m-net.org/4MAssociation

The special lens created by the MEC for Spire, the IR sensor for the Herschel Space Observatory. Looking ten billion light years into space. The School of Physics, Cardiff University using technology from the MEC

Some Results gained from the Welsh Assembly Government, DTI & ERDF funded programmes

Helping victims of Alzheimers disease using Daffodils grown in Wales. Atzeim Ltd. Talgarth, Powys, using expertise at the MEC

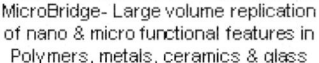

MicroBridge- Large volume replication of nano & micro functional features in Polymers, metals, ceramics & glass

Technically, the most advanced fly fishing line in the world. BVG Airflo, Brecon, Powys, using technology from the MEC

Manufacturing Engineering Centre at Cardiff University, winners of the Queen's Anniversary Prize and the DTI Prize for their work with industry.

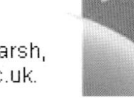

MEC, Cardiff University, Queen's Building, The Parade, Cardiff CF24 3AA. Contact: Frank Marsh, Marketing Director. Tel: 029 20 874641 Email: manufacturing@cf.ac.uk, web: www.mec.cf.ac.uk.

Micro EDM machined towers 120 x 120 microns

Enquiries: Dr Robert Hoyle at MicroBridge Services Ltd.

Tel: +44 (0)2920 870018   Email: hoylert@cf.ac.uk

www.microbridge.cf.ac.uk

Benefit from all the expertise and resources of the partners in the

4M Network of Excellence

via the

Research Advisory Service
(RAS)

Your single point of entry to a wealth of European skills and expertise in micro manufacturing technologies, processes and applications.

visit: www.4m-net.org

and begin searching for the solution to your micro manufacturing problems

 Sixth Framework Programme  MINAM